# 6th Annual International Conference on Material Science and Environmental Engineering (MSEE2018)

IOP Conference Series: Materials Science and Engineering
Volume 472

Chongqing, China
23 - 25 November 2018

**Editor:**

**Ke Wang**

ISBN: 978-1-5108-8220-1
ISSN: 1757-8981

**Printed from e-media with permission by:**

Curran Associates, Inc.
57 Morehouse Lane
Red Hook, NY 12571

**Some format issues inherent in the e-media version may also appear in this print version.**



Printed by Curran Associates, Inc. (2019)



Institute of Physics
Dirac House, Temple Back
Bristol BS1 6BE UK

Phone: 44 1 17 929 7481
Fax: 44 1 17 920 0979

techtracking@iop.org

**Additional copies of this publication are available from:**

Curran Associates, Inc.
57 Morehouse Lane
Red Hook, NY 12571 USA
Phone: 845-758-0400
Fax: 845-758-2633
Email: curran@proceedings.com
Web: www.proceedings.com

# TABLE OF CONTENTS

IOP Conf. Series: Materials Science and Engineering 472 (2019) 011001 doi:10.1088/1757-899X/472/1/011001

# Preface

6th annual 2018 International Conference on Material Science and Environmental Engineering [MSEE 2018] was held on November 23rd-25th, 2018, Chongqing, China. MSEE2018 aims to bring researchers, engineers, and students to the areas of Material Science and Environmental Engineering. MSEE2018 features unique mixed topics of Material Science and Advanced Materials, Material Engineering and Application, Environmental Protection and Sustainable Development, Renewable Energy and Building Energy Saving, Environmental Science and Engineering.

We received over 275 submissions from various parts of the world. The Technical Program Committee worked very hard to have all manuscripts reviewed before the review deadline. All the accepted papers have been submitted to strict peer-review by 2-4 referees, and selected based on originality, significance and clarity for the purpose of the conference. The conference program is extremely rich, profound and featuring high-impact presentations of selected papers and additional late-breaking contributions. We sincerely hope that the conference would not only show the participants a broad overview of the latest research results on related fields, but also provide them with a significant platform for academic connection and exchange. There is one keynote speaker and five invited sessions. The keynote speakers are internationally recognized leading experts in their research fields, who have demonstrated outstanding proficiency and have achieved distinction in their profession. The proceedings would be published by IOP Conference Series: Materials Science and Engineering (MSE).

We would like to express our sincere gratitude to all the members of Technical Program Committee and organizers for their enthusiasm, time, and expertise. Our deep thanks also go to many volunteers and staffs for the long hours and hard work they have generously given to MSEE2018. The last but not least, we would like to thank all the authors, speaker and participants for their great contributions to the success of MSEE2018.

MSEE2018 Organizing Committee
Prof. Ke Wang
East China University of Technology, China

General Chair
Dr. Klaus Doelle, State University of New York (SUNY), USA

Editor
Prof. Ke Wang, East China University of Technology, China

Technical Program Committee
Dr. Ray-Hwa Wong, Hwa-Hsia University of Technology, Taiwan
Dr. V.Preethi, Anna University, India
Dr. Elammaran Jayamani, University of Technology Sarawak Campus, Malaysia
Dr. Chih-Ta Tsai, National Cheng Kung University, Taiwan
Dr. Adnène Arbi, University of Kairouan, Tunisia
Dr. Ching-Tai NG, the University of Adelaide, Australia
Dr. Rzeszow University of Technology, Poland
Dr. Maciej Heneczkowski, Rzeszow University of Technology, Poland
Dr. Zaixing Huang, University of Wyoming, USA
Dr. Nesrin Horzum Polat, İzmir Kâtip Çelebi University, Turkey
Dr. Yanjun Li, Florida Atlantic University, USA
Dr. Hongli Chen, ZheJiang Sci-Tech University, China
Dr. Roshan Chitrakar, Nepal College of Information Technology, Nepal
Dr. Imran Memon, Zhejiang University, China
Dr. Teik-Thye Lim, Nanyang Technological University, Singapore
Dr. Qiao Lin, Columbia University, USA
Dr. Ming Liu, Tongji University, China
Dr. Khaled Benyounes, University M'Hamed Bougara of Boumerdes, Algeria
Dr. Jiangwei Liu, National Institute for Materials Science, Japan
Dr. Klaus Doelle, Dipl. Ing, Beijing University of Chemical Technology (BUCT), German
Dr. Arfan Bukhtiar, Beijing Institute of Technology, Pakistan
Dr. Changguo Wang, Harbin Institute of Technology, China
Dr. Malik Muhammad Akhtar,Balochistan University of Information Technology, Pakistan
Prof. Dan Dobrotă, Constantin Brâncusi University of Târgu Jiu, Romania
Prof. Rabah A. Khalil, University of Mosul, Iraqi
Dr. Bhat, Subrahmanya, Manipal University, India
Dr. Grzegorz Woroniak, Bialystok University of Technology, Poland
Dr. Udo Erich Schwingenschl¨ogl, KAUST, Saudi Arabia, German
Prof. João Ramos, Polytechnic University of Leiria, Portugal
Dr. Rajiv Kumar, Goswami Ganesh Dutt Sanatan Dharam College, India
Dr. Wangqiang Sun, Northwestern University, China
Prof. Andrey N. Dmitriev, Ural Federal University, Russia
Dr. José Luís Barroso de Aguiar, Universidade do Minho, Portugal
Dr. Zhengtao Zhu, South Dakota School of Mines & Technology, China
Dr. Hong Li, Northwestern Polytechnical University, China

IOP Conf. Series: Materials Science and Engineering 472 (2019) 011003 doi:10.1088/1757-899X/472/1/011003

# Peer review statement

All papers published in this volume of *IOP Conference Series: Materials Science and Engineering* have been peer reviewed through processes administered by the proceedings Editors. Reviews were conducted by expert referees to the professional and scientific standards expected of a proceedings journal published by IOP Publishing.

# Research on the Geo-Metrical and Radio-Metrical Topography Correction of the RADARSAT-2 Image

**Jiwen Zhu, Jiang Liu*, Jinling Yang, Xianglai Meng and He Zhang**
School of Surveying and Mapping Engineering, Heilongjiang Institute of Technology, Harbin 150050, P. R.China
Email: 121281531@qq.com

**Abstract.** Terrain geometry and radiation characteristics of radar images have strong influence. The correction of these effects becomes indispensable when quantitative image analysis is performed with respect to the derivation of geo- and biophysical parameters. Based on RD geo-location model and digital elevation model, an ortho-rectification and TRC implementation flow was introduced. The correctness and effectiveness of the method were qualitatively and quantitatively evaluated through several experimentations. By comparing two kinds of effective TRC methods, we found the project-angle based method is more effective than the local incidence-angle based method. The rationality of computing effective TRC factor using initial SAR geo-location model was also demonstrated.

## 1. Introduction

Terrain geometry and radiation characteristics of radar images have strong influence. In terms of geometric features, terrain causes obvious geometric distortion such as foreshortening, layover, shadow and so on in SAR range image, due to the SAR principle of side-view imaging. And in terms of radiation characteristics, Beaudoin [1] found that the error range of the backscattering coefficient reaches plus-minus 5 dB which is caused by the terrain in the region of the great terrain fluctuation. Such great error cannot be ignored when various parameters are extracted and the radar image is analyzed quantitatively. So the precise geometric and radiation rectification are required for the radar image before the parameter extraction and quantitative analysis, building correct the relationship between the ground and the image and generating orthophoto map which have its special reference coordinate system. And the error of the backscattering coefficient is corrected or compensated to produce SAR data independent of the terrain change.

Early the topography radiometric correction factor was used to correct the topography radiation, using the SAR simplified geometrical model by surface slope angle [2] or local incidence angle [3, 4, 5], when geocoding product is obtained directly. In spite of this it is inevitable to bring certain error when the topography radiometric correction factor is calculated because of the lack of the knowledge of the geometric relationship in the process of geocoding [6]. In order to overcome these disadvantages, in this paper based on the strict Range-Doppler (RD) geo-location model and SAR imaging simulation, the method is proposed to do the orthographical rectification and calculate topography radiometric correction factor simultaneously. Firstly, the method builds the geometric relationship between the SAR image and the ground based on the rigorous RD geo-location model. Secondly, local incidence angle, the angle between the normal of the imaging plane and the normal of object plane (projection angle) are found to calculate the terrain radiation correction factor. Finally, the orthographical correction for the SAR image and terrain radiation correction will be fulfilled after simulate the SAR image.

## 2. The Principle and Method

### 2.1. The RD Geo-Location Model

Due to the SAR principle of side-view imaging, terrain causes obvious geometric distortion such as foreshortening, layover, and shadow and so on in SAR range image. This causes object point, having the same distance from the subastral point, imaging different columns. And along the flight direction the relative speed will change with the terrain elevation between satellite sensors and object point, giving rise to Doppler shift. This brings about image distortion in the azimuth direction. The accurate geometric correction is carried on for SAR image to rebuild the correct imaging geometry relationship, which describes the relationship the image point and its corresponding object point. Based on the SAR imaging principle, RD geo-location model describes the geometrical relationship between the object and SAR image, which have been the standard model of the all successfully launched SAR satellite. RD model can be described in the following three equations:

$$\frac{X_t^2+Y_t^2}{(R_e+H_t)^2}+\frac{Z_t^2}{R_p^2}=1 \tag{1}$$

$$R=R(i,j) \tag{2}$$

$$f_d=-\frac{2dR}{\lambda.dt} \tag{3}$$

Where the equation (1) is the ellipsoid equation; (2) is the slant-distance equation; (3) is Doppler equation. $(X_t,Y_t,Z_t)$ are the coordinates of the object point in the geocentric rectangular coordinate system; $R_e$ is the earth ellipsoid equatorial radius; $R_p$ is the earth ellipsoid polar radius; $H_t$ is the ground point elevation; $H_t=H(X_t,Y_t,Z_t)$ can be expressed as the terrain equation; $R(i,j)$ is the slant distance of the corresponding point $(i,j)$ in the SAR image; and $\lambda$ is radar wavelength [7]

The required parameters solve the above the equations can be obtained from the header files of SAR data. Satellite orbit parameters such as vector data can be obtained through simple calculation. The RADARSAT-2 is a high-resolution SAR satellite; iterative localization method is adopted based on Doppler frequency to solve the RD geo-location model, where the location precision is better.

### 2.2. The Topography Radiometric Correction Model

The value of the backscatter coefficient is calculated by employing the calibration formula including the radiometric error from the terrain fluctuation when the terrain not flat, so topography radiometric correction factor is further computed to correct these errors based on RD geo-location model and DEM. Local incidence angle and projection angle is used as two kinds of topography radiometric correction factor. Where in the earth-fixed coordinate system (ECR) $O$ is as the earth's core, $S$ as the position of the satellite, $S'$ as the position of substellar point, $T$ as the object point (elevation is $H$ ) on the earth surface, $T'$ as its projection point on the earth ellipsoid surface, and $R_s$ ( $R_t$ )as the position vector in the ECR of $S$ ( $T$ ).So $R_{ts}=R_s-R_t$ describes the vector pointing $T$ to $S$ .Local incidence angle $\eta$ is calculated by equation 4 and projection angle by equation (5).

$$\cos\eta=\frac{\hat{n}.R_{ts}}{|\hat{n}||R_{ts}|} \tag{4}$$

$$\cos\psi=\frac{\hat{n}.|R_{ts}\times|R_{ts}\times R_t||}{|\hat{n}||R_{ts}\times|R_{ts}\times R_t||} \tag{5}$$

Where $\hat{n}$ is the normal vector of the earth surface cell located object point $T$ .The topography radiometric correction equation is obtained by equation (4) and (5) using incidence angle and projection angle:

$$\sigma^0 = \frac{\sigma^{0'}}{\sin\theta}\sin\eta \tag{6}$$

$$\sigma^0 = \frac{\sigma^{0'}}{\sin\theta}\cos\psi \tag{7}$$

Where $\theta$ is as the radar incident angle assuming that the earth surface is the ellipsoid surface, and $\sigma^{0'}$ as the backward scattering coefficient. It is noteworthy that the result of the backward scattering coefficient $\sigma^0(m^2m^{-2})$ with nonlinear transformation is different from $\sigma^0(dB)$ with linear transformation because of the nonlinear progress of topography radiometric correction. Doing topography radiometric correction, we should enter the backward scattering coefficient $\sigma^0(m^2m^{-2})$ into equation (6) and (7), then make its valve transform to dB valve.

## 3. Test Area and the Data

The study area of this paper is located in Zhayi Forest of Guizhou province, the size of study area is about 20 km*20 km. In study area the terrain ups and downs, the elevation range is 600~1 700m. This study area is conducive to verify the validity of the new method. The experimental data is Radarsat-2 Single Look Complex data, the resolution is 8m, and the specific parameters are shown in table 1. The DEM model data is standard 1:50 000 scale DEM data, which provided by National Administration of Surveying, Mapping and Geoinformation, the spatial resolution is 25 m.

According to the Radarsat-2 data technical documentation, the adjusted results of Radarsat-2 image with 8 m resolution should be sampled to 3.13 m, while the resolution of the DEM is 25 meters, and this will lead to the under sampling of simulated image, then it is necessary to carry out sampling process to DEM. The final DEM re-sampling rate is decided by formula (6) [4].

$$f \geq [\frac{\delta_{rg}}{\delta_r}\frac{\Delta S_{out}}{\sqrt{2}\Delta S_{dem}}]^{-1} \tag{8}$$

In formula, $\delta_{rg}$ is ground distance resolution of radar image, $\delta_r$ is oblique distance resolution of radar image, $\delta_{rg}/\delta_r = 1/\sin\theta$ , $\theta$ is incidence angle of radar image; $\Delta S_{dem}$ is the original size of DEM, $\Delta S_{out}$ is the size of the output pixel of the simulated image. According to formula (8) the final DEM will be sampled to 5m, so we can ensure that each pixel of the simulated image will be at least assigned once in the simulation process. In order to obtain the better effect after re-sampling from original DEM, the cubic convolution interpolation method was carried out on the original DEM data sampling step by step, this can ensure that in each step after re-sampling the before sampling resolution when divided by the sampling resolution is less than 3.

**Table 1.** System parameters of Radarsat-2 data

| Radar Satellite | Radarsat-2 |
|---|---|
| Level | SLC |
| Acquire date | 2009/2/8 |
| Ascending and Descending | Descending |
| Incidence angle (unit: °) | 40.34 |
| Azimuth resolution (unit: meters) | 5.12 |

## 4. Results and Discussion

### *4.1. Orthographic Correction Effect Assessment*

In order to quantitatively evaluate orthographic correction effect, this paper by using 1:1 0000 digital topographic map (DRG) for reference, analyzed the positioning accuracy of Radarsat-2 correction results, the results as shown in Table 2.

Through the analysis of the positioning accuracy, the positioning accuracy of Radarsat-2 orthographic correction result in east and west direction is about 15 meters and the north and south direction is about 45 meters, compared to the resolution of the Radarsat-2 itself, the positioning accuracy is high, and this proves that it is effective using the above methods for high resolution radar satellite Radarsat-2 data orthographic correction.

**Table 2.** Positioning accuracy analysis of orthographic correction results based on RD location model

| No. | X coordinate of DRG | Y coordinate of DRG | X coordinate of Radarsat-2 orthographic images | Y coordinate of Radarsat-2 orthographic images | dX(m) | dY(m) |
|---|---|---|---|---|---|---|
| 1 | 672770.864 | 2970497.183 | 672790.696 | 2970465.772 | -19.832 | 31.411 |
| 2 | 677316.567 | 2971458.724 | 677330.486 | 2971418.680 | -13.919 | 40.044 |
| 3 | 658260.804 | 2979256.470 | 658267.458 | 2979203.456 | -6.654 | 53.014 |
| 4 | 665861.075 | 2982663.367 | 665867.127 | 2982633.390 | -6.052 | 29.977 |
| 5 | 672895.287 | 2983721.548 | 672907.456 | 2983689.059 | -12.169 | 32.489 |
| 6 | 656679.879 | 2970421.380 | 656700.245 | 2970381.004 | -20.366 | 40.376 |
| 7 | 669675.319 | 2975956.899 | 669696.531 | 2975924.586 | -21.212 | 32.313 |
| 8 | 664509.917 | 2966713.910 | 664528.453 | 2966684.393 | -18.536 | 29.517 |
| 9 | 679009.950 | 2962922.148 | 678994.501 | 2962894.880 | 15.449 | 27.268 |
| 10 | 676583.453 | 2974709.069 | 676603.201 | 2974666.302 | -19.748 | 42.767 |

### *4.2. Topography Radiation Correction Effect Assessment*

After the orthographic process of radar images, the range image is converted into back scattering coefficient, and when put in formula (6) and (7), the SAR image contrast before and after the topography radiation correction as shown in Figure 1.

From visual effect, the geometric image distortion after terrain correction has been effectively suppressed, the difference between light and shade of toward and backward radar has been cut effectively, but the very bright area on original orthographic image is still not eliminate, these areas are mainly overlapping area, because the overlapping area is more than one feature point imaging into an image point, which cannot be eliminated through the topography radiation correction method based on single SAR image, and need to use more apparent SAR image method to eliminate. The shaded area due to no radar wave radiation to the region also cannot use the same method for radiation calibration.

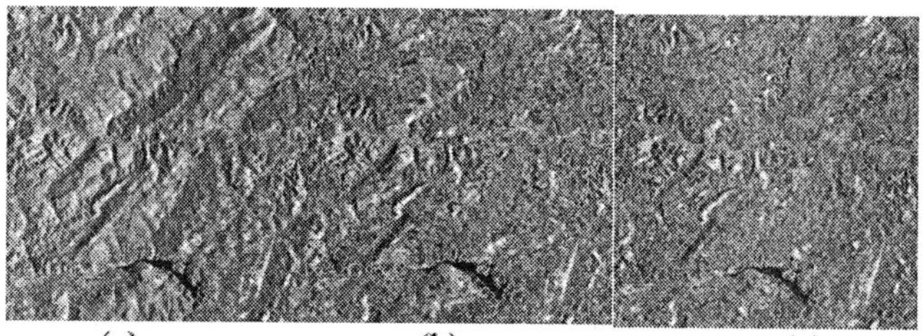

(a) (b) (c)

**Figure 1.** SAR image contrast before and after topography radiation correction (a) SAR image after orthographic correction; (b)SAR image after topography radiation correction based on local incidence; (c) SAR image after topography radiation correction based on project angle.

The important index that measures the effect of terrain on the backward scattering whether effectively be eliminated is image variance reduction percentage before and after correction, which is used by almost all researchers. The index $\alpha$ is expressed as:

$$\alpha = \frac{(s_1^2 - s_2^2)}{s_1^2} \times 100\% \tag{9}$$

**Table 3.** Statistics of variance of back scattering coefficients before and after topography radiation correction

| | $S0$ | $S1$ | $(S0-S1)/S0$ |
|---|---|---|---|
| Local incidence angle | 7.679721 | 7.094845 | 7.62% |
| Projection angle | 7.679721 | 7.005542 | 8.78% |

From the above chart we can see that, after topography radiation correction, image variance has been effectively reduced, which suggested that the topography radiation correction based on the projection angle and local incidence angle is effective. The variance of method based on projection angle reduces less than the method based on local incidence angle, which suggested that topography radiation correction method based on projection angle is more effective, that is in consistent with the conclusion of literature 6, in the practical application the projection angle is the recommended method for topography radiation correction processing.

### *4.3. The Rationality Based on the Initial Positioning Model Calculating Local Incidence Angle and Projection Angle*

Through statistics, we found that in this experiment for the initial positioning model the oblique distance error in distance direction is about 4.76 SLC pixel size, orientation error is about 16.6 SLC pixel size, equivalent to 22.99 m and 84.99m, respectively, but this is a small amount compared to the height of the satellite to the center of the earth (about 716 km) and the minimum slant distance from satellite to the features (about 1000 km), the local angle of incidence and angle of projection of the calculated value of change can be ignored. Therefore, even though the experiment SAR data has initial orbit error, but will not have a significant impact for local incidence angle and projection angle calculation, this shows that local incidence angle and projection angle were calculated based on initial positioning model are reasonable and effective.

## 5. Conclusion

In this paper, SAR orthographic correction and terrain radiant correction model based on strict SAR imaging geometric model and calibration formula are given, the experimental results show that the positioning accuracy is higher, orthographic correction can meet the requirements. The image variance significantly reduced after terrain radiant correction, and the terrain radiant correction image texture characteristics comparison before and after correction also shows that the influence of terrain has been effectively eliminated, and this demonstrates that the method of this paper is correct and effective; the quantitative evaluation of the two kinds of effective terrain radiant correction method showed that the method based on projection angle is more effective to remove the influence of terrain; this paper also analyse the terrain radiant correction factor calculation rationality based on initial positioning model, and found that the calculation results is not sensitive to initial positioning error model, the calculation of terrain radiant correction factor based on the initial positioning model is effective. This paper provides the SAR image processing a general and effective method.

## Acknowledgements

This work was financially supported by Natural Science Foundation Programs of Hei Longjiang Province (D201413) and educational bureau of Hei Longjiang Province (12541675).

**References**

[1] Beaudoin A, Troufleau D, Desbois N, Piet L and Deshayes M 1995 *Proc. IEEE Int. Geos. RS Symposium* On the use of ERS-1 SAR data over hilly terrain: Necessity of radiometric corrections for thematic applications **6** p10–14.

[2] van Zyl J J, Champman B D, Dubois P, et al 1993 *IEEE Transactions on Geoscience Remote Sensing* The Effect of Topography on SAR Calibration. **31** p1036-1043.

[3] Freeman A, Curlander J C, 1989 *Photogrammetric Engineering Remote Sensing* Radiometric Correction and Calibration of SAR images **55** p1295-1301.

[4] E. X. Chen, Z. Y. LI, X. J. Chen 2002 *High Technique Letters* Terrain Radiometric Correction Model and Its Validation for Space-Borne SAR Data **137** p10-14.

[5] Bernier M, Gauthier Y, Briand P, et al, 2002 *IEEE Int. Geos.RS Symposium* Radiometric Correction of RADARSAT-1 Images for Mapping the Snow Water Equivalent (SWE) in a Mountainous Environment Toronto **15** p127-132.

[6] Ulander L M, 1996 *IEEE Transactions on Geoscience and Remote Sensing* Radiometric slope correction of Synthetic Aperture Radar images **34** 1115-1122.

[7] E. X. Chen, 2004 *Chinese academy of agricultural sciences* Orthographic correction method research on space-borne synthetic aperture radar images **6** p26-32.

[8] J. P. Zhou, L. L. Tang, C. R. Li, 2002 *Journal of Remote Sensing* Two Practical R-D Location Model and Precision Comparison Between Them, **5** p191-197.

# Synthesis and Characterization of Ester Bond-Containing Polyethyleneimine Functionalized Carbon Nanotubes

**Qiao Zhang, Shuwen Huang, Fangzheng Yu, Huili Fu, Liang Li and Guoping Yan**

School of Materials Science and Engineering, Wuhan Institute of Technology, Wuhan, 430205, People's Republic of China.
E-mail: booyou@163.com

**Abstract**. Carbon nanotubes (CNTs) are emerging transporters for delivery of gene and drug. The degradable polycation functionalized CNTs (CNTs-HDDAPEI) were fabricated by grafting ester bond-containing polyethyleneimine (HDDAPEI). Fourier transform infrared spectroscopy, Raman spectrum, transmission electron microscopy and thermogravimetric analysis were used to characterize the degradable polyethyleneimine decorated CNTs. Compared with pristine CNTs, the CNTs-HDDAPEI displayed better dispersability and stability in water.

## 1. Introduction

Carbon nanotubes (CNTs) are one-dimensional nanomaterials and have received much attention for applications in various fields due to their unique mechanical, electrical and chemical properties [1,2]. They are nanometer scale diameter, high aspect ratio and mechanical strength, which make them an ideal reinforcing agent for high performance polymer composites [3]. Their physical and chemical stability as well as high surface area makes them can be functionalized to achieve improved biological properties and functions [4, 5]. Therefore, CNTs have been widely used in biotechnology and medicine in recent years [6, 7].

For biological applications, pristine CNTs usually need to be functionalized to achieve improved dispersion and biological properties in an aqueous environment. Recently, CNTs were studied to carry nucleic acids and penetrate mammalian cell membranes into the cytoplasm [8]. However, the hydrophobic character and lack of functional groups limit their function to condense DNA. To meet the increasing demands for better performances, the properties of CNTs should be further tailored and improved by modification and functionalization [9]. The most popular way is incorporation of cationic polymer such as polyethyleneimine (PEI) [10], polyamidoamine (PAMAM) [11] and chitosan [12]. The polycation functionalized CNTs have improved dispersion and are able to securely immobilize negatively charged DNA onto the surface of CNTs. The transfection efficiencies of functionalized CNTs are similar to or even several times higher than that of cationic polymer alone. The PEI functionalized CNTs were reported as gene vector with high transfection efficiency. However, due to the undegradable and high positive charge of PEI, they always displayed high cytotoxicity. Some biodegradable PEI derivatives with low cytotoxicity were synthesized to reduce the cytotoxicity.

In this study, the ester bond linked PEI (HDDAPEI) was synthesized by Michael addition between 1, 6-hexanediol diacrylate and low molecular weight branched 1.8 k Da PEI. Then carboxyl CNTs were functionalized through grafting of HDDAPEI to obtain CNTs-HDDAPEI. The product was characterized using Fourier transform infrared spectroscopy, Raman spectrum, transmission electron microscopy. In additional, the thermal stability and water dispersibility were examined.

## 2. Materials and Methods

### *2.1. Materials*
Carboxyl functionalized multi-wall carbon nanotubes (CNTs-COOH) with purity of 98 (wt%), containing 1.23 wt% carboxyl groups, having outer diameter of 20-30 nm, and length of 0.5-2 μm were obtained from Chinese Academy of Sciences Chengdu Organic Chemicals Co., LTD. BPEI (MW 1.8k, 25k) were purchased from Sigma-Aldrich. 1,6-hexanediol diacrylate, 1-Ethyl-3-(3-dimethylaminopropyl) carbodiimide hydrochloride (EDC), N-Hydroxysuccinimide (NHS), were purchased from Aladdin Co. (Shanghai, China). Acrylic acid, cyclohexane, thionyl chloride, triethylamine, dimethyl sulfoxide (DMSO), sodium bicarbonate and hydroquinone were purchased from Shanghai Chemical Reagent Company and used as received. Dialysis membrane (MWCO 3.5 kDa) was purchased from Biosharp.

### *2.2. Synthesis of 1,6-Hexanediol Diacrylate Linked PEI (HDDAPEI)*
The 1, 6-hexanediol diacrylate PEI were synthesized according to the modification of literature [13]. 1.8 g of branched PEI 1.8 k Da was dissolved in a mixture solution of 6 mL distilled water and 6 mL DMSO. Then the solution was added to the three-necked flask equipped with a condenser. To obtain HDDAPEI, 0.27 g 1,6-hexanediol diacrylate (HDDA) was dissolved in 10 mL DMSO and added to the PEI solution by dropping slowly. The reaction mixture was stirred for 48 h at 45°C under nitrogen atmosphere. After the reaction, the product was dialyzed against distilled water (MWCO: 3500) for 48 h to remove the unreacted 1.8 k Da PEI. Lastly, the solution was lyophilized for 48 h.

### *2.3. Synthesis of CNTs-HDDAPEI*
The HDDAPEI solution was prepared by dissolving certain HDDAPEI in 2 mL distilled water. CNTs-COOH (200 mg) was suspended in 20 mL distilled water and ultrasonic dispersed for 30 min at 50 °C. Then EDC (105 mg) and NHS (82.4 mg) were added to activate the carboxyl groups. Subsequently, the HDDAPEI solution was added and stirred for 24 h at room temperature. The weights of HDDAPEI used for functionalization process were 100 mg, 150 mg, and 200 mg. After the reaction, the product was centrifuged at 20,000 × g for 15 min, then redispersed in distilled water and dialyzed (MWCO:12000) for 48 h to remove the unreacted HDDAPEI and activating agent. Lastly, the solution was lyophilized for 48 h to obtain CNTs-HDDAPEI in a black powdered form.

### *2.4. Characterization of CNTs-HDDAPEI*
The Fourier transform infrared (FT-IR) spectra of CNTs-HDDAPEI was recorded on Nicolet Impact-420 Fourier transform infrared spectrometer in the range of 400 - 4000 $cm^{-1}$. Raman spectroscopy was conducted with a Kaiser optical system and a helium-neon laser of 532 nm wavelength. The morphology of CNTs-HDDAPEI was conducted by JEM-2100F field emission electron microscopy (TEM, JEOL, and Japan). Thermogravimetric analyzer (TGA) measurement of CNTs-HDDAPEI was performed by a TGA STA-409PC thermogravimetric analyzer under $N_2$ in the temperature range 30-600 °C with an increasing rate of 5 °C $min^{-1}$. The surface charge of CNTs-HDDAPEI was measured by a Nano-ZS ZEN3600 (Malvern Instruments).

### *2.5. Dispersibility Test*
Each of 4 mg CNTs, CNTs-COOH and three kinds of CNTs-HDDAPEI was ultrasonic dispersed in 8 mL water sufficiently. The dispersions were recorded after 48 h to evaluate the water dispersibility and stability of CNTs, CNTs-COOH and CNTs-HDDAPEI.

## 3. Results and Discussion

### *3.1. The Preparation of CNTs-HDDAPEI*
In this study, the ester bond-containing PEI (HDDAPEI) was synthesized by Michael addition between 1, 6-hexanediol diacrylate and low molecular weight branched 1.8 k Da PEI. The number

average molecular weight (Mn) of HDDAPEI is 8.15 kDa and polydispersity (Mw/Mn) is 1.36. Three CNTs-HDDAPEI were prepared by changing the weight ratio of carboxyl CNTs to HDDAPEI to study the influence of HDDAPEI content on the dispersion property of CNTs. The feed compositions of CNTs-HDDAPEI are listed in Table 1.

**Table 1.** Feed compositions of CNTs-HDDAPEI and grafting weight ratio of HDDAPEI to CNTs-COOH.

| Sample ID | CNTs-COOH [mg] | HDDAPEI [mg] | Grafting weight ratio[a] | Zeta potential [mV] |
|---|---|---|---|---|
| CNTs-HDDAPEI$_1$ | 200 | 100 | 25.76% | 31.6 |
| CNTs-HDDAPEI$_2$ | 200 | 150 | 30.72% | 36.7 |
| CNTs-HDDAPEI$_3$ | 200 | 200 | 36.17% | 39.0 |

[a] The grafting weight ratios of HDDAPEI to CNTs were calculated by thermogravimetric analysis.

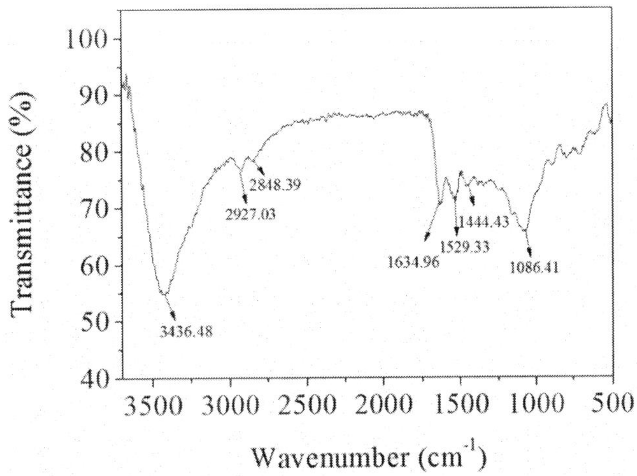


**Figure 1.** FTIR spectra of CNTs-HDDAPEI.

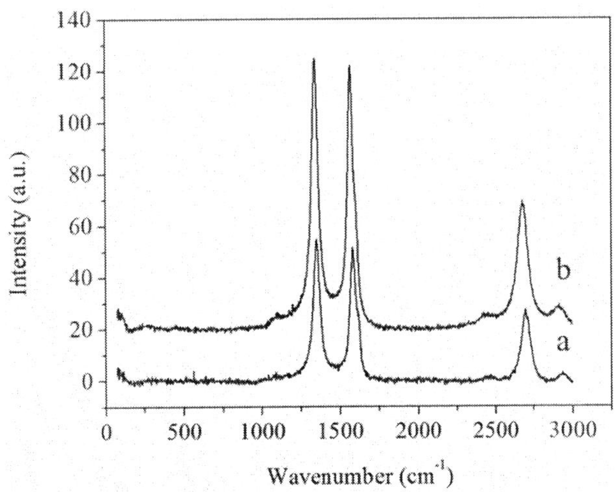


**Figure 2.** Raman spectra of carboxyl CNTs (a) and CNTs-HDDAPEI (b)

FTIR spectroscopy was employed to characterize the formation of CNTs-HDDAPEI. The typical FTIR spectrum of CNTs-HDDAPEI is shown in Figure 1. The broad and strong peak at 3436 cm$^{-1}$ is attributed to –NH stretching vibrations which indicates the amine groups on the surfaced of CNTs-

HDDAPEI [14]. The characteristic bands of –CO-NH- stretching are detected at 1634, 1529 and 1444 $cm^{-1}$, indicating the formation of CNTs-HDDAPEI [15]. The peak at 1634 $cm^{-1}$ is associated with the stretching of –C=O. The peaks at 1529 $cm^{-1}$ and 1444 $cm^{-1}$ refer to $-NH_2$ and –C-N stretching vibration, respectively. In addition, the peaks assigned to the H-C stretching vibrations at 2848 $cm^{-1}$ and 2927 $cm^{-1}$ are also detected. The peak at 1086 $cm^{-1}$ could be the stretching vibration of C-O-C [16].

Raman spectrum was used to investigate the structure of CNTs-HDDAPEI. Three raman bands at 1580 $cm^{-1}$ (G band), 1358 $cm^{-1}$ (D band) and 2711 $cm^{-1}$ (2D band) are observed for carboxylated CNTs and CNTs-HDDAPEI (Figure 2). The band at 1580 $cm^{-1}$ is a first-order raman mode attributed to vibrations of $sp^2$ carbon atoms. The band at 1358 $cm^{-1}$ is D band of a second-order scattering process, which related to the effect of defects or disordered structure in tubes. As we know that the ratio of D-band to G-band intensity ($I_D/I_G$) could be used to characterize the level of defect or disordered structure in CNTs. A higher value of $I_D/I_G$ indicates the CNTs with a higher level of defect structure [17]. As presented in Table 2, there is no noticeable change of $I_D/I_G$ values between the carboxyl CNTs and CNTs-HDDAPEI. This result shows that the functionalized process does not change the structure of carboxyl CNTs.

**Table 2.** $I_D$, $I_G$ and $I_D/I_G$ ratio of carboxyl CNTs and CNTs-HDDAPEI in Raman spectrum.

| | $I_D$ | $I_G$ | $I_D/I_G$ |
|---|---|---|---|
| CNTs-COOH | 104.46 | 102.19 | 1.02 |
| CNTs-HDDAPEI | 54.53 | 50.43 | 1.08 |

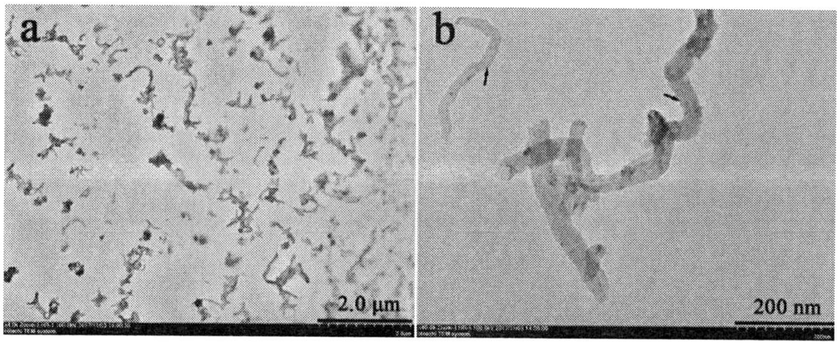


**Figure 3.** TEM images of CNTs-HDDAPEI.

The morphology and distribution of CNTs-HDDAPEI were explored by TEM. As shown in Figure 3a, the CNTs-HDDAPEI were dispersed into particles with the size of several hundreds nanometers. Figure 3b illustrates the morphology of single CNTs-HDDAPEI, which are the nanotubes with coating. The coating is attributed to the presence of HDDAPEI on the surface of CNTs. The length and dimensions are consistent with those given by the supplier.

### *3.2. Dispersion Study*

The dispersion states of carbon nanotubes, carboxyl CNTs, and three HDDAPEI functionalized CNTs were observed after sonication in water and maintained for 48 h. All the samples could be dispersed in water after sonication. However, as shown in Figure 4, pure carbon nanotubes were settled after 48 h. It is well known that CNTs usually agglomerate due to Van der Waals interactions and are difficult to disperse in water. Therefore, in order to improve dispersibility of CNTs, some water-soluble polymers were granted to CNTs [18, 19]. In this study, CNTs were decorated by HDDAPEI to increase the water dispersibility. As shown in the Figure 4b-e, the carboxylated CNTs and three HDDAPEI grafted CNTs did not settle and maintained homogeneous solution after 48 h. This enhanced dispersibility and stabilization in water is attributed to the hydrophilic groups of carboxyl on the carboxyl CNTs and amino on the CNTs-HDDAPEI.

**Figure 4.** The dispersibility of CNTs(a), carboxyl CNTs(b), CNTs-HDDAPEI$_1$ (c), CNTs-HDDAPEI$_2$ (d) and CNTs-HDDAPEI$_3$ (e) in water after ultrasounding and incubated for 48 h.

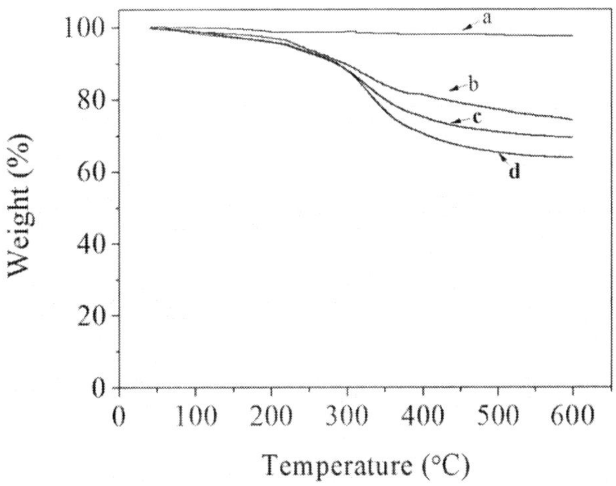


**Figure 5.** TGA profiles of carboxyl CNTs (a), CNTs-HDDAPEI$_1$ (b), CNTs-HDDAPEI$_2$(c) and CNTs-HDDAPEI$_3$ (d)

### *3.3. Thermogravimetric Analysis*

TGA was used to assay the HDDAPEI content in the CNTs-HDDAPEI. As shown in Figure 5, all the samples show small mass loss below 200 °C; this is due to the evaporation of moisture. Above 200 °C, carboxyl CNTs do not occur significant weight loss up to 600 °C because of their good thermal stability. However, the HDDAPEI grafted CNTs samples show a sharp weight decrease from 200 °C to 400 °C, which is mainly due to the thermal decomposition of HDDAPEI. There is no significant weight loss around 600 °C, indicating the functional groups of HDDAPEI are completely decomposed and the final residues are carbon nanotubes. The weight residues at 600 °C on TGA plots show the average weight residue of 74.24%, 69.38% and 63.83% for CNTs-HDDAPEI$_1$, CNTs-HDDAPEI$_2$ and CNTs-HDDAPEI$_3$, respectively. Therefore, the grafting weight ratios of CNT-HDDDAPEI were calculated and listed in Table 1. It has to be pointed out, compared with the feed ratios during synthesis, the grafted ratios are small. This is maybe because of the small amount carboxyl groups on the surface of CNTs (1.23 wt%). With the increasing of feed ratio of HDDAPEI to CNTs-COOH, the HDDPEI is excess. The zeta potential of CNTs-HDDAPEI is also listed in Table 1. The positive charge of surface demonstrated HDDAPEI was grafted to carbon nanotubes.

## 4. Conclusions

A novel ester bond-containing PEI decorated CNTs was prepared. The FTIR, Raman spectrum, TEM and TGA studies demonstrated that HDDAPEI was functionalized to CNTs successfully. The CNTs-HDDAPEI exhibited good dispersability and stability in water. These results indicated that CNTs-HDDAPEI would have promising potential applications in nanobiotechnology and nanomedicine.

**Acknowledgments**
This work was supported by the National Natural Science Foundation of China under Grant No.51503165, and Scientific research program of Hubei Provincial Department of Education, China under Grant No. Q20141509.

**References**
[1] Iijima S 1991 Nature. 354 56
[2] Sahoo N G, Rana S, Cho J W, Li L and Chan S H 2010 Prog. Polym. Sci. 35 837
[3] Kausar A, Rafique I and Muhammad B 2016 Polym-Plast Technol. 55 1167
[4] Sharma P, Mehra N K, Jain K and Jain N K 2016 Curr. Drug Deliv. 13 96
[5] Yan W, Pang D W, Wang S F and Zhao Y D 2005 Fuller. Nanotub. Car. N. 13 309
[6] Peretz S and Regev O 2012 Curr. Opin .Colloid. In. 17 360
[7] Kumar S, Rani R, Dilbaghi N, Tankeshwar K. and Kim K H 2017 Chem. Soc. Rev. 46 158
[8] Baates K and Kostarelos K 2013 Adv. Drug Deliver. Rev. 65 2023
[9] Caoduro C, Hervouet E, Girard-Thernier C, Gharbi T, Boulahdour H, Delage-Mourroux R and Pudlo M 2017 Acta. Biomater. 49 36
[10] Meng Y, Wang S S, Li C Y, Qian M, Yan X Y, Yao S G, Peng X Y, Wang Y and Huang R Q 2016 Biomaterials 100 134
[11] Liu M, Chen B, Xue Y N, Huang J, Zhang L M, Huang S W, Li Q W and Zhang Z J 2011 Bioconjugate Chem. 22 2237
[12] Liu X H, Zhang Y Y, Ma D M, Tang H, Tan L, Xie Q J and Yao S Z 2013 Colloid Surf. B. 111 224
[13] Yang L 2009 Petrochem. Technol. Appl. 27 508
[14] Moradiana H, Fasehee H, Keshvari H and Faghihi S 2014 Colloid. Surface. B. 122 115
[15] Pan B F, Cui D X, Xu P, Ozkan C, Feng G, Ozkan M, Huang T, Chu B F, Li Q and He R 2009 Nanotechnology. 20 183.
[16] Uc-Cayetano E G, Aviles F, Cauich-Rodriguez J V, Schonfelder R, Bachmatiuk A, Rummeli M H, Rubio F, GuteErrez-Amador M P and Cruz G J 2014 J. Nanopart. Res. 16 2192
[17] Dresselhaus M S, Dresselhaus G and Saito R 2005 Phys. Rep. 409 47
[18] Mehra N K, Mishra V and Jain N K 2014 Biomaterials. 35 1267
[19] Zhang Q, Lei M, Huang S W, Fu H L, Yan G P, Li L and Yan H Z 2017 Fuller. Nanotub. Car. N. 25 386

# Effect of C/S Ratio on Microstructure of Calcium Silicate Hydrates Synthesised By Solution Reaction Method

**Yazhou Wang[1,2], Qinglin Zhao[1,2], Shangqun Zhou[1] and Song Wang[1]**

1 Wuhan University of Technology, School of Materials Science and Engineering, Wuhan 430000, China.
2 State Key Laboratory of Silicate Materials for Architecture (Wuhan University of Technology), Wuhan 430000, China.
Email: 601424677@qq.com

**Abstract.** In this work, the calcium silicate hydrate was synthesized efficiently using the solution reaction of calcium nitrate ($Ca(NO_3)_2.4H_2O$) and sodium nitrate ($Na_2SiO_3.9H_2O$) at the C/S ratio of (0.8,1.2,1.5). The disciplinarian of C/S ratio with the calcium silicate hydrate was analyzed by XRD, SEM, IR, NMR. The experimental results showed that it could shorten the reaction time to 5 days, reduce the influence of the $CO_2$ in the air on the purity and ensure the quality of products controllable, when the calcium silicate hydrate was synthesized in nitrogen environment and with the ultrasonic dispersion; the d-spacing of (002),(101) and (200) decreased, the polymerization of silica tetrahedra of calcium silicate hydrate decreased 28.7%, the microstructure of calcium silicate hydrate varied from layer to irregular congeries and became denser as the C/S ratio increased from 0.8 to 1.5.

## 1. Introduction

The synthetic calcium silicate hydrate is regarded as a very promising new cement-based materials improver, but its performance is closely related to the synthesis process, research shows that the synthesis process, such as calcium-silicon ratio, reaction temperature, reaction time etc affects the composition and structure of calcium silicate hydrate [1, 9, 11]. Lv[1],He[2] synthesised the calcium silicate hydrate by the diatomite or fumed silica and calcium oxide and found that the polymerization of silicate tetrahedron of calcium silicate hydrate decreased as the calcium-silicon ratio increased. Zhao[14] synthesised the calcium silicate hydrate by solution reaction in the air and found that the microstructure of it transited from sheet to needle as the calcium-silicon ratio(1.0~1.7) increased. Zeng [10] synthesised the calcium silicate hydrate by silica and calcium oxide, found that its structure was more compact as the calcium-silicon ratio increased. calcium-silicon ratio affected synthesis of calcium silicate hydrate and its microstructure, previous studies more focused on hydrothermal synthesis of calcium silicate hydrate by calcium oxide and active silicon dioxide as raw material, but the process and product were complex, the minority synthesised the calcium silicate hydrate by solution reaction(7d) and investigated the effect of calcium-silicon ratio on the calcium silicate hydrate, with single testing method, longer synthetic cycle under air condition. Based on the related research, this paper synthesised the calcium silicate hydrate by solution reaction combined with ultrasonic instruments in the environment of nitrogen, in order to shorten the reaction time (5d), reduce the effect of $CO_2$ on the results and improve the controllability product quality etc. Combined with the XRD, SEM, IR, NMR test method to analyze the effect of products and calcium-silicon ratio on the microstructure of the calcium silicate hydrate, the relevant data can enrich the effect of calcium-silicon ratio on calcium silicate hydrate structure theories.

## 2. Raw Materials and Experimental Method

### *2.1. Raw Materials*
The calcium silicate hydrate was synthesised by solution reaction with the analytically pure calcium nitrate($Ca(NO_3)_2.4H_2O$), sodium nitrate($Na_2SiO_3.9H_2O$), sodium hydroxide(NaOH) and deionized water. Deionized water was prepared through the laboratory ultrapure water system.

### *2.2. Experimental Method*
Take appropriate calcium nitrate and sodium nitrate, and then prepare 1mol/L solutions respectively. To set the calcium-silicon ratio of 0.8, 1.2, 1.5, take the volume of the solution and deionized water accurately. The deionized water, calcium nitrate solution and sodium silicate solution were added to the container sequentially, added 1~2 mL sodium hydroxide solution, adjusting the solution pH>12, and then put into the water bath in 80°C, finally the mixed solution was dispersed by ultrasound instrument regularly. The synthetic calcium silicate hydrate named with CSH0.8, CSH1.2 and CSH1.5 respectively.

After the reaction was completed of, the products washed with deionized water and anhydrous alcohol (not less than 3 times), and then the products were dried in a vacuum desiccator (1~2d), finally the samples were analyzed using a variety of methods. The preparation of calcium silicate hydrate process was in figure 1.

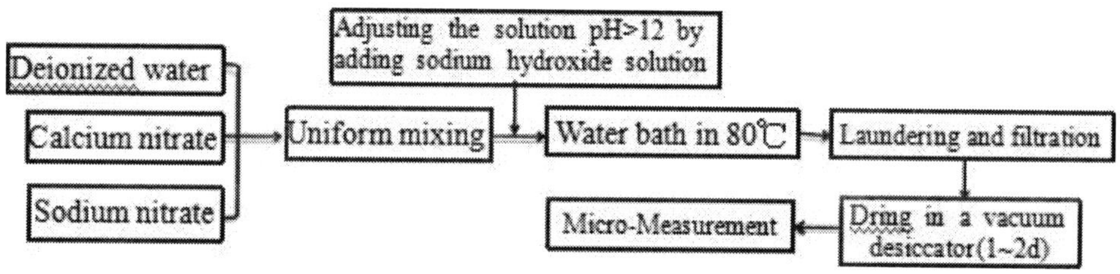


**Figure 1.** Process synthesis of Calcium Silicate Hydrate

## 3. Results and Analysis

### *3.1. XRD Analysis*
To verify the effect of calcium-silicon ratio on the purity and structure of calcium silicate hydrate, the calcium silicate hydrate at different of calcium-silicon ratio (0.8, 1.2, 1.5) were analyzed by XRD firstly, and observed the mineral composition and corresponding purity of synthetic product, the XRD results are given in figure 2.

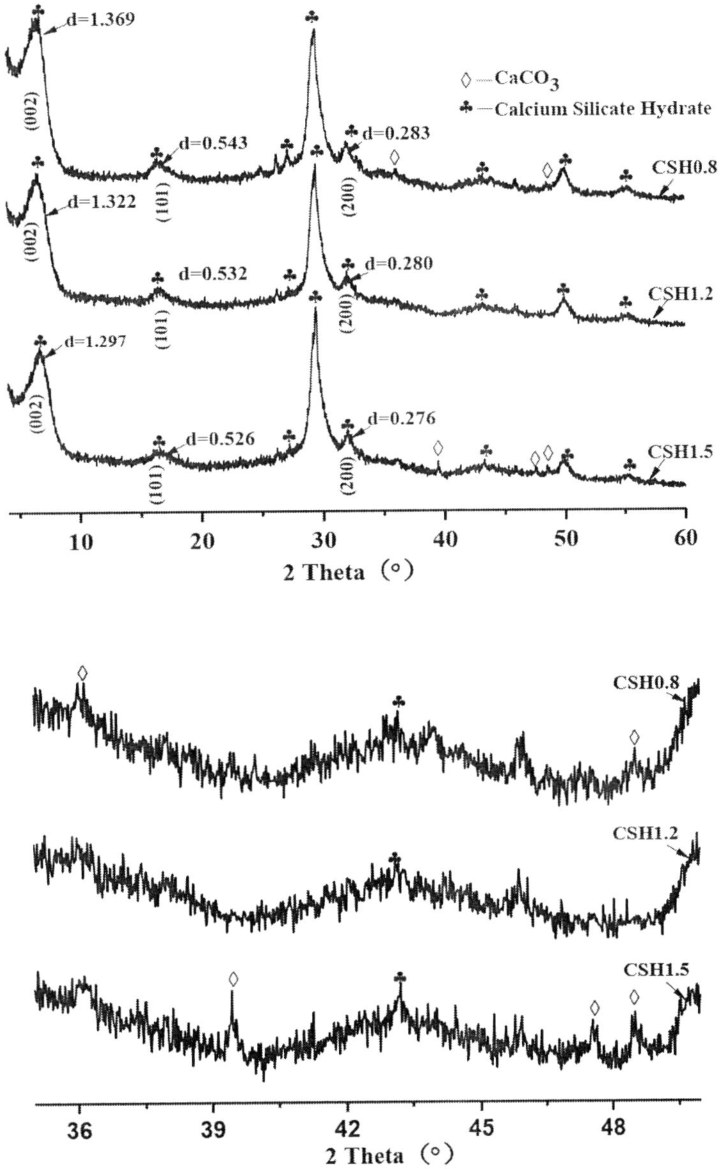


**Figure 2.** XRD patterns of Calcium Silicate Hydrate at different of C/S ratio.

By figure 2, the dispersion diffraction peaks at (7° and 17° nearby) and condyloma characteristics peaks at (29° and 31° nearby) in the XRD patterns showed that the synthesis of calcium silicate hydrate were materials with the crystalline and amorphous, the figure 2 (left) in 35~50° was amplified of the figure 2 (right). When the calcium-silicon ratio were 0.8 and 1.5, there was a small amount of $CaCO_3$ characteristic peaks in XRD patterns, there was no $CaCO_3$ characteristic peaks when the calcium-silicon ratio was 1.2; When the calcium-silicon ratio was 1.5, the $CaCO_3$ characteristic peak was more than compared to 1.2, this was because the high calcium-silicon ratio could accelerate the early carbide of calcium silicate hydrate, those were consistent with the research conclusion of Chang [19]. Besides, The figure 2 also showed that when the calcium-silicon ratio increased from 0.8 to 1.5, the d-spacing of (002) decreased by 0.072 nm, the d-spacing of (101) decreased by 0.007 nm, the d-spacing of (200) decreased by 0.007 nm, the d-spacing of (002) changed greatly compared to the d-spacing of (101) and (200), this might be the low angle diffraction peak was more sensitive than the high angle diffraction peak to the change of calcium-silicon ratio. With the increase of calcium-silicon ratio, the d-spacing of (002), (101) and (200) decreased, thus were identical with the research conclusion of He [2, 14], Kim [15] and Foley [17], namely with the increase of calcium-silicon ratio, the d-spacing of calcium silicate hydrate decreased.

### *3.2. IR Analysis*

The characteristic groups ($Q^1$, $Q^2$) of calcium silicate hydrate in the analysis of IR had specific wavenumber range (see table 1), when the structure of calcium silicate hydrate changed, its specific wavenumber would shift. The polymerization degree of silica tetrahedra of three different calcium silicate hydrate samples were analyzed by IR analysis method, in order to further clarify the effect of calcium-silicon ratio on the structural parameters of the samples.

**Table 1.** Formatting sections, subsections and subsubsections.

| Groups | O-Si-O | Si-O-Si | Si-O($Q^2$) | O-Ca-O | $H_2O$ | -OH |
|---|---|---|---|---|---|---|
| Wavenumber/$cm^{-1}$ | 450 around | 650 around | 970 around | 1445 around | 1640 around | 3440 around |

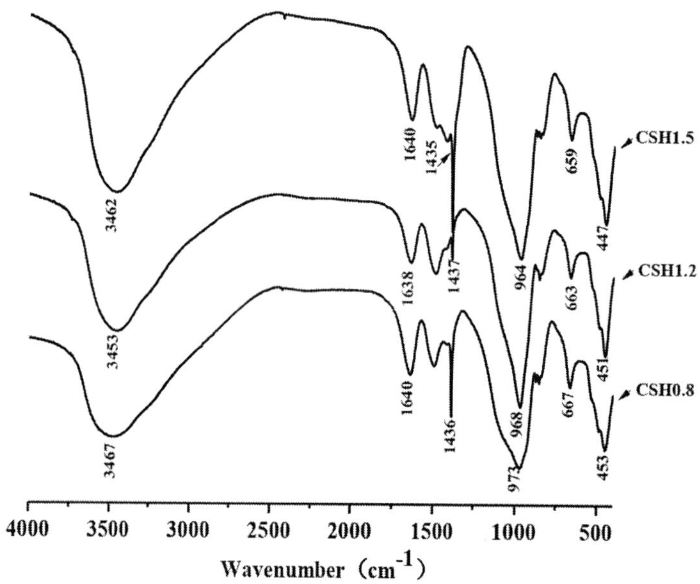


**Figure 3.** Infrared patterns of Calcium Silicate Hydrate at different of C/S ratio.

By figure 3, the characteristic vibration peaks position of the three kinds of calcium silicate hydrate samples were similar, and corresponding to various groups within the scope of wavenumber range, it showed that the characteristic group of the three kinds of calcium silicate hydrate by the solution reaction were basically the same. In figure 3, 970$cm^{-1}$, 450$cm^{-1}$ characteristic peaks were on behalf of Si-O-Si($Q^1$) bending vibration and Si-O($Q^2$) elastic vibration respectively[2]. Based on previous studies, the change of characteristic peaks position in the infrared spectrum represented the change of polymerization degree of silica tetrahedra of calcium silicate hydrate. When the calcium-silicon ratio increased from 0.8 to 1.5, the Si-O-Si($Q^1$) vibration wavenumber position from 453$cm^{-1}$ to 447$cm^{-1}$, and Si-O($Q^2$) vibration wavenumber position from 973$cm^{-1}$ to 964$cm^{-1}$, the position of characteristic peaks shifted to low wavenumber, showing that the polymerization degree of silica tetrahedra of calcium silicate hydrate decreased with the increase of calcium-silicon ratio, and thus were identical with the research conclusion of Yu [3],Wang [6],Yang [4], Beaudoin [16, 17], namely with the increase of calcium-silicon ratio, the polymerization degree of silica tetrahedra of calcium silicate hydrate decreased. Besides the 1436$cm^{-1}$, 1435$cm^{-1}$, 1437$cm^{-1}$ showed the vibration of O-Ca-O, and there were no obvious change of the wavenumber position from CSH0.8 to CSH1.5, but the peak intensity of CSH1.2 was weak, showing that the reactants could generate calcium silicate hydrate furthest. In figure 3, the vibration of $H_2O$ peak was near 1640cm$^{-1}$, this because it could not remove the interlayer water in the calcium silicate hydrate structure at 60°C under the condition of vacuum drying.

*3.3. $^{29}$Si NMR Analysis*

$^{29}$Si NMR was by measuring element chemical shift responding to element environment change. He[7], Feng [12] thought that chemical shift indicated the change of the polymerization degree of silica tetrahedra of calcium silicate hydrate, The average molecular chain length (degree of polymerization) could be calculated by $^{29}$Si NMR. In the silicate minerals, the common $^{29}$Si atoms chemical environment was used by $Q^n$, n=(0~4) showed the number of bridging oxygen in the silicate tetrahedron. The polymerization methods of silicon atom corresponded to different chemical shift range, and then the $^{29}$Si NMR test could indicate four polymerization status in the silicate tetrahedron, table 2 listed each state chemical shift range [12].

**Table 2.** The $^{29}$Si NMR Chemical shift of the structure unite of $Q^n$.

| Types of Si-O-X | Symbol | Range/ppm |
|---|---|---|
| Monomer | $Q^0$ | -68~-76 |
| Dimer& chain end groups | $Q^1$ | -76~-82 |
| Chain middle groups | $Q^2$ | -82~-88 |
| Layers & branched chain | $Q^3$ | -88~-98 |
| Three-dimensional networks | $Q^4$ | -98~-129 |

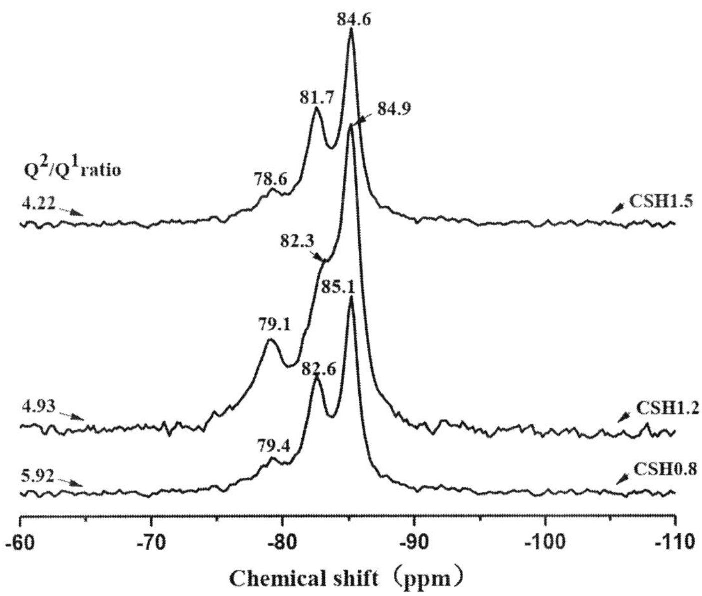


**Figure 4.** NMR patterns of Calcium Silicate Hydrate at different of C/S ratio.

The figure 4 showed that the characteristic peaks $Q^1$(-79ppm), $Q^2_{BT}$(-82ppm), $Q^2_{PT}$(-85ppm) were similar in three kinds of calcium silicate hydrate, there were no obvious $Q^0$,$Q^3$ and $Q^4$ characteristic peaks, indicating that the silicate tetrahedron polymerized long chain polymers mainly without few branched chain in three kinds of calcium silicate hydrate by solution reaction. Three samples of characteristic peaks $Q^2$ included two shoulder peaks ($Q^2_{BT}$, $Q^2_{PT}$), indicating that there were two different vibration forms in the silicate tetrahedron.

In figure 4, along with the increase of the calcium-silicon ratio, the characteristic peaks $Q^1$ (-79ppm), $Q^2_{BT}$ (-82ppm), $Q^2_{PT}$ (-85ppm) in three kinds of calcium silicate hydrate samples CSH0.8, CSH1.2, CSH1.5 shifted to positive wavenumber position slightly. The $^{29}$Si NMR tests showed that when calcium-silicon ratio was 0.8, $Q^2/Q^1$ was 5.92; when calcium-silicon ratio increased to 1.5, $Q^2/Q^1$ was 4.22, the ratio of $Q^2/Q^1$ decreased by 28.7%. Thus it could be seen with the increase of calcium-silicon ratio, the ratio of $Q^2/Q^1$ decreased and the characteristic peaks of $Q^1$, $Q^2$ shifted to positive wavenumber position slightly, indicating that the polymerization degree of silica tetrahedra decreased. The conclusion was in agreement with Feng [12], Lv [1], Kim [15] and He [2, 13] et al.

### *3.4. SEM Analysis*

Calcium-silicon ratio had an effect on micro-morphology and microstructure of calcium silicate hydrate. Taylor [20] found that the morphology of calcium silicate hydrate included four basic forms network-like large particles or iso-diameter particles, fibrous and layer. Figure 5 was the SEM pictures of CSH0.8, CSH1.2, CSH1.5 under 10000 times and 20000 times.

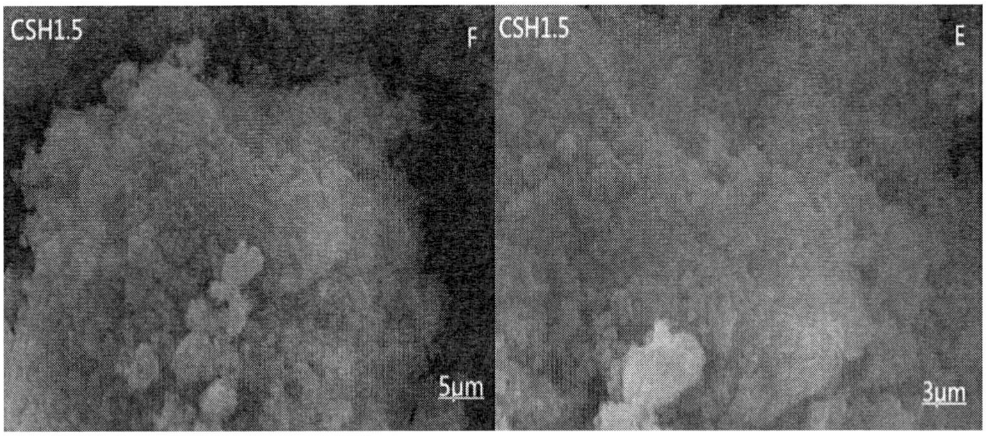


**Figure 5.** SEM patterns of Calcium Silicate Hydrate at different of C/S ratio.

By figure 5, the three samples CSH0.8, CSH1.2, CSH1.5 were gel state structure by random material. Under 10000 times, when the calcium-silicon ratio was 0.8, the morphology of calcium silicate hydrate was layer attached to a small number of irregular congeries; When the calcium-silicon ratio was 1.2, the morphology of calcium silicate hydrate was layer attached to a number of irregular congeries; When the calcium-silicon ratio was 1.5, the morphology of calcium silicate hydrate was irregular congeries primarily, the layer morphology disappeared nearly, and the structure more compact. Zeng [10] found that the calcium silicate hydrate by hydrothermal synthesis was random gel material, its structure was more compact with the increase of calcium-silicon ratio, Shi [9], Zhao [15] found that the morphology of calcium silicate hydrate changed from layer to honeycomb stacked by particles gradually.

Under 20000 times, the calcium silicate hydrate presented similar laws with the increase of calcium-silicon ratio (0.8~1.5). When the calcium-silicon ratio was 0.8, the morphology of calcium silicate hydrate was layer with few irregular congeries attached; When the calcium-silicon ratio was 1.2, the irregular congeries on the surface of calcium silicate hydrate increased and the layer morphology disappeared gradually; When the calcium-silicon ratio was 1.5, the morphology of calcium silicate hydrate was irregular congeries mainly with more compact structure.

## 4. Conclusions

(1) The calcium silicate hydrate was a material including crystalline and amorphous at the same time with high purity, it could be synthesised by solution method with calcium nitrate, sodium nitrate, sodium hydroxide and deionized water, with the increase of calcium-silicon ratio, the d-spacing of calcium silicate hydrate decreased.

(2) The calcium-silicon ratio had an effect on the polymerization degree of silica tetrahedra, with the increase of calcium-silicon ratio(0.8~1.2), the characteristic peaks ($450cm^{-1}$,$970\ cm^{-1}$) shifted to low wavenumber position in IR, showing that the polymerization degree of silica tetrahedra decreased with the calcium-silicon ratio than increased.

(3) The calcium-silicon ratio influenced on polymerization methods of silica tetrahedra, with the increase of calcium-silicon ratio(0.8~1.2), the ratio of $Q^2/Q^1$ reduced, the characteristic peaks $Q^1$and $Q^2$ shifted to positive wavenumber position slightly, when the calcium-silicon ratio increased from 0.8 to 1.5, the ratio of $Q^2/Q^1$decreased by 28.7%.

(4) The calcium-silicon ratio impact on calcium silicate hydrate micro-morphology, with the increase of calcium-silicon ratio (0.8~1.5), the morphology of calcium silicate hydrate changed from layer to irregular congeries with more compact structure.

**References**

[1] Lin Nv Lv, Xiao Gang Zhao, Yong Jia He, etc. Effect of C/S ratio on morphology and structure of calcium silicate hydrate[C]// Proceedings of the First Annual Meeting of the China Institute of Ceramics Cement Branch. 2009.

[2] Yong Jia He. Effect of C/S Ratio on Morphology and Structure of Hydrothermally Synthesized Calcium Silicate Hydrate [J]. Journal of Wuhan University of Technology, 2011, 26(4):770-773.

[3] Yu P, Kirkpatrick R J, Poe B, et al. Structure of Calcium Silicate Hydrate (C-S-H): Near-, Mid-, and Far-Infrared Spectroscopy [J]. Journal of the American Ceramic Society, 1999, 82(3):742-748.

[4] Nan Ru Yang. C-S-H gel and its research method [J]. Bulletin of the Chinese Ceramic Society, 2003, 22(2):46-52.

[5] Shu Mei Yu, Lin Nv Lv. Exploratory Study of Raman Spectroscopy for Polymerization of Calcified Calcium Silicate [J]. The World of Building Materials, 2011, 32(2):6-8.

[6] Lei Wang, Zhen He, Bo Zhang, etc. Revealing the mechanism of C-S-H polymerization based on infrared and nuclear magnetic resonance techniques [J]. Journal of Building Materials, 2011, 14(4):447-451.

[7] Yong Jia He, Shu Guang Hu. Application of 29Si Nuclear Magnetic Resonance in Research of Cement Chemistry [J]. Journal of Materials Science and Engineering, 2007, 25(1):147-153.

[8] Jian Min Xiao, Hui Mei Zhu, Feng Wu. Application of 29Si NMR Techniques of C-S-H Gel [J]. Bulletin of the Chinese Ceramic Society, 2016, 35(11):3594-3599.

[9] Lei Shi, Su Hua Ma, Wei Feng Li, etc. Hydrothermal synthesis of C-S-H and its properties [J]. Concrete, 2013(11):101-104.

[10] Lu Zeng, Mou He, Ding Mao. Influence of Hydrothermal Synthesis Conditions on Pore Structure of Calcium Silicate Hydrates [J]. Non-metallic minerals, 2017, 40(3):10- 13.

[11] Jin Wang, Wen Xu, Hui Li. Review on Effect of Different Conditions on the Structure and Performance of the Synthesized Calcium Silicate Hydrated [J]. Bulletin of the Chinese Ceramic Society, 2015, 34(5):1296-1301.

[12] Chun Hua Feng, Xi Jian Wang, Dong Xu Li. Progress in application of 29Si and 27Al NMR in cement-based materials [J]. Nuclear technology, 2014, 37(1):48-53.

[13] He Y, Lu L, Struble L J, et al. Effect of calcium–silicon ratio on microstructure and nanostructure of calcium silicate hydrate synthesized by reaction of fumed silica and calcium oxide at room temperature [J]. Materials & Structures, 2014, 47(1-2):311-322.

[14] Xiao Gang Zhao, Lin Nv Lv. Effect of C/S Ratios on Morphology of Calcium Silicate Hydrates Synthesized by Solution Reaction Method [J]. The World of Building Materials, 2010, 31(2):7-8.

[15] Kim J J, Foley E M, Taha M M R. Nano-mechanical characterization of synthetic calcium–silicate–hydrate (C–S–H) with varying CaO/SiO2, mixture ratios[J]. Cement & Concrete Composites, 2013, 36(1):65-70.

[16] Beaudoin J J, Raki L, Alizadeh R. A 29Si MAS NMR study of modified C–S–H nanostructures [J]. Cement & Concrete Composites, 2009, 31(8):585-590.

[17] Foley E M, Kim J J, Taha M M R. Synthesis and nano-mechanical characterization of calcium-silicate-hydrate (C-S-H) made with 1.5 CaO/SiO2, mixture [J]. Cement & Concrete Research, 2012, 42(9):1225-1232.

[18] Jun Chang, Yan Feng Fang, Yong Li. Effects of Calcium to Silicon Ratios on Accelerating Carbonation of Calcium Silicate Hydrate[J]. Bulletin of the Chinese Ceramic Society, 2014, 42(11):1377-1382.

[19] Taylor H F W. Cement Chemistry [J]. Chemistry for Engineers, 1992, 134.

# Mechanical Properties Analysis of Patch Repaired Composite Structures with Debonding Flaw

**Xia Guo[1,2,*], Hai Chi[1,2], Junzhi He[1,2], Weili Liu[1,2,3], Guanghui Hu[1,2] and Xia Gao[1,2,3]**

[1]Beijing Centre for Physical and Chemical Analysis, Beijing 100089, China;
[2]Beijing Key Laboratory of Organic Materials Testing Technology & Quality Evaluation, Beijing 100089, China;
[3]Beijing Academy of Science and Technology Key Laboratory of Analysis and Testing Technology, Beijing 10089, China.
Corresponding Email: guoxialihua@163.com, No. 7 Fengxian Middle Road, Haidian District, Beijing 100094

**Abstract.** Increasing use of composite structures in transport industry puts forward new requirements for its maintenance and repair. However, bonding repair process will produce inevitable debonding flaws, and there will also be debonding damage during use service. In this paper, mechanisms properties of compression failure of patch repaired composite structures were explored, and effects of flaws on the properties were analysed. The results showed that buckling of patch repaired structures first occurred in compression process and then failed. Patch repaired structures broke at the damage hole. Under compression load, debonding flaw of patch repaired composite structures would be expanded, which caused the patch and the parent laminates to be detached earlier, and compression bearing capacity became weaker.

## 1. Introduction

The application of composite materials on aircraft structures is increasing dramatically in the recent years, and it is becoming more important to develop bonding techniques for composite structures. Adhesively bonding technology provides advantages such as weight reduction over traditional mechanical joining methods. Adhesively bonding technology can reduce stress concentrations and thus increase fatigue and damage resistance of bonded structural assemblies. Bonding technology is usually applied to reinforcement and repair. Bonded repair methods include patch repair method, scarf repair method and injection repair. The advantage of patch repair method is that the process is simple and high-efficiency, and the damage to parent laminates is small.

Recently, the mechanical properties of ideal flaw free bonding structures and repair structures have caused widely public concern. Hu F Z et al. [1] examined the compressive behaviour of bonded external patch repairs. The problem of a laminate plate with an elliptical hole repaired by elliptical patches under in-plane and/or bending load is investigated by Engels H [2]. Liu X et al.[3] studied the tensile behavior of open-hole composite plates bonded with external composite patches, experimental tests and developed a 3-D progressive damage model. Three types of final failure modes of these repaired structures subjected to tensile loads are concluded, with each corresponding to a different damage progression process. Cheng P C et al. [4, 5] examined the tensile behaviour of composite structures repaired by bonding external patches. Guo X et al. [6] studied the tensile behavior of composite laminate containing penetrated damage and the optimization of repair designing parameters after being flush repaired on double-sided. Soutis C et al. [7] examined the compressive behaviour of bonded external patch repairs; the compression loading mode is more severe than the tensile mode owing to the instability of delaminated plies. Jiang Y P et al. [8] investigate the efficiency of repaired panels in this paper, and then modeling of the mechanical behavior of the repaired composite panel under compression static load is conducted by using of the finite element method.

However, bonding process will produce inevitable debonding flaws, and there will also be debonding damage during the use process. At the same time, many scholars have also studied the mechanical properties of two dimensional adhesive joints with flaws.

Kan H P et al. [9] developed a stress analysis technique for load transfer in metal-to-composite adhesively bonded step-lap joints with bondline flaws. Rossettosetal J N et al. [10] developed an analytical model to compare the effects of voids and debonds on the interfacial shear stresses between the adherends and the adhesive in simple lap joints. Cheuk P T et al. [11] analyzed the effect of cracks embedded in strap adherend on the lap shear joints. The failure loads and failure modes of specimens with different length of cracks are determined in the experimental study. de Moura M F S F [12] evaluated the influence of strip flaws on the mechanical behaviour of composite bonded joints. It was verified that specific strength of the joints was not affected by the size of the flaw. Wang C H [13, 14] conducted experiments on scarf joints containing disbonds of varying lengths. The results showed that the load-carrying capacity of scarf joints decreases with the size of the bondline flaw at a faster rate than the reduction in the effective bond area. The author of this article has carried out an investigation into the effects of debond flaws on the mechanical properties of adhesively bonded single lap joints and composite scarf joints [15, 16].

However, under the action of compression load, the research on the defect propagation mechanism of 3D bonded repair structure is very scarce. Therefore the compression characteristics of patch repaired composite structures were studied in this paper. The mechanism of compression damage expansion was analyzed and the relationship between strength and flaws was discussed.

## 2. Experiment Setup

### 2.1. Specimens

Specimens of patch repaired composite structures are fabricated with X850 prepreg material. Ply sequence of parent laminates is $[45/0/-45/90]_{2s}$. Diameter of the preset pierce hole is 20mm. Diameter of the repair patch is 80mm. Ply sequence of the patch is $[45°/0°/-45°/90°]_s$.

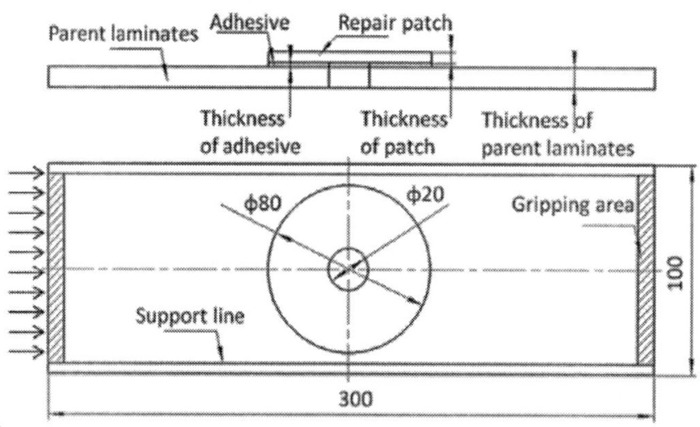


**Figure 1.** Schematic diagram of compressing specimen for patch repair structures (Unit: mm).

The parent laminates and the repair patch were bonded together with each other by J116B adhesive film. Schematic diagram of repair patch and parent laminates are shown in figure 1. In the sample processing stage, pieces of teflon films are inserted between the adhesive layer and the patch to form a preset debonding flaw. The flaws are located at the middle point of the long axis and the middle point of the short axis, as shown in figure 2.

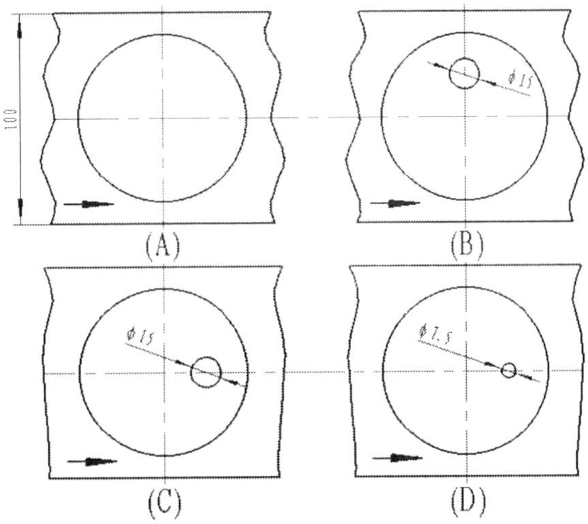


**Figure 2.** Schematic diagram of flaw location (Unit: mm).

*2.2. Test Equipments*
All the specimens were subjected to longitudinal compression loads on a 10 ton electric testing machine as shown in figure 3(a). Strain measurement and data acquisition were carried out by DH3821 static strain test system developed in Donghua as shown in figure 3(c).

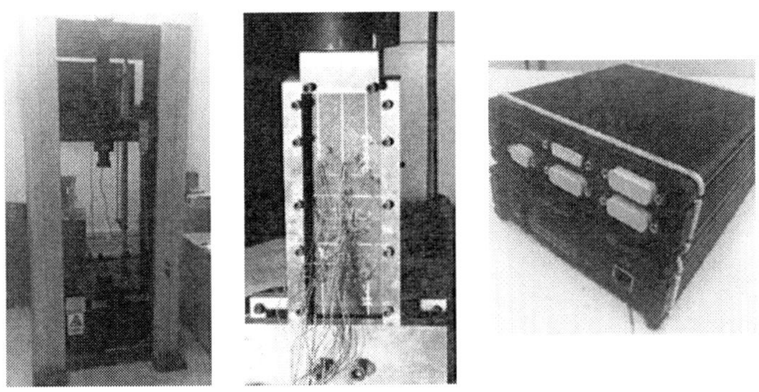

(a)WDW-100E (b) Test fixture (c)DH3821

**Figure 3.** Test equipment and fixture.

Compression test standards for adhesive bonded repair structures have not been established. Compression tests for composite bonded repair structures referred to the standard of compression test CAI for laminates with impact damage, which is ASTM D7137"Standard Test Method for Compressive Residual Strength Properties of Damaged Polymer Matrix Composite Plates"[17]. A test fixture was designed according to the compression test fixture after impact as shown in figure 3(b). The two sides of fixture were used to constrain the side deformation and prevent the overall buckling of the specimens.

## 3. Discussions and Results

### *3.1. Patch Repaired Composite Structures without Flaws*

The compression strength of patch repaired composite structures without flaws is 257.94MPa, and the dispersion coefficient is 6.02%. The failure modes are the repair patch peeled off, and first layer of the parent laminates stripped at first. Then the parent laminates crushed along the hole edge, as shown in figure 4.

The load displacement curve of the specimen is shown in figure 5. The slope of the load displacement curve is equivalent to the compression stiffness of the structure. In order to analyze the variation of the compression stiffness during loading, a straight line was drawn along the initial slope of the load displacement curve. It can be seen that the slope of the curve decreases obviously when the load reaches 58kN. The reduction of the structure stiffness means that the structure began to buckle from this point. When the load increased to 75kN after buckling, the whole structure failed. Post buckling stages of the structure was relatively short.

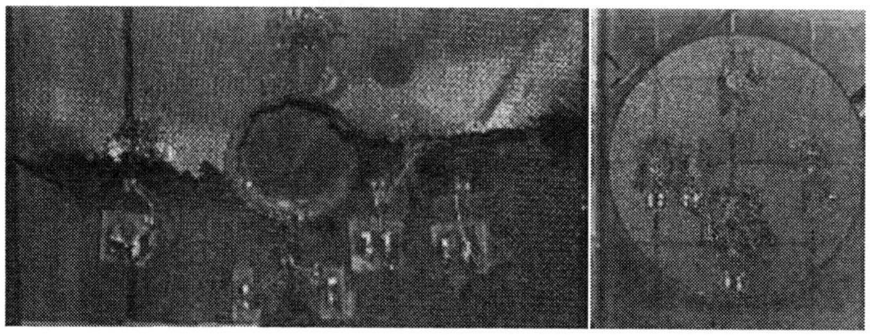

**Figure 4.** Failure modes.

In order to monitor the progressive damage process of the specimens, strain gauges were pasted at the corresponding positions on both sides of the specimens, as shown in figure 6. The positive and negative sides of the strain gauge were arranged symmetrically. The strain gauge x1 ~ x12 were arranged on the patch side of the specimens, and the strain gauge x101 ~ x112 were arranged on the other side of the specimens.

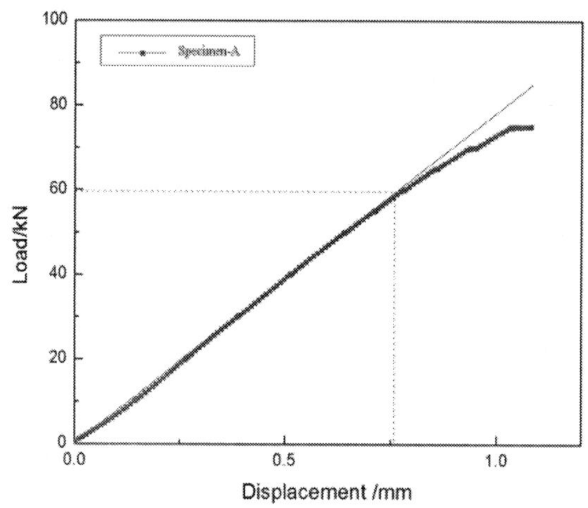


**Figure 5.** Load displacement curve of patch repaired composite specimen without flaws.

The strain gauges outside the repair area were monitored and used to adjust the load uniformity of the two sides to reduce the bending effect, as shown in figure 6(a). When the load was greater than 58kN, the strain difference of those strain gauges in figure 6(a) increased gradually. This indicates that the structure started to buckle, which caused local bending.

From figure. 6 (b), it can be observed that the compression strain values of 105# and 107# on the parent laminates side are not increased with the increase of 104# and 106# when the load is greater than 58kN. The compression strain values of 5# and 7# on the patch side are not increased with the increase of 4# and 6# when the load is greater than 58kN. At the same time, the value of strain gauge on the patch side was found to increase gradually to positive, while the value strain gauge on the other side continue to decrease, as before. The deformation on the long axis was smaller than that of the short axis. It can be explained that the buckling of the patch also occurred, and the external surface bears tensile action.

After buckling, when the load reached 70kN, the strain difference between 4#, 6# and 9# increased sharply, indicating that the patch and parent laminates were detached.

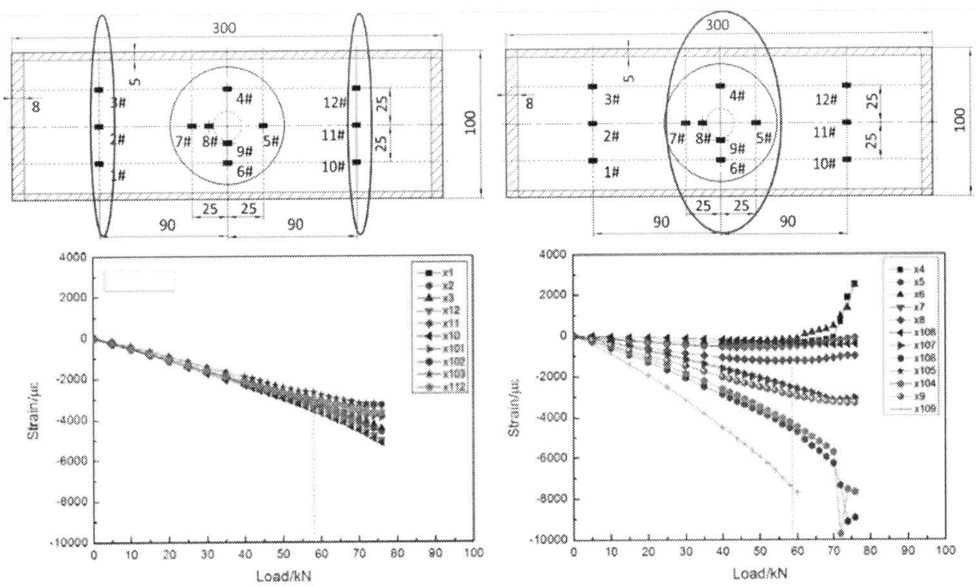

(a) Strain gauges outside the repair area (b) Strain gauges in the repair region

**Figure 6.** Strain load curve of patch repaired composite specimen without flaws.

The above results show that the structure buckled during the loading process of the repaired structure. The repair structure is weak in resistance to buckling, and then the overall failure occurs quickly.

### *3.2 Patch Repaired Composite Structures with Flaws*

The failure strength of all the compression specimens are shown in figure 7. When flaw is located at the long axis position, the flaw size has small effect on the compression properties of the structure. When the flaw is located at the short axis, the failure path of the structure will exceed the flaw, which will aggravate the failure of the structure.

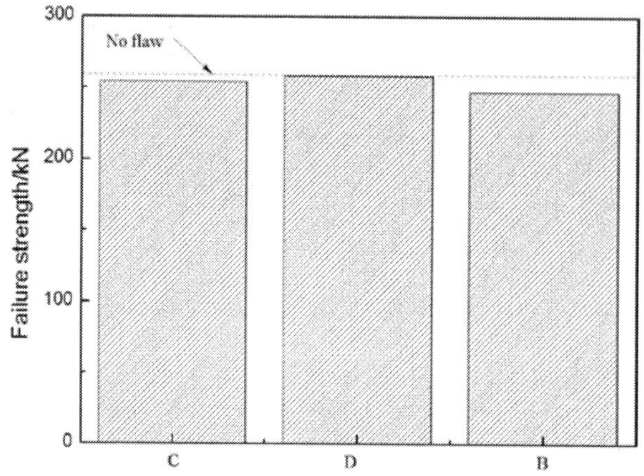


**Figure 7.** The failure strength of all the compression specimens.

The 4#, 5#, 6# and 7# strain differences between the parent laminates and the patch were taken to analyze the effect of the flaws, and the strain load curve was drawn as shown in figure 8. The strain differences values of 5# and 7# on the long axis are smaller than the strain differences values of 4# and 6# on the short axis. It is indicated that the strain value differences on the long axis is smaller than those on the short axis. After 70 kN, the strain difference on the short axis increased sharply, which indicates that the break onset position of the patch and the parent laminates was located next to the short axis. After 70 kN, the slope of the strain differences load curve on the long axis did not change much. This shows that there is no separation between the patch and the motherboard at the long axis.

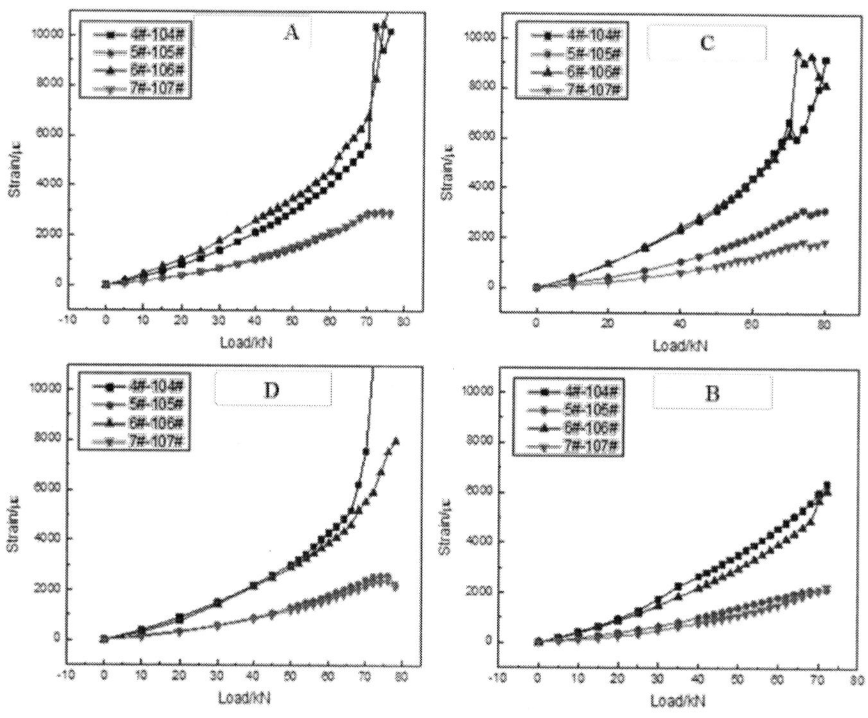


**Figure 8.** Strain loading curve for all the compression specimens.

## 4. Conclusions

In view of the compression properties of composite bonded repair structure, experimental study was carried out to explore the failure mechanism under compression load. At the same time, the influence of different locations of debonding flaws on the compression properties of patch repaired composite structures was studied, and the mechanism of flaw influence and the mechanism of flaw expansion were grasps. Flaw is located at the long axis position of the repair structure, and the debonding size has smaller effect on the compression properties of the structure. When the flaw is located at the short axis, the influence on the structure properties is greater. The flaw located at the path of damage extension speeded up the damage propagation of repaired structures. The debonding flaw may expand under the compression load, which causes the patch of the repair structure to be separated from the parent laminates earlier, and the compression bearing capacity becomes weaker.

## References

[1] Hu F Z and Soutis C 2000 *Compos. Sci. Technol.* **60** 1103
[2] Engels H and Becker W 2002 *Compos. Struct.* **56** 259
[3] Liu X and Wang G P 2007 *Compos. Struct.* **81** 331
[4] Cheng P C, Gong X J, Hearn D and Aivazzadeh S 2011 *Compos. Struct.***93** 582
[5] Cheng P C, Gong X J, Aivazzadeh S and Xiao X R 2014 *Polym. Test.* **34** 146
[6] Guo X, Guan Z D, Liu S, Liu J and Liu W P 2012 *Acta Materiae Compositae Sinica* **1** 176
[7] Soutis C, Duan D M and Goutas P 1999 *Compos. Struct.* **45** 289
[8] Jiang Y P and Yue Z F 2006 *Materialwiss. Werkstofftech.* **37** 597
[9] Kan H P and Ratwani M 1983 *J. Aircraft* **20** 848
[10] Rossettos J N, Lin P and Nayeb-Hashemi H 1994 *J. Eng. Mater. Technol.* **116** 533
[11] Cheuk P T, Tong L Y 2002 *Compos. Sci. Technol.* **62** 1079
[12] de Moura M F S F, Daniaud R, Magalhães A G 2006 *Int. J. Adhes. Adhes.* **26** 464
[13] Wang C H, Goh J Y, Ahamed J, Glynn A and Georgiadis S 2011 Jeju, Korea: *18TH International Conference on Composite Materials* 1
[14] Goh J Y, Georgiadis S, Orifici A C and Wang C H 2013 *Composites Part A* **55** 110
[15] Guo X, Guan Z D, Nie H C, Tan R M and Li Z S 2017 *J. Adhes.* **93** 216
[16] Guo X, Li Z S, Zhang W C, Tan R M and Guan Z D 2014 *Adv. Mater. Res.* **1016** 95
[17] ASTM D7137/D7137M-17 2017. America: *Annual Book of ASTM standards*, 2017.

# Effects of $AlCl_3$ on the Crystal Morphology of Calcium Sulfate Whisker Prepared from FGD Gypsum

**Xiao Wang[1,a], Biao Jin[1,a], Zhuoyue Xu[1,a], Xiaoting Zhang[1,a], Xueping Liu[1,a] and Liushuan Yang [1,2,a]**

[1] Department of Chemical & Material Engineering, Henan University of Urban Construction, Pingdingshan 467036, Henan Province, China
[2] School of Materials Science & Engineering, Henan University of Science and Technology, Luoyang 471003, Henan Province, China
[a]E-mail: wx407@126.com, [b]E-mail: liushuan@hncj.edu.cn

**Abstract.** Utilizing the purified flue gas desulfurization (FGD) gypsum as raw material and $AlCl_3$ as crystal modifier to prepare calcium sulfate whiskers (CSWs) by hydrothermal method, the effects of the $AlCl_3$ concentration and the pH of reaction slurry on the crystal morphology of CSW were discussed. The results show that the $AlCl_3$ concentration and the pH of reaction slurry had a significant effect on the crystal morphology of CSWs. CSWs grown in the $AlCl_3$-$H_2O$ slurry had diameters ranging from 2 to 6 um and aspect ratio above 80 μm, and their phase structure was calcium sulfate hemihydrate (HH).

## 1. Introduction

With the environmental protection, much attention has been given to the preparation of CSWs from flue gas desulfurization (FGD) [1-3]. However, compared with the CSWs prepared from chemical agents or high quality gypsum, the quality of CSWs prepared from FGD gypsum need still be improved. Because of the sparing solubility of FGD gypsum, it has no effect to control its crystallization by changing the temperature and super saturation of the reaction solution. According to reported studies, the crystal morphology of CSWs could be controlled by a proper crystal modifier, which maybe leaded to improve the quality of CSWs [4, 5]. Some metal ions, such as $Na^+$, $K^+$, $Mg^{2+}$, $Cu^{2+}$, etc. have a significant effect on the crystal morphology and size of CSWs [2, 6-9].

However, the effects of each metal ion on the crystal morphology and size of CSWs were completely different. Therefore, using $AlCl_3$ as crystal modifier, the influences of $AlCl_3$ concentrations and pH of reaction slurry on the crystal morphology were investigated to improve their quality in this paper.

## 2. Experimental

### *2.1. Sample Preparation*

The FGD gypsum used in this study was supplied by a power plant in Henan Province, China. Aluminum chloride ($AlCl_3$, >99 % purity), Sodium hydroxide (NaOH, >99 % purity) and sulfuric acid ($H_2SO_4$, density=1.84 g/cm$^3$) were provided by Luoyang Chemical Reagent Co. Ltd., China.

Based on our previous studies [2, 6, 8], the purified FGD gypsum was ball-milled and then shifted into an autoclave. Sulfuric acid was used to adjust pH, meanwhile, a scheduled amount of $AlCl_3$ crystal modifier was added into the slurry. The mixed slurry was heated at 120~140 °C for 90 min.

Finally, the hydrothermal products were rapidly filtered, washed and dried to constant weight at 80~200 °C.

### 2.2. Characterizations

The preliminary observations of samples were detected by a digital biological microscope (BA200, made by MOTIC China Group CO., LTD.). The morphology of samples was observed using a scanning electron microscope (SEM, QUANTA 450, made by FEI). The phase analysis of samples was determined using an X-ray diffracttometer with nickel-filtered Cu Kα radiation, λ=0.154 178 nm, 2θ=5°~80° (XRD, X'Pert PRO MPD, made by PANalytical B.V.).

## 3. Results and Discussion

Figure 1 shows the effect of $AlCl_3$ concentrations on the crystal morphology of CSWs prepared from the purified FGD gypsum by hydrothermal method.

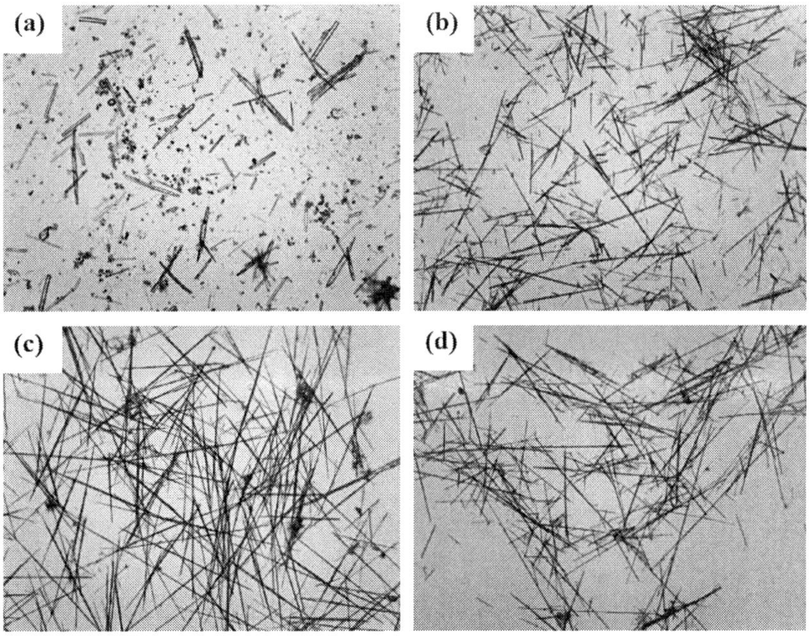


**Figure 1.** Microscopy images of the samples prepared at different $AlCl_3$ dosage (×100): (a) 0 %, (b) 2.0 %, (c) 4.0 %, (d) 6.0%.

Almost no CSWs could be obtained in the reaction slurry without the additive of $AlCl_3$, and a large of short columnar and gradual products could be easily observed from Figure 1(a). With the increment of $AlCl_3$ concentration, short columnar and gradual products were gradually reduced. At an $AlCl_3$ concentration of 2.0 %, the crystallization of products obviously changed from the short columnar and granular morphology to the needle morphology, but the crystal distributions in terms of diameter and length were not uniform. As the $AlCl_3$ dosage was 4.0 %, the obtained hydrothermal products were relatively best and mainly presented needle morphology, and its mean aspect ratio was about 80. However, with further increases of $AlCl_3$ concentration, the mean aspect ratio was reduced instead. This indicates that the $AlCl_3$ concentration has a significant effect on the crystal morphology of CSWs.

Due to the sparing solubility of gypsum, pH value has a significant effect on the solubility of the FGD gypsum, which would further affect the crystal morphology of hydrothermal products. Therefore, it was necessary to study the effects of pH on the crystal morphology of hydrothermal products. Figure 2 shows the influence of pH on the morphology of hydrothermal products.

As adding $10^{-5}$ M sulfuric acid or sodium hydroxide into the reaction slurry, the whiskers with a small diameter could be obtained, but their aspect ratio was relatively low. In both reaction systems, many short columnar and gradual products could still be easily found. Moreover, some fascicular crystals could be observed when added sodium hydroxide into the reaction slurry. Neither sulfuric acid nor sodium hydroxide was added into the reaction slurry, the obtained calcium sulfate whiskers had bigger lengths and smaller diameters, and the distribution of length and diameter was more uniform. Therefore, pH also had a significant effect on the crystal morphology of the hydrothermal products when using FGD gypsum as raw material.

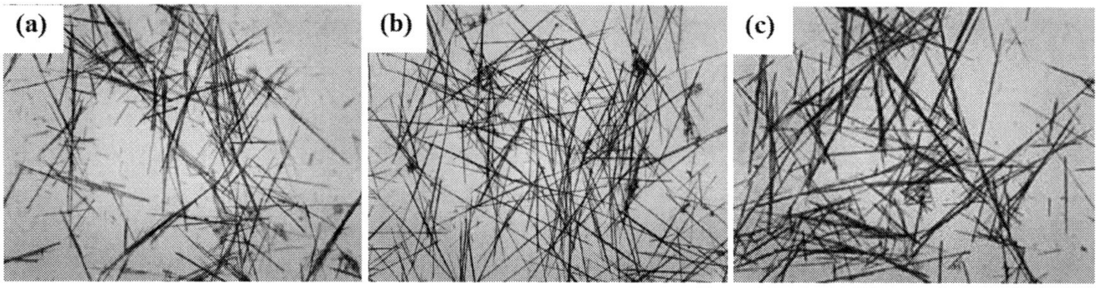


**Figure 2.** Microscopy images of the samples prepared at different pH and with $AlCl_3$ as crystal modifier (×100): (a) $H_2SO_4$, $10^{-5}$ M, (b) $H_2SO_4$, 0 M, (c) NaOH, $10^{-5}$ M.

Due to in the $AlCl_3$-$H_2O$ system, the obtained calcium sulfate whiskers have a relatively uniform crystal morphology. Therefore, it is necessary to detect its microstructure. Figure 3 is the XRD pattern of whisker sample prepared in the $AlCl_3$-$H_2O$ system conditions. It demonstrates that the phase of products is hemihydrate sulfate calcium and no second phase can be detected. The previous investigations had also reported that metal ions could markedly effect the crystal morphology of CSWs, but no effects on their phase, which could be testified by Figure 3 [2, 6-9].

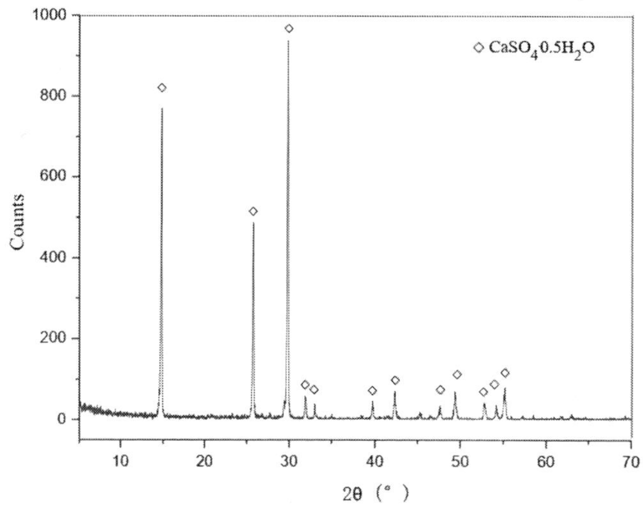


**Figure 3.** XRD pattern of the sample prepared in the $AlCl_3$-$H_2O$ system.

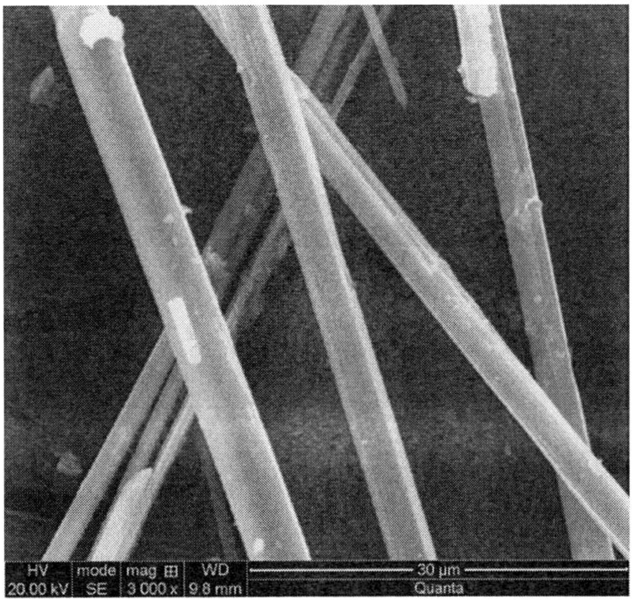


**Figure 4.** SEM images of the sample prepared in the $AlCl_3$-$H_2O$ system.

Figure 4 is the SEM images of CSWs prepared in the $AlCl_3$-$H_2O$ system conditions. Only a few of gradual products can be observed besides CSWs when 4.0 % $AlCl_3$ is added into the reaction slurry from Figure 4. Simultaneously, a small amount of precipitates on the surface of CSWs could also be detected, which may be caused by the precipitation of the impurities of FGD gypsum [2].

Combining the microscopy images in Figure 1, 2 and 4, it can be found that CSWs of a diameter about 2~6 um and aspect ratio above 80 can be obtained when 4.0 % $AlCl_3$ is added into the reaction slurry, which indicates that the CSWs prepared from FGD gypsum can be effectively improved when using $AlCl_3$ as crystal modifier.

## 4. Conclusions

Utilizing the purified FGD gypsum as raw material and $AlCl_3$ as crystal modifier, CSWs could be prepared. The $AlCl_3$ concentration and the pH of reaction slurry had a significant effect on the crystal morphology of CSWs. In the $AlCl_3$-$H_2O$ slurry, CSWs with relatively uniform crystal morphology could be obtained. The obtained CSWs had diameters ranging from 2 to 6 um and the aspect ratio above 80 μm, and their phase structure was calcium sulfate hemihydrate (HH).

## Acknowledgements

The authors would like to express thanks to the national nature Science Foundation of China (NSFC, 51674097) for the financial support. And the authors would like to appreciate financial supports of the Henan Provincial Natural Science Foundation (142102210387, 15210221009 and172102310729).

## References

[1] A Y Xu, H P Li, K B Luo and L Xiang, Res. Chem. Intermed. 37 (2011) 449-455.
[2] X Wang, L S Yang, X F Zhu and J K Yang, Particuology, 17 (2014) 42-48.
[3] C J Liu, Q Zhao, Y G Wang, P Y Shi and M F Jiang, Appl. Surf. Sci. 360 (2016) 263-269.
[4] S K Hamdona and U A Al Hadad, J. Cryst. Growth, 299 (2007) 146-151.
[5] S T Sargut , P Sayan and B Kiran, Chem. Eng. Technol. 33 (2010) 804-811.
[6] X Wang, L S Yang, X F Zhu and J K Yang, J. Synth. Cryst.(in Chinese), 42 (2013) 2661- 2667
[7] 9 Z T Yuan, X L Wang, Y X Han and W Z Yin, J. Northeast Univ. (Nat. Sci.)(In Chinese) 29 (2008) 573-576.
[8] X Wang, B Jin, L S Yang and X F Zhu. Cryst. Res. Technol. 8 (2015) 1-8.
[9] X L Mao, X F Song, G M Lu, Y X Xu, Y Z Sun and J G Yu, Chem. Engin. J. 278 (2015) 320-327.

# Study on Preparation and Photocurrent Response Properties of In/$In_2O_3$/$TiO_2$ Nanotubes Arrays Compound Heterojunction Semiconductor

**Yu Tan, Yunfei Xu, Kexin Liang and Shenghan Zhang**

Department of Environmental Science and Engineering, North China Electric Power University, Baoding 071003, Hebei Province, P. R. China
Email: 654526855@qq.com

**Abstract.** TNTs were prepared in ethylene glycol with 0.5wt%NH4F and 1.5Vol%$H_2O$ by the anode potential of 60V. In/$In_2O_3$/$TiO_2$ nanotube arrays (TNTs) heterojunction semiconductor was prepared by two electrochemical steps of reduction and oxidation in 0.01 mol/L $InCl_3$ ethanol solution. The element In and $In_2O_3$ nanofibers were distributed in and on the TNTs by SEM figures. By the photocurrent results, TNTs photocurrent response was reinforced by $In_2O_3$ and In modification in UV and visible light region. The monochromatic incident photon-to-electron conversion efficiency (IPCE) reached to 58% under the 350nm light radiation and was nearly 5 times that of TNTs. Two oxidation phases were got from the photocurrent data (the energy band gap Eg=3.08eV and Eg=2.2 eV), which was propitious to the absorption light red shift and response under the visible light. The flat band potentials shifted to electric positive direction for In/TNTs prepared by one step of reduction or In/$In_2O_3$/TNTs prepared by two steps of reduction and oxidation according to the Mott-Schottky curve. The donor density (ND) of the In/$In_2O_3$/TNTs was slightly higher than TNTs and the ND of In/TNTs was the highest for 14.6 times that of TNTs. The results show that heterojunction is formed between $In_2O_3$ and TNTs to promote visible light absorption and carrier separation. A great deal of element In makes the photo generated electrons and holes recombination to reduce the photocurrent response, but a small amount of element In can reduce the Eg of semiconductor and promote the absorption light to red shift for TNTs. The transfer mechanism of In and $In_2O_3$ modified by TNTs was discussed.

As a highly efficient photocatalyst, $TiO_2$ has received extensive attention, but its wide bandgap ($E_g$=3.2eV) limits the use of sunlight. Therefore, the modification research is widely carried out at present, and the modification mainly from two aspects. On the one hand, morphology modification of $TiO_2$ itself, and to form nano-particles, nano-wires, nano-clusters, nano-tubes, etc., increase its specific surface area and aspect ratio to increase photocatalytic efficiency, the $TiO_2$ nanotube array has become a research hotspot due to its neat tubular structure and a large specific surface area[1, 2]on the other hand, it is doped or added with other substances on $TiO_2$, such as deposition of metal elements or doped metal ions (Pt, Au, Cu, Pd, etc.) [3-6], doped non-metallic substances (C, N, etc.) [7,8], formation of heterojunctions with semiconductors such as metal oxides or sulfides[9-12], combined with organics [13] and other modification methods to improve the efficiency of light carrier separation or reduce the semiconductor band gap, enhance the utilization of visible light. In recent years, researchers have organically combined varieties of modification methods to prepare composite semiconductor catalysts [13-20], which can effectively promote the separation of photogenerated electrons and holes in $TiO_2$, and can also make the absorption light red-shift. Researchers composite semiconductor catalysts in order to effectively use the sun and play very good results.

$In_2O_3$ is a new n-type semiconductor functional material, with a bandgap width of Eg=2.8eV, a small resistivity, and a high catalytic activity, and has been widely used in the fields of optoelectronics, gas sensors, and photocatalysts [21]. The $In_2O_3$ particle size up to the nano level also has the surface effects of nanomaterials, quantum size effects, small size effects and macroscopic quantum tunneling effects in addition to the above functions. As an oxide semiconductor photocatalyst, $In_2O_3$ has similar photocatalytic properties as $TiO_2$, but its narrower band gap can also promote the absorption of light from the UV region to the visible region [22]. Liu et al.[23] prepared $In_2O_3$ nanorods/Ti-Ni-O nanomaterials Heterojunction photocatalyst can efficiently Photoelectrocatalytic pyrolysis of water.

In this paper, In/$In_2O_3$/$TiO_2$ nanotube array heterojunction composite semiconductor catalysts were prepared by electrochemical method. The photoelectrocatalytic performance of $TiO_2$ nanotube arrays is improved by the combination of metal In and oxide semiconductor $In_2O_3$.

## 1. Experiment

### *1.1. Preparation of $TiO_2$ Nanotube Arrays*

Titanium foil (10×10×0.2 mm) was ultrasonically cleaned with anhydrous ethanol and deionized water for 20 minutes respectively, then polished for 5 seconds with chemical polishing solution which the composition is V (HF): V ($HNO_3$): V ($H_2O$) =1:4:5, at last the treated sample was rinsed and dried with deionized water. Two-electrode anodization was used in the experiment, the titanium foil was an anode and platinum was a cathode. The ethylene glycol was added 0.5 wt% of $NH_4F$ and 1.5 Vol% of $H_2O$ as the electrolyte and magnetic stirring. Using phased voltage, the first stage uses a linear voltage, from 0V to 60V, for 60 seconds, the second stage is to maintain a constant voltage of 60V, and maintain the oxidation for 3600 seconds. Then it is annealed at 450°C for 2 hours, stored in a desiccator after natural cooling, prepared samples are referred to as TNTs.

### *1.2. Preparation of In/$In_2O_3$/$TiO_2$ Nanotube Array Heterojunction Semiconductors*

A certain amount of $InCl_3$ was dissolved in an absolute ethanol solution, and a solution with an $In^{3+}$ concentration of $1\times10^{-2}$ mol/L was configured. The $TiO_2$ nanotube array was placed in a sample holder, and the surface of the sample was sealed with an organic tape and leave 0.2826cm$^2$ of exposed area in contact with the solution. The sealed sample was placed in the holder and immersed in above solution, static adsorption 30min. Then, the $TiO_2$ nanotube array was used as the working electrode, the saturated calomel electrode (SCE) with capillary was used as the reference electrode, Pt was the negative electrode for indium deposition. The applied voltage was -6V and the electricity reached 0.01C. This sample was recorded as TNTs. -In. Subsequently TNTs-In was oxidized with an oxidation potential 1 V and oxidized charge of 0.001 C. The prepared samples were recorded as TNTs-InOx.

### *1.3. Morphologies Analysis*

Characterize the morphologies of $TiO_2$ nanotube arrays and their modified catalysts. Cold field emission scanning electron microscope (JSM-4800F) was used to characterize the composition of the sample. X-ray photoelectron spectroscopy (XPS, MT-500, Microtech) was used. Using the Al detector (C1s located at 284.8eV).

### *1.4. Photoelectric Response Performance and Mott-Schottky Test*

Photoelectric response performance test using 450W xenon lamp and CT110 monochromator, PARSTAT2273 potentiostat optocoupler test system, in a three-electrode electrolytic cell with a quartz window, the modified $TiO_2$ nanotube array electrode prepared above as a working electrode, the platinum film is an auxiliary electrode. The saturated calomel electrode connected by salt bridge is a reference electrode. Under the open circuit potential, a 0.1 mol•L$^{-1}$ $Na_2SO_4$ solution is used as the supporting electrolyte; the tests are performed at room temperature.

IOP Conf. Series: Materials Science and Engineering 472 (2019) 012006 doi:10.1088/1757-899X/472/1/012006

## 2. Results and Discussion

### *2.1. Characterization*

Figure 1 shows the $TiO_2$ nanotube array and the modified sample surface topography. It can be seen from Figure 1a that the $TiO_2$ nanotube array is well-arranged, the tube diameter is about 100 nm, and the tube wall thickness is about 10 nm. Figure 1b is the morphology of TNTs-In surface which after the reduced indium modification. It can be seen that the modified materials are distributed in the form of nanofibers on the surface of the $TiO_2$ nanotube array and in the $TiO_2$ tube. Figure 1c shows the surface morphology of indium-modified TNTs-$InO_x$. Compared with Figure 1b, it can be seen that the modified substances are also distributed on the surface of the $TiO_2$ nanotube array and in the $TiO_2$ tube after oxidation, but the nanofibers are shortened and contact status is not obvious. Figure 1d shows the lateral morphology of the damaged TNTs-$InO_x$ membranes. The film thickness is approximately 7-8 μm. The top of the membrane to the bottom of the membrane is in the form of a tube.

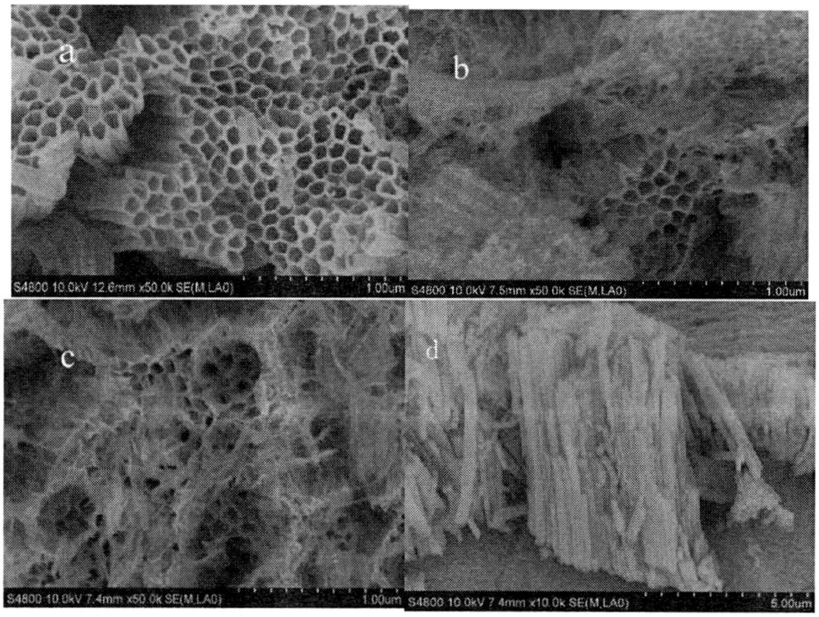


**Figure 1.** SEM images of the samples (a) TNTs (b) TNTs-In and (c) TNTs-$InO_x$(d)cross-section of TNTs-$InO_x$

Figure 2 shows XPS spectra of the surface composition of the TNTs-$InO_x$ samples. Figure 2a clearly shows that TNTs-$InO_x$ contains characteristic peaks for C1s, In3d, Ti2p, and O1s, and Figure 2b shows Ti2p3/2 and Ti2p1/2 characteristic peaks at 458.9 eV and 464.6 eV, respectively, these peaks show that Ti exists as a typical $TiO_2$ form [24]. Figure 2c shows the In3d5/2 characteristic peak. Obviously, this peak has a tailing phenomenon. Therefore, according to the literature fit the peak [23], the In3d5/2 binding energy of elemental In is 444.2 eV and the In3d5/2 binding energy of $In_2O_3$ is 445.1eV, Oxide-state indium modified substances containing two kinds of matters that are In and $In_2O_3$.The elemental In and $In_2O_3$ accounts for 9.4% and 90.6% respectively, based on the fitted peak area. Figure 2d is a fitting plot of the O1s map. Fitting peak 1 at 529.9 eV is the characteristic peak of lattice oxygen. Fitting peak 2 at 531.3 eV is the characteristic peak of adsorbed oxygen. This adsorbed oxygen may come from the air adsorbed on the sample surface like $O_2$ and $H_2O$ [25, 26].

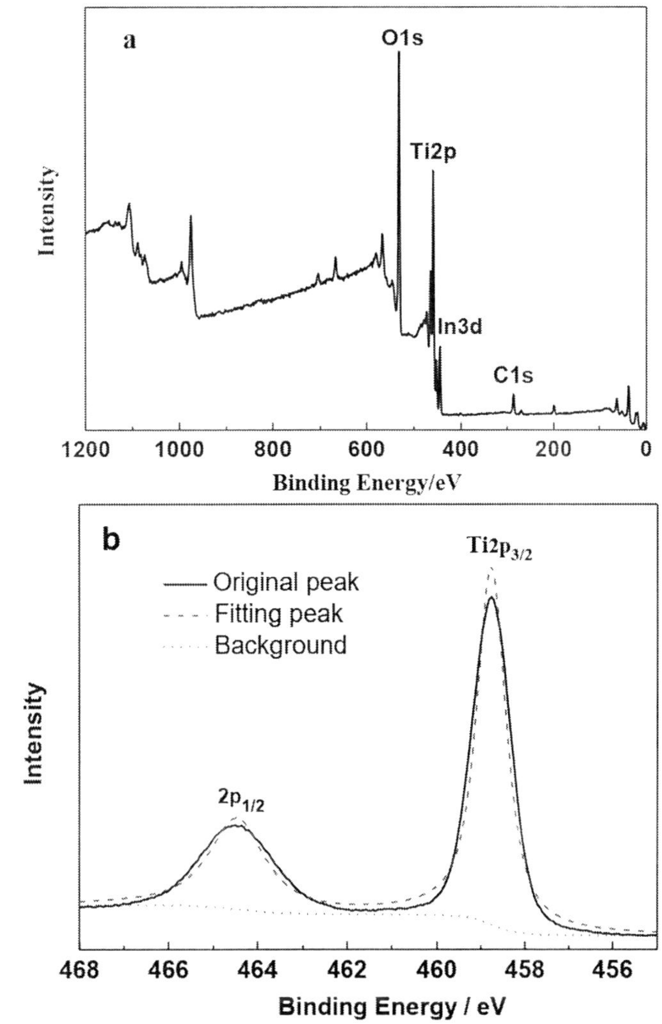
a
O1s
Ti2p
In3d
C1s
Intensity
1200
1000
800
600
400
200
0
Binding Energy/eV
b
Ti2p3/2
Original peak
Fitting peak
Background
2p1/2
Intensity
468
466
464
462
460
458
456
Binding Energy / eV

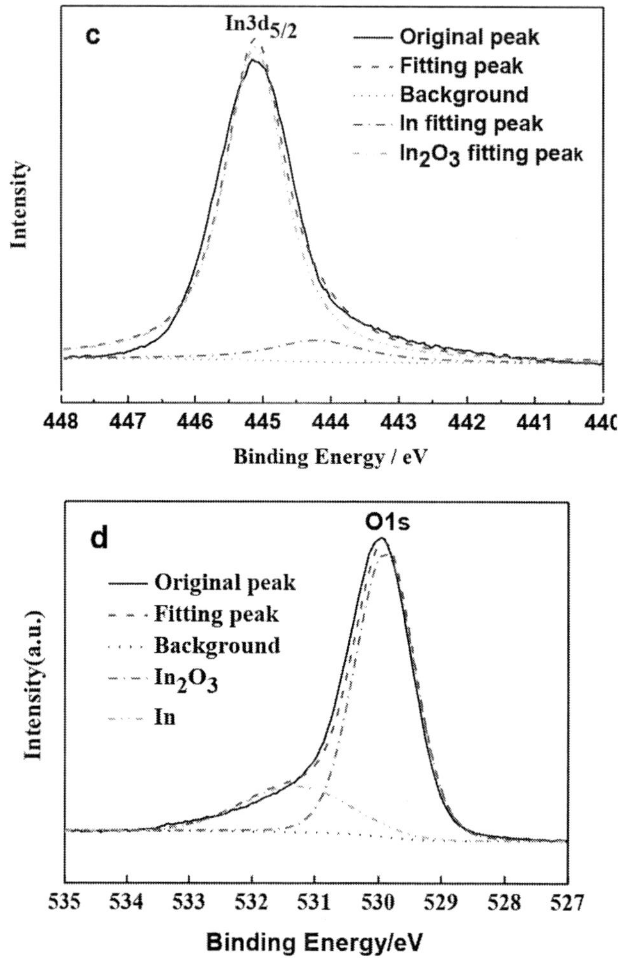


**Figure 2.** XPS spectra of (a) whole spectra (b) Ti2p (c) In3d and (d) O1s

*2.2. Photoelectric Response Performance Analysis*

Figure 3 shows the photocurrent spectrum of the modified $TiO_2$ nanotube array (λ= 250-600 nm), first in the ultraviolet (UV) part (λ= 250-400 nm), comparing the samples TNTs, the reducing In modified TNTs-In sample photocurrent response is weakened, but after the sample is oxidized, the photocurrent response of the TNTs-$InO_x$ in the UV part is significantly enhanced, and the wavelength of the highest photocurrent response is red-shifted.Figure 3 shows the photocurrent response spectrum of the visible part (λ= 400-600 nm). It can be seen from the figure that, compared with the $TiO_2$ nanotube array, the photocurrent response of the reduced indium modified TNTs-In is also not strengthened, but the photocurrent response has been enhanced after oxidation like the sample TNTs-InOx, this phenomenon indicates that the oxidized sample TNTs-$InO_x$ can effectively use visible light.

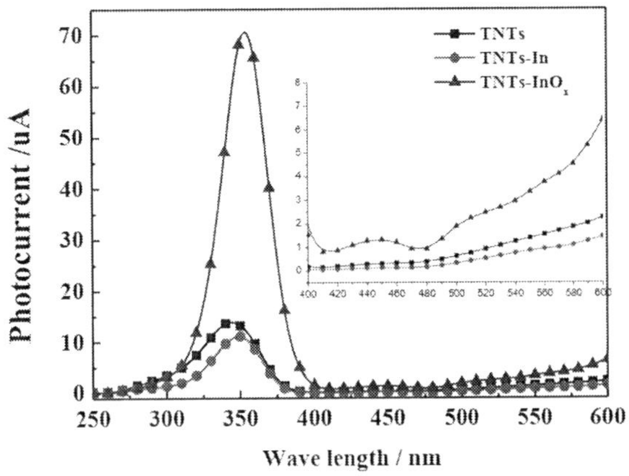


**Figure 3.** The photocurrent responses plots of $TiO_2$ nanotubes modified by In and its oxide

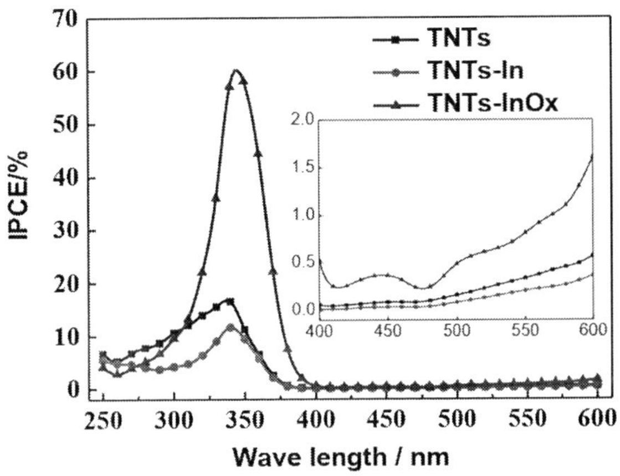


**Figure 4.** IPCE spectra of $TiO_2$NTs modified by In and its oxide

According to Figure 3, the monochromatic incident photon-to-electron conversion efficiency (IPCE) of the indium-modified $TiO_2$ nanotube array at each wavelength is converted. As shown in Figure 4, it can be seen that the $TiO_2$ nanotube array has the highest light quantum conversion efficiency of 11.7% at the wavelength of 340 nm. The $TiO_2$ nanotube array composite catalyst modified by In and $In_2O_3$ can achieve a photon conversion efficiency of up to 58% at a wavelength of 350 nm and five times that of a $TiO_2$ nanotube array. The photocatalyst conversion efficiency of this catalyst in the visible light region is also significantly enhanced.

According to the following formula (1) and the incident light intensity obtained by the Si photoelectrode, Figure 3 converts the relationship between the photocurrent response and the photon energy, as shown in Figure 5.

$$(I_{ph}h\nu / I_0)^{1/n} = A\,(h\nu - E_g) \qquad (1)$$

Where hv is the photon energy, $I_0$ is the light intensity of the incident light, and n is determined by the transition form of electrons between the valence band and the conduction band, for amorphous semiconductors, when the incident light photon energy is greater than its bandgap, the value is same as the indirect transition of the crystalline semiconductor, that is n=2, A is a constant, and $E_g$ is the bandgap width of the characteristic oxide phase. From the figure, it can be seen that TNTs and TNTs-In have only one oxidation phase energy band width. This is 3.18 eV, which corresponds to the band width of the anatase $TiO_2$, but after oxidation, the sample shows two oxidized phases of about 3.08 eV and 2.20 eV respectively.

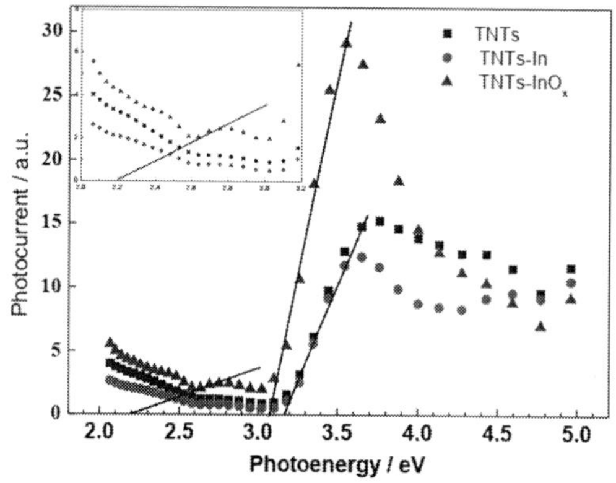


**Figure 5.** $(i_{ph}hv/I_0)^{1/2}$ vs. photon energy of TNTs modifying with In and its oxides

### *2.3. Analysis and Discussion*

The preparation process of the catalyst is shown in figure 6(a). A $TiO_2$ nanotube array film was prepared by electrochemical anodic oxidation. After annealing, In and $In_2O_3$ nanofibers were deposited in the film and on the surface and were formed $In/In_2O_3/TiO_2$ nanotube array heterojunction composite semiconductor catalyst.

Through the above experimental data analysis, the prepared reducibility In modified materials can not promote the photocurrent response of $TiO_2$ nanotube arrays, but instead reduce their photocurrent response. This may be because excessive elemental In has become a recombination center of photoelectrons and holes. The recombination center reduces the efficiency of photocarrier separation, resulting in a decrease in photoresponse. At the same time, these single In materials do not change the energy band gap width of $TiO_2$ nanotube arrays. However, after the oxidation, the photo-response is enhanced in both the UV and visible regions. On the one hand is due to semiconductor band theory: surface modification can form an additional energy level [27] in the original bandgap of $TiO_2$. The conduction band of $TiO_2$ is close to In doping level [28], and the electron transfer reaction mechanism is shown in figure 6(b). Through the above experimental data, it can be seen that the width of the semiconductor band after modification is reduced from 3.18 eV to 3.08 eV. It can be known from the XPS data that after the modified material is oxidized, the content of elemental In is smaller (9.4%), the introduction of a small amount of elemental In energy level in the band gap of $TiO_2$ enables photons with lower energy to excite electrons and holes which are trapped at the impurity level, causing red shifting of the absorption band edge of $TiO_2$, thereby broadening the absorption range of light by the semiconductor and at the same time the separation efficiency of photogenerated carriers is promoted, improve the utilization of photons; On the other hand, the reason is that oxidation generates $In_2O_3$ semiconductors (theoretical $E_g$=2.8 eV) can absorb visible light. The new energy band width is obtained from figure 4, which can calculate as $E_g$=2.2 eV. This may also be due to the effect of a small amount of elemental In on the $In_2O_3$ energy band, the $In_2O_3$ energy band width is reduced, and the absorbed light is further red shifted, and the sample is oxidized to form a heterojunction composite structure semiconductor catalyst $In/In_2O_3$/TNTs, which further improves the utilization of visible light and promotes photogenerated carrier separation.

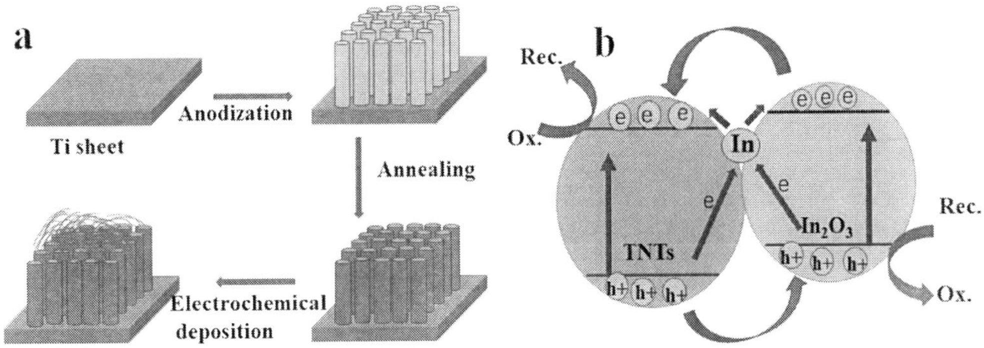


**Figure 6(a).** A schematic illustration of the synthesis process and structure of the catalyst and (b) a reaction mechanism diagramed for charge transfer in the $In/In_2O_3/TNTs$ photoanode.

## 3. Conclusion

Based on the above analysis, the $TiO_2$ nanotube array modified by reducing element In, although the carrier concentration is the highest, the single indium becomes the recombination center of photoelectrons and holes, it makes the photocurrent weakened; the $TiO_2$ nanotube arrays modified by the oxidation state of indium, form a heterojunction composite catalyst with $TiO_2$ nanotube arrays under the combined action of $In_2O_3$ and a small amount of elemental In, resulting in a reduction in the band width of the semiconductor catalyst and the photocurrent response enhancement in both the ultraviolet and visible regions.The modified material is conducive to light carrier separation and absorption of visible light. However, the modified materials make the flat band potential of the semiconducting catalyst become positive and reduce the reducibility of the photoelectron transition to the conduction band.

## References

[1] Paulose M, Mor G K, Varghese O K, Shankar K and Grimen C A 2006 *J.Photochem.Photobiol.,A* Visible light photoelectrochemical and water-photoelectrolysis properties of titania nanotube arrays **178** 8-15
[2] Gong J J, Lai Y K and Lin C J 2010 *Electrochim. Acta* **55** 4776-4782
[3] Chen H Y, Tang M N, Rui Z B, Wang X Y and Ji H B 2016 *Catal. Today* **264** 23-30
[4] Nam W; Han G Y 2007 *J. Chem. Eng. Jpn.* **40** 266-269
[5] Sakthivel S, Shankar M V, Palanichamy M, Arabindoo B, BahnemannD W and Murugesan V 2004 *Water Res.* **38** 3001-3008
[6] Wu N L, Lee M S. 2004 *Int. J. Hydrogen Energy* **29** 1601-1605
[7] Xu C K, Shaban Y A, Jr W B I, Khan U M. 2007 *Sol. Energy Mater. Sol. Cells* **91** 938-943
[8] Ohno T 2004 *Water Sci. Technol.* **265** 115-121
[9] Li C L, Yuan J, Han B Y, Jiang L and Shangguan W F 2010 *Int. J. Hydrogen Energy* **35** 7073-7079
[10] Cong Y Q, Li Z, Wang Q, Zhang Y, Xu Q and Fu F X 2012 *Acta Physico-Chimica Sinica* **28** 1489-1496
[11] Siripala W, Ivanovskaya A, Jaramillo T F, Baeck S-H and McFarland E W 2003 *Sol. Energy Mater. Sol. Cells* **77** 229-237
[12] Zhang Y, Liu Z Y and Zhai J 2011 *Acta Chim. Sinica* **71** 793-797
[13] Peng S Q, Zhang W, Li Y X, Lv G X and Li S B 2007 *J Funct Mater* **38** 362-365
[14] Rayalu S S, Dubey N, Labhsetwar N K, Kagne S and Devotta S 2007 *Int. J. Hydrogen Energy* **32** 2776-2783
[15] Dubey N, Labhsetwar N K, Devotta S and Rayalu S S 2007 *Catal. Today* **129** 428-434
[16] Li Y X, Xie C F, Peng S Q,Lu G X and Li S B 2008 *J. Mol. Catal. A: Chem.* **282** 117-123
[17] Yang H H, Guo L J, Yan W and Liu H T 2006 *J.Power Sources* **159** 1305-1309

[18] Sasikala R, Shirole A R, Sudarsan V, Jagannath, Sudakar C, Naik R, Rao R and Bharadwaj S R 2010 Appl. Catal. 377 47-54
[19] Ikeda M, Kusumoto Y, Yakushijin Y, Somekawa S, Ngweniform P and Ahmmad B 2001 Catal Comm. 8 1943-1946
[20] Gan Y P, Qin H P, Huang H, Tao X Y, Fang J W and Zhang W K 2013 Acta Phys-Chim Sin 29 403-410.
[21] Luo S, Feng J and Ng K M. 2013 Crysteng Comm. 16 9236-9244
[22] Hadia N M A, Mohamed H A. 2013 J. Alloys Compd. 547 63-67
[23] Liu Q, Ding D Y, Ning C Q and Wang X W 2016 Sol. Energy Mater. Sol. Cells 145 382-390
[24] Choi H J, Kang M 2007 Int. J. Hydrogen Energy 32 3841-3848
[25] You M, Kim T G, Sung Y 2009 Cryst. Growth Des. 10 983
[26] Wang C, Li Q, Wang F 2013 ACS Appl. Mater. Interfaces 6 1243
[27] Chen J T, Li X J, Yang Y, Wang L Y and He M X 2003 J Mater Scitechnol 21 67- 71.
[28] Long H J, Wang E J, Dong J Z, Wang L L, Cao Y Q, Yang W S and Cao S A 2009 Acta Chim Sinica 67 1533-1537.

# Exploring the Function Fatigue Behavior of Ferromagnetic Shape Memory Alloy

**Jia Ju[1,2,3], Yan Kang[1], Liguo Shuai[2*] and Chen Yan[4]**

1. Jiangsu Key Laboratory of Advanced Structural Materials and Application Technology, School of Materials Engineering, Nanjing Institute of Technology, Jiangsu, 211167, China
2. School of Mechanical Engineering, Southeast University, Nanjing, 211189, China
3. Hangzhou Kaierda electric welder CO., LTD, Hangzhou, 310000, China.
4. Zhongtian Alloy Technology CO., LTD, Nantong, 226010, China
Email: liguo. shuai@126.com

**Abstract.** In the industrial applications of ferromagnetic shape memory alloy, the large magnetic field induced strain and long function fatigue life are the essential conditions for their application. In this cases, the ferromagnetic shape memory effect and the function fatigue of $Ni_2MnGa$ under applied magnetic field has been investigated. The magnetic field induced strain appear typical nonlinearity increasing trend apparently with the applied magnetic field rising. Meanwhile, the magnetic field induced strain keep decreasing tendency as the applied magnetic field cycles increasing and the magnetic field induced strain kept a small value in a long cycle ranges of 400 to $10^7$. Furthermore, the sample appeared typical elastic deformation stage, twin martensite remigration stage and plastic deformation stage during compression test process.

## 1. Introduction

Since the first observation of the magnetic field induced strain in Ni2MnGa by Ullakko [1], the ferromagnetic shape memory alloys (FSMAs) have attracted a great attention as a kinds of intelligent material for actuator and sensor[2]. In past two decades, about 10% giant stains, biased on twin martensite variants remigration, were achieved in Ni2MnGa alloy [3]. However, the actuator and sensor made of Ni2MnGa has not getting a wide range of applications due to their function fatigue effect [4].

In industrial applications, the actuator and sensor must work a million times [5, 6]. In this case, the Ni2MnGa, as a drive material of actuator and sensor, should have a long function life (more than million cycles) [2, 7, 8]. Therefore, to clear the process of the magnetic field induced strain and to obtain a larger value of magnetic field induced strain is very significative. In this study, the ferromagnetic shape memory effect and the function fatigue of Ni2MnGa under applied magnetic field has been investigated.

## 2. Experimental

The alloy with a nominal composition of Ni2MnGa (at.%) was prepared by high purity nickel, manganese and gallium (>99.99 at.% ) in melting furnace with argon atmosphere. In order to make sure the homogenous, the sample was melted four times, then annealed at 1473 K for 6 hour and quenched into ice water.

The magnetic field induced strain and the function fatigue life of the sample was test by the homemade test system and the schematic was shown in Fig.1. Therein, the magnetic field induced strain was measured by HG-C1050 high precision laser rangefinder (Panasonic, Japan) in different applied magnetic field. Furthermore, the function fatigue life were test by the rotating magnetic field, which created by the rotating magnet in Fig.1. The compression tests with the size 5(length)×3(width)×3(thickness) of sample was performed on CMT5105 electronic universal testing machine at the cross-head displacement speed of 0.05 mm/min.

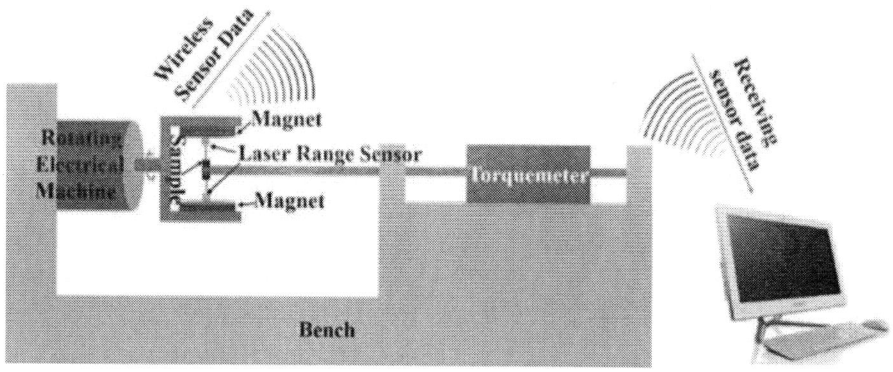


**Figure 1**. The schematic of homemade test system

## 3. Results and Discussion

### *3.1. Ferromagnetic Shape Memory Effect*

The ferromagnetic shape memory effect of FSMAs is the most basic function, which determines the area of their application. Therefore, to clear the process of the magnetic field induced strain and to obtain a larger value of magnetic field induced strain is very significative. Fig.2 shows the magnetic field induced strain value of sample changes in different applied magnetic field. One can see that the magnetic field induced strain of sample appear typical nonlinearity increasing trend apparently with the applied magnetic field rising. However, during applied magnetic field rising, the ascent rates of magnetic field induced strain demonstrate four different variations. The first step occurs at the beginning of applying magnetic field (about 0~350 Oe) and on obvious magnetic field induced strain observed in this stage when applied magnetic field increasing (step 1 in Fig.2). The second step occurs at a low applied magnetic field (about 350~1000 Oe), the magnetic field induced strain increase linearly when the applied magnetic field rising (step 2 in Fig.2). The third step occurs at moderate intensity applied magnetic field (about 1000~6400 Oe), the magnetic field induced strain increase very rapidly in this stage. The fourth step occurs at high applied magnetic field (more than 6400 Oe) and the increase trend of magnetic field induced strain become very gently. The largest magnetic field induced strain value (about 4.8%) was obtained near 1×104 Oe.

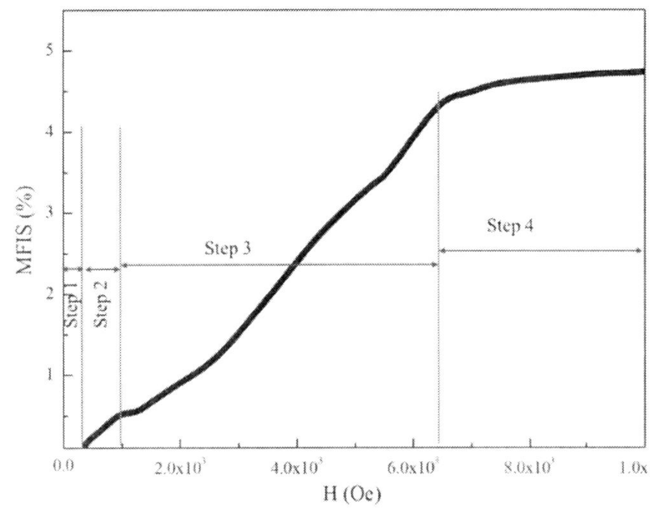


**Figure 2.** The magnetic field induced strain of sample changes in different applied magnetic field

*3.2. Function Fatigue*

The other factor to determine the application of FSMAs is the function fatigue. So, to study the function fatigue process and find out a way to improve the function life of FSMAs is benefit for industrial application of alloys. The magnetic field induced strain of sample depend on different magnetic field induced strain cycles was shown in Fig.3. As can be seen that the magnetic field induced strain value of the sample keep decreasing tendency as the applied magnetic field cycles increasing. The function fatigue curve in Fig.3 can be divided into four steps due to the different slope of the function fatigue curve. The first step is less than 40 cycles. In this stage, the magnetic field induced strain of the sample kept a high value (about 4.8%) and no obviously decrease as cycle increases. The second step is between 40 and 400 cycles. The magnetic field induced strain value obvious decreases form 4.8% to 1.2% as the cycle increasing. The third step occurs from 400 to 107 cycles. In this part, the magnetic field induced strain stable at a low value (1.2%) in a long cycle time. The fourth step is more than 107 cycles. The magnetic field induced strain of the sample decrease rapidly and then disappeared.

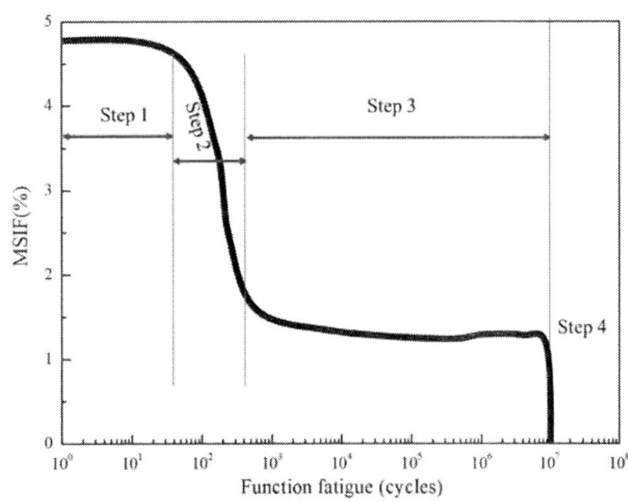


**Figure 3.** The magnetic field induced strain of sample depend on different magnetic field induced strain cycles

### *3.3. Discussions on Ferromagnetic Shape Memory Effect and Function Fatigue*

For FSMAs, the industrial applications of the alloy depend on their magnetic field induced strain and function fatigue life. In this study, the magnetic field induced strain value of sample was investigated and the result revealed that the strain appeared nonlinearity (4 step in curves) increasing with the applied magnetic field rising. At the first step, the applying magnetic field is too weak (about 0~350 Oe) to provide enough deformation driving force. So, the magnetic field induced strain of sample cannot be observed in this step. When the applied magnetic field strength become increase (about 350~1000 Oe), the strain of the sample appeared a linear increasing relation with applied magnetic field rising indicated that this strain is proportional to the magnetic force. Further increase the strength of applied magnetic field (about 1000~6400 Oe), the sample occurs an obvious strain due to a large magnetic force, which obtained by applied magnetic field, drove the twin martensite remigration and exhibits an obvious magnetic field induced strain in this stage. In the last step (more than 6400 Oe), the twin martensite remigration process finished and result in the magnetic field induced strain close to the maximum due to the magnetic force approaches the maximum value.

In order to demonstrate the analysis of the step of the magnetic field induced strain, the Stress-strain curve of the sample by compression test was shown in Fig.4. From Fig.4, the curve appeared three kinds stages due to their different slope. The first step is the typical elastic deformation region because of the linear relation between stress and strain. The second step appeared a large platform for strain increase. In this platform, the sample can be achieved a large strain with a small stress change as seen in Fig.4 (step 2). The reason for this platform can be ascribed into the twin martensite remigration by stress. The third step appeared nonlinear relation between stress and strain indicated that the sample become plastic deformation in this stage and then breakage as show in step 3. From Stress-strain curve analysis, the deformation stage, like elastic deformation, twin martensite remigration and plastic deformation, was consistent with the analysis of the step of the magnetic field induced strain.

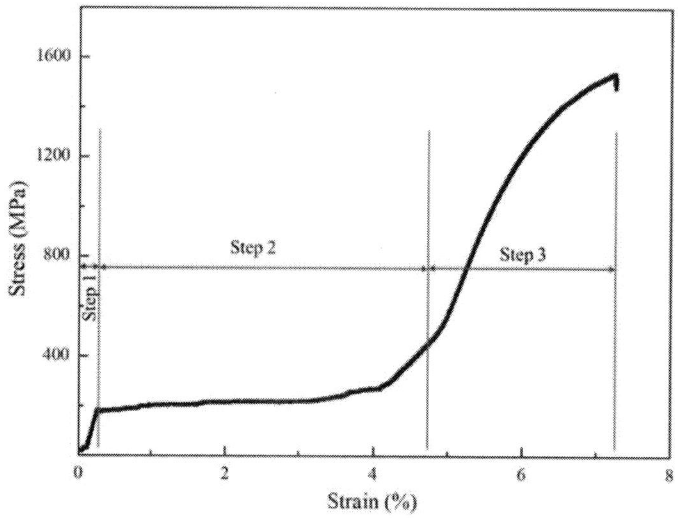


**Figure 4.** Stress-strain curve of the sample

Another main factor for industrial applications of FSMAs is function fatigue life. From the result of the function fatigue life analysis in Fig.3, the magnetic field induced strain of the sample keep decreasing tendency as the applied magnetic field cycles increasing indicated that the function fatigue life is very sensitive to the applied magnetic field cycles. During the function fatigue process, the magnetic field induced strain has no obvious decrease in low cycles (less than 40 cycles) due to the defects does not occur in low cycles. When the applied magnetic field cycles rises to 400 cycles, the defect, which introducing by strain cycles, become accrue in the sample and result in the magnetic field induced strain begin to decrease. With the strain cycles continue to increase (400 to 107 cycles), the magnetic field induced strain kept a small value in a long cycles range indicated that the number of defects curbed by the stain decrease. When the cycles reached a high value (more than 107 cycles), the magnetic field induced strain of sample disappeared due to the defects accumulation during long times cycles.

## 4. Conclusion

To improve the industrial applications of the Ni2MnGa shape memory alloy, the ferromagnetic shape memory effect and the function fatigue under applied magnetic field has been investigated. The main results were list as follows:

(1) The magnetic field induced strain of sample appear typical nonlinearity increasing trend apparently with the applied magnetic field rising.

(2) The magnetic field induced strain value of the sample keep decreasing tendency as the applied magnetic field cycles increasing and the magnetic field induced strain kept a small value in a long cycle's range of 400 to 107 cycles.

(3) The sample appeared typical elastic deformation stage, twin martensite remigration stage and plastic deformation stage by compression test process.

## Acknowledgments

This research is supported by the Natural Science Foundation of Jiangsu Province of China (Project No. BK20181475), the School Research Foundation of Nanjing Institute of Technology (CKJA201702), the postdoctoral research selected funding project of Zhejiang province, the Practice Innovation Program for College Students of Jiangsu (Project No. 201811276080H) and technological Innovation Team in Colleges and Universities of Jiangsu Province.

## References

[1] Ullakko K, Huang JK, Kantner C, O'Handley RC, Kokorin VV. Appl Phys Lett. 1996;69 pp 1966.
[2] Mendonca AA, Jurado JF, Stuard SJ, Silva LEL, Eslava GG, Cohen LF. J Alloys Compd. 2018; 738 pp 509-14.
[3] Lawrence T, Lindquist P, Ullakko K, Mullner P. Mater Sci Eng, A. 2016;654 pp 221-7.
[4] Kundu A, Ghosh S, Ghosh S. Physical Review B. 2017 pp 96.
[5] Waimann J, Junker P, Hackl K. Euro J Mecha a. 2016; 55 pp 120-1.
[6] Ramu P, Arul S. Estimating probabilistic fatigue of Nitinol with scarce samples. Int J Fatigue. 2016; 85 pp 31-9.
[7] Heczko O, Vertat P, Vronka M, Kopecky V, Perevertov O. Shape Memory and Superelasticity. 2016; 2 pp 272-80.
[8] Faran E, Shilo D. Smart Mater Struct. 2016; 25 pp 34-6.

IOP Conf. Series: Materials Science and Engineering 472 (2019) 012008 doi:10.1088/1757-899X/472/1/012008

# Poly (Vinylidene Fluoride)/Mg Doped $CaCu_3Ti_4O_{12}$ Composites with Improved Dielectric Properties

**Hanghang Wang, Huiwu Cai, Wenying Zhou and Shengnan Liu**

(College of Chemistry & Chemical Engineering, Xi'an University of Science and Technology, Xi'an, 710054 China)

**Abstract.** The $CaCu_{3-x}Mg_xTi_4O_{12}$ (CCMTO) particles were prepared by sol-gel method, ceramic/polymer composites made of CCTO and CCMTO particles as filler and polyvinylidene fluoride (PVDF) as matrix were prepared by powder mixing method. The microstructure and dielectric properties of the PVDF, CCTO, CCMTO particles and CCMTO/PVDF composites were investigated. The results indicate that the dielectric permittivity and the dielectric loss increased with increasing ceramics particles loading. The dielectric permittivity could reach 93 and dielectric loss was 0.2 when mass fraction of CCMTO is 60wt% at 100 Hz. SEM results show that CCTO and CCMTO particles dispersed well in the PVDF matrix. Doping Mg can improve dielectric permittivity of the CCTO/PVDF composites and slightly reduce the dielectric loss, compared raw CCTO/PVD composites.

## 1. Introduction

Microelectronic devices were inclined to miniaturization, lightness and high performance with the rapid development of the information technology. Dielectric materials with high dielectric constant, low dielectric loss and good temperature stability are highly desirable [1, 2]. Nowadays, polymer-based materials were widely used in electronic components especially in embedded circuits for their good thermal stability and process easily [3, 4]. However, polymer-based materials have low dielectric constant in range of 3-10, which limited their applications in electronic components [5]. Recently, ceramic-polymer composites were first introduced as a potential material for the application of electronic components because they combined the process ability of the polymer and the dielectric properties of the ceramics. Generally, polymer materials such as poly (polyvinylidene-trifluoroethylene) [6], polyetherssulfone [7], polyvinylidene fluoride (PVDF) and polyimide [8] had been used as the matrix. PVDF would be selected as polymer matrix due to its high permittivity($\varepsilon_\gamma$=8-13), and could be easily processed by injection melting process and better thermal stability [9]. Meanwhile, some high dielectric constant ceramics, such as Pb (Ti, Zr) [10], $BaTiO_3$ [11] and $Ba_{0.65}Sr_{0.35}TiO_3$[12], were selected as fillers. The $CaCu_3Ti_4O_{12}$ (CCTO) has attracted considerable attention in the past decade, due to its high and nearly temperature independent dielectric constant (Up to 10000 1kH at room temperature) over a wide temperature rang (100-600K) [13]. Therefore, CCTO should be good filler.

To further improve dielectric constant of CCTO, CCMTO particles were prepared by sol-gel method. And then the PVDF composites were prepared by powder mixing and hot-pressing. Dielectric properties of CCTO-PVDF and CCMTO-PVDF were investigated, respectively.

## 2. Experimental

### *2.1. Preparation of CCMTO Powers*

The sol-gel method was used to prepare CCTO and $CaCu_{3-x}Mg_xTi_4O_{12}$ (x=0.05; x=0.1; x=0.2) powers as follow, the $Cu(NO_3)_2.3H_2O$ (99.5% from Tianjin-Fuchen in Tianjin, China), $Ca(NO_3).4H_2O$ (99.0% from Tianjin-Fuchen in Tianjin, China) and $Mg(NO_3)_2.6H_2O$ (99.0% from Tianjin-Fuchen in Tianjin, China) in the stoichiometric ratio were dissolved in ethanol, and formed A solution, $C_{16}H_{36}O_4Ti$ (99.0% from Kermel in Tianjin, China) was dissolved in ethanol to form B solution. A solution, B solution and water were mixed with constant stirring, changed pH (PH=1-2) through adjusting the dosage of the glacial acetic acid and citric acid. The formed gels were dried at 180°C for 10h, and then were calcined at 700°C for 10h to form CCTO and CCMTO particles.

### *2.2. Preparation of CCMTO/PVDF Composite*

The prepared CCTO and CCMTO powders were thoroughly mixed with PVDF (From 3F in Shanghai, China) particles by melt mixing the mixtures, respectively. The mixtures were dissolved into N, N-dimethylformamide (from Sic-Tech Co, Ltd in Guangzhou, China) at room temperature to obtain a suspension, the obtained suspensions were stirred by magnetic stirring for 7 h. Then the suspensions was heated and stirred on an electric furnace until it became gelatinous, dried at 80°C in an over for 8 h and then molded by hot pressing at 220°C and 20MPa, the final disk shape sample has a thickness of about 2 mm and a diameter of 15 mm.

### *2.3. Characterization Techniques*

The phase structure of CCTO, CCMTO and CCMTO/PVDF composites were identified by X-ray diffraction technique (XRD D8 Advance). The surface morphology was measured by a scanning electron microscope (SEM JSM-7000F). The dielectric properties of CCTO/PVDF and CCMTO/PVDF composite were measured by an impedance analyzer (Agilent 4294A).

## 3. Results and Discussions

### *3.1. Structural Analysis*

X-ray diffraction patterns of CCTO, CCMTO PVDF and the composites with 40wt%, 50wt% and 60wt% $CaCu_{2.98}Mg_{0.02}Ti_4O_{12}$ in PVDF have been shown in Fig. 1(a-b). In the case of pure CCTO and CCMTO, the diffraction peaks corresponding to the planes (220), (310), (222), (321), (400), (422) and (440) confirm the formation of single phase compound [14]. There is no secondary phase in CCMTO. Pure PVDF crystallizes in a phase with characteristic peaks at $17.7^\circ$, $18.7^\circ$ and $19.9^\circ$ corresponding to (110), (020) and (111) reflections respectively [14]. Diffraction peaks corresponding to (220), (400) and (422) planes of CCTO and CCMTO are also present in the diffraction patterns of the composites. The intensities of the peaks pertaining to the PVDF are not significant. And with the higher content of ceramic powder, the weaker the characteristic peak of PVDF.

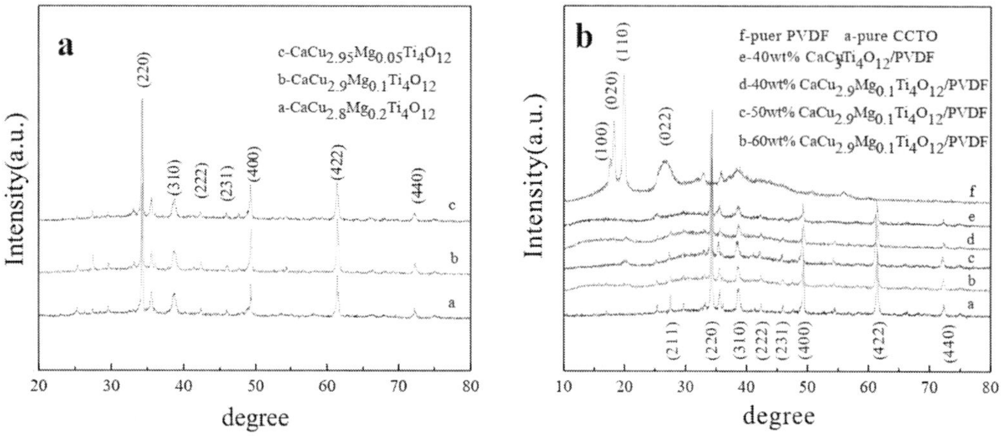


**Figure 1.** XRD diffraction patterns for CCTO, pure PVDF, CCMTO and the composites with 40wt%, 50wt% and 60wt% $CaCu_{2.98}Mg_{0.02}Ti_4O_{12}$ in PVDF

### *3.2. Morphology*

The SEM micrographs of composites containing various mass fractions of CCTO filler are shown in Fig. 2(a-c). It is seen that the CCTO fillers were homogeneously dispersed in the PVDF matrix without aggregation when the mass fraction of CCTO is 40wt%, and when the CCTO is 50wt% and 60wt%, the sample appears aggregation.

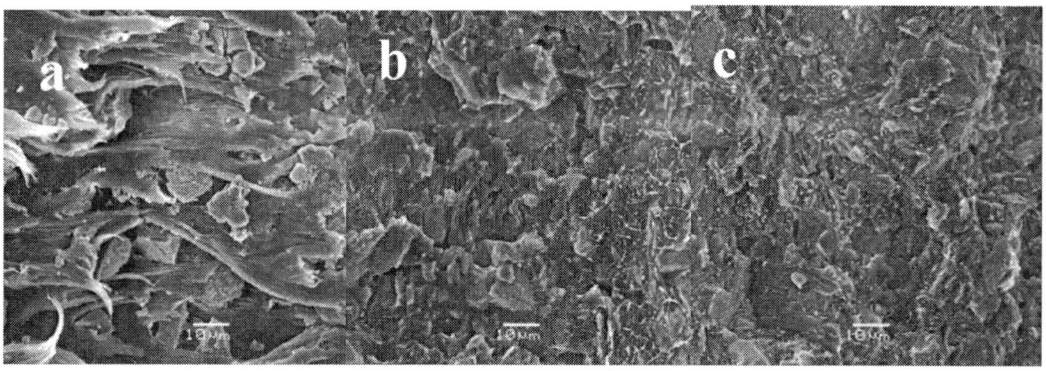


**Figure 2.** SEM images of fractured cross surfaces of CCTO/PVDF composites with various mass fraction of CCTO (a) 40wt% CCTO/PVDF, (b) 50wt% CCTO/PVDF, and (c) 60wt% CCTO/PVDF.

### *3.3. Dielectric Properties*

The frequency dependence of dielectric permittivity, dissipation factor and electric conductivity for PVDF, CCTO/PVDF and CCMTO/PVDF with various filler loading is shown in Fig. 3(a-h). As show in Fig. 3(a-d) the permittivity of CCTO/PVDF and CCMTO/PVDF composites were significantly enhanced with increasing the CCTO or CCMTO content. For the CCTO/PVDF composites, when the mass fraction of particles were 40wt%, 50wt% and 60wt%, the dielectric constant of the CCTO/PVDF composites were 32, 40 and 57 in the frequency range of 100Hz at room temperature, respectively. The CCMTO/PVDF (x=0.1) exhibits larger dielectric permittivity than that of CCTO/PVDF. The dielectric response in CCTO is due to formation of barrier layers at the interface between the grains and the grain boundaries because of difference in their conductivity. This gives rise to interfacial polarization. The giant permittivity depends on the microstructure [15]. Doping Mg significantly modifies the formation of IBLC by increasing the difference between the conductivity of the grains and grain boundaries, which increases the relative permittivity.

Doping Mg reduces the dielectric loss, for the CCTO/PVDF composites, when the mass fraction of particles were 40wt%, 50wt% and 60wt%, the dissipation factor of the CCTO/PVDF composites were

0.082, 0.126 and 0.198, respectively, in the frequency range of 100Hz at room temperature. The CCMTO/PVDF (x=0.1; x=0.2) exhibits low dissipation factor than that of CCTO/PVDF. The CCMTO/PVDF (x=0.1) exhibits low dielectric loss than that of CCTO/PVDF. The loss undergoes two relaxations; one in the low frequency region and the other at high frequencies. The relaxation beyond 1 MHz is related to the glass transition relaxation of PVDF and is denoted as $\alpha_a$ relaxation. Similarly, the relaxation peak that appears below $10^2$ Hz could be attributed to $\alpha_a$ relaxation associated with molecular motion in the crystalline regions of PVDF [16]. It is observed that (Fig. 3a-d) the dielectric loss decreases in the $10^2$ Hz to $10^4$ Hz frequency range and then subsequently increases in the $10^4$ to $10^7$ frequency range.

The frequency dependence of electric conductivity for PVDF, CCTO/PVDF and CCMTO/PVDF are shown in Fig. 3e-f. We can see that they all samples show typical insulator characteristics.

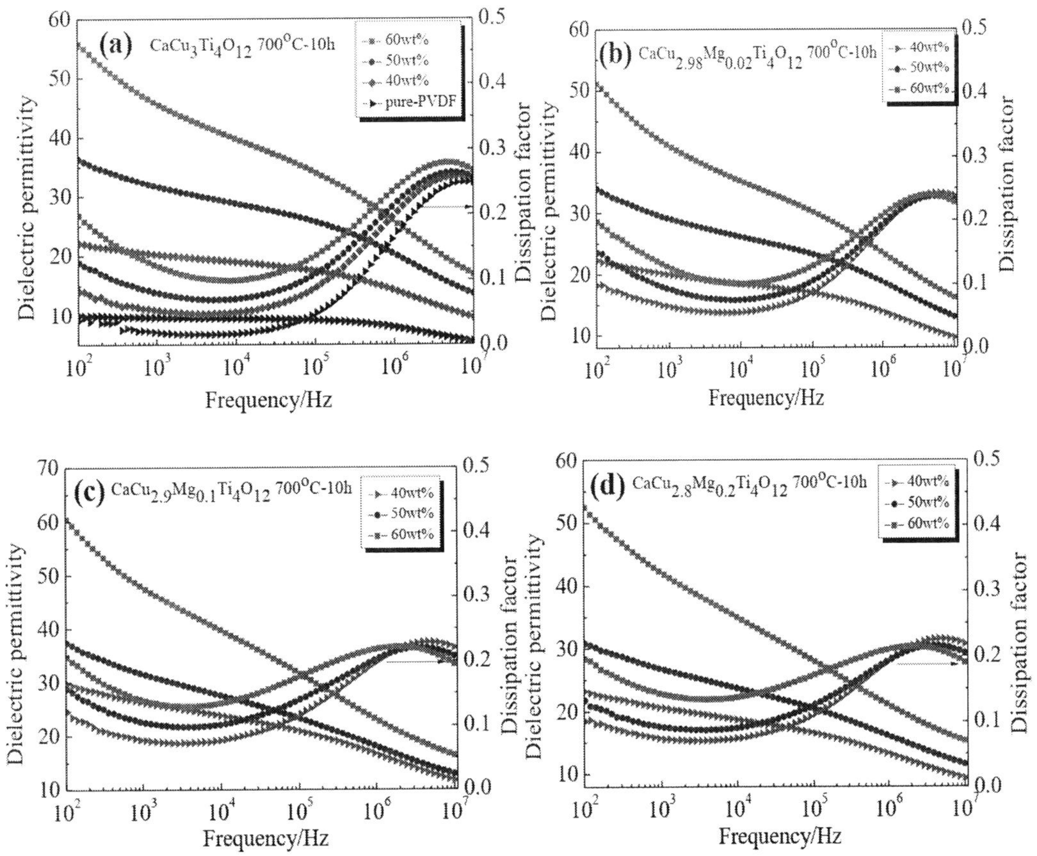

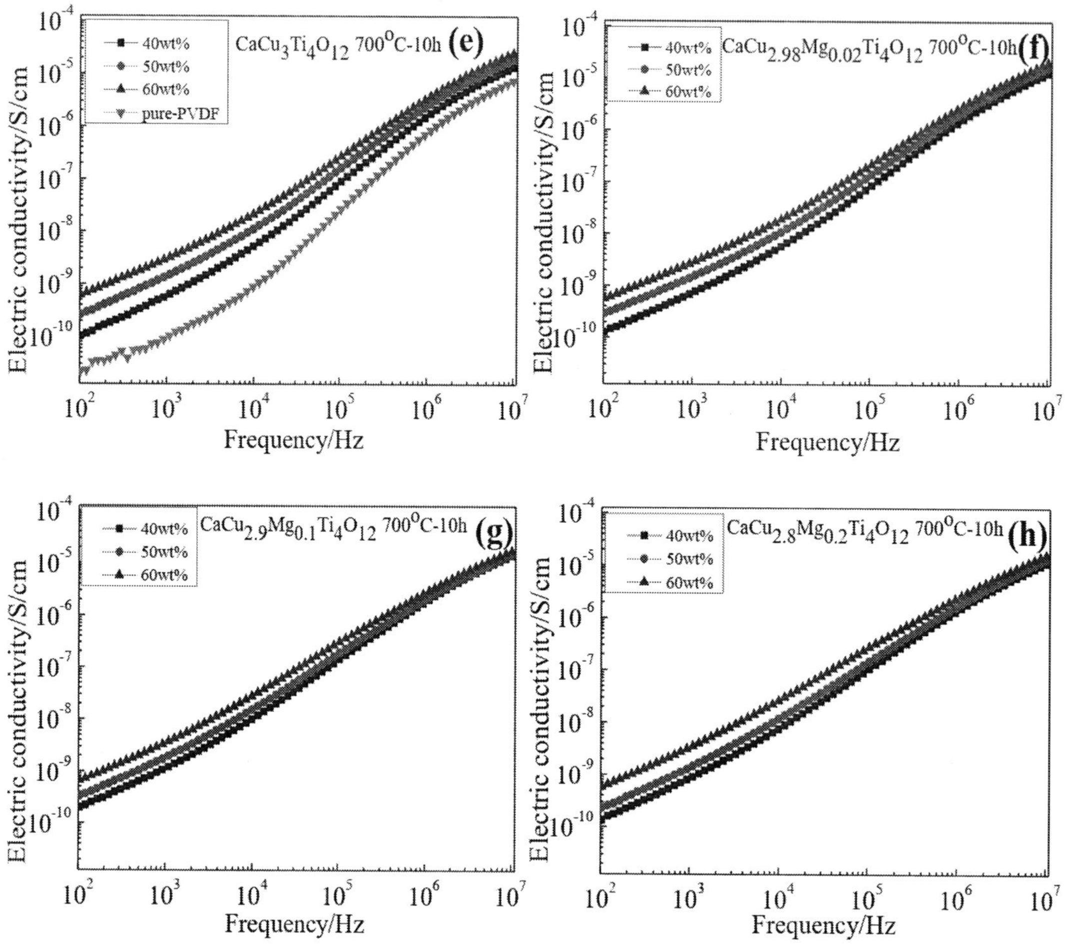


**Figure 3.** The frequency dependence of dielectric permittivity, dissipation factor ((a)CCTO composites and (b) CCMTO (X=0.05) composites and (c) CCMTO(X=0.1) composites and (d) CCMTO(X=0.2) composites)) and electric conductivity ((e)CCTO composites and (f) CCMTO(X=0.05) composites and (g) CCMTO(X=0.1) composites and (h) CCMTO(X=0.2))for pure PVDF, CCTO/PVDF and CCMTO/PVDF with various filler loading

## 4. Conclusion

In this study, CCMTO ceramics were successfully synthesized by the sol-method, and CCMTO/PVDF composites were prepared and their dielectric properties were investigated. The dielectric permittivity increases with the increasing of the addition of CCMTO as well as the dielectric loss increased. The dielectric permittivity of the composite is 93 when the mass fraction of CCMTO (x=0.1) is 60wt% which is higher than CCTO/PVDF (-55). The dielectric loss of the composite is 0.2 at the mass fraction of CCMTO (x=0.1) is 60wt% which has been below CCTO/PVDF (0.23). Doping Mg can improve dielectric properties of the CCTO/PVDF composites.

## Reference

[1] M. A. Subramanian, D. Li, N. Duan, B. A. Reiner, A. W. Sleight, J. Journal of Solid State Chemistry, 2000, 151(2):323-325.

[2] C. C. Homes, T. Vogt, S. M. Shapiro, S. Wakimoto, A. P. Ramirez, J. Science, 2001, 293(5530):673-676.

[3] M. Arbatti, X. Shan,Z. Y. Cheng, J. Advanced Materials, 2010, 19(10):1369-1372.

[4] M. A. Kakimoto, A. Takahashi, T. A. Tsurumi, J. Hao, L. Li, R. Kikuchi, T. Miwa, T. Oono, S. Yamad, J. Materials Science & Engineering B, 2006, 132(1–2):74-78.

[5] J. H. Yu, D. A. Chestakov, H. J. Eggink, J. Microelectronics Journal, 2013, 44(11):1025-1028.
[6] L. Zhang, X. B. Shan, P. X. Wu, Z. Y. Cheng, J. Applied Physics A, 2012, 107(3):597-602.
[7] F. J. Wang,W. Li, M. S. Xue, J. P. Yao, J. Composites Part B, 2011, 42(1):87-91.
[8] M. H. Chen, J. H. Yin, W. B. Bu, J. Applied Surface Science, 2012, 263(48):302-306.
[9] L. Zhang, Z. J. Tang, J. Phys.rev.b, 2004, 70(17):3352-3359.
[10] V. Pascariu, L. Padurariu, O. Avadanei, L. Mitoseriu, J. Journal of Alloys & Compounds, 2013, 574(48):591-599.
[11] R. K. Goyal, S. S. Katkade, D. M. Mule, J. Composites Part B Engineering, 2013, 44(1):128-132.
[12] S. U. Adikary, H. L. W. Chan, C. L. Choy, B. Sundarael, I. H. Wilson, J. Composites Science & Technology, 2002, 62(16):2161-2167.
[13] P. Ramirez, M. A. Subramanian, M. Gardel, G. Blumberg, D. Li, T. Vogt, S. M. Shapiro, J. Solid State Communications, 2000, 115(5):217-220.
[14] M. Arbatti, X. Shan, Z. Cheng, J. Advanced Materials, 2007, 19(10):1369-1372.
[15] M. Chen, J. Yin, D. Song, J. Polymers & Polymer Composites, 2014, 22(2):221-226.
[16] C. V. Chanmal, J. P. Jop, J. Express Polymer Letters, 2008, 2(4):294-301.

IOP Conf. Series: Materials Science and Engineering 472 (2019) 012009 doi:10.1088/1757-899X/472/1/012009

# Overview of Slope Monitoring Technology

**Panpan Yang [1], Nianqin Wang[1], Zhongyan Jiang[2], Yan Yang[2] and Heping Yan[2]**

[1]College of Geology and Environment,Xi'an University of Science and Technology, Xi'an, Shaanxi,710054,China

[2]Shaanxi 194 coalfield geology Co., Ltd., Tongchuan, Shaanxi, 727000, China

Email: xustypp@163.com

**Abstract.** The geological disaster caused by slope engineering has always been one of the most concerned issues in the society. In recent years, the rapid development of urbanization in China has caused relatively large changes in the geological environment, sometimes leading to landslides and mudslides, which pose a significant threat to human lives and properties. The monitoring technologies of slope deformation have been rapidly developed and played an increasingly important role in ensuring slope stability. This paper briefly introduces the conventional slope deformation monitoring technologies and their shortages. The latest development of slope deformation monitoring technologies is presented in detail, followed by analysis of the features of these technologies. Finally, some research suggestions are proposed for future development of slope deformation monitoring technologies.

## 1. Introduction

In the middle of the 20th century, due to the backward and limited level of science and technology, people generally used a simple macro-geological observation method to monitor the slope, that is, to determine the deformation characteristics of the surface and the anomaly of ground fissures by manual observation [1]. However, with the improvement of modern science and technology, there are more and more slope monitoring instruments and systems, which play an increasingly important role in quantitative monitoring of slope deformation.

With the rapid development of urbanization, road building and mining have increased the problem of slope. In the process of slope treatment, with the passage of time, the complexity of rock and soil mechanics inside the slope makes it impossible for us to find out the true mechanical effect inside the rock and soil body of the slope. By monitoring the deformation of the slope, the deformation mechanism of the slope can be fully understood and the slope supporting design can be optimized. This paper comprehensively summarizes and evaluates the research and application progress of slope deformation monitoring technology at home and abroad from the aspects of the principles, advantages and disadvantages of the new monitoring technology in recent years, and provides some suggestions for the development of slope deformation monitoring technology in the future [2].

## 2. Development History of Slope Monitoring Technology

Slope monitoring including construction safety monitoring, effect of treatment monitoring and dynamic long-term monitoring. The construction safety monitoring is to monitor every data in the construction period to guide construction. The effect of treatment monitoring is to test the effect of slope treatment design and construction, and to judge the stability of slope after treatment. The long-term monitoring of the slope will implement dynamic tracking on slope mass after the prevention and treatment engineering completed to realize the variation characteristics of stability of slope mass. The monitoring content should be considered in many aspects, as shown in Table 1[3].

**Table1.** Traditional slope monitoring technology and its shortages

| Monitoring Content | Monitoring method | Monitoring Displacement | shortage |
|---|---|---|---|
| Surface deformation monitoring | Geodetic method, seam measurement method | Theodolite, distance gauge, level gauge, steel tape measure, vernier caliper, etc. | The efficiency of realtime monitoring of surface displacement anges is low, and optical instruments are affected by environment. |
| Deep deformation monitoring | Borehole slanting method | Drilling inclinometer, multi-point displac ement meter, etc. | Long monitoring cycle and high equipment and instrument costs. |
| Structural stress monitoring | Monitoring geotechnical stress | Rock mass stress meter, earth pressure box, convergence meter, microvariation instrument, etc. | Vulnerable to environmental impacts such as humidity, strong acid, strong alkali.and rust. |
| Environmental factors | groundwater level,pore water pressure. | Soil moisture meter, water level gauge, pore pressure gauge, rain gauge, etc. | Low degree of automation, single monitoring param-eters, high instrument cost. |

## 3. New slope Deformation Monitoring Technology

In order to meet the requirements of modern slope engineering monitoring, the development direction of slope engineering at home and abroad has gradually reflected the following characteristics: automation, remote monitoring, wireless sensing, distributed and high precision. In recent years, a number of new types of slope deformation monitoring techniques and methods have emerged, which have effectively promoted the prevention and control of slopes.

### *3.1. Distributed Fiber Optic Strain Sensor (DFOSS)*

This technology of optical fiber sensing is mainly to measure the environmental parameters by measuring some parameters such as intensity and phase-related changes in optical fiber transmission light. Distributed optical fiber sensing technology is one of the most promising technologies in optical fiber sensing technology, mainly because it has the characteristics of reusability and long transmission distance. In addition, it is also the development trend of optical fiber sensing and monitoring technology. As for landslide monitoring, the effect of using distributed optical fiber sensing technology is obvious. For example, in Beichuan County West Slope Landslide Group Monitoring and Hong Kong New Territories Luk Keng Road Roadside Slope Monitoring, some optical fiber monitoring results are analyzed, and the experience of optical fiber monitoring technology in slope field application is summarized [4].

### *3.2. "3s"Technology*

Since the 20th century, the rapid development of global positioning system (GPS), remote sensing technology (RS), geographic information system (GIS) and other technologies has provided strong support for slope deformation monitoring. Compared with the traditional geodetic technology, this kind of technology has the advantages of high degree of automation, high positioning accuracy, short observation time, all-weather real-time monitoring of surface displacement, no need of mutual viewing between monitoring stations, and real-time processing of analytical data [5].

### *3.3. Close-Range Photogrammetry*

It is a process to put the close-range camera in two different fixed points, then take photographs of the observation point within the scope of slope to form stereo image pair, and to measure photographs for each point's 3D coordinates in the spatial position of a moment. Advantages are easy to photograph, relaxed in field surveying, can measure several point at the same time. However, the station is limited by the terrain, and the workload is large, so the application is not very extensive.

### *3.4. Terrestrial Laser Scanning (TLS)*

The ground laser scanning can directly obtain the high-precision, high-density spatial three-dimensional coordinates of the surface , that's point cloud data , without touching the object to be measured, and then processing it to constructing a high-precision, high-resolution three-dimensional digital model. Therefore, it is widely used in slope deformation monitoring. This paper introduced the ground 3D laser scanner in detail. It has the advantages of non - contact, fast data acquisition speed (50,000 points / second - hundreds of thousands of points / second), real - time, dynamic and dynamic, penetrability, high precision, high density, full automation ( schneider, 2006 ), and can be combined with GPS and external digital cameras, overcoming the one-sidedness of traditional point deformation monitoring and obtaining complete point cloud data in the monitoring area. Through processing and analyzing the point cloud data, the deformation characteristics of the landslide surface can be obtained [6].

### *3.5. Acoustic Emission (AE)*

It can deduce the morphological changes and inverse the failure mechanism of rock by the acoustic emission signals from the rock mass [7] compared with other methods, it has the advantages of low labor intensity and fewer human factors. And it has more obvious monitoring advantages at complex area and terrible weather. So it attracts more and more attention in recent years.

### *3.6. Time Domain Reflectometry (TDR)*

TDR is called the time domain reflection test technology. The method has been used in 1930s in the case of the United States for the detection of communication cable, and then applied to the geological industry. TDR system has many advantages, such as low price, short detection time, and data providing fast and high security. The shortage is that it can not be inclined to monitor, it can determine the shear plane, but can not determine the direction of displacement [8].

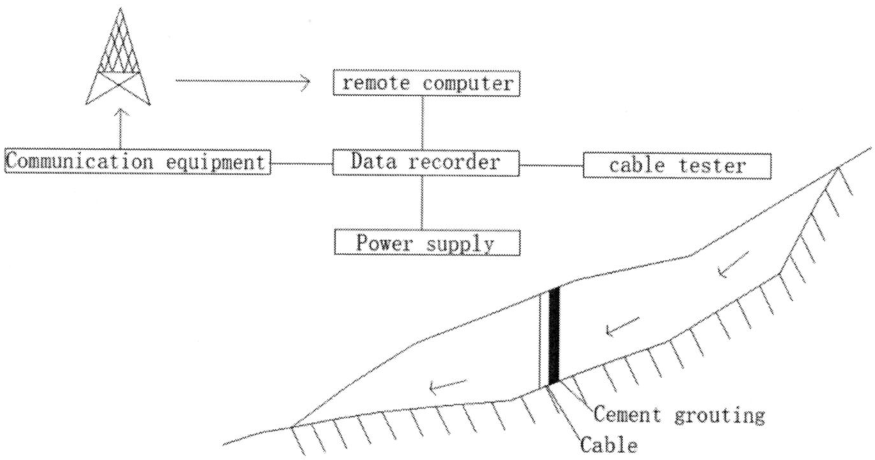


**Figure1.** TDR slope monitoring system (Tang et al. 2007)

### *3.7. Synthetic Aperture Radar (SAR)*

Synthetic Aperture Radar (SAR) is a microwave sensor. Synthetic Aperture Radar Interferometry (InSAR) combines synthetic aperture radar imaging technology and interferometric measurement technology to accurately measure the spatial position of any point on the surface by means of imaging geometry and sensor system parameters. And minor changes, the main application areas include: Digital elevation model (DEM) acquisition, map mapping, ground extraction and monitoring of surface deformation such as landslides and land subsidence [9]. Chinese expert Fan Qingsong has been working on the study of using this technology to monitor surface deformation [10].

## 4. Problems in New Engineering Monitoring Technology

In the past ten years, the application of slope monitoring technology at home and abroad has made considerable progress in engineering. However, due to the immature state of the art, the complexity of geological rock and soil, and the uncertainty of the slope deformation mechanism, these technologies also have various problems in the application process of engineering practice. Mainly reflected in the following aspects:

1) Technical application and on-site construction are difficult and the technical cost is high. echniques that monitor well under laboratory conditions, when actually applied to engineering, may encounter difficulties such as difficult construction, complex field conditions, and sensor damage.

2) Limitation of monitoring conditions. Existing deformation monitoring techniques may only be suitable for monitoring under certain conditions. This technology cannot accurately monitor deformations.due to vegetation cover, small deformation of slopes, and slow deformation. This is one of the problems that TLS technology needs to solve in the future.

3) Slope monitoring data analysis and theoretical model research are insufficient. The current monitoring technology may interfere with external factors, so scientific data preprocessing, the maximum access to the hidden information of monitoring values becomes a key point in the analysis of monitoring results.

4) Monitoring technology is poorly integrated and coordinated. The new monitoring system developed on the existing basis is generally based on its own programming. When combined with other technologies, there will be problems such as data cannot be read or read incorrectly, and monitoring results cannot be visualized.

## 5. Conclusion

In summary, with the development of science and technology, various advanced monitoring equipment and communication technologies are applied to the monitoring of slopes, and the comprehensive utilization of various means can more effectively understand the actual situation and stability of the slope and improve the monitoring. Accuracy and ease of operation, and provide reliable information on safety and design. However, the application of slope deformation monitoring technology still has the disadvantages of low automation of equipment and insufficient intelligence of data acquisition and transmission. To this end, key research can be carried out in the following aspects: 1) Develop more economical slope monitoring instruments and equipment. 2) Strengthen the integration and coordination of comprehensive application monitoring technology methods. 3) Strengthen the research on slope monitoring data analysis and theoretical models.

## Acknowledgements

This work was financially supported by the National Natural Science Foundation of China (41572287 40972174), Key project of national natural science funds (41130753), Shaanxi Province co-ordin-ator innovative engineering project (2013KTCL03-15) and Shaanxi Provincial Coal Geology Bureau of Sci ence and Technology Project (JB2014-4).

## References

[1] Yang Y B.A study on intelligentized methods of slope monitoring and prediction [D]. Wuhan: Graduate University of Chinese Academy of Sciences, 2005.

[2] Dong W W, Zhu H H, et al.Current status and new progress on slope deformation monitor ring technologys [J]. Journal of Engineering Geology, 2016, 24(6):1088-1095.

[3] Yingren Zheng, Zuyi Chen, et al.The slope and landslide engineering management [M]. Beijing: China communication press, 2010: 649 - 663.

[4] Chen Chi,Ke Xianjin,Cai Shnde, et al.Study on application of Distributed Optical Fiber Technology in Monitoring of Geological Flat Slope of Jinping I Hydropower Station [J]. Water conservancy and hydropower technology, 2009, 40 (3): 27-29..

[5] Zhang B L, Song S J, Feng W L,et al. Assessment and risk zonation of landslides in Panxi Area based on 3S technology[J]. Mountain Hazards, 2006, 11(4): 793-800.

[6] Liu J C. The research and application of 3D laser scanning technology in landslide monitoring

IOP Conf. Series: Materials Science and Engineering 472 (2019) 012009 doi:10.1088/1757-899X/472/1/012009

[D].Xi'an: Chang'an University, 2012.
[7] Wei Hu, Shulin Li. Study on AE technique in rock slope stability analysis [J]. Mining Research and Development, 2006, 26(5): 1-4.
[8] Wang N Q,Wang Y F,et al. The Review of China landslide prediction [J]. Geological review, 2008, 54(3): 355-361.
[9] Ker H A, Rosen P, Goldstein R,et al.On the derivation of coseismic displacement fields using differential radar interferometry: the landers earthquake [J]. Journal of Geophysical Research: Solid Earth, 1994, 99(B10): 19617-19634.
[10] Fan Q S,Tang C L, Chen Y,et al. Applications of GPS and In SAR in monitoring of landslide studies [J]. Science of Surveying and Mapping, 2006, 31(5): 60-62.

# Characterization of Oxide Scales on Super-Alloy HR120 at 1000 °C in Air and Steam Environments

**Xiumin Ren[1] and Zhiyuan Liang[2]**

1. Xijing University, Xi'an 710123, PR China. 2. Xi'an Jiaotong University, Xi'an 710049, PR China.

Email: 1172063626@qq.com

**Abstract.** Oxidation resistance of super-alloy HR120 in air and steam environments at 1000 °C was investigated. Scanning electric microscopy and X-ray diffraction analysis were used to characterize the corrosion products. The results show that weight gain in steam was much higher than that in air. Oxide scales formed in air were $MnCr_2O_4$ and $Cr_2O_3$, whereas oxides developed in steam were $NiFe_2O_4$, $(Cr, Fe)_2O_3$ and little internal Cr-Si oxides. Steam can increase the corrosion rate and fine the oxides formed on the sample's surface. Moreover, enhanced oxidation of HR120 in steam was mainly attributed to the evaporation of chromium oxyhydroxides.

## 1. Introduction

Super-alloy HR120 containing 25% Cr exhibits excellent corrosion resistance and high strength at elevated temperature. This new alloy has been used in carburizing and sulfidizing environments. HR120 is a potential heat-resistance alloy for super-heater in advanced power plants [1]. However, from the published literature its oxidation behaviour in a steam environment is not fully understood [2-5]. Ledouxet.al [3] investigated the oxidation kinetics using thermo-gravimetry under the air condition. A non-protective transient stage was found during test at 1050 °C. A multi-layered oxide scale (SiO2–Cr2O3–XCr2O4 with X = Mn and/or Fe, Ni) formed on the surface. Sarver and Tanzosh [5] found that in steam the oxide scale was Cr oxide above a thin Si containing layer at 800°C. There is no clear understanding on the effect of water vapour on the oxidation behaviour of HR 120, though most of the service conditions of super-alloy HR120 contain water vapour at high temperatures.

Therefore, the aim of this article is to investigate the influence of water vapor on the oxidation behaviour of super-alloy HR120 at high temperature.

## 2. Materials and Method

The oxidation tests were conducted in a horizontal furnace at 1000 °C for up to 120 h, as shown in a previous article [6]. The flow rates of air and steam were 90 mL/min and 134.4 mL/min, respectively. The nominal composition of super-alloy HR120 is listed in Table 1. Before oxidation tests, specimens with a size of 15 mm×15 mm×3.1 mm were ground up to 1000# SiC papers and then were ultrasonically cleaned in ethanol. The weight gain was measured using an electronic balance with an accuracy of 10-5 g. After tests, X-ray diffraction (XRD, X'pert PRO, Panalytical, Holland) and scanning electron microscope (SEM, SU3500, HITACHI, Japan) equipped with energy-dispersive X-ray spectroscopy were employed to characterize the oxidation products formed on the surface.

**Table 1.** Compositions of super-alloy HR120

| Ni | Fe | Cr | Co | W | Mo | Nb | Mn | Si | N | Al | C | B |
|---|---|---|---|---|---|---|---|---|---|---|---|---|
| 37 | 33 | 25 | 3 | 2.5 | 2.5 | 0.7 | 0.7 | 0.6 | 0.2 | 0.1 | 0.05 | 0.004 |

## 3. Results and Discussion

Figure 1 shows the weight gain of HR120 at 1000 °C under air and steam conditions. Both oxidation kinetics curves obeyed the power oxidation law. After 25 h test, the weight gain in steam was much higher than that in air, while weight gain under the two tested conditions was similar for the first 25 h. As chromium is the main beneficial element for oxidation resistance in super-alloy HR120, at the initial stage (<25 h) of exposure the uniform chromia film is formed at the samples' surface in both air and steam environments. However, for a relatively longer exposure time, the structure of uniform chromia layer was altered, which accounted for the higher weight gain in the presence of steam.

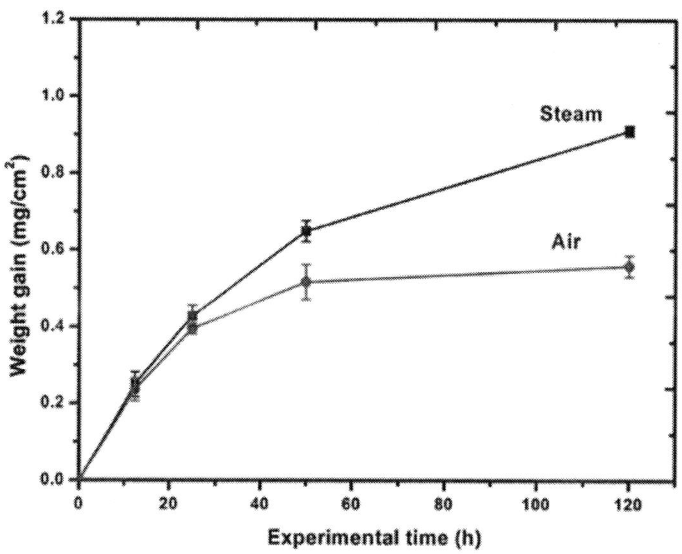


**Figure 1.** Weight gain versus oxidation time in the air and steam environments

Figure 2 displays that surface and cross sectional morphology of specimens exposing to air and steam for 120 h at 1000 °C. The spinel with smaller oxides crystal formed in steam were more porous than that formed in air, indicating the different morphologies of the oxides. The size of oxides also decreased in a steam environment, as showed in Figure 2(a) and Figure 2(b). In Figure 2(c-d), the average thickness of oxide scale due to the lattice diffusion was higher in steam than that in air at 1000 °C. What's more, the maximum penetration depth in steam due to grain boundary diffusion was found to be much larger than that in air. The thick oxide scale in the presence of steam is consistent with the weight gain data shown in Figure 1.

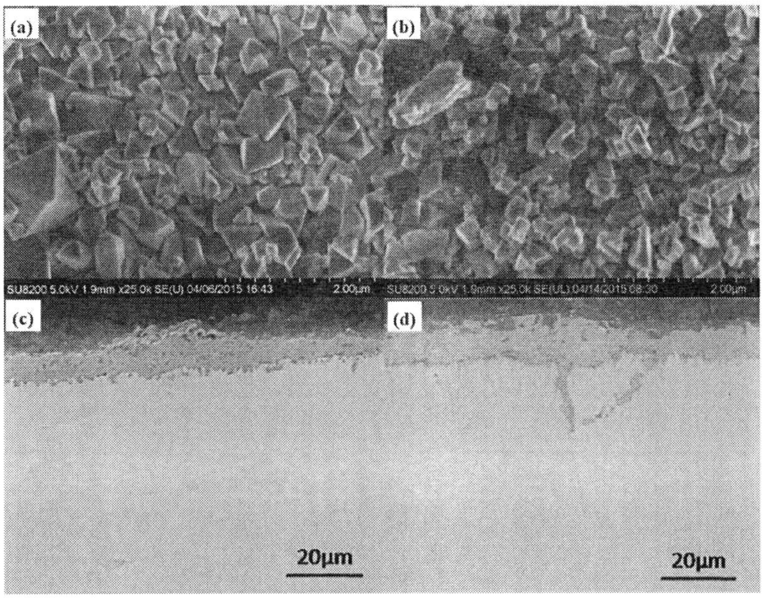


**Figure 2.** Surface and cross sectional morphology of specimens for 120 h. (a) Air, surface (b) steam, surface (c) cross section, air (d) cross section, steam

The element distribution of oxide scale gives a deeper understanding of oxidation behaviour of HR120, so EDS mapping analysis was done on the cross section of exposed samples, as showed in Figure 3 and Figure 4. From Figure 3, it is found that the oxide scale formed in air was mainly composed of Cr, O and Mn. However, in steam environment the oxide scale showed multi-layered structure including the outer layer of Fe-Ni-O, a middle layer with Cr-O and an internal oxidation layer with Cr-Si-O, as showed in Figure 4. In addition, some Nb-rich precipitates were observed underneath the oxide scale. From the results, it is concluded that the oxidation behaviour and metallic diffusion were greatly affected by the water vapour.

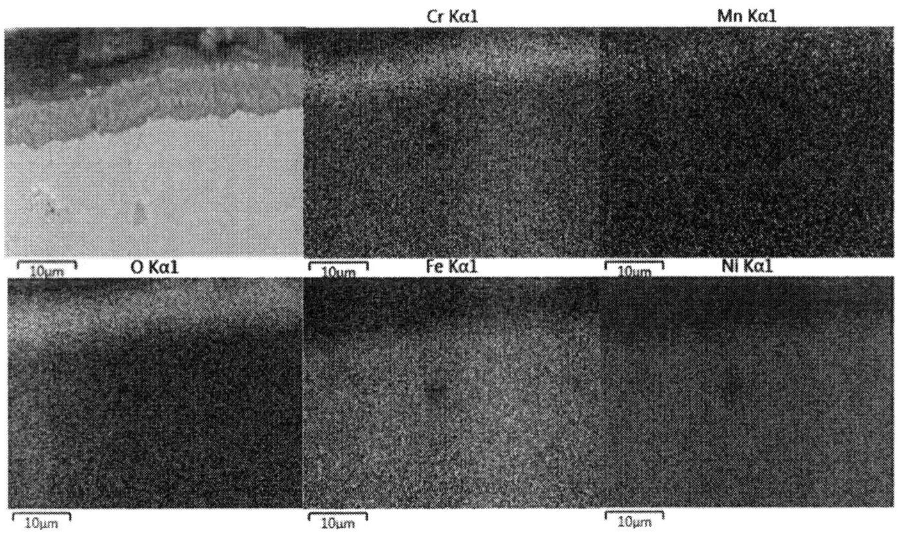


**Figure 3.** EDS mapping results of HR120 in air after 120 h at 1000 °C

The oxide scale was also characterized with XRD, as showed in Figure 5. XRD analysis shows that the oxides formed in the air consisted of $MnCr_2O_4$ and $Cr_2O_3$, while in steam the oxides were $NiFe_2O_4$

and (Cr, Fe) $_2O_3$.Moreover, intensity of alloy's peaks in steam was much lower than that in air, indicating a thicker oxide scale formed in the steam environment.

It is well known that at elevated temperatures alloys containing 25% or higher chromium content can develop a uniform protective chromia layer at the surface in a dry or wet environment [7-8]. However, it is not the case for super-alloy HR120 containing 25%Cr in the presence of steam. From the EDS and XRD results, it is confirmed that the oxide scales in steam were composed of $NiFe_2O_4$, (Cr, Fe) $_2O_3$ and internal Cr-Si-O oxides.

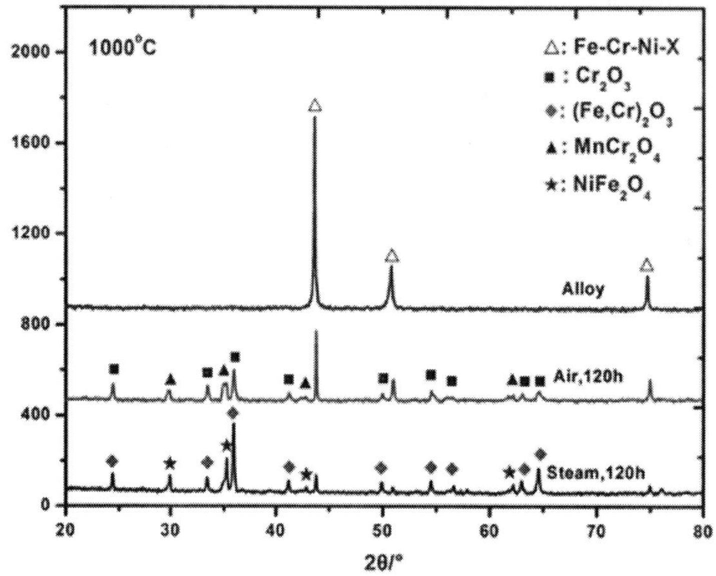


**Figure 4.** XRD results of specimens exposing to air and steam at 1000 °C

From the viewpoint of internal oxidation, the partial pressure of oxygen required for $SiO_2$ formation is lower than that for $Cr_2O_3$[10], so the partial pressure of oxygen at metal/oxide interface in steam is higher than the oxygen pressure at the same position in air. That is to say, the inward diffusion of oxygen in steam is faster than that in air. To further figure out the effect of water vapor, attention should be focused on the $Cr_2O_3$ layer in air and steam. From Figure 4, it is clear that the $Cr_2O_3$ layer was much thinner than that formed in air. This may potentially be due to the evaporation of volatile chromium oxyhydroxides, such as $CrO_2(OH)_2$.

$$2Cr_2O_3+4H_2O+3O_2=4CrO_2(OH)_2 \quad (1)$$

The above reaction indeed occurred even at lower temperature, which had been observed by Asteman et.al. Evaporation of volatile chromium oxyhydroxides changed the chemistry of formed oxides. As a result, the diffusion rates of Fe and Ni ions were significantly improved with the evaporation of chromium oxyhydroxides for providing more transportation paths. In addition, the partial pressure of oxygen in steam is sufficient for the development of NiO and $Fe_2O_3$ at 1000 °C. Therefore, $NiFe_2O_4$ and $Fe_2O_3$ formed above the $Cr_2O_3$ layer. Therefore, oxidation of super-alloy HR120 in steam was more serious than that in air gas at 1000 °C.

## 4. Conclusions

(1) The oxide scale formed on super-alloy HR120 in steam at 1000 °C was composed of $NiFe_2O_4$, (Cr, Fe)$_2O_3$ and small amount of internal Si-Cr-O oxides.

(2) The weight gain in steam is much higher, whereas the $Cr_2O_3$ layer is thinner than that formed in air at 1000 °C.

(3) Accelerated oxidation of super-alloy HR120 was mainly explained by the evaporation of chromium oxyhydroxides.

IOP Conf. Series: Materials Science and Engineering 472 (2019) 012010 doi:10.1088/1757-899X/472/1/012010

**References**

[1] Deodeshmukh VP 2013 *Oxid. Met.* **79** 567
[2] Viswanathan R, Sarver J and Tanzosh JM 2006 *J. Mater. Eng. Perform.* **15** 255
[3] Ledoux X, Mathieu S, Vilasi M, Wouter Y, Del-Gallo P and Wagner M 2013. *Oxid. Met.* **80** 25
[4] Deodeshmukh VP and Srivastava SK 2009 *JOM* **61** 56
[5] Philip JM, Shingledecker JP, Pint BA and Evans ND 2006 *J. turbomachinery 28* 814
[6] Liang Z, Singh PM, Zhao Q and Wang Y 2015 *Oxid. Met.* **84** 291
[7] Otsuka N 1992 *Sumitomo Met.* **44** 30
[8] Sundararajan T, Kuroda S, Kawakita J and Seal S 2006 *Surf. Coat. Tech.* **201** 2124

# Change in Plasticity of Copper under Weak Electrical Potentials

**Konovalov Sergey[1, 2, a], Shunqi Mei [1, b*], Osintsev Kirill[2,c], Filip'ev Roman[3,d] and Komissarova Irina[4,e]**

[1]Wuhan Textile University, Hubei Key Laboratory of Digital Textile Equipment, Fang Zhi Road 1, Wuhan, 430073, P. R. China
[2]Samara National Research University, 34, Moskovskoye Shosse, Samara, 443086, Russia
[3]Kuzbass Institute of the FPS of Russia, 49, Oktyabrsky Ave., Novokuznetsk, 654066, Russia
[4]Siberian State Industrial University, 42, Kirova Street, Novokuznetsk, 654007, Russia
[a]ksv@ssau.ru, [c]kirilloss@yandex.ru, [d]filipiev@live.ru, [e]i.r.i.ss@yandex.ru;
[b*]Corresponding author:1533876320@qq.com

**Abstract.** This paper examines the effect of weak electrical potentials on samples of commercially pure copper. Energy deposition was conducted in two ways. The first one involved alternating attachment of lead, titanium, chromium, aluminum, nickel and iron plates to copper samples. Within the second procedure a voltage from -3 to 3 V was applied to copper samples. The energy deposition lasted 30 min. When contacts and copper samples were disconnected, micro-hardness was measured and plasticity was calculated. As revealed, plasticity of copper drops in both cases. This value is 0.083 % for samples interacting with iron and nickel. The reaction of copper with chromium and aluminum resulted in 0.18 % and 0.10 % decrease of plasticity, respectively. The most significant change in plasticity was registered as a result of interaction between copper and lead or titanium – 0.21 %. It is found out that plasticity of copper samples drops sharply by 0.33 % as a result of connection with a source of power with a potential 0.2 to 1.5 V (in absolute value). The subsequent increase of voltage from 1.5 to 3.0 V hardly effects on plasticity, keeping it the same as at 1.5 V. A sign of an electrical potential is irrelevant for plasticity.

## 1. Introduction

Electrical contacts and various conductors are operated under external impacts like electrical and magnetic fields, electric potentials. Being insignificant, weak impacts aren't taken into consideration, however, ignoring these factors might cause failures in electrical circuits. Therefore, it is necessary to carry out research into the influence of weak energy deposition on different properties of metals. It is established that micro-hardness of aluminum alloys can change by almost 30% due to static and pulse magnetic fields up to 1 T in different conditions of thermal treatment [1, 2]. Electrical potentials and electroplasticity effect also can changes on the properties of the surface [3-5]. Data from studies [6] suggests that micro-hardness of aluminum rises by up to 0.16 % when it reacts with selenium, copper and zirconium. The impact of an electrical potential ±1 V initiates the increase in creep rate of commercially pure aluminum [7]. Consequently, surficial properties of metals might be changed even as a result of weak external energy impacts. Therefore, it is necessary to conduct more experiments because a lot of interaction mechanisms lack for an explanation. So, this work aims at research into the influence of weak electrical potential on copper plasticity.

## 2. Materials and Methods

The study is carried out at the ambient temperature. For the purpose of research samples of preliminary recrystallized commercially pure copper were used, the sizes were 20x20x10 mm. The sequence of the study is similar to one described in [8] and presented in Fig. 1.

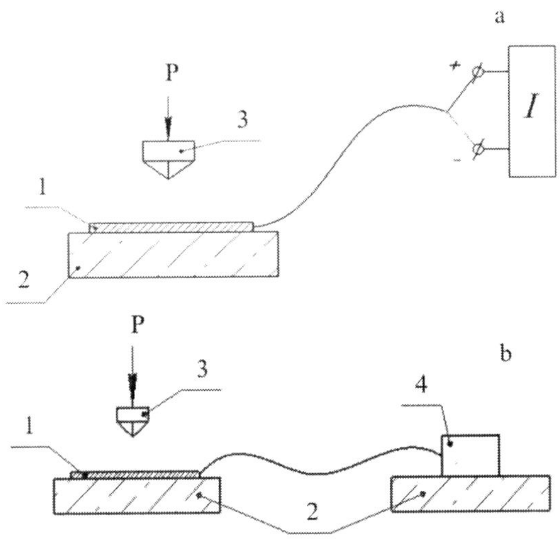


a – when connecting the metal; b – from the source of direct current
1 – sample, 2 – insulation layer, 3 – indenter, 4 – connected metal, P – load on the indenter,
I – source of direct current

**Figure 1.** Diagram of electric potential supply to the sample when measuring micro-hardness

Weak electrical potentials were generated in two ways. First, an electrical potential varying -3 to 3 V in increments of 0.2 V was supplied to copper samples from the source of stable voltage (Fig. 1, a). Within the second procedure different metals: lead, chromium, aluminum, titanium, nickel, and iron were connected with copper (Fig. 1, b). An electrical potential arose when a sample under consideration was connected with other metal with different work function. The experiment lasted 30 minutes. Work function for copper and other metals is selected from the data available [9, 10] and given in Table 1. Plasticity was studied indirectly: micro-hardness is related to plasticity as in [11]:

$$\delta = 1 - 14,3(1 - \upsilon_1 - 2\upsilon_1^2)\frac{HV}{E} \tag{1}$$

where δ – plasticity, HV – Vickers hardness, $\nu_1$ – Poisson coefficient of the material, E is Young's modulus;

Micro-hardness of metals *HV* was measured according to the Vickers method, using a micro-hardness measuring device HV-1000 with a load on indenter of 30 g, and deposition time of 18 seconds. Each sample was measured 30 times. The contact voltage difference was calculated according to the formula [12]:

$$\Delta\varphi = \frac{A_{Me} - A}{e} \tag{2}$$

where $A_{Me}$ – function work of a metal, A – function work of copper Cu, e – electron charge;

The percentage change of plasticity was estimated according to the formula:

$$Q = \frac{\bar{\delta}_E - \bar{\delta}_0}{\bar{\delta}_0} \cdot 100\% \quad (3)$$

where $\bar{\delta}_E$ and $\bar{\delta}_0$ - mean plasticity of copper samples when contacting with a metal and without the contact, respectively.

## 3. Results and Discussion

It is revealed that plasticity drops independently on the potential sign when connecting copper samples with different electrical potentials. The lowest plasticity was registered for a potential of 1.5 V, and the increase of electrical potential didn't change it. Plasticity of copper vs. electrical potential is shown in Fig. 2.

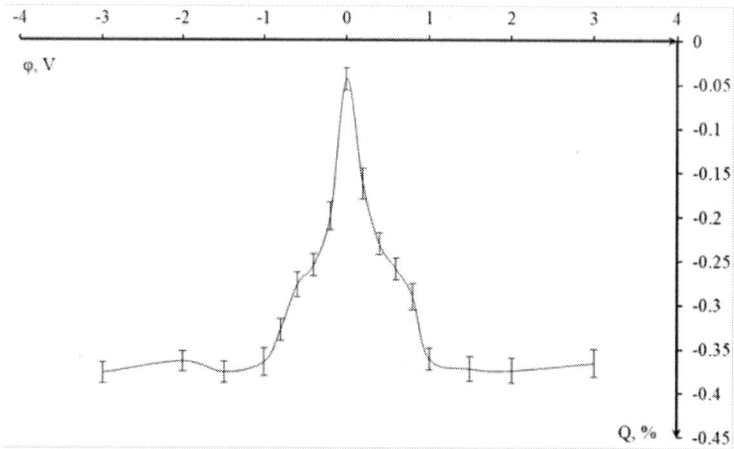


**Figure 2.** Copper plasticity vs. electric potential

The change in copper plasticity in conditions of the second procedure is presented in Table 1. When connecting copper samples with metals with different work function it was established that plasticity is in reverse proportion to the contact potential in absolute value (the higher potential, the lower plasticity), e.g. plasticity of copper decreased by 0.22 % and 0.21 % when samples reacted with lead and titanium, respectively. As a result of copper and chromium interaction plasticity drops by 0.19 %. Aluminum, nickel and iron caused the decrease in plasticity by less than 0.1 %.

**Table 1.** The change in plasticity as a consequence of the second experimental procedure

| Contact metal | $A_{Me}$, [eV] | $A_{Cu}$, [eV] | Δφ, [V] | $\bar{\delta}_0$ | $\bar{\delta}_E$ | $Q$, % |
|---|---|---|---|---|---|---|
| **Pb** | 4.25 | 5.10 | -0.40 | 0.97973 | 0.97762 | -0.22 |
| **Ti** | 4.33 | | -0.45 | | 0.97765 | -0.21 |
| **Cr** | 4.5 | | 0.18 | | 0.97790 | -0.19 |
| **Al** | 4.06 | | -0.15 | | 0.97871 | -0.10 |
| **Ni** | 5.04 | | 0.10 | | 0.97887 | -0.09 |
| **Fe** | 4.67 | | -0.09 | | 0.97891 | -0.08 |

It is suggested that the results above are possible, since an electric contact of diverse metals, generating contact potential difference, causes overcharging of double surficial layers [13, 14]. As a result, the density of surficial energy in the material under consideration changes, initiating the change in plasticity, respectively.

## 4. Conclusions

Experimental data suggest that electrical potential; generated both by the stable source of energy and due to contact potential difference of two metals, has an effect on mechanical properties of copper, e.g. on plasticity. A percentage drop of copper plasticity in comparison with the referential value is registered under contact electric potential. The maximal change was recorded when a sample under consideration was connected with lead and titanium contacts, the percentage plasticity was 0.22 % in this case. Exploring the influence of electrical potential on copper plasticity, it was determined that the sign of the supplied potential is irrelevant for plasticity, whereas its module is important. The decrease of plasticity by 0.3% was registered. The data obtained demonstrate that weak electrical potentials are important for mechanical properties of metals; this influence is to be considered when designing and modeling of complex systems and objects.

## References

[1] V.V. Shpeizman, P.N. Yakushev, Effect of a weak magnetic field on stepwise deformation of lead in the large strain region Phys. of the Solid St. 55 (2013) 1878-1883.
[2] G.-R Li, H.-M. Wang, P.-S. Li, L.-Z. Gao, C.-X. Peng, R. Zheng, Mechanism of dislocation kinetics under magnetoplastic effect Wuli Xuebao/Acta Physica Sinica 64 (2015) 10.
[3] A. Ghiotti, S. Bruschi, E. Simonetto, C. Gennari, I. Calliari, P. Bariani, Electroplastic effect on AA1050 aluminium alloy formability CIRP Annals 67 (2018) 289-292.
[4] B.J. Ruszkiewicz, T. Grimm, I. Ragai, L. Mears, J. T. Roth A Review of Electrically-assisted Manufacturing with Emphasis on Modeling and Understanding of the Electroplastic Effect. ASME J. of Manufact. Sci. and Eng. 139 (2017) 110801-110801-15.
[5] E.A.J. Sánchez, R.H.A. González, D.J. Celentano, J.A. Travieso-Rodríguez, L.I. Fuentes J., Electroplasticity-assisted bottom bending process, J. of Materials Proc. Tech., 214 (2014) 2261-2267.
[6] D. Orlova, L. Zuev, N. Ploskov, Variation in Mechanical Properties Due to The Effect of Electric Potential, IOP Conf. Series: Materials Science and Engineering 225 (2017) 012218.
[7] S.V. Konovalov, V.I. Danilov, L.B. Zuev, R.A. Filip'ev, V.E. Gromov, On the influence of the electrical potential on the creep rate of aluminum, Phys. of the Solid State, 49 (2007) 1457-1459.
[8] V.I. Danilov, L.B. Zuev, S.V. Konovalov, R.A. Filip'ev, B.S. Semukhin, On the effect of electric potential on resistance of metals' surface to microindentation, J. of Surf. Invest. 4 (2010) 157-161.
[9] M.W. Haynes, CRC Handbook of Chemistry and Physics, 96th Edition, CRC Press, Boca Raton, 2015.
[10] J. Hölzl, F. K. Schulte, Work function of metals Solid Surface Physics, Spring. Tracts in Modern Phys., 851979.
[11] Y.V. Milman, S.I. Chugunova, I.V. Goncharova Plasticity characteristic obtained by indentation Problems of Atomic Science and Technology, 4 (2011) 182-187.
[12] V.S. Bagotsky, Fundamentals of Electrochemistry, Willey Interscience, 2006.
[13] V.V. Emets, A.A. Mel'nikov, B.B. Damaskin, Electric double layer on renewable liquid (Cd–Ga) electrode in dimethylformamide solutions, Russian Journal of Electrochemistry, 53 (2017) 117-124
[14] V.V. Emets, A.A. Melnikov, B.B. Damaskin, Electric double layer in the solution of surface-inactive electrolyte and specific adsorption of Cl-, Brand I- ions on liquid (Cd-Ga) electrode in dimethylformamide J. of Electroanalytical Chem., 768 (2016) 145-154.

# Preparation and Properties of an Anti-Friction and Anti-Corrosive Radar Absorbing Material with Periodic Intermediate Coating

**Xingji Tang [1], Xianbin Li [1] and Hongguo Sun [2]**

[1] Academy of Naval Research, Liuliqiaobeili, Beijing, China
[2] Changchun Institute of Applied Chemstry, Renmin Street 565, Changchun, China
Email: Amani.11@163.com

**Abstract.** In this paper, an anti-friction and anti-corrosive radar absorbing material with periodic structure intermediate coating is studied. The preparation process is introduced and its electrical and mechanical properties are studied. The radar reflection loss of the material is less than -10 dB in the X-band and a tensile strength of 16 MPa, and the material has the characteristics of abrasion resistance and anticorrosion. The radar absorbing materials solve the problem of the existence of extended damage in the middle layer due to the continuity of the fiber cloth in the existing coating technology.

## 1. Introduction

Radar detection has the advantages of high detection accuracy and high precision. The development of new radars increases a growing threat to weapon equipment. Therefore, countries are striving to develop radar stealth technology [1]. Radar stealth technology includes shape stealth technology and radar absorbing materials stealth technology [2-3], and material stealth technology develops increasingly rapidly [4-5]. At present, coated absorbing materials with simple process, convenient use and strong ship adaptability have been widely used.

In this paper, an anti-friction and anti-corrosive radar absorbing material with periodic structure intermediate coating is studied. The preparation process is introduced and its electrical and mechanical properties are studied. As a result, the radar reflection loss of the material is less than -10 dB in the X-band and a tensile strength of 16 MPa, and the material has the characteristics of abrasion resistance and anticorrosion.

## 2. Preparation of Radar Absorbing Materials

### *2.1. Selection of the Material*

The material consists of a primer, an absorbing layer, intermediate coating and a top coating which are closely arranged from inside to outside. The primer has a thickness of 40-60 μm and is composed of an epoxy primer and an epoxy primer curing agent having a mass ratio of (6-8):1. The thickness of the absorbing layer is 1.5-2.0 mm, and is composed of an epoxy resin (one or two of E-44, E-20, 830), a low molecular polyamide curing agent (polyamide resin 650 or polyamide resin 651). A magnetic absorbent consisting of ferrite or carbonyl iron powder., a dispersing agent, a toughening agent, an antifoaming agent and a leveling agent. The intermediate coating has a thickness of 0.2-0.3 mm and has a periodic opening unit. The shape of the opening unit is square, diamond or rectangular, the side length of the opening unit is 20 mm-50 mm, and the spacing distance of the opening unit is not more than 2 mm. The intermediate coating is composed of an inner layer, an intermediate layer and an outer layer, and the inner layer and the outer layer are all a mixture of an epoxy resin, a polyamide resin curing agent, a toughening agent and a filler, and the intermediate layer is made of an impregnated epoxy resin. a glass fiber cloth of a mixture of a polyamide resin curing agent, a toughening agent and a filler. The top coating has a thickness of 0.20-0.40 mm and is composed of an abrasion resistant coating and an anticorrosive coating having a mass ratio of 1:1.

### *2.2. Preparation of the Material*

The preparation method of the above materials comprises the following steps:

Step one: preparation of primer.

The primer and the matching primer curing agent are uniformly mixed according to the mass ratio (6-8):1, and then sprayed or brushed on the metal substrate after roughening treatment and cleaning treatment, and the thickness is 40-60 μm. Curing at room temperature for 24 h or 60 ° C for more than 3 h, and the coating was obtained.

Step two: Preparation of absorption layer.

The epoxy resin, low molecular weight polyamide curing agent, dispersant, toughening agent, defoaming agent and leveling agent are stirred and mixed uniformly according to the proportion. The magnetic absorbent is added to mix and mix evenly. After 10-20 minutes of pre-curing at room temperature and pressure, uniform spraying or brushing was applied to the polished and cleaned bottom coatings for several times. After reaching the predetermined thickness, the absorbent layer was obtained after curing at room temperature for more than 24 hours.

Step three: Preparation of intermediate coating.

The epoxy resin, curing agent, toughening agent and filling agent were added into the solvent according to the proportion. After fully mixing and mixing, the mixture was prepared by pre-curing at room temperature and pressure for 10-20 minutes. A portion of the mixture to infiltrate the fiberglass cloth, and a portion of the mixture thinly and evenly sprayed or brushed on the polished and cleaned absorption layer, and the inner layer is obtained. Before the inner layer is cured, the fully infiltrated glass fiber cloth is coated on the inner layer surface, and the coating process is as tight as possible to ensure that there is no air bubble on the surface of the application, and the intermediate layer is obtained. Before the intermediate layer is cured, spraying or brushing a layer of the mixture on the surface of the intermediate layer, and removing the bubbles generated during the coating process to keep the glass fiber cloth flat, and spraying or brushing the mixture until the desired thickness is reached. The the outer layer is obtained. Curing at room temperature for 72h or more at 60°C for more than 8h, after curing the array is cut into a periodic structure according to a predetermined size by machining, and ensures that the fiber cloth between each cycle is completely cut during the processing process.

Step 4: preparation of top coating

The fluoroolefin-vinyl ether copolymer resin and the fluoroolefin-vinyl ether copolymer resin curing agent are thoroughly stirred and uniformly mixed according to the ratio to obtain an abrasion resistant coating. The solvent-based polyurethane rain-resistant resin and the solvent-based polyurethane resin curing agent are thoroughly stirred and uniformly mixed according to the ratio to obtain an anticorrosive coating. Mix the anti-corrosion coating and the anti-wear coating according to the ratio, add 3-5% of the total mass of the anti-corrosion coating and the anti-wear coating to the polyurethane solvent, stir well and mix evenly until the bubbles are eliminated. If necessary, add 0.1% defoamer. BYK-W980, then apply or spray on the middle coating after grinding and cleaning until the desired thickness is reached. If the organic processing demand is required, it is necessary to continue to increase the thickness of the top coating by 0.40mm or more, and reserve enough size for machining. Finally, curing at room temperature for more than 7 days, obtaining an absorbing coating with a periodic structure coating.

## 3. Properties of the Material

According to the absorbing coating with periodic structure coating proposed in this paper and its preparation method, two kinds of absorbing materials were fabricated according to different ratio schemes, and electrical performance test and material mechanics test were carried out.

### *3.1. Electrical Performance Test*

The electrical performance of the material was tested. The results are shown in Fig. 1. As can be seen from Fig. 1, the radar reflectivity loss of the material at the 8-12 GHz band can be less than -10 dB throughout the X-band.

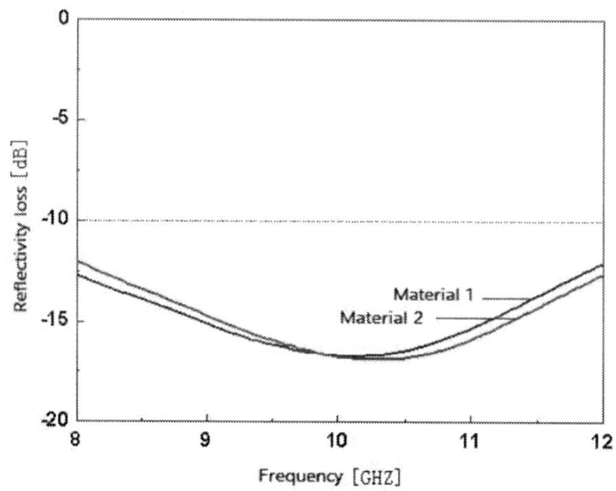


**Figure1.** Radar reflectivity loss

- 11 point Times or Times New Roman.
- The text should be set to single line spacing.
- Paragraphs should be justified.
- The first paragraph after a section or subsection heading should not be indented; subsequent paragraphs should be indented by 5 mm.

### *3.2. Mechanical Performance Test*

The coatings of Examples 1-2 were respectively subjected to rubbing resistance test, and 2,000 times of relative sliding friction tests with the metal surface were measured, and the coating wear thicknesses were 0.065 mm, 0.09 mm, and 0.06 mm, respectively. The coatings of example 1, example 2 and comparative to example 1 were tested for tensile strength according to GB/T 5210-2006. The results are shown in Table 1. As can be seen from Table 1, the material with periodic structure was better than the material with not a periodic structure.

**Table 1.** Comparison of tensile strength of material with or without the periodic structures

| Sample | destructive strength Non-periodic structure | Destructive strength Example 1 with periodic structure | Destructive strength Example 2 with periodic structure | The diameter of test column |
|---|---|---|---|---|
| 1 | 14.5573 | 15.5423 | 16.5423 | 20 |
| 2 | 14.1716 | 15.1261 | 17.1261 | 20 |
| 3 | 13.8107 | 16.2096 | 16.2096 | 20 |
| 4 | 14.1824 | 16.5693 | 15.5693 | 20 |
| 5 | 14.5012 | 16.7467 | 15.3467 | 20 |
| 6 | 13.9627 | 15.4702 | 15.7702 | 20 |

The absorbing coating of Example 2 was tested for peel strength in accordance with GB/T 2790-1995, and the results are shown in Table 2.

**Table 2.** Peel strength test

| Sample | Geometric shape | Width(mm) | Speed(mm/min) | Average peel(N/m) | Maximum peel(N/m) |
|---|---|---|---|---|---|
| 1 | 180degree | 25 | 100 | 4140.35 | 7647.01 |
| 2 | 180degree | 25 | 100 | 4426.28 | 6692.58 |
| 3 | 180degree | 25 | 100 | 3216.49 | 6157.77 |
| 4 | 180degree | 25 | 100 | 3805.72 | 5761.84 |
| 5 | 180degree | 25 | 100 | 5332.02 | 7029.96 |
| Average | | 25 | 100 | 4184.17 | 6657.83 |

Based on the above test description, the matrial has excellent corrosion resistance, wear resistance and impact resistance, and the environmental test results are consistent with the absorbing coating without periodic structure. The strength can reach above 16 MPa, which is higher than the mechanical strength of the material without periodic structure. The damage area is maintained within the structural unit of the periodic structure and does not result in a larger area expansion due to the fiber cloth.

## 4. Conclusion

The radar absorbing materials with periodic structure intermediate coatings studied in this paper have excellent electrical properties and ideal properties, and solve the problem of the existence of extended damage in the middle layer due to the continuity of the fiber cloth in the existing coating technology. The layers of the coating combine to provide excellent corrosion resistance, abrasion resistance and impact resistance.

## References

[1] Liu D. L., Liu P. A., Yang Q. S., Tang G. W., Zhao L. Y. , and Zeng F. C. *Research Status and Prospect of Wave Absorbing Materials*. J Material Review 74-78, 2013
[2] Yang D. Q. and Guo F. J. *Ship Appearance Optimal Design on RCS Reduction Using Response Surface Method and Genetic Algorithms*. J. Shanghai Jiaotong Univ. (Sci.) 336–342, 2008
[3] Gao Q, Yin Y, Yan D B, et al. *Application of meta-materials to ultra-thin radar-absorbing material design*. J Electron Lett. 936–937, 2005
[4] Li B. L. , Xie G. Z. , Song X. L. etal. *Progress and prospect of multi-layer radar absorbing coatings*. J. Journal of Hohai University 464-469, 2011
[5] Zhang X. Z, Sun W. *Three-layer microwave absorber using cement-based composites*. J magazine of concrete research 157-162, 2011

IOP Conf. Series: Materials Science and Engineering 472 (2019) 012013 doi:10.1088/1757-899X/472/1/012013

# Dynamic Systems: Passivity and Positivity Properties

**Manuel De La Sen**[1]

[1]Department of Electricity and Electronics, UPV-EHU, Campus of Leioa, 48080- Leioa, Spain

* E-mail: manuel.delasen@ehu.eus

**Abstract.** This paper is concerned with the discussion of passivity results and the characterization of the regenerative versus passive systems counterparts in dynamic systems. In particular, the various concepts of passivity as standard passivity, strict input passivity, strict output passivity and very strict passivity (i.e. joint strict input and output passivity) are given and related to the existence of a storage function and a dissipation function. The obtained results are related to external positivity of systems and positivity or strict positivity of the transfer matrices and transfer functions in the time- invariant case.

## 1. Introduction

This paper discusses certain aspects of passivity results in dynamic systems as well as the characterization of the regenerative versus passive systems counterparts. In particular, the various concepts of passivity as standard passivity, strict input passivity, strict output passivity and very strict passivity (i.e. joint strict input and output passivity) are given and related to the existence of a storage function and a dissipation function. Basic previous background concepts on passivity and positivity have been given in [1-5], in [6-9], and also in [10-12] and in some related references therein. The obtained results are linked to the properties of external positivity of dynamic systems and the positivity and the strict positivity of the transfer matrices and transfer functions in the time- invariant case. The way of proceeding in the case of passivity failing or how to eventually increase the passivity effects via linear feedback is also discussed to the light of the synthesis of the appropriate feed-forward or feedback controllers or, simply, by adding a positive parallel direct input-output matrix interconnection gain having a minimum positive lower-bounding threshold gain which is also an useful idea for asymptotic hyperstability of parallel disposals of systems, [10]. Finally, the concept of passivity is discussed for switched systems which can have both passive and non-passive configurations which become active governed by switching functions. The passivity property is guaranteed by the switching law under a minimum residence time at passive active configurations provided that the first active configuration of the switched disposal is active and that there are no two consecutive active non-passive configurations in operation.

## 2. Notation

$\boldsymbol{R}_{0+}=\boldsymbol{R}_{+}\cup\{0\}$, where $\boldsymbol{R}_{+}=\{r\in\boldsymbol{R}:r>0\}$, $\bar{p}=\{1,2,\dots,p\}$,

$\boldsymbol{Z}_{0+}=\boldsymbol{R}_{+}\cup\{0\}$, where $\boldsymbol{Z}_{+}=\{r\in\boldsymbol{R}:r>0\}$,

$D\succ 0$ denotes that the real matrix $D$ is positive definite while $D\succeq 0$ denotes that it is positive semidefinite, $\lambda_{min}(.)$and $\lambda_{max}(.)$ denote, respectively, the minimum and maximum eigenvalues of the real symmetric $(.)$-matrix, $\hat{G}\in\{PR\}$ denotes that the transfer matrix $\hat{G}(s)$ of a linear time-invariant system is positive real, i.e. $Re\,\hat{G}(s)\geq 0$ for all $Re\ s>0$ and $\hat{G}\in\{SPR\}$ denotes that it is strictly positive real, i.e. $Re\,\hat{G}(s)>0$ for all $Re\ s\geq 0$.

A dynamic system is positive (respectively, externally positive) if all the state components (respectively, if all the output components) are non-negative for all time $t\geq 0$ for any given non-negative initial conditions and non-negative input , $i=\sqrt{-1}$ is the complex unity, $I_m$ is the $m$-th identity matrix, the superscript $T$ stands for matrix transposition.

$\boldsymbol{H}_\infty$ is the Hardy space of all complex-valued functions $F(s)$ of a complex variable s which are analytic and bounded in the open right half-plane $Re\,s>0$ of norm $\|F\|_\infty = sup\{|F(s)| : Re\,s>0\} = sup\{|F(i\omega)| : \omega\in\boldsymbol{R}\}$( by the maximum modulus theorem) and $\boldsymbol{RH}_\infty$ is the sub set of real-rational functions of $\boldsymbol{H}_\infty$.

## 3. Basic Concepts and Results

Consider a dynamic system $G:\mathsf{H}_e\to\mathsf{H}_e$ with state $x\in\boldsymbol{R}^n$, input $u\in\boldsymbol{R}^m$ and output $y\in\boldsymbol{R}^m$, where $\mathsf{H}_e$ is the extended space of the Hilbert space $\mathsf{H}$ endowed with the inner product $\langle .,.\rangle$ from $\mathsf{H}_e\times\mathsf{H}_e$ to $\boldsymbol{R}$ consisting of the truncated functions $u_t(\tau)=u(\tau)$ for $\tau\in[0,t]$ and $u_t(\tau)=0$ ; $\forall t,\tau(>t)\in\boldsymbol{R}_{0+}$ and $u:\boldsymbol{R}_{0+}\to\boldsymbol{R}^m$. If $u\in\mathsf{H}_e$ then $\forall t\geq 0$.

Definitions [2].The above dynamic system is: *$L_2$-stable* if $u\in L_2^m$ implies $Gu\in L_2^m$.

*Nonexpansive* if $\exists\lambda$ and $\exists\gamma>0$ s. t. for all $u\in\mathsf{H}_e$

$\int_0^t (Gu)^T(\tau)u(\tau)d\tau \leq \lambda+\gamma^2\int_0^t u^T(\tau)u(\tau)d\tau$ ; $\forall t\geq 0$.

*Passive* if $\exists\varepsilon\geq 0$ such that $\int_0^t y^T(\tau)u(\tau)d\tau\geq -\varepsilon$ ; $\forall t\geq 0$.

*Strictly- input passive* if $\exists\varepsilon\geq 0$ and $\exists\varepsilon_u>0$ s. t.

$\int_0^t y^T(\tau)u(\tau)d\tau\geq -\varepsilon+\varepsilon_u\int_0^t y^T(\tau)u(\tau)d\tau$ ; $\forall t\geq 0$.

*Strictly- output passive* if $\exists\beta\geq 0$ and $\exists\varepsilon_y>0$ s. t.

$\int_0^t y^T(\tau)u(\tau)d\tau\geq -\varepsilon+\varepsilon_y\int_0^t y^T(\tau)y(\tau)d\tau$ ; $\forall t\geq 0$.

*Strictly input/output passive* (or *very strictly passive*) if $\exists\beta\geq 0$, $\exists\varepsilon_u>0$ and $\exists\varepsilon_y>0$ s. t. $\int_0^t y^T(\tau)u(\tau)d\tau\geq -\varepsilon+\varepsilon_u\int_0^t y^T(\tau)u(\tau)d\tau+\varepsilon_y\int_0^t y^T(\tau)y(\tau)d\tau$ ; $\forall t\geq 0$.

The constants $\varepsilon$, $\varepsilon_u$ and $\varepsilon_y$ are, respectively, referred to as the passivity, input passivity and output passivity constants.

Proposition 1. Consider a linear time-invariant SISO (i.e. m=1) system whose transfer function $\hat{G}\in\{PR\}$. Then, the following properties hold:

$\int_0^t y(\tau)u(\tau)d\tau\geq 0$ and $y(t)u(t)\geq 0$ ; $\forall t\geq 0$ and, furthermore, if $u\in L_2$ then $y\in L_2$ so that the system is passive.

Assume, in addition, that $\hat{G}\in\{SPR\}$. Then $\gamma\geq\int_0^t y(\tau)u(\tau)d\tau\geq\varepsilon_u\int_0^t|u(\tau)|^2 d\tau-\varepsilon$ for any $t\in(0,\infty]$ and some $\gamma,\varepsilon\in\boldsymbol{R}_{0+}$.

If, furthermore, the system is externally positive in the sense that the output is non-negative for all time if the initial consditions are non-negative and the input is non-negative for all time, then $\gamma\geq\int_0^t y(\tau)u(\tau)d\tau>0$ ; $\forall t\geq 0$ for any given non-negative initial conditions and non-negative input.

Define $R_{\hat{G}} = \left\| \frac{1-\hat{G}(i\omega)}{1+\hat{G}(i\omega)} \right\|_{\infty} = \sup_{\omega \in \mathbf{R}_{0+}} \left| \frac{1-\hat{G}(i\omega)}{1+\hat{G}(i\omega)} \right|$ as the relative passivity index of the transfer function $\hat{G}(s) = \frac{\hat{N}(s)}{\hat{D}(s)} \in \mathbf{RH}_{\infty}$ ( $\hat{N}(s)$ and $\hat{D}(s)$ being the numerator and denominator polynomials of $\hat{G}(s)$). Then, the constraint $a_G \le R_{\hat{G}} = \left\| \frac{\hat{D}_G(i\omega)-\hat{N}_G(i\omega)}{\hat{D}_G(i\omega)+\hat{N}_G(i\omega)} \right\|_{\infty} \le b_G$ is guaranteed for some $a_G, b_G(\ge a_G) \in \mathbf{R}_{0+}$ if

$$\frac{1-a_G^2}{2\left(1+a_G^2\right)}\left(Re^2\ \hat{D}_G(i\omega)+Re^2\ \hat{N}_G(i\omega)\right) \ge \frac{1-b_G^2}{2\left(1+b_G^2\right)}\left(Re^2\ \hat{D}_G(i\omega)+Re^2\ \hat{N}_G(i\omega)\right) \quad ; \quad \forall \omega \in \mathbf{R}_{0+}$$

.If $b_G \le 1$ (respectively, $b_G < 1$) then $\hat{G} \in \{PR\}$ (respectively, $\hat{G} \in \{SPR\}$).

Note that positivity is a very important property in some dynamic systems related to biological or epidemic-type models. See, for instance, [5-9] and references therein The generalization of Proposition 1 to the multi input multi-output (MIMO) case (i.e. $m > 1$) is direct by replacing the instantaneous power $y(t)u(t)$ by the scalar product $y^T(t)u(t)$ in the corresponding expressions. In particular, the subsequent two results discuss how the basic passivity property can become a stronger property as, for instance, strict-input passivity or very strict passivity, by incorporating to the input-output operator a suitable parallel static input-output interconnection structure.

## 4. Hyperstability and Passivity and Non-passivity of Dynamic Systems

It turns out that passive systems are intrinsically stable and either consume or dissipate energy for all time. However, unstable systems are essentially non-passive although some stable systems are also non-passive. Looking at Definition 3, we can give the next one:

Definition 7. A dynamic system is said to be *Non-passive* (or *Active* or, so-called, *Regenerative*) if $\int_0^{t_i} y^T(\tau)u(\tau)d\tau + \varepsilon_{t_i} < 0$ for some unbounded sequences $E = \{\varepsilon_{t_i}\} \subseteq \mathbf{R}_{0+}$, $T = \{t_i\} \subseteq \mathbf{R}_{0+}$ which satisfy the conditions:

$0 < \delta_{i-1} \le t_{i+1} - t_i \le \delta_i < \infty$ ; $\forall i \in \mathbf{Z}_{0+}$ for some positive bounded sequence $\Delta = \{\delta_i\}$,

$0 < \theta_{i-1} \le \tilde{\varepsilon}_{t_i} = \varepsilon_{t_{i+1}} - \varepsilon_{t_i} \le \theta_i < \infty$ ; $\forall i \in \mathbf{Z}_{0+}$ for some positive bounded sequence $\Theta = \{\delta_i\}$,

$\varepsilon_i, t_i \to +\infty$ as $i \to +\infty$.

The subsequent results follow:

Proposition 2. The following input output energy constraint is fulfilled by non-passive dynamic systems $\lim_{t \to \infty} \int_0^t y^T(\tau)u(\tau)d\tau = -\infty$.

Note that non-passive systems can reach an absolute infinity energy measure in finite time under certain atypical inputs as, for instance, a second-order impulsive Dirac input of appropriate component signs at some time instant $t_1 < \infty$ with $u(t) = 0$ for $t > t_1$. Then, $\int_0^{t_1} y^T(\tau)u(\tau)d\tau = \lim_{t \to \infty} \int_0^t y^T(\tau)u(\tau)d\tau = -\infty$.

Proposition 3. The following properties hold:

A passive system cannot be non-passive in any time sub-interval. A non-passive system in some time interval cannot be a passive system.

A passive system is always stable and also dissipative (i.e. the dissipative energy function takes non-negative values for all time) including the conservative particular case implying identically zero dissipation through time.

A non-passive system can be stable or unstable (so, stable systems are non- necessarily passive).

It is convenient to point out that, under certain rather weak assumptions about the switching operation, a switched system is passive for all time under a switching law $\sigma: \mathbf{R}_{0+} \times SW \to \{1, 2, \ldots, p, p+1, \ldots, q\}$, with at least one configuration being active, if:

a) The first active configuration of the switching law on $[t_0 = 0, t_1) \in SW_P$.

b) The switching law does not involve two consecutive active configurations being non-passive.

c) Each active passive configuration respects a minimum residence time, quantified in the proof, which can exceed a minimum common residence at such a configuration before next switching to another one to be marked as active.

A related property to passivity is that of the hyperstability. Such a property is a strong absolute stability property implying that the controlled system is closed-loop stable under a very general class of feedback controllers which satisfy a so-called Popov´s inequality. An asymptotically hyperstable system has a non-negative bounded input- output energy for all time which is also (strictly) positive within any nonzero time interval Consider the subsequent dynamic system under eventually time-delayed disturbances, whose nominal feed-forward part is linear and time-invariant while the feedback controller is nonlinear and, eventually, time-varying belonging to a class to be characterized later on through a Popov´s inequality:

$$\dot{x}(t)=Ax(t)+bu(t)+\eta(t) \ ; y(t)=c^T x(t)+du(t)+\eta_0(t) \tag{1}$$

$$u(t)=-\varphi(y(t),t) \tag{2}$$

where $x\in \boldsymbol{R}^n$ is the state vector and $u(t)$ and $y(t)$ are the scalar input and output, where $A$ is real square matrices of order $n$, $b$ and $c$ are real $n$- vectors and $d$ is a real scalar, which is subject to initial conditions $x(t)=\phi(t)$ for $t\in[-h(0),0]$ where $h:\boldsymbol{R}_{0+}\rightarrow[h_m, h_M]$ is a, in general, bounded piecewise-continuous time- varying internal delay and $\phi:[-h(0),0]\rightarrow \boldsymbol{R}^n$ is an absolutely continuous $n$-vector function with eventual finite isolated jump discontinuities, $x(0)=\phi(0)$, $\eta(t)=\eta(\bar{x}_h(t),u(t),t)$ and $\eta_0(t)=\eta(x(t),u(t),t)$ are, in general, unstructured real $n$-vector disturbance functions which take account of parametrical and unmodeled dynamics uncertainties and noise disturbances being eventually subjected to an internal delay $h(t)\geq h_m$, where $\bar{x}_h(t)$ is the strip $x:[t-h(t),t]\rightarrow \boldsymbol{R}^n$; $\forall t\in \boldsymbol{R}_0$.

It is assumed that the disturbance functions are subject to certain regularity conditions such as Lipschitz continuous with respect to all their arguments in order to ensure the existence and uniqueness of the solution for any bounded admissible initial conditions.

The nominal system is defined as the disturbance- free one, that is $\eta(t)=\eta(\bar{x}_h(t),u(t),t)\equiv 0$, $\eta_0(t)=\eta(x(t),u(t),t)\equiv 0$. Note that the nominal system is delay- free.

The function $\varphi:\boldsymbol{R}\times\boldsymbol{R}_{0+}\rightarrow\boldsymbol{R}$ defining the, in general, eventually non-linear and time-varying controller is any member of a class $\{\Phi\}$ satisfying a Popov´s -type integral inequality of the form:

$$\Gamma(t_0,t)=\int_{t_0}^{t} y(\tau)\varphi(y(\tau),\tau)d\tau\geq-\gamma_0 \tag{3}$$

for some finite $\gamma_0\in\boldsymbol{R}_+$; $\forall t_0\in\boldsymbol{R}_{0+}, t(\geq t_0)\in\boldsymbol{R}_+$. Such a class $\{\Phi\}$ is said to be a hyperstable class of controllers and any $\varphi\in\{\Phi\}$ is said to be $\Phi$-hyperstable.

If the closed-loop system is globally stable for any controller $\varphi\in\{\Phi\}$ with the forward- time invariant block having a positive real transfer function then it is said to be $\Phi$ *-hyperstable* so that the sate and the output are uniformly bounded for all time for any given finite initial conditions. If, in addition, the feed-forward block has a strictly positive real transfer function then the closed-loop system is $\Phi$ *-asymptotically hyperstable*, so that the state and output are uniformly bounded and converge asymptotically to zero as time tends to infinity, that is, it is globally asymptotically stable for any finite initial conditions and any controller of class $\{\Phi\}$. Thus:

1) the feed-forward controlled plant is *asymptotically hyperstable* (or *strictly positive*, or *strictly passive*) if its transfer function is strictly positive real,

2) the class $\Phi$ of feedback controllers is *hyperstable* if it satisfies the above Popov´s type inequlity,

3) any closed-loop configuration of the controlled $\Phi$ *-asymptotically hyperstable* (respectively, $\Phi$ *-hyperstable*) plant with a controller of class $\{\Phi\}$ (i.e. with a $\Phi$ -hyperstable controller) is $\Phi$ -asymptotically hyperstable (respectively, $\Phi$ - hyperstable).

Now, define truncate functions and $h$ -delay truncated functions for any $0 \le T \le \infty$, respectively, as

$$v_T(t) = \begin{cases} v(t) & \text{for } t \in [0, T] \\ 0 & \text{for } t > T \end{cases} ; \; v_{hT}(t) = \begin{cases} v(t) & \text{for } t \in [-h(t), T] \\ 0 & \text{for } t > T \end{cases} \tag{4}$$

It is assumed that $\eta : \boldsymbol{R}^{n+1} \times \boldsymbol{R}_{0+} \to \boldsymbol{R}^n$ , $\eta : \boldsymbol{R}^{n+1} \times \boldsymbol{R}_{0+} \to \boldsymbol{R}$ and $\varphi : \boldsymbol{R} \times \boldsymbol{R}_{0+} \to \boldsymbol{R}$ satisfy standard regularity conditions (so as to guarantee the uniqueness of the state and output trajectory solutions for any given finite initial conditions on $[-h(0), 0]$. Such solutions can be expressed by the concourse of the truncated relevant signals of the system by assuming that $u(-t) = 0$ , $\eta(-t - h(t)) = 0$ and $\eta_0(-t - h(t)) = 0$; $\forall t \in \boldsymbol{R}_+$ , as follows:

$$x(t) = 0; \; t \in (-\infty, -h(0)) \; ; \; x(t) = \phi(t); \; t \in [-h(0), 0)$$

$$x(t) = e^{At}\left(x(0) + \int_0^\infty e^{-A\tau}\left(bu_t(\tau) + \eta_{ht}(\tau)\right)d\tau\right); \; \forall t \in \boldsymbol{R}_{0+}$$

$$y(t) = c^T e^{At}\left(x(0) + \int_0^\infty e^{-A\tau}\left(bu_t(\tau) + \eta_{ht}(\tau)\right)d\tau\right) + du(t) + \eta_0(t)$$

$$= y_f(t) + c^T e^{At}\left(x(0) + \int_{-\infty}^\infty e^{-A\tau}\eta_{ht}(\tau)\,d\tau\right) + \eta_0(t); \; \forall t \in \boldsymbol{R}_{0+}$$

where $y_f(t)$ is the forced output. It is well- known that the Fourier transform exist of any absolutely-integrable vector real function on $(-\infty, +\infty)$. Truncated functions with no escape time instants over finite time intervals always fulfil this property so that for any finite $t \in \boldsymbol{R}_{0+}$ , the Fourier transforms $\hat{u}_t(i\omega)$, $\hat{x}_t(i\omega)$, $\hat{y}(i\omega)$, $\hat{\eta}_{ht}(i\omega)$ and $\hat{\eta}_0(i\omega)$ exist for all $\omega \in \boldsymbol{R}$. From Parseval's theorem, it follows that the following input-output energy measure $E(t)$ on the time interval $[0, t]$ fulfils the subsequent associated relationships for any $t \in \boldsymbol{R}_{0+}$ :

$$E(t) = \int_0^t y(\tau)u(\tau)d\tau = \int_{-\infty}^\infty y_t(\tau)\,u_t(\tau)d\tau$$

$$= \frac{1}{2\pi}\int_{-\infty}^\infty \hat{y}_t(i\omega)\hat{u}_t(-i\omega)d\omega = E_0(t) + \tilde{E}(t) \; ; \; \forall t \in \boldsymbol{R}_{0+}$$

where the nominal value and its incremental term generated by the combination of the unforced output response and the contributions of the uncertainties to the output $E_0(t)$ and $\tilde{E}(t)$

The main relevant result whose proof is omitted follows.

Proposition 4. Assume that

a) the dynamic system is under feedback control generated by any controller $\varphi \in \{\Phi\}$, where the class $\{\Phi\}$ is defined by the Popov's -type inequality ,

b) the nominal transfer function, namely, that when $\eta(t) = \eta(\bar{x}_h(t), u(t), t) \equiv 0$ and $\eta_0(t) = \eta(x(t), u(t), t) \equiv 0$ , satisfies $\hat{g} \in \{SSPR\}$ such that $\min_{\omega \in \boldsymbol{R}_{0+}} \hat{g}(i\omega) \ge \lambda d$ for some real constant $\lambda \in (0, 1]$, where the direct input-output interconnection gain $d$ is positive,

c) $\tilde{E}(t) \le \tilde{\gamma}(t) \le +\infty$ ; $\forall t \in \boldsymbol{R}_{0+}$ for some real function $\tilde{\gamma} : \boldsymbol{R}_{0+} \to \boldsymbol{R}$ subject to the constraint:

$\tilde{\gamma}(t) \le \tilde{\varepsilon}_\gamma \int_0^t u^2(\tau)d\tau$; $\forall t \in \boldsymbol{R}_{0+}$

Thus, the following properties hold:

(i) $u \in L_2$ (the space of square-integrable functions on $[0, \infty])$, $u(t) \to 0$ as $t \to \infty$, $\operatorname{ess\,sup}_{t \in \boldsymbol{R}_{0+}} |u(t)| < +\infty$, $\sup_{t \in \boldsymbol{R}_{0+}} |\tilde{E}(t)| < +\infty$, $0 < E(t) < +\infty$ and $0 < E_0(t) < +\infty$; $\forall t \in \boldsymbol{R}_+$ for any non-identically zero control on a time interval $[0, t_0)$ of nonzero measure if $\tilde{\varepsilon}_\gamma < \lambda d$. Also, $x(t)$ and $y(t)$ are uniformly bounded for all time for any given admissible finite initial conditions.

(ii) If $\|\eta\|_2$ and $\eta_0$ are in $L_\infty \cup L_1 \cup L_2$ then the state $x:[0,+\infty]\times([-h(0),0]\times \mathbf{R}^n)\to \mathbf{R}^n$ and output $y:[0,+\infty]\times([-h(0),0]\times \mathbf{R}^n)\to \mathbf{R}$ are uniformly bounded for each given vector function of initial conditions. As a result, the closed-loop system is $\Phi$ - hyperstable for the class $\{\Phi\}$ of output feedback controllers defined by the Popov's inequality. Furthermore, if those unstructured functions describing the uncertainties are identically zero then the closed-loop system is $\Phi$ -asymptotically hyperstable for any finite initial conditions, i.e. globally asymptotically stable for any member of the class $\{\Phi\}$ of output feedback controllers being defined by the Popov's inequality.

Proposition 5. A switched system is passive for all time under a switching law with at least one of the inviolved configurations is active, if for $\{t_i\}$ ; $i\in \mathbf{Z}_{0+}$ being the sequence of switching time instants:

a) The first active configuration of the switching law on $[t_0=0\,,t_1)$ is passive

b) The switching law does not involve two consecutive active configurations being non-passive.

c) Each active passive configuration respects a minimum residence time, quantified in the proof, which can exceed a minimum common residence time which depends on the parameterizations and can be explicitly calculated.

## 5. Conclusions

Some relationships between positivity, passivity and hyperstability in dynamic systems, as well as and the characterization of the regenerative versus passive systems counterparts, have been described and mutually inter-linked in a simple way.

## Acknowledgement

This research was financially supported by the Spanish Government and by the European Fund of Regional Development FEDER through Grant DPI2015-64766-R and by UPV/EHU by Grant PGC 17/33.

## References

[1] G. Zames, IEEE Transactions on Automatic Control 11 pp. 228-238 (1966)
[2] N. Kottenstette and P.J. Antsaklis, 2010 American Control Conference, Baltimore 1, pp. 409-416 (2010)
[3] B.A. Francis, A Course in $H_\infty$ Control Theory, Lecture Notes in Control and Information Sciences No. 88, Springer-Verlag, Heidelberg (1992)
[4] P. Dorato , L. Fortuna and G. Muscato, Robust Control for Unstructured Perturbations- An Introduction, Lecture Notes in Control and Information Sciences No. 168, Springer-Verlag, Heidelberg (1992)
[5] M. De la Sen and A. Ibeas , Discrete Dynamics in Nature and Society 2005, pp. 31-67 (2005)
[6] M. De la Sen, Positivity 6, pp. 31-45 (2002)
[7] M. De la Sen M , Applied Mathematical Modelling 32, pp. 2312-2328 (2008)
[8] M. De la Sen M and S. Alonso-Quesada, 2008 Discrete Dynamics in Nature and Society 2008, Article ID 793512, 21 pages (2008)
[9] M. De la Sen and S. Alonso-Quesada, Applied Mathematics and Computation 215, pp. 2616-2633 (2009)
[10] M. De la Sen, International Journal of Control 44, pp. 1769-1775(1986)
[11] Z.Shen and S.B. Andersson , 49th IEEE Conference on Decision and Control 1, pp. 4819-4824 (2010)
[12] A. Fradkov, European Journal of Control 2003, pp. 577-586 (2003)

IOP Conf. Series: Materials Science and Engineering 472 (2019) 012014 doi:10.1088/1757-899X/472/1/012014

# Design and Implementation of Intelligent Monitoring System for Brake Shearing Machine

**Yong Wang[1,a], Jiamin Zhu[2,b,*], Sheng Chen[3,c], Jun Yu[4,d] and Dabing Chen[5,e]**

[1,2,3] School of Mechanical Engineering, HeFei University of Technology, HeFei, 230009,China

[4,5]MaAnshan ZhongYa Machine Tools Manufacturing Co. Ltd., MaAnshan, 243131, China

E-mail: [a]ywang9868@163.com, [b]1192457501@qq.com, [c]1905058587@qq.com,[d]631686773@qq.com, [e]249471281@qq.com

**Abstract.** In order to obtain the working state of the brake shearing machine, the paper proposed a monitoring system designed for the key data of the shearing machine such as the throat deformation, the temperature of the shearing cylinder and the vibration of the upper tool holder. The selection and connection of the hardware and the method of setting up the wired and wireless networks of the monitoring system are specified first in this paper. Then the methods of data acquisition and processing are put forward. Finally the functions of the host computer and the cloud server as well as the structure of the software are illustrated in detail. The monitoring system is suitable for the production line which contains many sheet metal machine tools, and it is of reference significance to establish a larger range of machine tool monitoring system.

## 1. Introduction

As the basic industrial equipment of sheet metal industry, shearing machines are widely used in many fields. *The intelligent monitoring system* of them will be an important part of the future sheet metal intelligent factory. But at present, there are few theoretical researches or engineering applications in this field [1].This paper presents a monitoring system which can collect, store and transmit the key data during the operation of the brake shearing machines, so that the intelligence degree of the traditional brake shearing machines can be improved. On the one hand, by analyzing the monitoring data, the shearing machine manufacturing enterprises can obtain the first-hand operating parameters so that they can make the machine tools more responsive to customer needs, on the other hand, machine tool production and application enterprises can better understand the use of equipment through the monitoring system, and more effectively supervise the equipment.

## 2. The Monitoring System Overall Design

The monitoring system is mainly composed of three modules: the data acquisition module, the cloud storage and management module, and the information display and processing module. *The data acquisition module* mainly relies on the sensors of temperature, acceleration, strain, etc., to obtain the operating status information of the brake shearing machines via wired or wireless networks. *The cloud storage and management module* uses the internet to send information collected by sensors to cloud servers for classification and storage, and thereafter for data sharing. *The information display and processing module* mainly relies on the on-site client software and cloud server software of the brake shearing machines. Figure.1 shows the architecture design of the monitoring system.

The monitoring system uses Wireless Sensor Network (WSN) to achieve sensor data acquisition and transmission. The mating hardware includes sensors, gateways, and controllers, etc. WSN is composed of spatially distributed routers and terminal nodes. A typical WSN system consists of routers, terminal nodes, and gateways. The distributed measurement nodes communicate with the central gateway wirelessly. The central gateway is responsible for the node authentication, message buffering, and network coordination from the IEEE 802.15.4 wireless network to the wired ethernet bridge. In this way, the measurement data collected by the sensors can be collected, processed, analyzed and displayed via the WSN network, as shown in figure 2.

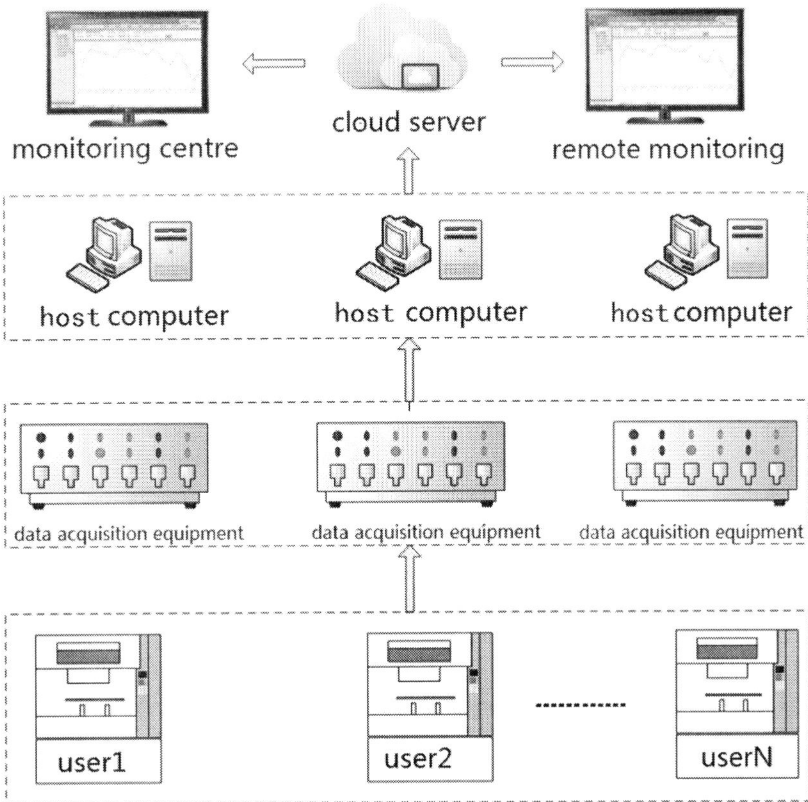


**Figure 1.** Architecture design of the monitoring system

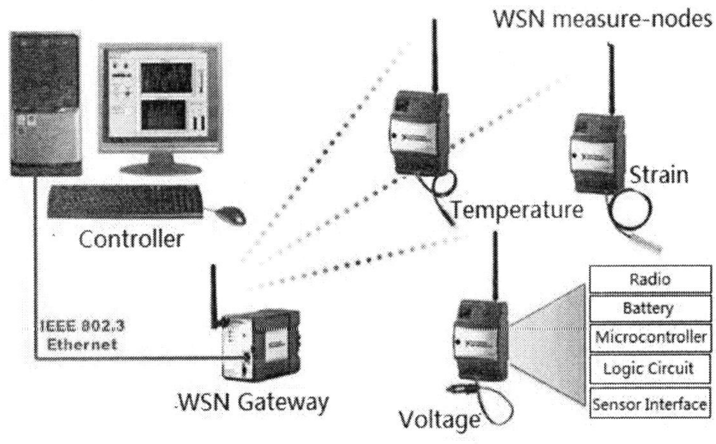


**Figure 2.** WSN network monitoring system

The WSN measurement nodes [2] can be configured as terminal nodes or routers using the NI Measurement & Automation Explorer tool. In the both configurations, nodes can collect measurement data from sensors, control their DIO channels, or program through the LabVIEW WSN for some more advanced functionality. To reduce power consumption, at the most of time, the terminal nodes are in the low-power sleep mode and are only awake when doing sampling and transmitting data and other interior information processing. However, the router nodes must always be awake to relay data from other nodes back to the gateway, extending the distance and reliability of the network.

## 3. Hardware Selection of the System

### *3.1. Data Measurement*

The throat strain measurement: The Wheatstone bridge has a high reliability and accuracy when measuring micro-resistance and temperature changes [3]. The strain gauge configuration of the half-bridge (type I) takes into account the influence of the wire resistance on the measurement, and can simultaneously measure the axial and bending strains with high sensitivity. Therefore, the Wheatstone half-bridge circuit is used to measure the strain of the gate type plate shearing machine in this paper.

During monitoring the strain on the brake shearing machines, NI WSN-3214 module is used. WSN-3214 module has 4 analog input channels. Each channel can be individually configured corresponding to Wheatstone bridge single arm bridge, half-arm bridge, full bridge circuit settings, and they can be used specifically for strain measurement. Figure 3 shows half bridge connection circuit of WSN-3214 and the strain gauge. The sampling frequency, waveform size, and time interval can be defined according to the actual needs of the monitoring. The WSN-3214 can also be used as a mesh network router to access more nodes or extend network distance. When it is used as a gateway, the WSN-3214 can connect up to 8 terminals or 36 measurement nodes, supporting a 300-meter sight-distance range.

The temperature measurement of the hydraulic system: In order to avoid the high hydraulic cylinder temperature in shearing machines under high load conditions, it is needed to measure the hydraulic cylinder temperature. A platinum resistance temperature detector (RTD) is used for temperature detection in this monitoring system. Taking into account the influence of the wire resistance during measurement, a three-wire connection is used. A thin-film RTD with a nominal temperature resistance of 100 Ω and a TCR value of 3851 is selected to monitor the temperatures of the cylinders at both ends of the upper tool carrier of the brake shearing machine. The NI WSN-3226 module can directly measure platinum RTD with nominal temperature resistance of 100Ω or 1000Ω. Because of the non-strict linear relationship between resistance and temperature, the NI WSN-3226 also incorporates a Callendar-Van Dusen linearization algorithm to fit the curve.

Vibration measurement: in order to avoid forced vibration near the natural frequency when the gate type plate shearing machine works, which will affect the machine service life and processing quality, it is particularly important to monitor the vibration of the gate type plate shearing machine [4]. When the shearing machines are in operation, the blade fixed on the tool carrier is sheared in the Z-axis direction. The vibration of the tool holder along the X-axis and the Z-axis will affect the straightness and parallelism of the cut plate, so it needs to consider the vibration detection in X and Z directions of the tool carrier.

When monitoring the vibration of the tool carrier along the two axial directions, NI C Series 9304 module and a 780985-01 industrial accelerometer with BNC terminals are used. The NI 9234 module has 4 channels for high-speed, high-accuracy acquisition of dynamic signals, which are suitable for IEPE sensors [5]. The dynamic measurement range of the NI 9234 module is 102dB, and a 2 mA constant current IEPE signal conditioning circuit is built in, which can be used for accelerometer and microphone data acquisition. The 4 channels can achieve an acquisition speed up to 51.2 kS/s, and the built-in anti-aliasing filter automatically adjusts the sampling rate. It is compatible with single-module USB box, NI CompactDAQ, and NI CompactRIO hardware and is ideal for the detection of vibration and noise. Shown in figure 4 is the grounding connection circuit of NI9234

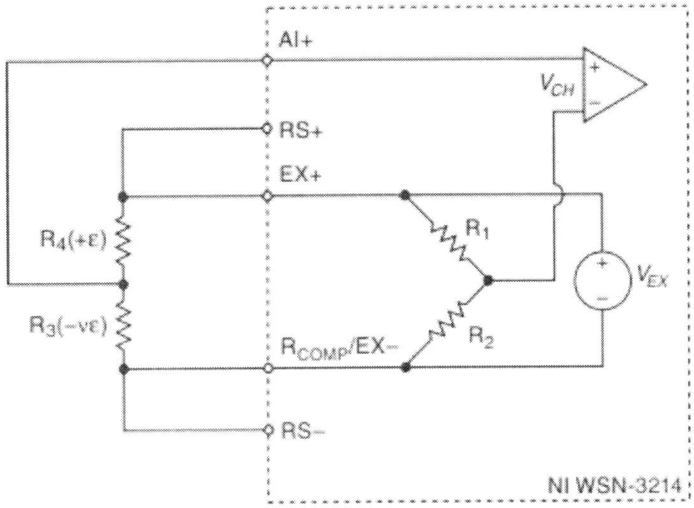


**Figure 3.** WSN-3214 Half-bridge circuit

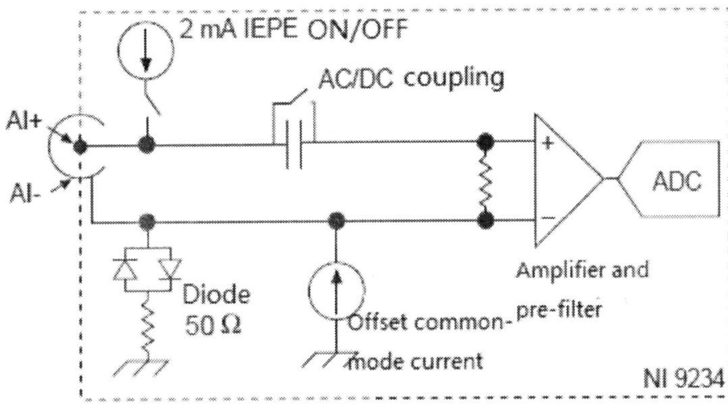


**Figure 4.** Grounding connection circuit of NI9234

*3.2. Gateway and Controllers*

The combination of wireless and wired sensor detection network constitutes the hardware system for data acquisition of the entire monitoring system. The WSN system also needs to configure the gateway and controller, and the controller module should be compatible with the NI 9234 module for wired detection. Therefore, the NI-9795 C series wireless gateway and the CompactRIO-based cRIO-9039 controller module are selected.

The NI 9795 wireless gateway and the NI WSN-32xx node are combined to form a wireless sensor network that can communicate with up to 36 wireless sensor network nodes. The CompactRIO-based cRIO-9039 controller module combines an open embedded architecture. It is with small size, ruggedness and includes custom timing, triggering, FPGA processing and modular I/O ports.

*3.3. Hardware Architecture and Configuration*

The monitoring system can be used for a lot of different sheet metal processing equipment to achieve the distributed monitoring of machine tools; meanwhile it is available for multi-state data acquisition by adding multiple types of sensors.

Figure 5 shows the overall hardware architecture configuration, including the accelerometer, strain gauge, three RTD sensors, and the corresponding three detection data acquisition modules: NI-9234, WSN-3214, and WSN-3226. The two main wireless acquisition nodes communicate with the NI-9795 wireless gateway to form a wireless detection network. The NI-9234 high-speed dynamic acquisition separately constitutes a wired detection network and is inserted into the CRIO-9039 controller together with the wireless gateway NI-9795. The controller communicates with the host computer to collect commands and send and receive data.

The sensor installation position is shown in figure 6.When measuring the dynamic strain of the throat, the strain rosette arrangement is adopted for each side wall plate, and multi-channel data can be acquired; the middle part of the upper tool carrier is subjected to relatively great stress and deformation during the shearing process, so the accelerometer is arranged on the middle position of the upper tool carrier; The temperature sensors are arranged on the both sides of the shear cylinders.

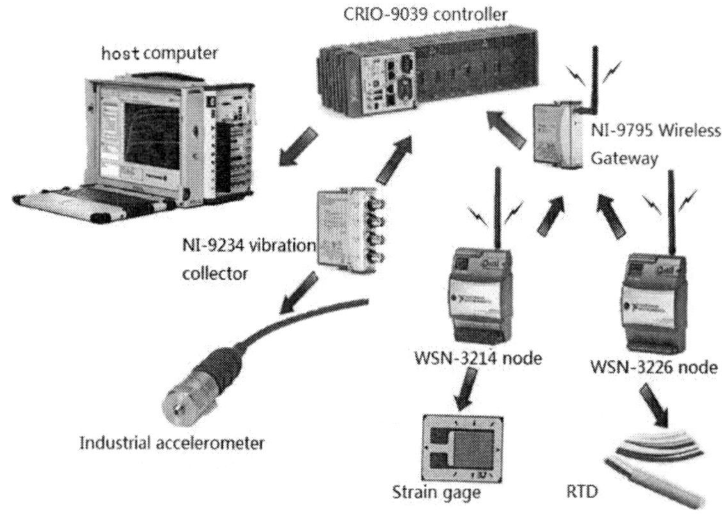


**Figure 5.** Hardware architecture

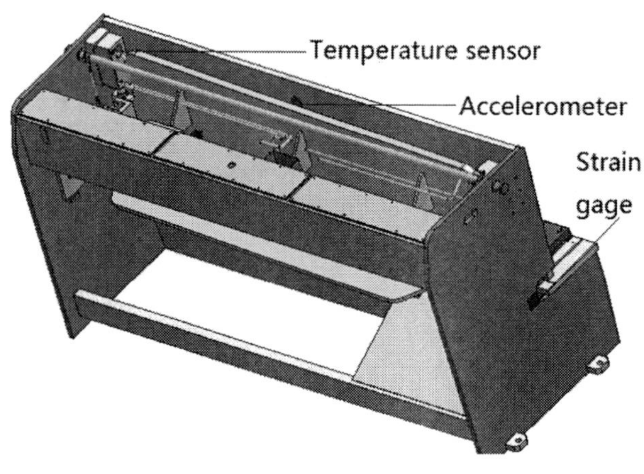


**Figure 6.** Sensor installation position

## 4. Data Transmission and Management

### *4.1. Acquisition of Sensor Signals*

When the signal is collected, the working status of the sensor is detected by the host computer or the control system, the operating time of the brake shearing machine is obtained, and the sensor signal is collected through the short distance wireless and wired methods. For the wired method, it is directly connected to the controller. The signal processing module in the controller processes the sensor signal and then connects the CDAQ or the computer. The computer or the CDAQ transmits the data to the database through the Internet or the wireless network [6].

Taking into account the stability of data transmission, transmission distance and multi-sensor information collection, the short-range wireless data transmission ZigBee has advantages like long transmission distance, low power consumption and high reliability. Therefore, it is more appropriate for the establishment of wireless sensor networks. ZigBee supports mesh network topology, so it has greater network flexibility and can route packets from the terminal nodes to the gateway through the shortest path. The gateway can be connected to DAQ (Data Acquisition Equipment) or computer equipment via Ethernet, USB, and serial ports.

### *4.2. Data Processing on the Server*

*4.2.1. Data transmission.* The information collected by the sensors is transmitted through the domain server, that is, the users of the brake shearing machines local area network server. And the VPN (virtual private) network is constructed by the same structure, which is used to assess the information of the work status of the gate type plate shearing machine issued in the company's local area network. The real-time monitoring and downloading of the shearing machine operating status parameters could be realized, which is shown in Figure 7.

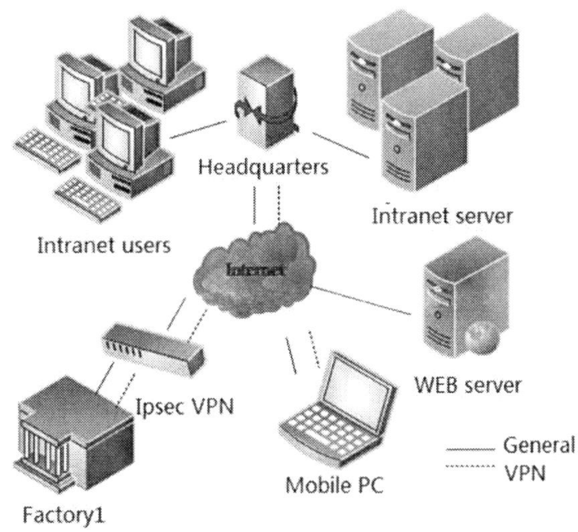


**Figure 7.** VPN application topology

The required configuration of the domain server itself is very low, and the occupation of network bandwidth is also insignificant. Under normal circumstances, the domain controller will not be deployed to the external network because it is related to the security of the entire domain organization. If the user wishes to log in to the enterprise domain on the external network, it needs to set the gateway VPN function, which not only can guarantee the security of account password transmission, but also as easy access to network resources as the LAN.

*4.2.2. WEB remote publishing.* Remote monitoring generally includes B/S (Brower/Server) and C/S (Client/Server) modes. In C/S mode, each client has full authority to operate the server database, the server responds at the same time, which is easy to cause server lack of resource. Its system security is poor, and performance is low. The B/S mode conforms to the TCP/IP conventions. In the modern WEB-centric enterprise network, HTTP is used as the transmission protocol. In this mode, the client accesses the web server and the background database through the browser. The client only needs to install the corresponding browser. The server installs the WEB server software. The client sends a data upload request to the server through the browser. After the server responds, it can complete the indirect access to the server database.

The intelligent monitoring system based on the gate type plate shearing machine needs to integrate the operation data of company's multiple shearing machines to realize remote monitoring. Therefore, the WEB release function of this system selects the remote monitoring method based on B/S mode.

## 5. Software Function

Combine the functions should be realized in this system with the functions required by software. The functions of the monitoring system software are determined as follows:

(1)Display the working status of the server, including IP address, port number, working mode and so on.

(2)The receiving status of data can be controlled and the relevant parameters can be modified artificially.

(3)Display wireless network module login, including module ID, phone number, current dynamic IP, and so on.

(4)Remote monitoring center and work data can be updated in time to ensure data consistency.

(5)Display the operation state of the brake shears, record abnormal data and give an alarm..

(6)The data can be stored reliably in the corresponding tables of the server database, and the historical data can be queried as needed.

## 6. Software Architecture

According to the software system requirements, the intelligent monitoring system based on brake shearing machine is divided into three subsystems: data collection, information management and on-line state monitoring.

Shown in figure 8 is the work flow chart of the intelligent monitoring software system of the CNC brake Shearing machine. The data collection system will store the data of stress, temperature and vibration acquiring from the sensor in the local disk and send it to the remote cloud server according to the agreed transport protocol.

The on-line condition monitoring system is the core of the software. The system uses graph to process the data of the sensor and displays the working state of the gate shearing machine used by field worker or remote worker

Information management system is the core of the whole system, with functions of manage information of the machine tool model, enterprise name, machine tool operation status, sensor installation. It is the link between the users and online status monitoring system.

Shown in figure 9 is the user interface of *Online Status Monitoring System* designed in this paper.

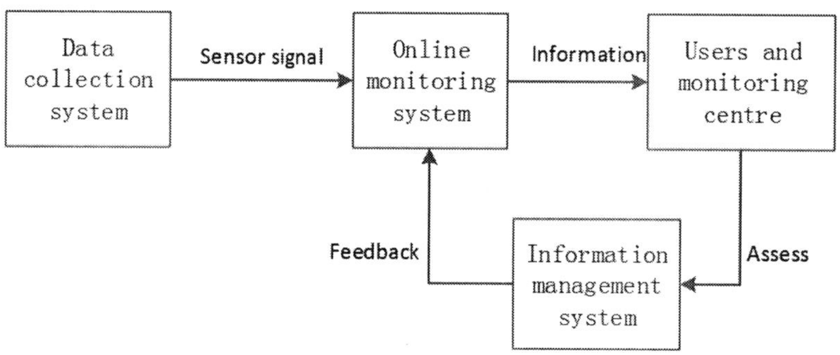


**Figure 8.** Work flow chart of software system

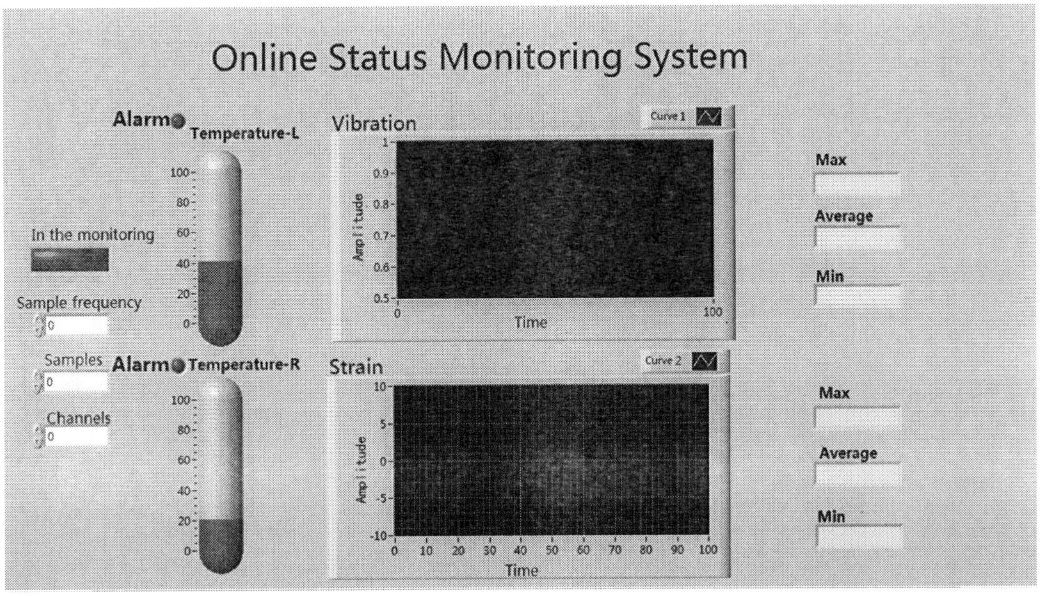


**Figure 9.** User interface of Online Status Monitoring

## 7. Conclusion

Aiming at the phenomenon of low intelligence of traditional brake shearing machine, this paper designs an intelligent monitoring system for the on-line monitoring of the throat deformation, shearing cylinder temperature and vibration of the upper tool holder, expounds the collection method of the key data and the transmission and processing method of the system data. The function and system software architecture of host computer and cloud server are expounded. The monitoring system is suitable for the production line which contains many sheet metal machine tools, and it is of reference significance to establish a larger range of machine tool monitoring system.

## Acknowledgement

This research was financially supported by the Anhui Provincial Major Research and Development Project < JZ2017AKKZ0987>, Anhui Provincial Digital Design and Manufacturing Key Laboratory Performance Subsidy Project < JZ2016AKSY1100>.

## References

[1] ZHAO Fei, MEI Xuesong, LI Guangdong, TAO Tao, JIANG Gedong. Machining Error Online Monitoring and Compensation of Numerical Control Forming GearGrinding Machine [J]. Journal of Mechanical Engineering, 2013, 49(1): 171-177.

[2] JIN Min, LUO Enze, ZHOU Xiang. Wireless Communication Protocol for Remote Intelligent Monitoring of Construction Machinery [J].China Mechanical Engineering, 2011, 22(19): 2316-2324.

[3] H Jiang, JG Vogel, S Nihtianov. A Power-Efficient Readout for Wheatstone-Bridge Sensors with COTS Components [J].IEEE Sensors Journal, 2017, PP (99):1-1.

[4] ZHU Shaowei, Li Weidong, Tang Liming, Du Li.ARTIS Tool Monitoring System Using in Milling of Aerospace Structural Components[J]. China Mechanical Engineering, 2016, 27(15): 2040-2043.

[5] YANG Zhe, CAO Lifang, WANG Yutian, HOU Peiguo, LI Hongjin, CHENG Pengfei, Pan Zhao. Analysis of Circuit Noise in Integral Electronics Piezoelectric Accelerometer [J]. Journal of Vibration, Measurement & Diagnosis, 2016, 36(05): 948-953+1026-1027.

[6] ZHANG Qiang, WANG HaiJian, LI liYing, LIU Zhiheng. Online Monitoring of Shearer's Pick Wear Based on ANFIS Fuzzy Information Fusion [J]. Chinese Journal of Scientific Instrument, 2016, 27(19): 2607-2614.

# Optimization of Chainsaw Setting Angle in Pruning Robot

**Haruhisa Kawasaki**[1,a,*], **Mouri Tetuya**[1,a], **Toshitake Yanagawa**[2,b] **and Haruo Nagayama**[2,b]

[1]1-1 Gifu University, Faculty of Engineering, 1-1 Yanagido, Gifu, Japan
[2]Leonix Co., Ltd, 3-10-7 Irifune, Chuuou-Ku, Tokyo, Japan
E-mail: [a]{[*]h_kawasa, mouri}@gifu-u.ac.jpmail, [b]{yanagawa, nagayama}@leonix.jp

**Abstract.** This paper presents an optimal mechanical design for a chainsaw setting angle to reduce the driving force of wheels in a pruning robot that cuts branches while spiraling up. Analysis and experimental results demonstrate the effectiveness of the optimal mechanical design for the chainsaw setting angle.

## 1. Introduction

Pruning is necessary in forests to produce high-quality wood and to maintain the forest environment. Pruning is, however, very dangerous work because a forestry worker must climb high trees, and support his body with one hand while cutting branches with the other gripping a pruning device. Thus, several engine-driven automatic pruning machines have been reported [1]; their use is limited, however, because of their heavy weight and the high frequency of branch bites. Pruning robots driven by servomotors have been developed, including a wheel-type [2, 3] and an inchworm-type [4]. Most of these robots are equipped with a strong pressing mechanism to protect the device from slipping downward due to its own weight. No practical and lighter weight pruning robot driven by servomotor has yet been realized.

We have developed the world's first climbing robot that uses its own weight [5], and is thus effective for weight saving. This robot can remain on a tree using its own weight without energy consumption, since it is driven by four active wheels through servomotors, worm-wheel reduction mechanisms with non-back-drivability, and a center of mass located outside of the tree. Other climbing robots using their own weight have been reported [6, 7]; they have one active wheel, two passive wheels and a low climbing speed, since the robot slips off the pole when the active wheel slips.

Power saving is a major challenge for such pruning devices, because the pruning robots are battery driven, and pruning consumes a fairly large amount of energy. When a prunng robot cuts a branch with a chainsaw while spiraling up a tree, it must then cut open a cut section of the branch to turn around the tree, which requires a high driving force of wheels. An optimization design for the chainsaw setting angle could contribute to reducing the driving force of the wheels.

## 2. Developed Pruning Robot

As shown in Figure 1, our pruning robot is equipped with four active wheels. Two wheels are located above and two below the pruning device. Each wheel is placed at the same angle, $\pm \pi / 4$ rad relative to the x-axis, when viewed from the top of the cylinder. The active wheel is driven by a DC servomotor via a worm-wheel reduction mechanism, which has non-back-drivability and allows the robot to be at rest without any energy expenditure. The robot's center of mass is located outside of the cylinder due to the controller being located by the down-side active wheels. Each of the four active wheels has a steering system to control the steering angle. The steering system is also driven by DC servomotors via a worm-wheel reduction mechanism.

The pruning device consits of a chainsaw with a pose-adjustment mechanism. When a branch bite occurs during pruning, work efficiency decreases considerably, because the operator must remove the robot/pruning device. To the best of our knowledge, no robot with a pruning mechanism that includes a branch-bite prevention function has yet been proposed. We previously presented a chainsaw prototype with a branch-bite prevention function [8]. In the prototype, a movable bar was mounted on the guide bar of the chainsaw so that it could move on the guide bar even if bitten by a branch. The robot could then continue to prune and advance, because the guide bar was not constrained by the branch. A movable bar is attached to the guide bar by a spring to enable planar motion of the movable bar.

(a)Visual appearance

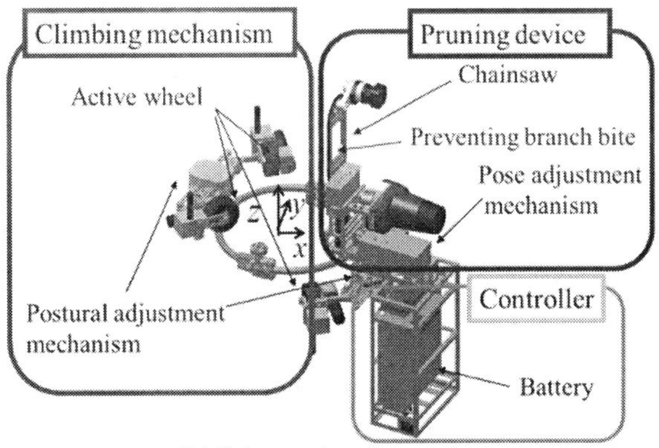


(b)Schematic of the robot

**Figure 1.** Developed pruning robot

## 3. Optimization of Chainsaw Setting Angle

When the robot prunes branchs during the straight motion that is normal in the use of a chainsaw device, the power to open up a cut section of branch is not needed. However, such power is needed through the guide bar of a chainsaw when the robot prunes branchs while moving up in a spiral around the tree, as shown in Fig. 2. In the figure, the orthogonal coordinate system is set at the center of the tree, points $A$ / $B$ are the right / left points of the chainsaw guide bar, point $C$ is the right-side point of the cutting section of a branch, point $C^*$ is a transfer point of $C$, $D_t$ is the diameter of the tree, $D_b$ is the diameter of the branch, $W$ is the width of the chainsaw, $l$ is the displacement of the chainsaw along the y-axis, $\theta$ is the rotation angle of the chainsaw device around the tree, $\theta_d$ is the cut angle of the branch, and $\xi$ is the setting angle of the chainsaw guide bar. When the chainsaw cuts the branch at displacement $l$, point $C$ moves to point $C^*$, which means that the guide bar spreads the cutting section with $\Delta x_c$. If $\xi$ is reasonably set, it is expected to reduce $\Delta x_c$. On the other hand, an interference between the cut surface of the branch and the left surface of the guide bar, $\Delta x_p$, is revealed. This means that the guide bar must compress the cut surface of the branch. This relation shows that an optimal setting angle exists for the pruning device.

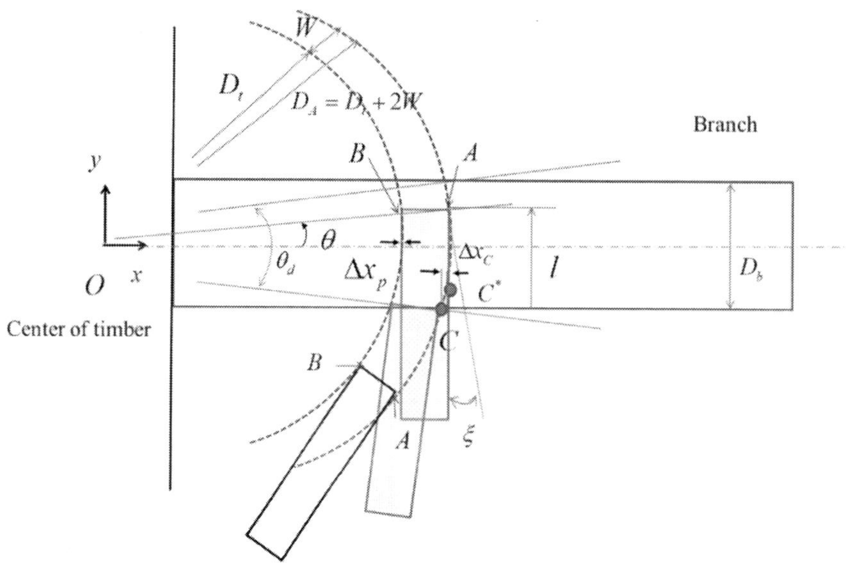


**Figure 2.** Geometric model of pruning by chainsaw device

The opening force to spread the cutting section, $F_o$, is derived by the relationship between the bending moment and the deflection as follows:

$$F_O = \frac{2EI_z\phi_A}{W((D_b + l)\cos\gamma + W\sin\gamma)} \tag{1}$$

where $E$ is the Young's modulus of the branch, $I_z$ is the moment of inertia of the area, $\phi_A$ is the deflection angle, and $\gamma(=\theta_d/2+\theta+\xi)$ is the slope angle of the guide bar. When the displacement $l$ and the setting angle $\xi$ are given, then $\Delta x_c$, $I_z$, $\phi_A$ and $\gamma$ are obtained geometrically, and $F_o$ is computed using these values. The drive force of the wheel is needed to overcome the opening force and friction force between the guide bar and the cut surface of the branch, which is caused by the opening force.

The interference $\Delta x_p$ varies with the position of the guide bar along the y-axis, and its maximum is derived by

$$(\Delta x_p)_{\max} = \frac{D_t}{2}(1-\cos\xi). \tag{2}$$

This shows the maximum of $\Delta x_p$ does not depend on the displacement $l$ of the chainsaw. From the relationship between the compressive force and the deformation, the compression force $F_P$ is derived by

$$F_P = \int_{y_{p1}}^{y_{p2}} \frac{\Delta x_P}{D_t} E\sqrt{D_t^2 - 4y^2}\,dy \tag{3}$$

where $y_1$ and $y_2$ are both ends of the interference interval along the y-axis. The drive force of the wheel is needed to overcome the compression force and friction force between the guide bar and cut surface of the branch caused by the compression force. This varies with $\xi$. Therefore, it is desirable that the total of the opening force and compression force given by

$$F_{total} = F_o + F_p. \tag{4}$$

be kept as small as possible by designing the optimal setting angle for the chainsaw device.

**4. Numerical Analysis**

To evaluate the effect of the setting angle of the chainsaw device, numerical analysis was conducted on the condition of $E = 9kN/mm^2$, which is the Young's modulus of Japanese cypress. Figure 3 shows the opening displacement $\Delta x_c$ and the opening forces $F_o$ with the condition $D_b = 40, 50, 60\,[mm]$, $D_t = 250[mm]$, $W = 6\,[mm]$ and $\xi = 0[rad]$. This shows that the curve of $\Delta x_c$ does not vary with $D_b$, while $F_o$ varies with $D_b$ remarkably.

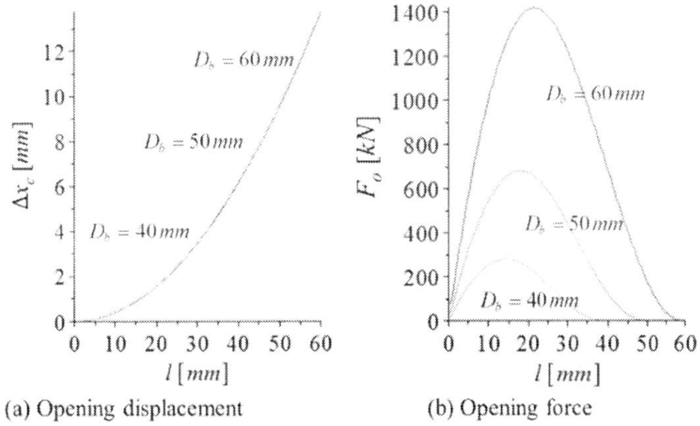


**Figure 3.** Opening displacement and opening forces.

Figure 4 shows $F_o$, $F_p$ and $F_{total}$ with the conditions $D_t = 250[mm]$, $D_b = 50[mm]$, and $\xi$=-0.105 [rad] (-6 [deg]). $F_o$ becames drastically smaller than that of $\xi$=0 [rad], $F_p$ increases, and $F_{total}$ is not so large. The maximum of $F_{total}$ (= $F_{\max}$) should be as small as possible.

Figure 5 shows $F_{\max}$ at $D_t = 150, 200, 250[mm]$. It shows that the minimum $F_{\max}$ varies with $D_t$ and $\xi$. Optimum $\xi$ minimizing $F_{\max}$ is -0.163, -0.128, -0.098 [rad] at $D_t = 150, 200, 250[mm]$, respectively. For a simpler mechanical design, it is better that $\xi$ is constant. This leads to an optimum $\xi$ of about -0.142 [$rad$] (8.1 [$deg$]), which is the cross point of the curves of $D_t = 150[mm]$ and $D_t = 250[mm]$. Then $F_{total}$ is 280 [$kN$], which is about 25% in the case of $D_t = 150[mm]$ and $\xi = 0[rad]$.

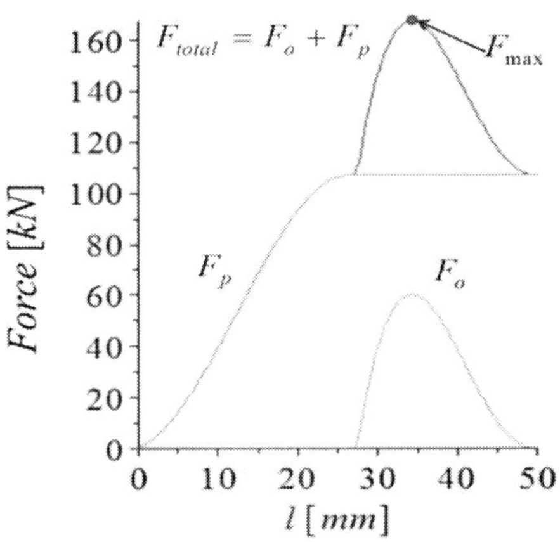


**Figure 4.** Force $F_o$, $F_p$ and $F_{total}$

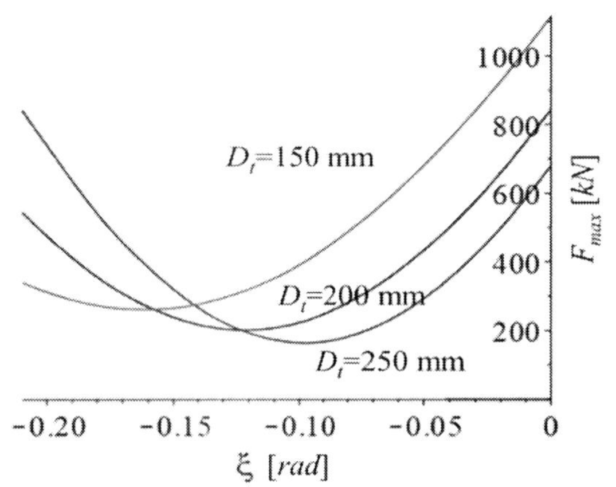


**Figure 5.** Maximum of $F_{total}$

## 5. Experiment

The designe method was evaluated experimentally using the circular experimental unit shown in Figure 6. Two types of chainsaw guide bars were used, as shown in Figure 7. One was our developed chainsaw guide bar [8], which has the branch-bite protection function; the other was a commercial guide bar.

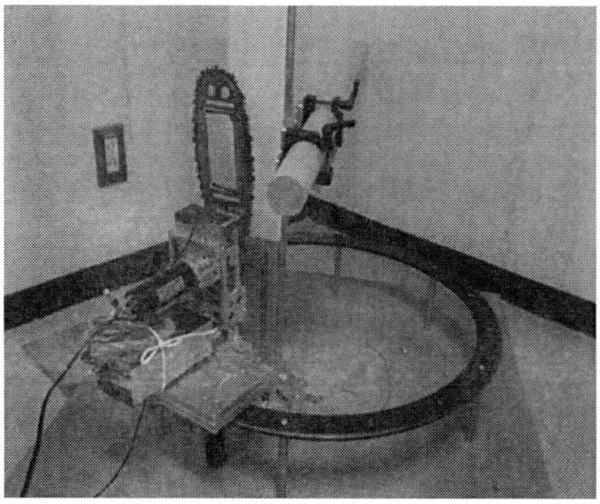

**Figure 6.** Circular experiment unit

(a) Developed guide bar Type (a) (b) Commercial guide bar Type (b)

**Figure 7.** Two types of guide bar

Table 1 shows power consumption and maximum torque of the circular motor drive system. The branch was a drying round bar, which was harder than a living branch of Japanese hinoki. The power consumption was evaluated during cutting time. Type (b) was evaluated at carriage velocity 3 [mm/s], because it could not cut the round bar at carriage velocity 9 [mm/s]. The power consumpution and maximum motor torque of both types (a) and (b) were largely reduced by setting the angle from 0 to 9 [deg]. The large effect of type (a) was due to the tilt mechanism of the movable plate on the guide bar [8].

**Table 1.** Power consumption and maximum torque of circular motor drive system

| Type of guide bar | Evaluation item | Setting angle $\xi$ [deg] | | Ratio [%] |
|---|---|---|---|---|
| | | 0 | -9 | |
| Type (a) at 9 [mm/s] | Power consumption [ws] | 22.6 | 0.56 | 2.5 |
| | Maximum motor torque [mNm] | 53.9 | 1.75 | 3.2 |
| Type (b) at 3 [mm/s] | Power consumption [ws] | 5.1 | 0.93 | 18.2 |
| | Maximum motor torque [mNm] | 1.16 | 0.69 | 59.4 |

## 6. Conclusion

We demonstrated that the optimization of the setting angle of a chainsaw device contributed to reduce power consumption and maximum motor torque of the carrying unit substantially in pruning during rotational motion. In future work, we plan to apply this optimum setting angle for the chainsaw device to the developed pruning robot and to evaluate it in a field setting.

**Acknowledgements**

This research is supported by a Grant-in-Aid for Scientific Research from JSPS, Japan ((C) 18K04017) and NEDO project P14033.

## References

[1] https://www.youtube.com/watch?v=eZdZTOxiR6s

[2] P. Gui, L. Tang, S. Mukhopadhyay, A Novel Design of Anti-falling Mechanism for Tree Pruning Robot, Proc. of IEEE 10th Conf. on ICIEA, pp.812-816, 2015
[3] G. Fu, X. Liu, Y. Chen, J. Yuan, Fast-growing forest pruning robot structure design and climbing control, Advances in Manufacturing, 3(2), pp.166-172, 2015,
[4] B.C. Widanagamage, T. N. Gallege, S. Salgado, J. Wijayakulasooriya, "TREEBOT: An Autonomous Tree Climbing Robot Utilizing Four Bar Linkage System," Proc. of SAITM Research Sympo. On Engineering Advancements 2014, pp.181-185, 2014,
[5] H. Kawasaki1, S. Murakami, H. Kachi, and S. Ueki, Novel Climbing Method of Pruning Robot, Proc. of SICE Annual Conference 2008, pp. 160-163, 2008
[6] W. Fischer, R. Siegwart, Wheeled Pole-Climbing-Robot With High Payload Capacity, Using Clamping Mechanism Which Is Inspired By The Rope-Clamps in Human Climbing, Proc. of CLAWAR'2010, pp.399-406, 2010
[7] A. Sadeghi, H. Moradi, and M. N. Ahmadabadi, Analysis, Simulation, and Implementation of a Human-Inspired Pole Climbing Robot, Journal of Robotica, Vol. 30, pp.279-287, 2011
[8] Y. Ishigure, H. Kachi, Y. Mori, H. Kawasaki, "Pruning Machine with A Mechanism for Preventing Branch Bite," Proc. of FORMEC 2010, pp. 1-9, 2010

IOP Conf. Series: Materials Science and Engineering 472 (2019) 012016 doi:10.1088/1757-899X/472/1/012016

# Speed and SOC Planning-based Control Strategy for Hybrid Electric Vehicle

**Qingyun He[1], Benhan Yang[2], Daming Cai[1] and Zongxiao Yang[1,*]**

[1]Henan University of Science and Technology, No.48 Xiyuan Road, Luoyang 471003, China

[2]Northeastern University, No.3-11 Wenhua Road, Shenyang 110819, China

*Corresponding author Email: zxyang@haust.edu.cn

**Abstract.** For solving the problem of low engine fuel economy caused by the low average driving speed in the running process of the hybrid power city bus, a control strategy based on the speed and the state of charge (SOC) rule extraction is proposed for hybrid electric vehicle (HEV) in this paper. Combining different vehicle speeds and battery SOC feedback to find the best fuel economy engine and motor status based on the traditional power-based control strategy, and to enable the battery pack to prolong life. The control strategy model is built in MATLAB/SIMULINK software in order to verify the effectiveness of the control strategy, and imported into the Cruise software to carry out co-simulation with the built vehicle power system model. The simulation results show that the proposed control strategy has good working condition adaptability compared with other rule based control strategies and the balance control of SOC is within 5%. It is shown that the design of the hybrid electric bus control strategy not only improves the fuel economy of the engine, but also prolongs the battery life.

## 1. Introduction

With the improvement of living standards, the per capita ownership of traditional fuel vehicles is increasing gradually. However, the emission of automobile exhaust has caused serious environmental pollution. Based on the existing industrial foundation, HEV and electric vehicles have become one of the best solutions at this stage, Plug-in hybrid vehicle have the advantages of ordinary hybrid and pure electric vehicles, and the state regulations that the production of a plug-in hybrid vehicle will be subsidized by at least a 50 Km[1]. When the battery power is sufficient, the electric mode runs; when the battery power is insufficient, the general hybrid mode is adopted.

As the core of the control system, the energy management control strategy of HEV can be divided into rule-based and optimization-based control strategy, and the rule based control strategy is widely used because of its simplicity. Cordiner et al. propose a control strategy to optimize the power consumption stage [2]. Capata et al. establish the four stage control strategy on the basis of the original charge depleting-charge sustaining control strategy [3]. Wang proposes a multi-stage and multi-objective control strategy based on the vehicle's different state of charge and power requirements [4]. Padmarajan et al. set the control strategy with the target mileage and battery SOC to select the best way to use electric energy [5]. Lin based on the state planning of power battery charge state, the control strategy of hybrid vehicle is formulated, and the balance of SOC is realized by using the torque allocation method referring to the SOC trajectory [6]. The above control strategy is mainly aimed at the planning control strategy of the battery SOC, but little consideration is given to the excessive charge and discharge of the power battery which will reduce its service life. For HEV, a control strategy that accords with its working conditions is formulated.

In order to improve the fuel economy of the whole vehicle and prolong the service life of the power battery, the energy management control strategy based on the battery SOC and the driving speed planning is put forward. So that make the engine and motor and battery pack work in the best condition.

## 2. Speed & SOC Planning-based Control Strategy

### *2.1. Topology Structure of Hybrid Power Transmission System*

The power system structure of HEV is mainly divided into four types: series, parallel, parallel-series and power split. Although power split hybrid power occupies a dominant position in the passenger car market, the parallel-series hybrid power is widely used in the commercial bus field because of its simple configuration and adaptation to the characteristics of various complex conditions [7]. The power system mainly includes engine, battery, Integrated Starter and Generator (ISG), driving motor and motor controller shown in Figure 1.

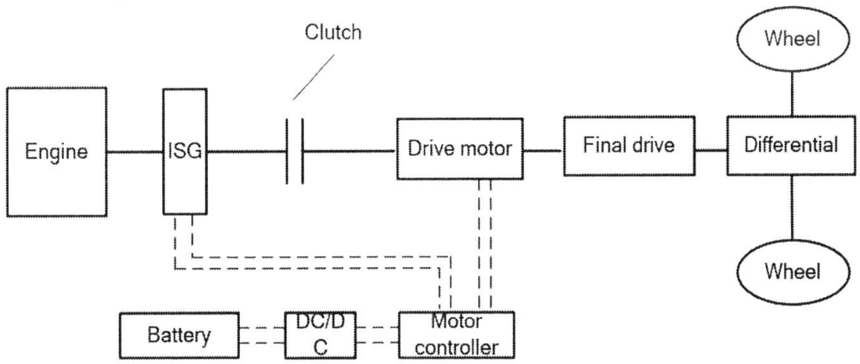


**Figure 1.** Topology diagram of hybrid power transmission system.

### *2.2. Construction of Control Strategy Based on Speed and SOC Planning*

Based on the torque control strategy, combined with the vehicle speed and the SOC state of the power battery, a control strategy based on vehicle speed and SOC planning is formulated. Because the bus running speed is in the middle and low speed period most of the time, the best working mode can be designed according to the different speed range, and the best efficient working interval of the power battery is at 40%~60%, which can combine the demand power of the whole vehicle. The specific hybrid power control strategy is designed, and $V_{min}$ shows the speed when the engine starts at the lowest speed when the hybrid drive is driven when the speed is in the 0~$V_{min}$, the whole vehicle is in a pure electric running mode. The $V_{max}$ indicates the corresponding motor speed when the engine is driven by the engine, and the P indicates the demand power in the driving process, and the $SOC_{min}$ indicates the minimum value of the power charge state of the power battery. The specific control strategy for vehicle driving is shown in Table 1 below.

**Table 1.** Control strategy of hybrid power bus

| Vehicle speed | Power | SOC | |
|---|---|---|---|
| | | $SOC<SOC_{min}$ | $SOC>SOC_{min}$ |
| $V<V_{min}$ | $P_{min}<P<P_{max}$ | Extended range mode | Motor drive |
| $V_{min}<V<V_{max}$ | $P<P_{min}$ | Vehicle charging | Motor drive |
| | $P_{min}<P<P_{max}$ | Vehicle charging | Engine drive |
| | $P>P_{max}$ | Vehicle charging | Joint drive |
| $V>V_{max}$ | $P_{min}<P<P_{max}$ | Engine drive | Engine drive |

### 2.3. The Working Mode of Hybrid Power System

The switching of the working mode of the hybrid bus is as follows: the power demand of the vehicle controller according to the current driver's operation, such as the acquisition of the acceleration pedal signal or the brake pedal signal, then calculates the corresponding demand torque, and combines the current working mode of the current demand of the power state of the power battery shown in Figure 2. According to the current speed signal and the power demand signal, the current mode of the output is judged by the power state of the battery.

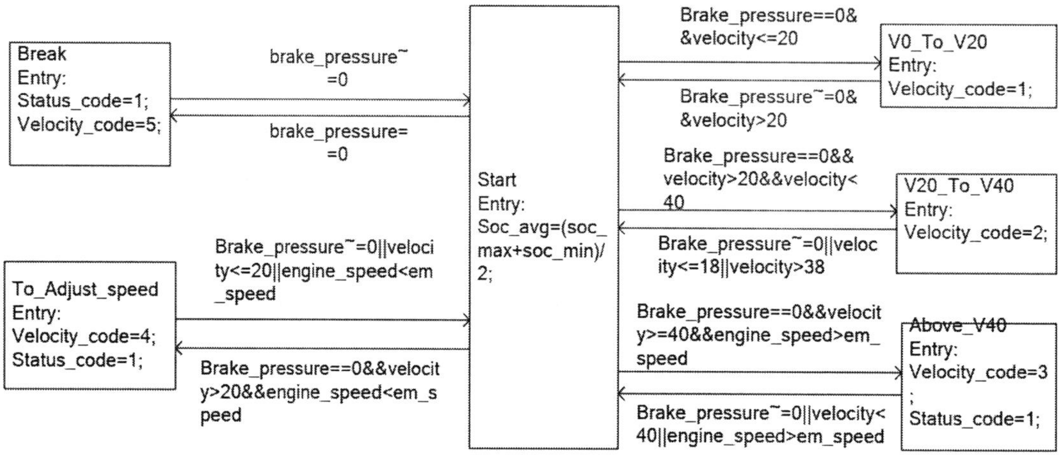


**Figure 2.** Stateflow working pattern judgment diagram.

## 3. Modelling and Simulation of Control Strategy

### 3.1. Parameter Selection of Hybrid Power Vehicle

In the simulation of vehicle control strategy, vehicle basic parameters, engine parameters, main reducer parameters and so on should be provided. The power of the vehicle is calculated according to the maximum speed, the maximum climbing requirement and the acceleration time, and the power of the motor is calculated with the different energy management strategies [8]. The power selection of the power battery pack is related to the motor to meet the power demand of the motor [9].

$$P_{max1} = \frac{v_{max}}{3600n_t}\left(m \cdot g \cdot f + \frac{C_D \cdot A \cdot v_{max}^2}{21.25}\right) \tag{1}$$

where, $v_{max}$ is the highest speed for the best pure electric driving; $\eta_t$ is the transmission efficiency for the power system; $f$ is the rolling resistance coefficient; $C_D$ is the air drag coefficient.

$$P_{max2} = \frac{v_i}{3600n_t}\left(m \cdot g \cdot f\cos\alpha + m \cdot g \cdot \sin\alpha + \frac{C_D \cdot A \cdot v_i^2}{21.25}\right) \tag{2}$$

where, $v_i$ is the speed for a car to climb at a certain speed; $A$ is the Frontal area.

$$P_{max3} = \frac{v_i}{3600}\left(m \cdot g \cdot f + \frac{C_D \cdot A \cdot v_i^2}{21.25} + \delta \cdot m \cdot \frac{dv_i}{dt}\right) \tag{3}$$

where, $\delta$ is the car rotating mass rotation factor.

$$P_e \geq \max\{P_{max1}, P_{max2}, P_{max3}\} \tag{4}$$

where, $P_e$ is the rated power of the drive motor and must meet the requirements of the car's maximum speed, acceleration time, and climbability.

$$P_{b\_max} \geq \frac{P_{m\_peak}}{n_b n_{m\_peak}} \tag{5}$$

where, $P_{b\_max}$ is the maximum power that can be provided for the power battery pack; $P_{m\_peak}$ is the peak power of the motor; $\eta_b$ is the efficiency of the power battery pack; and $\eta_{m\_peak}$ is the efficiency of the motor at the maximum power.

**Table 2.** Basic parameters of the vehicle.

| Parameters | Value |
|---|---|
| Curb weight /kg | 12500 |
| Full weight /kg | 18000 |
| Frontal area/m$^2$ | 7.5 |
| Drag coefficient | 0.65 |
| Rolling resistance coefficient | 0.018 |
| Engine power/kW | 147 |
| ISG power/ kW | 80 |
| Drive motor power/ kW | 144 |
| Battery capacity / (kW·h·V$^{-1}$) | 12.96/460 |

*3.2. Establishing the Simulink Model of Control Strategy*

The dynamic and fuel economical simulation of the SOC control strategy based on speed and power battery group is designed. First, a hybrid control strategy model is built in Simulink to deal with the signals transmitted from the Cruise software. According to the current speed signal and SOC signal, Simulink determines the working mode of the next step, establishes the Stateflow driving model judgment model, and then carries out the engine and motor torque distribution in the drive execution model.

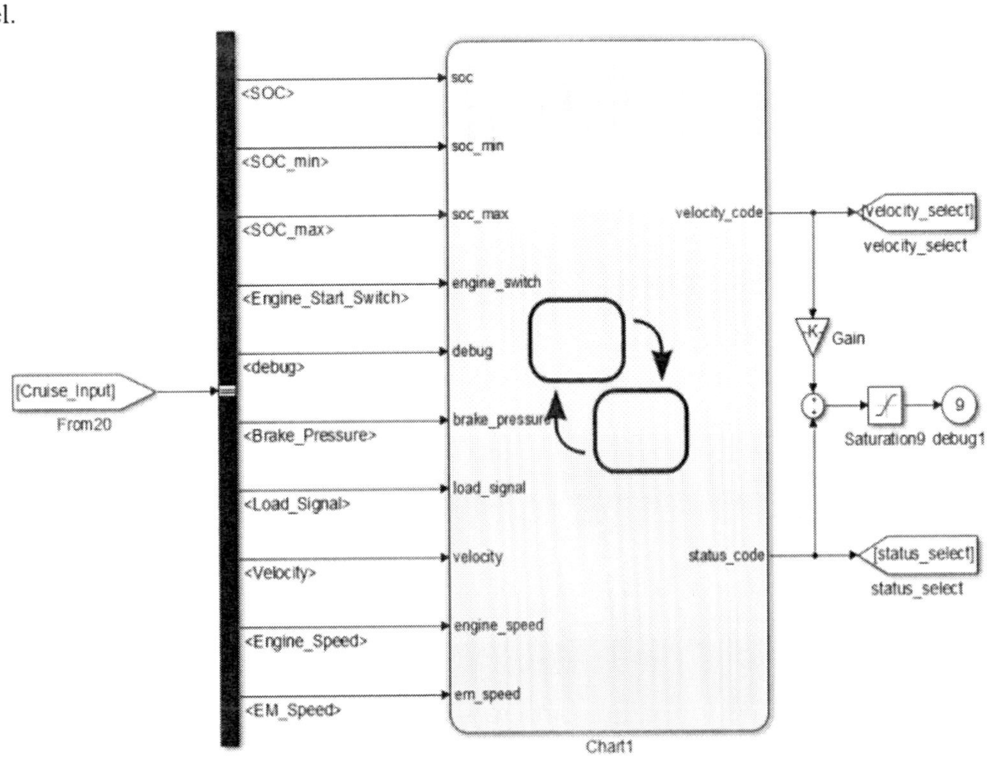


**Figure 3.** Diagram of stateflow pattern output.

IOP Conf. Series: Materials Science and Engineering 472 (2019) 012016 doi:10.1088/1757-899X/472/1/012016

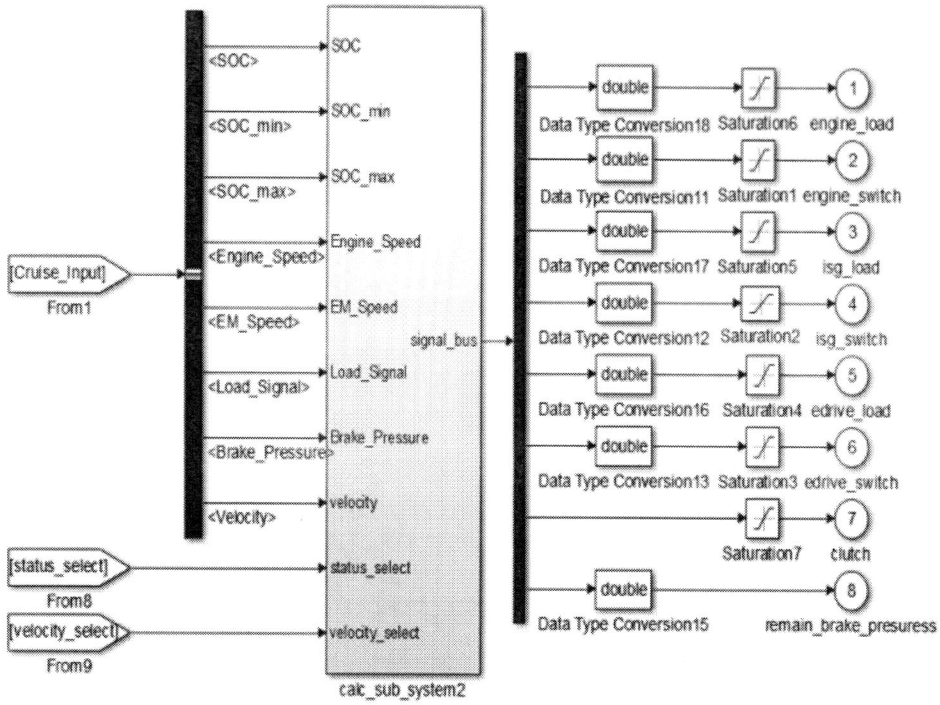


**Figure 4.** Diagram of drive pattern execution .

### *3.3. Building a Vehicle Cruise Model and Joint Simulation*

The whole vehicle model is built in the Cruise software according to the dynamic requirement of the whole vehicle, and the necessary parameters in each module are input and the connection of mechanical and electrical of the parts are conducted [10]. Communicate through the CAN-BUS, transmit the speed and load of cruise to Simulink, and judge the working mode through the drive calculator in Simulink, and transmit the engine signal and motor signal to the cruise model. The specific process is shown in the following figure 5 and 6.

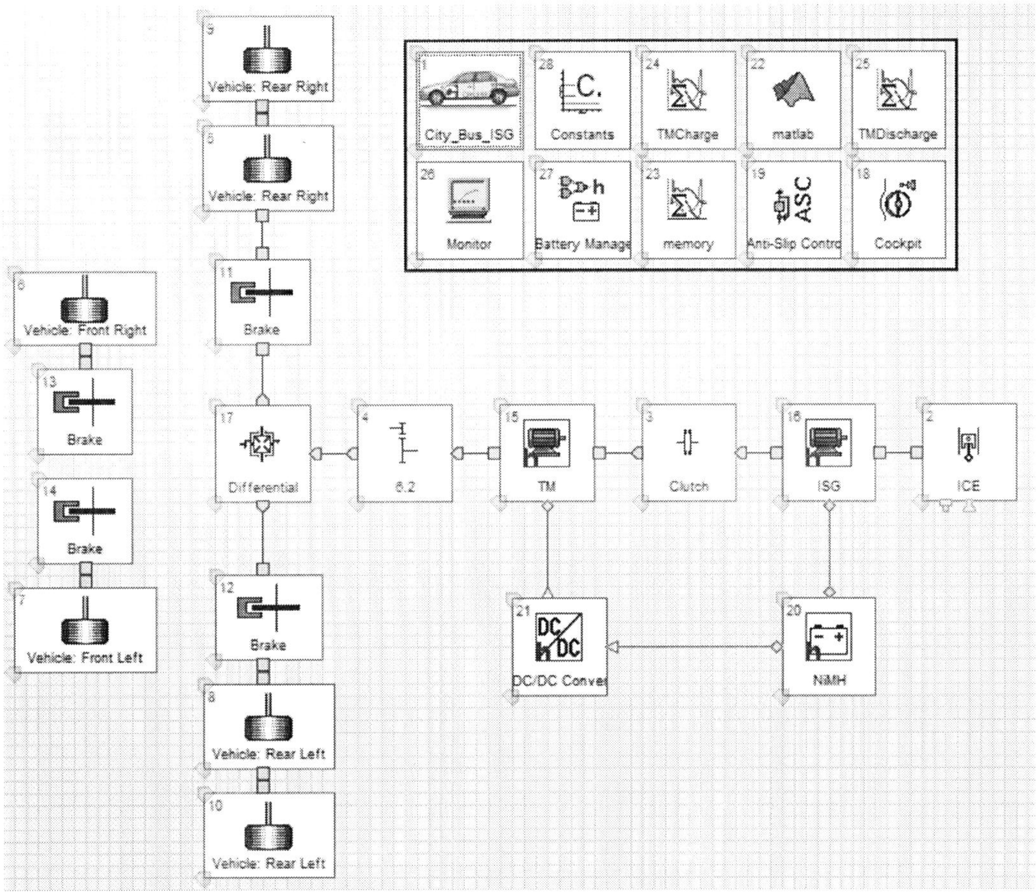


**Figure 5.** Vehicle mode of Cruise.

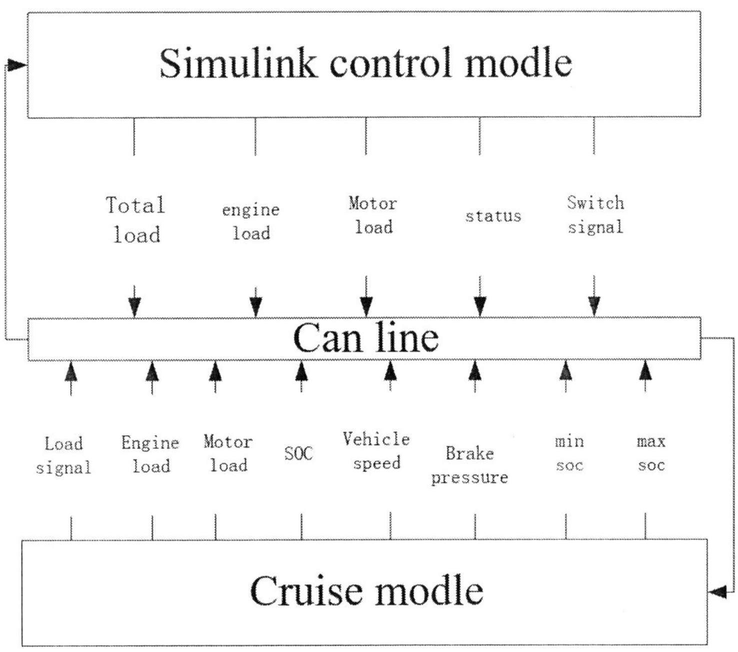


**Figure 6.** Diagram of joint simulation.

## 4. Simulation and Result Analysis

### *4.1. Analysis of Fuel Economy*

The fuel economy simulation is carried out in the typical city bus condition (CCBC) in China [11]. According to the control strategy mentioned above, we select electric mode when $0<V<20km/h$. When $20<V<40km/h$, this is a hybrid driving stage, when $V>40km/h$, the motor torque is taken into account. As the speed increases, the engine is driven independently. The results are shown in the following table.

**Table 3.** Fuel economy of different control parameters.

| $V_{min}$ | $V_{max}$ | Inital SOC (%) | Fuel (L/100Km) | Electric (KW·h/100Km) | Total (L/100Km) |
|---|---|---|---|---|---|
| 40 | 20 | 50 | 20.42 | -9.68 | 17.20 |
| 40 | 20 | 60 | 20.01 | -7.35 | 17.56 |
| 40 | 20 | 70 | 16.34 | 5.02 | 18.02 |
| 40 | 20 | 80 | 9.81 | 26.57 | 18.67 |

From the above table, the overall fuel consumption increases with the increase of the initial value of SOC. The main reason is that when the initial value of SOC becomes larger, according to the set control strategy, the driving mode is mostly pure electric working mode and the power consumption of power battery is more, and the less the corresponding braking energy recovery is.

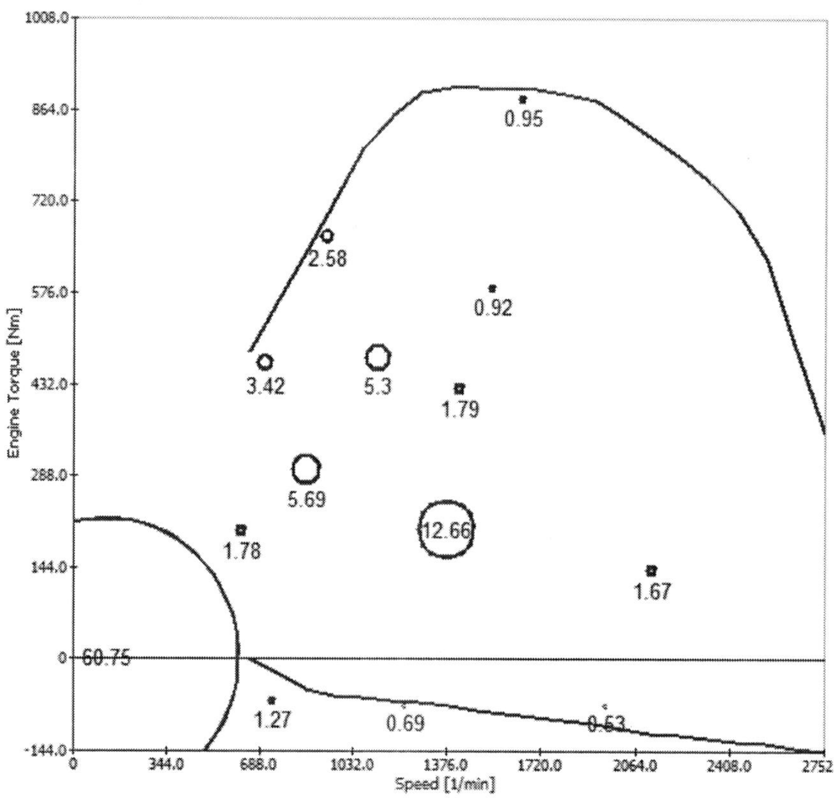


**Figure 7.** Inital soc is 40%

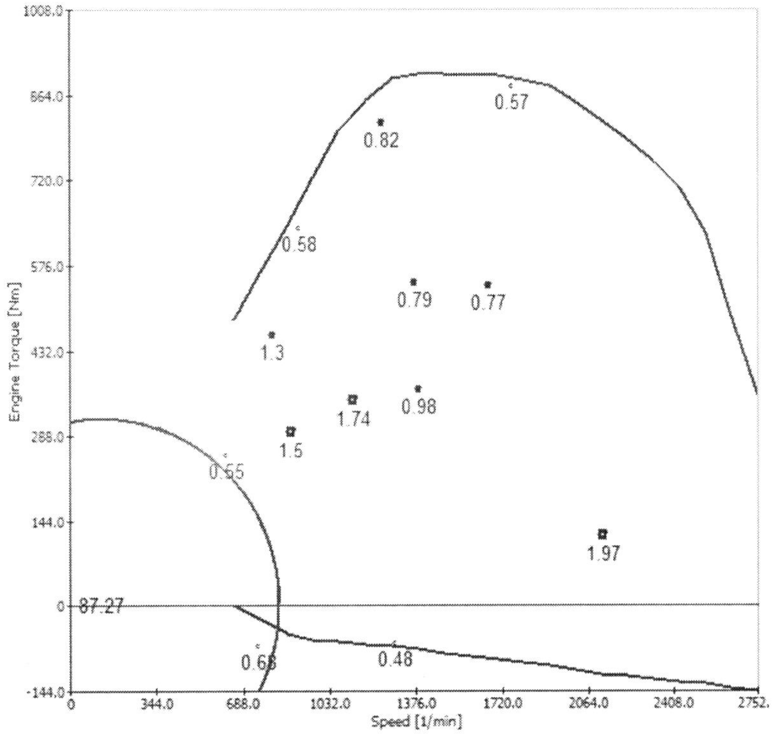


**Figure 8.** Inital soc is 80%.

*4.2. Effect of Initial SOC on Engine Working Point*

When the power battery state selects different initial values, the corresponding comprehensive fuel consumption will change. When the initial charge state of the battery is 40%, the power system will choose the pure electric mode first, so the engine fuel consumption is reduced in the whole simulation condition and the consumption of electricity will increase accordingly. When the initial charge state of the battery is 80%, the vehicle controller controls the engine to work on the optimal efficiency curve, so the engine working point is close to the engine's optimal efficiency curve.

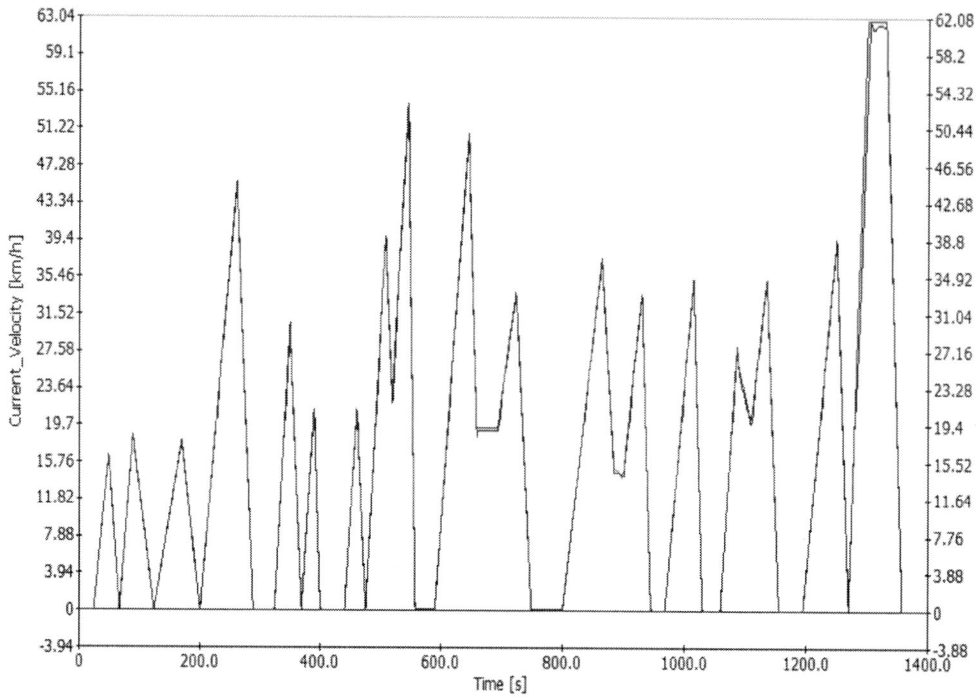


**Figure 9.** Target speed and current speed.

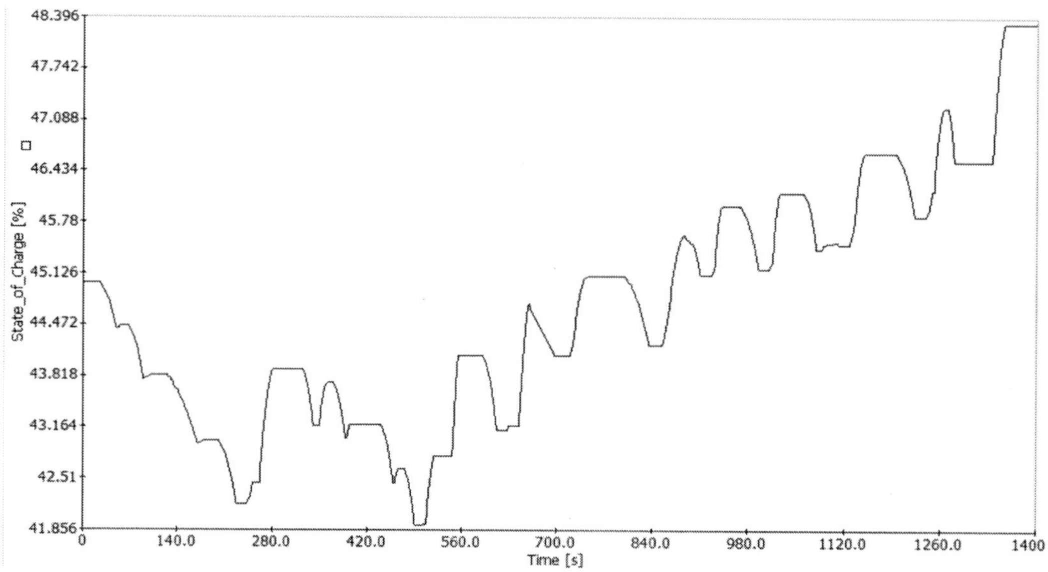


**Figure 10.** SOC curve changes with time.

### *4.3. Analysis of the Speed Following and SOC Balance*

As shown in Figure 9.In the whole simulation process, the current speed is in good agreement with the target speed, which indicates that the hybrid electric vehicle is simulated with the standard condition of CCBC. In the process of charging and discharging, the high efficiency area of the power battery is generally located at between 40%~60% [12], The SOC balance is mainly to ensure that the power battery should not change too much during the process of simulation. From the Figure 10, the dynamic battery SOC changes from the initial 45% to the end of 48.3%, and the SOC balance is kept within 5%, so the control strategy is designed reasonably.

## 5. Conclusion

The control strategy based on speed and SOC planning is designed in this paper. The simulation analysis shows that the fuel economy of the whole vehicle is better when the initial value of SOC is 45%~55%, and the fuel economy is less than 3% under different simulation conditions, and the balance of SOC is within 5%, which validates the effectiveness of the control strategy.

## Acknowledgement

The financial support from the National Natural Science Foundation of China (NSFC) under the grant No. 70671035 and No. 71071078 is acknowledged gratefully.

## References

[1] Wu G, Boriboonsomsin K, Barth M J. Development and Evaluation of an Intelligent Energy-Management Strategy for Plug-in Hybrid Electric Vehicles [J]. IEEE Transactions on Intelligent Transportation Systems, 2014, 15(3):1091-1100.

[2] Cordiner S, Galeotti M, Mulone V, et al. Trip-based SOC management for a plugin hybrid electric vehicle[J]. Applied Energy, 2016, 164:891-905.

[3] Capata R, Bella G, Sciubba E, et al. A Real Time Energy Management Strategy for Plug-in Hybrid Electric Vehicles based on Optimal Control Theory[C]// ENERGY PROCEDIA. 2014:949-958.

[4] Wang J, Wang J, Wang Q, et al. Control rules extraction and parameters optimization of energy management for bus series-parallel AMT hybrid powertrain [J]. Journal of the Franklin Institute, 2018, 355(5):2283-2312.

[5] Padmarajan B V, Mcgordon A, Jennings P A. Blended Rule-Based Energy Management for PHEV: System Structure and Strategy [J]. IEEE Transactions on Vehicular Technology, 2016, 65(10):8757-8762.

[6] Lin X, Ivanco A, Filipi Z. Optimization of Rule-Based Control Strategy for a Hydraulic-Electric Hybrid Light Urban Vehicle Based on Dynamic Programming[J]. Sae International Journal of Alternative Powertrains, 2012, 1(1):249-259.

[7] Banvait H, Anwar S, Chen Y. A rule-based energy management strategy for Plug-in Hybrid Electric Vehicle (PHEV) [C]// American Control Conference. IEEE, 2009:3938-3943.

[8] Škugor B, Cipek M, Deur J, et al. Design of a power-split hybrid electric vehicle control system utilizing a rule-based controller and an equivalent consumption minimization strategy [J]. Proceedings of the Institution of Mechanical Engineers Part D Journal of Automobile Engineering, 2014, 228(228):631-648.

[9] Lee H, Jeong J, Park Y I, et al. Energy management strategy of hybrid electric vehicle using battery state of charge trajectory information [J]. International Journal of Precision Engineering and Manufacturing-Green Technology, 2017, 4(1):79-86.

[10] Chen C. The development of hybrid electric vehicle control strategy based on GT-SUITE and Simulink [J]. Atlantis Press, 2015.

[11] Arata J P I. Simulation and control strategy development of power-split hybrid-electric vehicles [J]. Georgia Institute of Technology, 2011.

[12] Millo F, Rolando L, Fuso R, et al. Analysis of Different Energy Management Strategies for Complex Hybrid Electric Vehicles [J]. Computer-Aided Design and Applications, 2014, 11(sup1):S1-S10.

IOP Conf. Series: Materials Science and Engineering 472 (2019) 012017 doi:10.1088/1757-899X/472/1/012017

# Basic Experiment of LIDAR Sensor Measurement Directional Instability for Moving and Vibrating Object

**Shinya Hama[1,*] and Hideki Toda[1]**

1Dept. of Electric and Electronic Engineering, University of Toyama
Gofuku Campus, 3190 Gofuku, Toyama, 930-8555, Japan
*Corresponding author's e-mail: m1871027@ems.u-toyama.ac.jp

**Abstract.** In the present paper, a basic study of LIDAR sensor measurement data directional instability for moving straight object such as wall and vibrating object such as human or trees trembling by wind was confirmed. To realize autonomous car control without human, the correct current position estimation (localization) is an important factor, and it would be measured by LIDAR device generally. Since the LIDAR sensor used time of flight information of laser to measure the distance from the target, the micro or nano second order time difference would be disturbed by a few physical factors - sensor movement / vibration, target shape complexity and target movement / vibration. Especially, if the LIDAR was used for object recognition, it is necessary to know what kind of environment. Object and movement factors disturb the sensor signal. By using a developed artificial tree that was influenced by the wind and moving stand attached the LIDAR, the effect of the sensor self position movement and the object shape vibration were measured in this paper. The experimental result of 0.4×0.2 m straight shape board movement shows that there are large directional dependencies of moving straight shape board about the rotation angle estimating process by using ICP algorithm. In addition, the experimental result of 0.4×0.2 m artificial tree attached 40×200 mm green ribbon shows that the estimated rotation angle was continuously departing from the true value (zero degree) especially in few minutes experiment. Our experimental result shows that there is a weak direction to move the LIDAR sensor itself and it is hard process to estimate the rotation angle by ICP when the object is vibrating or moving.

## 1. Introduction

In this paper, a basic study of LIDAR sensor measurement data directional instability for straight shape moving object such as wall and vibrating object such as human or trees trembling by wind was confirmed by developing artificial tree attached ribbons, and the evaluation was performed by estimating the object existing angle by simple ICP algorithm. Government and companies cooperate each other and the technology development of automatic driving is progressing actively for the aim of practical application by 2020 [1]. To realize autonomous car control without human, the correct current position estimation (localization) is an important factor, and it would be measured by LIDAR device generally [2].LIDAR (Light Detection and Ranging) is used as an important device for recognizing surrounding environment like human eyes and ears. Since the LIDAR sensor used time of flight (TOF) information of laser to measure the distance from the target, the micro or nano second order time difference would be disturbed by some few physical factors - sensor movement / vibration, target shape complexity and target movement / vibration (Fig. 1). Especially, if the LIDAR was used for object recognition (wall, tree, human and so on), it is necessary to know what kind of environment, object and movement factors disturb the sensor signal.

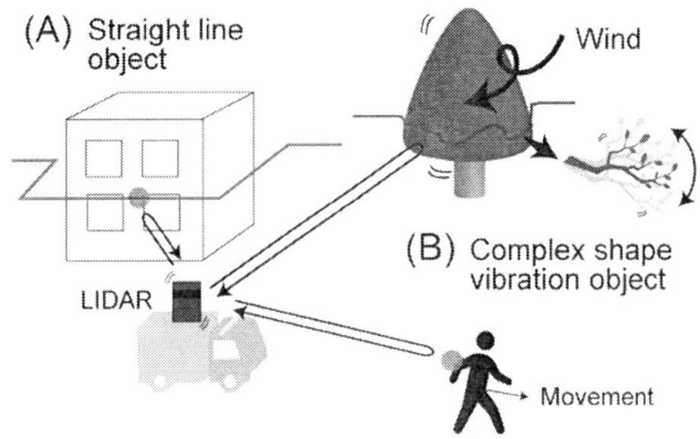


(A) Straight line shape object such as wall or building. (B) Complex shape and movement / vibration object.

**Figure 1.** Some factors of noise appearing on the LIDAR sensor.

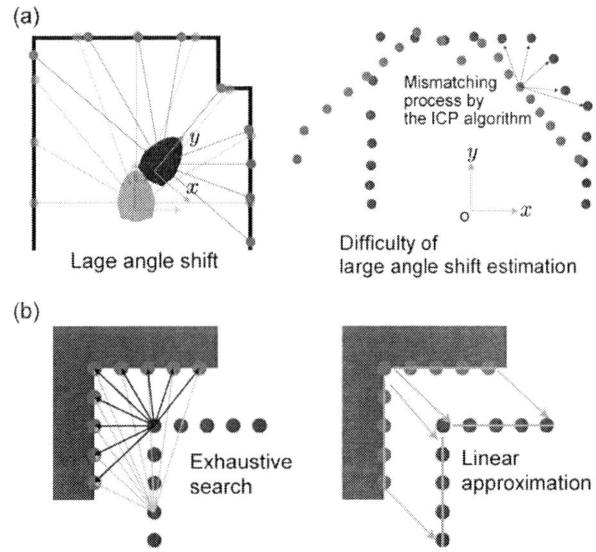


**Figure 2.** Difficulty of rotation angle estimation process by using ICP algorithm.

In previous study, the localization methods which are robust against the influence of sensor noise and past estimation errors by using the improved scan matching techniques and a probability estimation that take into account the errors have been proposed and developed [2]. In those studies, the LIDAR sensor output's disturbance have been treated as uniform "noise", and it have not been taken into account the factor of movement direction of the sensor itself for example.

The aim of this study is firstly to measure the dependency of the target object movement direction by the rotation angle θ estimation calculated by ICP (Iterative Closest Point) algorithm. The reason of using the rotation angle θ as the evaluation function is that the variable with the large influence on the accuracy of self position estimation is the rotation angle θ [3]. As shown in Fig. 2a, even if there is a small positional movement of the LIDAR, once the rotational angle would be dispersed from the original angle, re-estimation of the shifted angle becomes a difficult problem in the ICP algorithm that performs the corresponding point search of the closest point.

In addition as shown in Fig. 2b, to reduce the computing explosion problem of the ICP, many previous study used that linear approximation was performed to try to improve the efficiency of calculation cost by grouping points [2]. However, if the points would be grouped as "line", the grouping algorithm does not operate when there is a complicated shape object such as trees in the environment.

In this paper, by using the rotation angle estimation based by ICP as an evaluation function, the performance of the LIDAR was confirmed for the moving a straight line board and an object whose shape changes temporally like trees.

## 2. Method

Figure 3 shows the experimental setup devices, (a) LIDAR sensor (RPLIDAR A1, SLAMTEC Corp., 0.15-6 m and 360deg ranging, precision 1%, 2000 samples/sec) with a stand, (b) 0.4×0.2 m straight shape board object, and (c) 0.4×0.2 m vibrating object by using many green ribbons. Object (c) was developed as a model of tree leaves in natural environment, and the ribbons were vibrated by a wind flow (2.0 m/sec) while the experiment.

As a rotation angle θ estimation, the evaluation function E(θ) was used,

$$E(\theta) = \sum_{i=1}^{m} \left\| \vec{b_i} - R(\theta)\vec{a_i} \right\|^2 \tag{1}$$

$$R(\theta) = \begin{pmatrix} \cos\theta & -\sin\theta \\ \sin\theta & \cos\theta \end{pmatrix} \tag{2}$$

Where $\vec{a_i}$ is the previous measured point set represented as (x,y) axis, and $\vec{b_i}$ is the present measured point data set. R means the rotation matrix. The rotation angle θ was estimating by minimize the evaluation function E by simple Euler method.

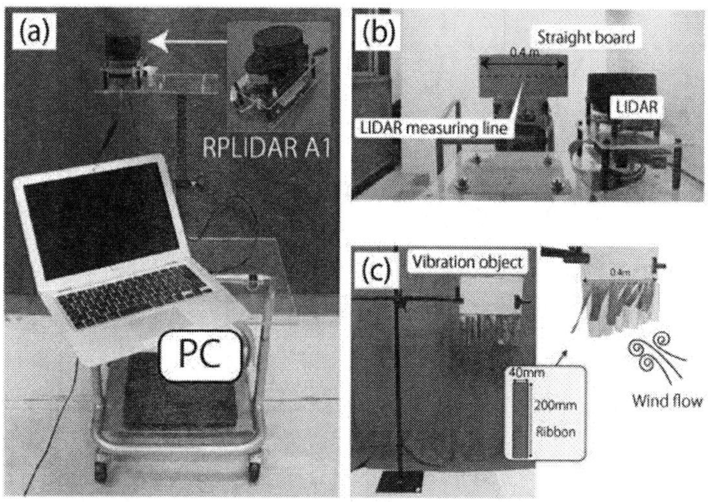


(a) LIDAR sensor (RPLIDAR A1, SLAMTEC Corp). (b) Straight line shape board object (0.40.2 m). (c) 0.4×0.2 m artificial tree attached 40×200 mm green ribbons

**Figure 3.** Experimental setup devices.

## 3. Experiment

Figure 4a shows the straight board movement experimental setup (experiment 1). The 0.4×0.2 m straight board was positioned on the 1.5 m front of the LIDAR sensor, and the board was moved 1 m distance from the original position with 0.25 m/sec to (L) left, (R) right, (B) backward and (F) front direction. The area of the measurement was limited at a black line square 3×3 m area to reject another object except for the board.

Next experiment 2, the experimental setup was shown in Fig.4b. The 0.4×0.2 m green ribbon vibration object was positioned on the 1 m front of the LIDAR sensor, the ribbon object was vibrating by 2 m/sec wind flow generated from an electric fan. The standard deviation (S.D.) of the ribbon vibration was measured as 4.6 cm in the case of 2 m/sec wind flow.

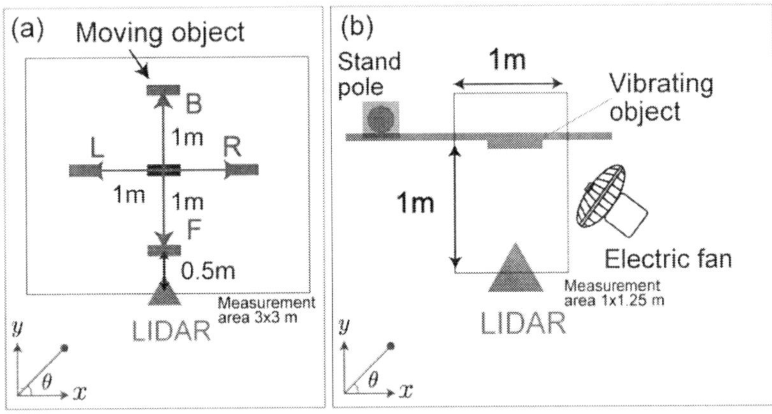


(a) Straight shape board objects movement strategy in experiment 1. (b) Artificial tree vibrating object setup of experiment 2.

**Figure 4.** Experimental setup of experiment 1 and 2.

## 4. Result

Figure 5 shows the rotation angle θ estimation result of the straight board object movement. In the case of (R) right or (L) left direction movement, the estimated θ (Blue is (R), Green is (L)) were shifting about 10 degree while the 4 sec board movement. Average number of distance measurement points of the straight board was about 7 points while the experiment. Next, in the case of (F), the estimated θ (Red) was transitioned to over 40 degree and the was changing unstably while 7 sec period, and it was settling to about 10 degree close to (L) case. Although this phenomenon does not always occur, it would affect the position / direction estimation accuracy. Last, in the case of (B), there is large unstable θ estimation error about 120 degree, and the estimation could not recover after the 4 sec movement. In the (B) back movement case, the average number of measurement points was about 3.5 after the 1 m movement. Experiment 1 result clearly shows that there was movement direction dependency even if the target was straight board shape.

Figure 6 shows the rotation angle θ estimation result of the vibrating ribbon object by wind. In the without wind (0 m/sec), the θ was changing within small range (5.09±2.5 degree, average ± S.D.). However, the θ was shifted to 40 degree after over 80 degree estimation within 1 min when the wind flow start, and the shift could not recover to zero degree and it kept maintaining around 50 degree (-48.9±10.5 degree). Important point is that the distance movement of the ribbon was about 4.6 cm, and this small vibration also affects to the θ estimation.

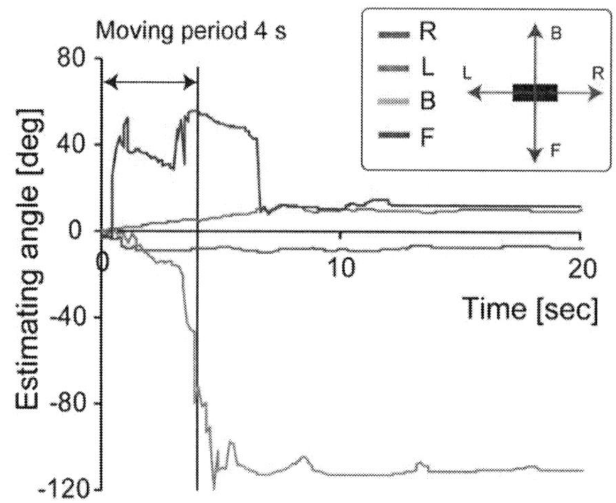


**Figure 5.** Result of experiment 1. 0.4×0.2 m straight shape board movement case.

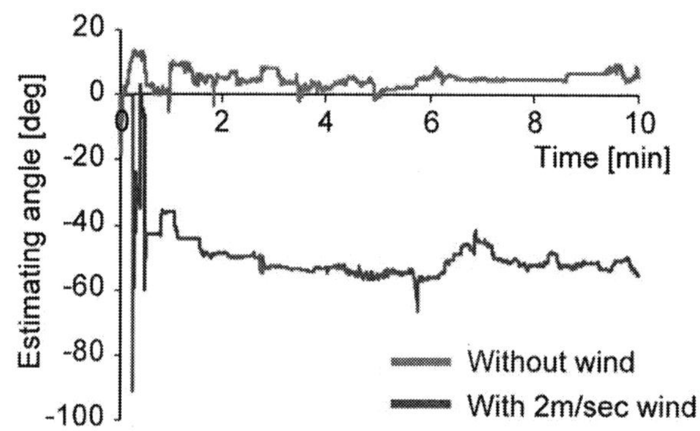


**Figure 6.** Result of experiment 2. 0.4×0.2 m artificial tree vibrating objects case.

## 5. Conclusions

In this paper, a basic study of LIDAR sensor measurement data directional instability for moving straight object such as wall and vibrating object such as human or trees trembling by wind was confirmed. The experimental result of experiment 1 shows that even if the object was straight shape board, there was directional dependency of object movement for angle estimation based by ICP algorithm. There was large effect to (B) backward direction movement, and it was related to reducing the measurement points of the LIDAR. In the case of vibration object made with green ribbons, there were large estimation angle shifting about 40 degree (max 80 degree) when the wind flow start, and about S.D. 10 degree vibration was occurred while the wind flow period. Above experimental result shows that (1) in the SLAM algorithm, it is necessary to consider the directional information and (2) the rotation angle estimation would be largely affected by the small distance change generated by small wind flow.

## References

[1] MILT, "Main government policies concerning automatic driving", URL: https://www.mlit.go.jp/jidosha/jidousha_tk7000018.html, (accessed on 2018 March in Japanese).
[2] Shinya IWASHINA, Atsushi YAMASHITA, Toru KANEKO, "3-D SLAM in Dynamic Environments by a Mobile Robot Equipped with two Laser Range Finders", RSJ 2008, 2008.
[3] P.J. Besl, Neil D. McKay, "Method for registration of 3-D shapes", IEEE Transactions on Pattern Analysis and Machine Intelligence, Vol. 14, Issue 2, pp. 239-256, Feb. 1992.

IOP Conf. Series: Materials Science and Engineering 472 (2019) 012018 doi:10.1088/1757-899X/472/1/012018

# Development of Autonomous 4 Rotor Helicopter 50 M-range Movement Control Method by Using Single Search Light and the Evaluation

**Naoki Kasamatsu**[1,*] **and Hideki Toda**[1]

[1]Dept. of Electric and Electronic Engineering, University of Toyama
Gofuku Campus, 3190 Gofuku, Toyama, 930-8555, Japan
E-mail: m1871010@ems.u-toyama.ac.jp

**Abstract.** In the present study, 50 m-range autonomous 4 rotor helicopter movement control method by using single search lights was proposed, and the autonomous flight test evaluation was performed. To realize over 50 m long distance flight of the drone under unstable GPS signal situations such as under the bridge or inside tunnels for the periodic inspection, the correct self-position measurement and improvement of the control method are indispensable for stable control. The aim of this study is to compare three control methods of over 50 m range autonomous 4 rotor helicopter movement using a high power 100V AC single search light as InfraRed sources, and the searchlight is positioned at near the investigation target such as bridge handrail. By comparing the standard deviation (S.D.) from the central orbit divided by the moving distance, a proposed control method using 2nd order velocity factor and a control method using boundary condition of both sides represents high performance against general optimized P-D control. The proposed 2nd order velocity factor control and the control method using boundary condition realized 12% and 45% small S.D. comparing with the optimized PD control respectively.

## 1. Introduction

In this paper, 50 m-range autonomous 4 rotor helicopter movement control method by using single search lights was proposed, and the autonomous flight test evaluation was performed. To realize over 50 m long distance flight of the drone under unstable GPS signal situations such as under the bridge, inside tunnels and buildings for the periodic inspection, easy and correct self position measurement method and the control theory (strategy) of the drone are indispensable. Even if the drone would be controlled by human being, the flight by visual confirmation of man is difficult in the over 10 m long distance flight. In that situation, other supporting mechanisms of the position measurement and the control method would be necessary [1]. The aim of this study is to compare three control methods of over 50 m range autonomous 4 rotor helicopter movement using a high power 100V AC single search light as InfraRed sources, and the searchlight is positioned at near the investigation target such as bridge handrail (Fig. 1). By comparing the standard deviation (S.D.) from the central orbit divided by the moving distance, a proposed control method using 2nd order velocity factor and a control method using boundary condition of both sides represents high performance against general optimized P-D control.

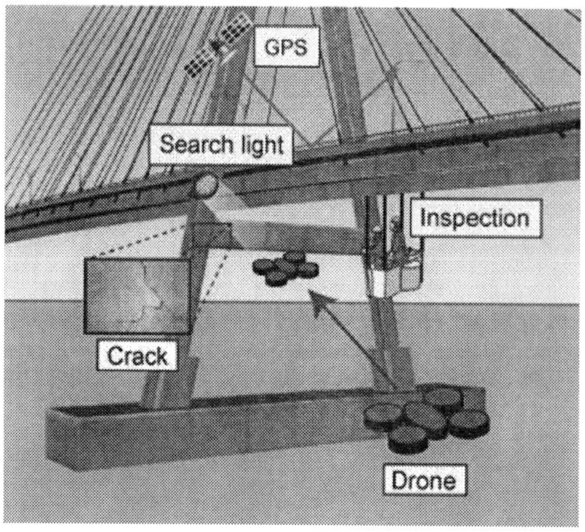


**Figure 1.** Proposed over 50 m range autonomous four rotor helicopter movement control using high power 100V AC single search light.

## 2. Previous Study

Four rotor helicopter would not be included autonomous position controlling program as itself, and the positioning system is necessary in the space using InfraRed 3D cameras or GPS sensor system at least to control the drone position [1]. In the case of the InfraRed 3D camera, the precision of the position measurement is 1 mm order, however, the area of the using this method is within 10 m and indoor situation only. On the other hand, if the drone would be controlled in the outdoor situation, the GPS signal can be used in the situation that there is almost no obstacles upper direction (sky) and movement direction. However, in order to use the drone for the periodic inspection of under the bridge or inside tunnels, the two approaches could not be adopted. In addition, attitude estimation and autive for the position control of the drones. In this study, we used three types oftonomous flight control are an important topics in the study field of the drone system. It is not clear that what kind of the flight control law (algorithm) is effec the controller design (1) optimizing gain adjusted PD control, (2) a method using 2nd order velocity factor and (3) a control method using boundary condition of both sides. Especially, the control strategy (2) has been discussed in previous our studies [2].

## 3. Method

### *3.1. Experimental Devices and the Setup*

Figure (fig2) shows the experimental setup devices, (1) AR Drone 2.0 quad rotor helicopter (Parrot Corp.), (2) same AR Drone with InfraRed filter (IR76, FUJI FILTER, FUJIFILM Corp.) attached on the front camera, (3) fluorescent red color sheet (0.29×0.24 m, 5065 fluorescent sheet red, Myst Corp.), and (4) high power search light (Stage Evolution, PAR56SBG and SYLVANIA light, 16 cm diameter light PAR56 300W, SOUND HOUSE Corp.) was used for the InfraRed light source. The camera (320×240 image with 10 Hz) at the front of the drone was used to find the fluorescent marker or search light position, and the drone was controlled by the fluorescent marker or search light position on the camera image (using the center of gravity of the point). The drone was controlled by ARDroneForP5 library by Y.Shigeo.

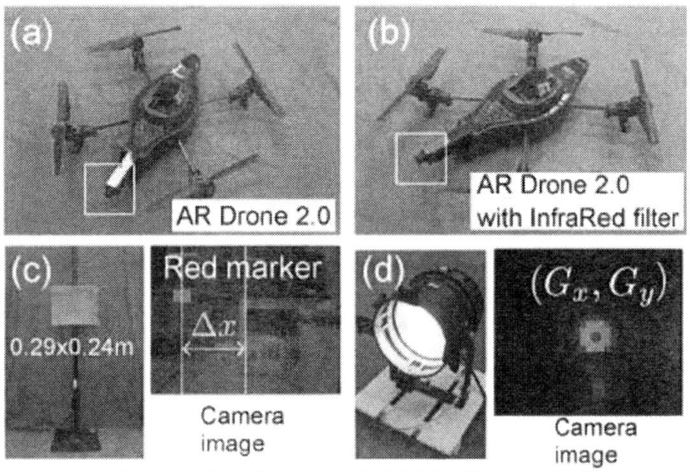


(1) AR Drone 2.0 (parrot Corp.) and (2) with InfraRed (IR76) filter at the front camera. (3) Red colored target marker (size 0.29×0.24 m). (4) Search light (combination of PAR56SBG and PAR56 300W).

**Figure 2.** Experimental setup devices.

*3.2. Three Control Methods*

The drone was controlled by three different control strategy in this study. When the center of the gravity of the red target object was $\left(G_x, G_y\right)$, the drone's movement roll and pitch speed commands $F^r$,$F^p$were described as,

$$F^r = -\gamma_r(160 - G_x) - \eta_r V_r \quad (1)$$

$$F^p = \gamma_p \quad (2)$$

where the meaning of $\gamma_r$, $\eta_r$, $\gamma_p$ were constant, and $V_r$ was drone roll velocity. Roll direction used simple P-D feedback control, and pitch direction was feed forward control process. The feedback parameters $\gamma_r$ , $\eta_r$ were determined by continuously measured 30 sec flight control experiments (N=over 50) in order to minimize the S.D. $\sigma_x$. The speed of the control command transfer was same with the camera frame rate (10 Hz). It is the first control strategy (it denoted as P control) of the three methods denoted in Figure 3a.

Next control strategy was represented as below control equations:

$$F^r = -\eta_r V_r |V_r| \quad (3)$$

The parameters $\eta_r$was determined same with the above P control method, and the control command of pitch direction was also same with Eq.2.It is second control strategy denoted as Method (1).

Last control strategy was represented as below control equations:

$$F^r = \begin{cases} -\gamma_r(\Delta x + d) & (\Delta x < -d) \\ 0 & (|\Delta x| < d) \\ -\gamma_r(\Delta x - d) & (\Delta x > +d) \end{cases} \quad (4)$$

$\gamma_r$ was determined same with the P-D control, and $d$ was determined by the 1/4 distance of the aisle. It is thrid control strategy denoted as Method (2).

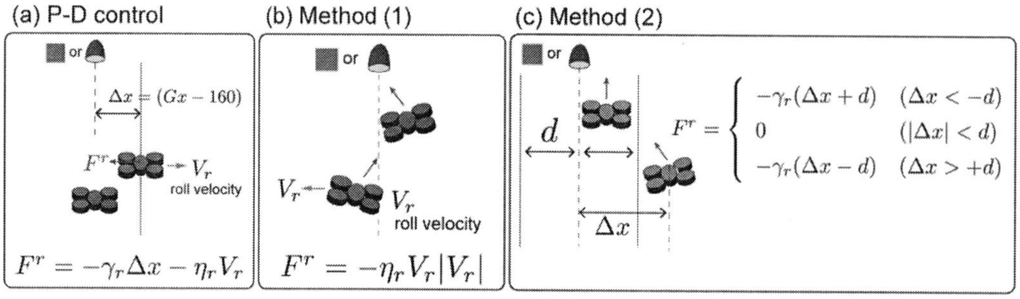


**Figure 3.** Three control strategy using simple PD control, a proposed control method using 2nd order velocity factor, and a control method using boundary condition of both sides.

## 4. Experiment

Experiment 1 performs over 50 m distance autonomous back and forth control stability by using the red marker (Fig. 4).The drone was controlled by the red marker's position on the front camera image (A of Fig. 4), and it was moved back and forth within 2.0 m about 30 turns by changing the control parameter of pitch direction $\gamma_p$ (Ep.2) positive or negative. The position of the drone was measured by the camera attached on the ceil (B of Fig. 4), and the $x$ and y position were continuously measuring by the image. Axis $x$ and y correspond to the roll and pitch direction of the drone respectively. In this situation, three control methods (P-D control, Method (1), Method (2)) were performed.

Experiment 2 performs the long distance autonomous control stability of the corridor surrounded by wall. Width of the corridor was 2.0 m, and the search light was positioned 35 m distance from the start point. The drone was controlled by the search light position on the front camera image (IR75 InfraRed filter was attached on the camera), and it was moved back and forth within 15 m by changing $\gamma_p$ parameter. The position of the drone was calculated by the search light's center of the gravity and area of the image. In this situation, three control methods (P-D control, Method (1), Method (2)) were performed.

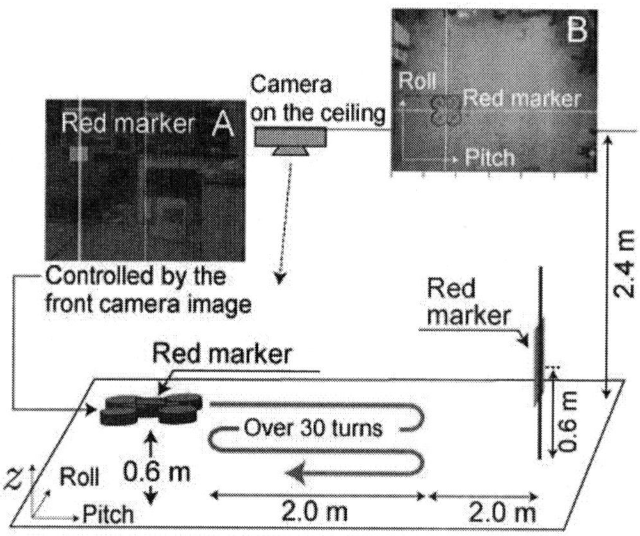


**Figure 4.** Experimental setup of over 50 m distance autonomous go and back control experiment by using red marker in 6×6 m room.

## 5. Result

Figure 5 left represents the result of roll and pitch position transition over 30 turns within 2.0 m distance. The trajectory of Method (1) and Method (2) were shown in Fig. 5a and Fig. 5b respectively. In the Method (1) result, a total length of the trajectory was measured about 60 m, and the standard deviation

(S.D.) of the roll movement was $\sigma$=0.2 m. Values of the $\sigma$ at the point of start position and end position were calculated as $\sigma_s$=0.23 m and $\sigma_e$=0.12 m respectively, and it represents that if the drone's position is close to the red marker, the position control performance becomes to increase.

In the Method (2), the boundary condition limitation $d$ was fixed to 0.5 m. The total $\sigma$ equals to 0.31 m, and it was 1.55 (=0.31/0.2) times larger than the Method (1). Values of the $\sigma$ at the point of start position and end position depended on the distance from the red marker, and it was same feature of the Method (1).

In Fig. 5 right, the $\sigma$ divided by the total flight distance (it was denoted as $\sigma/l$) were compared in the three control methods. The graph and the error bar (S.D.) shows the result of five time experiments (N=5) of the three control methods. Optimized P-D control (denoted as P control) got worst value ($\sigma/l \times 10^3 = 25.7 \pm 6.9$) comparing to the remaining two methods, and the Method (1) took best value ($\sigma/l \times 10^3 = 3.1 \pm 1.0$) in three methods. Though the control method of the Method (2) ($\sigma/l \times 10^3 = 11.6 \pm 3.7$) is simple, the control performance was good comparing to the P-D control. Since the Method (2) is just controlling the drone when the boundary condition $d$ is exceeded, the battery or energy performance would be good and it would be difficult for the control to become unstable. On the other hand, the Method (1) have been studied and analyzed in our previous studies [2], and it could realize a good performance in a specific condition. However, by the reason that the controller design is complex (2nd order term is used) and the real time control is indispensable, the battery or energy performance would be bad, and it would be easy to become the control unstable.

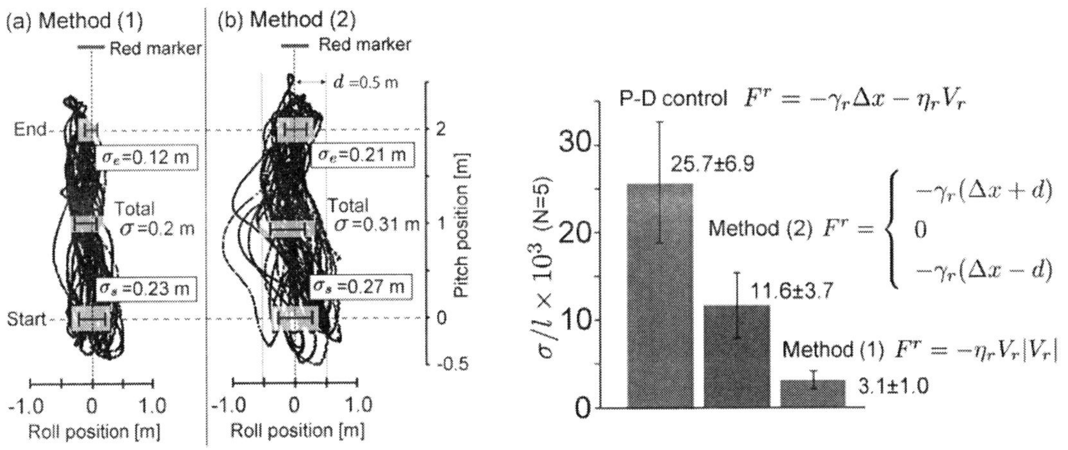


**Figure 5.** Result of experiment 1 roll and pitch position transition over 30 turns within 2.0 m distance. The controller Method (1) and Method (2) are compared.

Experiment 2 performs the control to go and back 15 m distance corridor (width 2 m) surrounded by wall (Fig. 6). The blue / red mean the going and returning trajectories. The boundary condition $d$ set as 0.25 m in the Method (2). The S.D. of the go / back area in the Method (1) took 0.23 / 0.20 respectively. On the other hand, the S.D. of the go / back in the Method (2) took 0.18 / 0.17 respectively, and this value is smaller than the Method (1). This means that the Method (1)'s non linear instability would be appeared in the long range control condition, and at least in the 15 m range control, simple controller design Method (2) is useful.

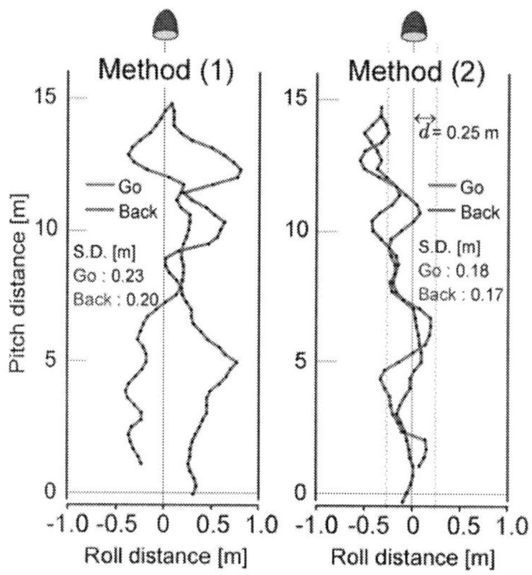


**Figure 6.** Result of the experiment 2 roll and pitch position transition one turn 15 m distance. The controller Method (1) and Method (2) are compared.

## 6. Conclusion

In this paper, 50 m-range autonomous 4 rotor helicopter movement control method by using single search light was proposed and the autonomous flight test evaluation was performed. In order to realize over 50 m long distance flight of the drone under unstable GPS signal situations such as under the bridge or inside tunnels for the periodic inspection, the correct self position measurement and the controller design are indispensable for stable control. In the 50 m distance long flight control experiment of the room, a proposed control method using 2nd order velocity factor (Method (1)) and a control method using boundary condition of both sides (Method (2)) represented good performance comparing with the optimized P-D control. Especially in the small room condition, the Method (1) took most good performance. On the contrary, more practical situation, 2 m width 15 m range corridor surrounded by wall long distance flight control showed a good performance in the case of Method (2). It represents that the controller design depends on the flight situation, and there would be effective method even by a simple controller. Our results would be useful to design an autonomous drone controller in the situation that there is no skilled the drone control operator and the flight by visual confirmation of man are hard conditions.

## References

[1] K.Nonami, F.Kendoul, S.Suzuki, W.Wang and D.Nakazawa, "Autonomous flying robots : unmanned aerial vehicles and micro aerial vehicles ", 2010, Springer Japan.

[2] Hideki Toda , Syuichi Honda , Hiroaki Takano, "Effectiveness of Fast Speed Yaw and Roll Control Switching Instead of Normal Roll Control for AR Drone 4 rotor Helicopter", International Journal of Innovative Research in Advanced Engineering, Vol. 3, Issue 7, Paper ID JYAE10081, doi:10.17632/y52bmbgtkg.1, July 2016

IOP Conf. Series: Materials Science and Engineering 472 (2019) 012019 doi:10.1088/1757-899X/472/1/012019

# Development of Rough Terrain Mobile Robot Control Strategy Capable of Position and Direction Estimation by a Single Marker

**Yuki Kurahashi**[1,*] **and Hideki Toda**[1]

[1]Dept. of Electric and Electronic Engineering, University of Toyama
Gofuku Campus, 3190 Gofuku, Toyama, 930-8555, Japan
E-mail: m1871014@ems.u-toyama.ac.jp

**Abstract.** In the present study, a mobile robot control strategy of rough terrain capable of position and direction estimation by a single marker was proposed, and the effect was performed. To realize wheel / caterpillar mobile robot movement on the rough terrain such as house farm, two major problems (1) the precise position estimation and (2) the direction estimation are important before the development of the control theory. Generally, the position / direction have been estimated by the wheel rotation (odometry) and gyro or acceleration sensor's output integration, but there are many problems such as many impacts from the ground and the wheel slipping on the rough ground to the precise estimation. The aim of this study is to develop a method (algorithm) to estimate the mobile robot direction using a single marker placed at the top of the robot by changing the front movement and the target direction movement periodically. The reason using a single marker is that it is difficult process to find and analyze positions of the many markers attached on the robot in the camera image from a long distance, and the positions of many markers have been unstable by the tilting of the robot, agricultural product's shadow, getting dirty of the body and so on.
Our proposed method would be effective in the difficult situation to find or analyze the marker position of the robot surfaces by the reason of camera position and the environment, and it could realize wheels / caterpillar type mobile robot movement on the muddy situation.

## 1. Introduction

In this paper, a mobile robot control strategy of rough terrain capable of position and direction estimation by a single marker was proposed, and the effect was performed. Current agriculture in Japan has a problem of aging of agricultural workers, and the substitution of agricultural work by robots is becoming increasingly important socially [1]. To realize wheel / caterpillar mobile robot movement on the rough terrain such as house farm, two major problems (1) the precise position estimation and (2) the direction estimation are important before the development of the control theory.

Figure 1a shows the major situation of the mobile robot movement environment for application in the house farm. Previously proposed robot application in house farm used multi aluminum / stainless rails on the ground to realize correct position / direction movement of the robot, however many cost is necessary for the equipment and the rails becomes a factor hindering effective use of agricultural ground. Even if the rail is set up on the farm, the problem of measurement and movement of the mobile robot's current position is caused by dirt or slippage on the rail. The need for long-term maintenance has not yet been discussed. As another solution of the mobile robot localization and control, colored markers attached on the mobile robot (ex. Fig. 1b) could be used, and the position would be measured by camera attached on the ceil. But the extracting position of the markers are generally difficult by the reason of mixing with environment colors, marker shape (area) instability, the distance from the camera to the markers, and multi markers geometric position combination problem. To estimate the position and the direction of the robot, two markers are necessary at least in 2D map. But it is not always possible to detect the two markers in the rough terrain field in the house farm by above some reasons; especially a marker fusion is a serious problem (Fig.1d) in this situation. These causes disable robot direction detection. In addition, since the wheel / caterpillar rotations affected by the slip on the dirt (Fig.1c), the position estimation by the wheel / caterpillar rotations (odometry) was failed generally.

Our proposed method uses one marker on the mobile robot to increase the detection performance of the robot by the camera on the ceil, and the direction of the robot is estimating by a novel two phase movement control algorithm. In this paper, basic examination of the position estimation / movement control of a caterpillar mobile robot in rough terrain situation were performed.

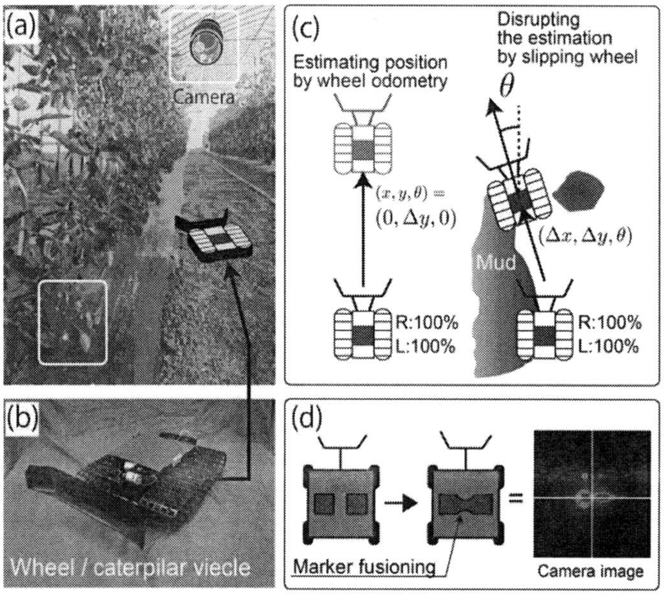


**Figure 1.** Difficulty of the position estimation of the wheel / caterpillar type mobile robot platform in the rough terrain house farm

## 2. Previous study

Auto-runner[2] is a previously proposed unmanned pesticide spreader in house farm environment. It could realize the pesticide spread by moving the all area on the house railed with metal rail. Since the robot does not estimate its own position, it cannot manage with slippage on the rail, and it is expensive setup. In University of Shimane[3], it represents omnidirectional mobile robot to convey objects in a complicated and narrow environment by using a special wheel called a omni-wheel, it can move in all directions. It have constructed an algorithm that estimate the position and orientation of the mobile robot with high accuracy acceleration and gyro sensors mounted on the robot, and the rotation angle of the driving wheel. Since the wheel usually slipped on the farm, the acceleration and gyro sensor values could not correspond to the real wheel rotation angle information and it is impossible to estimate the correct position and direction.

## 3. Method

### *3.1. Experimental Devices and the Setup*

Figure 2 shows the experimental setup devices, (a) Caterpillar mobile robot (BLIZZARD SR, Kyosyo Corp., 35×30×8cm) attached red marker (0.18×0.18m), (b) Setup of the experiment. Camera (640×480 pixels, view angle 130 deg, 30 fps) was attached on the ceiling, and two markers were used as the robot and target position respectively. The mobile robot was controlled by two channel motor driver, Arduino microcomputer and middle spec PC (AMD A10-7860K, 8GB ram).

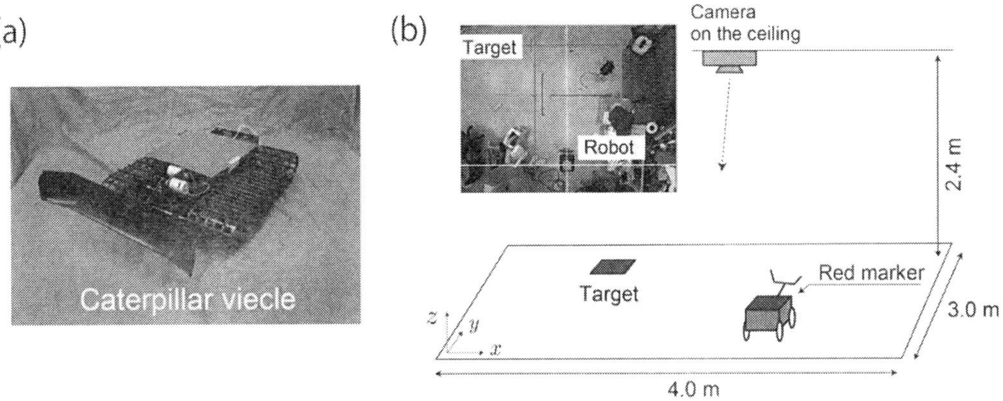


(a) Caterpillar mobile robot (BLIZZARD SR, Kyosyo Corp., 35×30×8 cm) attached red marker (0.18×0.18m). (b) Setup of the experiment. Camera (640×480 pixels, 30 fps) was attached on the ceiling and two markers (the robot and target) were found by red color image analysis.

**Figure 2.** Experimental setup devices.

### *3.2. Control Algorithm*

The mobile robot was controlled by two different control strategy in this study. When the center of the gravity of the red markers were measured as $\vec{x}_1$ (robot) and $\vec{x}_2$ (target), the direction of the target was defined $\vec{b} = \vec{x}_2 - \vec{x}_1$ (Fig. 3a). Normal vector $\vec{a}$ was defined as estimated robot direction, and it was calculated from Method (1) the robot caterpillar rotations and (2) a 300 msec period "forward" movement of the robot denoted by A in Fig. 3b. Method (1) was previously proposed general direction estimation algorithm and the caterpillar rotations only were used to estimate the direction in this study. On the other hand, Method (2) was a proposed strategy, and there is two control phases in the strategy - (A) Forward movement phase to estimate which is the head direction of the robot, and (B) Approach phase to control the robot to the target position. In the Forward phase, the R and L channel power balances of the caterpillar robot were set same values eg.100% while 300 msec period, and the red marker position movement was measured and set as $\vec{a}$ in the 300 msec phase. Next Approach phase use $f_R$ and $f_L$ as R and L channel powers of the robot defined below in the period of 300 msec,

$$Cross_z = \vec{a} \times \vec{b} \tag{1}$$

$$f_R = \alpha Cross_z + \beta_R \tag{2}$$

$$f_L = \alpha Cross_z + \beta_L \tag{3}$$

where α is constant, $\beta_R$ and $\beta_L$ are power bias parameters depended on the machine.

The parameters were set to move the robot forward direction if the cross product value $Cross_z$ is zero. In the Method (1) control, the Eq.1 to Eq.3 are used.

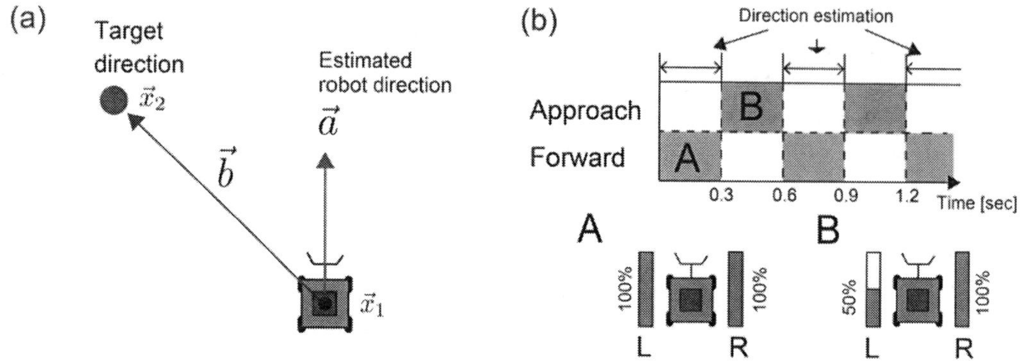


**Figure 3.** (a) Definition of the vector $\vec{a}$ and $\vec{b}$. (b) Proposed control strategy two phase method.

## 4. Experiment

Experiment 1 performs basic forward movement test when the right caterpillar power changed 40 to 100% and the left power was fixed 100% (power supply was 12 V DC, the power was changed by 1 kHz PWM and the ON time ratio). Experimental time was 500 msec. From this experiment, the power bias parameters $\beta_R$ and $\beta_L$ were determined. The ground condition of the room was flat (polished concrete), but slippy surface for the plastic caterpillar.

Experiment 2 performs movement control to the target position on the slippy surface of the polished concrete when Method (1) and (2) control strategy. The area of the experiment was 4×3 m field, and the target position was set (-0.4, 2.3) m from the start position.

## 5. Result

Figure 4 represents the trajectory plotting result of experiment 1. The robot was started from (0,0) position. Since 100% plot is tilting to the left side (x,y)=(-0.13,0.45), the movement power of the right motor is large comparing with the left. Even if the power input to the motors were completely same, the real robot movement would be affected by another factor such as caterpillar surface friction, gear friction loss and so on. From this result, the robot was moved to almost straight ahead when the power of R:L = 70%:100%. As a result, we set $\beta_R$=35% and $\beta_L$=50% respectively in later experiment.

Figure 5 represents the trajectory plotting result of experiment 2. Red and green circle were start and target position respectively. Blue colored plot was the result of Method (1): oddmetry. The trajectory largely deviated to the left side, and the front direction of the robot was different from the target direction (black arrow). Red colored plot was the result of our proposed Method (2). 300 msec periodical "forward" and "approach" switching control could realize the movement control to the target direction even though in the slippy surface of the polished concrete, and the estimated direction $\vec{a}$ was estimating correctly in all the situation.The mechanism of the Method (2) was summarized in Fig. 6 especially in the slipper situation. If the left caterpillar was slipped in a situation (Fig. 6A), even though the power balance was 50:50, the robot position was moved left top side and the direction also changed at the same time.

In the situation B of Fig. 6, the robot performs "Forward" phase (Fig. 3), and the vector $\vec{a}$ would be obtained. In a small area of the slippy environment, the direction of the $\vec{a}$ would be a kind of pseudo "straight" direction of the robot. Next in the approach phase, the robot would calculate the wheel power on the condition that the vector $\vec{a}$ is regarded as "straight" direction in the small slippy area. This periodical control process was assumed in the proposed algorithm, and it could realize the performance of rough terrain movement.

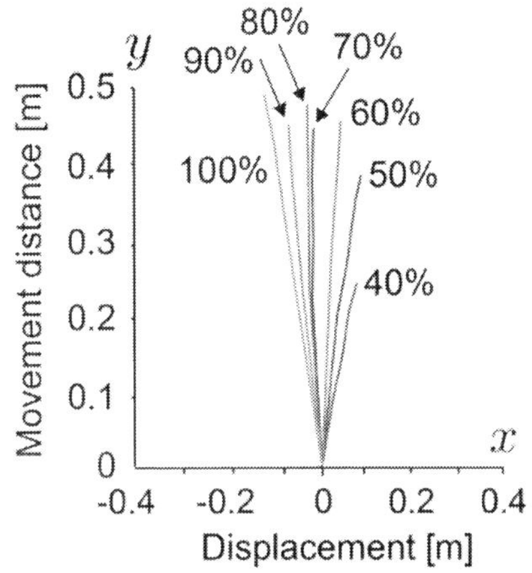


**Figure 4.** Result of experiment 1 basic forward movement test dependency of the right caterpillar power.

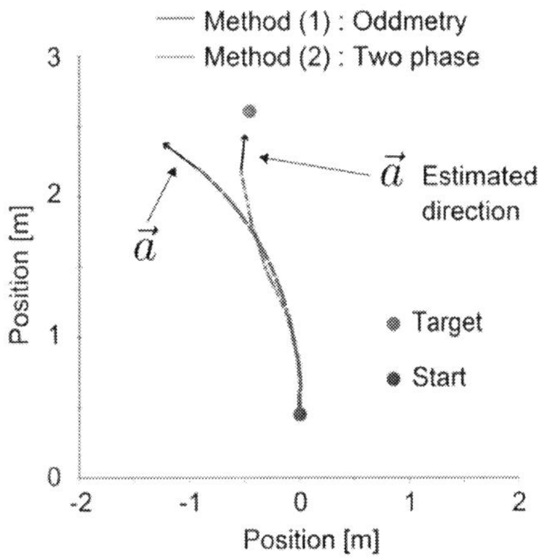


**Figure 5.** Result of experiment 2: Difference of the trajectory of Method (1) and Method (2) control strategy.

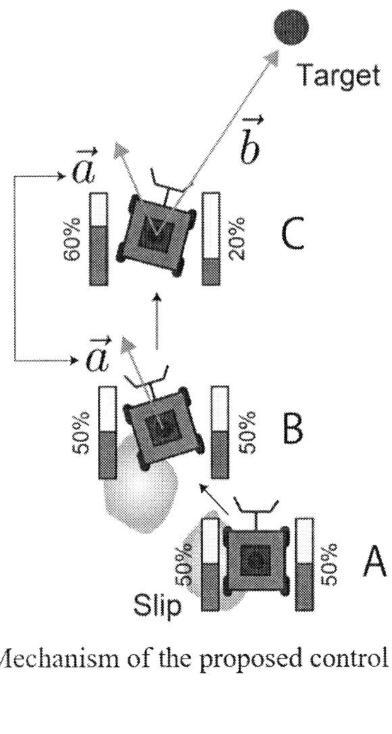


**Figure 6.** Mechanism of the proposed control Method (2).

## 6. Conclusion

In this paper, a mobile robot control strategy of rough terrain capable of position and direction estimation by a single marker was proposed, and the effect was performed.

To realize wheel / caterpillar mobile robot movement on the rough terrain such as house farm, it is necessary to solve the precise position estimation and the direction estimation are important before the development of the control theory. This study developed a method to estimate the mobile robot direction using a single marker placed at the top of the robot by changing the front movement and the target direction movement periodically. Experimental result shows that even if there is slip while the forward phase, obtained vector $\vec{a}$ could treat as a kind of pseudo "straight" direction of the robot, and cross product between $\vec{a}$ and $\vec{b}$ would be useful for R and L channel power control. Our proposed method would be effective in the difficult situation to find or analyze the marker position of the robot surfaces by the reason of camera position and the environment, and it could realize wheels / caterpillar type mobile robot movement on the muddy situation.

## 7. References

[1] Hiroaki KINOSHITA, kenjiro TADAKUMA, Keiji NAGATANI, Kazuya YOSHIDA, "Traversal Experiments on rough terrain for a Leg-track Hybrid Robot in Mt.Asama", SICE System Integration 2010, Sendai, 2010.
[2] Arimitsu K.K., URL: http://www.arimitsu.co.jp/ghc/04/02/index.html, (accessed on March, 2018).
[3] RoboticsLab.inShinameUniv.,URL:http://www.ecs.shimane-u.ac.jp/~hamaguchi/robotics/study.html, (accessed on March, 2018).

# Movable Ankle Joint Stretching Device by Stabilization of Ankle Position Using Tension Rod Structure

**Shin Sugihara**[1,a,*], **Hideki Toda**[1,b] **and Takahiro Yamaguchi**[1,c]

[1] Dept. of Electric and Electronic Engineering, University of Toyama Gohuku Campus, 3190 Gohuku, Toyama, 930-0855, Japan
E-mail: [a,*]m1771023@ems.u-toyama.ac.jp, [b]toda@eng.u-toyama.ac.jp, [c]m1771041@ems.u-toyama.ac.jp

**Abstract.** In this paper, we proposed a tension rod mechanism for fixing ankle rehabilitation device, which is formed of the structure tightly stretched between the patients' sitting chair and floor, and the frictional force of the both ends of it keeps the position of the device during the ankle joint stretching. Generally, physical therapists treat their patients to prevent contracture and to improve walking function, but sufficient rehabilitation therapy does not conducted demanding their labor-tasks because of increase elderly population and shortage of personnel of physical therapists. Therefore rehabilitation robots have been proposed in order to help physical therapist to treat their patients. However, since patients have to transfer their body to the rehabilitation device in order to take mechanical therapy, caregivers every time supports their transfer which cause low back pain when the person who needs ankle rehabilitation uses it. Proposed mechanism allows the device to switch the state of fix/non-fix, so that it does not require transfer assistance by caregiver. Result of experiment based on developed device, proposed system had the function to stabilize within approximately 6 mm the device position during 60 N pushing dorsiflexion stretching.

## 1. Introduction

The human ankle joint plays a major role in daily activities such as walking, sitting and running. In the case of injury or disease, the ankle joint causes contracture, shortening of a muscle or joint, due to excessive immobility of the joint [1]. Once contracture occur, it is difficult to stretch these ankle joint again, so that it lower the quality and quantity of daily activity after the treatment. Therefore, rehabilitations by physical therapists are helping their patients to prevent contracture and to recover their gait function. One of the traditionally therapy by physical therapists is stretching patient's ankle joint by pushing patient's foot with hand. These treatments, however, does not conducted sufficient because most of physical therapist have high demanding labor tasks.

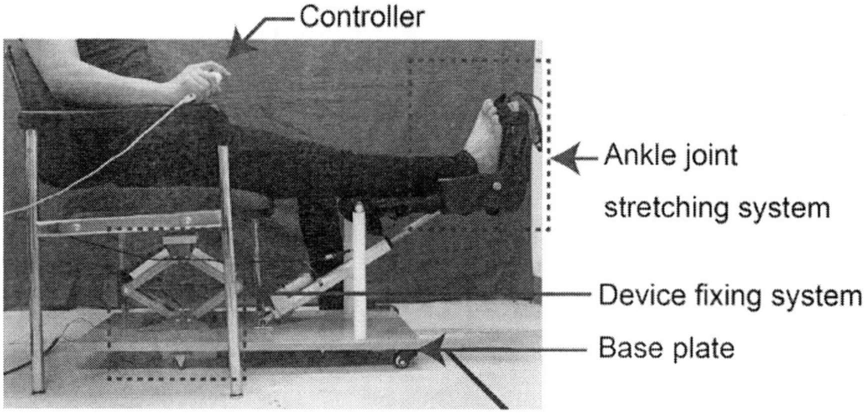


**Figure 1.** Developed ankle joint stretching device with tension rod structure.

Thus, many research groups have been developed robot-aid rehabilitation device in order to support physical therapist tasks [2][3]. However, most of the previous ankle rehabilitation devices have a large mechanical body, which requires extensive setup space and transfer assistance by caregivers. This large mechanical structure is because the ankle stretching devices must have a mechanical stiffness against the reaction force generated during foot sole pushing process in order to prevent deviation of the device rotation axis [2]. Thus these devices connect ankle pushing mechanism to bed or chair in order to fixed ankle pushing parts, so that this integrated device require large space except for the ankle pushing function. In addition, more significant problem is transfer assistance which cause low back pain. This is because caregivers take heavy physical burden to support the patient's weight to move the patients to the rehabilitation device. In fact, some research reported that the risk of back pain is increased due to transfer assistance in nursing home care facilities [4].

In this study, to address these problems, we proposed tension rod structure for fixing ankle rehabilitation device, which does not require transfer assistance.

Section 2 explains the developed ankle stretching device with proposed tension rod structure. In section 3 and 4, we confirm the function to stabilize the device position by developed equipment, and report results, respectively. Finally, we conclude in section 5.

## 2. Method

### *2.1. Developed Ankle Joint Stretching Device.*

Figure 1 shows developed ankle joint stretching device. The mechanical structure of this device is functionally divided into ankle joint pushing system and the device fixing system. Both pushing of the ankle joint and fixing of the device system can be moved by the controller with patient's hand.

### *2.2. Ankle Joint Pushing Mechanism.*

Figure 2 shows ankle joint pushing mechanism, which is composed of (a) a footplate, (b) two DC motors (Changzhou DAIWEI Motor Co. Ltd, Model: JSX 1622-31 ZY) and (c) a footrest. (a) The footplate can be rotated dorsiflexion/planter flexion direction by (b) the DC motors which mounted on both sides of the footplate rotation axis. (c) The footrest is made of soft sponge, and the foot sole of the patient placed on the footrest is pushed by rotated (a) the footplate. One load cell is embedded in (a) the foot plate in order to measure the pressure applied to the foot sole.

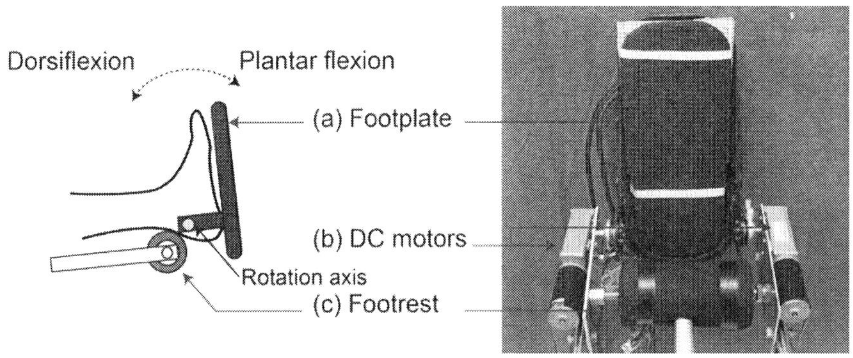


**Figure 2.** Mechanism of the ankle joint pushing.

The foot sole of the patient is pushed by the foot plate rotated by DC motors.

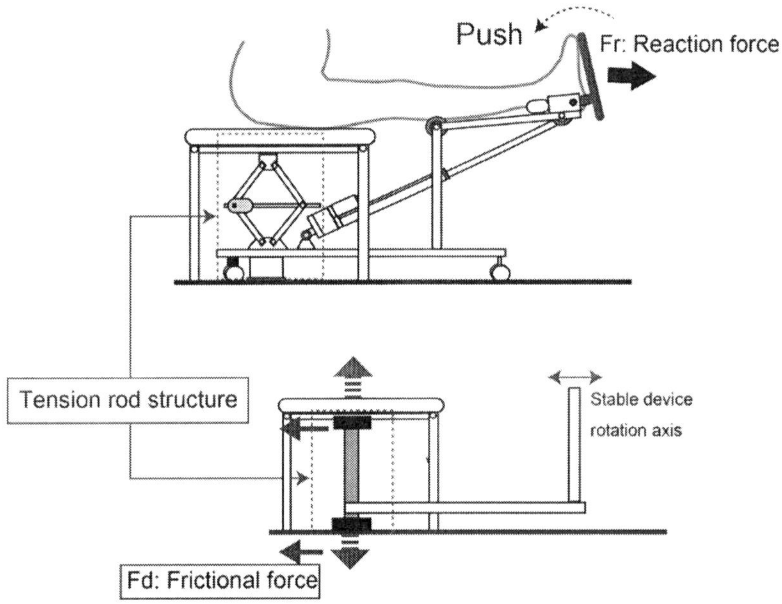


**Figure 3.** Mechanism of the proposed tension rod structure which stable the device rotation axis by the structure between patient's sitting chair and floor.

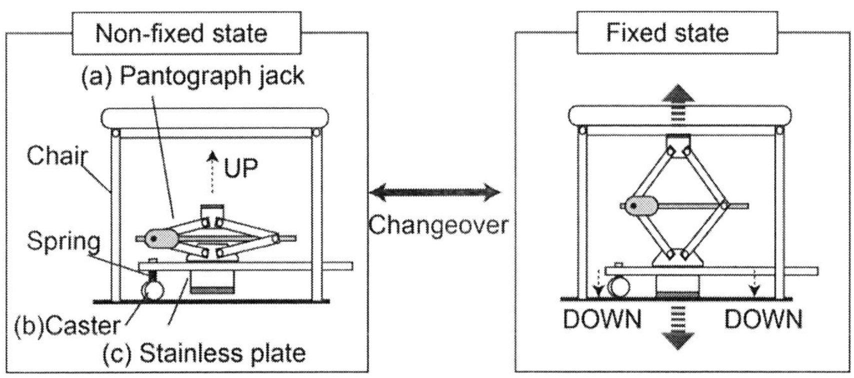


**Figure 4.** Mechanical movements of the tension rod structure. This movement allow the device to switch the state of fix and non-fix.

*2.3. Proposed Tension Rod Mechanism.*

Figure 3 shows the overview of the tension rod structure developed as fixed system for the ankle joint stretching device. This system is formed of a bar tightly stretched between the chair and floor (highlighted in green in fig3), and the frictional force Fd of the both ends of the bar prevents the position of the device from being displaced by reaction force Fr generated by dorsiflexion stretching.

Fixing system movement of the developed device is shown in figure 4. This system is consisted of (a) an electric pantograph jack, (b) two casters with spring and (c) a stainless plate.

The tension rod structure of the developed device is constructed as follows; First, (a) the pantograph jack adjusts the length to the height of the chair which patient sitting, and then transmits the pressure from the lower part of the seating face of the chair to the base plate of the device. When pressure is applied, the position of the base plate is down because (b) the spring of the caster is shrunk. Finally, (c) the stainless plate which attached lower part of the device lower part is pressed the floor, so that tension rod stricture formed on the patient's sitting chair and the floor. As show in figure 4, since the proposed system can change the state of fix and non-fix by controlled (a) the jack, this system does not require to transfer to the stretching device before/after using it. In addition, this system can adjusts the force of the ends of tension rod structure (Fig.4, red arrow), which force is measured by load cell embedded in the top of the structure.

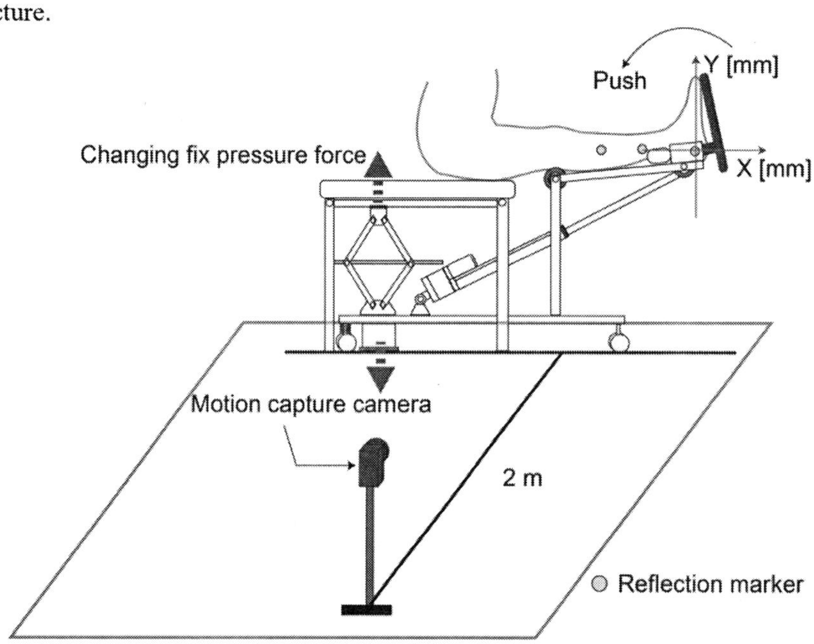


**Figure 5.** Set up of the motion capture system to measure patient's ankle joint position of the subject's and ankle rotation axis of the device.

## 3. Experiment

Experiment 1: The positional relationship between the ankle joint and device rotation axis during ankle joint stretching process. In this experiment, the ankle joint and device rotation axis movement trajectory during the foot sole pushing process is confirmed by using the proposed system with/without tension rod structure. One subject (age 24, adult man) is participated in the experiment. As shown in Figure 5 the trajectories were measured by motion capture camera (Flex 3, 0.3 MP (640 × 480, 100 fps), OptiTrack) which film reflective markers attached subject's ankle joint and device rotation axis. The pressure force of the end of fixed part is coordinated 150 N. The maximum pressure force of the foot is controlled under 60 N during the dorsiflexion stretching process, and if reach 60 N, foot plate is stopped 20 sec and then it is rotated planter flexion.

Experiment 2: The deviation of the device position against fixed pressure. Experiment 2 is conducted for checking the stable fixation of the device against the pressure force of the both ends of tension rod structure during the dorsiflexion stretching. One subject (age 24, adult man) is participated in this experiment. The movement of the device rotation axis is measured as well as experiment 1. The deviation of the device is measured 5 times adjusting pressure force of the end of fixed part to 0, 50, 100 and 150 N, respectively.

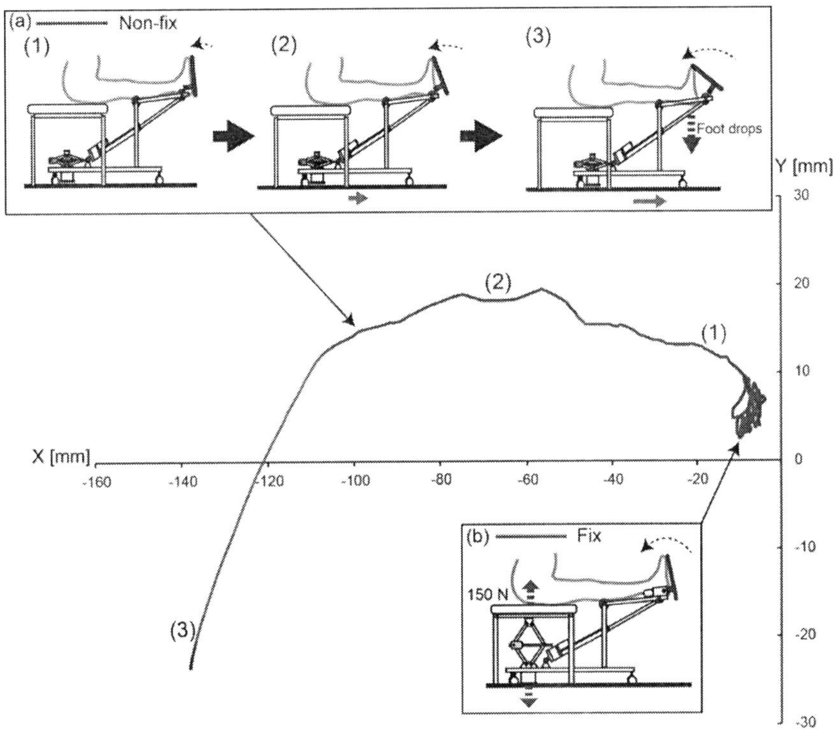


**Figure 6.** Result of experiment 1: Transition of the positional relationship between ankle joint and device rotation axis during the ankle joint stretching process.

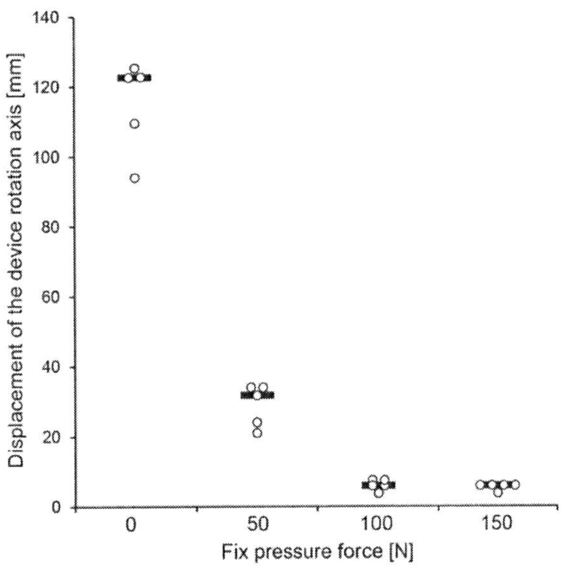


**Figure 7.** Result of experiment 2: The deviation of the device rotation axis against the pressure force of the fixed part (N = 5)

## 4. Result

Result of experiment 1. Figure 6 shows the positional relationship between ankle joint and device rotation axis with/without the tension rod structure during the foot sole pushing process by developed device. Without the tension rod structure, the subject's foot drops from footrest as the device is leaved horizontal direction from the subject. Contrary to the without fixing, with the proposed tension rod structure, the positional relationships was stable at (x, y) = (-7.5± 2.4, 5.9 ± 2.5) mm. Since the displacement between the subject's ankle joint and device rotation axis was approximately only 2.5 mm, this result indicate the proposed tension rod structure has the capability of the fixed function of the stretching device.

Result of experiment 2. Figure 7 shows the displacement of the device from the initial position against the pressure force of the top of device fixed part. Less than 100 N of the fixing force, the displacement and its variation was related fixing force. On the other hand, 100 N or more, the displacement was the approximately 5.8 mm, and its variation does not depend on fixing force. This stationary displacement is considered caused by mechanical distortion of the devise, not the low pressure force of the tension rod structure. This result also supports the fixed function of the tension rod structure as well as result of experiment 1. In addition, it reveals the fixed function is appeared when the fixing pressure is over 100 N.

## 5. Conclusion

In this research, we have developed ankle joint stretching device by using tension rod structure, which mechanism allow the device position to fix by the frictional force of the structure between the floor and chair. In addition, this mechanism is able to switch the state of fix/non-fix, so that patient's does not require transfer to the stretching device. The experiment 1 and 2 based on motion analysis has indicated that the tension rod structure is functioning to fix the position of the device against the reaction force Fr generated ankle pushing process. The proposed tension rod mechanism will be useful for developing a device that does not require transfer assistance to add a large force in order to the ankle pushing.

## References

[1] Suzuki Rie, Ohta Yuji, Causes and preventions of Articular Contracture, Journal of Human Environmental Engineering, Vol. 9, No.1, pp. 142-143 (2007)

[2] H. Toda, T. Matumoto, R. Tanizaki, T. Imaeda, Ankle joint pushing mechanism by Stabilization of ankle position using a brace structure, Journal of Advanced Mechanical Design Systems and Manufacturing, Vol. 10, No. 1, pp. 1-11 (2016)

[3] M. Boiro, A new concept of parallel robot for rehabilitation and fitness: the Lambda, IEEE international conference on Robotics and Biomimetics, pp.2503-2508 (2009)

[4] K. Iwasaki, M Takahashi, M, Sotoyama, X. Liu and A. Koda, Low back pain among workers in care facilities for elderly after introducing welfare equipment, Sangyo Eiseigaku Zassi, pp. 130-142, (2016)

IOP Conf. Series: Materials Science and Engineering 472 (2019) 012021 doi:10.1088/1757-899X/472/1/012021

# Study on the Change of Incident Direction and Incoming Flow Velocity to the Reattachment Point and the Center Position of the Large Vortex

**Yaonan Li[1,a,*], Xing Zhou[2,b] and Wentao Jiang[3,c]**

[1,2,3]School of Mechanical Engineering, Ningxia University, in Chuan 750021 China
*Corresponding author E-mail: [a,b,c]601154336@qq.com

**Abstract.** This paper establishes a model with the variation of the standard angle with the height of the step. Numerical simulation is carried out to study the influence of the change of the incident direction and the incoming velocity on the large eddy structure and the length of the recirculation zone in the reattached shear layer under different standard angles. Finding that with the increase of the incident angle β, the length of the recirculation zone tends to increase firstly and then decrease. The large vortex center position Y/H firstly maintains a small range of fluctuations; when the incident angle β reaches a certain value, the large vortex center position suddenly rises sharply to a certain value. The center position of the large vortex decreases again when β continues to increase. When the Reynolds number increases from 150 to 300, the value of Xr/H will suddenly increase sharply and the position of the large vortex will have a significant downward trend. When the Reynolds number is greater than 300, the length of the recirculation zone will remain unchanged from the center of the large vortex. That is the recirculation zone length and the Y/H value of the center position of the large vortex is not affected by the Reynolds number.

## 1. Introduction

As the simplest and most typical flow in the two-dimensional separation and reattachment movement, the backward step flow has been generally valued.

Roos et al. [1] pointed out through experiments that large-scale vortex structures exist in both laminar and turbulent separations, as the shear turbulence increases, the large-scale structures become more and more irregular. Zhou Huiliang et al. [2] had used flow field display technology to find that the scale near the reattachment point is comparable to the step height.

The study of the large eddy structure in the above literature is carried out in the case where the incident direction is perpendicular to the step, that is, the incident angle is $\beta=0°$ according to the model presented in Fig 1. H. Iwai et al. [3] studied the influence of fluid gravity on the backward step flow when the tilt angle and roll angle change with time and found that the reattachment point moves forward with the increase of the tilt angle which within a certain range from 0° to 180°. The influence of different inflow velocity on the large eddy structure is analyzed by numerical simulation in this paper. It is found that with the increase of the incident angle β, the length of the recirculation zone shows a trend of increasing firstly and then decreasing and the value Y/H of center of the large vortex firstly keeps fluctuations in a small range; when the incident angle β reaches a certain value, the center position of the large vortex suddenly rises sharply to a certain value and it decreases again when β continues to increase. When the Reynolds number increases from 150 to 300, the value of Xr/H will suddenly increase sharply and the position of the large vortex will have a significant downward trend. The recirculation zone length and the Y/H value of center position of the large vortex is not affected by the Reynolds number as it grows greater than 300.

## 2. Model Establishment and Verification

The backward step flow studied in this paper is shown in Figure 1, where θ represents a standard angle, which is related to the step height H, β is the incident angle and V is the incoming flow velocity.

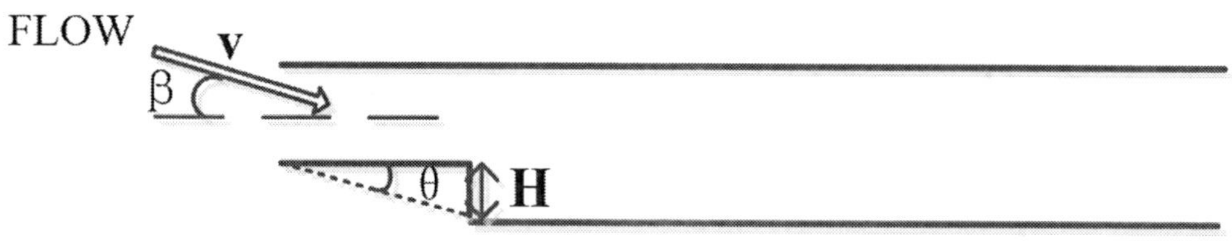


**Figure 1.** Schematic diagram of the flow backward step

In order to check the accuracy of the construction of this model, the calculation parameter is H=0.049m and the height of the flow before the step is 0.052m according to the literature [4]. The standard angle is θ=30°. The calculation grid is 98×490 under the grid noncorrelation test. Setting the fluid medium to water and the outlet pressure to 0 in Fluent. To study the response of the recirculation zone length with the variation Reynolds number in the case of low Reynolds number. The specific verification method is to determine whether the length of recirculation zone changes with the Reynolds number is consistent with the literature [4] when the standard angle is equal to the incident angle, ie β=θ=30 °, under the Reynolds number is changed from 100-1500. The results obtained in this experiment and the results obtained in the literature [4] are shown in Figure 2. Xr/H represents the ratio of the length of the recirculation zone to the height of the step in the figure. It is found through analysis that the length of the recirculation zone becomes longer as the Reynolds number increases, however, the length of the recirculation zone is reduced and kept fluctuating up and down within a certain range as the Reynolds number higher than a certain value shows in literature [4]. We obtained through numerical simulation that the length of the recirculation zone increases with the increases of the Reynolds number in the low Reynolds number range, but this trend continues for a short time only and it fluctuates with a small range in the range of the Reynolds number.

In terms of the overall trend, the Reynolds number equals 290 where the length of recirculation zone reached limit in this paper as compared to the Reynolds number at the maximum length of recirculation zone in the literature [4] is 1070. It can be considered that the construction of grid in this paper is more reasonable.

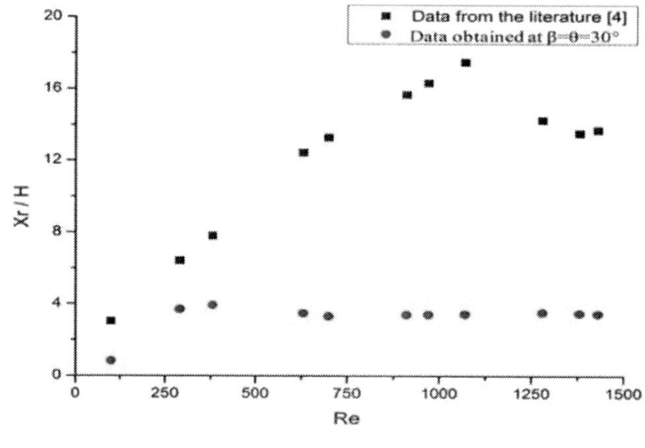


**Figure 2.** The relationship between the lengths of recirculation zone to Reynolds number

## 3. Calculation Results and Analysis

### *3.1. Influence of Incident Angle on Flow Field Vortex Structure*

A backward step model with a standard angle θ of 15, 30, 45 and 60 degrees is constructed successively to study how the influence of the change of the incident angle on the position of the reattachment point and the influence of the position of the large vortex center based on the above-mentioned verified mesh model. In each of the backward step models, the incident angle value is changed and satisfies 0° ≤ β ≤ 60°. The obtained result is processed by the software GetData to obtain the data as shown in Table 1 and the influence of the change of the incident angle on the length of the recirculation zone and the position of the large vortex at different standard angles is respectively extracted by software Oringin 8.5, as shown in Figure 3,4, where the XY coordinates established in the figure are the origin of the coordinates of the bottom of the step.

We found that the length of the recirculation zone is always increased when 15° ≤ θ ≤ 30° by changing the height of the step whatever the β is. However, the length of the recirculation zone has been reduced when 30° ≤ θ ≤ 60° by changing the height of the step, which is most obvious when β = 0°. We found there is a phenomenon of symmetrical change of θ = 30°. The length of the recirculation zone also changes as β changes when θ keeps constant under each model. It can be seen from Fig. 3 that the length of the recirculation zone shows a tendency to increase firstly and then decrease as the β increases. This trend is easy to see from θ = 30°, θ = 60°, while θ = 15°, θ = 45° only shows a tendency to increase the length of the recirculation zone however the downward trend has not yet appeared.

Analysis of Figure 4 shows that the center position of the large vortex is increasing when 15 ° ≤ θ ≤ 30 ° by changing the height of the step under the β = 0°. The center position of the large vortex shows a downward trend when 30° ≤ θ ≤ 60°. The Y/H value of center position of the large vortex seems to satisfy a trend that firstly maintains a small range of fluctuations as the incident angle increases in the backward step model structure with different standard angles. The position of the center of the large vortex suddenly rises sharply to a certain value when the incident angle β reaches a certain value. The center position of the large vortex decreases again when β continues to increase. However, the trend is not obvious and the center of the large vortex shows a tendency to rise slowly when the standard angle is 30°.

**Table 1** The effect of incident angle on reattachment point and the center position of large vortex

| | | | | | | | | | | | | |
|---|---|---|---|---|---|---|---|---|---|---|---|---|
| θ=15°, Re=150 | | | | | | | | | | | | |
| β | 0 | 10 | 15 | 20 | 25 | 30 | 35 | 40 | 45 | 50 | 55 | 60 |
| $X_r$/H | 0.7 | 0.7 | 0.8 | 0.7 | 0.7 | 0.8 | 0.8 | 0.8 | 0.8 | 0.8 | 0.8 | 1.0 |
| Y/H | 0.3 | 0.3 | 0.3 | 0.3 | 0.7 | 0.7 | 0.7 | 0.7 | 0.7 | 0.7 | 0.7 | 0.6 |
| θ=30°, Re=150 | | | | | | | | | | | | |
| β | 0 | 10 | 15 | 20 | 25 | 30 | 35 | 40 | 45 | 50 | 55 | 60 |
| $X_r$/H | 3.2 | 3.3 | 3.4 | 3.2 | 3.4 | 3.4 | 3.3 | 3.1 | 3.1 | 3.0 | 2.8 | 2.5 |
| Y/H | 0.5 | 0.5 | 0.5 | 0.5 | 0.5 | 0.5 | 0.5 | 0.5 | 0.5 | 0.5 | 0.6 | 0.7 |
| θ=45°, Re=150 | | | | | | | | | | | | |
| β | 0 | 10 | 15 | 20 | 25 | 30 | 35 | 40 | 45 | 50 | 55 | 60 |
| $X_r$/H | 0.7 | 0.7 | 0.8 | 0.9 | 1.0 | 1.0 | 1.1 | 1.2 | 1.3 | 1.4 | 1.5 | 1.8 |
| Y/H | 0.3 | 0.4 | 0.3 | 0.4 | 0.4 | 0.4 | 0.4 | 0.7 | 0.7 | 0.7 | 0.7 | 0.7 |
| θ=60°, Re=150 | | | | | | | | | | | | |
| β | 0 | 10 | 15 | 20 | 25 | 30 | 35 | 40 | 45 | 50 | 55 | 60 |
| $X_r$/H | 0.8 | 0.8 | 0.9 | 1.1 | 1.1 | 1.2 | 1.2 | 1.3 | 2.2 | 2.4 | 1.9 | 1.8 |
| Y/H | 0.3 | 0.3 | 0.7 | 0.7 | 0.7 | 0.7 | 0.6 | 0.6 | 0.5 | 0.5 | 0.6 | 0.6 |

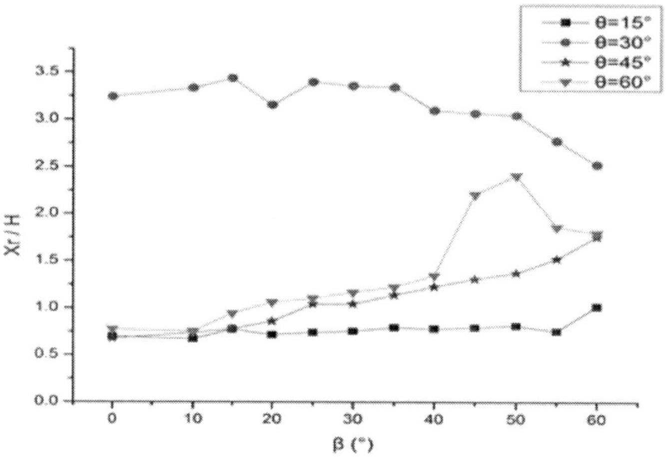


**Figure 3.** The effect of incident angle change on the length of the recirculation zone at different standard angles

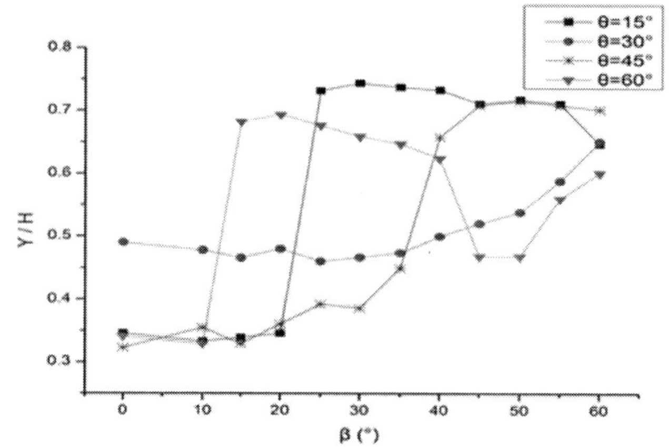


**Figure 4.** The effect of incident angle change on the center position of large vortex at different standard angles

### *3.2. Influence of Reynolds Number on Vortex Structure of Flow Field*

We obtained the Reynolds number $150 \leq Re \leq 5000$ by changing the magnitude of the flow velocity V in this paper to study the influence of the Reynolds number on the reattachment point and the center position of the large vortex when the standard angles are equal to the incident angle and the data as shown in Table 2.

It is found that the change of the length of the recirculation zone is similar under four different standard angles with the increasing Reynolds number. That is said that the value of Xr/H will suddenly increase sharply when the Reynolds number increases from 150 to 300 at a low Reynolds number. Then the length of the recirculation zone can be considered to remain unchanged, that is, not affected by the Reynolds number. However, there will be a small range of fluctuations in the graph because an error in the data acquisition from GetData and the error range is in the range of 0.2-0.4. Considering this small range of fluctuations in the figure 5 can be considered to be consistent. It can also be clearly seen from the figure that the standard angle satisfies $15° \leq \theta \leq 60°$ by changing the height of the step and the value of Xr/H decreases continuously as the standard angle θ increases.

From the analysis of Fig. 6, it is found that the center position of the large vortex has a significant downward trend when the Reynolds number is $150 \leq Re \leq 300$. The center position of the large vortex remains unchanged when the Reynolds number satisfies $300 \leq Re \leq 5000$, that is, The Y/H value of the center position of the large vortex center is not affected by the Reynolds number. Moreover, it can be analyzed from the figure that the position of the large vortex center is not affected by the change of the step height, that is, the center position of the large vortex remains unchanged when the standard angle θ is increased from 15° to 60°.

**Table 2.** The effect of Reynolds number on reattachment point and the center position of large vortex

| | | | | | | | | | | | | |
|---|---|---|---|---|---|---|---|---|---|---|---|---|
| | | | | | | β=θ=15° | | | | | | |
| Re | 150 | 300 | 600 | 1000 | 1500 | 2000 | 2500 | 3000 | 3500 | 4000 | 4500 | 5000 |
| $X_r$/H | 0.8 | 4.7 | 4.3 | 4.2 | 4.4 | 4.4 | 4.4 | 4.6 | 4.5 | 4.6 | 4.6 | 4.6 |
| Y/H | 0.3 | 0.4 | 0.4 | 0.5 | 0.5 | 0.5 | 0.5 | 0.5 | 0.5 | 0.5 | 0.5 | 0.5 |
| | | | | | | β=θ=30° | | | | | | |
| Re | 150 | 300 | 600 | 1000 | 1500 | 2000 | 2500 | 3000 | 3500 | 4000 | 4500 | 5000 |
| $X_r$/H | 3.4 | 3.9 | 3.6 | 3.6 | 3.6 | 3.6 | 3.6 | 3.6 | 3.7 | 3.7 | 3.6 | 3.6 |
| Y/H | 0.5 | 0.4 | 0.5 | 0.5 | 0.5 | 0.5 | 0.5 | 0.5 | 0.5 | 0.5 | 0.5 | 0.5 |
| | | | | | | β=θ=45° | | | | | | |
| Re | 150 | 300 | 600 | 1000 | 1500 | 2000 | 2500 | 3000 | 3500 | 4000 | 4500 | 5000 |
| $X_r$/H | 1.3 | 3.2 | 3.0 | 3.0 | 3.1 | 3.1 | 3.1 | 3.1 | 3.1 | 3.1 | 3.1 | 3.0 |
| Y/H | 0.7 | 0.4 | 0.5 | 0.4 | 0.4 | 0.5 | 0.5 | 0.5 | 0.5 | 0.5 | 0.5 | 0.5 |
| | | | | | | β=θ=60° | | | | | | |
| Re | 150 | 300 | 600 | 1000 | 1500 | 2000 | 2500 | 3000 | 3500 | 4000 | 4500 | 5000 |
| $X_r$/H | 1.8 | 3.0 | 2.8 | 2.9 | 3.0 | 2.5 | 2.5 | 2.4 | 2.3 | 2.2 | 2.2 | 2.2 |
| Y/H | 0.6 | 0.4 | 0.4 | 0.4 | 0.4 | 0.4 | 0.4 | 0.4 | 0.4 | 0.4 | 0.4 | 0.4 |

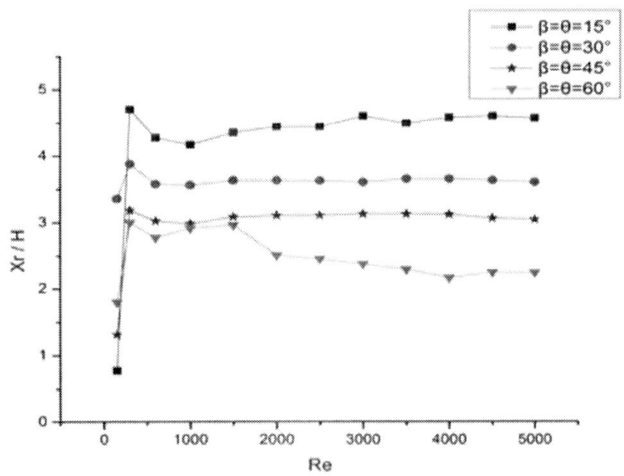


**Figure 5.** The effect of Reynolds number change on the length of the recirculation zone at different standard angles

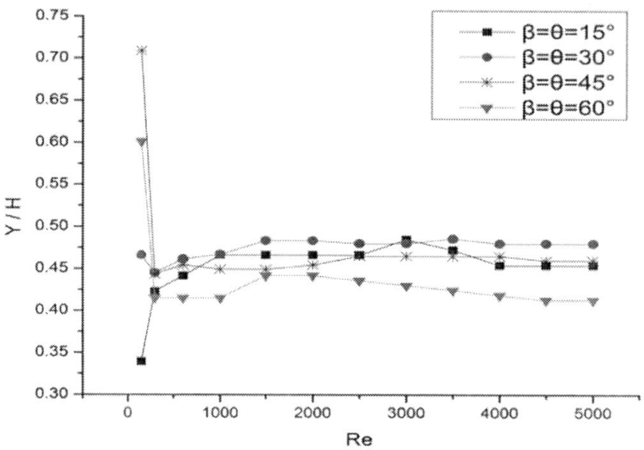


**Figure 6.** The effect of Reynolds number change on the center position of large vortex at different standard angles

## 4. Conclusions

The length of the recirculation zone Xr/H and the center position of the large vortex Y/H are studied by changing the incident angle β and the flow velocity V in this paper through numerical simulation method and the following conclusions are obtained:

Firstly: The length of the recirculation zone exhibits a tendency to increase firstly and then decrease as the incident angle β increases. This may have a lot to do with the structure we are studying; the reattachment point achieves the maximum value when the incident angle increases to a certain value, which is also obvious in practice. Similarly, similar results were obtained by changing the pitch angle of the duct in the article by Iwai H et al in literature [3].

Secondly: The value of Xr/H will suddenly increase when the Reynolds number increases from 150 to 300. The length of the recirculation zone can be considered to remain unchanged when the Reynolds number is greater than 300, that is, not affected by the Reynolds number. The conclusions we get here are consistent with the conclusions obtained by Zhou Ning et al. [6], however, our study reached saturation at a Reynolds number of 300 because our study is carried out under the model of the incident angle equal to the standard angle ($\beta = \theta$) as compared to the incident angle of the literature [6] is always zero $\beta$=0.

Thirdly: The center position of large vortex firstly maintains a small range of fluctuation as the incident angle increases; The center position of large vortex suddenly rises sharply to a certain value when the incident angle β reaches a certain value and the position of the center of the large vortex is decreasing again with β continues to increase. The conclusions of our research here are inconsistent with the conclusion that the scale of the vortex near the reattachment point is comparable to the height of the step proposed in literature [2] in which the incident angle is $\beta$=0. The conclusion of this paper is more realistic since the incident angle is changeable.

Lastly: The center position of the large vortex has a significant downward trend when the Reynolds number $150 \leq Re \leq 300$ and the center position of the large vortex remains unchanged when the Reynolds number satisfies $300 \leq Re \leq 5000$, that is, the Y /H value of the center position of large vortex is not affected by the Reynolds number. It can be clearly seen from Fig. 6 that the center position of the large vortex is maintained at about 0.5 which means that the scale of the large eddy is equivalent to the height of the step, regardless of which model of different standard angle we are studying. This conclusion is consistent with the literature [4].

The shortcoming of this paper is that there will be a human factor error in the range of 0.2-0.4 when taking value of the recirculation zone length Xr and the large vortex center coordinate Y.

**Acknowledgement**
This research was financially supported by the Reform Project of Higher Education in Ningxia (NXBJG2018011).

**References**

[1] Roos F W, Kegelman J T. Control of coherent structures in reattaching laminar and turbulent shear layers [J]. Aiaa Journal, 2014, 24(12):1956-1963.
[2] Zhou Huiliang, Xu Jingmin, Cao Wei, et al. Large eddy structure in reattachment shear layer [J]. Hydrodynamics Research and Progress: 2015(1):23-29
[3] Iwai H, Nakabe K, Suzuki K, et al. The effects of duct inclination angle on laminar mixed convective flows over a backward-facing step [J]. International Journal of Heat & Mass Transfer, 2016, 43(3):473-485.
[4] Wang Bing, Zhang Huiqiang, Qi Jianfeng, et al. Numerical Simulation of Large Vortex Structure Evolution in Separated Flows of Back Steps [J]. Chinese Quarterly of Mechanics, 2013, 24(2): 166-173.
[5] Qi Erong, Huang Minghai, Li Wei, et al. Experimental Study on Flow Characteristics of Two-Dimensional Backward Step Flow [J]. Laboratory Mechanics, 2016, 21(2):225-232.
[6] Zhou Ning, Pei Cheng, Pan Wei. Effect of Reynolds Number on Transition Characteristics of Back Step Shear Layer [J]. Journal of Beijing University of Aeronautics and Astronautics, 2014, 40(7): 978-983.

IOP Conf. Series: Materials Science and Engineering 472 (2019) 012022 doi:10.1088/1757-899X/472/1/012022

# Assignment of Weights to Indexes in Evaluating Usability of Child Safety Chairs Based on AEMV Synthetic Weighting Method

**Honglei Dong [1,a], Yan Wang [1], Lingyun Xiao [1,*] and Jinhuan Zhang [2]**

[1] 1.China National Institute of Standardization, China

[2] Department of Automotive Engineering Tsinghua University, China

E-mail: [a] donghl@dpac.gov.cn

**Abstract.** Usability is a decisive factor of any child safety restraint (child-safety chair) in assessing its functions, making purchase decision, and using it properly and effectively. The article takes advantage of well-developed child chair evaluation regulations and processes worldwide, itemizes and discusses every correlation between usability evaluation index and misuse, and thus formulates usability evaluation indexes system consisting of symbol usability, usability of user manual, usability of child protection and usability of chair installation; builds AEMV synthetic weight assignment modeling combining AHP method, Entropy method and Mean Variance method, on account of usability evaluation that can be subjectively biased, and index weighting that can not be decided properly, in consideration of expertise and objective test data; and calculates any index weight assigned to child chair usability evaluation based on the present synthetic weight assignment modeling, and justifies the scientific strength and objectivity of the weights assignment method in question by means of result comparison and analysis.

## 1. Introduction

Thanks to increasing numbers of purchased vehicles and enhanced awareness of consumers about safety, child-safety restraint systems have drawn greater attention than ever. In developed nations, child safety restraint systems are legally compulsory and as a result wide applied. The systems are equipped on about 90% of vehicles in the US, Canada, Australia and Germany equip with the system [1-4]. In contrast to this situation, laws and regulations on compulsory use of child safety chairs for child occupants are only locally constituted in Shanghai, Shenzhen and Shandong. The legislation is poorly implemented due to limited enforcement. In 2015, a survey conducted by the AQSIQ Defective Product Administrative Center showed that of the 4,900 parent who replied inquiries in Beijing, Shanghai, Guangzhou, Jinan and Chongqing, 60.2% have his/her child/children travel in car every week, while simply 19.7% use child safety restraint systems. It can be a huge gap. According to statistics, correct use of child safety chairs reduces fatal injuries of children in traffic accidents by 71%, and serious injuries by 67%[5]. On the contrary, unqualified child safety chairs or misuse will result in greater fatal risk, rather than expected protection [6]. Certain measures are adopted worldwide to ensure effective protection by child safety chairs. Continuous improvement and modifications of technical regulations and standards for child safety restraint systems on the one hand motivate manufacturers to enhance constantly product quality by means of product certification and callback campaigns[7]; and performance evaluation and misuse studies on child safety restraint systems on the other hand are implemented to practically guide consumers to correctly select and use child safety chairs[8-9]. GB27887-2011 Restraining devices for child occupants of power-driven vehicles came into effect as of 1st July 2012. On 22nd January 2014, the AQISQ put child-safety chairs on China Compulsory Certification list. According to survey data released in 2013 by China Automotive Technology & Research Center CATARC, child safety chairs made in China satisfy international standards in terms of technical design, manufacturing processes and quality. Products are most exports. Therefore, performance evaluation and possible callback campaigns of child safety chairs would be major concerns in child occupants' protection in China.

A study project team was organized in 2011 and occupied in study on usability evaluation system of child-safety chair in response to market features and status quo of child-safety chair products in China. Members of the joint team came from the AQSIQ Defective Product Administrative Center, Tsinghua University, National Center of Supervision and Inspection on Motor Vehicle Products Quality (Shanghai). In May 2014, the Usability Evaluation Regulations and Processes of Child Safety Chair in China were published as periodical study outcome. Usability evaluation of child safety chair serves rational consumption and prevents misuse of child safety chairs, so as to ensure on-board safety of child occupants; it works also as a set *of* guidelines for manufacturers in designing and producing products, which also help improve usability with better built child safety chair products. Weights assigned to evaluation indexes have major impact on evaluation results. Usability evaluation indexes are most qualitative, the present regulations and processes adopt Analysis Hierarchy Process (AHP) for assignment. AHP depends on expert experience and can be subjective. Its results could become less precise somewhat [10]. Instead of single weighting method with poor outcomes, various synthetic weighting methods are getting popular. For instance, literature [11-12] combine AHP and Entropy methods to decide weights to be assigned to evaluation indexes; literature [13] combine AHP and CPA methods to decide weights to be assigned to synthetic indexes for evaluating operation of storage tank zone; literature [14] combine Entropy method, Max Dispersion method and Inter-Class Standard Variance method and CRITIC method to decide weights to be assigned to categorical features of stored-grain insects based on extension theory. Synthetic weighting method is an amalgamation of expertise and objective test data. The present article integrates AHP, Entropy and Mean Variance methods together into original *AEMV* synthetic weight assignment modeling for index weighting of usability evaluation of child-safety chairs. It is an approach with more scientifically objective results than AHP alone.

## 2. Usability Evaluation Index System of Child Safety Chair

There are globally 2 types of regulations for performance evaluation of child-safety chairs: synthetic evaluation regulations, including dynamic performance evaluation and usability; and usability evaluation regulations. Usability is a concept derived by Jakob Nielsen. It indicates if a product is performance friendly [15]. A child-safety chair is well protective as long as its safety performance is logically designed and manufactured as well as it is correctly and properly used. Usability of any child-safety chair therefore shall be equally composed of operation-friendly or comfort-oriented equipment, and easy and smart design to keep it from misuse.

The article takes advantage of international studies on child chair evaluation systems, especially US research results, itemizes and analyses correlation between usability evaluation index and misuse, and thus formulates a usability evaluation index system in response to market situation and status quo of child-safety chair products in China. Refer to Figure 1 for details. The system consists of 4 indexes, i.e. symbol usability, usability of user manual, usability of child protection and usability of chair installation. Symbol usability refers to completeness, preciseness and legibility of any information delivered by a symbol; usability of user manual refers to completeness, preciseness and legibility of any information delivered by a user manual; usability of child protection refers to ensured safety and easy operation when a child is fixed to a child-safety chair; and usability of chair installation refers to ensured safety and easy operation when a child-safety chair is fixed to vehicle seat.

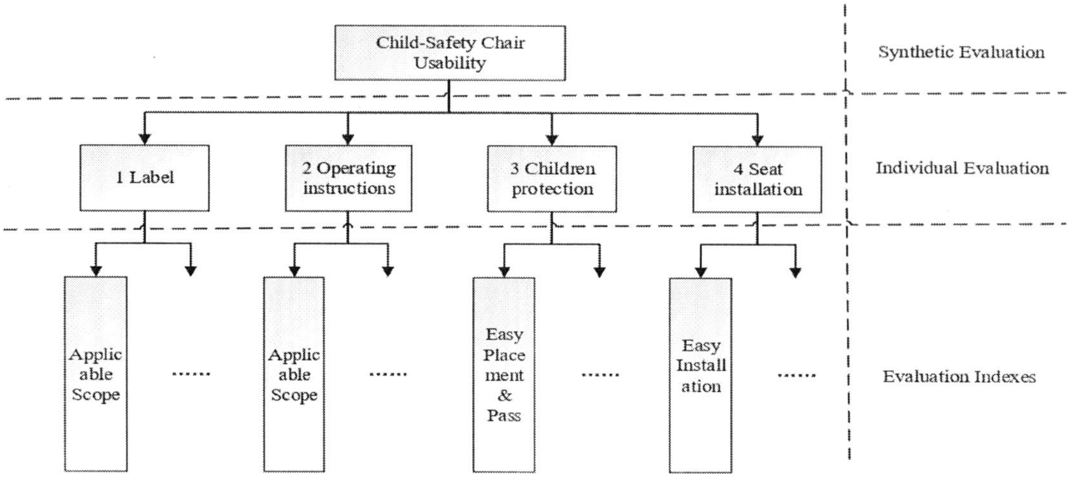


**Figure 1.** Usability evaluation index system of child safety chair

## 3. AEMV Synthetic Weight Assignment Modeling

As for an index-based evaluation, weights that indicate respectively significance shall be assigned to fit indexes. A weight decides the contribution of every assigned index. Determination of each weight lays consequently the ground for evaluation. AHP, developed by American operation-focused primary theoretician Saaty, is an understandable weighting method widely adopted in various fields[16]. Basically, it decomposes complex problems into sub-problems and elements at hierarchies. Indexes are compared to each other two at a time by professional hands with respect to their impact on any element relevant to decision making. A matrix of judgement is then structured. Maximum eigenvalue and associated eigenvector of the matrix are calculated, so as to decide a weight of any factor with certain significance. AHP method comes with risks when making decisions due to uncertain experience and subjectivity of an operator. Dispersion of evaluation data is introduced here as countermeasure to decide respective weight, wider the scatter of any data, greater the weights assigned[17]. Entropy measures uncertainty. Less an uncertainty, less an entropy value[18]. Mean variance is also possible for measuring data scatter degrees. Greater standard deviation means greater significance an index value has in synthetic evaluation, and therefore greater weight. In consideration of expertise and evaluation data at the same time, AHP method, Entropy method and Mean Variance method are applied jointly. AEMV synthetic weight assignment modeling is thus formulated:

$$\begin{bmatrix} w_1 \\ w_2 \\ \vdots \\ w_n \end{bmatrix} = \begin{bmatrix} I_1 & A & E+Mv & \\ I_2 & A & E+Mv & ; \quad R \\ \vdots & A & E+Mv & \\ I_n & A & E+Mv & \end{bmatrix} \tag{1}$$

Of the coefficients, $[I_1 \quad I_2 \quad \cdots \quad I_n]$ is $n$ as many as N indexes; $A$ is AHP method; $E$ is Entropy method; $Mv$ is Mean Variance method; $R$ is correlation among the present 3 weighting methods; $[w_1 \quad w_2 \quad \cdots \quad w_n]$ is weight values gained jointly thanks to the 3 weighting methods.

Given $W_S$ is the weight vector based on AHP method, $W_{O1}$ is the weight vector based on Entropy method, and $W_{O2}$ is the weight vector based on Mean Variance method. $W_{O1}$ and $W_{O2}$ are gained based on evaluation index measurements and have geometric mean values: $W_O^{'} = \sqrt[p]{\prod_{i=1}^{p} W_{Oi}}$

$i=1,2,\cdots,p$ ; $W_O^{'}$ can be normalized: $W_O=[w_{Oi}=\frac{w_{Oi}^{'}}{\sum_{i=1}^{p} w_{Oi}^{'}}]$ $i=1,2,\cdots,p$ ; have $W_S$ and $W_O$ integrated by means of additive ensemble method: $W=\alpha W_S+\beta W_O$; wherein $\alpha$ and $\beta$ are both undetermined coefficients for synthesis, and $\alpha>0$, $\beta>0$, $\alpha+\beta=1$, and values can be calculated via coefficient variation method: $\alpha=\frac{n}{n-1}T_S$, and $\beta=1-\alpha$; wherein $T_S$ serves as variation coefficient of individual weight of set $W_S$, $T_S=\frac{2}{n}(1P_1+2P_2+\cdots+nP_n)-\frac{n+1}{n}$; and $P_1,P_2,\cdots,P_n$ indicate sequencing of $W_O$ vectors from lesser values to greater values.

## 4. Verification by Application

Provided the usability evaluation index system of child-safety chairs in Figure 1, $M_1,M_2,M_3,M_4$ indicate 4 indexes respectively as symbol usability, usability of user manual, usability of child protection and usability of chair installation. AEMV synthetic weight assignment modeling is used to decide individual weight.

### *4.1. AHP-based Weighting.*

Matrix should be constructed based on vectorization, which means all raw data shall be converted into values in standardized format for direct comparison, according to defined measurement scale system[19] Saaty adopted reciprocal 1-9 scale of measurement for AHP vectorization[20] Elements of present hierarchy in AHP modeling are compared to an element of higher hierarchy. Pairwise comparison is commonly used [21]. As for comparison to an element of higher hierarchy, a criterion $z_{ij}$ indicates the comparative importance between element $i$ and element $j$ in the present hierarchy. $z_{ij}$ is usually a positive integer from 1~9 and the reciprocal thereof $z_{ij}$ composes so-called comparison matrix:

$$Z=(z_{ij})_{n\times n}=\begin{bmatrix} z_{11} & z_{12} & \cdots & z_{1n} \\ z_{21} & z_{22} & \cdots & z_{2n} \\ \vdots & \vdots & \vdots & \vdots \\ z_{n1} & z_{n2} & \cdots & z_{nn} \end{bmatrix} \tag{2}$$

If $i=j$, $z_{ij}=1$ $z_{ij}$ ; other values are defined with reciprocal 1-9 scale. Refer to Table 1.

**Table 1.** Reciprocal 1-9 scale of measurement

| $Z_{ij}$ | Value selection |
|---|---|
| 1 | element $i$ and element $j$ equally important |
| 3 | in comparison between element $i$ and element $j$, $i$ is slightly more important than $j$ |
| 5 | in comparison between element $i$ and element $j$, $i$ is obviously more important than $j$ |
| 7 | in comparison between element $i$ and element $j$, $i$ is strongly more important than $j$ |
| 9 | in comparison between element $i$ and element $j$, $i$ is extremely more important than $j$ |
| 1/3 | element $i$ is less important in comparison with element $j$ |
| 1/5 | element $i$ is obviously less important in comparison with element $j$ |
| 1/7 | element $i$ is strongly less important in comparison with element $j$ |
| 1/9 | element $i$ is extremely less important in comparison with element $j$ |

Pairwise comparison is adopted for judgement matrix $Z$. Its maximum eigenvalue $\lambda_{\max}$ and associated eigenvector $P_m$ can be calculated. $P_m$ is normalized to present the weight vector of various elements at same hierarchy. Conformity of the comparison-driven judgement matrix $Z$ need be verified by calculating conformity rate $C.R.$. Verification is positive in case $C.R. < 0.1$, otherwise reconstruction is required. $C.R.$ formulation is shown below: $C.R. = C.I./R.I.$

Of the coefficients: $C.I.$ is conformity index, $C.I. = (\lambda_{\max} - n)/(n-1)$; $n$ means the order of the comparison-driven judgement matrix $Z$; $R.I.$ is random index, with its value selection shown in Table 2.

**Table 2.** Random index $R.I.$ value selection

| $n$ | 1 | 2 | 3 | 4 | 5 | 6 | 7 | 8 | 9 |
|---|---|---|---|---|---|---|---|---|---|
| $R.I.$ | 0 | 0 | 0.58 | 0.90 | 1.12 | 1.24 | 1.32 | 1.41 | 1.45 |

Scale is given in Table 1, judgement matrix is then acquired shown as Table 3 below, thanks to expertise. Calculated maximum eigenvalue is $\lambda_{\max} = 3.0092$; $C.I. = 0.0046$ is derived according to conformity index formulation; $C.R. = 0.0088 << 0.1$ is derived according to conformity rate formulation, indicating qualified conformity verification of judgement matrix. Eigenvector values are normalized to obtain weight vector $W_S = \{0.1091 \quad 0.1891 \quad 0.3509 \quad 0.3509\}$ based on AHP method.

**Table 3.** AHP-based judgement matrix

| $Z$ | $M_1$ | $M_2$ | $M_3$ | $M_4$ |
|---|---|---|---|---|
| $M_1$ | 1 | 1/2 | 1/3 | 1/3 |
| $M_2$ | 2 | 1 | 1/2 | 1/2 |
| $M_3$ | 3 | 2 | 1 | 1 |
| $M_4$ | 4 | 2 | 1 | 1 |

*4.2. Entropy-based Weighting.*

Literature [22] displays usability evaluation results of 6 child-safety chairs. Matrix composed of evaluation scores is shown in Table 4.

**Table 4.** Matrix usability indexes evaluation values

| | $M_1$ | $M_2$ | $M_3$ | $M_4$ |
|---|---|---|---|---|
| CSR-1 | 2.283 | 2.861 | 2.321 | 1.738 |
| CSR-2 | 2.501 | 2.112 | 2.298 | 2.825 |
| CSR-3 | 1.595 | 1.882 | 2.356 | 2.437 |
| CSR-5 | 1.867 | 2.282 | 2.623 | 1.896 |
| CSR-6 | 1.595 | 1.876 | 2.155 | 1.257 |

When applying Entropy method, index $j$ has its entropy value $e_j$ calculated:

$e_j = -k\sum_{i=1}^{n} P(x_{ij})\ln P(x_{ij})$, $e_j > 0$, $k > 0$. Wherein $x_{ij}$ is evaluation score of CSR $i$'s index $j$;

$P(x_{ij})$ is eigenratio of CSR $i$ under index $j$: $P(x_{ij}) = x_{ij} / \sum_{i=1}^{n} x_{ij}(n!/r!(n-r)!)$.

With regard to the present evaluation, $n = 6$, while $k = 1/\ln 6 = 0.5581$. Calculation is:

$$E = \{e_1 \quad e_2 \quad e_3 \quad e_4\} = \{0.9917 \quad 0.9937 \quad 0.9965 \quad 0.9791\} \tag{3}$$

Index $j$ has coefficient variation $g_j$ calculated: $g_j = 1 - e_j$, then the calculation is:

$$G = \{g_1 \quad g_2 \quad g_3 \quad g_4\} = \{0.0083 \quad 0.0063 \quad 0.0035 \quad 0.0209\} \tag{4}$$

Provided the normalization formulation $w_j = g_j / \sum_{j=1}^{n} g_j$, weight vector values shall be decided by Entropy calculation:

$$W_{O1} = \{w_1 \quad w_2 \quad w_3 \quad w_4\} = \{0.2131 \quad 0.1609 \quad 0.0892 \quad 0.5367\} \tag{5}$$

### 4.3. *Mean-Variance-based Weighting.*

Refer to the usability evaluation values given in Figure 4. Calculation of mathematically expected value is performed as $\overline{x_j} = \frac{1}{n}\sum_{i=1}^{n} x_{ij}$, and get values

$\overline{X} = \{\overline{x_1} \quad \overline{x_2} \quad \overline{x_3} \quad \overline{x_4}\} = \{2.0065 \quad 2.1790 \quad 2.4588 \quad 2.1630\}$; Calculation of mean variance is performed as: $v_j = \sqrt{\frac{1}{n-1}\sum_{i=1}^{n}\left(x_{ij} - \overline{x_j}\right)^2}$, and get values:

$V = \{v_1 \quad v_2 \quad v_3 \quad v_4\} = \{0.3784 \quad 0.3672 \quad 0.3058 \quad 0.6363\}$

Provided the normalization formulation $w_j = v_j / \sum_{j=1}^{n} v_j$, weight vector values shall be decided by Mean Variance calculation:

$$W_{O2} = \{w_1 \quad w_2 \quad w_3 \quad w_4\} = \{0.2242 \quad 0.2176 \quad 0.1812 \quad 0.3770\} \tag{6}$$

### 4.4. *Synthesis of Weight Values and Result Analysis.*

Entropy-based $W_{O1}$ and Mean-Variance-based $W_{O2}$ are synthesized to get individual weight vector $W_O'$:

$$W_O' = \{w_{O1}' \quad w_{O2}' \quad w_{O2}' \quad w_{O2}'\} = \{0.2186 \quad 0.1871 \quad 0.1271 \quad 0.4498\} \tag{7}$$

$W_O^{'}$ is later normalized to get $W_O = \{0.2225 \quad 0.1904 \quad 0.1294 \quad 0.4578\}$

Vector set $W_S$ is organized from the minimum to the maximum: $\{0.1091 \quad 0.1891 \quad 0.3509 \quad 0.3509\}$, and calculate coefficient variation of individual weight $W_S$:

$$T_S = \frac{2}{4}(1\times0.1091 + 2\times0.1891 + 3\times0.3509 + 4\times0.3509) - \frac{4+1}{4} = 0.2218 \tag{8}$$

Calculation of coefficients as $\alpha = \frac{n}{n-1}T_S = \frac{4}{4-1}\times0.2218 = 0.3$; $\beta = 1-\alpha = 1\text{-}0.2957 = 0.7$;

$W_S$ and $W_O$ are now synthesized via additive ensemble into weight vector set $W$:

$$\begin{aligned} W &= 0.3\{0.1091 \quad 0.1891 \quad 0.3509 \quad 0.3509\} + 0.7\{0.2225 \quad 0.1904 \quad 0.1294 \quad 0.4578\} \\ &= \{0.1885 \quad 0.1900 \quad 0.1959 \quad 0.4257\} \end{aligned} \tag{9}$$

According to weight values shown in Figure 2, chair installation usability $M_4$ has the greatest weight, regardless of adopted weighting methods. Weighting both expertise and test data, AEMV synthetic approach produces a more logical distribution of weights, thanks to experience and preference of experts in decision making as well as correlation and features of objective data series.

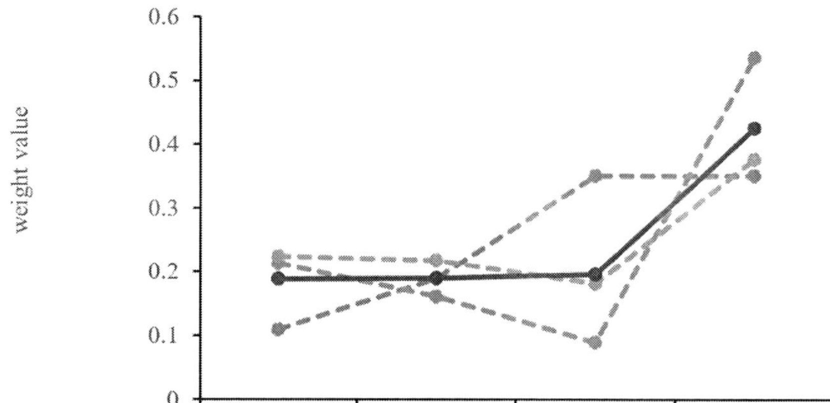


| | w1 | w2 | w3 | w4 |
|---|---|---|---|---|
| AHP method | 0.1091 | 0.1891 | 0.3509 | 0.3509 |
| Entropy method | 0.2131 | 0.1609 | 0.0892 | 0.5367 |
| Mean Variance method | 0.2242 | 0.2176 | 0.1812 | 0.377 |
| AEMv synthetic method | 0.1885 | 0.19 | 0.1959 | 0.4257 |

**Figure 2.** Weight values based on various weighting methods

## 5. Conclusions

Child-safety chairs are perceptibly less applied in China in comparison to the situation in developed countries. Child-safety chair evaluation serves rational and correct purchase and use of every consumer. Usability is a decisive factor of any child safety chair in assessing its functions, making purchase decision, and using it properly and effectively. AHP method brings about less proper results due to intrinsic limitation when deciding index weights of usability evaluation of any child-safety chair products. AEMV synthetic weight assignment modeling as a combined method is whereas derived on the basis of AHP method, Entropy method and Mean Variance method. Expertise and objective test data are also reckoned in. Comparative study on results demonstrates the weight assignment mechanism of the present synthetic weighting as a more scientifically objective approach, with more logical outcomes of usability evaluation.

**Acknowledgments**
This work was financially supported by National Fund for Fundamental Research (282017Y-5303, 282018Y-5973).

**Reference**

[1] Decina L E, Lococo K H. Misuse of child restraints [C]//SAE Tech Paper, 2004.US DOT /NHTSA No. DOT HS 809 671.
[2] Snowdon A W, Hussein A, Ahmed E. Canadian national survey on child restraint use 2010 [R/OL]. 2010-12-07.
[3] Koppel S. Child restraint system misuse and /or inappropriate use in Australia [J]. Traffic Injury Prevention, 2009, 10: 302-307.
[4] German Insurance Association. Misuse of child restraint systems: A 2008 observation study in Germany [R/OL]. 2009-01-01.
[5] Kahane C J. An evaluation of child passenger safety: The effectiveness and benefits of safety Seats [R /OL].2010-01-01.
[6] HU Jia,ZHONG Zhihua,Koji Mizuno.Frontal Crash Sled Test Simulations of Child Seat in Both Correct Use and Misuse Condition[J].Automotive Engineering,2009,31(1):10-17.
[7] ZHANG Jinhuan,YANG Fan,ZHOU Yong,etc.Lateral impact sled test system for child restraint system based on ECE R129[J].J Automotive Safety and Energy,2015,6(3):237-244.
[8] ZHANG Jinhuan,LI Hao,ZHOU Yong,etc.Influence of support leg misuse on the safety performance of child restraint system ISOFIX[J].J Automotive Safety and Energy,2015,6(4):360-364.
[9] Catherine S.Gotschall, Eichelberger Martin R, Morrissey J R, et al. Injury patterns associated with child restraint misuse [J]. Society of Automotive Engineers.1997:197-193.
[10] Bruce L.Golden, Edward A.Wasil, Patrick T.Harker. The analytic hierarchy process [M]. McGraw Hill Inc. NY, 1989.
[11] GUO Jinwei,PU Xuqiang,GAO Xiang,etc.Improved method on weight determination of indexes in multi-objective decision[J].JOURNAL OF XIDIAN UNIVERSITY, 2014, 41(6):118-125.
[12] LIU Jie,LI Chaofeng,LI Xiaopeng,etc.Study and Application of Generalized Quality Evaluation of Products by Combination Weights and Fuzzy Matter Element[J].Journal of Northeastern University(Natural Science),2008,29(9):1314-1317.
[13] JIA Liangjun,WANG Guangshe,SUN Jie,etc.Determining weights of comprehensive evaluation indexes for irrigation district operation performance by method of game theory[J].Journal of Drainage and Irrigation Machinery Engineering,2014,32(9):808-813.
[14] ZHANG Hongtao,MAO Hanping.Comparison of four methods for deciding objective weights of features for classifying stored-grain insects based on extension theory[J]. Transactions of the CSAE, 2009, 25(1):132-136.
[15] Jakob Nielsen. Usability engineering [M]. Morgan Kaufmann, 1993.
[16] PENG Hongying,ZHANG Li,WEN Bangchun.AHP -Based Assignment of Weight in Product Design Quality[J].Journal of Northeastern University(Natural Science),2009,30(5):712-715.
[17] XU Baoli,LI Rongli,WEI Huoming,etc.Entropy Weight Fuzzy Comprehensive Evaluation on Failure Mode Criticality of Integrated Transmission System[J].ACTA ARMAMENTARH, 2013, 34 (11):1380-1386.
[18] FU Yu,WU Xiaoping,YE Qing,etc.An Approach for Information Systems Security Risk Assessment on Fuzzy Set and Entropy-Weight[J].ACTA ELECTYONICA SINICA,2010,38(7):1489-1494.
[19] WANG Guipeng,JIA Yazhou,ZHOU Guangwen.Evaluation Method and Application of CNC Machine Tool's Green Degree Based on Fuzzy-EAHP[J].JOURNAL OF MECHANICAL ENGINEERING,2010,46(3):14-147.
[20] Satty T L. The analytic hierarchy process [M]. New York: Mc Craw-Hill, 1980.
[21] DONG Honglei, CHEN Zhongzheng Jinhuan.Defect identifcation of automotive product based on fuzzy risk matrix [J]. J Automotive Safety and Energy, 2016, 7(4):377-381.

IOP Conf. Series: Materials Science and Engineering 472 (2019) 012022 doi:10.1088/1757-899X/472/1/012022

[22] ZHANG Xiong.Research on Ease of Use Rating System for Vehicle Child Restraint System [D]. Tsinghua University, 2014.

IOP Conf. Series: Materials Science and Engineering 472 (2019) 012023 doi:10.1088/1757-899X/472/1/012023

# Analysis of the Influence of Mass Imbalance on the Performance of Photoelectric Stabilization Platform

**Hui Guo**[1, a] **and Huajie Hong**[1,b,*]

[1]College of Artificial Intelligence, National University of Defense Technology, Chang Sha, Hu Nan, 410073, China

E-mail: [a]1518467261@qq.com, [b,] *opalqq@163.com

**Abstract.** Photoelectric stabilization platform is an optoelectronic imaging tracking system widely used in aircraft, vehicles and other equipment. It can effectively isolate the disturbance of the carrier and ensure the stability of the platform. In practical applications, there are many factors that affect the stability of the platform, including friction, disturbance, and mass imbalance. The purpose of this paper is to use Simulink to model and simulate the integrated three-axis photoelectric stability platform, and to explore the influence of mass imbalance torque caused by different factors on the accuracy and stability of the photoelectric stability platform. The results show that the unbalanced moment generated by the eccentricity of the frame has a great influence on the accuracy of the platform, and the unbalanced torque generated by the uneven mass distribution has little or no negligible influence on the accuracy of the platform.

## 1. Introduction

The photoelectric stability platform is an optoelectronic imaging tracking system consisting of three parts: photoelectric sensor system, signal processing and control system and tracking stabilization system [1]. It is generally a multi-axis, multi-frame spherical or hemispherical structure that can carry high-precision photoelectric loads such as visible light, infrared, and laser [2]. Since the photoelectric stabilization platform can not only effectively isolate the disturbance of the carrier, but also achieve fast and accurate capture, tracking and aiming of the moving target. Therefore, it has been widely used in modern weapons and equipment such as airplanes and ships.

In the analysis and design of the photoelectric stability platform system, it is very important to ensure its stability. There are many factors affecting the stability of the photoelectric stability platform, including friction, frame coupling, carrier disturbance and unbalanced disturbance. Yu Shuang et al. used sensors to test the frame, and derived the frame unbalanced moment formula in dynamic environment and established the inertial platform control model based on unbalanced torque compensation [3]. Wu Hailong et al. established a simulation model of mass static and dynamic imbalance for the two-frame stable platform, and studied its influence on the platform system [4]. Zhu Huazheng et al. used the two-frame seeker servo mechanism as the object to analyze the influence of mass static imbalance on the performance of the platform stability loop [5].

Although the above-mentioned references are all aimed at the modeling and simulation of the factors affecting the stability of the photoelectric platform, most of them ignore the modeling and simulation analysis of the unbalanced torque caused by the uneven distribution of the frame mass. Combined with the actual situation, this paper will model the simulation of the integrated three-axis three-frame photoelectric stability platform under the condition that the frame mass eccentricity and uneven mass distribution cause unbalanced torque, and study the influence of mass imbalance on the stability of the platform.

## 2. Kinematics Analysis

### 2.1. Coordinate System Definition

The three-axis stabilization platform consists of three frames, the structure of which is shown in Fig.1.

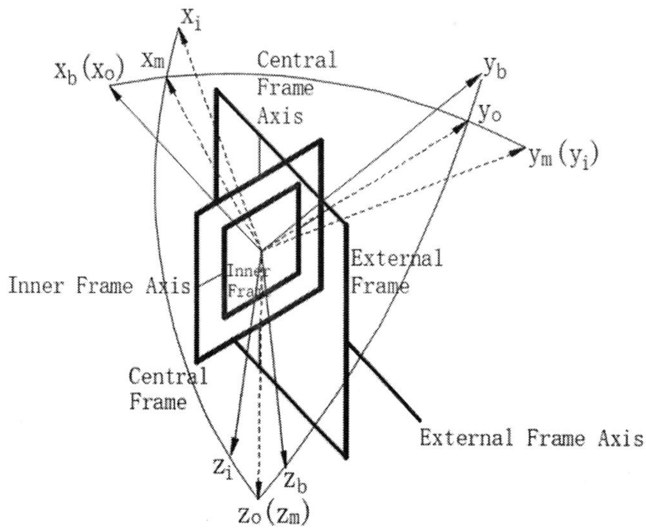


**Figure 1.** Schematic diagram of three-axis stable platform structure and coordinate system.

In Figure 1, $o_b x_b y_b z_b$ is the pedestal coordinate system, and the origin is at the intersection of the three frame axes of the stable platform. $o_o x_o y_o z_o$ is the outer frame (rolling frame) coordinate system, the roll axis $o_o x_o$ is in the same direction as the $o_b x_b$, and the rotation produces a roll angle of $\alpha$. $o_m x_m y_m z_m$ is the middle frame (azimuth frame) coordinate system, the azimuth axis $o_m z_m$ is in the same direction as $o_o z_o$, and the rotation produces an azimuth angle of $\beta$. $o_i x_i y_i z_i$ is the inner frame (pitch frame) coordinate system, the pitch axis $o_i y_i$ is in the same direction as $o_m y_m$, and the rotation produces the pitch angle $\varepsilon$.

### 2.2. Analysis of Frame Motion Coupling Relationship

The coordinate transformation matrices from the base to the outer frame, the outer frame to the middle frame, and the middle frame to the inner frame are $R_{ob}, R_{mo}, R_{im}$ , respectively. Let $\omega_b=[\omega_{bx} \quad \omega_{by} \quad \omega_{bz}]^T$ be the angular velocity of the pedestal coordinate system with respect to the inertial space coordinate system. $\dot{\alpha}, \dot{\beta}, \dot{\varepsilon}$ are the angular velocities of the outer frame, the middle frame, and the inner frame around the axis of rotation. According to the coupling relationship between the three frames of the platform, the angular velocity of the outer frame, middle frame, inner frame and detector's boresight in the inertial space coordinate system can be obtained as follows:

$$\omega_o = R_{ob}\omega_b + [\dot{\alpha} \quad 0 \quad 0]^T \tag{1}$$

$$\omega_m = \omega_{mo} + \dot{\beta} = R_{mo} R_{ob}\omega_b + R_{mo}\dot{\alpha} + \dot{\beta} \tag{2}$$

$$\omega_i = \omega_{im} + \dot{\varepsilon} = R_{im} R_{mo} R_{ob}\omega_b + R_{im}R_{mo}\dot{\alpha} + R_{im}\dot{\beta} + \dot{\varepsilon} \tag{3}$$

$$\omega_g = \omega_i \tag{4}$$

## 3. Quality Imblance Model Analysis

Due to manufacturing and assembly errors, the quality imbalance of stable platforms is widespread. Generally, it consists of two parts: one part is due to the mass eccentricity of the frame, so that the frame produces an unbalanced torque in the presence of the carrier overload acceleration; the other part is due to the uneven distribution of the frame mass, that is, the product of inertia exists in the inertia tensor matrix. This produces an unbalanced torque.

### *3.1. Unbalanced Torque Generated by the Mass Eccentricity of the Frame*

Define the overload acceleration at the origin of the stable platform base coordinate system as $a_b=[a_{bx} \quad a_{by} \quad a_{bz}]^T$. According to the coordinate transformation relationship between frames, the acceleration of the outer frame, middle frame and inner frame are:

$$a_o=R_{ob}a_b \tag{5}$$

$$a_m=R_{mo}a_o=R_{mo}R_{ob}a_b \tag{6}$$

$$a_i=R_{im}a_m=R_{im}R_{mo}R_{ob}a_b \tag{7}$$

The projections of the outer frame, the middle frame and the inner frame centroid in the respective frame coordinate systems are $[r_{ox} \quad r_{oy} \quad r_{oz}]$, $[r_{mx} \quad r_{my} \quad r_{mz}]$, $[r_{ix} \quad r_{iy} \quad r_{iz}]$.The torque due to the force generated by the overload acceleration at the eccentricity can be expressed as:

$$M=m\cdot a\cdot r \tag{8}$$

Where m is the frame quality; $a$ is the overload acceleration; r is the eccentricity.

Therefore, the mass imbalance torques on the outer frame axis, the middle frame axis and the inner frame axis are:

$$M_{diso}= m\cdot (a_{oz}\cdot r_{oy}-a_{oy}\cdot r_{oz}) \tag{9}$$

$$M_{dism}= m\cdot (a_{my}\cdot r_{mx}-a_{mx}\cdot r_{my}) \tag{10}$$

$$M_{disi}= m\cdot (a_{ix}\cdot r_{iz}-a_{iz}\cdot r_{ix}) \tag{11}$$

### *3.2. Unbalanced Moment Generated by Uneven Distribution of Frame Mass*

Define $J_i$, $J_m$, and $J_o$ as the inertia matrix of the inner frame, the middle frame, and the outer frame, respectively. $\omega_i$, $\omega_m$, $\omega_o$ are the angular velocities of the inner frame, the middle frame, and the outer frame relative to the inertial space, respectively.

According to the momentum moment theorem, the dynamic model of the inner frame, the middle frame and the outer frame are:

$$M_i=J_i\dot{\omega}_i+\omega_i\times J_i\omega_i \tag{12}$$

$$M_m=J_m\dot{\omega}_m+\omega_m\times J_m\omega_m+[M_i]_m=J_m\dot{\omega}_m+\omega_m\times J_m\omega_m+R_{mi}M_i \tag{13}$$

$$M_o=J_o\dot{\omega}_o+\omega_o\times J_o\omega_o+[M_m]_o = J_o\dot{\omega}_o+\omega_o\times J_o\omega_o+R_{om}M_m \tag{14}$$

Where $[M_i]_m$ is the projection of the reaction torque of the inner frame to the middle frame in the middle frame coordinate system.$[M_m]_o$ is the projection of the reaction torque of the outer frame of the middle frame in the outer frame coordinate system.

In the formula, the term of the inertial product is the unbalanced torque due to the uneven distribution of the mass of the frame.

## 4. Simulation Analusis

The three-axis stabilized platform servo control system uses speed-current dual-loop control, and the controller uses a classic controller(PID). The stable platform simulation model built by Simulink in Matlab is shown in Figure 2.

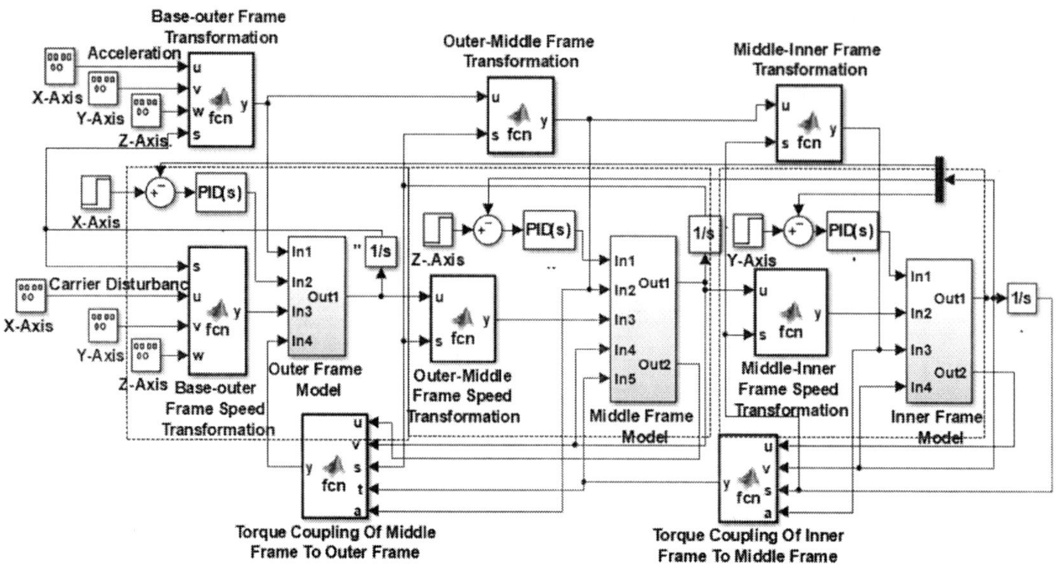


**Figure 2.** Simulation model of photoelectric stability platform Simulink.

The solid line box above Figure 2 is the system acceleration input and transfer model(See 3.1 for details). The three dashed boxes in Figure 2 are the outer frame model, the middle frame model and the inner frame model from left to right. There is an angular velocity transfer relationship between the frame models(See 2.2 for details).

### *4.1. Simulation without Mass Imbalance Torque*

When there is no frame mass eccentricity and the inertia product term in each frame inertia tensor matrix is ignored. Set the controller parameters and set the carrier speed disturbance to be a sinusoidal signal with amplitude 1°/s and frequency 2Hz. Obtaining the boresight angle, that is, the system response is shown in Figure 3.

It can be seen from Fig. 3 that there is also a certain oscillation of the boresight angle due to the influence of the carrier disturbance. The magnitudes of the x, y, and z directions are $6.1\times10^{-5}$,$8.2\times10^{-5}$,$15\times10^{-5}$(rad). The frequency is consistent with the carrier disturbance frequency.

### *4.2. Simulation in the Presence of Unbalanced Moment Due to the Eccentricity of the Frame*

The inertial product term in each frame inertia tensor matrix is ignored. Let the eccentricity of each frame be $[0.005 \quad 0.005 \quad 0.005]$ m, and the carrier velocity perturbation is a sinusoidal signal with amplitude of 1°/s and frequency of 2Hz.

(1) Refer to the helicopter equipment parameters, only take the overload acceleration amplitude of 1.5g (gravity acceleration), the frequency is 10Hz in the x direction, and the other settings are unchanged. The system response is shown in Figure 4.

It can be seen from Figure 4 that under the interference of the unbalanced torque, the angles in the x, y, and z directions of the boresight are respectively signals of a peak value of $0.7\times10^{-4}$,$2.9\times10^{-4}$,$4.9\times10^{-4}$(rad).

(2) Only take the overload acceleration amplitude of 1.5g in the y direction, the frequency is 10Hz.The other settings are unchanged. The system response is shown in Figure 5.

It can be seen from Figure 5 that under the disturbance of the unbalanced torque, the angles in the x, y, and z directions of the boresight are respectively signals of a peak of $1.4\times10^{-4}$,$0.8\times10^{-4}$,$2.5\times10^{-4}$(rad).

(3) Take only the overload acceleration amplitude of 1.5g in the z direction, the frequency is 10Hz, and the other settings are unchanged. The system response obtained by simulation is shown in Figure 6.

It can be seen from Figure 6 that under the interference of the unbalanced torque, the angles in the x, y, and z directions of the boresight are respectively signals of a peak value of $1.4\times10^{-4}$,$2\times10^{-4}$,$3.2\times10^{-4}$(rad).

In summary, in the presence of frame eccentricity and overload acceleration, the angular error of the boresight is significantly increased compared to the result of no unbalanced torque added, which has a great impact on the accuracy and stability of the system. More serious will also cause system instability.

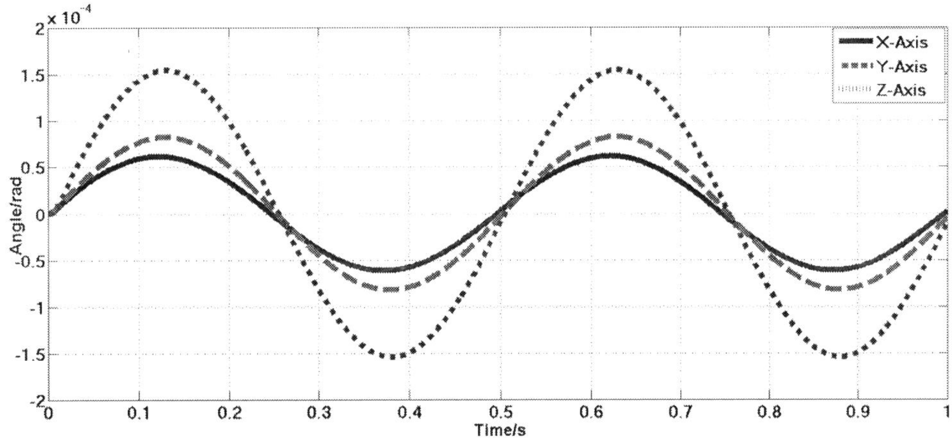


**Figure 3.** The angular response of the boresight under carrier disturbance.

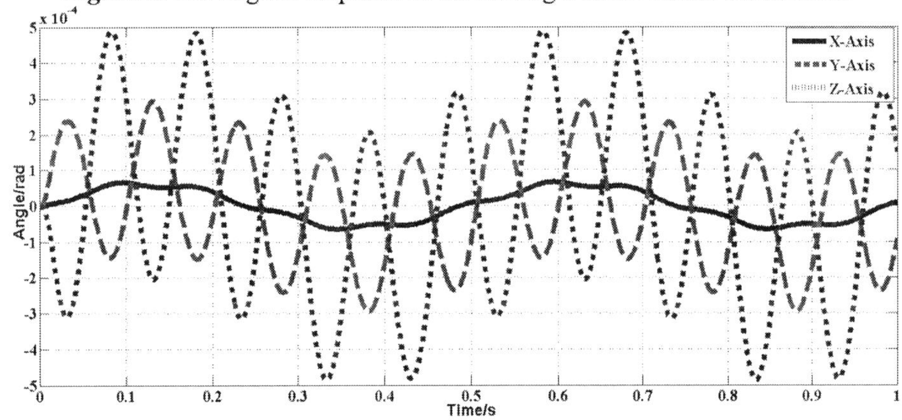


**Figure 4.** Angle response of the boresight under disturbance and unbalanced torque.

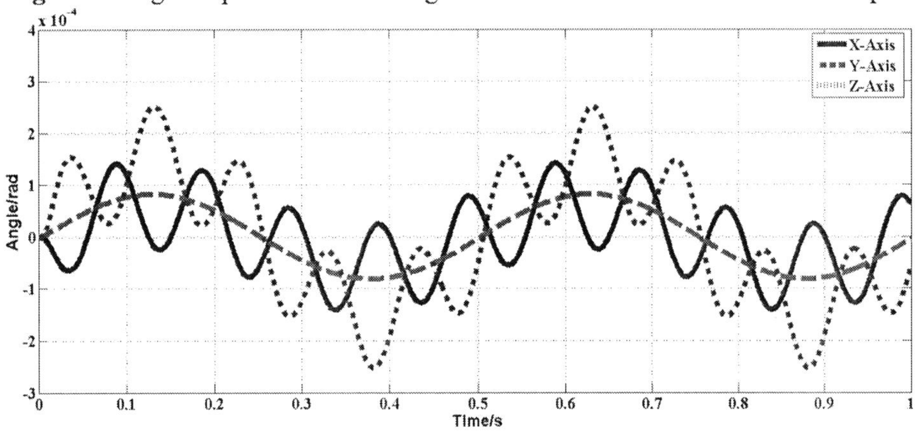


**Figure 5.** Angle response of the boresight under carrier and unbalanced torque.

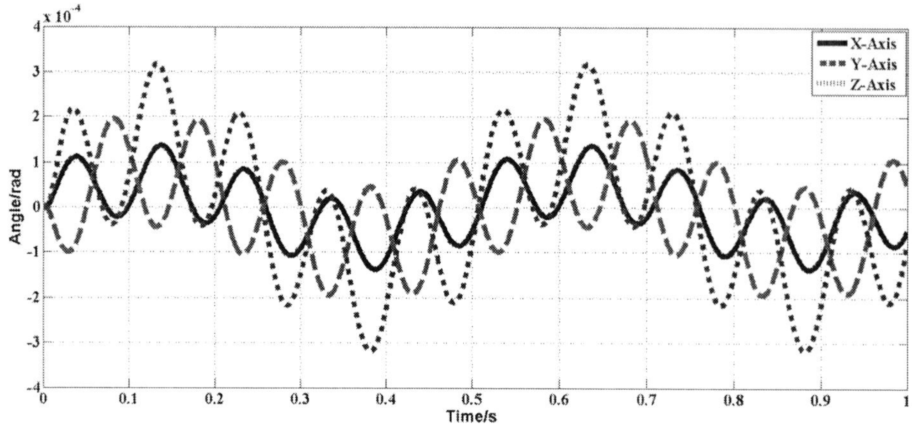


**Figure 6.** Angle response of the boresight under carrier and unbalanced torque.

### *4.3. Simulation in the Presence of Unbalanced Torque Due to Uneven Distribution of Frame Mass*
It is assumed that there is no mass eccentricity in each frame, but there is an inertial product term in each frame inertia tensor matrix. The carrier rate perturbations are sinusoidal signals with amplitude 1°/s and frequency 2Hz, and other settings are unchanged. The system response obtained by simulation is shown in Figure 7.

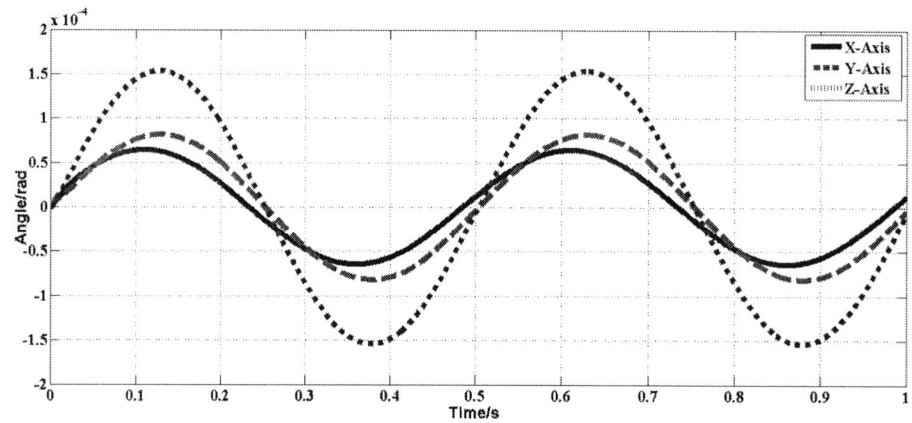


**Figure 7.** Angle response of the boresight under carrier and unbalanced torque.

It can be seen from Figure 7 that under the disturbance of the unbalanced torque, the angles of the boresight x, y, and z directions are respectively $6.4\times10^{-5}$,$8.2\times10^{-5}$,$15\times10^{-4}$ (rad), and the frequency is consistent with the carrier disturbance frequency.

Compared with the boresight angle without adding unbalanced torque interference, it can be seen that the unbalanced torque generated by the frame due to uneven mass distribution has little influence on the boresight angleTherefore, to a certain extent, the unbalanced torque of the frame due to uneven mass distribution can be ignored.

## 5. Conclusions
In this paper, Simulink is used to model the simulation of the integrated three-axis three-frame photoelectric stability platform to explore the impact of mass imbalance on the accuracy and stability of the platform. The simulation results show that:

(1) When the platform is only subjected to the mass imbalance torque due to the eccentricity, the boresight angle has a large oscillation, which affects the accuracy and stability of the platform.

(2) When the platform is only subjected to mass imbalance torque due to uneven mass distribution, the boresight angle is less affected by it and has less influence on the accuracy and stability of the platform.

Therefore, the use of frame eccentricity should be avoided during the use of photoelectric stabilization platforms. However, in the actual manufacturing assembly process, the eccentricity cannot be avoided, and the position of the center of mass needs to be adjusted by the counterweight to eliminate the existence of the eccentricity and improve the accuracy and stability of the stable platform.

**References**

[1] Y.Y. Li, Research of three axis tracking-pointing platform servo control System, Master Thesis, Henan University of Science and Technology, Henan China, 2015.

[2] R.H. Xie, The Research of Stabilization and Tracking Control Techniques on Airborne Opto-electric Platform Servo System, Doctoral thesis, University of Chinese Academy of Sciences, Beijing China, 2017.

[3] S. Yu, Z. Fu, H. Zhao, Y.Z. Zhao, Measurement Method and Compensation Control for Unbalanced Moment of Inertial Platform. JOURNAL OF SHANGHAI JIAOTONG UNIVERSITY. 2008(10):1692-1696.

[4] H.L. Wu, H.G. Jia, Q. Wei, H.H. Jiang, R.F. Zhu, Analysis of Mass Imbalance for Roll-Pitch Intertial Stabilized Platform. JOURNAL OF XI'AN JIAOTONG UNIVERSITY. 49(05):108-115.

[5] H.Z. Zhu, D.P. Fan, W.B. Zhang, D.X. Ma, Influence analysis of the mass imbalance torque on the performance of seeker servo mechanism. Infrared and Laser Engineering. 38(05):767-772.

[6] P. Zhang, X.M. Dong, K.S. Fu, F.Y. Deng, 2011. Modeling and Control of Airborne/Missile-borne Vision-guidance Stabilized Platform. Beijing: National Defense Industry Press.

[7] McKerley, Charles W. A model for a two degree of freedom coupled seeker with mass imbalance.Conference Proceedings- IEEE Southeastcon. Tampa,FL, USA: IEEE, 1996: 84-87.

IOP Conf. Series: Materials Science and Engineering 472 (2019) 012024 doi:10.1088/1757-899X/472/1/012024

# Experimental Exploration of Macro-fiber Composite Morphing Wing

**Xiangying Guo***, **Shuaibo Wang and Wei Zhang**

Beijing Key Laboratory of Nonlinear Vibrations and Strength of Mechanical Structures, College of Mechanical Engineering, Beijing University of Technology, Beijing,100124, China

*E-mail:eagle2008guo@yeah.net

**Abstract.** In this paper, a experimental model of morphing swept-wing with carbon fiber is designed and driven by Macro-Fiber Composites (MFC). Dynamic analysis of the wing structure is carried out. The maximum twist angle of the carbon fiber structure driven by the MFC in the experiment is 10.72°. External excitation can be applied to the wing structure for its dynamic analysis and the modal performance of the structure is analyzed.

## 1. Introduction

The dream of every aeronautical engineer is to design the best possible wing which is able to smoothly run in all operating flight conditions. Morphing wing, as a type of shape-adaptive structures, are enabled to improve performance with fight condition and smooth geometric changes. Therefore, morphing wing becomes a hotspot in research nowadays. Morphing swept angle for airfoils is a classical morphing type and good at reducing the shock wave resistance and improving the flight dynamics. Moreover, the development of smart materials and structure offers a great potential in the field of shape control in different fields of application, especially in the aviation field.

Some studies have explored the use of such actuators to change or control the shape of aerodynamic bodies. Bilgen et al. investigated the use of MFCs to change the wing camber for roll and pitch control of a remotely piloted micro-air-vehicle (MAV) [1]. At Virginia tech in 2004, Neal and others designed a kind of adaptive UAV model, of which the wing can turn 20°, drone sweepback can vary from 0° to 40°, wing exhibition can change 17.3% and long tail wing can compress closer to 11%. The effectiveness of the MAV model under different deformation was verified by wind tunnel test [2]. In 2011, the national research council of Canada's school of aeronautical research designed a wing structure with variable curvature, in which MFC was used as the driver, and torsional angle of the wing was 24.2°from the theoretical calculation. In one of the experiment analyses, the wing torsion angle of micro-air-vehicles could arrive to 27.3° with 40N axial force and 2000V driving voltage [3].

In this paper, a preliminary exploration of a new design of morphing swept-wing is studied, which is driven by MFC, and the dynamic response of the smart structure is analyzed by using the experimental methods.

## 2. Experiment Results

### *2.1. Driving Performance Testing of MFC Actuator*

Firstly, consider a single cantilever carbon fiber composite plate with MFC actuator added at the free end as shown in Figure 1. The geometries of the plate are 150*40*0.2 mm and the dimensions of MFC actuators are 85*28*0.3 mm [4]. The other material properties of the components are as follows:

cross-ply carbon fiber composite: E=117GPa, μ= 0.44, ρ = 1800 kg/m$^3$

MFC: $E_x$=30.3GPa, $E_y$=15.9GPa, $E_z$=15.9GPa, $G_{xz}$=10.7GPa, $G_{yz}$=10.7GPa, $G_{xz}$=5.7GPa, $d_{33}$=4.18*10$^{-10}$m/V, $d_{32}$=$d_{31}$=-1.98*10$^{-10}$m/V, ρ = 5440 kg/m$^3$

Here, $E$ is Young's modulus, $G$ is the shear modulus, μ is Poisson's ratio, $\rho$ is the density and $d_{ij}$ is the effective piezoelectric constant.

The torsion angle of the composite plate varies with the increasing voltages are shown in Table 1. And the proportional relationship of these two variables is shown in Figure 2, where the maximum deflection angle is 3.32°.

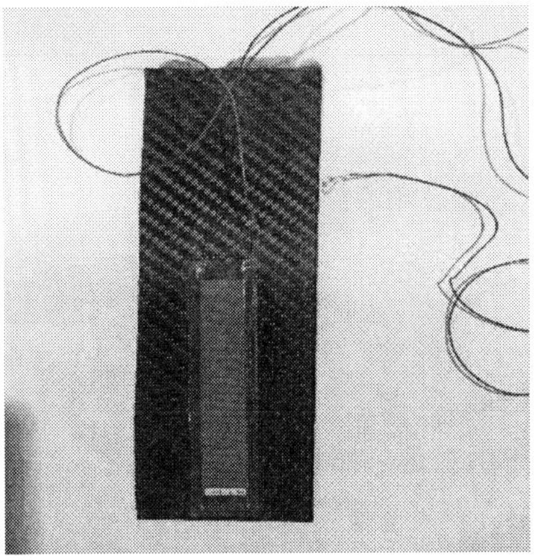

**Figure 1.**The MFC actuator

**Table 1.** Deflection angle and displacement at different voltages

| | Experiment | |
|---|---|---|
| Voltage (V) | Displacement (mm) | Deflection Angle (°) |
| 300 | 2.08 | 0.66 |
| 600 | 4.53 | 1.44 |
| 900 | 6.65 | 2.12 |
| 1200 | 8.27 | 2.63 |
| 1500 | 10.41 | 3.32 |

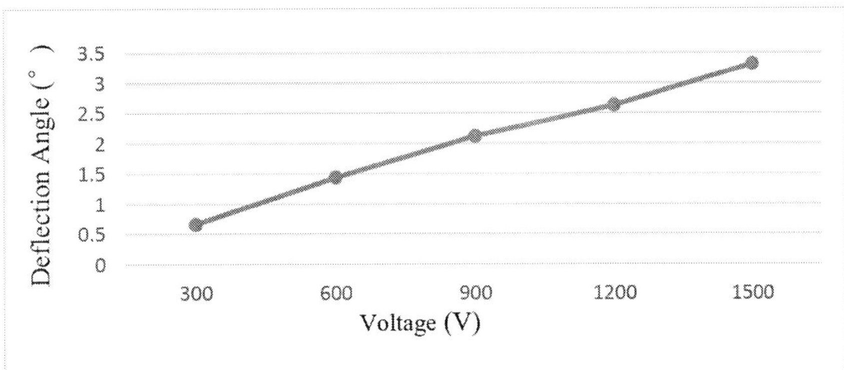


**Figure 2.** Structural deflection angle under different driving voltage

*2.2. The Modal Analysis of Carbon Fiber Wing Ribs*

Then, a now kind of the experimental model of morphing swept wing are design and the dynamical characteristics are analyzed in this part. In order to optimize the structure of swept wings and minimize the weight, we design a full carbon fiber composite wing rib structure as shown in Figure 3.

it contains four cantilever fin plates with certain radians and connected by the slots board, the geometrical parameters are: The total length is 410mm, the maximum height is 110mm, and the thickness is 0.2mm. The MFC actuators are fixed at the root of the fixed end in the two middle fin plates to drive the deformation as what has been finished in the section 2.1

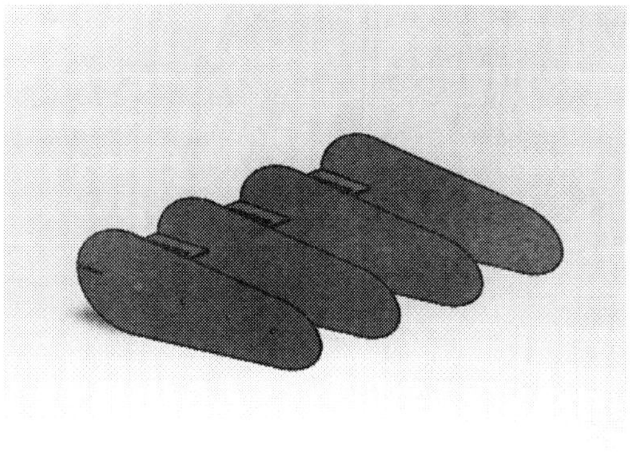

**Figure 3.**The three dimensional wing rib model

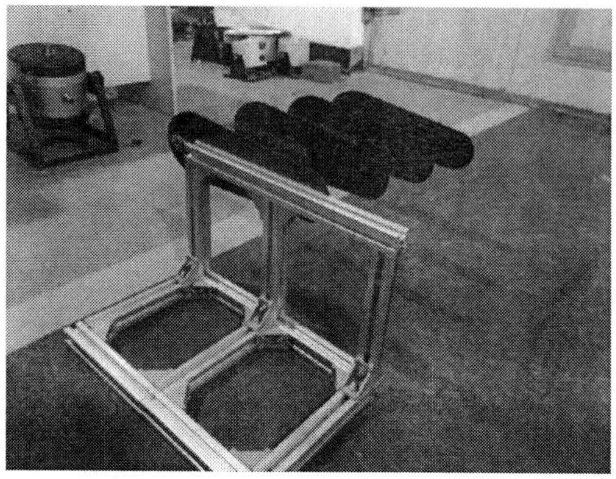

**Figure 4.** Physical model of wing rib

JZK-5T are applied at the root of the experimental model to simulate the aerodynamic disturbance, and The B&K 4517 acceleration sensor is mounted at the rear of the wing rib. Firstly, the first and second natural frequencies of the structure are obtained as 9Hz and 28Hz, respectively, through sweep test of LMS test lab [5]. Then, the dynamical behavior of the morphing wing in the resonant case is studied. Set the frequencies of excitation forces be 9Hz and 28Hz, respectively and increase the amplitude of the forces though change the voltage of signal generator which can change the amplitude forces of the excitation. At the same time, the extra voltage 1500V is added to the wing rib. And the maximum swept angle is 10.72°.

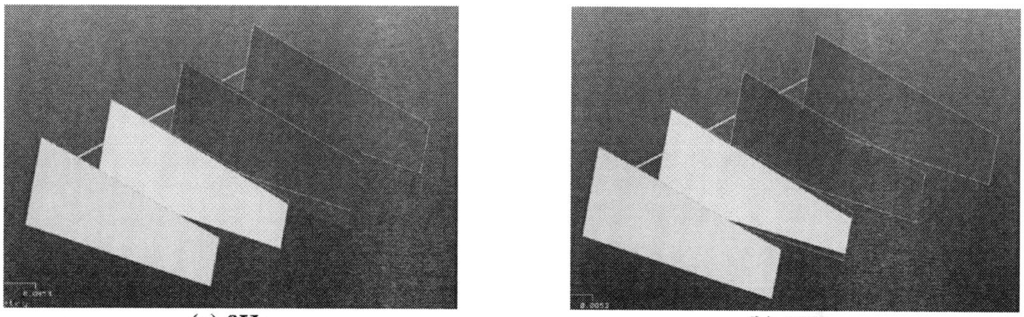

(a) 9Hz (b) 28Hz

**Figure 5.**Some mode shapes corresponding to the natural frequencies of the model

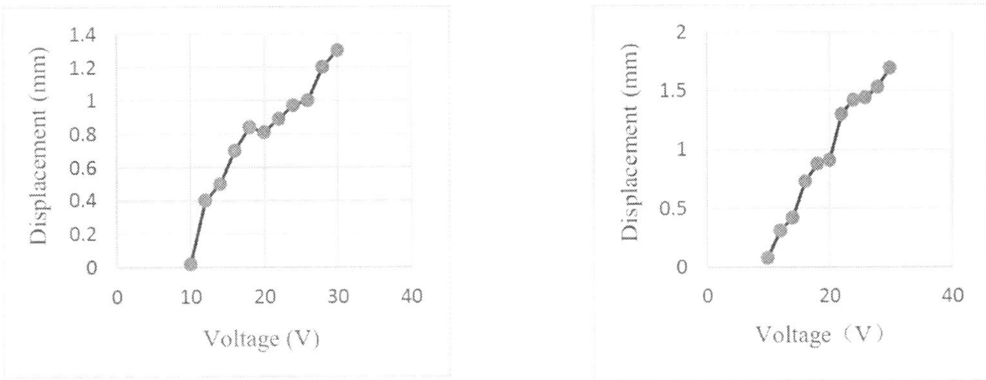


(a) 9Hz (b) 28Hz

**Figure 6.**The amplitude of the carbon fiber wing model at different drive voltages

The figure 5 shows the vibration modes of the first and second natural frequencies. And the varies of the amplitudes of the wing with voltage are shown in Figure 6. It is concluded that the displacements of middle two wing ribs with MFC are relatively synchronous at the first frequency, which have large amplitude at the root of the ribs. In the second frequency, the vibration of the outer flank ribs is more violent and they are out of sync.

## 3. Conclusions

In this paper, a new type of driving material MFC is proposed, and a deformable machine rib model is designed. The feasibility of this method is demonstrated through preliminary experimental exploration. The performance of composite plate with MFC actuators are measured Firstly. Deflection angle is proportional to the driving voltage and the maximum deflection angle is 10.72°. Secondly, a kind of intelligent fin structure with MFC as driver is designed, and the modal performance of the structure is analyzed. The vibration of the two wing ribs is more synchronous under the first resonant frequency and they are different under the second resonant frequency. This study lays a foundation for the feasibility of applying the morphing swept-back airfoil with MFC.

## References

[1] O. Bilgen, K. Kochersberge, E.C. Diggs, A.J. Kurdila, D.J Inman, Morphing Wing Micro-Air-Vehicles via Macro-Fiber-Composite Actuators, Structural Dynamics, and Materials Conference, Honolulu, Hawaii, 2007.

[2] D. Neal, M, Good, C. Johnston, Design and wind tunnel analysis of a fully adaptive aircraft configuration, Proc of the 45th AIAA/ASME/ASCE/AHS/ASC Structures, Structural Dynamics and Materials Conference, Palm Springs, California, 2004,

[3] V. Wickramasinghe, Y. Chen, M. Martinez, F. Wong, R. Kernaghan, Design and verification of a smart wing for an extreme-agility micro-air-vehicle, Smart Materials and Structures 2011,20(12)

[4] Information on http:// www.smart-materials.com,

[5] LMS Test. Lab, Modal Impact User Manual, Beijing, 2009

IOP Conf. Series: Materials Science and Engineering 472 (2019) 012025 doi:10.1088/1757-899X/472/1/012025

# GA Based Real Frequency Technique for Antenna Broadband Matching

**Zhixun Huang[a, *] and Hu Yang[b]**

National University of Defense Technology, Changsha, Hunan, China
E-mail: [a, *]huangzhixun12@nudt.edu.cn, [b]yanghu@nudt.edu.cn

**Abstract.** In this paper, the genetic algorithm is used as the optimization algorithm in real frequency technique for antenna wideband matching. A broadband matching network for an inductive whip antenna is designed by using genetic algorithm to optimize the dividing point impedance function, and the dividing point admittance function is optimized for a capacitive monopole. The achievable matching circuits are simulated by ADS and the results agree well with the optimized ones. Simulated results show that the matching networks increase the bandwidth of the antennas, verifying the effectiveness of the technique.

## 1. Introduction

Wideband impedance matching technique are widely used in RF and microwave systems which can greatly increase transducer power of the system. Within all wideband impedance matching technique, the real frequency technique has the advantages of high precision, good convergence and easy to implement in engineering.

Unlike other impedance matching technique, the real frequency technique does not require a default network topology which may not be the optimal topology. In this technique, the network topology is synthesized by its dividing point function. The same dividing point function can be synthesized by different networks, so the matching network can be designed to be an easy-to-implement structure in engineering. Despite these advantages, the matching performance of the technique is not so effective at first and it has been improving.

The real frequency technique was first proposed by H.J. Carlin in 1977 for wideband passive matching [1]. After that, the technique has been widely used in matching network design [2], [3]. Nowadays, real frequency technique is simplified and applied in antenna design for wideband operation [4]- [6]. In order to pursue better performance, some researchers begin to use different optimization methods in the technique [7].

This paper improved the real frequency technique by applying genetic algorithm (GA) to optimize the dividing point function. The GA based real frequency technique was implemented on antenna broadband matching. Two examples of designing matching network using the dividing point impedance function and the dividing point admittance function were given.

## 2. GA Based Real Frequency Technique

Provided the impedance of the antenna to be matched is $Z_l$, and the impedance of the dividing point from the antenna to the matching network is $Z_q$, then the transducer power gain (TPG) is

$$TPG=\frac{4R_q(w)R_l(w)}{\left[R_q(w)+R_l(w)\right]^2+\left[X_q(w)+X_l(w)\right]^2} \tag{1}$$

Where $R_l$, $X_l$ is the real part and imaginary part of $Z_l$, and $R_q$, $X_q$ is the real part and imaginary part of $Z_q$. If the real frequency data $Z_l$ is given, TPG is determined by dividing point impedance $Z_q$ or admittance $Y_q$. In order to promise the optimized matching network can be realized by ladder network which is easy to implement in engineering, only the real part of the dividing point function is optimized and the imaginary part can be obtained by the real part using the method of Gewertz.

For wideband matching of antenna, it is necessary to extract the sampling points in a wide frequency band, and optimize the matching characteristics in these frequency points. The least squares method was used to optimize the TPG before, but it was found that genetic algorithm had better performance in this technique as it searched for the global optimal solution rather than the local optimal solution. Moreover, GA cost less time in computing and its convergence was better.

Because the sampling frequency points are discrete, it is generally necessary to use a rational function that can be realized to fit the discrete optimized dividing point function. The rational function of the real part of the dividing point function used in this paper is

$$R_q(w) = \frac{A_0 w^{2k}}{1 + B_1 w^2 + \cdots + B_n w^{2n_z}} \tag{2}$$

With $R_q$ obtained, according to the properties of analytic functions, the imaginary part of $Z_q$ can be found using the method of Gewertz [8]. After deriving the expression of the dividing point function $Z_q(s)$, it can be regarded as a two-port lossless network terminated with a 1-ohm resistance, which can be transformed into a continuous fraction by rolling division. The topology structure of the $n$ order ladder network terminated with 1-ohm source is shown in Fig. 1. After performing anti-normalization on these parameters, the actual parameters of the matching network can be obtained.

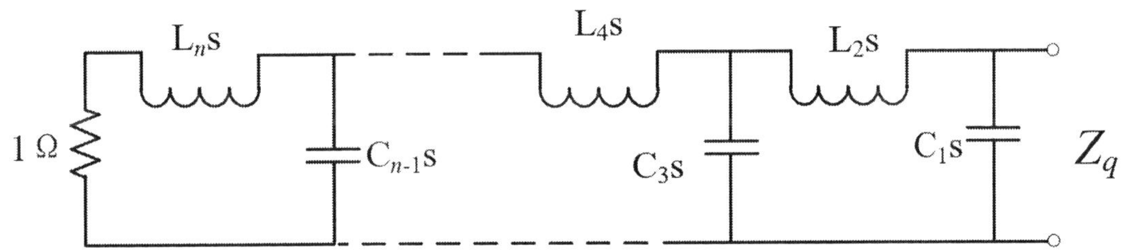


**Figure 1.** Ladder network of *Zq*.

## 3. Wideband Antenna Matching Network Design

### *3.1. Impedance Function Matching*

In this section, dividing point impedance function and the dividing point admittance function are used respectively to optimize the TPG of two antennas. When the equivalent impedance of the antenna is capacitive, it is better to use the dividing point admittance function. Conversely, when the equivalent impedance of the antenna is inductive, dividing point impedance function is more suitable.

The first antenna to be match is a dual band UHF whip antenna with its reflection coefficient less than -10 dB in 28.3~30.5 MHz and 37.5~40.6 MHz. The GA based real frequency technique was implemented to optimize the TPG of the antenna from 27.5 MHz to 39.5 MHz. The GA population size was set as 50 and the generation was set as 500.The TPG of the whip antenna is shown in Fig.2, where the blue dash line is the TPG before matching and the red line is the optimized TPG by real frequency technique. It can be seen that the TPG of the matched antenna has been significantly improved in 32~39MHz. The high frequency band of the antenna moves closer to the low frequency band, and the match performance between the two bands becomes better. The optimized dividing point function of the matching network is

$$Z_q = \frac{2.2391s^2 + 1.1585s + 1}{4.1247s^3 + 2.1341s^2 + 3.7749s + 1} \tag{3}$$

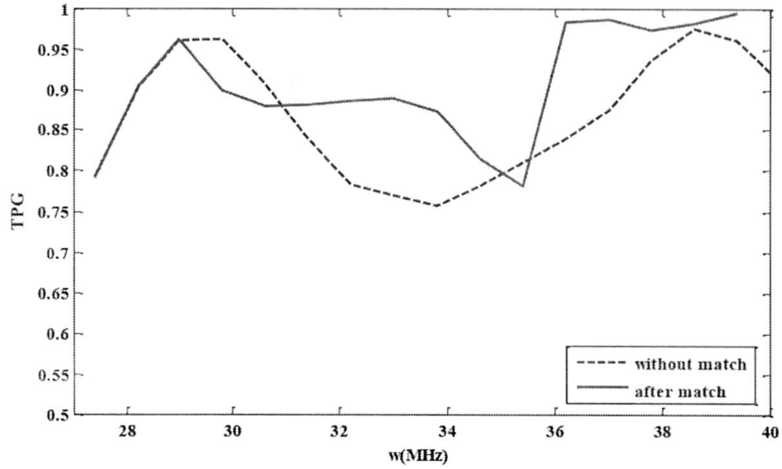


**Figure 2.** TPG of the whip antenna.

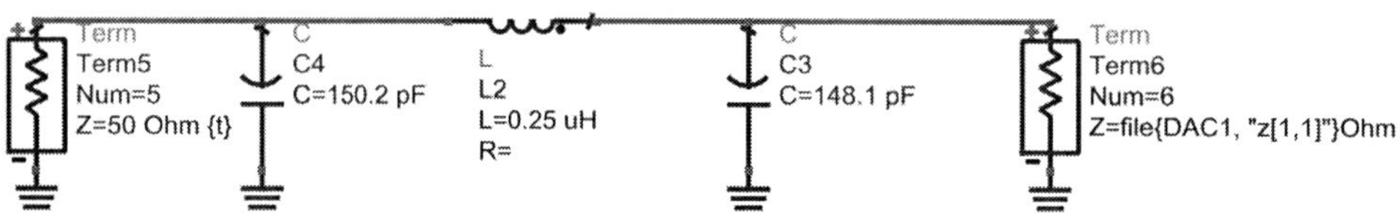


**Figure 3.** Schematic of the matching network.

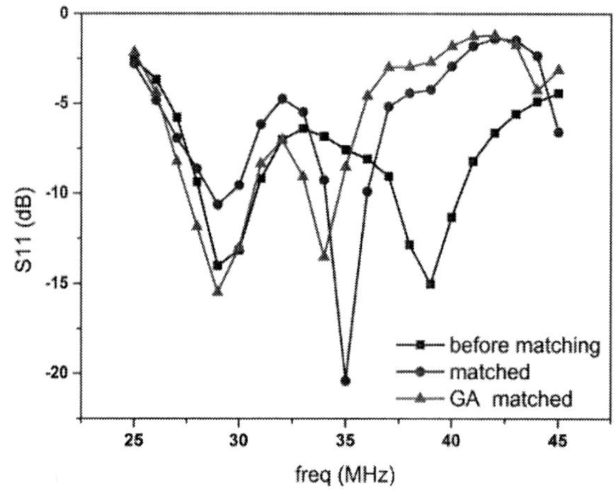


**Figure 4.** Reflection coefficient of the antenna before and after matching.

By applying Eq.3 and Eq.4 to Eq.5, the parameters of the matching network can be obtained. The achievable network simulated via ADS is shown in Fig.3, where the right port terminates with the antenna and the left term is 50-ohm term. The simulated results of the unmatched antenna, least squares based technique matched one and GA based technique matched one are shown in Fig.4, indicating that the GA based real frequency technique improves the matching performance in broadband. The results are consistent with those optimized by real frequency technique, which verifies the effectiveness of the technique.

### *3.2. Admittance Function Matching*

In section 3.1, a whip antenna was matched by means of real frequency technique using dividing point impedance function as the antenna was inductive. In this section, a capacitive monopole was matched using dividing point admittance function. The monopole was 1.75 meters high and its ground radius was 0.85 meters, whose resonant frequency was 198 MHz with 30 MHz bandwidth.

Applying real frequency technique, the whole frequency band was dividing into 5 bands from 0 to 500 MHz, and the passband was set between 50~450 MHz. 21 frequency points were selected in the passband for optimization, and the network complexity $n_s$ was set as 3. The TPGs of the monopole before and after matching are shown in Fig.5. It indicates that the optimized TPG has been improved in almost the whole passband. The optimized dividing point function of the matching network is

$$Y_q = \frac{2.5443s^2 + 1.3835s + 1}{5.6518s^3 + 3.0733s^2 + 4.0604s + 1} \tag{1}$$

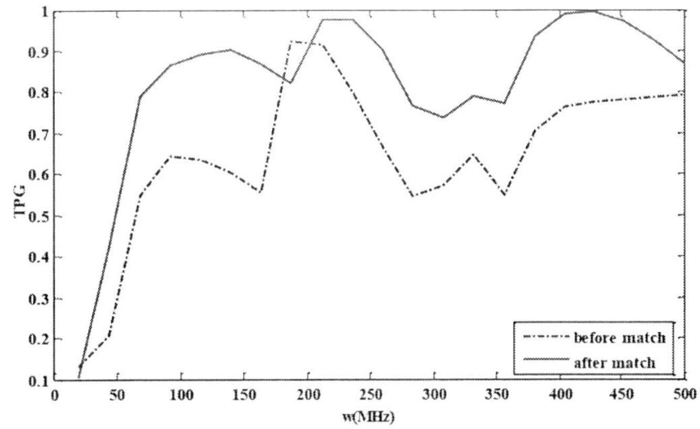


**Figure 5.** TPG of the monopole.

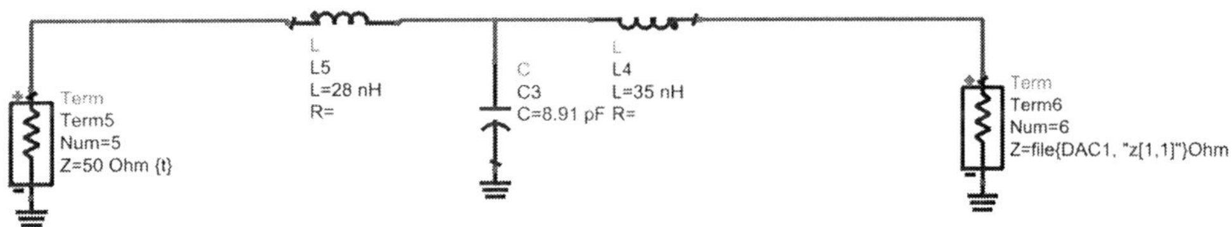


**Figure 6.** Schematic of the matching network.

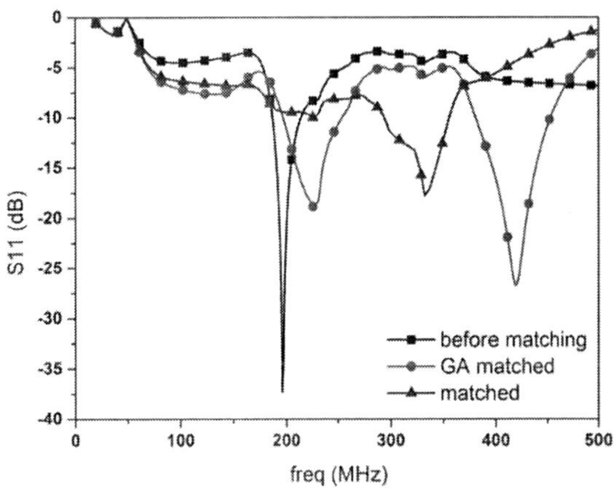


**Figure 7.** Reflection coefficient of the antenna before and after matching.

The achievable network simulated by ADS is shown in Fig.6 and the simulated reflection coefficient is shown in Fig.7. It indicates that the matching network optimized by the GA based real frequency technique not only improves the matching performance by introducing a new resonance point at high frequency, but also widens the original resonant bandwidth. The matching network designed by GA based real frequency technique is effective. The least square based real frequency technique also improves the matching performance in broadband, but it is not as well as the GA based one. The performance of the frequencies below the original resonant frequency have also been improved, even it is not obvious due to Bode-Fano limitation [9].

## 4. Conclusions

This paper applies GA in real frequency technique for antenna wideband matching network design. According to the impedance of the antennas, dividing point impedance function and admittance function are used to design matching network of a whip antenna and a monopole respectively. Simulated result shows that the TPG of the whip antenna after matching network improves in 32~39MHz and the TPG of the matched monopole improves in 50~470 MHz. The matching networks obtained by the technique were simulated through ADS, and the results are consistent with the optimized ones. The correctness of the GA based real frequency technique and the practicability of its application in antenna broadband matching are verified.

## References

[1] Yarman, B. S., and H. J. Carlin. "A Simplified "Real Frequency" Technique Applied to Broad-Band Multistage Microwave Amplifiers." IEEE Transactions on Microwave Theory & Techniques 30.12(1982):2216-2222.

[2] Dai Z, He S, Pang J, Peng J, Huang C, You F. Sub-optimal matching method for dual-band class-J power amplifier using real frequency technique. IET Microw Antennas Propag. 2017; 11(9):1218–1226.no.pp.

[3] Kilinc, Sedat, and B. S. Yarman. "A novel method to design broadband flat gain and sufficiently efficient power amplifiers using real frequency technique." IEEE International Conference on Electronics, Circuits and Systems IEEE, 2017:30-33.

[4] Wang, Xiaojun, Z. Niu, and Y. Liu. "Matching network design for a double band PIFA antenna via simplified real frequency technique." International Conference on Microwave Technology and Computational Electromagnetics IET, 2010:92-94.

[5] Aksen, Ahmet, et al. "A frequency transformation based real frequency design approach for dual-band matching." International Symposium on Signals, Circuits and Systems 2017:1-4.

[6] Yildiz, Serkan, A. Aksen, and S. B. Yarman. "Quad-band matching network design with real frequency technique employing frequency transformation." IEEE First Ukraine Conference on Electrical and Computer Engineering IEEE, 2017:143-147.

[7] Wang, Fei, and C. To. "A generic broadband matching network synthesizer using real frequency technique." International Journal of Rf & Microwave Computer - aided Engineering 9(2017):e21227.

[8] Shiye Guo, Jingye Yan, and Ji Wu. "Numerical error study on parametric approach and Gewertz procedure in designing broadband matching network." Electronic Design Engineering, 24. (2016):119-121.

[9] Fralich, Russell, J. Wang, and J. Litva. "An optimized impedance-matching technique for increasing the bandwidth of microstrip antenna." Antenna Technology and Applied Electromagnetics, 1990. ANTEM 1990. Symposium on IEEE, 2017.

IOP Conf. Series: Materials Science and Engineering 472 (2019) 012026 doi:10.1088/1757-899X/472/1/012026

# Design and Simulation of the Key Parameters of Speed Control Valve Based on MATLAB

**Guangyong Huang**[a] **and Dongyuan Ge**[b,]

College of Mechanical Engineering, Guangxi University of Science and Technology, Liuzhou, Guangxi 545006, China

E-mail: [a]hguangyong@foxmail.com, [b]gordon399@163.com

**Abstract.** MATLAB was used to realize the characteristic curve of inlet and outlet pressure difference and flow rate, the characteristic curve of hydraulic force and spring force in speed control valve, and obtain key parameter values. SIMULINK was used to model speed control valve, and the parameter values were substituted into the modeling for simulation verification.

## 1. Introduction

Speed control valve is an important component in hydraulic system; its dynamic performance plays an important role in the speed stability of the working element in hydraulic circuit. The design process of the speed control valve involved many parameters and the selection of parameter valves had a great impact on the working performance, especially the design of the main parameters is more critical. The main parameters affecting the dynamic performance of the speed control valve were introduced in the literature [1]. A manual method was introduced to obtain the spring stiffness, initial compression amount, maximum/minimum load on the spring, and other main parameters in the literature [2]. In this paper, the main design parameters were obtained through MATLAB, and the parameter values were substituted into the modeling of speed control valve for simulation verification, reduced the design work [3-6].

## 2. Working Principle of Speed Control Valve

The speed control valve is mainly composed of a throttle valve and a constant differential pressure reducing valve, including valve body, adjustable port, and spring and valve core of reducing valve. Its structure is shown in the Figure 1.

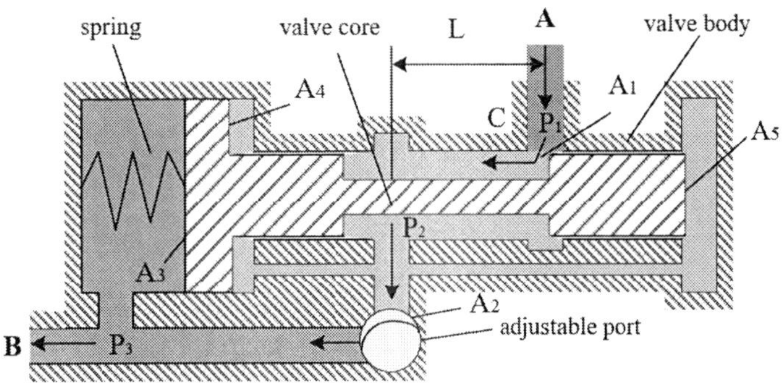


**Figure 1.** The working principle of speed control valve

The throttle valve of the speed control valve through the flow,

$$Q_2 = C_{d2} A_2 \sqrt{\frac{2}{\rho}(P_2 - P_3)} \tag{1}$$

When the inlet pressure $P_1$ or outlet pressure $P_2$ changes, the valve core of speed control valve moves to the left or right. At this point, there will be new changes in spring force and steady-state hydraulic force, and the force balance equation of the valve core at a new balance position was as follows,

$$P_2 - P_3 = \frac{(F_t \pm \Delta F_t) - (F_{RS} \pm \Delta F_{RS})}{A_3} = \frac{(F_t - F_{RS}) + (\pm \Delta F_t \mp \Delta F_{RS})}{A_3} \tag{2}$$

Trying to make the spring force $F_t$, steady-state hydraulic force $F_{RS}$ as close as possible, so that the pressure difference remains unchanged.

## 3. Design Actual Example

### *3.1. Characteristic Curve*

Give an example, the working pressure of the speed control valve was 15MPa and the working flow speed control valve was 25L/min. The speed control valve inlet and outlet pressure difference and flow rate characteristic curve of MATLAB procedures as follows,

```
posion=1e-6:1e-6:0.5852e-3;
for i=1:size(posion,2)
    x=posion(i);
    Lam1=Cd2*A2*sqrt(2*K*x0/ro/A3);
    Lam2=1/x0;
    Lam3=2*Cd2^2*0.3583679495*A2^2/(A3*Cd1);
    Lam4=ro/(2*Cd2^2*A2^2);
    Lam5=Cd2^2*A2^2/Cd1^2;
    q=Lam1*sqrt((1-Lam2*x)/(1+Lam3/(w1*x)));
    p=Lam4*q^2*(1+Lam5/(w1*x)^2);
    Q(i)=q;
    p1p3(i)=p;
end
[p1p3,sequence]=sort(p1p3);
Q=Q(sequence);
[value,sequence2]=min(abs(p1p3-deltapmin));
[value,sequence3]=min(abs(p1p3-pg));
Qmax=max(Q(sequence2:sequence3));
Qmin=min(Q(sequence2:sequence3));
eta=2*(Qmax-Qmin)/(Qmax+Qmin);
plot(p1p3(1:sequence3),Q(1:sequence3));
xlabel('pressure difference p1-p3(Pa)'),ylabel('flow rate Q(m^3/s)');
fprintf('flow rate variation ratio=%d\n',eta);
```

The flow rate variation ratio was 4.3%,and the speed control valve inlet and outlet pressure difference and flow rate characteristic curve was as follows,

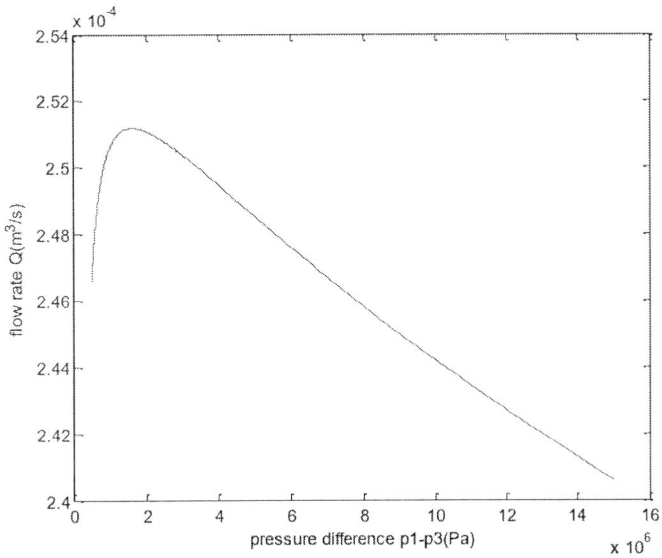


**Figure 2.** The characteristic curve of flow rate

The procedure for calculating the spring stiffness and initial compression amount was as follows,

```
x=1e-6:1e-6:0.5852e-3;
for i=1:size(x,2)
   Fw(i)=ro*Qg^2*0.3583679495/(Cd1*w1*x(i));
end
Fmin=Qg*Cv1*sqrt(2*ro*(15e6-0.2e6))*0.3583679495;
[Fmin2, poi]=min (abs (Fw-Fmin));
Fy=0.2e6*A3;
Ft=Fw(poi:end)+Fy;
p=polyfit(x(poi:end),Ft,1);
K=-p(1);
xt=p(2)/K-0.5852e-3;
line=-K*x(poi:end)+K*(xt+0.5852e-3);
plot(x(poi:end),Ft,'-k',x(poi:end),line,':b');
xlabel('displacement of valve core /m'),ylabel('force/N');legend(' steady-state hydraulic force + hydraulic force','spring force');
```

The spring stiffness was 16879N/m and initial compression amount was 0.0045m, and characteristic curve of hydraulic force and spring force in speed control valve was as follows,

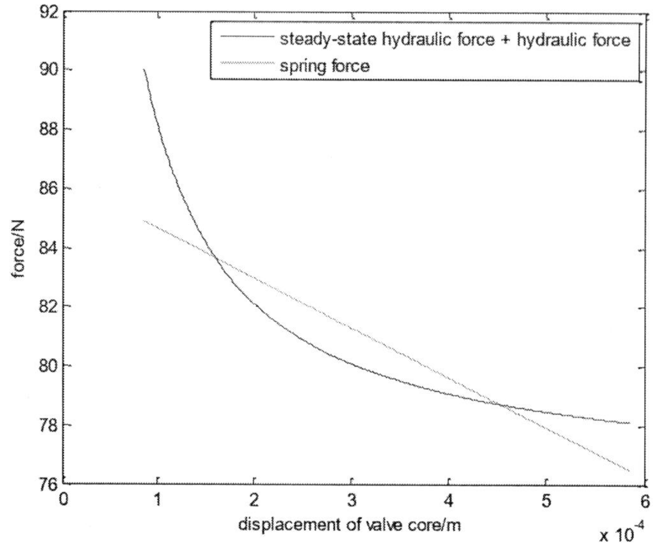


**Figure 3.** The characteristic curve of hydraulic force and spring force

*3.2. Simulation Verification*

When inlet pressure or outlet pressure was changed, the dynamic characteristic analysis of speed control valve was mainly to analyze the dynamic balancing process of valve core of reducing valve. When valve core of reducing valve did not reach the new balance position, it was subject to the pressure, spring force and steady-state hydraulic force, and also transient-state hydraulic force and viscous resistance of hydraulic oil.

Steady-state hydraulic force,

$$F_{RS} = 2C_{d1}C_v w_1 x(P_1 - P_2)\cos\theta \tag{3}$$

Transient-state hydraulic force,

$$F_{Rt} = -LC_{d1}w_1\sqrt{2\rho(P_1 - P_2)}\frac{dx}{dt} \tag{4}$$

Viscous resistance of hydraulic oil,

$$F_s = B\frac{dx}{dt} \tag{5}$$

Spring force,

$$F_t = K(x_0 - x) \tag{6}$$

Stress condition of valve core,

$$M\frac{d^2x}{dt^2} = K(x_0 - x) + P_3A_3 - P_2(A_4 + A_5) - 2C_{d1}C_{v1}\omega_1 x(P_1 - P_2)cos\theta - LC_{d1}\omega_1\sqrt{2\rho(P_1 - P_2)}\frac{dx}{dt} - B\frac{dx}{dt} \tag{7}$$

and $P_2$ pressure value,

$$P_2 = \frac{C_{d1}^2\omega_1^2 x_1^2 P_1 + C_{d2}^2 A_2^2 P_3}{C_{d1}^2\omega_1^2 x_1^2 A_1^2 + C_{d2}^2 A_2^2} \tag{8}$$

Eq.8 generation into the equation Eq.7, so we had,

$$M\frac{d^2x}{dt^2}=K(x_0-x)+P_3A_3-\frac{C_{d2}^2A_2^2P_3+C_{d1}^2\omega_1^2x^2P_1}{C_{d2}^2A_2^2+C_{d1}^2\omega_1^2x^2}(A_4+A_5)-2C_{d1}C_{v1}\omega_1x\frac{C_{d2}^2A_2^2}{C_{d2}^2A_2^2+C_{d1}^2\omega_1^2x^2}(P_1-P_3)cos\theta$$
$$+LC_{d1}\omega_1\sqrt{\frac{2\rho\rho_{d2}^2A_2^2}{C_{d2}^2A_2^2+C_{d1}^2\omega_1^2x^2}(P_1-P_3)}\frac{dx}{dt}-B\frac{dx}{dt} \tag{9}$$

The dynamic model of speed control valve was established by SIMULINK, as can be seen from Figure 4,

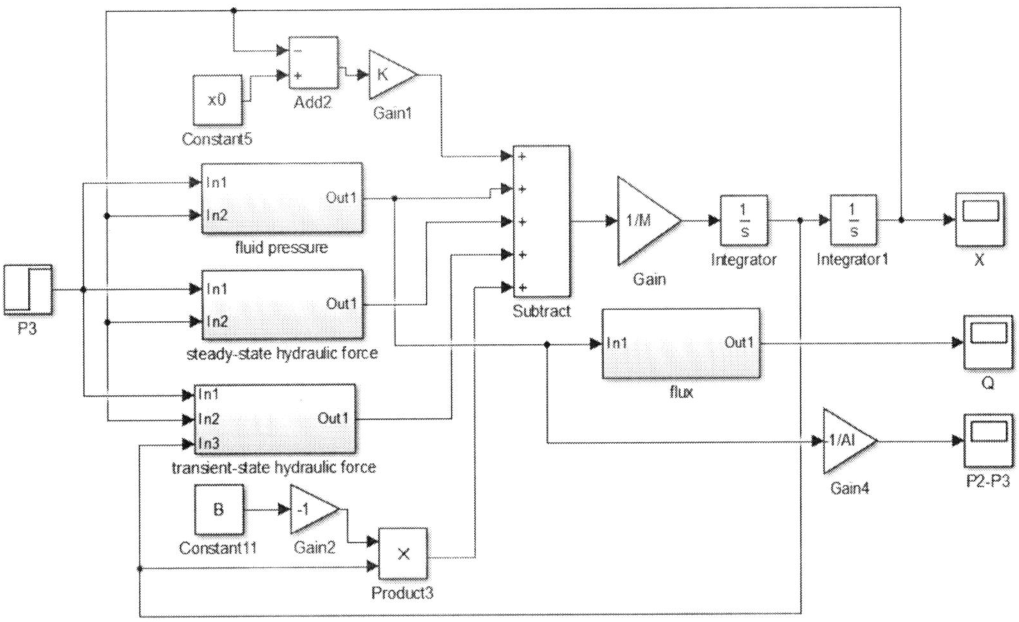


**Figure 4.** The dynamic model of speed control valve

The dynamic model of speed control valve contained fluid pressure, steady-state hydraulic force, transient-state hydraulic force and flux.

The previous parameter values were substituted into the model, and the simulation results were shown in the Figure 5.

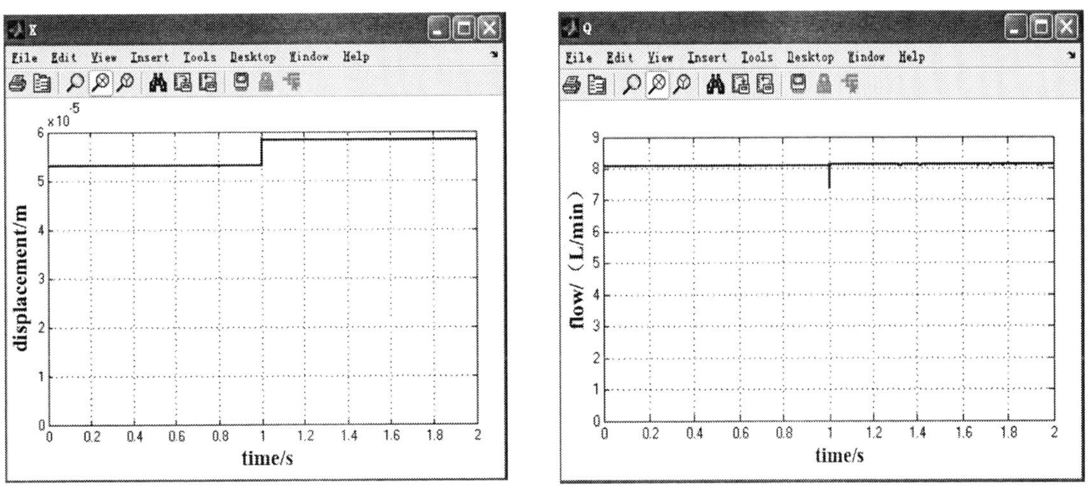


**Figure 5.** The displacement curve of pressure reducing valve valve core and flow

According to the simulation results, when the inlet pressure $P_1$ remains unchanged and the outlet pressure $P_3$ rised, the valve core was finally in a new equilibrium position through the servo feedback of the speed control valve itself. During the dynamic change of valve core, the flow rate was basically maintained at 8.2L/min. When the outlet pressure $P_3$ rised suddenly, the flow rate rised after a sudden drop due to hydraulic shock.

## 4. Conclusions

Through MATLAB programming, it was easy to obtain the characteristic curve of the pressure difference between the inlet and outlet of the speed control valve and the flow rate, flow rate variation ratio was 4.3%. MATLAB was used to realize the characteristic curve of hydraulic force and spring force in speed control valve, and obtain key parameter values, for example the spring stiffness was 16879N/m and initial compression amount was 0.0045m. SIMULINK was used to model speed control valve, and the parameter values were substituted into the modeling for simulation verification. This paper provided a graphical method for designing the structural parameters of the speed regulating valve.

## Acknowledgement

This research was financially supported by by the Guangxi University Young and Middle-aged Teacher Basic Ability Enhancement Project under grant No. 2017KY0356, and National Natural Science Foundation of China under grant No. 51765007, and the Guangxi Provincial Natural Science Foundation of China under grant No. 2016GXNSFAA380111, and Guangxi University of Science and Technology Natural Science Foundation under grant No.174518.

## References

[1] Zongqi Tan, Cancan Li, Min You, Huijun Ye. Orthogonal Design and Analysis of Key Parameters of a Speed Control Value. Journal of China Three Gorges University (Natural Sciences), 2011, 33(3):78-80.
[2] Zhi-an Song, Shanshan Zheng ,Qingrong Guo. Computer Aided Graphical Method on Designing Parameters of Pressure-relief Valve Spring in Speed Regulating Valve. MACHINE TOOL & HYDRAULICS.2009, 37(7):227-229.
[3] Pingge Zhang, Jinghan Huang, Yongzhao Peng, Zhimin Liu. Application of MATLAB in the Hydraulic System Parameters Optimization. MACHINE TOOL & HYDRAULICS.2013, 44(22):81-82.
[4] Wenbo Li. Computer Simulation of Hydraulic System Based on Matlab. Coal Mine Machinery.2013, 34(11):209-211.
[5] Liu Tao, Xinxing Li ,Yun Wang , Xiong Lei, Zhongli Li , Jing Wang, Simulation Study of Throttle Velocity Modulation Circuit Based on AMESim. Coal Mine Machinery, 2016, 37(7):66-68.
[6] Yao Yao, Jiadong Han, Xiaoguang Yu, Pengfe Wang, Jie Huang. Hydraulic System Dynamic Simulation Based on Bond Graphs Modeling and MATLAB /Simulink. Chinese Hydraulics & Pneumatics, 2013(12):103-106.

# On-Line Fault Diagnosis Method for Power Transformer Based on Missing Data Repair

**Xiansi Lou[1], Weihan Liao[1], Jianbo Xin[2], Qiukuan Zhou[2], Chen Kang[2], Shiying Ma[3] and Dunwen Song[3]**

1. College of Electrical Engineering, Zhejiang University, Hangzhou 310027, China
Email: louxiansi@zju.edu.cn, 449057978@qq.com
2. Electric Power Research Institute of State Grid Jiangxi Electric Power Company, Nanchang 330006, China
Email: mandyzhuhai@163.com, zqk0791@163.com, kangchen-123@163.com
3. China Electric Power Research Institute, Beijing 100192, China
Email: abde@epri.sgcc.com.cn, 719528304@qq.com

**Abstract.** Data quality is an important factor affecting the accuracy of transformer fault diagnosis. In order to reduce the impact of missing data, an on-line fault diagnosis method using a loop iterations of improved k-Nearest Neighbour (kNN) and multi-class SVMs based on the missing data repair is proposed in this paper. In the kNN method, the improved Manhattan distance weighted by the negative exponent of the correlation coefficient is designed to measure the distance between samples. On one hand, the influence of the strong correlation indicators on the missing data can be highlighted to improve the accuracy of data repair. On the other hand, the improved Manhattan distance is suitable for an efficient search strategy based on the k-d tree which can achieve the fast search for massive historical data and meet the real-time demand of on-line diagnosis. Diagnosis test results show that the proposed method can keep the high diagnostic accuracy on the incomplete data and realize the efficient on-line fault diagnosis for transformers.

## 1. Introduction

The transformer is an important equipment and core component of the entire substation system and even the power grid which is responsible for the conversion and transmission of electrical energy. Once the transformer especially the oil-immersed main transformer fails, it will affect the safe and stable operation of the power grid and lead to severe economic losses. Dissolved gas analysis (DGA) is the widely recognized method to detect latent faults of transformers [1-3]. With the development of smart substations, oil chromatographic online monitoring devices have become widely applied in power systems [4]. However, due to the constraints of objective conditions such as network topology, geographical environment, and economic factors, the most transformer on-line monitoring systems still use low-quality communication modes like power distribution carrier, Zigbee wireless technology, and industrial distribution. These low-cost channels often occur broken codes under the attack of over-voltage, high-current and other strong electromagnetic interference resulting the loss of monitoring data [5]. The incompleteness of oil chromatographic data will seriously affect the accuracy of transformer fault diagnosis, and even lead to unavailability of traditional methods. Therefore, how to use the incomplete data to achieve accurate and fast online diagnosis for transformer faults has attracted wide attention in academic and industrial circles.

At present the fault diagnosis of transformers based on the incomplete data has two main ideas. The first idea is to clean the bad data and then use existing models such as neural network [6-8], Bayes classifiers [9-10] and support vector machines [11-14] to classify the failure type. There are some research results on the methods of cleaning transformer online monitoring data. The literature [15] treats the missing values as one type of outliers and a time-series analysis method is utilized to identify the data anomaly patterns. Then the autoregressive integrated moving aver-age model (ARIMA) model is adopted to fit the data curve. Literature [5] proposes a method based on Pearson correlation coefficient and regression model to achieve the recovery of missing data. In literature [16] the wavelet neural network model is used to predict the missing data and correct the error data. And the effect of data cleaning can be improved by modifying the wavelet neural network parameters and combined prediction. The above data cleaning methods are all based on the time series analysis. When the missing data is in the middle of the time series, they will have satisfying performance. However, in the actual situation, the missing data is generally at the end of the time series. In addition, there are few literatures discussing the computational efficiency of the cleaning method under the large sample of oil chromatographic data.

Another idea is to select some flexible methods to deal with the missing data in the monitoring results. The rough set theory is a typical method. The literature [17] uses the rough set to reduce the transformer state information, and diagnoses transformer faults based on the reduction set. A transformer fault diagnosis method combining with Bayesian network classifier and rough set is proposed in [18]. This method has the fault-tolerance characteristics, and its performance is obviously better than using Bayesian classifier or rough set separately. The advantages of the rough set theory is that it can effectively analyse and deal with the inaccurate and incomplete data and extract implicit knowledge from a large amount of information. Nevertheless, when there is the lack of key characteristic indicators, the accuracy of such fault diagnosis method will rapidly decline.

Aiming at the deficiencies in the above studies, this paper proposes an online fault diagnosis method for the incomplete oil chromatographic data based on an improved k-Nearest Neighbour (kNN) and multi-class SVMs. This method which designs a loop iteration mechanism for data repair and fault diagnosis inherits the first idea. The reminder of this paper is structured as follow: section 2 introduces a data repair method based on improved kNN, fast search strategy is described in section 3, a cycle iterative diagnosis method is presented in section 4, case study on 309 sets of real oil chromatographic data is shown in section 5, followed by conclusions in section 6.

## 2. Data Repair Method based on Improved k-Nearest Neighbour

The k-Nearest Neighbour (kNN) algorithm is a supervised learning method for finding samples with the similar data pattern. This paper is based on a basic idea that the same transformer fault has a similar data pattern. An improved kNN method is proposed to repair the missing transformer oil chromatographic data. The working mechanism of this method is as follows: 1) input the incomplete test sample, 2) find the $k$ nearest training samples in the history oil chromatographic database, 3) synthesize the $k$ samples' information to estimate the missing data. In order to highlight the correlation between different oil chromatographic indicators, a modified Manhattan distance using the negative exponent of the correlation coefficient as the weight coefficient is adopted to measure the distance between two samples.

Assume that there are $N$ sets of transformer oil chromatographic history data as training samples, and each sample $x_i = (x_{i1}, x_{i2}, ..., x_{im})$ has $m$ characteristic gas indicators. The correlation coefficient $\rho_{pq}$ between indicator $p$ and $q$ can be calculated as follow.

$$\rho_{pq} = \frac{\frac{1}{N-1}\sum_{i=1}^{N}(x_{ip} - \mu_p)(x_{iq} - \mu_q)}{\sqrt{\frac{1}{N-1}\sum_{i=1}^{N}(x_{ip} - \mu_p)^2}\sqrt{\frac{1}{N-1}\sum_{i=1}^{N}(x_{iq} - \mu_q)^2}} \quad (1)$$

Where $\mu$ is the mean value of the corresponding characteristic gas indicators. The range of the correlation coefficient is [-1, 1] where -1 (1) denotes strong and negative (positive) correlations between the two characteristic gas indicators and 0 means no correlation. If two characteristic gas indicators have the close correlation, it is more credible to use one of them to estimate the other.

When the test sample $x_j$ lacks the $p$th ($1 \le p \le m$) gas indicator $x_{jp}$, the improved Manhattan distance between the test sample $j$ and the training sample $i$ is formulated as follow:

$$d_{ij|p} = \sum_{q=1}^{m} e^{-|\rho_{pq}|} \bullet |x_{iq} - x_{jq}| \quad q \neq p \tag{2}$$

By weighting the spatial distance of the gas indicators with the negative exponent of correlation coefficient, the correlation between the gas indicators can be reflected. Therefore, the test sample is more likely to approach the training sample which is closed on the strong correlation indicator. Select the $k$ nearest training samples and calculate the mean of the corresponding gas indicators as the estimated value of the missing data:

$$\hat{x}_{jp} = \frac{\sum_{i=1}^{k} x_{ip}}{k} \tag{3}$$

Compared with the traditional stochastic filling or time series data repair methods, the improved kNN method considers the similarity of the oil chromatographic data pattern for the same transformer fault type. And due to breaking the temporal correlation of oil chromatographic indicators, it is suitable for the repair of missing data in the event of rapid changes during the development of faults.

## 3. Fast Sample Search Strategy Based on K-D tree

With the wide utilization of online monitoring devices for transformer oil chromatography, the volume of the monitoring data has increased dramatically. Taking a provincial power grid for example, the number of oil-immersed transformers above 220 kV is over 2,000. Millions of oil chromatographic data are produced every year, reaching the GB level of data volume. If the traditional linear scanning method is still used to find the k nearest training samples from the massive historical database, the calculation efficiency and operating speed will not meet the on-line requirements. The improved Manhattan distance presented in section 2 is suitable for a fast sample search strategy based on k-d tree.

### *3.1. Construction of K-D tree*

Each oil chromatographic sample usually contains 5-8 indicators, so an oil chromatographic sample set corresponds to a high-dimensional data space. The k-d tree, as a data structure for partitioning k-dimensional data space, can be applied to the search of key data in high-dimensional space. Figure 1 illustrates the structure and generation process of a k-d tree with $k = 3$.

The generation process of a k-d tree includes the following three main steps:

Step1: Calculate variances of samples for each dimension. Select the dimension corresponding to the maximum value of the variance as the split field for effectively guaranteeing the balance of the tree.

Step 2: Sort the sample according to the value of the split dimension, and the intermediate data point is selected as the pivot point which is represented in bold in Figure 1.

Step3: The samples are divided by the pivot point to form the left and right k-d subtrees. When the size of the subtree is smaller than the acceptable number of leaf nodes, the operation of generating the k-d tree is terminated otherwise return to Step 1. In Figure 1, the acceptable number of leaf nodes is 2.

When this method is applied to the structured storage for transformer oil chromatographic data, each sample is stored in one node of the k-d tree containing $k$ indicators.

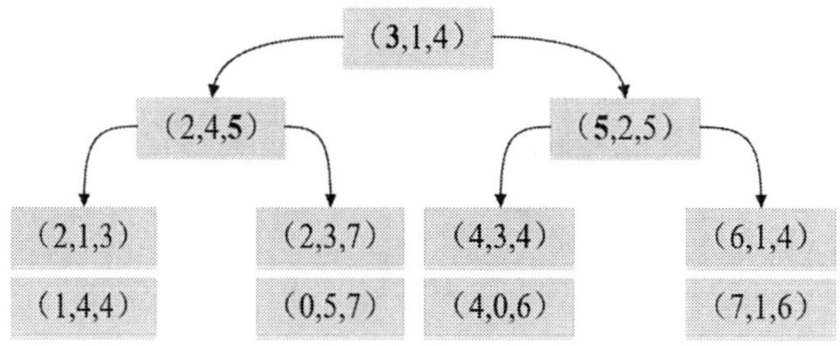


**Figure 1.** The structure of a three-dimensional k-d tree

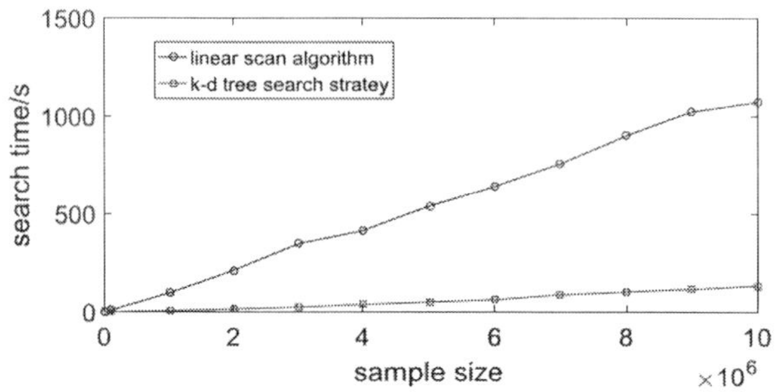


**Figure 2.** The comparison of algorithm performances for different search strategy

### *3.2. Neighbour Search Strategy*

The data structure based on k-d tree can achieve efficient neighbour search for oil chromatographic sample data. The literature [19] has proved that the complexity of neighbour search using k-d tree is O (log *n*), where *n* is the number of samples. When applied to the search of massive high-dimensional samples, it has significant advantages compared with the linear scan algorithm with the complexity of O(n). The neighbour search strategy based on k-d tree structure contains the following three steps.

Step1: Find the leaf node where the test sample is located along the k-d tree, get the nearest sample, and add all of the nodes on the search path to a queue.

Step2: Make a hypersphere around the nearest sample with radius of the distance between the test sample and the nearest sample. Verify whether the sample in the queue is within this hypersphere. If not, remove the sample from the queue and repeat Step 2 until the queue is empty, otherwise go to Step 3.

Step3: Update the nearest sample, add its child node in the queue, and return to Step 2.

Figure 2 presents the relationship between search time and sample size of the oil chromatographic data containing five indicators using the linear scan algorithm and the k-d tree search strategy for the improved Manhattan distance defined in section 2. The test results show that when the k-d tree search strategy is used, the search time for neighbour samples increases slowly with the expansion of the sample size, which has a significant advantage comparing with the linear scan algorithm.

## 4. Cycle Iterative Diagnosis Method Based on kNN and SVMs

### *4.1. Design of Multi-class SVMs*

The basic idea of the SVM algorithm is mapping the linearly inseparable sample data to high-dimensional feature space through a kernel function, which may be constructed as a linearly separable problem. By solving a convex optimization problem, the optimal hyper plane is obtained which can be utilized to achieve data classification. Since the selection of the optimal hyper plane only depends on support vectors, this method is robust and has a good generalization performance when applied to the transformer fault diagnosis. A single SVM can only solve a binary classification problem. In this paper, a multi-classifier containing 8 SVMs is designed to achieve accurate diagnosis for 6 main fault types. The structure of multi-class SVM classifier is shown in the figure 3.

### *4.2. Cycle Iteration Fault Diagnosis Process*

In the transformer fault diagnosis method based on incomplete data, the kNN method is used for the repair of missing data, and the multi-class SVMs classifier performs fault diagnosis based on the repaired sample. According to the results of the diagnosis, the sample space is reduced. And the missing data is re-estimated using the new *k* nearest neighbour samples. This iteration is cycled until the fault type of the neighbour sample is consistent with the fault diagnosis result of the test sample. The fault diagnosis process based on the missing data repair is shown in Figure 4, which specifically includes the following five steps:

Step1: Input the oil chromatographic data. If there is a missing indicator, apply the kNN method to repair.

Step2: Use the multi-class SVMs model for fault diagnosis.

Step3: Judge whether the diagnosis result is consistent with the fault types of the *k* nearest training samples. If there is any inconsistency, go to Step 4, otherwise go to Step 5.

Step4: Select the historical samples which have the same fault type with the multi-class SVMs's diagnostic results as the new sample space. And re-use the kNN method to estimate the missing data. Go back to Step 2.

Step5: Obtain the fault diagnosis result, store the repaired data and diagnostic result in the historical database, and retrain the SVM multiple classifier periodically.

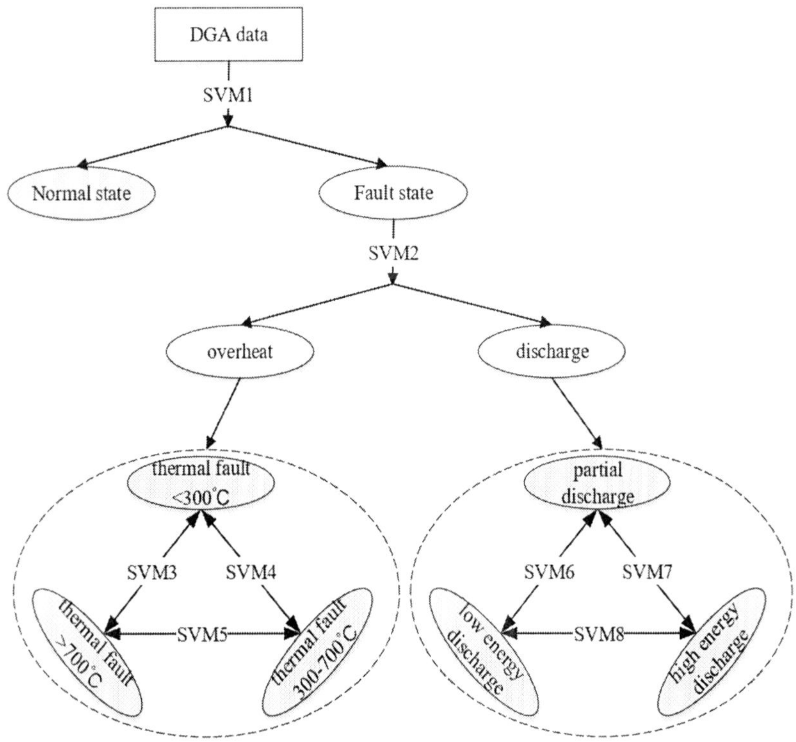


**Figure 3.** The fault diagnosis model for transformer based on multi-class SVMs

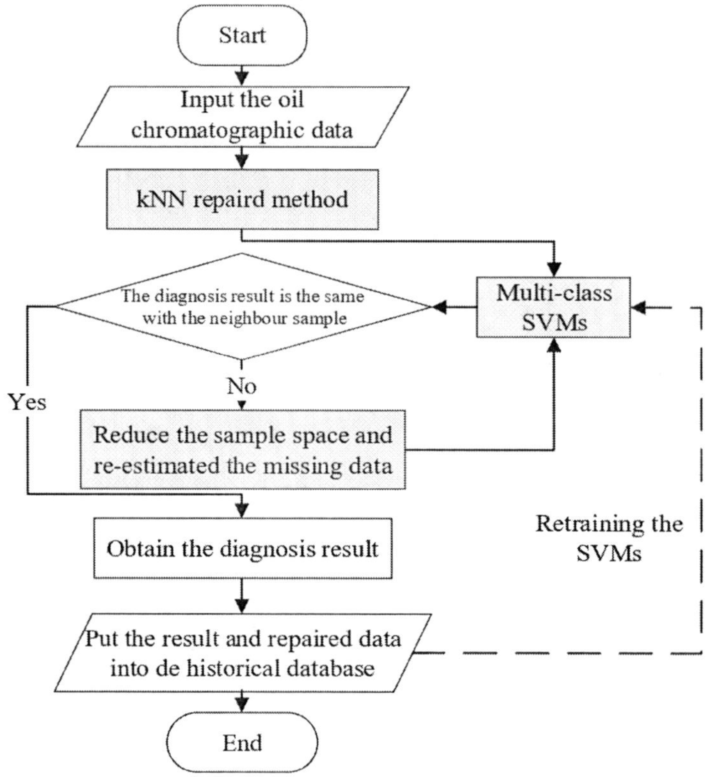


**Figure 4.** The fault diagnosis flow for transformer based on missing data repair

## 5. Case Study

In the case study, 309 sets of complete transformer oil chromatographic sample with the identified fault type, including thermal fault at <300°C (T1), thermal fault at 300°C-700°C (T2), thermal fault at >700°C (T3), partial discharge (PD), low energy discharge D1, and high energy discharge (D2), has been collected. And the incomplete data samples are constructed by random deletion. To facilitate results comparison, the leave-one method (LOOM) [20] is used to calculate the diagnostic accuracy rate for three different methods: kNN+ multi-class SVMs (proposed in this paper), ARIMA+ multi-class SVMs and rough set method.

**Table 1.** Comparison of fault diagnosis accuracy by different methods

| Fault type | Number of training samples | Multi-class SVMs (complete data) | kNN+SVMs (incomplete data) | ARIMA+SVMs (incomplete data) | Rough set (complete data) | Rough set (incomplete data) |
|---|---|---|---|---|---|---|
| T1 | 47 | 85.11% | 80.00% | 74.47% | 72.34% | 59.58% |
| T2 | 39 | 87.18% | 91.79% | 82.05% | 79.49% | 68.21% |
| T3 | 51 | 76.47% | 74.51% | 70.59% | 58.82% | 44.71% |
| PD | 47 | 74.47% | 71.49% | 68.09% | 63.83% | 62.13% |
| D1 | 49 | 87.76% | 75.10% | 77.55% | 71.43% | 55.92% |
| D2 | 34 | 79.41% | 72.35% | 58.82% | 70.59% | 72.35% |
| Normal | 42 | 92.86% | 87.14% | 85.71% | 71.43% | 57.14% |
| total | 309 | 83.17% | 78.64% | 74.11% | 69.26% | 59.09% |

From Table 1, it can be seen that the diagnostic accuracy of the multi-class SVMs and the rough set method based on the complete oil chromatographic data is 83.17% and 69.26% respectively, which has the equal classifier performance in the literature [21]. When the rough set method is applied to the incomplete data, the diagnostic accuracy is reduced to 59.09%. This is because the gas indicator is determinant attributes which seriously affect the accuracy of fault diagnosis.

Compared with the 3rd, 4th and 5th column data, the diagnostic accuracy dropped from 83.17% to 74.11% by using the ARIMA+ SVMs method. When using the kNN+ SVMs method proposed in this paper, the accuracy rate is 78.64% with only 4.53% reduced. Therefore, the high accuracy of the multi-class SVMs can still be maintained. A specific example is illustrated in the Table 2 and Figure 5 for the further analysis between the ARIMA+ SVMs and kNN+ SVMs.

**Table 2.** Missing data repair using kNN

| Incomplete data sample | Neighbour samples | Actual value | Estimated value | Relative error |
|---|---|---|---|---|
| (*,12.85,3.65,2.6,0) | (83,13.05,2.3,1.65,0)<br>(85.34,11.05,2.82,0.79,0)<br>(84,14.17,0.98,0.85,0)<br>(87.41,10.13,2.3,0.16,0)<br>(74.63,13.73,7.58,4.06,0) | 80.9 | 82.876 | 2.44% |

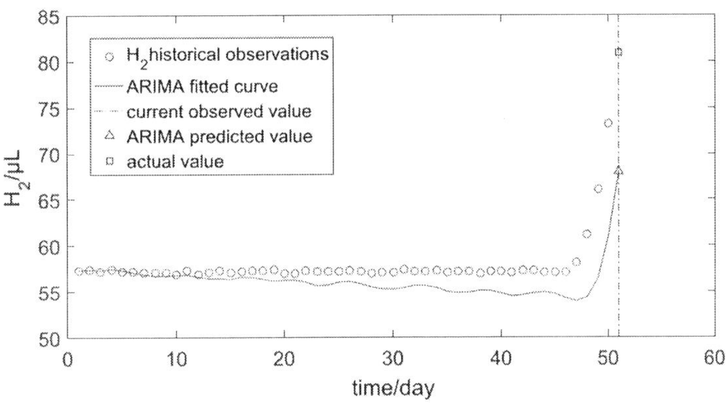


**Figure 5.** Missing data predict using ARIMA

(*, 12.85, 3.65, 2.6, 0) is an oil chromatographic sample with five indicators which corresponds to the gas of $H_2$, $CH_4$, $C_2H_6$, $C_2H_4$, and $C_2H_2$ respectively. * denotes the missing data of $H_2$, and the actual value is 80.9μL. The final five closest data samples searched in the historical database by using the kNN method is shown in Table 2. The estimated value of $H_2$ is 82.867μL with the relative error of 2.44%. The result of the fault diagnosis result using the repaired sample data is thermal fault at <300°C (T1) which is consistent with the actual situation. When the ARIMA method is adopted to fit the history data of the $H_2$, the fitted curve is presented in Figure 5 and the predicted value is 68.02μL. Based on the predicted data, the result of the fault diagnosis is thermal fault at 300°C-700°C (T2) which is incorrect.

It can be seen from the Figure 5 that due to the rapid increase of $H_2$ in the fault development course, the ARIMA method overfits the historical data under the normal condition leading to a conservative final predicted result. Compared to the predicted method based on the timing characteristics, the kNN method is a global related search strategy that breaks the mandatory association relationship of oil chromatographic data from the perspective of time. Therefore, it has an obvious advantage when applied to the rapid development of faults.

## 6. Conclusion

With the popularization and application of the transformer on-line monitoring system, the quality of monitoring data has gradually attracted people's attention. For the incompleteness of oil chromatographic monitoring data, the traditional transformer fault diagnosis method has a rapid decline in accuracy or even cannot be applied. In this paper, an online transformer fault diagnosis method based on missing data repair is presented and the main innovations are as follows:

1) A fault diagnosis method based on the loop iteration of kNN and multi-class SVMs is proposed, which improves the accuracy of transformer fault diagnosis based on incomplete data.

2) An improved Manhattan distance with the negative exponent of the correlation coefficient as the weight is used to measure the distance between samples. Therefore, the influence of the strong correlation indicators can be highlighted.

3) The k-d tree-based data structure is applied to speed up the search in massive historical samples to meet the actual needs of transformer on-line fault diagnosis.

The results of the case study show that the proposed method can realize on-line transformer fault diagnosis based on incomplete data and adapt to the trend of power transmission equipment big data. Compared with the methods based on rough set theory or time series data prediction, the proposed method still has a higher diagnostic accuracy rate in the incompleteness of the key indicator or the situation of rapid development of faults. But there are also some limitations. This method is only applicable to the repair of missing data, but it does not have the ability to identify the wrong data. At the same time, more monitoring data and maintenance records should be used to evaluate the dynamic state of transformers which is also the direction of work for further research.

**Acknowledgement**
Project Supported by Electric Power Research Institute of State Grid Jiangxi Electric Power Company (52182016001J) and China Electric Power Research Institute (XT71-16-053).

**References**

[1] GAO Jun, HE Junjia. Application of quantum genetic ANNs in transformer dissolved gas-in-oil analysis 2010 *J. Proceedings of the CSEE* **30(30)** pp 121-127.
[2] ZHANG Dong-bo, XU Yu, WANG Yao-nan. Neural network ensemble method and its application in DGA fault diagnosis of power transformer on the basis of active diverse learning 2010 *J. Processdings of the CSEE* **30(22)** pp 64-70.
[3] Zheng H B, Liao R J, Grzybowski S, et al. Fault diagnosis of power transformers using multi-class least square support vector machines classifiers with particle swarm optimisation 2011 *J. IET Electr. Power Appl* **5(9)** pp 691-696.
[4] LIANG Yongliang, LI Kejun, ZHAO Jianguo, et al. Research on the dynamic monitoring cycle adjustment strategy of transformer chromatography on-line monitoring devices 2014 *J. Processdings of the CSEE* **34(9)** pp 1446-1453.
[5] LIU Yuankun, LUAN Wenpeng, XU Yu, et al. Data cleaning method for distribution transformer 2017 *J. Power System Technology* **41(3)** pp1008-1014.
[6] LIU Na, GAO Wensheng, TAN Kexiong. Fault diagnosis of power transformer using a combinatorial neural network 2003 *J. Transactions of China Electrotechnical Society* **18(12)** pp 11-16.
[7] ZHANG Jianguang, ZHOU Hao, XIANG Canfang. Application of super SAB ANN model for transformer fault diagnosis 2004 *J. Trans-actions of China Electrotechnical Society* **19 (7)** pp 49-52.
[8] ZANG Hongzhi, HU Yuhua, YU Xiaodong. Integrated ANN based on radial basis function applied in transformer fault diagnosis 2003 *J.Proceedings of the EPSA* **15(1)** pp 51-53.
[9] WU Lizeng, ZHU Yongli, YUAN Jinsha. Novel mothed for trans-former fault integrated diagnosis based on Bayesian network classifier 2005 *J. Transactions of China Electrotechnical Society* **20(4)** pp 45-51.
[10] WANG Yongqiang, LU Langcheng, li heming. Intelligent fault diagnosis for power transformer based on bayesian network and DGA 2004 *J. High Voltage Engineering* **30(5)** pp 12-13.
[11] TAO Xinmin, LI Zhen, LIU Furong, et al. Fault detection method for power transformer based on SVM using reduced vector set 2016 *J. High Voltage Engineering* **42(10)** pp 3199-3206.
[12] WU Xiaohui, LIU Jiong, LIANG Yongchun, et al. Application of support vector machine in transformer fault diagnosis 2007 *J. Journal of Xi'an JiaoTong University* **41(6)** pp 457-457.

[13] BI Jianquan, LU Mingming, GUO Chuangxin, et al. A transformer fault diagnosis method based on multi-classified probability output 2015 *J. Automation Electric Power System* **39(5)** pp 88-93.
[14] GUO Chuangxin, ZHU Chengzhi, ZHANG Lin, et al. A fault diagnosis method for power transformer based on multiclass multiple-kernel learning support vector machine 2010 *J. Proceedings of the CSEE* **30(13)** pp 128-134.
[15] YAN Yingjie, SHENG Gehao, CHEN Yufeng, et al. Cleaning method for big data of power transmission and transformation equipment state based on time sequence analysis 2015 *J. Automation Electric Power System* **7** pp 138-144.
[16] LIN Jun, YAN Yingjie, SHENG Gehao, et al. Online monitoring data cleaning of transformer considering time series correlation 2017 *J. Power System Technology* **41(11)** pp 3733-3740.
[17] MO Juan, WANG Xue, DONG Xing, et al. Diagnosis model of insulation faults in power equipment based on rough set theory 2004 *J. Proceedings of CSEE* **24(7)** pp 162-167.
[18] ZHU Yongli, WU Lizeng, LI Xueyu. Synthesized diagnosis on transformer faults based on Bayesian classifier and rough set 2005 *J. Proceedings of the CSEE* **25(10)** pp 159-165.
[19] Bentley, J. L. Multidimensional binary search trees used for associative searching 1975 *J. Communications of the ACM* **18(9)** pp 509-517.
[20] Chang C S, Jin J,Chang C,et al.Online source recognition of partial discharge for gas insulated substations using independent component analysis 2006 *J. IEEE Transactions on Dielectrics and Electrical Insulation* **13(4)** pp 892-1002.
[21] WANG Ke, LI Jinzhong, ZHANG Shuqi, et al. New features derived from dissolved gas analysis for fault diagnosis of power transformers 2016 *J. Proceedings of the CSEE* **36(23)** pp 6570-6579.

# Numerical Study on Effect of the Slippage on the Cross-Section Size and Profile for Inverse Extrusion Forming of Plastic Micro-Tubes

**Zhong Ren[a*] and Xingyuan Huang[b]**

[a]Key Laboratory of Optic-Electronic and Communication, Jiangxi Science and Technology Normal University, Nanchang, 330038, China
[b]School of Mechanical and Electrical Engineering, Nanchang University, Nanchang, 330031, China
*Corresponding author: renzhong0921@163.com

**Abstract.** We studied the effect of slippage on the size and profile of outlet cross-section for the extrusion forming of plastic micro-tubes by means of finite element software Polyflow. The same material and process parameters were imposed on the established geometric model and mesh. The inverse extrusion forming method was used in the numerical simulation. The different slip coefficients were imposed on the surface of the inner and outer wall of plastic micro-tube. Numerical results show that with the increasing of the slip coefficient, the size of inner and outer diameter for the outlet cross-section is smaller and the profile is gradually distorted, which will increase the difficulty of the design and manufacture for the outlet cross-section of extrusion die. The filed distributions of melt with different slip coefficients were also obtained. The results show that the increasing of X flow velocity, nonuniform Z velocity and the increasing of the shear rate at the outlet face induced by the increasing of the slip coefficient are the main factors of changing the size and profile for the outlet cross-section of plastic micro-tube extrusion forming.

## 1. Introduction

Plastic micro-tubes have a great deal of economic and application values because it can be used in many fields including the medical diagnosis [1], optical communication, automobile oil or gas tube, precision instruments etc. The main manufacture way of the plastic micro-tubes is the extrusion forming [2, 3] of molten melt by means of the extruder, die, cooling device, tractor, etc. In the process of the extrusion forming of the plastic micro-tubes, the design and manufacture of the extrusion die [4] is one of most important work. At the same time, the size and profile of the cross-section for the outlet of the die can directly impact the extrudate quality of the plastic micro-tubes. Except for the material properties, structure of die, process parameters, the interface situations between the melt and the surface of die's channel can also impact the extrusion quality of plastic micro-tubes. As we know, the interface between the melt and the surface of die's channel can be different when the shear stress of melt is changed at the interfaces, which will impact the flow behaviors of melt and extrudate qualities. Therefore, during the design of the outlet cross-section of the die, the slippage between the melt and the surface of the die's channel should be considered. In this paper, the effect of the slippage on the cross-section profile of die for extrusion forming of plastic micro-tubes was numerically studied by using the finite element software Polyflow [5]. The change laws of slippage on the size and profile of outlet cross-section for the plastic micro-tubes were obtained under the fixed process and material parameters.

## 2. Numerical Simulations

### 2.1. Geometric Models

The geometric model of plastic micro-tube was established, which is shown in Figure 1(a). The cross section is shown in Figure 1(b). The outer diameter and the wall thickness of plastic micro-tube are 3mm and 0.5mm, respectively. The lengths of plastic micro-tube's melt inside and outside die are all 10mm.

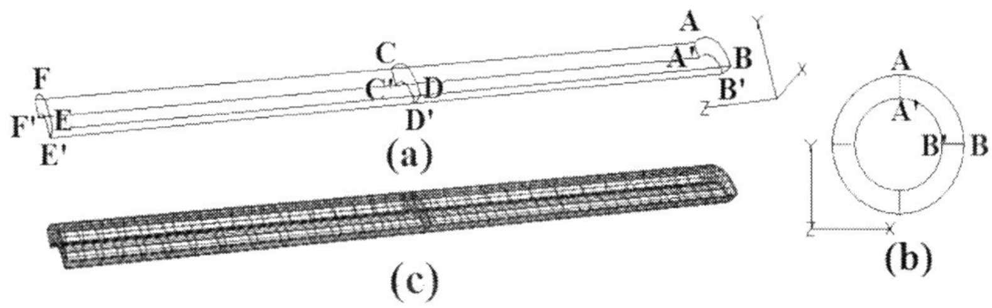


**Figure 1.** The geometric model of plastic micro-tubes (a), the cross section (b), and the finite element mesh (c).

The finite element mesh is shown in Figure 1(c). To improve the numerical precision, the mesh was refined near the surface and interface of the boundaries. The finite element mesh number is 1920.

### 2.2. Governing Equations

The melt of plastic micro-tubes is regarded as the iso-thermal, steady, non-Newtonian, and laminar fluid. At the same time, the gravity and inertia force were neglected due to the properties of high viscosity and low flow velocity. Based on these above reasonable hypotheses, the governing equations are given as follows,

$$\nabla \cdot v = 0 \tag{1}$$

$$\nabla p - \nabla \cdot \tau = 0 \tag{2}$$

where, $\nabla$ is Hamilton operator, which is equal to $\nabla = \frac{\partial}{\partial x}\vec{i} + \frac{\partial}{\partial y}\vec{j} + \frac{\partial}{\partial z}\vec{k}$. $v$ is the flow velocity, $p$ is pressure drop, $\tau$ is the extra stress.

In the numerical simulation, the constitutive model should be used to describe the rheological properties of plastic micro-tube's melt. In this paper, Phan-Thien–Tanner (PTT) constitutive model [6] was used, which is shown as follows,

$$\exp\left[\frac{\varepsilon\lambda}{(1-\eta_r)} tr(\tau)\right]\tau + \lambda\left[\left(1-\frac{\xi}{2}\right)\overset{\nabla}{\tau} + \frac{\xi}{2}\overset{\Delta}{\tau}\right] = 2(1-\eta_r)\eta D \tag{3}$$

where, $\varepsilon$ and $\xi$ are the parameters correlated with the characteristics of tensile and the shear, respectively. $\lambda$ is the relaxation time of melt, $\overset{\nabla}{\tau}$ and $\overset{\Delta}{\tau}$ are the upper and lower convected derivative of the extra stress tensor $\tau$, respectively. $\eta$ is the total viscosity of melt, $\eta_r=\eta_2/\eta_1$ is the viscosity ratio, $\eta_1$ is the Non-Newtonian component viscosity of the melt, $\eta_2$ is the Newtonian viscosity component of the melt. $D$ is the strain-rate of the tensor.

### 2.3. Boundary Conditions

(1) inlet face: In Figure 1(a), ABA'B' is the inlet face. Supposed that the melt flow has already full-developed, the following relationship should satisfy, i.e., $\partial v_z/z=0$, $v_x=v_y=0$, where, $v_x$, $v_y$, and $v_z$ are the flow velocities of melt at $x$, $y$, and $z$ direction.

(2) wall face: ABCD and A'B'C'D' are the outer wall and inner wall of plastic micro-tube,

respectively. In this paper, the different slip conditions are considered, the slip equation is given as follows, i.e., $f_s=-F_{slip}\times(v-v_{wall})^{eslip}$, where, $F_{slip}$ is the slip coefficient, for the no slip flow, $F_{slip}=10^9$. For the full slip flow, $F_{slip}=0$. $e_{slip}$ is equal to 1.

(3) free face: CDEF and C'D'E'F' are the free faces. The following relationships are satisfied, i.e., $f_n=f_s=0$.

(4) exit face: EFE'F' is the exit face. Supposed that the traction velocity of 0.04m/s was imposed on the exit face, the following relationship should be satisfied, i.e., $v_n$=-0.04m/s, $v_s$=0.

*2.4. Rheological Parameters*

In the simulations, the rheological parameters of PTT constitutive model is given in Table 1.

**Table 1.** The rheological parameters of PTT constructive model

| **Parameters** | $\eta$(Pa.s) | $\lambda$(s) | $\varepsilon$ | $\xi$ | $\eta_r$ |
|---|---|---|---|---|---|
| **Values** | 2700 | 0.2 | 0.23 | 0.18 | 0.12 |

## 3. Numerical Results and Analysis

*3.1. Effect of Slippage on the Size and Profile of Outlet Cross-Section*

The volume rate of 40mm$^3$/s was imposed on the inlet face, the traction velocity of 0.04m/s was imposed on the exit face. By means of the finite element computing for inverse extrusion of plastic micro-tube, the outlet cross-section profiles and sizes of the plastic micro-tubes with different slip coefficients of die's channels were obtained, which are shown in Figure 2(a), and (b).

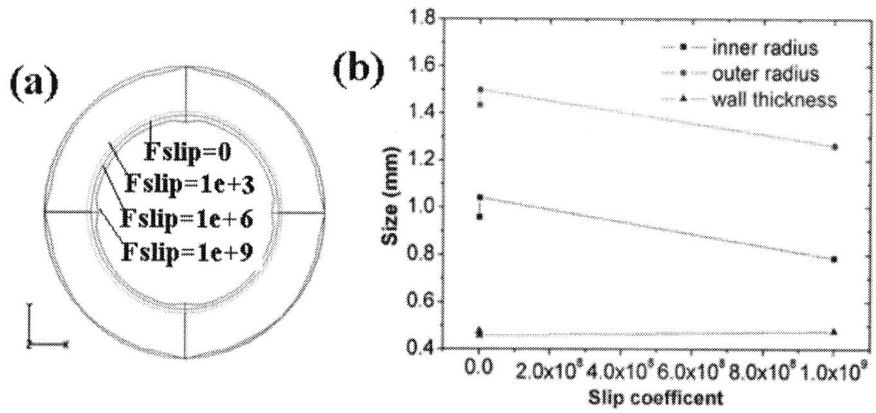


**Figure 2.** Effect of slippage on the size and profile of outlet cross-section. (a) outlet cross-section profiles, (b) sizes of the plastic micro-tubes with different slip coefficients of die's channels

From Figure 2, it can be seen that the sizes and profiles of plastic micro-tube are nearly same when the slip coefficient is less than $1\times10^{+3}$. However, when the slip coefficient is larger than $1\times10^{+3}$, it can be seen that the inner radius of cross-section at the outlet of die decreases, and the outer radius of cross-section at the outlet increases, which result in the increasing of the wall thickness, i.e., the annular width of die's channel was increased with the increasing of the slip coefficient when the process and material parameters are same. At the same time, from Figure 2(a), it can be found that the profile of outlet cross-section is distorted when the slip coefficient is equal to $1\times10^{+9}$, i.e., the extrusion forming of plastic micro-tube with no slippage, which will increase the difficulty of die designing and process stability controlling.

*3.2. Distribution of Flow Velocity*

To ascertain the effect of slippage on the size and profile of plastic micro-tube, the field distributions of X and Z flow velocity are obtained, which are shown in Figure 3(a), and (b).

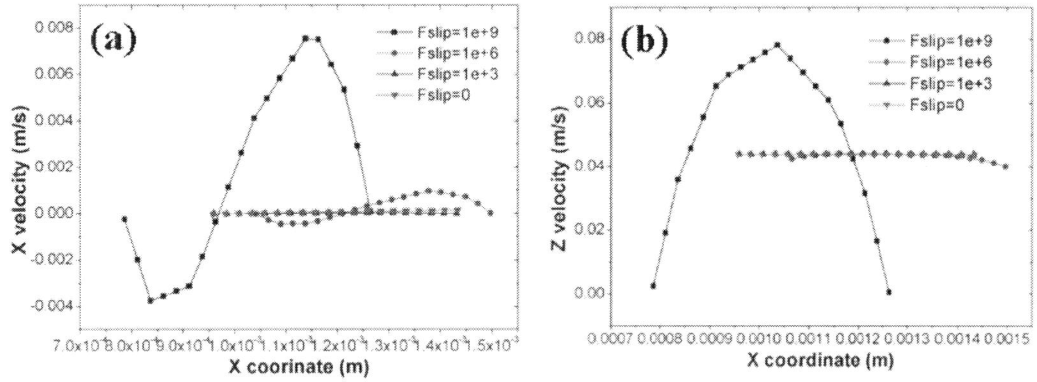


**Figure 3.** Effect of slippage on the flow velocity. (a) X velocity; (b)Z velocity.

From Figure 3(a), it can be seen that when the slip coefficient is less than $1\times10^{+3}$, the field distributions of X and Z flow velocities are same. Moreover, for the slip coefficient is less than $1\times10^{+3}$, the X flow velocity is equal to 0. However, with the increasing of the slip coefficient, the amplitudes of X flow velocity increase. When the slip coefficient reach the maximum value, i.e., the extrusion forming of no slip, it can be seen that the X flow velocity obviously occurs at the outlet cross-section of die, which show that the extrudate swell phenomenon is generated at the time of traditional extrusion forming with no slip, and the extrusion forming with full slip can greatly eliminate the extrudate swell of plastic micro-tube. From Figure 3(b), it can be seen that when slip coefficient is less than $1\times10^{+3}$, the field distribution of Z velocity is uniform. However, with the increasing of the slip coefficient, the Z velocity near wall will be decreased, but the Z velocity in the middle of the channel is increased, which illustrates that with the increasing of slip coefficient, non-uniformity of the Z flow velocity will result in the extrusion distortion (See Figure 2(a)). At the same time, from Figure 2, it can be seen that the position at X coordinate and size of outlet cross-section is different for the different slippages.

### *3.3. Distribution of the Shear Stress*

Then, the field distributions of shear stress at the outlet cross-section of die with different slip coefficients were obtained, which are shown in Figure 4.

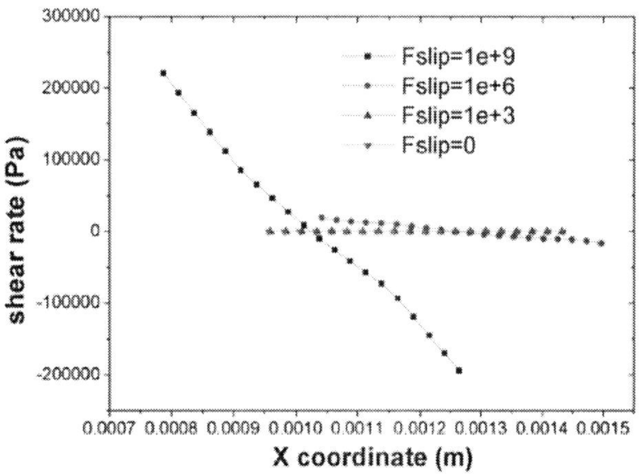


**Figure 4.** Distributions of shear stress with different slip coefficients

From Figure 4, it can be seen that the shear stress is very large especially near the surface of wall for the slip coefficient is equal to $1\times10^{+9}$, i.e., the traditional extrusion forming with no slip. With the decreasing of the slip coefficient, the shear stress of melt decreases, the shear stress of melt is equal to 0 when the slip coefficient decrease to 0, i.e., extrusion forming with full slip. As we know, the shear stress is one of most important factors of not only storing the elastic energy of melt molecular to generate the extrudate swell phenomenon, but also generating the extrudate fracture when the shear stress reaches the inherent tensile resistance. Therefore, the effect of slippage on the size and profile of outlet cross-section is also induced by the melt suffered from different shear stress effects in the channel of die.

## 4. Conclusion

In this paper, the effect of slippage on the size and profile of outlet cross-section of die for extrusion forming of plastic micro-tube were numerically studied by using the finite element method. Based on the same geometric model, material parameters and process parameters, the inverse extrusion forming method was used. The different slip coefficients were imposed on the boundary of the die's channel, i.e., the inner and outer wall of melt. Numerical results show that the inner and outer radius of outlet cross-section decrease but the annular width increases with increasing of the slip coefficient. Moreover, the profile of outlet cross-section is distorted when the no slip effect occurs on the surface of channel wall. The field distributions results illustrate that the slippage effect results in the different distributions of flow velocities and shear stress.

## Acknowledgments

This work was supported by the National Natural Science Foundation of China (51763011), 2018 Natural Science outstanding youth fund project of Jiangxi Province (2018ACB21006), and Doctor start-up foundation project of JXSTNU (2017BSQD021).

## References

[1] Jain P., Parada J. P., David A., et al. 1995 *Arch. Intern. Med.* **155** 1425
[2]Ha Y. S., Cho J. R., Kim T. H., et al. 2008 *J. Mater. Process. Tech.* **201** 168
[3] Ren Z., Huang X. Y., Liu H. S., et al. 2015 J. Appl. Polym. Sci., **132** 1
[4] Carneiro O. S., Nóbrega J. M., Pinho F. T., et al. 2001 *J. Mater. Process. Tech.* **114** 75
[5] ANSYS Inc., 2012 ANSYS POLYFLOW 14.5 User's Guide (ANSYS. Inc., Southpointe)
[6] Thien N. P., Tanner R. I., 1977 *J.Non-Newton. Fluid Mech.* **2** 353

# The Properties and Latest Application of Geopolymers

**Yanguang Wu, Bowen Lu, Zhuangcheng Yi, Feipeng Du and Yunfei Zhang**

School of Materials Science and Engineering, Wuhan Institute of Technology, LiuFang Campus, No.206, Guanggu 1st road, Donghu New & High Technology Development Zone, Wuhan, Hubei Province, P.R. China.
Email: wygddxyz@163.com

**Abstract.** With unique and favorable performances, geopolymer have attracted extensive attention in recent years. It combines some characteristics of organic polymers,cements and ceramics for its special poly-condensed network structure. This paper introduced the synthesis and performance of geopolymer, and mainly reviewed its application in recent years.We can see that the geopolymer has a great development future in many fields like civil engineering,innovative geopolymer binder, rapidly-repair and dealing with poisonous and radioactive waste material, geopolymer-type coatings, novel ceramic materials and so on. In addition, the existing problems and several suggestions according to personal researches are put forward.

## 1. Introduction

Geopolymer (GP) is a kind of new cementitious material with high-temperature ceramics, fire-resistant coatings, radioactive and toxic waste encapsulation, medicinal applications and environmental friendliness [1–4]. And the application of geopolymer (GPs) involves many scientific fields: geology, mineralogy, colloid science, physical chemistry, modern inorganic chemistry and other engineering process technologies. In recent years, extensive studies have been done on geopolymerization chemistry, geopolymeric applications as binders and cements [5–7].

There are many literatures about the study of GPs and the reaction mechanism is still controversial, which brought a certain degree of complexity to the late-stage study. Therefore, the preparation, properties and applications of GP are systematically reviewed, and the future research is prospected in this paper.

## 2. Synthesis and Properties of Geopolymers

### *2.1. Geopolymers Synthesis*

GP materials are usually synthesized using an aluminosilicate raw material and an activating solution which is mainly composed of alkalis of sodium or potassium and waterglass [8]. A GP contains a variety of semicrystalline or amorphous phases and it is a three-dimensional aluminosilicate mineral polymer, futher more, the network structure of GP consists of $Si(OH)_4^-$ and $Al(OH)_4^-$ linked alternately by sharing oxygen [9,10].

In recent years, researchers have done a lot of research on the GP material synthesis for further improving its performance. J.J. Trochez used a spent fluid catalytic cracking catalyst as precursor to prepare GPs, and the study found that increased $SiO_2/Al_2O_3$ was beneficial to the formation of GPs, but the mechanical strength was reduced [11]. In addition, there are many researches on the preparation process of GPs. Jean Noël Yankwa Djobo studied the effects of different preparation process conditions on the reactivity of volcanic ash [12]. Results showed that the reactivity of volcanic ash in alkaline solution is lower than that of other aluminosilicate, but the reaction activity increases slightly with the increase of concentration and curing temperature. Prinya Chindaprasirt has found that the use of outdoor heat exposure (OHE) for curing polypropylene (PP) fiber-reinforced GP composite can achieve the purpose of saving energy [13]. Solar energy is a sustainable and renewable energy for OHE-curing of GP composites leading to strong bridging properties between the PP fiber and the matrix, and strong composites.

### *2.2. The Properties of Geopolymers*

*2.2.1. Environmental friendliness.* GPs, an environmental protective silicon aluminate inorganic material, whose products can significantly reduce the carbon footprint in their preparation and application process. And in the nature world, various containing silicon and aluminum oxide can be used to synthesize GP. For instance, GP concrete is a building material with high application potential. It can reduce the amount of $CO_2$ emissions to a large extent, and the maximum can be 86% lower than that of standard ordinary portland cement [14].

*2.2.2. Rapid hardening and high strength.* GP has the characteristics of high early strength and fast setting time. In recent years, researchers have done a lot of research on GP owing to its high mechaincal strength. Du revealed that using SiC can obviously enhance the compressive strength of GP composites. When the content of SiC is 10 wt %, the compressive strength of GP composites can reach 155 MPa, which is equivalent to increasing 100% on the basis of unfilled GP [2].

*2.2.3. High temperature resistance and heat insulation.* The oxidation reaction and decomposition reaction of the GP products between 1000-1200 °C are not possible, and they are stable under high temperature. The linear shrinkage at 400 °C and 800 °C are 0.2% to 1% and 0.2% to 2%, respectively, and the original strength above 60% can be maintained [15]. This shows better mechanical strength at elevated temperature.

*2.2.4. Good corrosion resistance and durability.* GPs show good stability in acidic solutions and in various organic solvents. In 5 wt.% hydrochloric acid and sulfuric acid solution, the decomposition rate of GP were only 1/13 and 1/12 of it of portland cement, respectively. GP can form compact structure with high strength and excellent impermeability.

*2.2.5. Strong interfacial bonding ability.* At the interface between the traditional portland cement and the aggregate, the transition zone of calcium hydroxide enrichment and preferred orientation is easy to occur, resulting in weak interfacial bonding force. And for GP, the transition zone does not appear due to the reason that the hydration reaction of calcium silicate does not exist, and the final product is a three-dimensional network of gel, which combined with the aggregate interface.

*2.2.6. Immobilizing toxic metal ions.* Mallowc and others believe that metal ions are also involved in the formation of GP structures, so that metal ions in the system can be more effectively fixed .

## 3. The Application of GPs

The excellent properties of GPs determine their broad application and development prospects, and the main applications are as follows.

### *3.1. Innovative GP Binder and Sustainable Repair Material*

Utilization of GP binder has been extensively investigated for its low shrinkage, high early strength, sulfate attack and corrosion, high resistance to thawing and freezing. Used in civil engineering, it can greatly shorten the stripping time, speed up the template operation cycle, and improve the construction speed. One-part GP mixtures is a new type of geopolymeric binders, which can be used to simplify the process of activating GPs in silicate solution. At the temperature of 950°C, Peng has performed thermal treatment of low-quality kaolin, and the strength of the corresponding binders is up to 47MP, which could reach comparable strength with two-part mix systems [16]. Nasvi revealed that GP cement is suitable for storage wells under deep down-hole stress conditions and carbon capture owing to its good acid resistance, high mechanical strength, durability and low permeability [17]. GP cement is regarded as potential replacement of portland cement. This is mainly based on the properties and sustainability criteria.

### *3.2. GP-Type Coatings*
The GP can be connected into the network structure of film material through the aluminum tetrahedral and silicon tetrahedral, so as a kind of inorganic coating, with non-toxic environmental protection, low cost, fire proof and water proof etc. Various proportions of β-SiC were used to prepare the GP-type paints by A. Khosravanihaghighi, and the results show that, by increasing the proportion of nano-SiC, the porosity is reduced, and the corrosion resistance and wear resistance are improved [18]. In addition, reflective heat insulation GP coating was synthesized from metakaolin and sodium silicate solutions as main initial raw material, and talcum powder, hollow glass microspheres and sericite powder are added as fillers [19]. The properties of two hybrid type GP concrete composed of alkaline-activated metakaolin and fly ash has been studied by Ana MaríaAguirre-Guerrero, and it was treated as protective coatings against chloride-induced corrosion in its application [20].

### *3.3. Novel Ceramic Materials*
Compared with the hydrothermal and the sol-gel methods, GP technology provides a new method for the direct synthesis of final structural ceramics. GPs can form a structure comparable to those of high temperature sintered ceramics, and the low temperature GP ceramic is convenient for forming products of various complex shapes and has good overall performance.

### *3.4. Sealing Materials for Industrial Toxic Waste Residue*
Harmful heavy metals can be produced from modern industry, agriculture, disposal of waste and others. GP technology can transform industrial solid waste containing aluminosilicates into useful products thanks to the flexibility and capability to immobilize and stabilize the wastes inside the GP network. Metals such as Co, Cu, Pb, Cd, Ni, Zn, Pd, As, Ra and U can be incorporated in the 3 dimensional GP network decreasing the mobility of the heavy ions through metal hydroxide precipitation, ion substitution or physical encapsulation [21].

### *3.5. GP Based Self-Cleaning Concrete Materials*
Severe air pollution is mainly caused by industrial production, natural disasters, engineering construction and automobile exhaust, etc. Thus building facilities will become dirty when exposed to the pollute air. This problem has forced the industries to develop alternative materials for supporting the sustainable development. GP based self-cleaning concrete material is green alternative for solving the problem owing to their unique properties. And the self-cleaning performance of GP based concrete are studied by introduction of photocatalytic materials such as ZnO and $TiO_2$ [22].

### *3.6. GP Biological Materials*
GP composite has the advantages of non-toxic, biocompatibility, good drug carrying capacity and sufficient mechanical strength, etc., which can be used as oral drug delivery carrier. Metakaolin-based Na GP has been reported as an appropriate carrier for the powerful synthetic the sedative drug zolpidem and opioid analgesic drug fentanyl [23]. Further more, the ion exchange property of GP has been developed into germicidal material with potential application prospect of water purification, and it has a great application prospect in water-scarce countries.

### *3.7. The Applications of Other Fields*
GPs have a wide variety of applications in other fields owing to its excellent properties. For example, there are some of the following:
- Thermal shock refractories;
- Fire-resistant materials, thermal insulation, foams;

- Environmentally friendly adhesive for wood based panels and modern biotechnology;
- High-tech carbon fiber materials used inside and outside of aircraft;
- Dielectric materials for electronic packaging and molecular sieve catalysts material.

## 4. Conclusion and Recommendations

GPs have many advantages, such as rich raw materials, simple process, saving resources and energy, and they also have excellent properties of organic polymers, ceramics and cement, so that more and more attention has been paid to them. In this paper, the properties and applications of GP have been comprehensively reviewed. At the same time, with the development of GP research, the sources of sustainable raw materials will be expanded, the performance of GPs will continue to be enhanced, and some new potential applications will be found.

Based on this review, some recommendations on future research can be provided. On the basis of the existing literature review, the future research of GP should be focused on the following aspects. On the one hand, the research time of GPs is relatively short, and many properties required long time monitoring are not clear, such as volume stability, long-term durability and so on. In addition, toughening modification and mechanism of GP should be studied further owing to its brittleness. On the other hand, NaOH and $Na_2SiO_3$ and other alkaline activator resources are limited, the price is expensive, and may pollute the environment, which affected the popularization and application of this material to some degree.

## Acknowledgment

This work was supported by the National Natural Science Foundation of China [grant numbers 51603156]; and the Wuhan Institute of Technology Scientific Research Foundation of China [grant number k 201704].

## References

[1] Masi G, Rickard WDA, Bignozzi MC, Riessen AV 2015 *Compos Part B Eng.* **76** 218
[2] Du F P, Xie S S, Zhang F et al 2016 *Composites Part B Eng.* **105** 93
[3] Menna C, Asprone D, Ferone C et al 2013 *Compos Part B Eng.* **45** 1667
[4] Nematollahi B, Sanjayan J, Shaikh F U A 2016 *Compos Part B Eng.* **89** 253
[5] Ke X, Bernal S A, Ye N, Provis J L, Yang J 2015 *J Am Ceram Soc.* **98** 5
[6] Juenger M C G, Winnefeld F, Provis J L, Ideker J H 2011 *Cement Concrete Res.* **41**1232
[7] Khan M I, Azizli K, Sufian S et al 2015 *Scientific Conference of Microscopy Society Malaysia.***1669** 299
[8] Duxson P, Mallicoat S W, Lukey G C 2015 *Colloids Surf. A Physicochem. Eng. Asp.* **292** 8
[9] Balcikanli M, Ozbay E 2016 *Compos Part B Eng.* **91** 243
[10] Izquierdo M, Querol X, Davidovits J, Antenucci D, Nugteren H, Fernández-Pereira C 2009 *J. Hazard. Mater* **166** 561
[11] Trochez J J, Gutiérrez R M, Rivera J, Bernal S A 2015 *Materiales de Construcción* **65** 46
[12] Djobo J N Y, Elimbi A, Tchakouté H K, Kumar S 2016 *J Mater Sci* **51** 10301
[13] Chindaprasirt P, Rattanasak U 2017 *Clean Techn Environ Policy* **19** 1985
[14] McLellan B C, Williams R P, Lay J, Riessen A V, Corder G D 2011 *J. Clean. Prod* **19**1080
[15] Wang E, Wen N, Sun H 2005 *Multipurpose Utilization of Mineral Resources* **2** 30
[16] Peng M X, Wang Z H, Shen S H, Xiao Q G 2015 *Mater. Struct* **48** 699
[17] Nasvi M C M, Rathnaweera T D, Padmanabhan E 2016 *Geomech. Geophys. Geo-energ. Geo-resour* **2** 245
[18] Khosravanihaghighi A, Pakshi M 2017 *J Aust Ceram Soc* **53** 909
[19] Zhang Z T, Wang K T, Mo B H, Li X F, Cui X M 2015 *Energy and Buildings* **87** 220
[20] MaríaAguirre-Guerrero A M, Robayo-Salazar R A, Gutiérrez R M D 2017 *Appl Clay Sci* **135** 437
[21] Guo X, Hu W, Shi H 2014 *Constr. Build. Mater* **56** 81
[22] Norsaffirah Zailan S, Mahmed N, Bakri Abdullah M M A, Sandu AV 2016 *Mater Sci Eng* **133** 012026.
[23] Jämstorp E, Strømme M, Bredenberg S 2012 *J. Pharm. Sci* **101** 1803

# Study on Fractal Model for Predicting Permeability and Effective Thermal Conductivity to Fibres of Cut-Tobacco

**Zhicai Jiang, Xue Ding, Hua Huang, Wei Zhou, Ting Fang and Qi Sun**

China Tobacco Zhejiang Industrial Co., Ltd. Hangzhou, China
Email: snowdingxue@163.com

**Abstract.** The permeability and the effective thermal conductivity of cut-tobacco were studied using two-dimensional fractal theory. The porous structure of cut-tobacco samples on their cross sections was analysed and their box-counting dimensions were calculated. The permeability process of gas through cut-tobacco has been analysed and fractal model of the process has been established. The heat conduction process of a single hole was studied and the thermal conductivity of cut-tobacco was obtained by the thermal resistance method. The results show that the fractal dimension of porous structure of cut-tobacco approximately is equal to 1.3. With the increasing of the fractal dimension, high porosity cut-tobacco is in favor of permeability process. Oppositely, with the decreasing of the fractal dimension, the porosity of cut-tobacco decreases corresponding to the component of cut-tobacco increasing which lead to the strength of the thermal conductivity coefficient. Comparison between theoretical and experimental results, the results show that the fractal model can effectively predict the permeability and effective thermal conductivity to fibres of cut-tobacco.

## 1. Introduction

Cigarettes are tobacco products made by wrapping rolling paper around cut tobacco. They constitute a substantial market in the world. During smoking, mainstream smoke is emitted from the end of a lit cigarette. Numerous chemical changes take place in the anoxic hydrogen-rich environment as the burning tobacco produces carbon monoxide, carbon dioxide, hydrogen, and volatile hydrocarbons at different temperatures. To facilitate smoking cessation and reduce the harm of smoking, the market has launched various novel tobacco products that mainly consist of three types—smokeless tobacco, heat-not-burn tobacco, and electronic cigarettes [1]. While looking like traditional cigarettes, heat-not-burn products heat the tobacco at under 500 °C, just enough to release its flavour without burning it. Related studies show that traditional cigarettes produce an enormous amount of harmful substances at high temperatures of 350 °C to 600 °C. Meanwhile, low-temperature e-cigarettes discharge much fewer of these substances because they are heated under 300 °C. In addition, heat-not-burn products result in only 5% of the substances and 14% of the cytotoxicity produced by traditional cigarettes. Compared with other novel cigarettes, their flavour can better satisfy consumers' needs and provide them with an authentic tobacco taste [2].

Inside heat-not-burn cigarettes is a porous material made of cut tobacco. Its structure, composition, and crafting affect its porosity, density, and, consequently, the air permeability and heat transfer during inhalation. Studies on the permeability and heat transfer coefficient are useful for developing a pyrolysis model of cut tobacco, which has a complex structure of randomness and irregularity. A substantial number of studies have indicated that the microstructure, pore size, and pore distribution of fibrous porous media made of cut tobacco carry fractal properties and that the transport and heat transfer in such media can be analysed with the fractal theory and method [3].

The fractal theory was devised by Mandelbrot during his study on the irregular and self-similar geometry of coastlines. In recent years, the theory and its applications have been developing so rapidly that it has become one of the most important topics in nonlinear science. Katz and Thompson [4] were the first to provide experimental data to justify that the pore spaces of sandstone samples follow a self-similar and fractal scaling law. The fractal theory has become an effective approach for studying the structure and transport of porous media. Adler [5] computed the thermal conductivity of the Sierpinski carpet, a fractal porous medium, with certainty. Based on the fractal geometric theory, Yu et al. [6, 7] proposed a scaling law followed by the cumulative pore-size distribution in porous media; based on the exact self-similar Sierpinski carpets, Yu and Li [6] obtained the correlation between the porosity and fractal dimensions of porous media. Liu [8] used a fractal model to describe the hydrate saturation and pore-scale behavior dependent permeability of hydrate-bearing sediments. Yu et al. [9] adopted the fractal theory to study the porous structure of wood sections perpendicular to fibres and obtained the effective thermal conductivity perpendicular to the grain. Yu et al. [6] employed the fractal theory to study the cut-tobacco size distribution and found its fractal properties and statistically significant self-similarity.

This study investigated the porous microstructure of cut tobacco using the fractal theory. First, it used the photomicrograph of a section of cut tobacco to analyse the fractal properties of the fine-cut tobacco in a cigarette and its porous structure. It also used the box-counting dimension to compute the fractal dimension of the porous tobacco. The effective permeability of the fractal porous media was then calculated with both the Hagen–Poiseulle equation and Darcy's law. Then, the energy equation was employed to reveal the correlation between the thermal diffusivity and porosity of the cigarette. Eventually, this study compared the results of the fractal model and the experiment to prove the effectiveness of the fractal theory in predicting the permeability and heat transfer of cut tobacco.

## 2. Internal Microstructure of Cigarettes

A cigarette is a typical fibrous medium with a complicated inside porous structure. The experiment in this study selected a novel heat-not-burn cigarette available in the market as the target sample, the central part of which was observed with the HD4i VED tobacco test system manufactured by Rinco Asia PTE Ltd. at 10X magnification. The photomicrograph is shown in Figure 1.

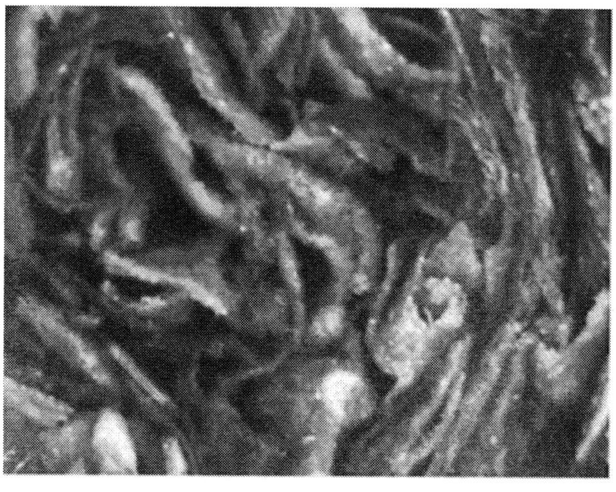

**Figure 1.** Sectional view of cut tobacco

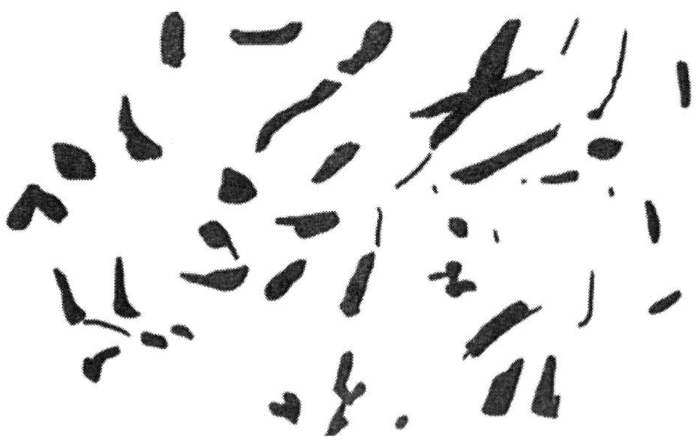

**Figure 2.** Binary image of pore distribution

From the photomicrograph, it is evident that the pores between the tobacco fibres display intricate and messy boundaries, random distribution, and great discrepancies in diameter. Irregularity and self-similarity are manifest in the porous structure.

Analysis of the porous structure reveals the difficulty in using conventional Euclidean geometry to describe its characteristics: no shapes with regular dimensions suffice to characterize the complexity of the porous structure. With reference to the properties of fractal figures, the porous medium exhibits a random coherent structure and a certain level of self-similarity. Hence, the porous structure of cut-tobacco fibres can be categorized as a fractal structure [10] and thus examined with the fractal theory.

## 3. Computation of Fractal Dimension

Fractal dimension is an important parameter for describing fractals and representing their basic properties. It is a tool for comparing fractals objectively and distinguishing one from another based on their irregularity. Its most distinctive characteristic is that it accepts fractional dimensions.

The fractal dimension is expressed as follows [11]:

$$D = \frac{\ln(N)}{\ln(s)} \tag{1}$$

In this equation, $D$ denotes the fractal dimension and $N$ is the number of parts in the scaled down version of set $A$ by ratio $s$ that does not cover set $A$.

The commonly used fractal dimensions include the Housedoff dimension, the self-similarity dimension, and the box-counting dimension. Among them, the box-counting dimension can perform approximation through experiments and has wide applications [12]. Chaudhury and Sarkar [13] suggested the calculation of fractal dimensions by differential box-counting. The computational thinking is: consider A as an image with $M\times M$ pixels and S a grid with $s\times s$ pixels; A is partitioned by S, and $N(s)$ denotes the smallest number of boxes on S needed to cover A. The box-counting dimension $D(A)$ is expressed as follows:

$$D(A) = \lim_{s\to 0} \frac{\ln N(s)}{-\ln s} \tag{2}$$

During the experiment, the first step of using the box-counting dimension to calculate the fractal dimension of a two-dimensional object was to pre-process and convert the digital image into a black-and-white binary image. The next step was to cover the fractal image with a grid consisting of square boxes measuring on each side and then count the number of boxes that contain image pixels, $N(s)$, which represent the black parts of the pore spaces. The same process was repeated by decreasing the box size of grids until the smallest square covering the pixels is found. Then, a linear regression line was fit to the log-log plot ($\ln N(s)$-$\ln(1/s)$) of the obtained data. The slope of the linear fit is the fractal dimension of the image. From equation (2), we can get:

$$N(s) \sim (s)^{D(A)} \tag{3}$$

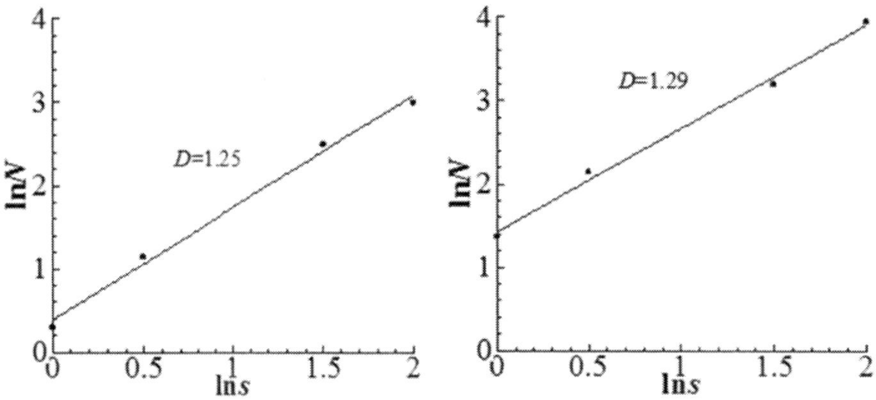


(a) Sample 1 (b) Sample 2

**Figure 3.** Curve fit of the box-counting dimension

An image of the porous structure is taken from the photomicrograph produced by the tobacco test system and then converted into a black-and-white binary image as shown in Figure 2. Analysis of the fractals in Figure 2 returns Figure 3, the curve fit of the box-counting dimension of the porous structure. Results show that the box-counting dimension demonstrated good linearity. The slope $D$ of the linear fit represents the required box-counting dimension. The fractal dimension is smaller than 2 because the pores do not fill the entire plane. It is also relatively small because of the low porosity of the section. Hence, the value is between 1 and 1.5.

## 4. Fractal Model of Cigarette Permeability

The volume of airflow $q$ passing through the capillaries in the porous structure satisfies the Hangen–Poiseulle equation [14]:

$$q = \frac{\pi}{128} \frac{\Delta p}{L(\lambda)} \frac{\lambda^4}{\mu} \tag{4}$$

In this equation, $L(\lambda)$ is the length of capillary with pore size as $\lambda$, $\mu$ is the viscosity of air, and $\Delta p$ is the pressure difference between the two ends of a capillary.

Based on the fractal model of the capillaries in the cut tobacco, assuming the air passage resembles a straight line, the total volume of airflow $Q$ can be obtained from the integrals in equation (4), $l$ is the length of the capillary cell.

$$Q = \int_{\lambda_{\min}}^{\lambda_{\max}} q(\lambda) dN(\lambda) = \frac{\pi}{128\mu} \frac{\Delta p}{l} \frac{D}{4-D} \lambda_{\max}^4 \tag{5}$$

Integrating Darcy's law, we obtain the effective permeability of the fractal porous medium:

$$\zeta = \frac{\mu l Q}{\Delta p A} = \frac{\pi}{128} \frac{1}{A} \frac{D}{4-D} \lambda_{\max}^4 \tag{6}$$

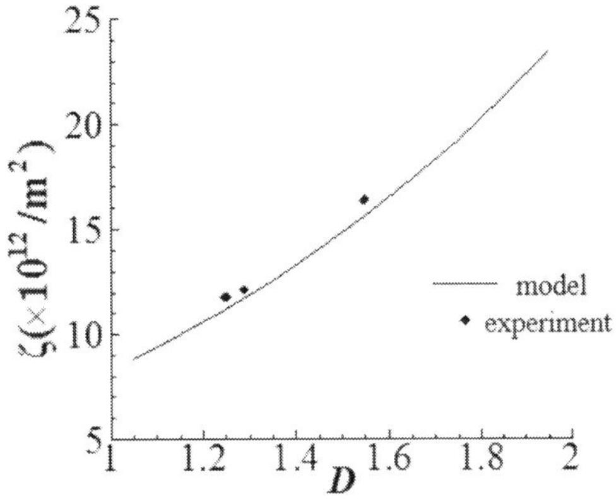


**Figure 4.** Correlation between fractal dimension and permeability

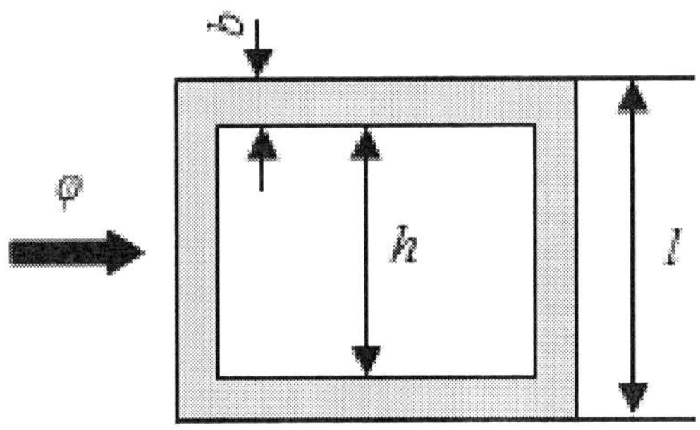


**Figure 5.** Simplified unit structure

Based on the structural parameters of the cigarette, namely the sectional area $A$, length $l$, the maximum pore size $\lambda$max, and the fractal dimension $D$, the permeability was computed as shown in Figure 4. The experimental results were in line with the theoretical results of the fractal model. The permeability increased with the fractal dimension because a higher fractal dimension leads to higher porosity, which facilitates the effusion of air. When the fractal dimension equal to 2, maximum permeability was achieved.

## 5. Fractal Model of Effective Thermal Conductivity

During inhalation, the heat-not-burn cigarette transfers heat through the thermal conductivity of the cut tobacco. It is essential to build a two-dimensional model of how the effective thermal conductivity changes with the fractal dimension. The unit structure of the cut tobacco can be simplified as in Figure 5. The convection and radiation in the pore spaces are ignored, and heat transfer is assumed a pure process of thermal conduction. Assume a uniform thickness of cut tobacco and a transfer of heat flux $\varphi$ in the $x$-direction. The thermal resistance network of the porous structure is illustrated in Figure 6: the thermal resistance in each part can be expressed as follows:

$$r_1 = r_5 = l/(bk_w) \tag{7}$$

$$r_2 = r_4 = b/(hk_w) \tag{8}$$

$$r_3 = (l-2b)/(hk_a) \tag{9}$$

The total thermal resistance is:

$$r = l/[(h+2b)k_t] = \left(\frac{2}{r_1} + \frac{1}{2r_2 + r_3}\right)^{-1} \tag{10}$$

Thus, the effective thermal conductivity is:

$$k_t = \frac{(4b^2 + hl)k_w k_a + 2b(l-2b)k_w^2}{2b(h+2b)k_a + (h+2b)(l-2b)k_w} \tag{11}$$

Substitute porosity $\varepsilon$ into the equation. The above equation is transformed to:

$$k_t = \frac{0.5(1+\varepsilon)^2 k_w k_a + \varepsilon(1-\varepsilon)k_w^2}{\varepsilon(1-\varepsilon)k_a + 2\varepsilon k_w} \tag{12}$$

where

$$\varepsilon = (\lambda_{min}/\lambda_{max})^{2-D} \tag{13}$$

Through substitution of the above equation, we obtain the effective thermal conductivity of the cigarette.

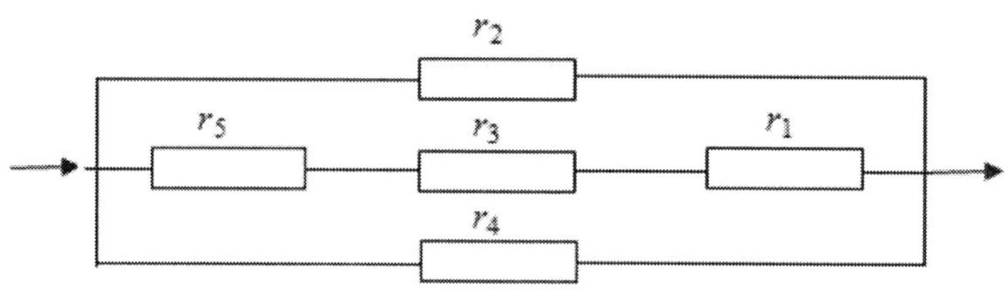


**Figure 6.** Illustration of thermal resistance network.

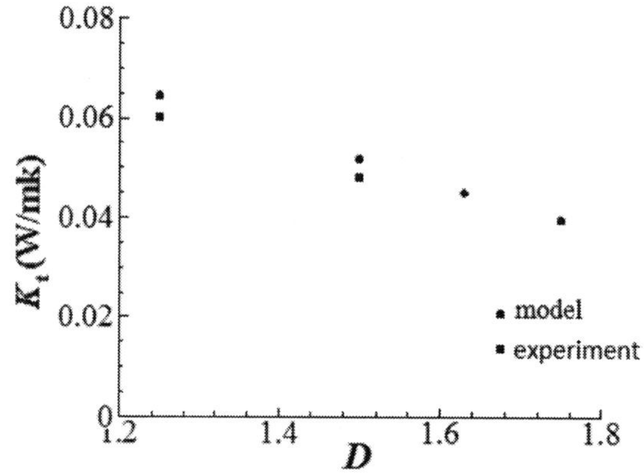


**Figure 7.** Correlation between fractal dimension and effective thermal conductivity

The thermal conductivity *kw* of solid tobacco is 0.086 W/mk. Assuming the pores are filled with air, we obtain the thermal conductivity of *ka* = 0.024 W/mk. The computational results are shown in Figure 7. To test the reliability of the fractal model, the thermal conductivity of different pores was measured with an apparatus for measuring the thermophysical properties based on the transient plane source method manufactured by Hot Disk AB. During measurement, the temperature distribution in the sample constitutes an unsteady temperature field that is constantly changing over time. The rate of temperature change is used to measure the heat diffusivity of the sample and, in turn, the thermal conductivity of the material. This method is suitable for measuring a wide range of thermal conductivity and has a short measurement time and high measurement accuracy. Figure 7 shows that the results of experimental and computational are consistent which means the computational results of the fractal model are reliable. The effective thermal conductivity declined as the fractal dimension increased. The lower the fractal dimension, the smaller the porosity, which means a higher concentration of tobacco. Since the cut tobacco has a higher thermal conductivity than air, the porous medium has greater thermal conductivity. However, the closer the fractal dimension is to 2, the greater the porosity, which means a higher concentration of air and, consequently, lower thermal conductivity.

## 6. Conclusion

This study used the fractal theory to build two-dimensional computational models of the fractal dimension, fractal permeability, and effective thermal conductivity of the porous medium of cut tobacco. According to the comparison of the model outcomes and the experimental results, the fractal model was effective in predicting the permeability and heat transfer of the cut tobacco. The fractal dimension of porous structure of cut-tobacco approximately is equal to 1.3 The permeability increases and the effective thermal conductivity declines as the fractal dimension increasing. This provides a significant theoretical foundation and research method for follow-up studies.

**Acknowledgments**

The project supported by ZJTI Science and technology project" Research on electronic cigarette device and heating characteristic (ZJZY2015A008)

## References

[1] Shen Y 2012 Trends in Research and development of a new generation of tobacco substitutes, *Tob. Sci. Tech.***3** 38-40.
[2] Ruprecht A A, Marco C D, Saffari A, et al.2017 Environmental pollution and emission factors of electronic cigarettes, heat-not-burn tobacco products, and conventional cigarettes, *Aerosol Sci Tech.* **51**(6):674-684.
[3] Yu N, Shen X F, Xu D Y, Li H J and Chang M B 2012 Characterization of cut tobacco size distribution based on fractal theory, *Tob. Sci. Tech.* **4** 5–8
[4] Katz A J and Thompson A H 1985 Fractal sandstone pores: implications for conductivity and pore formation, *Phys. Rev. Let.* **54**(12) 1325
[5] Adler P M 1996 Transports in fractal porous media, *J. Hydrol.* **187**(1-2) 195-213
[6] Yu B and Li J 2001 some fractal characters of porous media. *Fract.* **9**(3) 365-372.
[7] Yu B and Cheng P 2002 A fractal permeability model for bi-dispersed porous media, *Int.J. Heat Mass Trans.* **45**(14), 2983-93
[8] Liu L L,Zhang Z,Ning F L,Li C F,Cai J C,Wang D G, Liu C L, Wu N Y. 2019 A fractal model for the relative permeability prediction of hydrate-bearing sediments, *Scientia Sinica.***49**(3):034614
[9] Yu Z T, Hu Y C, Tian T and Fan L W 2007 Fractal model for predicting effective thermal conductivity perpendicular to fibres of wood, *J.Zhejiang Univ.***41**(2) 351–5
[10] [10] Shen H M, Ye Q and Meng G X 2017 Anisotropic fractal model for the effective thermal conductivity of random metal fiber porous media with high porosity, *Phys. Let.* A. **381**(37) 3193-96
[11] Pia G and Sanna U 2014 An intermingled fractal unit's model to evaluate pore size distribution influence on thermal conductivity values in porous materials, *Appl.Therm.Eng.* **65**(1-2) 330-6
[12] Yu M C, Rao J P and Xie Y P 2011 Moisture and temperature distribution of MDF mat after microwave preheating.*J. Northeast Forestry Univ.* **39**(6) 47-9
[13] Chaudhuri B B and Sarkar N 1992 an efficient approaches to compute fractal dimension in texture image, *Proc. 11th IAPR Int.Conf. on. IEEE* pp358-61
[14] Xu P , Li C H ,Liu H C, Qiu S X and Yu B M 2017 Fractal features of the effective gas transport coefficient for multiscale porous media, *Earth Sci.* **42**(8)1373–78

IOP Conf. Series: Materials Science and Engineering 472 (2019) 012031 doi:10.1088/1757-899X/472/1/012031

# Nonlinear Finite Element Simulation on Seismic Behaviour of Steel Frame-Central Brace with Ring Damper

**Xiaotong Peng[1], Chen Lin[2], Yongming Cao[1] and Wenxu Duan[1]**

[1] School of Civil Engineering and Architecture, University of Jinan, Jinan 250022, China.

[2] School of Architecture and Landscape Design, Shandong University of Art & Design, Jinan 250014, China.

Email: pengxito@163.com.

**Abstract.** The central brace combined with a ring damper is used as lateral resistant to improve the lateral stiffness and seismic performance of steel frame. In order to reveal the working mechanism of the new system, three nonlinear finite element models: steel frame with central brace, steel frame with eccentric brace, and steel frame infill central brace with steel ring damper, were established. The results indicate that compared with other two structural systems, the steel frame infill central brace with steel ring damper has the best bearing capacity, energy dissipation ability and seismic behaviour.

## 1. Introduction

Using of the brace system could increase the lateral stiffness of steel frames dramatically, and reduce the seismic demands on steel frames. Previous researches show that different brace systems have different effects on the performance and energy consumption of steel frames. Chong ping Bai [1-4] carried out finite element simulation analysis on steel frames with various supporting system, such as “X-shaped” and “herringbone type”. It was found that the supporting steel frame structure can reduce the interlayer displacement, but the support rods are easily destabilized under compression. Jie Leng etc. [5-9] conducted a cyclic loading test on the self-repairing support steel frame. The results indicated that the structure had residential deformation after loading and a poor permanent in energy dissipation. In order to study the reinforcement brace system, a new type of supporting brace system for steel frame was produced. Based on that, an analysis model using ABAQUS software was established to study it’s lateral resistant and energy dissipation ability.

## 2. Steel Frame--Central Brace with Ring Damper

The calculating example is based on a 6-storey frame-brace structure. A horizontal composite steel frame in the central area was selected as the research object (Figure. 1). Based on that, a steel frame--central brace with ring damper (SBRD) was simplified for analysis (Figure 2).

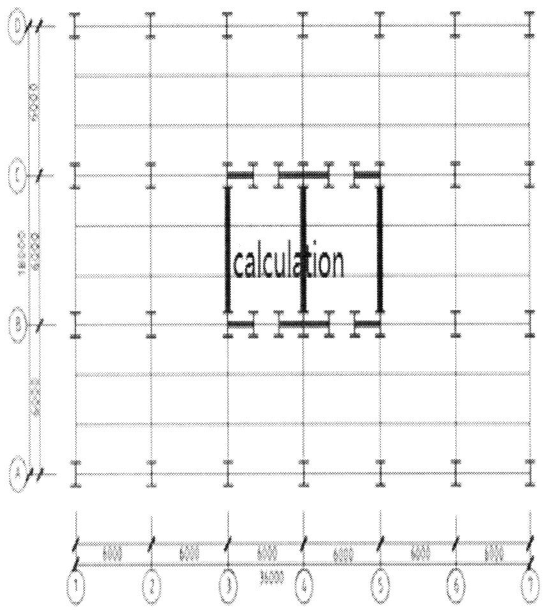


**Figure 1.** SBRD Plan

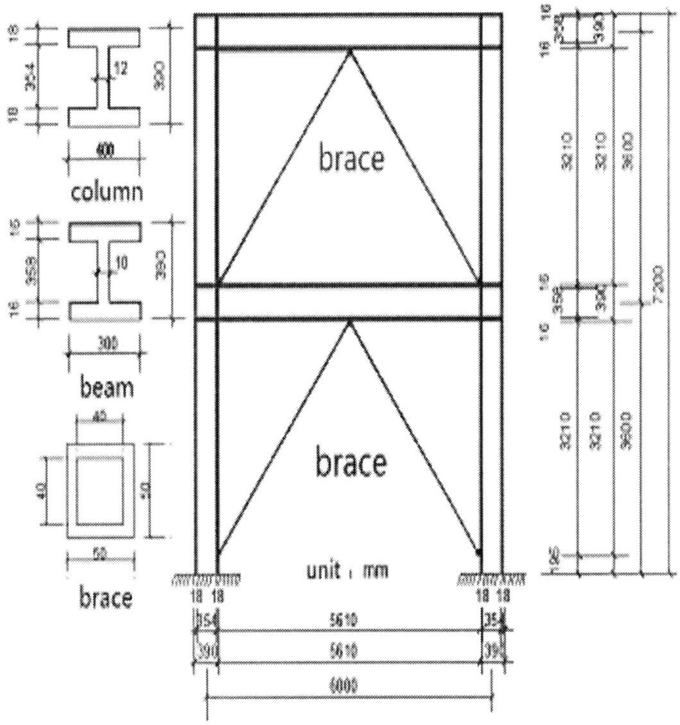


**Figure 2.** Simplified Model for SBRD

### 3. Nonlinear Finite Element Analysis

In order to compare the seismic behavior of steel frame with central brace, steel frame with eccentric brace and SBRD, the same frame element type, mesh division, boundary conditions and loading methods are adopted.

IOP Conf. Series: Materials Science and Engineering 472 (2019) 012031 doi:10.1088/1757-899X/472/1/012031

### 3.1. Finite Element Models Establishment

In order to simulate the Steel Frame-central brace with ring damper accurately, S4R shell unit is adopted in beam and column, C3D8R hexahedron solid unit is adopted in brace and ring damper. Specimens were selected according to ASTM-A370-05 for material properties. The beams and columns are made by Q345B steel, while the braces and ring dampers are made by Q235B steel. Material mechanical properties data is shown in Table 1.

**Table1.** Mechanical Properties of Materials

| | fy[a]/MPa | $f_u$[b]/MPa | E[c]/GPa |
|---|---|---|---|
| Steel frame(Q345) | 300 | 460.5 | 204 |
| Brace and damper (Q235) | 200 | 336.2 | 203 |

[a] $f_y$ is the yield strength.
[b] $f_u$ is the ultimate strength.
[c] E is the elastic modulus.

To ensure the accuracy and efficiency of calculation, structural meshing is adopted in all elements, as shown in Figure 3.

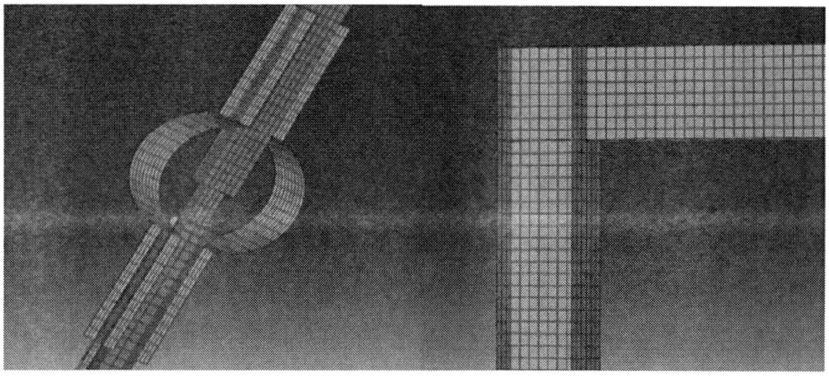

**Figure 3.** Meshing Grid

To meet the requirements of nodal rigidity and seism, the braces are articulated connect to beams and columns, and beams and columns are connected by bolt and weld. Ring dampers are bolted to braces to transfer the shear (Figure. 4).

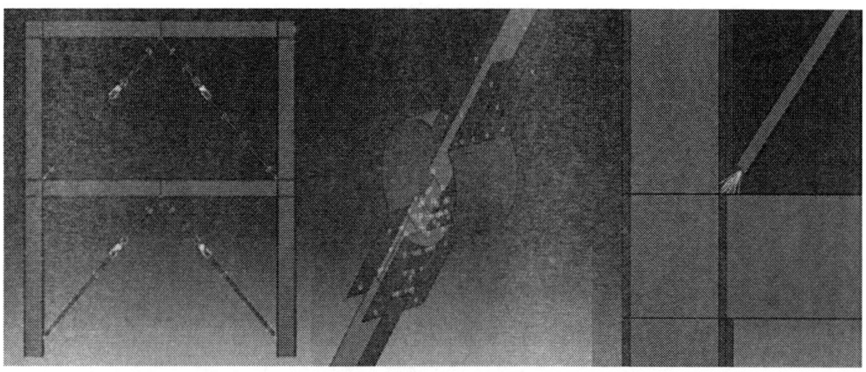

**Figure 4.** Interaction

The boundary conditions are set as follows: all bottom nodes of the brace and column are fixed completely. A lateral constraint is arranged in the middle of the beams and columns web to prevent the out-of-plane overturning of the steel frame. In order to avoid the stress concentration and local buckling at the loading point, the beam-column section is coupled to the hinged joint, on which lateral displacement is applied. The boundary conditions and load arrangement are shown in Figure. 5.

**Figure 5.** Boundary Conditions

The monotonic load and cyclic loading scheme are controlled by displacement and the load step increases 2mm each stage. Each stage of the load moves only once. The load-displacement performance and hysteretic behaviour are recorded under the loading scheme.

### *3.2. Finite Element Results*

The displacements of three structures increase with the increase of monotonic load (Figure 6). The steel frame with a central support has a desirable initial stiffness in the initial elastic stage (Table 2). In the elastic-plastic stage, although the stiffness of each structure decreases gradually. The ultimate bearing capacity of SBRD is 915.2kN and the ultimate displacement is about 310mm. It has satisfactory bearing capacity and strong deformation ability under lateral loads.

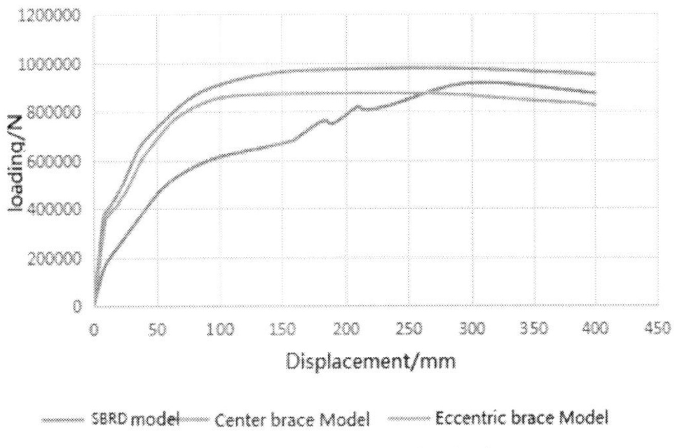


**Figure 6.** Comparison of Three Load-Displacement Curves

**Table 2.** Comparison of Bearing Capacity and Initial Stiffness

| Model | Bearing capacity | Initial Stiffness |
|---|---|---|
| SBRD | 915.2KN | 22.4KN/mm |
| Steel Frame with Central Brace | 976.0KN | 50.8KN/mm |
| Steel Frame with Eccentric Brace | 874.2KN | 37.8KN/mm |

Three hysteresis curves of three structures under the cyclic loads are listed in Figure 7 respectively. Two of them are all shuttle shape, while the SBRD hysteresis curve is much plumper showing that it has better deformation performance and energy dissipation ability.

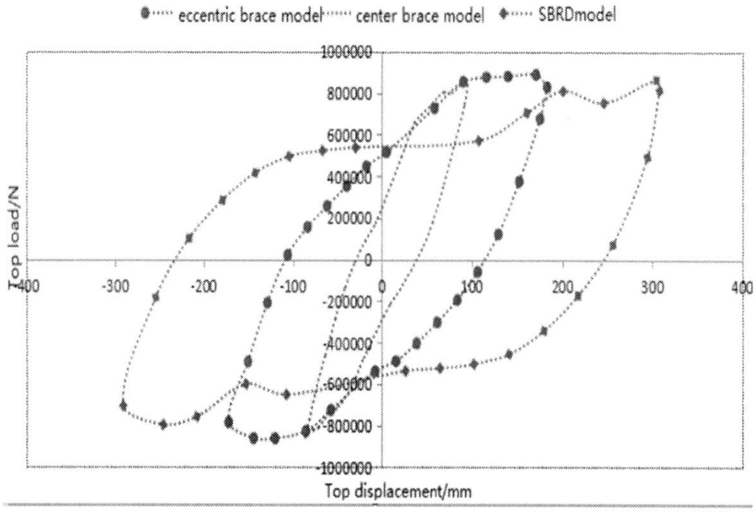


**Figure 7.** Hysteresis Curves of Three Models

The Stress nephograms of three structures are shown in Figure 8-10. It indicates that the forces in steel frame with central brace and SBRD are mainly applies on the joints, beams and braces. The SBRD delays compression buckling of braces through deformation of the ring damper to realize better energy dissipation performance. The forces in the steel structure with eccentrically brace are mainly applies on the joints, beams and energy-consuming beams. Once the energy-consuming beams are damaged in the loading procedure, it may cause premature failure of the structure.

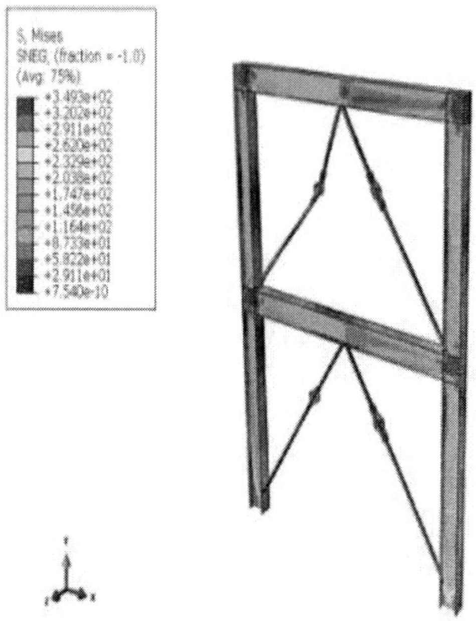

**Figure 8.** SBRD

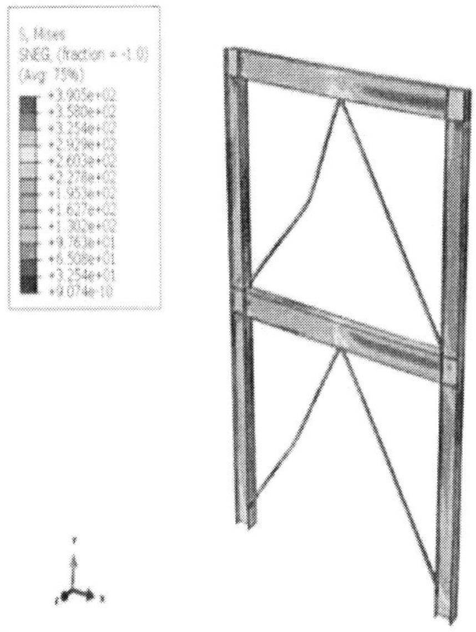

**Figure 9.** Steel Frame with Central Brace

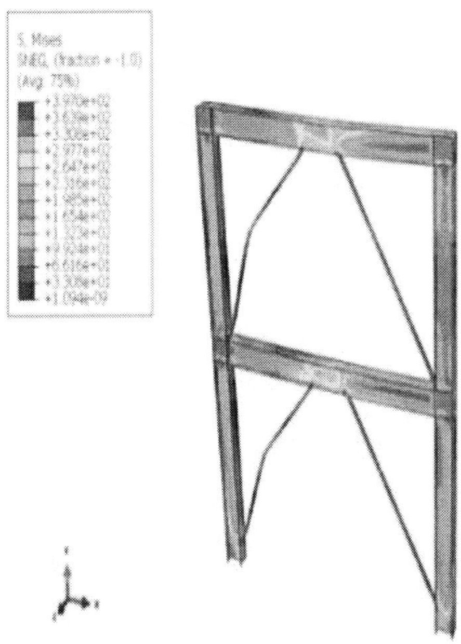

**Figure 10.** Steel Frame with Eccentric Brace

## 4. Conclusion

According to the performance of three structural models under monotonic loads and cyclic loads, conclusions can be drawn as follows:

(1) The SBRD has the best deformation ability under the monotonic loads, and thus has better performance under large deformation conditions. Also the SBRD has desirable lateral stiffness and bearing capacity

(2) The SBRD has the best seismic behaviour under cyclic loads. It dissipates the energy through a ring damper.

(3) SBRD could not only delay the buckling of the brace, but also avoid the structural damage caused by the failure of energy-consuming beams.

**Acknowledgements**

The work was sponsored by Key Research and Development Program of Shandong Province (2015GSF122003; 2016GSF122012). The writers gratefully acknowledge the support provided and a Project of Shandong Province Higher Educational Science and Technology Program (J17KB048; J18KA208).

**References**

[1] Bai C 2016 *Discussion on seismic design of X-shaped centre bracing of steel structure* (G ansu Science and Technology).p15
[2] Li Y, Cai K, Tang Y 2010 *Design and Research of Central Support-Frame Steel Structure* (Industrial Architecture). p11
[3] Lian Z, Zhang Y 2005 *Study Status, Inadequacy and Improvement of Double Support System with Steel Support and Frame-centre Support.*(Earthquake Engineering and Engineering Vibration).p3
[4] Tian C. 2016 *Seismic performance analysis of steel frame-cruciform buckling restraint support structure* (Harbin Engineering University).
[5] Leng J, Liu W. 2016 *Self-returning hysteretic behaviour of K-eccentrically braced steel frame structures* (Journal of Civil Engineering and Management).p73-79.
[6] Liu W, Sun G, Leng J 2015 *Hysteretic behaviour of a new type of self-return Y-eccentrically braced steel frame structure* (Earthquake Resistant Engineering and Retrofitting). p51-57
[7] Lian M, Su M. 2016 *Seismic performance of Y-shaped eccentrically supported high-strength steel frame structure* (Civil Engineering and Environmental Engineering). p86-95
[8] Yao Z. 2016 *Seismic performance analysis of self-reset energy-consuming supporting steel frame structure* (Lanzhou University of Technology).
[9] LIU W,WU B. 2016 *Analysis of vibration reduction effect of self-resetting buckling-supported steel frame*(Journal of Building Structures).p4

# Nonlinear Finite Element Analysis on Seismic Performance of Steel Corrugated Shear Wall

**Xiaotong Peng[1], Chen Lin[2], Tingting Zhang[1] and Xu Zhang[1]**

[1] School of Civil Engineering and Architecture, University of Jinan, Jinan 250022, China.

[2] School of Architecture and Landscape Design, Shandong University of Art & Design, Jinan 250014, China.

Email: pengxito@163.com

**Abstract.** In order to study the seismic behavior of corrugated steel shear wall, a three dimensional nonlinear FEM of steel corrugated shear wall was established by use of ABQUS software. Based on that, the bearing capacity of the corrugated steel shear wall under the monotonic loads and its hysteretic behavior under the cyclic loads were analyzed. The results indicate that the corrugated steel shear wall could decrease storey drift and improve lateral bearing capacity and stiffness of steel frames dramatically; the corrugated steel shear wall has better seismic performance than steel plate shear wall.

## 1. Introduction

Steel plate shear walls (SPSWs) are widely used and efficient lateral force resistant system. However, it is prone to buckle under lateral load. Although stiffened steel shear walls and composite steel plate shear walls are produced by Alinia and Shafaei respectively [1-2] to delay the buckling of steel plates, the effect is not satisfactory. The corrugated steel plate is quailed as shear walls in steel frame for its large initial stiffness and shear yield strength. It can reduce the out-of-plane deformation of shear walls and improve lateral stiffness of steel frames. Thus, tanking corrugated steel plates as shear walls was proposed by Hossain and Wright [3].Then the buckling performance of steel corrugated plate shear walls (SCSWs) was studied by Li Feng et al [4-6]. Li Guo qiang [5] studied the influence of waveform on SCSWs shear capacity, and Zhu Wei [6] found that the shear wall has the best lateral resistant performance when corrugations placed horizontally. Non-linear finite element analysis of SCSWs was conducted by Liu Ying [7] to study its hysteretic behavior.

Based on the research above, two models of SCSW and SPSW were produced taking a 6-story office building as a prototype. ABAQUS 6.14 software was used to analyze the seismic behavior of SCSWs and SPSWs, and its lateral resistant behavior and hysteretic performance under monotonic and cyclic loading were compared.

## 2. Corrugated Steel Plate and Flat Steel Plate

Two models of SCSW and SPSW was designed based on the 3-4 stories in the central area of the 6-story office building prototype (Figure. 1). The span of the model is 7800mm, the storey height is 3600mm, the section size of the beam, column and ground beam is 400×8×280×18mm, 500×14×320×18mm, and 1200×75×900×75mm respectively. The corrugated steel plate in SCSWs adopts type YX-51-250-750 with 1 mm in thickness. It is composed of two parallel-connected trapezoidal corrugated steel plates formed honeycomb shape. Flat steel plate used in SPSW is 2mm in thickness. The beams and columns are connected rigidly. The infill steel plate is separated from the columns with a distance of 30 mm. whereas the infill steel plates are welded to flanges of beams by connection plates.

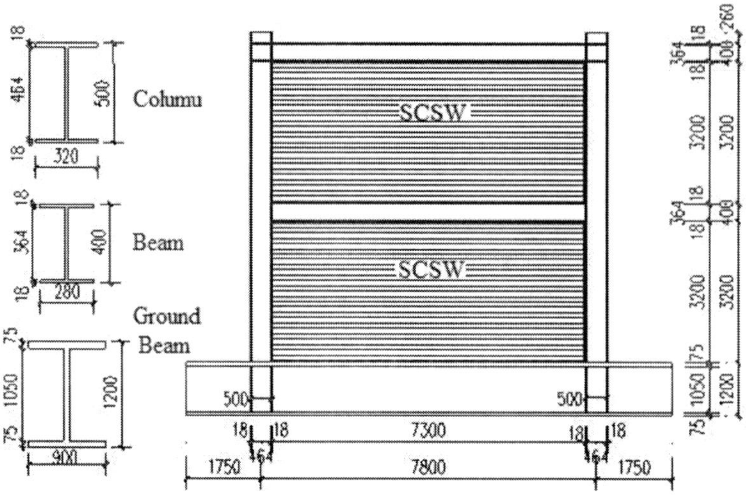


(a) SCSW structure

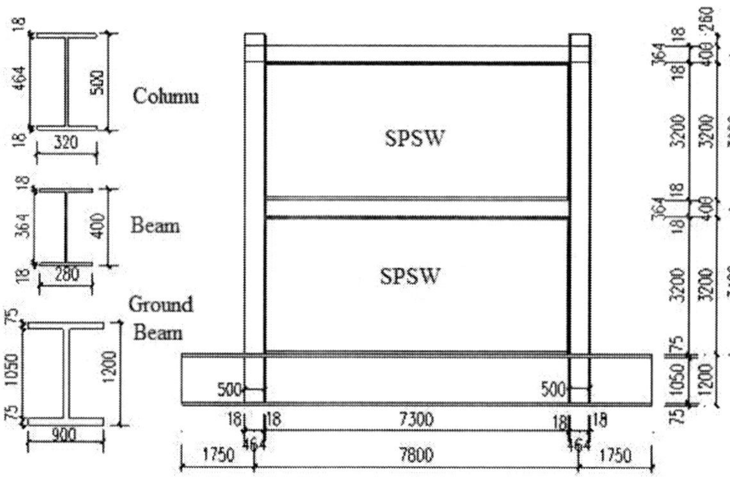


(b) SPSW structure

**Figure 1.** Model Dimensions

## 3. Nonlinear Finite Element Analysis

In order to analysis the seismic behavior of steel corrugated plate shear walls, SCSW and SPSW were compared. Besides the infill steel plate, the modeling methods for frame size, element type, meshing, boundary conditions and loading history used in the two models is the same.

### *3.1. Finite Element Models*

S4R shell elements are adopted to simulate beams, columns, and flat steel plate or corrugated steel plate, which can realize the buckling deformation under stress accurately. The rigid joints are achieved by the 'merge' option in assembly function module. The welded connections are achieved by 'tie' option. The steel frame and the steel plate and connection used Q345 and Q235 steel respectively (Table. 1).

**Table 1.** Material Properties

| | $E^a$(N/mm$^2$) | $f_y^b$(N/mm$^2$) | $f_u^c$(N/mm$^2$) |
|---|---|---|---|
| Steel frame(Q345) | 204000 | 310 | 460.3 |
| Infill plate(Q235) | 203000 | 210 | 323.2 |

[a] E is the elastic modulus.
[b] $f_y$ is the yield strength.
[c] $f_u$ is the ultimate strength.

The models are meshed in a simple and regular way using free meshing with quadrilateral element (Figure. 2). The accuracy and calculating time are influenced by meshing density directly. To ensure accuracy of calculating results and save computer time, the meshing size is 50mm for beams, and 100mm for columns, ground beams, flat steel plates, corrugated steel plates and fishplates.

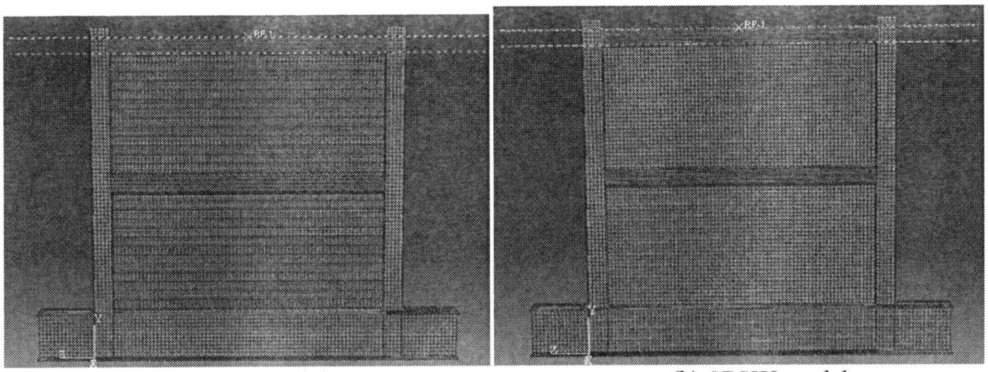
(a) SCSW model (b) SPSW model

**Figure 2.** Free Meshing Partition

The boundary conditions are set as follows: the nodes of the ground beam lower flanges are constrained fully to realize the fixed end. The load is applied in the Z-axis direction with X and Y axes degrees constrained (Figure. 3). In order to avoid stress concentration and local buckling at the loading point, an infinite rigid plane is placed on the loading point to ensure the transfer lateral load uniformly.

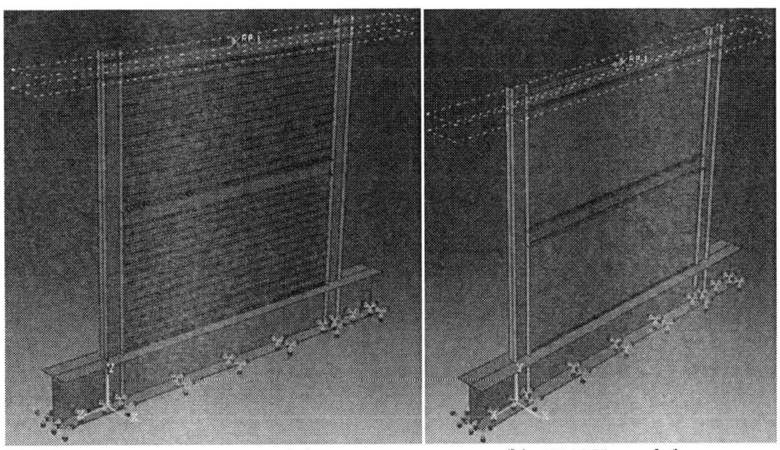
(a) SCSW model (b) SPSW model

**Figure 3.** Boundary Condition

The static and cyclic loading scheme is controlled by peak displacement, which is applied along the Z-axis direction. The amplitudes of elastic loading history at each level are 2mm, 4mm, 6mm, 8mm, 10mm, 12mm, and 14mm, with each loading level recycled only once (Figure. 4).

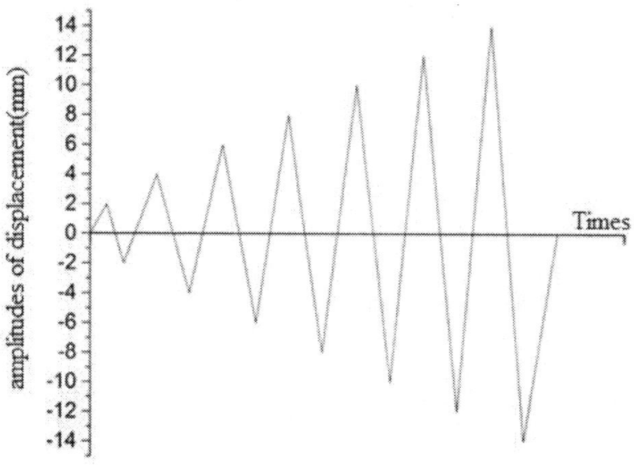


**Figure 4.** Loading History

*3.2. Finite Element Analysis*

The lateral deformation of both structures increased gradually under the monotonic loads (Figure. 5). The end of elastic stage of the SPSW is represented by horizontal load of 4550 kN, while that of the SCSW is 5564 kN. The ultimate bearing capacity of the SPSW is 6858KN with a displacement of 66.36mm, while the bearing capacity of the SCSW is 7490KN. It indicates that within the elastic stage, the peak drift of the SCSW is smaller but its ultimate bearing capacity is 9.2% higher than that of the SPSW.

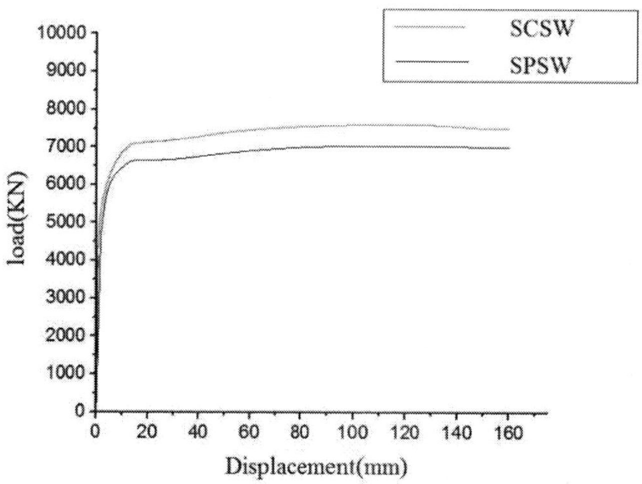


**Figure 5.** Load-displacement Curves

In order to study the hysteretic behavior in the elastic stage, the cycling loading is applied in elastic stage with displacement in 14mm. Figure. 6 shows the hysteretic curves of the SCSW and the SPSW under cyclic loading. They are both spindle shaped and plump indicating both models have good energy dissipation performance. Moreover, the energy dissipation performance of the SCSW is better than that of the SPSW, since hysteretic loop areas of the SCSW is obviously larger than that of the SPSW.

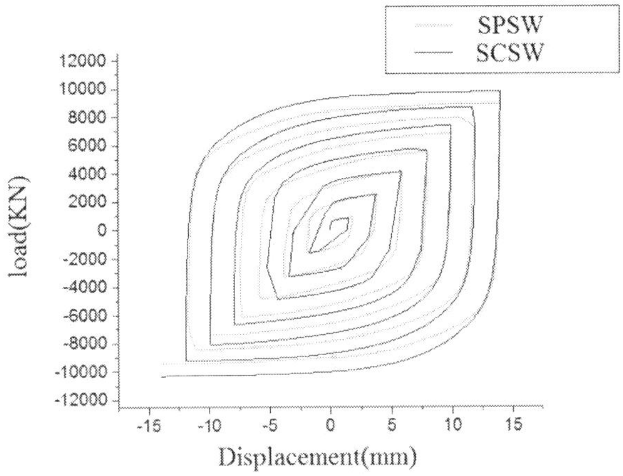


**Figure 6.** Hysteretic Curves

Within the elastic stage, buckling waves are formed diagonally in first and second floors of SPSW (Figure. 7). The steel frame mainly yields around joints, and the out-of-plane deformation occurs in the flange of columns. As for the SCSW, buckling waves are formed in steel plate area near the upper and lower beams on second floor and no buckling waves are observed on the steel plate of first floor. Only a small part of the steel around the connection reaches ultimate strength, no deformation is observed on the flange of columns.

There is a large part of steel plate around the top of column in SPSW yields, while only a small range on the top of the column in SCSW yields. The corrugated steel plate of the SCSW absorbs most parts of the lateral loads of the structure, and shares the lateral loads that columns subjected to.

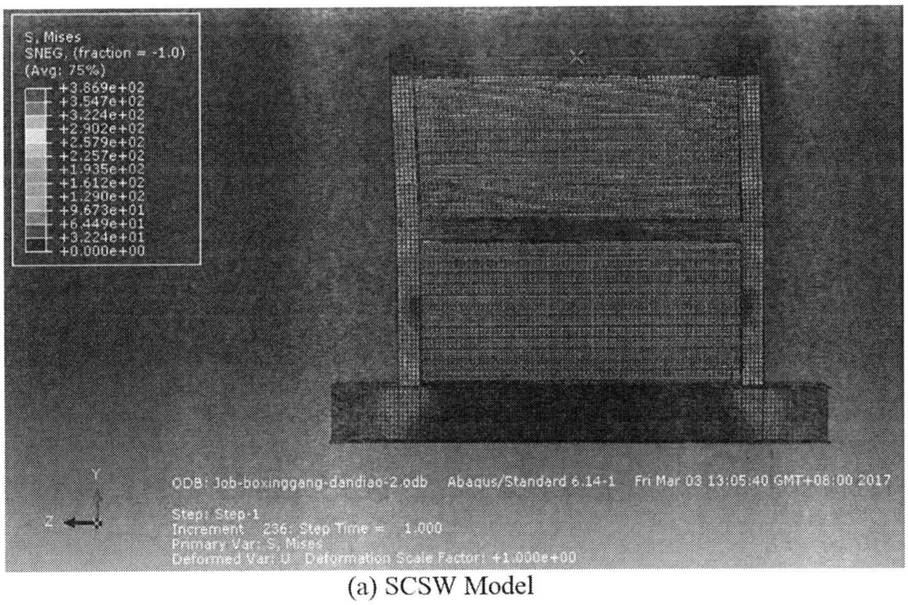


(a) SCSW Model

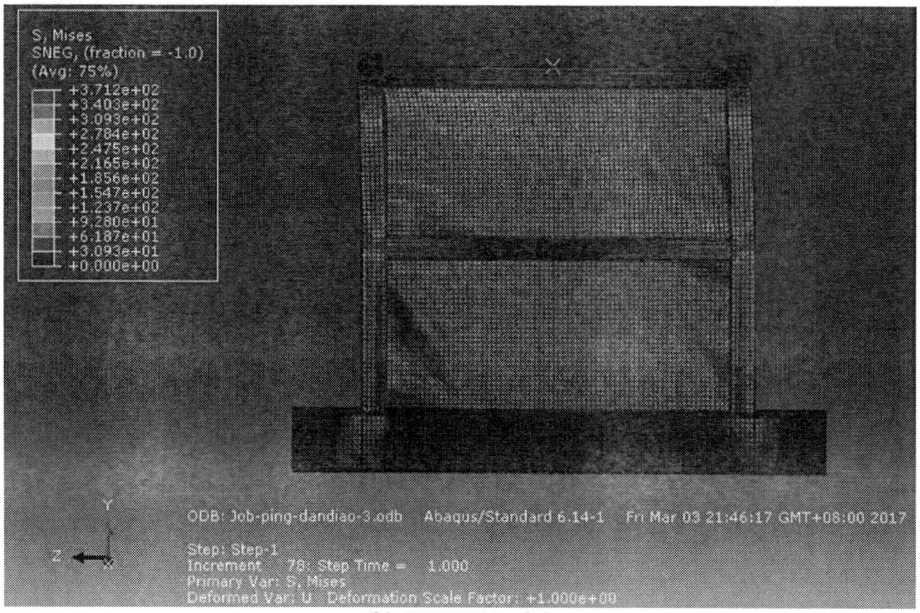


(b) SPSW Model

**Figure 7.** Stress Nephogram

## 4. Conclusions

(1)Under the monotonic loads, compared to the SPSW, the SCSW has better lateral stiffness and bearing capacity.

(2)Under the cyclic loads, the SCSW has a better hysteretic behavior in the range of displacement amplitude (14mm). The corrugated steel plate has made great contributions in increasing lateral stiffness, improving energy dissipation capacity and delaying the premature buckling of steel plates.

## Acknowledgments

The work was sponsored by Key Research and Development Program of Shandong Province (2015GSF122003; 2016GSF122012). The writers gratefully acknowledge the support provided and a Project of Shandong Province Higher Educational Science and Technology Program(J17KB048; J18KA208).

## References

[1] Alinia M.M and Shirazi R.S On the design of stiffeners in steel plate shear walls 2009 *Journal of Constructional Steel Research* **65** 2069-77.

[2] Shafaei S, Ayazi and Farahbod F *T*he effect of concrete panel thickness upon composite steel plate shear walls 2016 *Journal of Constructional Steel Research* **117** 81-90.

[3] Chan C.L, Khalid Y.A and Sahari B.B Finite element analysis of corrugated web beams under bending 2002 *Journal of Constructional Steel Research* **58** 1391-1406

[4] Li F and Zhao L Elastic Bucking Analysis of Transverse Honeycomb Shaped Section Steel Plate Shear Wall 2014*Steel Construction* **29** 59-62.

[5] J.W. Berman and M. Bruneau. Experimental Investigation of Light-Gauge Steel Plate Shear Walls 2005 *Journal of Structural Engineering* **131** 259-267.

[6] Yanlin Guo, Qinglin Zhang and Xiao'an Wang. A Theoretical and Experimental Study of the Shear Strength of H-Shaped Members with Sinusoidal Corrugated Webs 2010 *China Civil Engineering Journal* **10** 45-52.

[7] Li G, Zhang Z and Sun F Shear Strength of H-beam with Corrugated Webs 2009 *Journal of Tongji UniversityNatural Science* **37** 709-14.

[8] Zhu W and Mei H Analysis of Lateral Force Performance of Corrugated Steel Plate Shear Wall 2013 *Development orientation of building materials* **5** 20-2.
[9] Liu Y 2012 Nonlinear Finite Element Analysis of Semi-rigid Composite Steel Frame with Profiling Steel Plate Infill Wall *China Steel construction Society Lodge of Structural stability and Fatigue*
[10] Yu A, Zhao J, Jiang A 2006 Ductility Steel Folded Plate Shear Wall *China Steel construction Society Lodge of Structural stability and Fatigu*

IOP Conf. Series: Materials Science and Engineering 472 (2019) 012033 doi:10.1088/1757-899X/472/1/012033

# Three-Point Bending Fracture Properties of Multilayer Metal Hot Forging Die Specimen

**Huajun Wang[1], Qingyang Liu[2], Nian Han, Liang Yao and Chundong Zhu**

School of Materials Science and Engineering, Wuhan University of Technology, Wuhan 430070 China

Email: [1] wanghuajunhb@163.com; [2] 376368471@qq.com

**Abstract.** Multilayer metal hot forging dies deposited by the plasma transferred arc welding presents high wear resistance, oxidation resistance and high temperature stability. In the process of hot forging, bending load made the cavity of multilayer metal hot forging die easily failure. We investigated the bending performance of multilayer metal hot forging die by three-point bending test. The model of bending specimen used H13 as the substrate, transition layer was W6Mo5Cr4V2 and outer layer was Ni60A. Mechanical properties of H13 homogeneous specimens and multilayer metal specimens were tested and compared. ABAQUS finite element method was used to simulate the process of three-point bending under temperature gradient. The results show that crack appears firstly in the H13 substrate and Ni60A layer, expands and meets in the transition layer gradually, results in specimen fractures. At service temperature, bending performance of the multilayer metal specimens is better than homogeneous specimens.

## 1. Introduction

Hot forging dies have very poor service conditions. The process of fast heating and cooling leads to thermal stress and organizational change, then the chamber surface will occur wear, thermal fatigue, and soften deformation failure [1, 2]. It restricts the service life of hot forging die. Several techniques, such as the PTAW, carburizing, boronizing, nitrocarburized, thermal diffusion technology, electroless plating, CVD and PVD, are use for hardfacing [3-5]. One or more layers of high performance material coatings are formed on the surface of hot forging die, then a special multilayer structure obtained. Due to its high bonding strength, wear resistance, corrosion resistance and impact fatigue resistance at high temperature. This method is widely used in hot die repair and life prolonged.

Wang and co-workers [6] proposed that a continuously running homogeneous hot forging die. Its temperature field is divided into three regions: temperature fluctuation zone, temperature equilibrium zone and temperature drop zone. According to the temperature distribution characteristics of hot forging die, required high temperature alloy coatings can be prepared in different thickness region. Its hardness is characterized by the gradient distribution. Kashani [7] prepared Inconel625, Stellite6 and Stellite21 coatings on the surface of the H11 material by TIG welding. Through pin-on-disk testing at different temperature, he found that alloy coating improved the high temperature wear resistance. Xie[8] proposed a composite surfacing repair method by adding transition layer between substrate and outer layer. It improved the bonding property of substrate and coating. Liu [9] compared the organization and performance of the coating structure, between wear resistant layer surfacing and transition layer + wear resistant layer gradient overlay welding. He indicated that transition layer was beneficial to rational transition of alloying elements and microstructure, improved the bonding strength. Imran [10] successfully prepared multilayer metal die casting dies on the surface of copper substrate. The transition layer is 41C stainless steel, coating is H13 steel.

In the process of hot forging, the cavity of the multilayer metal hot forging model is non-uniform stressed. The surface is subjected to bending load inevitably and the coating materials are usually hard and brittle. Therefore, Deng[11] prepared specimens with different thickness by PTAW, researched the effect of thickness on bending strength by three-point bending test at room and high temperature. Through tensile and impact toughness test, Imran [12] held that multilayer metal hot forging die specimen impact toughness is fine. However, the fracture mode of the multilayer metal die at service temperature has not been investigated intensively.

This paper prepared the specimens of multilayer metal hot forging die by PTAW. Three-point bending test and mechanical property test of the multilayer metal hot forging die were carried out by the universal tensile testing machine. The deformation mechanism of multilayer metal hot die fracture at room temperature is explained by FEM, and predicted the mechanical properties of the specimens under the service temperature.

## 2. Experimental Procedure

### *2.1. Materials and Specimens Preparation*

The temperature field of hot forging die in work can be divided into three regions: temperature fluctuation zone (600-700 °C), temperature equilibrium zone (400-600 °C) and temperature falling zone (200-400 °C). According to the thermal physical parameters of material, the Ni60A alloy is suitable for the temperature fluctuation zone, the W6Mo5Cr4V2 high speed steel as temperature equilibrium zone and H13 hot forging die steel as temperature drop zone. Thermal physical properties of the above three materials in respective service temperature range as shown in Table 1.

Multilayer metal blanks were prepared by PTAW. Firstly,W6Mo5Cr4V2 high speed steel transition layer was prepared by spray welding on the surface of H13 substrate. Heating at 500 °C for 30 minute in furnace and polished the weld layer to 1mm. Secondly, prepared the Ni60A coatings by welding, annealing at 500 °C for 2 hours to avoid weld cracks. Finally, prepared the specimens according to the experimental design by wire cutting, as shown in Figure 1b. Parameters were shown in Table 2.

The specimen surface quality is fine and there are no weld cracks or shrinkage aging pore defects.

**Table 1**. Physical property parameters of materials in the service temperature range.

| Material | Thermal expansion coefficient ($10^{-6}K^{-1}$) | Thermal conductivity ($W \cdot m^{-1} \cdot K^{-1}$) | Specific heat capacity ($J \cdot kg^{-1} \cdot K^{-1}$) | Elastic modulus (GPa) |
|---|---|---|---|---|
| 4Cr5MoSiV1 | 10.3-12.2 | 31.2-28.6 | 506-587 | 185-190 |
| W6Mo5Cr4V2 | 10.8-11.27 | 27.1-26.5 | — | 230-250 |
| Ni60A | 13.9-14.4 | 26.34-25.56 | — | 180-195 |

**Table 2**. Plasma-transferred arc welding parameters

| Parameter | Ni60A | W6Mo5Cr4V2 |
|---|---|---|
| Current[A] | 70 | 80 |
| Voltage[V] | 30 | 30 |
| Powder feeding rate[g/min] | 10 | 10 |
| Welding speed[mm/min] | 50 | 50 |
| Arc oscillation width[mm] | 12 | 12 |
| Standoff distance[mm] | 8 | 9 |
| Preheating temperature[°C] | 500 | 500 |
| Overlap ratio[%] | 40 | 40 |

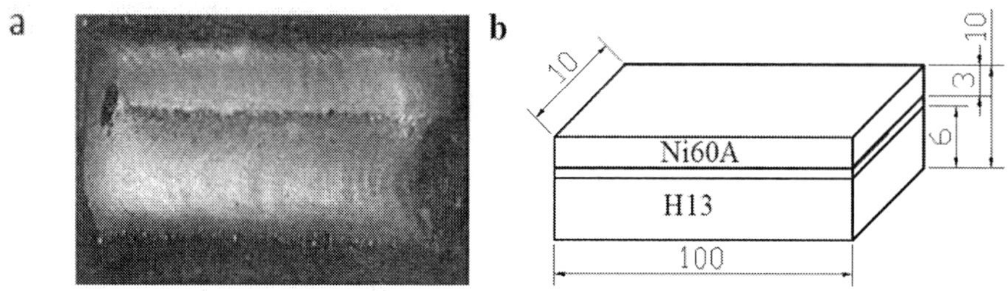


**Figure 1.** Preparation of multilayer metal specimens. (a) blank appearance; (b) size of specimens.

*2.2. Three-Point Bending Tests*

The three-point bending test was carried out on the universal tensile test machine (Figure 2b). According to the experiment, the load-displacement curve between the bending load and the bending deformation of the specimen was plotted. The flexural strength ($\sigma_{bb}$) and flexural deflection ($f_{bb}$) of the specimens were calculated from data. The test span (Ls) was 80mm, compression speed was 1mm/min, and support roll diameter was 10 mm.

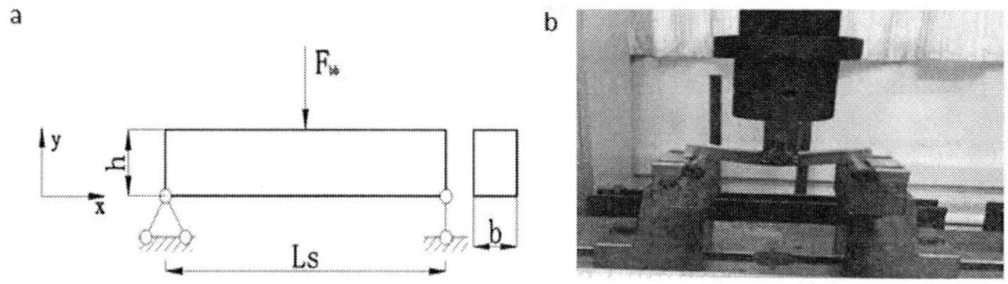


**Figure 2.** Three-point bending tests. (a) schematic diagram; (b) test machine.

*2.3. FEM Simulation*

Three-point bending finite element model was established by ABAQUS. The bending process was simulated at room and service temperature. The characteristic model of homogeneous and multilayer metal specimens was established by model segmentation and optional material (Figure 3a). Contact conditions were: no friction between the three-point bending specimen, the support roller and the loading roller. Two support rollers were fixed point constraints. Restriction of X axis on the left and right symmetry axis of the specimen. Due to the possibility of ductile fracture and it needed large mesh distortion to simulate large strain deformation, we selected ductile damage as fracture criterion. The type of mesh unit was CPE4RT, which is used to simulate thick structures.

During the hot forging, the near-surface temperature of die exhibited gradient changes. The three-point bending experiment and the gradient temperature distribution were difficult to realized under high temperature. So the finite element model was used to simulate the three-point bending process of the hot forging die specimen under the temperature gradient. According to the temperature range of the hot forging die partition. The temperature of fluctuation zone, equilibrium zone and falling zone were set to 600°C, 500°C and 400°C respectively(Figure 3b).

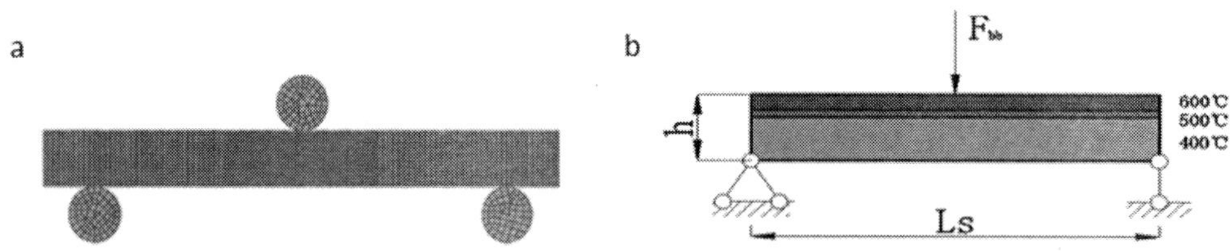


**Figure 3.** Finite element model. (a) three-point bending experiment ; (b) gradient temperature diagram of hot forging die specimen.

## 3. Results and Discussions

### 3.1. Load-Displacement Curves

Load-displacement curves of homogeneous and multilayer metal specimens during three-point bending tests are plotted in Figure 4a. According to the formula (1) and (2), the bending strength of homogeneous and multilayer metal hot forging die specimen can be calculated (Table 3).

$$\sigma_{bb} = \frac{3F_{bb}L_S}{2bh^2} \quad (1)$$

$$\sigma_{bb} = \frac{3EF_{bb}L_S h}{E_s[bh^3-b(2h_s-h)^3]+E_t[bh^3+b(2h_s-h)^3]+E_c[bh^3-b(2h_s-h)^3]} \quad (2)$$

In which, $h_s$, $h_t$, $h_c$ is the thickness of the substrate, transition layer, cladding. $E_s$, $E_t$, $E_c$ is the elastic modulus of the substrate, transition layer, cladding.

**Table 3.** Mechanical properties of three-point bending specimen

| Specimen | Maximum bending load (N) | Maximum deflection (mm) | Bending strength (MPa) |
|---|---|---|---|
| Homogeneous | 29984.2 | 8.9 | 3598.1 |
| Multilayer | 25250.5 | 9.2 | 3030.1 |

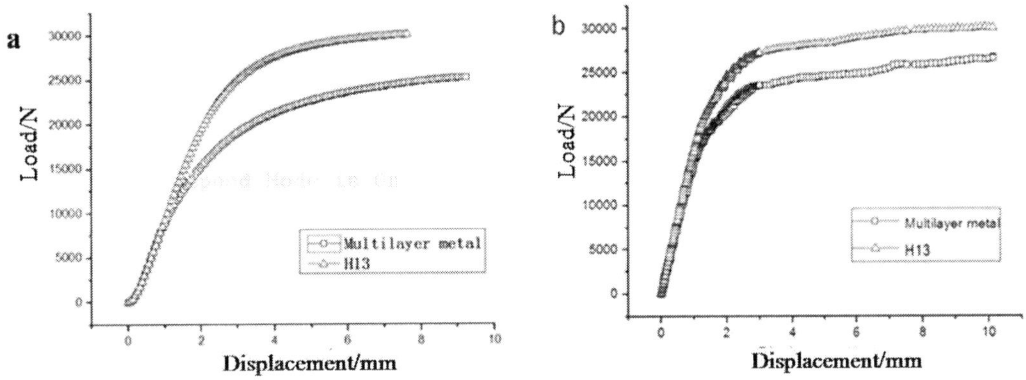


**Figure 4.** Load- displacement curve under room temperature. (a) experiment; (b) simulation.

The bending strength of homogeneous hot forging die specimens is 18.7% higher than multilayer ones. That the strength of H13 is higher than Ni60A cladding at room temperature may be the answer. By comparing the simulated curves (Figure 4) with the experimental curves, the peak and slope are equal. This verifies the validity of the finite element calculation in this experiment.

### 3.2. Fracture Mode

Figure 5a shows the multilayer metal specimen stress distribution in the three-point bending under gradient temperature. The tensile stress on the substrate side is at the maximum value and crack is easily occurred. On the other hand, the stress in the cladding area is high; two high-stress channels appear on both sides of the loading roller and parallel to the loading direction. So the cladding side is also easy to crack. Due to the hardness and brittleness of the Ni60A coating, the crack propagation is promoted. The initial crack appeared on the Ni60A coating and the surface of the substrate, than crack expands and meets in the transition layer gradually, results in specimen fracture .According to the previous simulation, the fracture mode of the multilayer metal specimens at service temperature is nearly the same with the result in room temperature. The experimental results (Figure 5b) show the fracture mode of the multilayer metal specimen is consistent with the finite element.

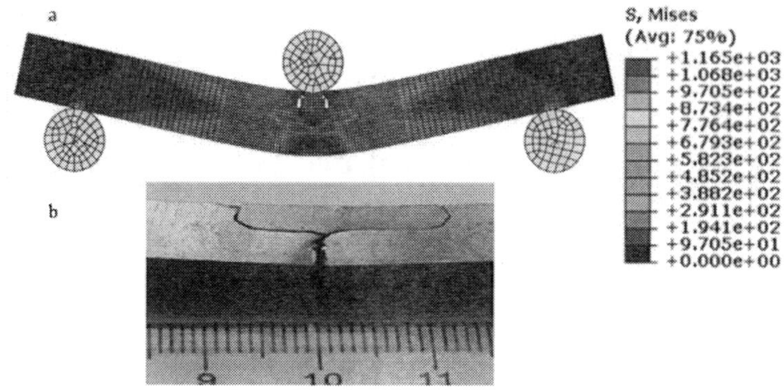


**Figure 5.** Three-point bending result. (a) Bending stress distribution simulation; (b) Experimental.

### *3.3. Microscopic Fracture Morphology Analysis*

Figure 6 is the fracture cross-section scanning electron microscope morphology of the multilayer metal bending specimen. It showed a fiber-like morphology (Fig.6a) and there are shallow dimples in the substrate. Therefore, from the fracture morphology the fracture mode of the substrate is ductile fracture. The brittle cleavage of hard particles observed in fracture of the transition layer in Figure 6b, the hard brittle property of the W6Mo5Cr4V2 steel leads transition layer to brittle fracture. Figure 6c is the fracture morphology of the Ni60A coating; hard particles have large brittle fracture on the fracture surface, so the fracture mode is brittle fracture too. During the three-point bending experiment, tensile stresses in the substrate, the stress is always at maximum value.

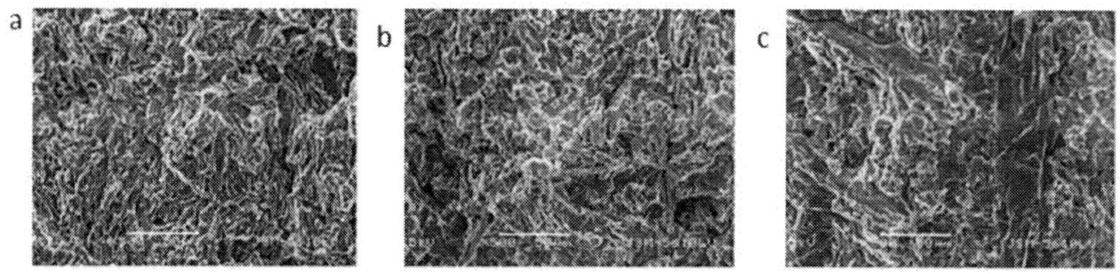


**Figure 6.** Fracture morphology of multilayer metal hot forging die bending specimen (a) substrate 500×; (b) transition layer 500×; (c) outer layer 500×.

### *3.4. Analysis of Bending Properties under Temperature Gradient*

The displacement load curves of homogeneous and multilayer metal bending specimens under temperature gradient are shown in Figure 7. When the bending deflection is 8mm, the bending load of H13 specimen is about 20000N, the bending load of multilayer metal hot forging die specimen is 22000N. This shows that the bending load of the multilayer metal specimen is 10% larger than the homogeneous specimen. Ni60A is a self-fluxing and high temperature red hardness alloy powder. Its thermoset physical properties are better than H13. Therefore, the thermal bending strength of multilayer hot forging dies is better than that of homogeneous ones.

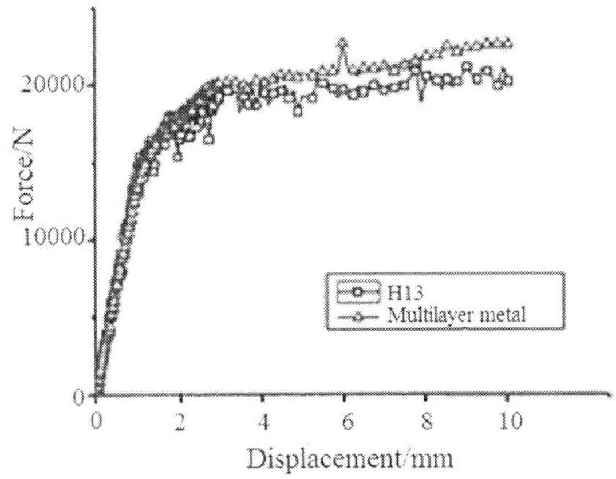


**Figure 7.** Load-displacement curves under temperature gradient.

## 4. Conclusions

(1) At room temperature, the bending strength of homogeneous hot forging die specimens is 18.7% higher than multilayer metal. The bending load of the multilayer metal specimen is 10% larger than the homogeneous specimen under temperature gradient.

(2) At room and service temperature, the fracture modes of multilayer metal specimen are similar: high-stress channel exists on both sides of loading roller; crack appears near the Ni60A cladding layer and extends to the transition layer. The tensile stress on the substrate side is always the maximum. Crack easily germinates and spreads to the transition layer. The cracks on both sides of the transition layer meet and finally fracture.

(3) The fracture cross-section micrographs of three-point bending test show that the fracture mode of substrate (H13) is ductile fracture, the transition layer (W6Mo5Cr4V2) and coating layer (Ni60A) are brittle fracture.

## Acknowledgements

This work was financially supported by the Natural Science Foundation of China (51475346), Natural Science Foundation of China (51741506) and Wuhan University of Technology Graduate Outstanding Thesis Cultivation Project (2017-YS-006). The authors gratefully acknowledge Ms. Bing Xie for her assistance in experiment.

## References

[1] J H Kang, I W Park, J S Jae and S S Kang 1998 A study on die wear model of warm and hot forgings *Metals & Materials* vol 4 pp 477-83

[2] K Venkatesan, C Subramanian, and E Summerville 1997 Three-body abrasion of surface engineered die steel at elevated temperatures *Wears* vol 203–204 pp 129–38

[3] B Navinšek, P Panjan, and F Gorenjak 2001 Improvement of hot forging manufacturing with pvd and duplex coatings. *Surface & Coatings Technology* vol 137 pp 255-64

[4] C. M. D Starling and J R T Branco 1997. Thermal fatigue of hot work tool steel with hard coatings. *Thin Solid Films* vol 308–309 pp 436-42

[5] W Roberts and B Johansson 1989 *Tool steel EP* US 4863515 vol 15 pp 557-9

[6] Wang H J, et al. 2011 Design method and verification for long life hot forging die. *Material Research Innovations* vol 15 pp 377-80.

[7] H Kashani, A Amadeh and H M Ghasemi 2007 Room and high temperature wear behaviors of nickel and cobalt base weld overlay coatings on hot forging dies *Wear* vol 262 pp 800-6

[8] Xie Bing, Luo Jian, Li Ainong and Hu Fang 2006 Study on microstructure and properties of gradient hardfacing layer for mould repairing *China Mechanical Engineering* vol 17 pp 338-43

[9] Y Liu, J Zhou, BI Huan, Y Ding and LU Shun 2011 Study on microstructure and mechanical properties of low alloy cast steel by surfacing with transition layer and anti-wear layer *Hot Working Technology* vol523 pp 881-92
[10] M K Imran, SH Masood, M Brandt, S Bhattacharya and J Mazumder 2011 Direct metal deposition (dmd) of h13 tool steel on copper alloy substrate: evaluation of mechanical properties *Materials Science & Engineering A* vol 528 pp 3342-9
[11] H Deng H Shi and S Tsuruoka 2012 Influence of coating thickness and temperature on mechanical properties of steel deposited with co-based alloy hardfacing coating. *Surface & Coatings Technology* vol 35 pp 63-70
[12] MK Imran, SH Masood, M Brandt, S Bhattacharya, S Gulizia, M Jahedid and J Mazumderc 2012 Thermal fatigue behavior of direct metal deposited h13 tool steel coating on copper alloy substrate Surface & Coatings Technology vol 206 pp 2572-80

# Heaviside-Based Morphological Filters for Topology Optimization

**Longfei Qie[1], Shikai Jing[1, 2] and Ruichao Lian[1]**

[1] School of Mechanical Engineering, Beijing Institute of Technology, 5 South Zhongguancun Street, Haidian District, Beijing, China
[2] Corresponding author
Email: corauthor@163.com

**Abstract.** To ensure manufacturability and mesh independence in density-based topology optimization schemes, it is imperative to use restriction methods. This paper proposes a family of Heaviside-based morphological filters based on the morphology-based restriction schemes, namely open, close, open-close and close-open filters. In the proposed morphological filters, Heaviside projection filter plays the role of dilation filter and modified Heaviside projection filter acts as the erosion filter. They work as principle operators and then applied sequentially to form the close, open, close-open, and open-close filters. The proposed morphological filters are tested by the compliance minimization problem. Test results show that the proposed four filters are volume preserving, and the characteristics of each filter are discussed.

## 1. Introduction

Topology optimization is a design tool for identifying optimal distributions of materials within a domain [1-3]. The fundamental concept of the density-based topology optimization method is to represent the geometry as a digital image where the colour of each pixel corresponds to the value of density [4]. It is by now well-known that the use of restriction methods in density-based topology optimization schemes is critical because they facilitate checkerboard free and mesh independence [5].

Density filtering method [6] is one of the commonly used restriction method in density-based topology optimization schemes. It can be used either on its own or in conjunction with other filters to eliminate the checkerboard patterns and ensure mesh independence. The morphology-based filters proposed by Sigmund [7] can be regarded as extensions of the density filtering method. They include a family of regularization schemes that are based on image morphology operators, namely dilation filter, erosion filter, open filter, close filter, open-close filter and close-open filter. The dilation and erosion filters work as the principle operators. The close filter is defined as a dilation followed by an erosion, the open filter as an erosion followed by a dilation, the close-open filter as opening followed by closing, and the open-close filter as closing followed by opening [8].

These morphological filters have been spread to solve other topology optimization problems since its introduction in the seminal paper by Sigmund [7], including robust design [9] and minimum length scale control [10], etc. Based on the principles of morphology-based filters, a series of Heaviside-based morphological filters are proposed in this paper. In the proposed filters, the Heaviside projection filter [11] plays the role of dilation filter, and the modified Heaviside projection filter [7] works as the erosion filter. The other filters are obtained by the embedded combinations of these principle filters. The next section will introduce the construction procedure of the proposed Heaviside-based filters.

## 2. Construction of Heaviside-Based Morphological Filters

### *2.1. Heaviside Projection Filter*

Heaviside projection filter uses an independent design variable field that is associated with a single-phase. When a design variable magnitude is greater than zero, the associated phase is projected from its location onto nearby element space. The Heaviside step function is defined as follows:

$$\bar{x}_e = H(\tilde{x}_e) = \begin{Bmatrix} 1 \text{ if } \tilde{x}_e > 0, \\ 0 \text{ if } \tilde{x}_e = 0. \end{Bmatrix} \tag{1}$$

where $\tilde{x}_e$ represent the densities obtained from the density filter, which is defined as:

$$\tilde{x}_e = \frac{1}{\sum_{i \in N_e} H_{ei}} \sum_{i \in N_e} H_{ei} x_i \tag{2}$$

where $x_i$ represent the design variable densities, $N_e$ is the neighbourhood of element $e$, which is a set of elements $i$ for which the centre-to-centre distance $\Delta(e,i)$ to element $e$ is smaller than the filter radius $r_{\min}$, $H_{ei}$ is a weight factor defined as:

$$H_{ei} = r_{\min} - \Delta(e,i) \tag{3}$$

In order to allow for the use of a gradient-based optimization scheme, the Heaviside step function is replaced with the following smooth function:

$$\bar{x}_e = 1 - e^{-\beta \tilde{x}_e} + \tilde{x}_e e^{-\beta} \tag{4}$$

where the parameter $\beta$ controls the smoothness of the approximation: for $\beta$ equals to zero, the Heaviside filter is identical to the original density filter; for $\beta$ approaching infinity, the approximation approaches the Heaviside step function.

### *2.2. Modified Heaviside Projection Filter*

By the definition of Heaviside projection filter, any design variable with a value greater than zero will be projected to one. So the Heaviside projection filter is a max filter, that is, the volume of filtered densities is larger than the volume of original densities. While in the modified Heaviside projection filter [7], the volume of filtered densities is smaller than the volume of original densities, i.e. the modified Heaviside projection filter is a min filter. Therefore, in the proposed Heaviside-based morphological filters, the Heaviside projection filter plays the role of dilation filter and the modified Heaviside projection filter acts as the erosion filter. The modified Heaviside projection filter is defined as [7]:

$$\hat{x}_e = e^{-\beta(1-\tilde{x}_e)} - (1 - \tilde{x}_e) e^{-\beta} \tag{5}$$

### *2.3. Close-Open and Open-Close Filter*

Based on the principles of morphology-based filter, the max and min filters are applied sequentially to form the close, open, close-open, and open-close filters. The close filter is defined as a max followed by a min, the open filter as a min followed by a max, the close-open filter as opening followed by closing, and the open-close filter as closing followed by opening [7-8].

As an example, the open-close filter is discussed. The filtered densities obtained with the open-close filter are denoted by $\bar{\bar{\bar{x}}}_e$. They are computed in four steps:

$$\bar{x}_e = 1 - e^{-\beta \tilde{x}_e} + \tilde{x}_e e^{-\beta} \tag{6}$$

$$\bar{\bar{x}}_e = e^{-\beta(1-\bar{x}_e)} - (1 - \bar{x}_e) e^{-\beta} \tag{7}$$

$$\hat{\tilde{\bar{x}}}_e = e^{-\beta(1-\tilde{\bar{x}}_e)} - (1-\tilde{\bar{x}}_e)e^{-\beta} \tag{8}$$

$$\bar{\hat{\tilde{\bar{x}}}}_e = 1 - e^{-\beta\hat{\tilde{\bar{x}}}_e} + \hat{\tilde{\bar{x}}}_e e^{-\beta} \tag{9}$$

The sensitivity $\frac{\partial f}{\partial x_k}$ of a function $f(\bar{\hat{\tilde{\bar{x}}}}_e)$ with respect to the $i$-th design variable $x_i$ is obtained by applying the chain rule:

$$\frac{\partial f}{\partial x_i} = \sum_e \sum_p \sum_q \sum_r \sum_s \frac{\partial f}{\partial \bar{\hat{\tilde{\bar{x}}}}_s} \frac{\partial \bar{\hat{\tilde{\bar{x}}}}_s}{\partial \hat{\tilde{\bar{x}}}_r} \frac{\partial \hat{\tilde{\bar{x}}}_r}{\partial \tilde{\bar{x}}_q} \frac{\partial \tilde{\bar{x}}_q}{\partial \bar{x}_p} \frac{\partial \bar{x}_p}{\partial \tilde{x}_e} \frac{\partial \tilde{x}_e}{\partial x_i} \tag{10}$$

## 3. Case study

### *3.1. Problem Formulation*

The classic compliance minimization problem is investigated in this section. The MBB beam is employed as research object, and the settings are shown in figure 1. The load is applied vertically in the upper left corner and there is symmetric boundary condition along the left edge $\Gamma_D$, the structure is supported horizontally in the lower right corner $\Gamma_E$.

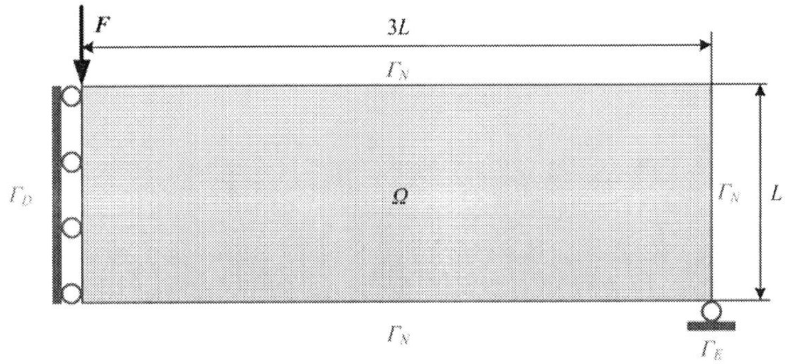


**Figure 1.** Settings of MBB beam.

The optimization problem for the MBB beam with the SIMP scheme may be written as:

$$\begin{aligned} &\min_{\rho} : f(\boldsymbol{x}) = \mathbf{U}^T\mathbf{K}\mathbf{U} = \sum_{e=1}^{N} \mathbf{u}_e^T \mathbf{k}_e \mathbf{u}_e \\ &s.t.: \mathbf{K}\mathbf{U} = \mathbf{F} \\ &\quad : g = V(\boldsymbol{x})/V^* - 1 = \sum_{e=1}^{N} v_e x_e / V^* - 1 \le 0 \\ &\quad : \mathbf{0} \le \boldsymbol{x} \le \mathbf{1} \end{aligned} \tag{11}$$

where $f$ is the compliance, $\mathbf{K}$, $\mathbf{U}$ and $\mathbf{F}$ are the global stiffness matrix, displacement vector and force vector, respectively, $\mathbf{u}_e$ is the element displacement vector, $\mathbf{k}_e = x_e^p \mathbf{k}_0$, is the element stiffness matrix, $\mathbf{k}_0$ is the element stiffness matrix for unit Young's modulus, $p$ is the penalty factor, $N$ is the number of elements used to discretize the design domain $\Omega$, $V$ is the material volume and $V^*$ is the material resource constraint, $v_e$ is the volume of element $e$.

### *3.2. Test Examples*

For filters that apply continuation schemes to update filter parameters, volume preserving is critical because this property helps to avoid oscillations in optimization iterations when using the continuation scheme, thereby improving the stability and convergence of the optimization process [12].

As mentioned above, the Heaviside projection filter is a max filter and the modified Heaviside projection filter is a min filter, thus both of them are not volume preserving. While the embedded filter, i.e. the open, close, open-close and close-open filter are volume preserving. Figure 2 displays the optimization results of MBB beam. Parameter settings are as follow: the design domain $\Omega$ is discretized with 120 by 40 bi-linear quadrilaterals, the material resource constraint is $V^* = 0.5$, the penalty power is $p = 3$, and the filter radius is $r_{min} = 3.5$.

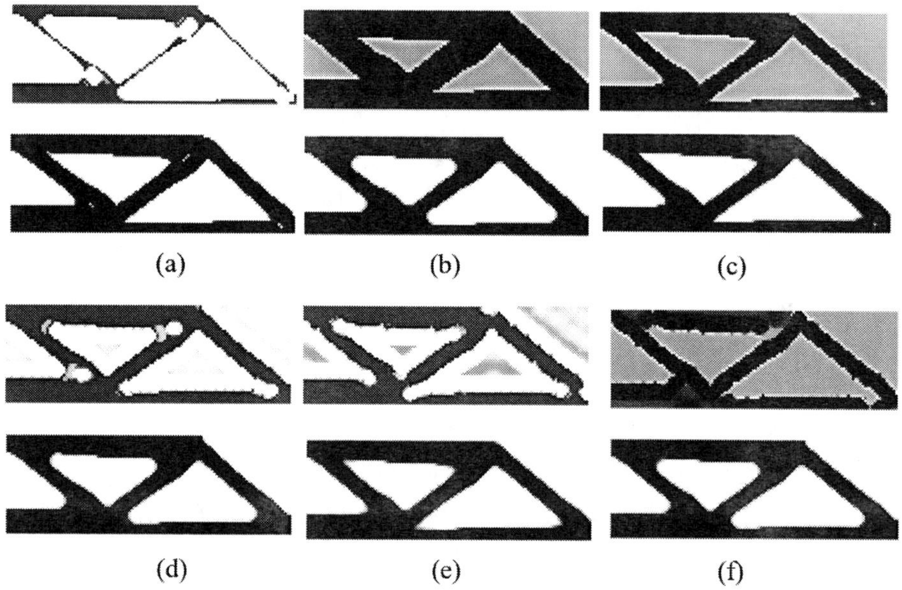


**Figure 2.** Results for the MBB test example with Heaviside-based morphological filters. For each filter, two images are shown. The upper one shows the original design variable field and the lower one shows the filtered density field. (a) Heaviside projection filter, (b) modified Heaviside projection filter, (c) open filter, (d) close filter, (e) open-close filter and (f) close-open filter.

Figure 2(a) and (b) display the optimization results of the Heaviside projection filter and modified Heaviside projection filter, respectively. As seen from the figures, the filtered volumes are different from the original density volumes, these two filters are not volume preserving. The remaining four filters, open filter, close filter, open-close filter and close-open filter (shown in figure 2(c) to figure 2(f)), however, are volume preserving. This is the advantage of the morphological filters.

In the Heaviside projection filter, some small-scale holes appear in the result, which may make the end design unmanufacturable, and the topologies exhibit sharp corners that may cause stress concentrations. In the modified Heaviside projection filter, the topologies exhibit round corners and no small-sized holes appear in the optimization result.

The open filter exhibits the characteristics of Heaviside projection filter, that is, the topologies formed with sharp corners, and a small hole appear in the lower right corner of the structure. The close filter keeps the characteristics of modified Heaviside projection filter. In both the open-close and close-open filters, no small-scale holes appear in the final topologies. The open-close filter exhibits the characteristics of open filter and the close-open filter exhibits the characteristics of close filter.

## 4. Conclusion

Based on the principles of morphology-based restriction schemes, this paper proposes a series of Heaviside-based morphological filters. In the proposed morphological filters, the Heaviside and modified Heaviside filter work as the principle operators. The open, close, open-close and close-open filters are obtained by applying the two principle operators sequentially.

Test results show that the open, close, open-close and close-open filter are volume preserving. This property facilitates stability and convergence as it helps to avoid oscillations in optimization iterations when using continuation scheme. The analysis indicates that the open filter exhibits the characteristics of Heaviside projection filter, and the close filter keeps the characteristics of modified Heaviside projection filter. The open-close filter exhibits the characteristics of open filter and the close-open filter exhibits the characteristics of close filter.

**Acknowledgment**
The authors gratefully acknowledge the fund support from National Key Research and Development Program of China [grant numbers 2017YFB1102804]; National High-tech R&D Program (863Program) [grant numbers 2015AA042505]. We thank the anonymous reviewers for their valuable comments.

**Reference**

[1] Rozvany G I N. A critical review of established methods of structural topology optimization. *Structural and Multidisciplinary Optimization*, 2009, 37(3):217-237.
[2] Liu J, Ma Y. A survey of manufacturing oriented topology optimization methods. *Advances in Engineering Software*, 2016, 100: 161-175.
[3] Liu J, Gaynor A T, Chen S, et al. Current and future trends in topology optimization for additive manufacturing. *Structural and Multidisciplinary Optimization*, 2018, 57(6):2457-2483.
[4] Gersborg A R, Andreasen C S. An explicit parameterization for casting constraints in gradient driven topology optimization. *Structural and Multidisciplinary Optimization*, 2011, 44(6): 875-881.
[5] Sigmund O, Maute K. Topology optimization approaches. *Structural and Multidisciplinary Optimization*, 2013, 48(6): 1031-1055.
[6] Bruns T E, Tortorelli D A. Topology optimization of non-linear elastic structures and compliant mechanisms. *Computer Methods in Applied Mechanics and Engineering*, 2001, 190(26):3443-3459.
[7] Sigmund O. Morphology-based black and white filters for topology optimization. *Structural and Multidisciplinary Optimization*, 2007, 33(4-5):401-424.
[8] Schevenels M, Sigmund O. On the implementation and effectiveness of morphological close-open and open-close filters for topology optimization. *Structural and Multidisciplinary Optimization*, 2016, 54(1): 15-21.
[9] Sigmund O. Manufacturing tolerant topology optimization. *Acta Mechanica Sinica*, 2009, 25(2):227-239.
[10] Hägg L, Wadbro E. On minimum length scale control in density based topology optimization. *Structural and Multidisciplinary Optimization*, 2018:1-18.
[11] Guest J K, Prévost J H, Belytschko T. Achieving minimum length scale in topology optimization using nodal design variables and projection functions. *International journal for numerical methods in engineering*, 2004, 61(2): 238-254.
[12] Li L, Khandelwal K. Volume preserving projection filters and continuation methods in topology optimization. *Engineering Structures*, 2015, 85: 144-61.

# Research on High Temperature Friction and Wear Characteristics of H13-$Cr_2C_3$-$CaF_2$ Self-lubricating Materials

**Runyin Rao[1], Huajun Wang[1*], Menglu Li[1], Zhenhua Yao[1], Qingyang Liu[1] and Songshan Yan[2]**

[1] School of Materials Science and Engineering, Wuhan University of Technology, Wuhan 430070 China

[2] School of Mechanical and Electronic Engineering, Wuhan University of Technology, Wuhan 430070 China

Email of the corresponding author: * wanghuajunhb@163.com

**Abstract.** Based on the service environment and requirements of hot working die,we proposed a H13-$Cr_2C_3$-$CaF_2$ self-lubricating material formula system and powder metallurgy process parameters, and prepared the high temperature composite materials. The phase composition and distribution of H13-$Cr_2C_3$-$CaF_2$ samples were analyzed by energy dispersive spectrometer. The friction coefficient, wear rate, wear morphology and spectrum of samples with different contents of $CaF_2$ were tested and compared by ball-on-disc high temperature tribometer under the conditions of test temperature 600 °C, disk rotation speed 200 r/min and friction for 15 min,and the mechanism of high temperature friction and wear was described. It will support the green manufacturing and sustainable development of forging industry.

## 1. Introduction

Die is an important process equipment in the parts forming manufacturing industry, which directly affects the product's qualification rate and production cost. Friction and wear are the key factors affecting the life of the die, and lubrication is the most effective measure to reduce friction and improve die life [1]. Lubricating oil or grease is usually used to reduce friction and wear. The working temperature of lubricating oil and grease is generally lower than 180 °C, and the working temperature of high-temperature grease is not more than 300 °C [2].

The solid lubricant has good stability and lubrication property under high temperature and pressure [3]. Based on the low temperature utilization of liquid lubricants, researchers have proposed a solution for solid lubricants [4]. Self-lubricating materials are one of the important methods of solid lubrication. Solid self-lubricating materials have been applied in aerospace industry in the United States [5], [6]. In order to protect the natural environment and sustainable social development, the preparation and application of self-lubricating hot work die materials are proposed [7]. Using powder metallurgy technology, the pure metal or alloy is used as the matrix phase and solid lubricating material is added to make the compound antifriction material, so as to achieve the purpose of reducing the friction of the lubrication under the condition of high temperature and high pressure [8].

In this paper, H13 steel metal powder is used as the material matrix, $CaF_2$ is used as the lubricating phase, and $Cr_2C_3$ is used as the hard phase to prepare the high temperature self-lubricating material. As a commonly used hot work die material, H13 steel has good comprehensive mechanical properties [9]. $CaF_2$ has good thermal stability and can lubricate at 900 °C [10]. $Cr_2C_3$ has strong oxidation resistance and wear resistance [11].

The friction coefficient, wear rate, wear surface morphology and surface composition of H13-$Cr_2C_3$-$CaF_2$ materials were tested. The wear mechanism of H13-$Cr_2C_3$-$CaF_2$ composites was analyzed. The effect of $CaF_2$ on the high temperature wear properties of H13-$Cr_2C_3$-$CaF_2$ was studied.

## 2. Experimental Materials and Methods

The test materials were H13 steel spheroidized powder, $CaF_2$ solid lubricant, $Cr_2C_3$ ceramic powder, zinc stearate, and the particle size was -200+500 mesh. The content of $Cr_2C_3$ is 10wt%, and the addition amount of $CaF_2$ is 0%, 5wt%, 7.5wt%, 10wt%, 12.5wt%, and content of zinc stearate is 2wt% of the total mass of the powder. The powder was dry-mixed by ball milling in a pure argon atmosphere. After mixing for 6-8 hours, it was sieved through a 200-mesh sieve, and 22 g of the powder was weighed and compressed at 600 MPa for 1 min. After demolding, the sample was sintered at 1250 ° C for 2h.

The high-temperature wear characteristics of the composite were tested by a ball-on-disc tester for high temperature friction and wear. The selected abrasive material was 45 steel with a diameter of 3.5 mm, friction and wear environment temperature was 600 °C, the friction time was 15 min, and the disc speed was 200r/min and the friction radius is 2mm. The friction coefficient, wear rate, wear surface morphology and surface composition of H13-$Cr_2C_3$-$CaF_2$ materials were tested.

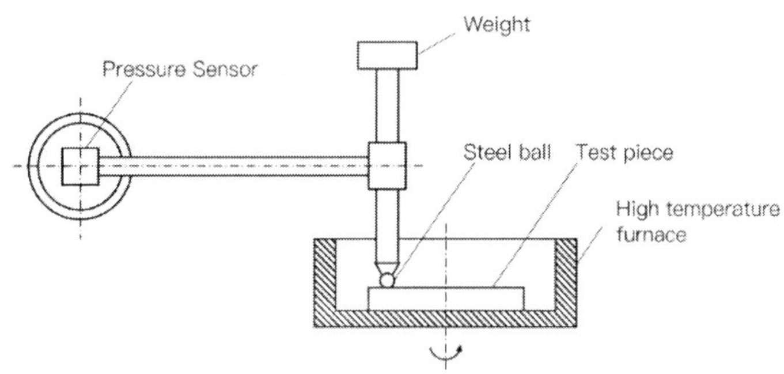


**Figure 1.** Ball-on-disc tribometer for high temperature friction and wear

## 3. Results and Discussions

### *3.1. Composition Distribution*

Figure 2 is micrograph of a sintered sample adding 7.5wt% $CaF_2$. By analyzing the EDS data by the dot analysis, it can be judged that the large-area circular shape is the H13 matrix, and a small amount of $Cr_2C_3$ and $CaF_2$ exist in the area, more $Cr_2C_3$ and $CaF_2$ exist in the pores between round particles

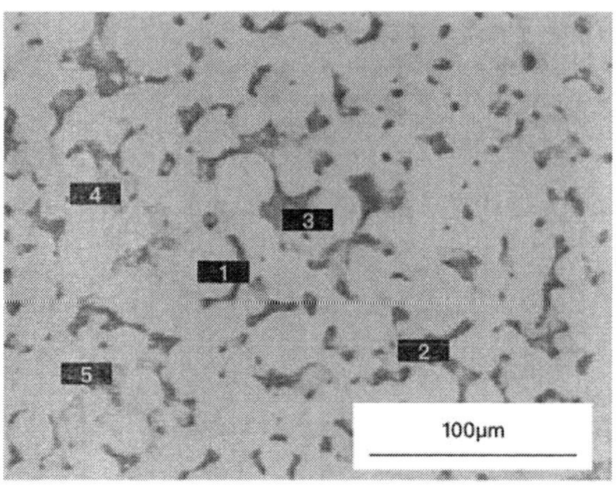


**Figure 2.** Micrograph of sample adding 7.5wt% $CaF_2$

### 3.2. *Coefficient Of friction*

Figure 3 is a graph of time versus friction coefficient for samples with different $CaF_2$ contents. In Figure 3, the average friction coefficient of H13-$Cr_2C_3$ composite material is 0.5479, which is the highest in the samples. The friction coefficient increases with time. For samples adding 5wt% and 7wt% $CaF_2$, the average friction coefficients were 0.4411 and 0.4194, respectively. The friction coefficient is low in the early stage of the experiment, the medium-term coefficient fluctuates greatly, and it remains stable after a certain increase in the later stage. The composite adding 10wt% $CaF_2$ has the lowest average friction coefficient of 0.3907 in the samples, which is relatively stable throughout the experiment. The friction coefficient of the composite adding 12.5wt% $CaF_2$ is low in the early stage, and it rises steadily in the later stage, and the average friction coefficient is 0.4088.

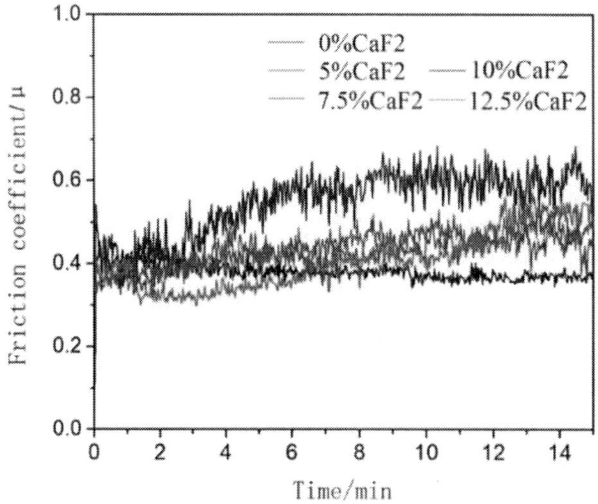


**Figure 3.** Friction coefficient curves of samples with different $CaF_2$ contents

At the beginning of the test, the contact area of the friction pair is small, and the contact point has a stress concentration phenomenon, convex portion of surface is deformed, and the friction coefficient increases rapidly. When the content of $CaF_2$ is low, the precipitation of the lubricant during the high-temperature friction is insufficient to form a complete lubricating film, and the abrasive grains generated by the friction increase the friction coefficient of the material. With $CaF_2$ contents increasing, sample precipitates more lubricant during the friction process, forming a continuous lubricating film on the friction surface, which reduces the friction coefficient and wear rate. When the content of $CaF_2$ exceeds a certain range, the bonding force of the sintered sample matrix and the hardness is reduced, the hard phase in the composite material is stripped, the lubricating film is destroyed and the friction coefficient is increased.

### 3.3. *Sample Wear Rate*

Figure 4 shows the variation of the wear rate of the samples with different $CaF_2$ contents. The H13-$Cr_2C_3$ composite material with no lubricating phase has a high friction coefficient. The wear rate of the H13-$Cr_2C_3$-$CaF_2$ composite decreased first and then increased with the increase of $CaF_2$ content. As the content of CaF2 increases, the area of the lubricating film increases, and the friction coefficient of the sample decreases, so that the wear rate decreases. When the content of $CaF_2$ exceeds 10wt%, the compactness of the sample is decreased and the pores are increased, the crack generated by the friction is rapidly expanded, and the hard phase and the lubricating phase of the sample are peeled off and rubbed against the friction surface, so that the wear rate increases again.

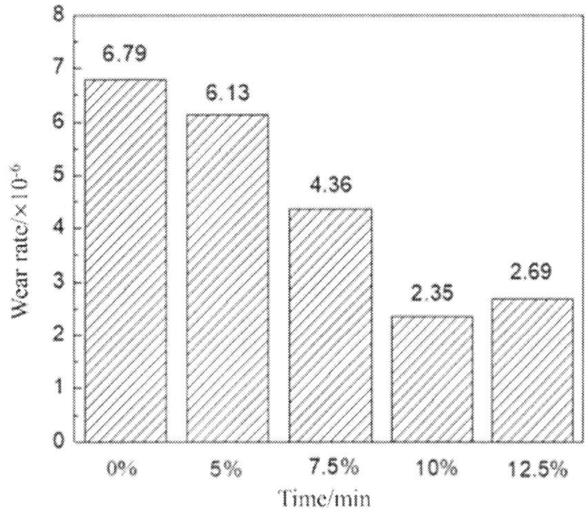


**Figure 4.** Wear rate histogram of samples with different $CaF_2$ contents

### *3.4. Wear Mechanism*

Figure 5 is topography of the high temperature friction and wear electron probe of the sample with different $CaF_2$ contents. Figure 5a is 200-fold backscattered electron image of the high-temperature friction and wear profile of a sample containing no $CaF_2$, Figures 5b~5e are high temperature friction and wear morphology of samples with $CaF_2$ addition of 5wt%, 7.5wt%, 10wt%, and 12.5wt%, respectively. Due to adhesive wear, the material on the wear area is partially peeled off, and a large amount of abrasive grains are produced, and a bright white peeling area appears (In Figure 5a). The main wear form of H13-$Cr_2C_3$ composites is oxidative wear, followed by abrasive wear and adhesive wear. H13-$Cr_2C_3$ composite material will promote the formation of a large number of oxides under the high temperature environment of 600 °C. Some of these oxides form an oxide film during the rubbing process, which protects the grinding surface; some oxides form hard abrasive grains, so the oxide film and the matrix material on the surface are plastically deformed by the abrasive grains and then fall off [12].

In Figure 5b~5e, there is ploughing phenomenon in the wear area. The main wear forms of the composite are slight oxidative wear and abrasive wear. Under the environment of high temperature and frictional stress, the lubricating phase $CaF_2$ with lower hardness is separated from the composite material and precipitates out of the friction surface. When the content of the lubricating phase $CaF_2$ is increased, the hardness of the material decreases, and the hard phase $Cr_2C_3$ is worn out to the surface. In Figure 5b~5c, the lubricating phase $CaF_2$ content is low, the lubricating film formed on the friction surface is discontinuous. When the lubricating phase $CaF_2$ content is increased, a relatively continuous lubricating film can be formed, and the friction coefficient can be further reduced. When the content of $CaF_2$ exceeds a certain range, it will decrease the bonding force of the matrix particles, which leads to the decrease of mechanical properties and bearing capacity of the composite material [13].

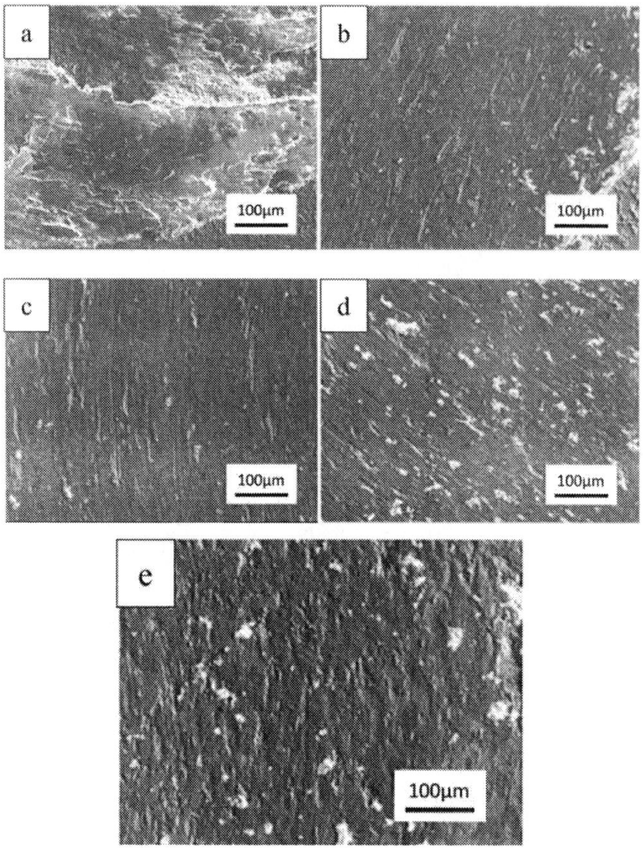


**Figure 5.** The surface morphology of samples with (a) 0% (b) 5wt% (c) 7.5wt% (d) 10wt% (e) 12.5wt% $CaF_2$ content after high temperature friction

## 4. Conclusions

(1) This paper proposed H13-$CaF_2$-$Cr_2C_3$ self-lubricating material formula system and preparation process.

(2) In the microstructure of H13-$Cr_2C_3$-$CaF_2$ self-lubricating material, there is mainly H13 matrix and a small amount of $Cr_2C_3$ and $CaF_2$ in the large-area circular shape. In the pores between circular shape particles, there are $Cr_2C_3$ and $CaF_2$.

(3) The friction coefficient and wear rate decrease first and then increase with the increase of $CaF_2$ content. When the CaF2 content is 10wt%, friction coefficient and wears rate reach the minimum value.

(4) In the high temperature friction and wear test, the main wear form of H13-$Cr_2C_3$ composite is oxidative wear, accompanied by abrasive wear and adhesive wear. The main wear forms of H13-$Cr_2C_3$-$CaF_2$ composite are slight oxidative wear and abrasive wear when load is small. When load is large, the dominant wear mechanism of H13-$Cr_2C_3$-$CaF_2$ composite is oxidative and adhesive wear.

## Acknowledgements

This work was financially supported by Wuhan University of Technology Graduate Outstanding Thesis Cultivation Project (2017-YS-006). The authors gratefully acknowledge Ms. Jiaqi Li for her assistance in experiment.

## References

[1] Tung S C and Mcmillan M L 2004 Automotive tribology overview of current advances and challenges for the future Tribology International vol 37(7) pp 517-36.

[2] Xue Q J and Lv J J 1999 Research status and developing trend of solid lubrication at high

temperatures Tribology vol 19(1) pp 91-6.
[3] Liu W M and Guo Z G 2014 Colorful friction, wear and lubricants Chinese Journal of Nature vol 36(04) pp 241-7.
[4] Wang C C, Wang R C, Peng C Q, Feng Y and Wei X F 2012 Research progress of metallic solid self-lubricating composites The Chinese Journal of Nonferrous Metals vol 22(07) pp 1945-55.
[5] Sun R L, Sun S W, Guo L X and Yang D Z 1999 Application of solid lubrication technique for space mechanism Aerospace Materials & Technology vol 29(1) pp 17-22.
[6] Cai Z Y, Li M Z and Feng Z H 2001 Theory and method of optimum path forming for sheet metal Chinese Journal of Aeronautic vol 14(2) pp 118-22.
[7] Wang H J, Li M L, Yan S S, Zhou C Y, Yao Z H and Gan K K 2017 High-temperature bionic self-lubricating hot work die material and preparation method China Patent 201710735974.8
[8] German R M, Hens K F and Johnson J L 1994 Powder metallurgy processing of thermal management materials for microelectronic applications International Journal of Powder Metallurgy vol 30(2) pp 205-15.
[9] Genel K 2006 Boriding kinetics of h13 steel Vacuum vol 80(5) pp 451-7.
[10] Øgaard B 2001 CaF2 formation: cariostatic properties and factors of enhancing the effect Caries Research vol 35(Suppl.1) pp 40-4.
[11] Sadeghimeresht E, Markocsan N, Nylén P and Björklund S 2016 Corrosion performance of bi-layer Ni/Cr2C3-NiCr HVAF thermal spray coating Applied Surface Science vol 369 pp470-81.
[12] Tyagi R, Xiong D S, Li J and Dai J 2010 Elevated temperature tribological behavior of Ni based composites containing nano-silver and hBN Wear vol 269(11) pp 884-890.
[13] Deng J C, Cao T K, Yang X F and Liu J H 2006 Self-lubrication of sintered ceramic tools with CaF2 additions in dry cutting International Journal of Machine Tools & Manufacture vol 46(9) pp 957-63.

IOP Conf. Series: Materials Science and Engineering 472 (2019) 012036 doi:10.1088/1757-899X/472/1/012036

# Numerical Simulation Method Study of Rock Fracture Based on Strain Energy Density Theory

**Tengfei Ma**[1], **Shuchen Li**[1, 2] **and Qian Sun**[1]

[1] Geotechnical and Structural Engineering Research Center, Shandong University, Jinan, Shandong 250061, China

[2] Corresponding author. Email: 316159025@qq.com

**Abstract.** Many numerical methods are carried out to study the nonlinear failure behaviors of the rock; however, the numerical simulation methods for the failed rock are still in the research stage. This paper establishes the damage constitutive equation by combining the bilinear strain softening constitutive model with energy dissipation principles, as well as the energy failure criterion of mesoscopic elements based on the strain energy density theory. When the strain energy stored by an element exceeds a fixed value, the element enters the damage state and the damage degree increases with increasing energy dissipation. Simultaneously, the material properties of the damaged element change until it becomes an element with certain residual strength. As the load increases, the damage degree of an element increases. When the strain energy stored by an element exceeds the established value of the energy criterion, the element is defined to be failed. As the number of failed elements constantly increases, failed elements interconnect and form macrocracks. The rock fracture calculation program on the basis of the preceding algorithm is successfully applied to the fracture simulation process in Brazilian splitting and intermediate crack tensile tests. This method completes the nonlinear calculation process with linear calculation, avoids singularity of numerical calculation in element fracture, and simulates the rock post-peak fracture behaviors. The simulation results agree well with the test results, indicating the accuracy and feasibility of this method for simulating the rock fracture process.

## 1. Introduction

As a product of geological movements, the rock is a heterogeneous material with complex mechanical properties. The rock medium usually shows strong nonlinear characteristics [1] in the deformation process. Physical and mechanical properties will be irreversible during the failure process, and this irreversible process will cause various forms of energy dissipation. According to the law of thermodynamics, energy conversion is the essential feature during the material physical process, and material failure is a kind of instability driven by energy. Therefore, studying the energy change rules and establishing the relation between energy changes and strength as well as structural failure during the rock failure process will be more conducive to reflecting strength changes and essential characteristics [2] of structural failure of the rock under the action of external load.

Since Huber firstly introduced the potential energy concept to define material damage, many scholars at home and abroad have described rock deformation behaviors through energy analysis and achieved tremendous progress [3-8]. Most work focused on tests and theoretical research and macroscopic failure behaviors of the rock were obtained through energy analysis. According to their studies, material failures are mainly caused by irreversible internal energy dissipation; meanwhile, the energy criterion is generally significant for determining the rock failure.

The rock fracture is an entire process from damage, material progressive degradation, microcrack generation, expansion, and till penetration. Therefore, studying rock fracture from the micromechanics perspective, analysing the damage rules of tiny rock elements, and studying element failure through element energy dissipation and energy criterion can systematically show the entire rock fracture process. Currently, the most representative strain energy failure criterion is distortional strain energy density theory (fourth strength theory) [9]. This is a better strength theory for plastic materials; however, it is applicable only to plastic materials with the same tension and compression properties rather than triaxial equivalent tension. Therefore, besides the distortional strain energy, the volume deformation energy also needs to be considered [10]. The application of strain energy density theory [11] can comprehensively consider the preceding problems and take the strain energy density that is the sum of the volume deformation energy density and shape change energy density as the criterion of material failure. The advantage of this theory is that it can be well applied to complex geometry, loading conditions, and development situations of mixed cracks [12].

According to the preceding analysis, nonlinear failure behaviours of the rock are considered to simulate the rock partial failure and the entire process from microcrack generation, expansion, and complete fracture. In addition, the bilinear strain softening constitutive model is adopted by referring to document [13], and the energy criteria for the damage and failure of mesoscopic rock elements according to the strain energy density theory and energy dissipation principle. When the strain energy stored by an element exceeds a fixed value, the element enters the damage status. The damage degree of the element is determined based on the strain energy density and the material properties of the damaged element change until it becomes an element with certain residual strength. For rock elements entering the damage state, the energy failure criterion of strain energy density is used to determine whether the element is damaged. As the load increases, the number of failed elements gradually grows, and the failed elements interconnect and form macrocracks, causing the structural failure of the rock specimen. During the numerical simulation process, the elastic modulus reduction of damaged elements after reaching the stress extreme value is discretized. The preceding method completes the nonlinear calculation process with linear calculation, avoids singularity of numerical calculation in element fracture, and simulates the rock post-peak fracture behaviours. In this calculation method, the rock fracture calculation program is developed with Fish language in the Flac. It is successfully applied to the fracture simulation process in Brazilian splitting and intermediate crack tensile tests, indicating the accuracy and feasibility of this method for simulating the rock fracture process.

## 2. Strain Energy Density Theory

The rock fracture mode is affected by many factors such as the loading type, geometry, and material properties. However, the strain energy density theory can comprehensively consider these factors. The rock is assumed to be a continuum composed of many tiny structural elements and each element contains per unit volume of the material. If the element deforms under the action of external force, strain energy will be stored inside the element. In this way, the energy stored by each element is called strain energy density $(dW/dV)$.

### *2.1. Determination of the Strain Energy Density Function Considering Strain Softening*

According to the strain energy density theory, the strain energy density function of each element in the isothermal condition can be expressed as:

$$\frac{dW}{dV}=\int_0^{\varepsilon_{ij}}\sigma_{ij}d\varepsilon_{ij} \tag{1}$$

According to the preceding formula, the density of strain energy stored in an element is determined by its stress $\sigma_{ij}$ and the deformation history of strain increment $d\varepsilon_{ij}$. This theory determines the yield failure of the material element based on strain energy density $(dW/dV)$ and stipulates that the element fails when all strain energy absorbed by the element is equal to the energy absorbed during fracture under uniaxial tension $(dW/dV)_c$.

According to this theory, before the rock texture shows global instability under the action of load, partial failure and crack propagation already occur and seriously affect the macroscopic failure behaviors of the rock texture. The steady propagation of microcracks inside the rock will finally cause rock macroscopic fracture.

The deformation process of each rock element is accompanied with energy dissipation that will cause material progressive damage, property deterioration, and strength loss [2]. The material mechanical damage can be described by strain softening. The bilinear strain softening constitutive model of the rock under uniaxial tension in document [13] is considered, that is, the uniaxial tension stress-strain relation of the rock is simplified, as shown in Figure.1.

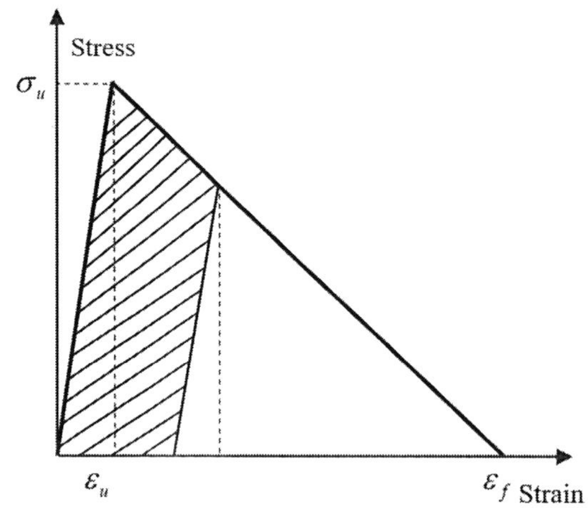


**Figure 1.** Bilinear strain softening stress-strain curve

Under the action of external force, the rock first shows elastic deformation and does not fail immediately after reaching the stress limit point U. Instead, it enters the strain softening stage and the material starts to damage. As shown in the figure, the density $(dW/dV)$ of strain energy absorbed by the material element at point A is area OUAC surrounded by the stress-strain curve. If the element is unloaded at point A, the unloading path will be along line AB and the new loading path will be along line BAF. In the process of unloading and reloading, energy represented by area OUAB is dissipated. Therefore, the density of strain energy absorbed by the material element is composed of the following two parts:

$$\frac{dW}{dV} = (\frac{dW}{dV})_d + (\frac{dW}{dV})_e \tag{2}$$

In the preceding formula, $(dW/dV)_d$ is the density of strain energy dissipated OUAB while $(dW/dV)_e$ is the density of strain energy recoverable BAC.

If point A is on line OU in the elastic stage, the unloading path will be along the loading path and the material will not damage. For the material element without damage, its critical strain energy density $(dW/dV)_c$ is equal to area OUF. However, energy dissipation occurs after the material element damages. The residual critical strain energy density after damage $(dW/dV)_c^*$ is the density of strain energy regained after the element is unloaded (that is area BAF) and can be expressed as the following:

$$(\frac{dW}{dV})_c^* = (\frac{dW}{dV})_c - (\frac{dW}{dV})_d \tag{3}$$

The preceding formula indicates that a higher density of strain energy dissipated by the material element means more serious element damage and lower density of critical strain energy that can be borne.

According to the preceding analysis, larger deformation of the material element under the action of external force means a higher density of strain energy absorbed $(dW/dV)$ and lower density of critical strain energy $(dW/dV)_c^*$. When $(dW/dV) \geq (dW/dV)_c^*$, cracks start to be generated. When $(dW/dV)$ is equal to the initial critical strain energy density $(dW/dV)_c$ (that is, area OUF) of the material element, the element is completely fractured and cannot bear any load.

The bilinear constitutive relation of the preceding rock element can be simply obtained through uniaxial tensile tests. The rock texture is usually under the action of complex external force and the internal rock elements are under the combined action of tension, compression, and shear stress. The strain energy density can be obtained based on the loading history of the rock element and it can comprehensively reflect the loading status of the element. The damage status of the rock element under complex stress conditions can be determined by comparing it with the strain energy density in different stages under uniaxial tension.

### *2.2. Energy Dissipation and Damage Constitutive Model*

According to energy opinions, after the rock shows inelastic deformation, the inelastic deformation energy the rock can bear has been significantly reduced, that is, the rock constitutive energy has decreased. This is also an expression of rock performance deterioration caused by the changes of rock microstructure [15]. After the rock element reaches the peak point of the tensile stress, it enters the strain softening stage and shows inelastic deformation and decreased material strength. The strength decrease of the material element is defined as the elastic modulus reduction and is expressed by equivalent elastic modulus $E^*$. As shown in Figure.2, as the energy loses, the unloading peak strength decreases from point U to points G, H, I ... and the equivalent elastic modulus is $E_1^*$, $E_2^*$, $E_3^*$, ...and $E_n^*$ respectively.

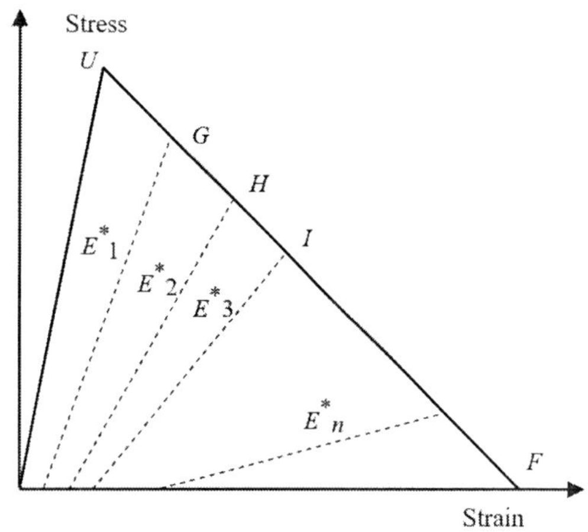


**Figure 2.** Reduction of elastic modulus after energy dissipation

For the sake of calculation, the equivalent elastic modulus is discretized into 20 different values:

$$E^*(n) = \frac{(21-n)}{20}E \text{, n=1, 2, ..., 20} \tag{4}$$

The following can be obtained according to the constitutive relation of continuous medium damage mechanics theory [16]:

$$\sigma = E\varepsilon(1-D) \tag{5}$$

E is the elastic modulus of the rock without damage.

The damage parameter $D$ of the rock element in this model is:

$$D = \frac{E - E^*(n)}{E} \tag{6}$$

n refers to the damage degree of the material element.

For the discretization of the equivalent elastic modulus, a larger value of n means more accurate simulation results but remarkably lower calculation efficiency. n is set to 20 by taking both factors into consideration, which meets related requirements.

According to the energy dissipation principle, energy dissipation is directly related to the damage and strength loss. The dissipation quantity reflects the reduction of the initial strength [2]. According to Figure.1 and formula (2), both the strain energy density and the energy dissipation increase after the rock element reaches the stress limit point. Therefore, the strain energy density is used as the criterion to determine the rock element damage and failure.

(1) When $(dW/dV) < (dW/dV)_u = 1/2(\sigma_u \varepsilon_u)$, the material element is in the elastic stage; no damage occurs; both the equivalent elastic modulus and critical strain energy density are the initial values of the element, that is, $E^* = E$, $(dW/dV)_c^* = (dW/dV)_c = 1/2(\sigma_u \varepsilon_f)$.

(2) When $(dW/dV) >= (dW/dV)_u$, the material element enters the damage stage. The discretized equivalent elastic modulus $E^*(n)$ is regarded as the elastic modulus of the material element after damage. n refers to the damage degree of the material element ( $0 \le n \le 20)$ ) and its value is determined by the strain energy density. It also can be seen from Figure.1 that the critical strain energy density $(dW/dV)_c^*$ decreases with increasing $(dW/dV)$ after the material element enters the damage stage.

(3) When $(dW/dV) >= (dW/dV)_c^*$, the element fails, marking the beginning of crack generation inside the material.

(4) When $(dW/dV) >= (dW/dV)_c = 1/2(\sigma_u \varepsilon_f)$, according to formula (4), the damage degree of the material element reaches the largest, that is, $n = 20$; both the equivalent elastic modulus and critical strain energy density of the element change to zero; the element is completely fractured and loses the bearing capacity. To maintain the integrity of the entire structure calculation model and element continuity, the element completely fractured will be given a very small residual modulus $E_c^* = 0.05E$ instead of being dug out.

### *2.3. Numerical Implementation*

The FLAC2D finite difference numerical software is adopted to establish the rock element damage equation based on the preceding theory and develop a rock fracture calculation program with Fish language. Model loading adopts the load control stepwise loading method. In particular, equivalent tiny loads are accumulatively added to the model in turn. After calculation is balanced, the strain energy density of each element is calculated. Then, the preceding strain energy density criterion is used to determine the element damage degree and failure. After the first loading step is performed, all elements are in the elastic stage due to the small load and the strain energy density of each element is:

$$\frac{dW}{dV} = \frac{1}{2E}[\sigma_x^2 + \sigma_y^2 + 2\tau_{xy}^2 - 2v(\sigma_x\sigma_y - \tau_{xy}^2)] \tag{7}$$

After loading step i is performed, the strain energy density of each element is:

$$(\frac{dW}{dV})_i = (\frac{dW}{dV})_{(i-1)} + \Delta(\frac{dW}{dV}) = (\frac{dW}{dV})_{(i-1)} + \frac{1}{2}(\sigma_x{}^i + \sigma_x{}^{i-1})(\varepsilon_x{}^i - \varepsilon_x{}^{i-1}) + \frac{1}{2}(\sigma_y{}^i + \sigma_y{}^{i-1})(\varepsilon_y{}^i - \varepsilon_y{}^{i-1}) + \frac{1}{2}(\tau_{xy}{}^i + \tau_{xy}{}^{i-1})(\gamma_{xy}{}^i - \gamma_{xy}{}^{i-1}) \quad (8)$$

In the formula, i≥2

$\sigma_x{}^i$, $\sigma_y{}^i$, and $\tau_{xy}{}^i$ are the element stress in loading step i while $\sigma_x{}^{i-1}$, $\sigma_y{}^{i-1}$, and $\tau_{xy}{}^{i-1}$ are the element stress in the last loading step. Similarly, the strain has the same expression way.

When the strain energy density of an element $(dW/dV) > (dW/dV)_u = 1/2(\sigma_u \varepsilon_u)$, the damage degree n is:

$$n = \frac{(dW/dV) - (dW/dV)_u}{\frac{1}{20}[\,(dW/dV)_c - (dW/dV)_u\,]} = 20\frac{2(dW/dV) - \sigma_u \varepsilon_u}{\sigma_u \varepsilon_f - \sigma_u \varepsilon_u} \quad (9)$$

n is set to the integer part of the calculation results on the right of the equation and ranges from 0 to 20. When $n = 0$, the element is still in the linear elastic stage and no damage occurs; when $n = 20$, the damage value is the maximum, indicating that the element is completely fractured and loses the bearing capacity. Meanwhile, the element elastic modulus $E^*$ is obtained by using formula (4).

The density of strain energy dissipated by the element is:

$$(dW/dV)_d = (dW/dV) - \frac{1}{2E^*}[\sigma_x{}^2 + \sigma_y{}^2 + 2\tau_{xy}{}^2 - 2v(\sigma_x \sigma_y - \tau_{xy}{}^2)] \quad (10)$$

Therefore, the critical strain energy density of the element is:

$$(dW/dV)_c^* = (dW/dV)_c - (dW/dV)_d = \frac{1}{2}\sigma_u \varepsilon_f - (dW/dV) + \frac{1}{2E^*}[\sigma_x{}^2 + \sigma_y{}^2 + 2\tau_{xy}{}^2 - 2v(\sigma_x \sigma_y - \tau_{xy}{}^2)] \quad (11)$$

Figure.3 shows the calculation process of the damage constitutive calculation model based on strain energy density.

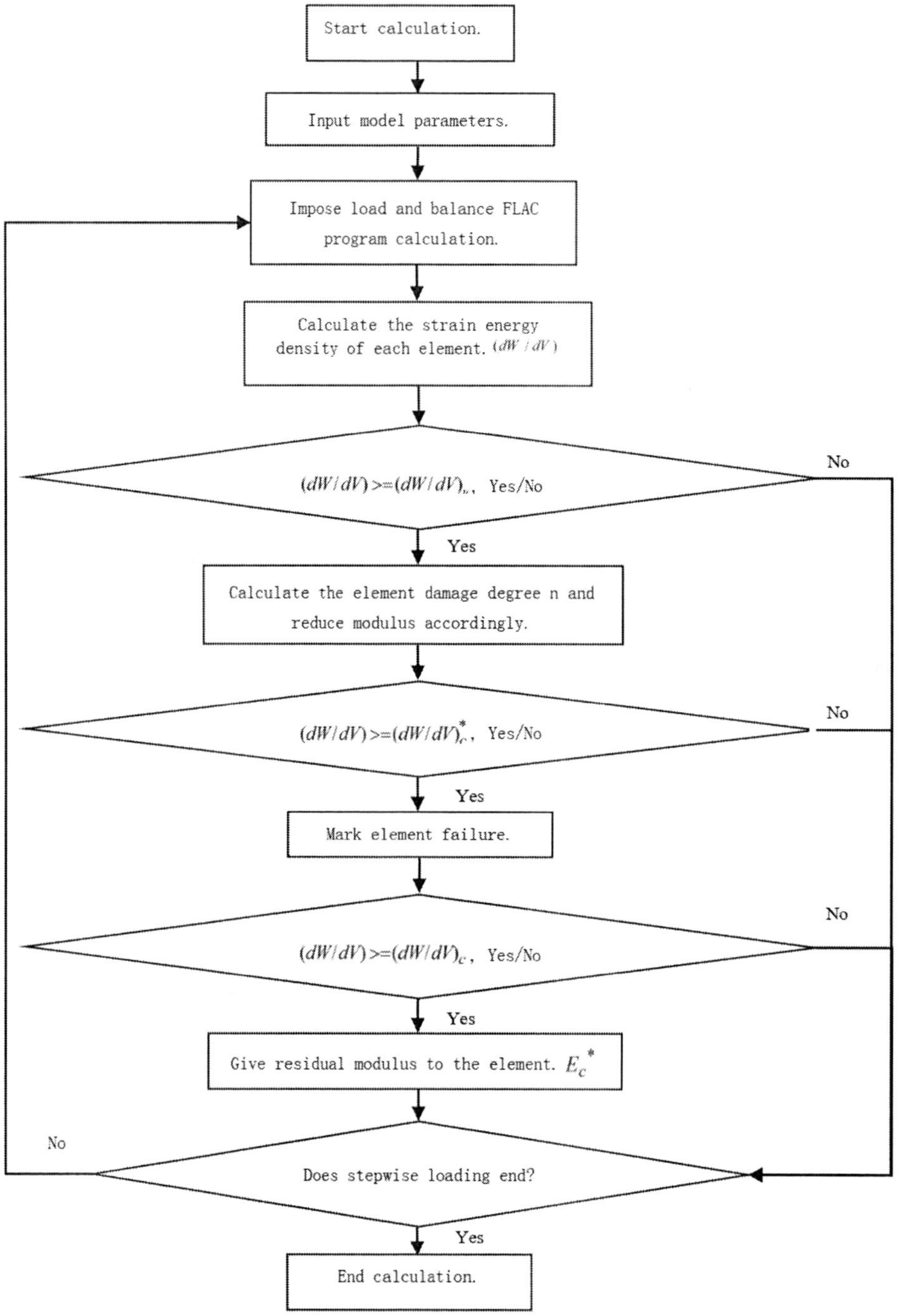


**Figure 3.** Numerical simulation flow of rock failure based on strain energy density theory

## 3. Numerical Simulation of Rock Fracture

### *3.1. Brazilian Splitting Numerical Simulation*

This case takes the Brazilian splitting with lateral concentrated load in document [17] as an example. Figure.4 shows the calculation model. The disk diameter is 50mm and it is divided into 7860 elements. The model material adopts the marble in document [17] with the elastic modulus $E = 66.8GPa$ ,

Poisson's ratio $v = 0.33$, density $\rho = 2620 Kg/m^3$, element tension ultimate stress $\sigma_u = 20MPa$, strain under ultimate stress $\varepsilon_u = 1.3 \times 10^{-5}$, and strain when fracture occurs $\varepsilon_f = 1 \times 10^{-4}$. The model top and bottom are imposed with load increased by 2KN in the negative y direction and positive y direction respectively.

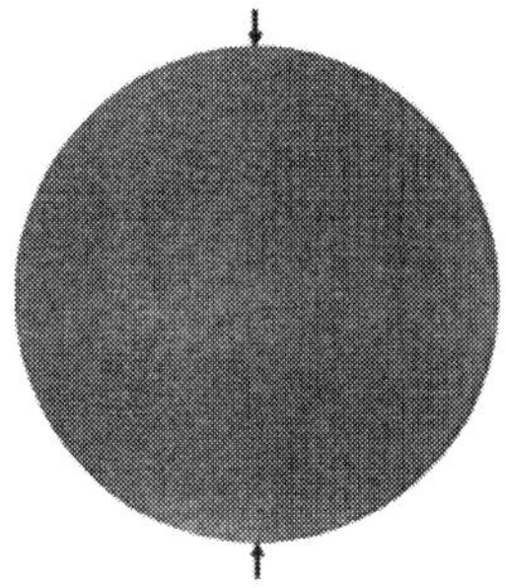

**Figure 4.** Numerical calculation model of Brazil splitting

After the first 2KN load is imposed, no damage occurs inside the disk; each element is in the linear elastic stage; Figure.5 shows the distribution of the stress inside the Brazilian disk. The results of comparison of distribution rules and theoretical results of stress distribution inside the disk in document [17] show that the simulated stress distribution rules basically agree well with the theoretical distribution.

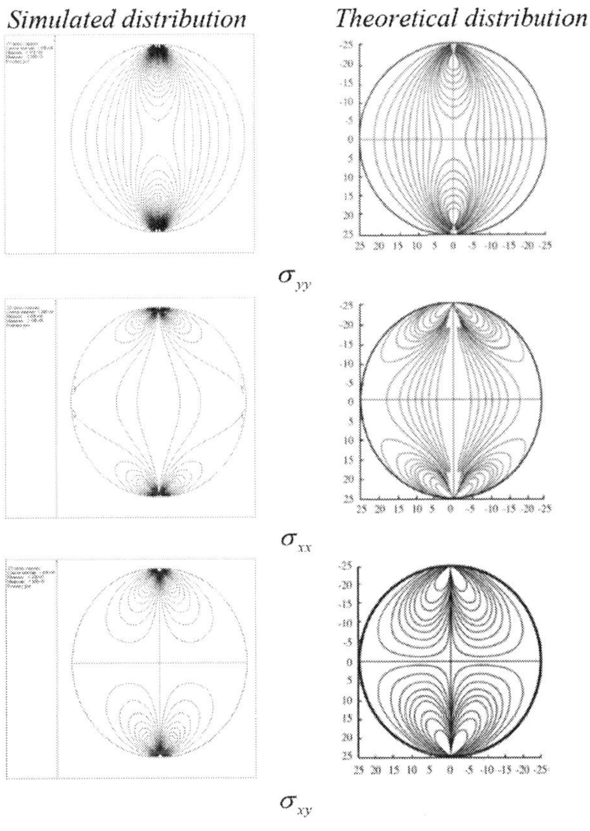


**Figure 5.** Comparison between theoretical and numerical results of stress distribution in Brazil disc

As the load increases, the strain energy absorbed by each element gradually increases. The energy change process of elements in the middle of the disk top is recorded, as shown in Figure.6. The strain energy density absorbed by the element $(dW/dV)$ gradually increases from zero while the critical strain energy density $(dW/dV)_c^*$ gradually decreases from $(dW/dV)_c = 1000$. When they two intersect, the element fails.

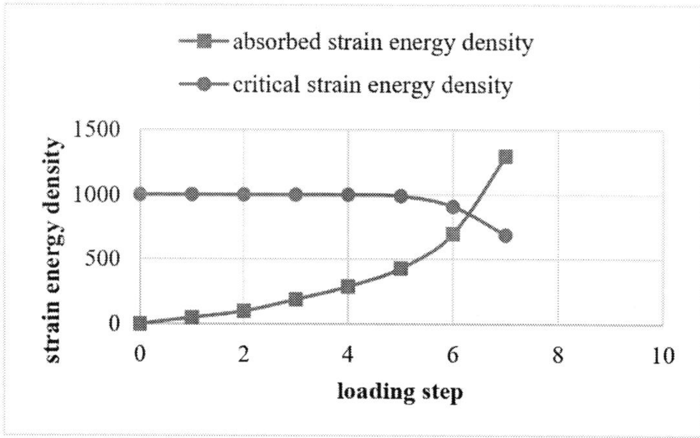


**Figure 6.** Curve of element strain energy density versus load change

It can be seen from Figure.6 that when calculation reaches the 7th loading step, that is, 14KN load is imposed, elements in the middle of the disk top have $(dW/dV) \geq (dW/dV)_c^*$, indicating that elements at both ends of the disk start to fail. When calculation reaches the 15th loading step, that is, 30KN load is imposed, partial damage at both ends is suddenly developed into complete penetration in the middle of the specimen; meanwhile, there are partial failure areas on both ends of the disk. Figure.7 shows the failure of the specimen when it is loaded to loading step 7, 12, 14, and 15.

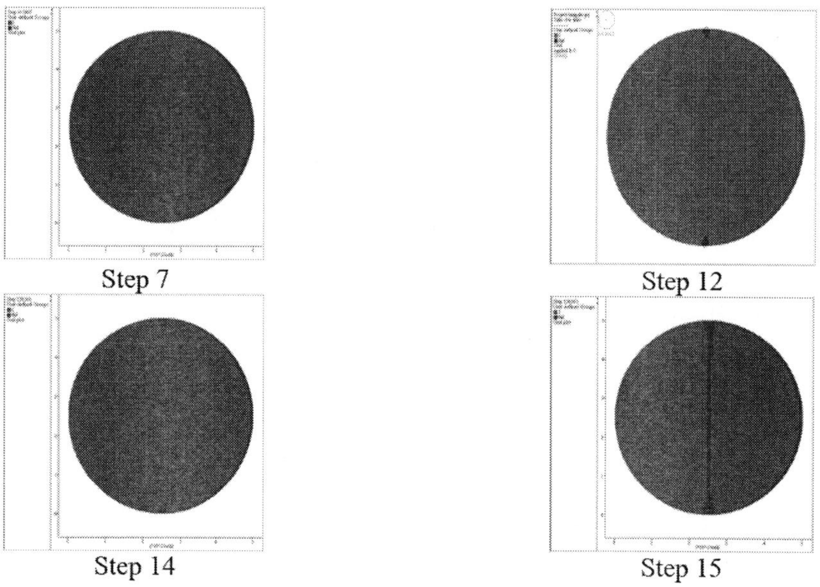


**Figure 7.** Element failure process of Brazil splitting under different loading steps

Figure.8 shows the damage status of each element when the specimen fails. The damage value n is expressed by the built-in variable ex_5 of the element and ranges from 1 to 20. It can be seen from the damage cloud chart of the element that the damage value of each element at the penetration point in the middle of the disk reaches the maximum value 20, indicating that elements in the middle of the disk are completely fractured and lose the bearing capacity. Elements on both ends of the disk except a small number of elements are completely fractured, and surrounding elements also show different levels of damage.

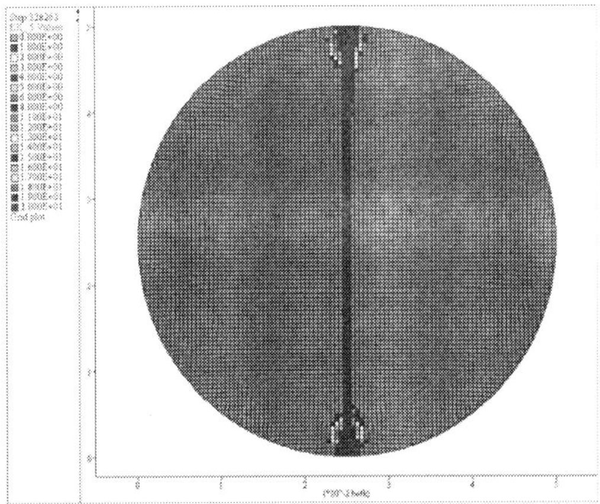

**Figure 8.** Element damage status in Brazilian disc in numerical calculation

Documents [18-20] all conducted experimental studies on Brazilian disk imposed with lateral load. The results show that both ends of the disk firstly fail due to the stress concentration caused at the loading point during the splitting process of the brittle rock. As the load continues to increase, penetration occurs in the middle of the disk and the failure pie section appears on the loading points at both ends of the disk. Figure.9 shows the sketch of splitting failure of the marble under different stepping bars in document [18]. Comparison of Figure.7 and Figure.9 shows that simulated results of numerical calculation in this paper agree well with test results in the document, indicating the feasibility of this method for simulating the rock fracture.

**Figure 9.** Test results of Brazilian splitting for marble in document [18]

*3.2. Numerical Simulation of Tensile Test with Built-in Crack*

This case takes the tensile test with built-in crack in document [21] as an example. Figure.10 shows the calculation model. The specimen size is $50mm \times 100mm$; the length of the built-in crack is $20mm$; the specimen is divided into 9772 elements. The specimen material adopts the rock mortar with the elastic modulus $E = 5.17GPa$, Poisson's ratio $\nu = 0.192$, density $\rho = 2300Kg/m^3$, element ultimate tensile stress $\sigma_u = 2.7MPa$, strain under ultimate stress $\varepsilon_u = 5.8 \times 10^{-4}$, and strain when fracture occurs $\varepsilon_f = 7 \times 10^{-4}$. The model top and bottom are imposed with uniform tensile load increased by 100N respectively.

**Figure 10.** Numerical calculation model of uniaxial tensile test with built-in crack

The following calculation results are obtained: under the action of tensile load, stress concentration occurs on the crack tip; elements at both ends of the crack start to fail when calculation reaches the fourth loading step, that is, 400N load. Figure.11 shows the strain energy density of the first element at the left end of the crack with changing load.

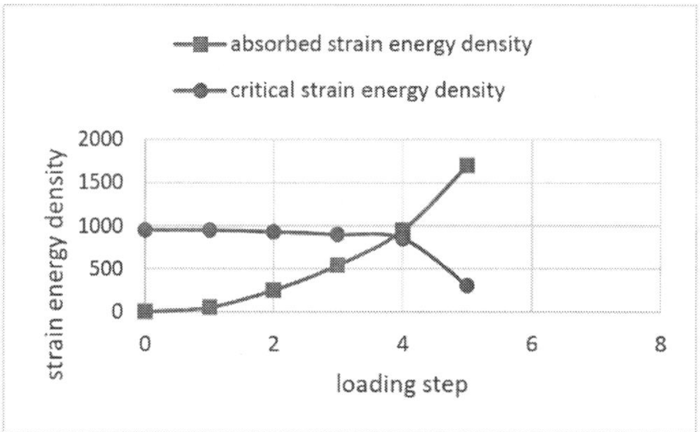


**Figure 11.** Curve of element strain energy density versus load change

When calculation reaches the fifth loading step, that is, 500N load, penetration occurs in the middle of the specimen in the crack direction. Figure.12 shows the failure status of the model when calculation reaches loading steps 4 and 5.

Step 4 Step 5

**Figure 12.** Element failure process of uniaxial tensile test under different loading steps

Figure.13 shows the damage status of each element when the specimen fails. It can be seen from the damage cloud chart of the element that each element surrounding the crack tip shows different levels of damage due to stress concentration and that the damage value of a series of elements from both ends of the crack to the model edge reaches 20, indicating that the specimen is completely fractured along the cross section of the crack.

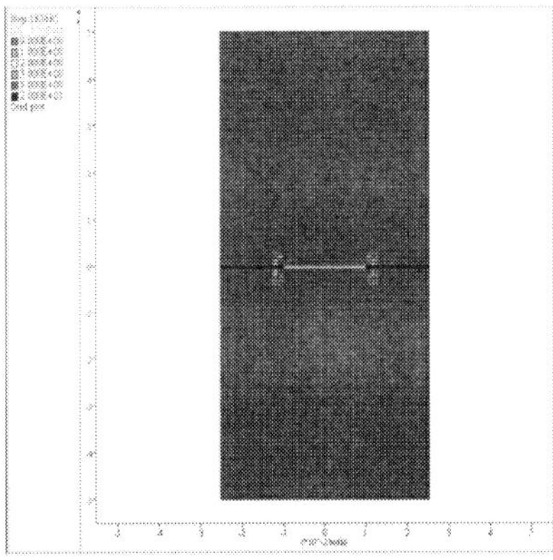

**Figure 13.** Element damage status in tensile test with built-in crack

Document [21] conducted uniaxial tensile tests on mortar materials with different prefabricated crack dips on the rigidity servo testing machine. The test results show that when the tensile stress reaches a specific value, the prefabricated crack starts to initiate and then rapidly expands, causing the structural failure of the specimen. When the crack is horizontal (that is, crack dip $\alpha = 90^\circ$), wing crack is generated at the front edge of the crack and expanded in the direction of the original crack until the specimen is fractured. Finally, the fracture trace is perpendicular to the axial tensile stress direction and the fracture surface is flat. Figure.14 shows the tensile failure of horizontal crack in document [21]. Comparison of Figure.12 and Figure.14 shows that the simulated failure process and damage status in the built-in crack tensile tests of numerical calculation in this paper basically agrees with test results, indicating the accuracy and feasibility of this method in this paper for simulating the rock fracture.

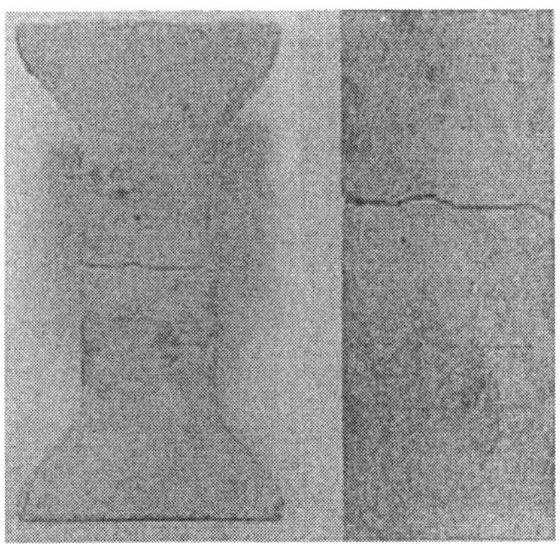

**Figure 14.** Test results of uniaxial tensile test in document [21]

## 4. Conclusions

This numerical simulation method of rock fracture based on strain energy density theory completes the nonlinear calculation process with linear calculation, avoids singularity of numerical calculation in element fracture, and simulates the rock post-peak fracture behaviors.

(1) The tensile bilinear strain softening constitutive model is adopted to analyse the energy and damage evolution process after the rock element reaches the stress limit point from the micromechanics perspective. In addition, the criterion is created to determine element failure based on strain energy density theory. This method can well simulate the entire process of rock damage.

(2) The strain energy density of the rock element is obtained by accumulating the energy absorbed by the element after the load is imposed each time. Meanwhile, the modulus reduction of the damaged element is discretized to complete nonlinear calculation with linear calculation. A very small residual modulus is given to elements (regarded to be completely fractured) with the maximum damage values. Continuous calculation is used to implement discontinuous fracture behaviours of the rock to avoid singularity of numerical calculation in element fracture.

(3) Numerical simulation of Brazilian splitting and built-in crack tensile tests shows that the preceding method can well simulate the entire process from partial failure to structural fracture of the rock and obtain the damage evolution process of each element during the failure process.

(4) Studying rock fracture from the energy perspective can comprehensively reflect the rock strength changes and essential characteristics of the structural failure. This paper develops a rock damage numerical calculation program with Fish language in the FLAC. Compared with the theory and analytic method, this program can directly resolve various complex problems and therefore has a wider application prospect.

## Acknowledgments

The authors would like to thank the support of the National Basic Research Program of China (Grant No. 2010CB732002), the National Natural Science Foundation of China (Grant No. 51179098、51379113).

## References

[1] Cai M 2002 *Rock Mechanics and Engineering* (Beijing: Science Press)
[2] Xie H P, Ju Y and Li L 2005 *Chinese J of Rock Mech. and Eng.* **24** 3003-10
[3] Mikhalyuk A V, Zakharov V V 1997 *Appl. Mech. and Tech. Phys.* **38(2)** 312-18
[4] Steffler E D, Epstein J S and Conley E G 2003 *Int. J. Fracture* **120** 563-80
[5] You M Q, Hua A Z 2012 *Chinese J. of Rock Mech. and Eng.* **21(6)** 778-81
[6] Xie H P, Peng R D and Ju Y 2014 *Chinese J. of Rock Mech. and Eng.* **23(21)** 3565–70
[7] Yang S Q, Xu W Y and Su C D 2006 *Acta Mech. Solida Sinica* **27(2)** 213-16
[8] Carloni and C Nobile L 2002 *Theoretical and Appl. Fracture Mech.* **(38)** 109-19
[9] Liu H W 1999 *Mechanics of Materials (I)* (Beijing: High Education Press)
[10] Yu X Y 2012 *J. of Guangxi University(Natural Science)* **27** 63-9
[11] Li Q M 2011 *Int. J. of Solids and Structures* **(38)** 6997-7013
[12] Gdoutos E E 1984 *Problems of Mixed Mode Crack Propagation* (Martinus Nijhoff Publ)
[13] Wang X B, Hai L and Huang M 2014 *Chinese J. of Rock Mech. and Eng.* **23(9)** 1446–49
[14] Jeong D Y, Orringer O and Sih G C 1994 *Theoretical and Appl. Fracture Mech.* **22(2)** 127-37
[15] Jin F N, Jiang M R and Gao X L 2014 *Chinese J. of Rock Mech. and Eng.* **23(12)** 1976–80
[16] Li Z X 2012 *Damage Mechanics and Its Application* (Beijing: Science Press)
[17] Ye J H, Wu F Q, Sun J Z 2013 *Int. J. of Rock Mech. & Mining Sci.* **46** 568-76.
[18] Yang T, Wang B X and Sun L 2012 *Site Investigation Sci. and Tech.* **(01)** 3-7
[19] Hudson J A, Brown E T and Rummel F 1972 *Int. J. Rock Mech. Mining Sci.* **9** 241–48
[20] Markides C F, Pazis D N and Kourkoulis S K 2010 *Int. J. of Rock Mech. & Mining Sci.* **47** 227-37.
[21] Yang L 2012 *Experimental and Numerical Simulation Study on the Tensile Fracture Mech. of 3D Flaws in Fractured Rock Mass* (Ji Nan: Shandong University)

# Superfine Ce Oxide-Induced Precipitation Behavior of Copper in Steel

**Zhongyuan Li [1,2], Changrong Li [1,2]*, Peng Wang [1,2], Shixin Zhang [1,2], Longhai Cheng [1,2] and Wenhui Du [1,2]**

[1] College of Materials and Metallurgy, Guizhou University, Guiyang 550025, China
[2] Guizhou Province Key Laboratory for Metallurgical Engineering and Process Energy Saving, Guiyang 550025, China
*Email address: cr263@163.com (Changrong Li)

**Abstract.** Residual copper is gradually enriched in short steelmaking and produces copper crunch. To solve this problem, this study explores the influence of rare earth element Ce on copper precipitation behavior in steel. Chemical composition analysis, scanning electron microscopy, and EDS are used to characterize the phases. Results show that the inclusions MnS, $Al_2O_3$, $Ce_2O_3$, and $Ce_2O_2S$ in steel play a role in inducing the heterogeneous nucleation and precipitation of copper elements. The amount of segregation at the grain boundaries is significantly reduced, and the inclusions remain in the matrix and are uniformly dispersed to reduce the copper crunch problem.

Scrap is the main metal raw material for short-process steelmaking. The recycling of scrap steel introduces harmful elements such as copper, tin, arsenic, and antimony. These residual elements constantly accumulate in steel and lead to scrap "garbage" [1-4]. The increase in copper in scrap is mostly pronounced, and residual copper is gradually enriched in steel, leading to hot brittleness during hot process [5-7] and easy segregation at the austernite grain boundary. Liquid copper-rich phase infiltrates along the grain boundary and forms copper oxide (cuprous oxide) with oxide, which is reduced to copper by the iron matrix, permeating oxidation grain boundaries to destroy the matrix [8-10]. Copper severely deteriorates the hot working properties of steel [11] and reduces the yield of materials. Hence, studies on the effect of typical inclusions MnS and $Al_2O_3$ and rare earth inclusions on heterogeneous nucleation of copper in steel for controlling the precipitation of copper and reducing thermal embrittlement are crucial.

## 1. Theoretical Analysis

Bramfitt [12] and Turnbull proposed a 2D mismatch based on a 1D mismatch, which can reflect the lattice matching of the substrate and the crystal nucleus better when the lattice structure is different.

$$\delta_{(\mathrm{hkl})_n}^{(hkl)_s} = \sum_{i=1}^{3} \left[ \left| d[_{uvw}]_s \cos\theta - d[_{uvw}]_n \right| / d[_{uvw}]_n \right] / 3 \times 100\% \tag{1}$$

In the following form, $(hkl)_s$ is a low surface index of the base, $(hkl)_n$ is a low exponential surface for the crystalline phase, $[uvw]_s$ is a low-index direction on the $(hkl)_s$ class of crystal faces, $[uvw]_n$ is a low exponential direction on the crystal surface $(hkl)_n$, $d_{[uvw]s}$ is the atomic spacing along the $[uvw]_s$ direction, $d_{[uvw]n}$ is the atomic spacing along the $[uvw]_n$ direction, and $\theta$ is the angle between $[uvw]_s$ and $[uvw]_n$. According to Bramfitt [12], in heterogeneous nucleation, the core at $\delta$<6% is the most effective, that at $\delta$=6%–12% is slightly effective, and that at $\delta$>12% is invalid.

Given the strong affinity of rare earth elements with oxygen and sulfur, the existing forms of rare earth elements in steel are rare earth oxides, oxygen sulfides, and sulfides, whereas the form of copper in steel is mainly ε-Cu[13-15]. On the basis of MnS, $Al_2O_3$, $Ce_2O_3$, CeS, and $Ce_2O_2S$ as nucleation basements and ε-Cu and $Cu_2S$ as nucleation phases, the validity of the method is calculated. The use of the abovementioned inclusions as particles for the heterogeneous nucleation of copper in steel is determined, and the calculation results are shown in Table 1.

**Table 1.** Misfit calculation results (%)

| | MnS | $Al_2O_3$ | $Ce_2O_3$ | $Ce_2O_2S$ | CeS |
|---|---|---|---|---|---|
| ε-Cu | 2.4 | 7.5 | 6.7 | 8.8 | 27.2 |
| $Cu_2S$ | 6.7 | 19.2 | 31.3 | 29.9 | 20.8 |

Table 1 shows that MnS, $Al_2O_3$, $Ce_2O_3$, and $Ce_2O_2S$ can be used as effective nucleation particles for ε-Cu. However, MnS is only the effective nucleation particle of $Cu_2S$. Moreover, the ability of the other inclusions to become $Cu_2S$ nucleation particles is weaker than that of MnS.

## 2. Experimental Process and Research Method

Experimental raw materials for industrial use include pure iron, electrolytic pure manganese powder, pure copper, pig iron (carbon content 2.94 wt.%), electrolytic aluminum tablets, rare earth cerium, and experimental steel melted in medium-frequency induction furnace. The main chemical constituents are shown in Table 2.

**Table 2.** Main chemical composition of samples (wt.%)

| Sample | C | S | Cu | Mn | Ce | Fe |
|---|---|---|---|---|---|---|
| S1 | 0.053 | 0.034 | 0.820 | 0.360 | 0.000 | balanced |
| S2 | 0.230 | 0.024 | 0.620 | 0.560 | 0.010 | balanced |
| S3 | 0.310 | 0.015 | 0.720 | 0.550 | 0.023 | balanced |
| S4 | 0.320 | 0.005 | 0.720 | 0.580 | 0.069 | balanced |

After completing casting, demolding, and sampling under an initial forging temperature of 1050 °C and final forging temperature of 850 °C, the steel ingot is forged into a Φ30 mm × 300 mm steel bar. The metallographic sample is prepared by wire spark cutting, inlaying, sanding, polishing, and corroding with 4% nitric alcohol solution. The ZEISS-SUPRA55 scanning electron microscope and energy spectrometer are used for observation and analysis.

## 3. Result Analysis and Discussion

### 3.1. *Effect of Rare Earth Ce Elements on the Inclusions in Steel*

After the addition of rare earth Ce element, the inclusions in the steel are obviously refined, made spherical, and increased in quantity [16]. To study the effect of rare earth elements on the distribution of inclusions, 50 different fields of view are randomly selected from two samples to observe their main distribution state, quantity, and form. The statistical results are shown in Figure 1.

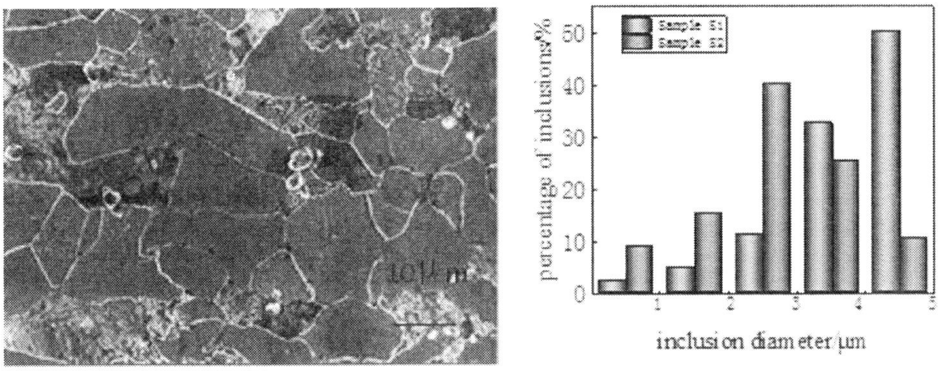


**Figure 1.** Distribution and size distribution of inclusions

In Figure 1, the inclusions in S1 samples without adding Ce are generally large, and about 50% or more of them have a particle diameter of more than 4 μm. In the S2 sample with Ce, the inclusion size is smaller than that of the S1 sample and concentrated in the 2–4μm interval. After adding Ce, the inclusions in steel are obviously refined, spherical, and high in quantity. The particle size of nearly half of the inclusions is at 2–3 μm, providing a large number of nucleation particles for the heterogeneous nucleation of residual copper elements in steel.

*3.2. Induction of Heterogeneous Nucleation of Copper in Steel by Ultrafine Ce Oxide*

The validity of the method is calculated based on MnS, $Al_2O_3$, $Ce_2O_3$, CeS, and $Ce_2O_2S$ as nucleation basement and ε-Cu and $Cu_2S$ as nucleation phase. The results show that MnS, $Al_2O_3$, $Ce_2O_3$, and $Ce_2O_2S$ can be used as nucleation sites for ε-Cu. However, only MnS can become the effective nucleation particle of $Cu_2S$.

The EDS analysis in Figure 2 shows that spectrum 10 is $Al_2O_3$ and the copper content is 5.8%, which is far larger than the average content of 0.72% in the steel matrix. The matrix continues to cool down after the solidification of molten steel. When the temperature drops to the melting point of copper, the copper element begins to grow in the $Al_2O_3$ inclusion, where there is no obvious stratification between the two. On spectrum 11, its composition is analyzed as $Al_2O_3$ and MnS, and the content of copper is 9.9%, which shows that $Al_2O_3$ and MnS are mixed and precipitated together during solidification. When the temperature continues to drop to the melting point, copper solidifies and nucleation precipitation occurs.

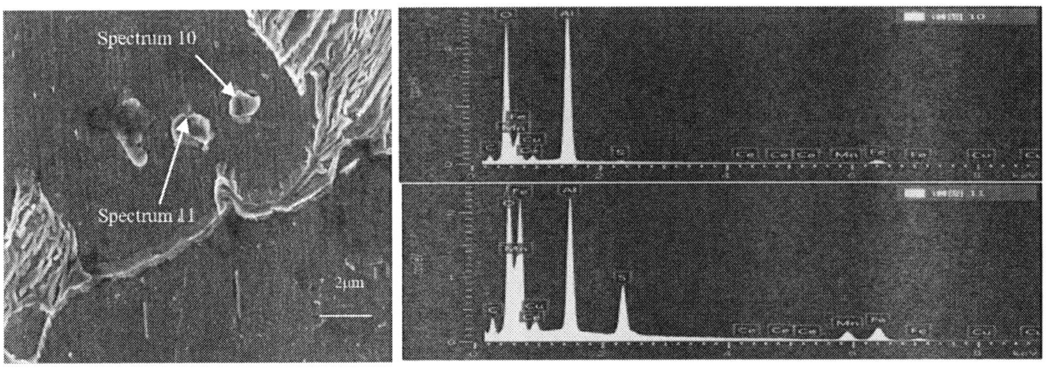


**Figure 2.** Morphology and energy spectrum analysis of heterogeneous nucleation of inclusions

Figure 3 shows that the contents of Ce, O, and S in point 36 are the highest. However, the atomic percentage is not strictly in accordance with $Ce_2O_2S$ (i.e., 2:2:1) probably because the phase composition of each point is a mixed inclusion of $Ce_2O_3$ and $Ce_2O_2S$. Its content is 7.10%, which is far higher than the average copper content in sample 3 (0.72%). Thus, $Ce_2O_3$ and $Ce_2O_2S$ of rare earth inclusions in steel induce the precipitation of copper heterogeneous nucleation. During the solidification of molten steel, copper precipitates around the rare earth inclusions, and a common mixture of rare earth inclusions and copper is formed.

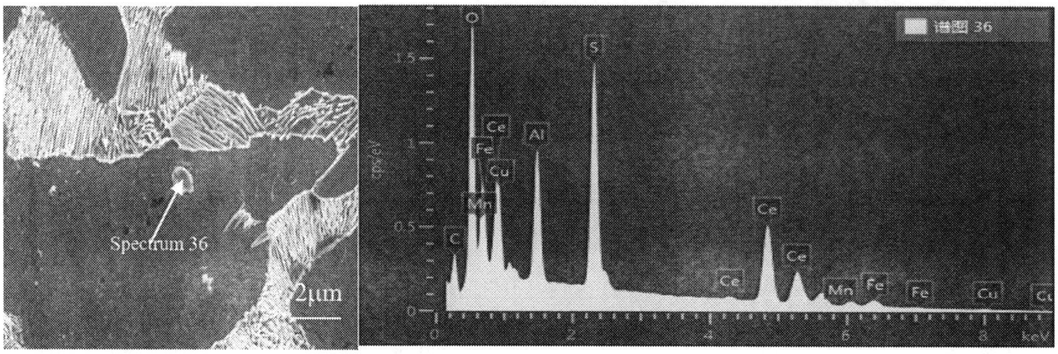


**Figure 3.** Morphology and energy spectrum analysis of the heterogeneous nucleation of inclusions

### *3.3. Effect of Superfine Ce Oxide on the Distribution of Copper in Steel*

Copper is highly prone to segregation in steel and easily precipitates at the grain boundary [17]. Given the redox reaction, iron copper oxide occurs in the matrix redox reaction during the solidification of molten steel. Copper and oxygen react to form copper oxide, which is reduced by iron to form copper. As the reaction continues and the grain boundary erodes, the matrix is destroyed. With ten points randomly selected for detection and the average value obtained to analyze the copper content, the data are shown in Table 3.

**Table 3.** Copper content at different positions in four groups of samples (wt%)

| Sample | Average | Intragranular | Grain boundary |
|---|---|---|---|
| S1 | 0.82 | 1.11 | 6.72 |
| S2 | 0.62 | 0.90 | 2.92 |
| S3 | 0.72 | 0.87 | 1.39 |
| S4 | 0.72 | 0.76 | 0.92 |

Table 3 shows that the Ce content increases with decreasing copper content in the grain boundary. When the Ce content in sample S4 is 0.72%, the copper content at the grain boundary is 0.92%. Compared with the grain boundary of sample S1, the copper content is reduced by 86.3%.

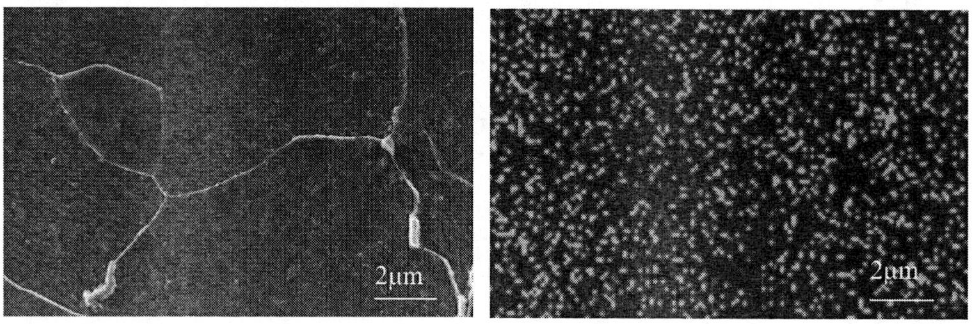


**Figure 4.** SEM morphology and EDS surface sweep distribution of copper in sample S4

Figure 4 shows that the randomly selected area of S4 sample is employed to test the distribution of copper, which is basically homogenized. Increasing Ce content results in a state of dispersion homogenization.

### *3.4. Effect of Ultrafine Ce Oxide on the Distribution of Copper in Hot Working Process*

The hot working temperature of steel is determined by its composition. It ranges from 1000 °C to 1200 °C during which copper-containing steel is prone to cracks during hot working process [18-20]. Cracks are formed when the temperature reaches the melting point of copper, which forms a copper-rich phase between the substrate and oxide layer. Moreover, the redox reaction of iron, copper, and oxygen makes the grain boundary embrittlement crack.

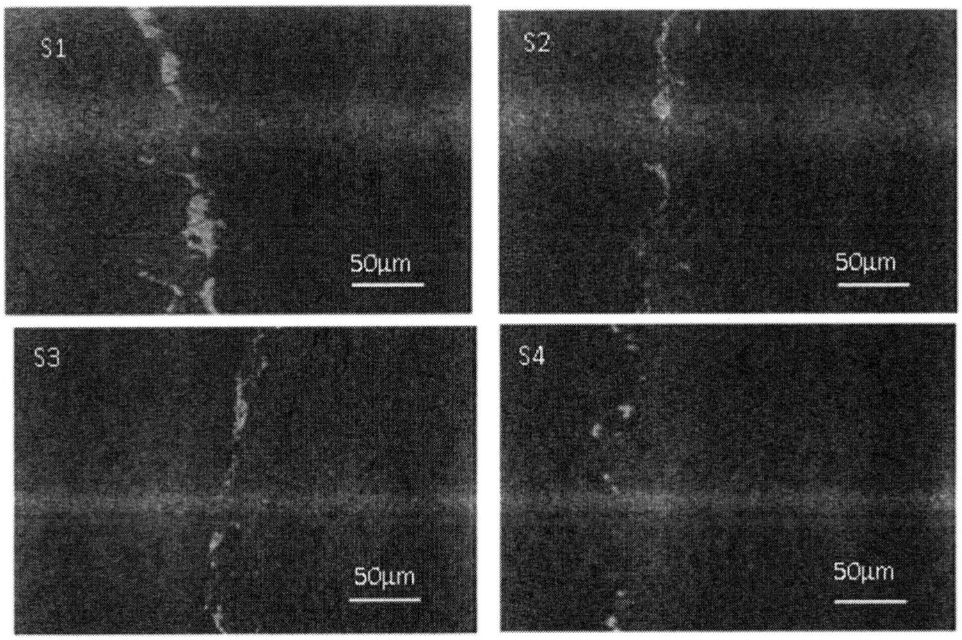


**Figure 5.** Distribution of copper for each sample at 1150 °C

In this experiment design, four sets of samples are heat treated at 1050 °C, 1100 °C, and 1150 °C for 1 h to observe the distribution of copper at high temperature. Figure 5 shows the distribution of copper at 1150 °C for each sample. For all samples, as the rare earth content increases, the copper distribution is gradually reduced on the boundary of the matrix and decreases in content.

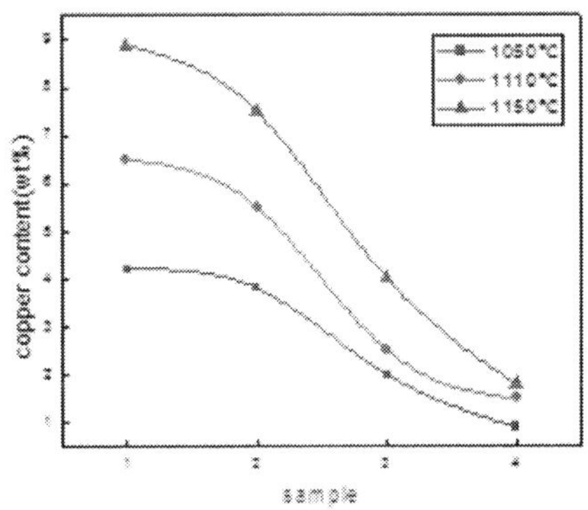


**Figure 6.** Distribution of copper content for each sample at different temperatures

Figure 6 shows the distribution of copper content at different temperatures for each sample. As the temperature rises, copper is heavily enriched on the boundary. When the temperature reaches the melting point of copper, copper is diffused to the boundary to form a copper-rich phase. At the same temperature, as the content of rare earth elements increases, the copper content on the boundary is greatly reduced. The copper content on sample S4 at the boundary under hot working treatment is the lowest at 1.8%, and its hot brittle tendency is also the lowest among the four samples. As a result of grain refinement and increased amount of grains, the number of grain boundaries increases, and the copper content decreases in the unit grain boundary to increase the solubility of copper in steel. Simultaneously, after the addition of rare earth elements, numerous fine inclusions in steel form, which induce the precipitation of copper to decrease its diffusion to the boundary?

## 4. Conclusion

(1) Theory of 2D lattice mismatch calculation shows that MnS, $Al_2O_3$, $Ce_2O_3$, and $Ce_2O_2S$ can all become nucleation particles of residual copper in steel. Experiments reveal that the inclusions MnS, $Al_2O_3$, $Ce_2O_3$, and $Ce_2O_2S$ in steel induce the heterogeneous nucleation and precipitation of copper. Moreover, they limit its segregation to the grain boundary and cause dispersed homogenization in the matrix.

(2) Rare earth inclusions produced by rare earth elements in steel provide a large number of nucleation particles for the heterogeneous nucleation of residual copper in steel, thereby inducing the precipitation of copper.

(3) After the addition of rare earth elements, the precipitation of copper changes. The precipitation of copper at the grain boundary is greatly reduced, resulting in distributed dispersion. The copper brittleness of steel can be alleviated during hot working processes.

## Acknowledgments

This work was financially supported by the National Natural Science Foundation of China (51364006).

## References

[1] A. P. Xian, D. Zhang Y. K. Wang, Residual Elements in Steel and its Effect on Properties of Steel, Iron and Steel 1999, 34 (1999) 64-68.

[2] F. Y. Liu, Surface Enrichment and Grain Boundary Oxidation of Residual Elements-Discussion on the Mechanism of Hot Brittleness in Low Carbon Steel, Acta Metall. Sin. 14 (1978) 310-347.

[3] K. Noro, M. Takeuchi, Y. Mizukami, Necessity of Scrap Reclamation Technologies and Present Conditions of Technical Development, ISIJ Int. 37 (2007) 198-206.

[4] L. F. Li, Z. H. Jiang, Method for Removing Copper from Scrap Steel, J. Iron Steel Res. 12 (2000) 77-80.

[5] L. G. Garza, C. J. V. Tyne, Surface Hot-shortness of 1045 Forging Steel With Residual Copper, J. Mater. Process. Technol. 159 (2005) 169-180.

[6] G. Sahoo, M. Deepa, B. Singh, A. Saxena, Hot Ductility and Hot-shortness of Steel and Measurement Techniques: A Review, JOM. 26 (2016) 1-6.

[7] G. D. Zhang, Basic Research on Removing Harmful Residual Elements in Scrap Steel-Removal of Copper in Scrap Steel, Northeastern University, Shenyang, 2000.

[8] H. Peng, the Role of Silicon in Hot Shortness Amelioration of Steel Containing Copper and Tin, Oxid. Met. 85 (2016) 1-12.

[9] M. E. Fine, D. Isheim, Origin of Copper Precipitation Strengthening in Steel Revisited, Scr. Mater. 53 (2016) 115-118.

[10] R. Y. Chen, W. Y. D. Yuen, Copper Enrichment Behaviours of Copper-containing SteelsIn Simulated Thin-slab Casting Processes, ISIJ Int. 45 (2006) 807-816.

[11] H. Matsuoka, K. Osawa, M. Ono, M. Ohmura, Effects of Cu and Other Tramp Elements on Steel Properties. Influence of Cu and Sn on Hot Ductility of Steels with Various C Content, ISIJ Int. 37 (1997) 255-262.

[12] B. L. Bramfitt, the Effect of Carbide and Nitride Additions on the Heterogeneous Nucleation

Behavior of liquid Iron, Metall. Trans. 1 (1970) 2958-2958.
[13] J. Zhu, H. Y. Huang, J. X. Xie, Recent Progress and New Ideas for Accelerating Research in Rare Earth Steel, J. Iron Steel Res. Int. 29 (2017) 513-529.
[14] R. Tuttle, K. Song, Characterization of Rare Earth Inclusions from Cast 1010 Steel, Int. J. Metall. 9 (2015) 23-31.
[15] P. Wang, C. R. Li, R. Liu, S. Shi, Calculation of Misfits of ε-Cu Precipitation in Steel Induced by Rare Earth Inclusions, J. Chin. Rare Earth Soc. 36 (2018) 316-318.
[16] R. B. Tuttle, S. R. Kottala, Population Size Distribution of Rare Earth and Non-metallic Inclusions in 4130 and 8630 Steels, Int. J. Metall. 6 (2018) 1-13.
[17] C. R. Li, F. S. Zhang, Q. Zhang, L. Zhang, Crystallographic Analysis of the Second Phase Induced Copper Precipitation in Steel, J. Iron Steel Res. 28 (2016) 63-66.
[18] W. Xin, B. Song, Z. Yang, L. Li, Effect of Arsenic and Copper + Arsenic on High Temperature Oxidation and Hot Shortness Behavior of C–Mn Steel, ISIJ Int. 56 (2016) 1232-1240.
[19] C. F. Yang, H. Su, L. Li, Y. Q. Zhang, Effect of Heating Process on Surface Oxidation of Copper-Containing Steel, J. Iron Steel Res. 19 (2007) 48-52.
[20] M. Hatano, K. Kunishige, Mechanism for Suppression of Surface Hot-shortness in Cu Containing Ferritic Stainless Steel, Tetsu-to-Hagane. 90 (2009) 134-140.

# Study on Performance of Polyvinyl Chloride Treated by Plasma

**Runjin Gui and Zhiwen Wang***

Guang Xi University for Nationalities, Nanning 530007 China.

* E-mails: 310484815@qq.com

**Abstract.** Treating polyvinyl chloride material with plasma, and the contact angle and shear strength, then study the changes of wettability and surface cohesiveness of the polyvinyl chloride material after modification. Measuring the background mass spectrometry and the mass spectrum of the intermediate product in the treatment of PVC, then we investigated the changes in gas composition during plasma treatment of PVC. Based on the analysis of contact angle and we found that the contact angle of PVC treated by plasma changed and the surface wettability was obviously improved. We also found that the shear strength of PVC treated by plasma showed a significant increase in surface bonding performance. The analysis of mass spectrometry shown that the composition of the air after plasma treatment of PVC changed a lot.

## 1. Introduction

Polyvinyl chloride (PVC) is a high-molecular material, which is a white powder, and its original color is faint yellow, and the degree of transparency is translucent. Because PVC is wear-resistant, flame-retardant, and not easily corroded, and its cost is relatively low compared to other polymer materials, it has been widely used in industry and agriculture [1]. Polyvinyl chloride has poor surface wettability and is hydrophobic and lipophilic, thus limiting its application in oil-water separation membranes, industrial printing and food packaging [2]. In many cases, PVC plastic products need to be subjected to secondary processing such as bonding when applied. Therefore, in many fields, the PVC plastic sheet should be improved as much as possible to make it more firm when bonded. An effective way to improve the surface wettability of PVC [3] and increase the surface roughness [4] is to treat polyvinyl chloride with plasma surface modification technology. The plasma surface treatment does not require wet water treatment, as long as it can be carried out in a dry environment, and there is no need to dry the material during the treatment again. Compared with other materials, the advantages of energy saving and high efficiency, while eliminating the waste water treatment, have protected the environment to a certain extent. The plasma surface treatment is modified only in the range of several nm to several hundred nm on the surface of the material, so that the surface properties of the material are effectively improved while ensuring that the material is not affected. In addition, the plasma surface treatment effect is relatively long-lasting, with simple operation, relatively low cost, and convenient for continuous automated production [5]. Modified surface plasma treatment of chemically inert polymers can lead to chemically reactive groups those result in functionalized polymer surfaces [6]. In this paper, polyvinyl chloride was treated by RF glow discharge plasma to study the change of contact angle of polyvinyl chloride plastic surface after plasma treatment and the bond strength of polyvinyl chloride plastic after adhesive treatment of polyurethane adhesive, and to analyze the reasons for the influence of plasma treatment on the bonding strength of PVC.

## 2. Experiment and Test

### 2.1. Specimens and Instruments
Main samples and instruments: polyvinyl chloride plastic sheet, polyurethane adhesive, ultrasonic cleaner, contact angle measuring instrument and HD-2 cold plasma modification equipment.

### 2.2. Sample Pretreatment
(1) Sanding the edges of the sample to be tested for easy marking and bonding.

(2) Putting the test sample (PVC sheet) in an ultrasonic cleaner for 3 minutes, and then takeing out the sample to dry.

(3) Radio frequency glow discharge: A solid glass rod was used as a support part of the sample to be processed in a vacuum chamber.

### 2.3. Low Temperature Plasma Treatment
The HD-2 cold plasma modification equipment uses a capacitive coupled RF glow discharge plasma device with a frequency of 13.56 MHz. Putting the pretreated PVC plastic sheet in the vacuum chamber of the plasma modification equipment: (1) Evacuating the vacuum chamber to 20 Pa, and plasma discharge treatment was performed on four groups of PVC, the discharge power was 50 W, and the treatment time was 3 min. , 5 min, 10 min and 15 min respectively ; mass spectrometry was measured simultaneously. (2) The vacuum chamber was evacuated to 20 Pa, and four sets of PVC samples were treated with plasma for 5 min, and the discharge powers were 30 W, 50 W, 100 W and 150 W, respectively.

## 3. Measurement of Contact Angle of PVC Plastic Sheet
In the measurement experiment of the contact angle, in order to reduce the experimental error, we decide to take four points for each PVC board, and the average of the four points was taken as the value of the last contact angle. The values of the contact angles of PVC (P9, P10) without plasma treatment are shown in Table 1.

**Table 1.** Contact angle of PVC (P23, P25) without plasma treatment

| Measuring position | | 1 | 2 | 3 | 4 | average value |
|---|---|---|---|---|---|---|
| Contact | Left | 91 | 87.75 | 85.5 | 87.5 | 87.94 |
| angle | Right | 89.25 | 85.75 | 85 | 85.25 | 86.31 |

According to the table, the difference in position at the measurement of the contact angle in the table is ±6. The generation of this error may be related to the difference in temperature, the impurities on the surface of PVC sheets and the operation at the time of dropping. Therefore, taking the average value can obtain the value of the contact angle more accurately.

A comparison of the values of the contact angles of plasma treated PVC sheets at the same power (50 W) and at different times is shown in Table 2.

**Table 2.** Same power (50W), contact angle of plasma treated PVC sheet at different times

| Processing time | | Unprocessed | 3(min) | 5(min) | 10(min) | 15(min) |
|---|---|---|---|---|---|---|
| Contact | Left | 89.19 | 61.63 | 62.25 | 59.13 | 60.13 |
| angle | Right | 87.5 | 59.63 | 60.25 | 58 | 57.75 |

A comparison of the values of the contact angles of the plasma treated PVC sheets at different powers at the same time (5 min) is shown in Table 3.

**Table 3.** Contact of plasma treated PVC sheets at different powers at the same time (5 min)

| Processing power | | Unprocessed | 30(W) | 50(W) | 100(W) | 150(W) |
|---|---|---|---|---|---|---|
| Contact | Left | 87.94 | 57.75 | 58.94 | 57.75 | 58.56 |
| angle | Right | 86.31 | 55.44 | 56.69 | 55.44 | 56.56 |

Analysis of the above table shows that the PVC surface contact angle changes significantly after plasma treatment of PVC, and the contact angle after treatment is reduced by about 30° than that treatment. The smaller the contact angle is, the better the wettability of the PVC surface is. Therefore, the apparent wetting effect of the PVC surface after plasma treatment is improved, which will help to improve the bonding performance.

Further analysis can also show that the plasma treatment of PVC under the same power, different time, the contact angle have not changed much; plasma treatment of PVC at the same time, under different power effects, the contact angle have not changed much, the difference is Between ±3°and ±2°.

Factors such as temperature, water purity, and adsorption of impurities on the surface of the PVC sheet during the experiment may affect the measured data, but it is in the error range. We can consider that time and power has some effect on the plasma treatment of PVC.

## 4. Measurement and Data Analysis of Bond Shear Strength

### *4.1. Test Unit and Test Method*

The polyvinyl chloride is bonded using wet-solid one-component polyurethane. Test basis and method: test temperature 20 ° C, tensile speed 2 mm / min, clip spacing 91 mm, shear strength = shear force / bonding area, unit MPa, bonding area are 625 $mm^2$ (25 mm × 25 mm), calculation results All 3 significant digits are reserved.

### *4.2. Results*

Table 4 is the shear strength data of the bonded body of the PVC sheet after the RF glow discharge plasma treatment, the electric power is 50 W, and the treatment time is 3 min, 5 min, 10 min and 15 min, respectively. We can see that the PVC sheet treated by the nitrogen glow discharge plasma has the best treatment effect when the discharge power is 50W, and the best shear strength after bonding is 1.110MPa.

Table 5 shows the shear strength data of the bonded sheets of the PVC sheets after the RF glow discharge plasma treatment. The treatment time was 5 min, and the discharge powers were 30 W, 50 W, 100 W and 150 W, respectively. We can see that the PVC sheet treated by the nitrogen glow discharge plasma has the best treatment effect when the treatment time is 5 min and the discharge power is 150 W.

**Table 4.** Maximum shear strength of PVC sheets at different treatment times when the treatment power is 50W

| Processing time/min | Unprocessed | 3(min) | 5(min) | 10(min) | 15(min) |
|---|---|---|---|---|---|
| Maximum shear strength/kN | 0.490 | 0.605 | 0.644 | 0.696 | 0.685 |
| Shear strength/MPa | 0.780 | 0.970 | 1.030 | 1.110 | 1.100 |

**Table 5.** Maximum shear strength of PVC sheets at different treatment powers at a treatment time of 5 min

| Processing power | Unprocessed | 30(W) | 50(W) | 100(W) | 150(W) |
|---|---|---|---|---|---|
| Maximum shear strength | 0.490 | 0.638 | 0.655 | 0.693 | 0.763 |
| Shear strength | 0.780 | 1.020 | 1.050 | 1.110 | 1.220 |

## 5. Mass Spectrometry

In this experiment, we used a quadrupole mass spectrometer to analyze the mass spectrum of the vacuum residual gas, the background mass spectrum of the PVC sample, and the mass spectrum when the PVC sample was discharged. We can determine the composition of the gas qualitatively and semi-quantitatively based on the mass-to-charge ratio and the relative intensity value. According to Fig. 1, among the residual gases, there are: m/e=28 gas such as $N_2$ or $C_2H_4$; m/e=16 determines the presence of $O_2$; m/e=17, 18 determines the presence of $H_2O$; m/e=44 Gases such as $C_3H_6$ and $CO_2$ are present. The components in the air are nitrogen, oxygen, water vapor, carbon dioxide, and other gas components. The gas components obtained from Fig. 2 and Fig. 3 substantially correspond to the air components.

By comparing the mass spectrometer background mass spectrum and the PVC sample outgassing background mass spectrum in the air environment, we can find that when the PVC sample is placed, as shown in Fig. 2, m/e = 14, 16, 28, 32, 40 The relative intensity of the ion peaks is significantly higher, that is, the content of $N_2$, $O_2$, Ar, and $CO_2$ is increased. The reason for this phenomenon is that in the air environment, PVC itself absorbs the components of the gas in the air. When the PVC is placed in a vacuum environment, the air gas absorbed inside is released, so the relative value of each ion peak will be higher. Among them, the existence of Ar came from experimental residue.

By comparing the mass spectrum of the PVC sample treated by discharge (Fig. 3) and the mass spectrum of the PVC sample without discharge (Fig. 2), we can find that when discharging PVC, m/e = 2, 14, 16, 17, 18, 19, 20, 29, 30, 32, 40, 44, and 45The relative intensities of the ion peaks of are significantly higher, that is, the contents of $H_2$, $N_2$, $CH_4$, $H_2O$, $C_2H_4$, Ar, and $CO_2$ increased. The reason may be that when the PVC is discharged, the air gas absorbed inside the PVC is affected by the discharge power and time when released, and is released more than when not discharged, and the release speed is fast.

We can draw conclusion that the discharge treatment can increase the gas released inside the material and improve the treatment effect, especially when the power is large.

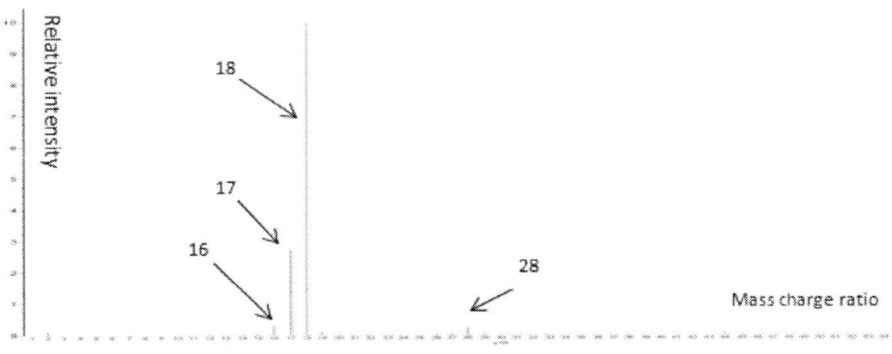


**Figure 1.** Background spectrum of residual gas in a vacuum chamber

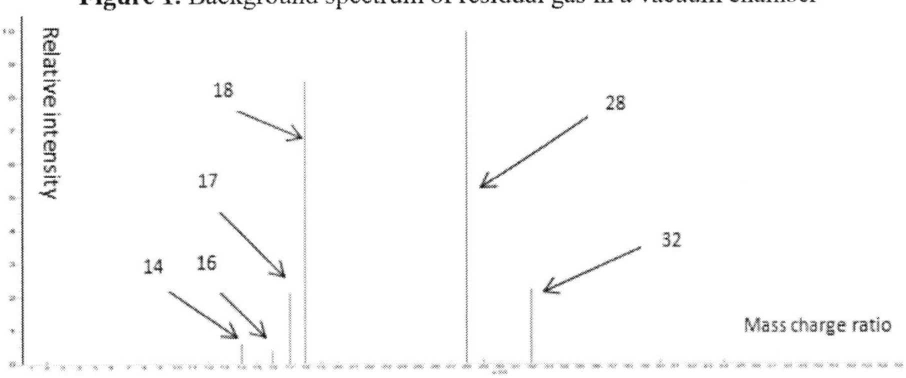


**Figure 2.** Gas output spectrum after placing a PVC sample

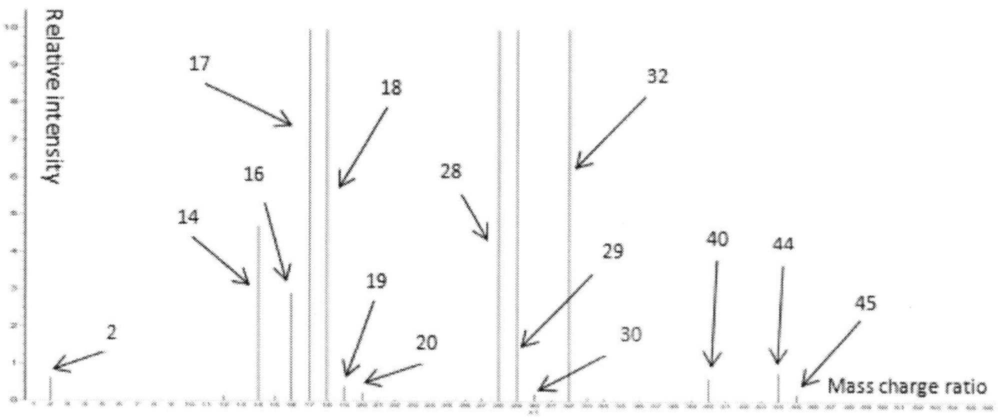


**Figure 3.** Mass spectrum of discharge treated PVC samples

## 6. Mechanism Analysis

After the treatment of plasma, the surface energy of the PVC plastic sheet was significantly improved. The contact angle was reduced, whilethe wettability and hydrophilicity of the plastic surface were greatly improved. When high energy particles in the plasma hit the surface of the material, the electron energy is about 5-10eV, the plasma high-energy particles transfer their energy to the surface molecules, causing the molecular bonds in the polymer to break, and then undergo a series of changes such as etching, cross-linking, and chemical reaction, thereby changed the surface chemical composition and properties of plastic sheet [8].

In addition, when the polymer treated by $O_2$ and $N_2$ plasma is in contact with air, the active radicals on the surface of the polymer will bring in oxygen-containing or nitrogen-containing polar group such as -COOH, -OH, -C=O and $-NH_3$ on its surface[3,7]. They have a strong reactive activity, so they can react with other compounds through these groups, and introduce the desired functional groups to functionalize the polymer surface, thereby greatly improving the bonding performance. In addition, after measuring and analyzing the mass spectrum, it is found that the PVC sample adsorbs certain air gas components in the air environment, so when it is in a vacuum environment, the absorbed air gas is released, so that the relative intensity of the ion peak of the gas component is remarkably increased in the background mass spectrum.

## 7. Conclusion

Polyvinyl chloride has poor wettability and the shear strength when using polyurethane directly for mixing and bonding sometimes fails to meet the requirements of industry and agriculture. In order to make a breakthrough in this respect, this paper uses RF glow discharge plasma to treat the surface of PVC plastic, the surface roughness of plastic increased, and a large number of polar groups were introduced on the surface of plastic. After the plasma treatment, the surface of the plastic is activated, the wettability of the plastic surface is remarkably improved, and the shear strength of the PVC plastic bonded body is remarkably improved.Based on the research of He Yanhe et al., "Adhesive of Polypropylene and Polyethylene Treated by Plasma", we have carried out further research on polyvinyl chloride (PVC) materials. We have found that the shear strength of the bond for PVC is not as high as that of polypropylene (PP) and polyethylene (PE). The reason may be that the hardness of the polyvinyl chloride material is larger than that of polypropylene and polyethylene. The stability is better and the surface energy is relatively smaller, so the shear strength of the bond is not as large as the other two, which requires further research.

## Acknowledgement

We are grateful to the project 113000101200012000300006 which have supported us all the time.

**References**

[1] Yang Liting. Polyvinyl chloride modification and formulation [M]. Beijing: Chemical Industry Press, 2011. P1-P5.
[2] Zuo Xiuqin, Ye Zhiyin, Zhang Baixin, Liu Shuping. Study on Surface Wettability of Rack of Polyvinyl Chloride (RPVC) [J].Plastic, 2004(06):50-53.
[3] He Yanhe, Li Xuemei, Wang Zhiwen, et al. Bonding of plasma treated polypropylene and polyethylene [J]. Intense Laser and Particle Beams, 2012, 24(9): 2099-2102.
[4] Zhao Huaqiao. Plasma Chemistry and Technology [M]. Beijing: University of Science and Technology of China Press, 1993. P299-P301.
[5] Zhao Qing, Liu Shuzhang, Tong Honghui. Plasma Technology and Application [M]. Beijing: Chemical Industry Press, 2006. P1-P5.
[6] Gui-qiu Ma, Ya-peng Liu, Shu-xian Wei, Jing Sheng, et al. Surface modification of polypropylene by ethylene plasma and its induced β-form in polypropylene. Chinese Journal of Polymer Science, 2015, Vol. 33 (5), pp.669-673
[7] MARYAM A, SIAMAK M, MOJTABA M, et al. investigating the effect of power/time in the wettability of Ar and O2 gas plasma-treated low-density polyethylene. Progress in Organic Coatings, 2009, 64(4): 482- 488
[8] Lin Lizhong. Principle and application of plasma surface treatment machine for plastic products [J]. Plastics Industry, 1999 (06): 19 - 21.

IOP Conf. Series: Materials Science and Engineering 472 (2019) 012039 doi:10.1088/1757-899X/472/1/012039

# Failure Analysis of Fractured Cylinder Bolt

**Xiaodan Zhang, Nan Li, Zhiqiang Luo, Jikang Li and Long Liu**

Central Iron and Steel Research Institute, College South road 76th. Haidian District, Beijing, China
Email: zhangxiaodan@nercast.com

**Abstract.** Failure analysis of 42CrMo cylinder bolt had been done through the chemical component analysis, tensile test, metallographic test and fracture morphology observation. It was identified brittle fracture. The crack was mainly intergranular and propagated to fracture. The results show that it is hydrogen induced delayed fracture.

## 1. Introduction

The materials of fractured cylinder bolts which used on cars were 42CrMo. When the car was running more than 1000 kilometers, it was found that the bolt broke. The fracture was cracked at the tail of bolt without thread. And there was no plastic deformation, so it was identified brittle fracture. The manufacturing process was cold forging, heat treatment, hardness test, grinding, flaw detection, rolling thread, bluing and finished product test. The macro-photos of fractured bolts were shown in figure1.

**Figure 1.** Macro-photos of bolt

## 2. Test results and Analysis

### *2.1. Chemical Analysis*

A sample taken from bolt was analyzed seen Table 1. The chemical composition is fulfilled requirements.

**Table 1.** Chemical composition of the bolt (wt, %)

| | C | Si | Mn | P | S | Cr | Ni | Cu | Mo |
|---|---|---|---|---|---|---|---|---|---|
| Bolt | 0.44 | 0.24 | 0.76 | 0.023 | 0.0031 | 0.95 | 0.013 | 0.017 | 0.19 |
| GB /T 3077 | 0.38-0.45 | 0.17-0.37 | 0.5-0.8 | ≤0.03 | ≤0.03 | 0.9-1.2 | ≤0.30 | ≤0.3 | 0.15-0.25 |

*2.2. Tensile Test at Room Temperature*

The samples taken from bolt were proceeded tensile experiment. The results seen Table 2 is fulfilled requirements.

**Table 2.** Mechanical behavior of the bolt

| | | Tensile strength Rm, Mpa | Yield strength Rp0.2 ,Mpa | Elongation A, % | Shrinkage Z, % |
|---|---|---|---|---|---|
| Bolt | Sample 1 | 1383 | 1204 | 12.0 | 44 |
| | Sample 2 | 1384 | 1242 | 12.5 | 48 |
| GB/T 3098.1 | | ≥1200 | ≥1080 | ≥8 | ≥44 |

*2.3. Metallographic Investigation*

Using metallographic microscope we investigated the microstructure of bolt, as shown in figure 2 to figure 3. The microstructure of bolt in intact section and cracking section were all tempered sorbite. It could find no abnormal microstructure. At the edge of fracture it could be observed secondary crack along the grain boundary.

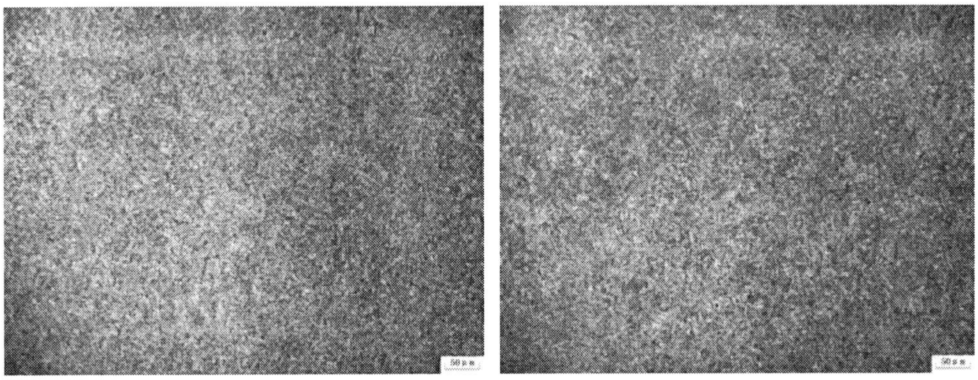

**Figure 2.** Microstructure of bolt (intact section)

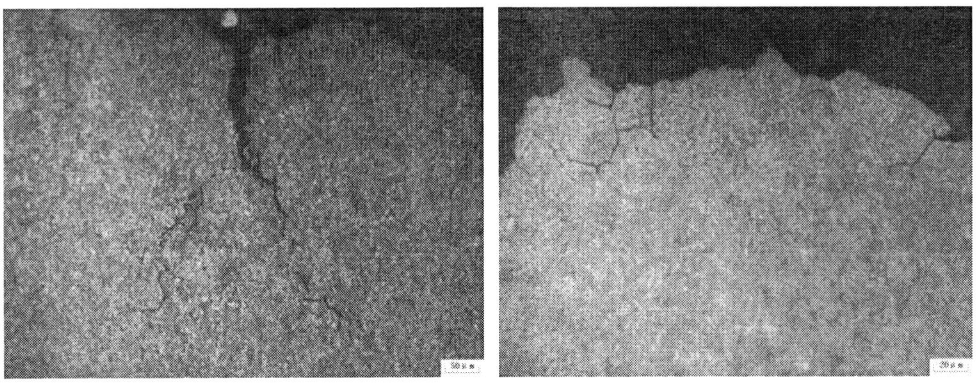

**Figure 3.** Microstructure of bolt (cracking section)

### *2.4. Fracture Analysis*

Using scanning electron microscope, the appearance of the microscopic fracture was investigated, as shown in figure 4 (a)-(d). It could be observed crystallographic features at large area. Along the crystallographic surface "chicken claw mark" can be seen which meets the characteristics of hydrogen induced delayed fracture. The microscopic morphology of the final break zone is the dimple.

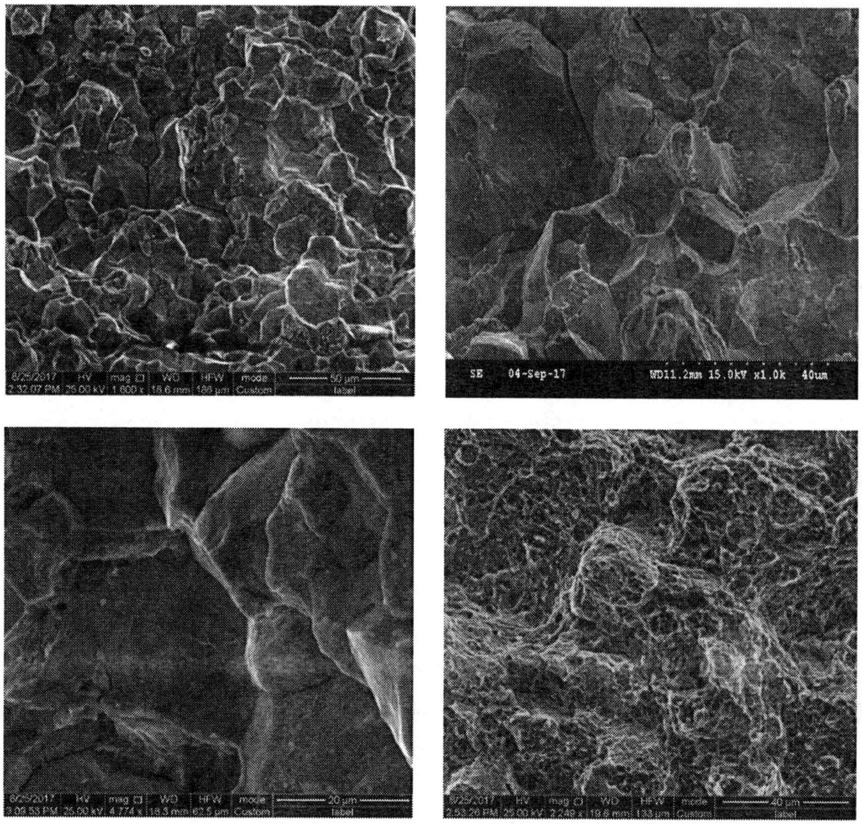


**Figure 4.** Micro-photos of fracture (a)-(c) fracture initiation (d) final break

## 3. Discussion

Analyzing the fractured cylinder bolts comprehensively, it could find that the chemical composition and mechanical behavior were all fulfilled requirements. Observing the fracture, it could be found crystallographic features at large area. Along the crystallographic surface "chicken claw mark" can be seen which meets the characteristics of hydrogen induced delayed fracture. The workpiece may be invaded by hydrogen atoms during smelting, pickling, electroplating, welding, etc [1]. It is understood that there are two links to be invaded hydrogen. One is quenching, the other is bluing [2]. The first case is using methanol medium to prevent decarburization during heat treatment. And then quench after volatilization of methanol. Methanol can be thermally decomposed to form carbon monoxide and hydrogen. The other case is pickling off the rust on the surface of bolts before bluing. The bolts reacted with acid to form hydrogen. Hydrogen invaded in bolts with diffusion and aggregation [3]. Then due to the stress being hydrogen induced delayed fracture.

## 4. Conclusion

(1) Analysis results showed that the chemical composition and mechanical behavior were all fulfilled requirements.

(2) The crack is mainly intergranular and "chicken claw mark" can be seen on the fracture.

(3) The results of comprehensive analysis show that it is hydrogen induced delayed fracture.

**Acknowledgment**
I would like to express my gratitude to all those who helped me during the writing of this article. I gratefully acknowledge the help of my colleague, Ms. Li, who has offered me valuable suggestions in the studies.

**References**
[1] Yuan g, Jao y, Wang c and Feng y Failure analysis of hub bolt of main land gearing ICAMCS 2017 (China: Henan) p 1024
[2] Hui w et al Influence of cold deformation and annealing hydrogen embrittlement of cold hardening bainitic steel for high strength bolts Materials Science & Engineering A 2016 662 p 528
[3] Dong l et al Metal magnetic memory testing for early damage assessment in ferromagnetic materials Journal of Central South University of Technology 2015 12 p 103

IOP Conf. Series: Materials Science and Engineering 472 (2019) 012040 doi:10.1088/1757-899X/472/1/012040

# Effect of Dealumination of Citric Acid-Modified Hβ Molecular Sieve on Hydroisomerization Performance of n-Hexadecane

**Xuefeng Bai[1,2], Xiaomeng Wei[2], Yang Liu[1] and Wei Wu[2*]**

[1] Institute of Petrochemistry, Heilongjiang Academy of Sciences, Harbin 150040, China

[2] School of Chemistry and Material Sciences, Heilongjiang University, Harbin 150080, China

Corresponding author and e-mail: W Wu, marywu64@126.com

**Abstract.** Hβ molecular sieves with $SiO_2/Al_2O_3$ ratio of 40 were modified by citric acid dealumination (Hβ-Cn) and Hβ-Cn loading Pd (0.5 wt%) bifunctional catalysts(Pd/Hβ-Cn) were prepared by hydrothermal synthesis and incipient wetness impregnation. The structure, morphology, pore and acid properties of Hβ-Cn and Pd dispersion of Pd/Hβ-Cn were characterized using XRD, SEM, $N_2$ physical adsorption, $NH_3$-TPD and $H_2$ chemical absorption. The skeleton structure of all Hβ-Cn was not damaged and their relative crystallinity increased somewhat after the treatment of citric acid. With the increase of citric acid concentration, the change of morphology,structure and pore properties of Hβ-Cn was very little, but the amount of weak acid, strong acid and total acid of Hβ-Cn, and Pd dispersion for Pd/Hβ-Cn decreased significantly. It was known from hydroisomerization perforamce of n-$C_{16}H_{34}$ that the hydroisomerization reaction temperature over Pd/Hβ-Cn slightly rises under the condition of the same conversion of *n*-$C_{16}H_{34}$, but the amount of multi-branched *iso*-$C_{16}H_{34}$ over Pd/Hβ-Cn, especially Pd/Hβ-C2 (60.51%) and Pd/Hβ-C3 (57.84%) was much higher than that over Pd/Hβ (36.99%). This could be due to the remove of partial skeleton aluminum species with strong acid sites to effectively inhibit the further cracking of the muti-branched isomerized products.

## 1. Introduction

Biodiesel is a kind of renewable, efficient and safe clean energy, which has been greatly developed in recent years [1]. The hydroisomerization of n-alkanes is the key to produce all kinds of high quality fuel oil [2]. In hydroisomerization reaction process, the cracking reaction of the long carbon chain alkanes will produce light hydrocarbon component content,which reduce the selectivity of alkane isomer and liquid yield of diesel fuel [3]. The development of efficient hydroisomerization bifunctional catalysts for skeletal branching of long n-alkanes without cracking is core research topic [4]. Beta molecular sieves have been widely used in the fields of catalytic cracking, hydroisomerization, alkylation because of thier strong acidity and unique three-dimensional pore structure, good thermal stability and hydrothermal stability. But the strong acidity and more number of acidic sites of Hβ aggravate the cracking reaction in the hydroisomerization of n-hexadecane [7]. In this paper, Hβ were modified by citric acid dealumination to reduce appropriately acid strength and number of acids sites and investigted the performance of the hydroisomerization reaction of n-hexadecane.

## 2. Experimental Section

### *2.1. Synthesis of Hβ Molecular Sieve and its Dealumination by Citric Acid*

Hβ molecular sieves were hydrothermally synthesized using $NaAlO_2$ (CP, Sinopharm chemical reagent co. LTD), ludox (26.85 wt% $SiO_2$, Qingdao Yumin Company, China) and tetraethylammonium hydroxide (CP, Sinopharm chemical reagent Co. LTD) according to $SiO_2/Al_2O_3$ ratio of 40 at 145°C for 5 days. After crystallization products were separated by centrifugation, dried at 110°C, calcined at 560°C to remove the template, treated by 0.5 mol/L $NH_4NO_3$ solution, then dried and roasted to obtain Hβ molecular sieves. As-prepared Hβ was mixed with citric acid solutions with concentrations of 0.5mol/L, 1.0mol/L and 1.5mol/L respectively in a solid-liquid mass ratio of 1:20 at 80°C, stirred on 500 prm for 7 h. The dealuminated Hβ molecular sieves were obtained by centrifugation and drying, labeled **Hβ**-Cn (n=1, 2, 3, represent three concentrations respectively).

### *2.2. Dealuminated Hβ Molecular Sieves Loading Palladium Bifunctional Catalysts*

The bifunctional catalysts were prepared by incipient wetness impregnation with a Pd $(NO_3)_2$ solution according to Pd loading amount of 0.5 wt%. Then, these samples were dried at room temperature overnight and calcined at 600 °C for 4 h in air, and referred to as Pd/**Hβ**-Cn.

### *2.3. Characterization of Hβ-Cn and Pd/ Hβ-Cn Catalysts*

XRD patterns of Hβ and Hβ-Cn were measured on a Bruker D8 Advance diffractometer with a Cu Ka radiation (λ= 0.154 nm). The morphology of Hβ and Hβ-Cn was performed by SEM on a FEI Sirion microscope. The $N_2$ absorption and desorption isotherms were carried out using an Autosorb-iQ2 apparatus. The surface area and pore volume were calculated by BET method and the t-plot method, respectively. The acid sites of Hβ and Hβ-Cn was measured by $NH_3$-TPD on a conventional set-up equipped with TCD. The Pd dispersion of Pd/Hβ and Pd/Hβ-Cn was determined by $H_2$ chemical absorption analysis using an AutoChem II 2920 Chemisorption Analyzer.

### *2.4. Catalytic Hydroisomerization of n-Hexadecane*

The catalytic hydroisomerization of n-hexadecane (n-$C_{16}H_{34}$) were carried out in a fixed-bed reactor. Firstly, 1.0 g Pd/Hβ and Pd/Hβ-Cn were reduced in situ with $H_2$ at 400°C for 1 h. Then, the catalyst was cooled to the reaction temperature, and $H_2$/n-$C_{16}H_{34}$ were fed into the reactor at a pressure of 2.0 MPa, a $H_2$/n-$C_{16}H_{34}$ of 8.72 (molar ratio) and WHSV of 3.70 $h^{-1}$. The composition of products after reacting for 2 h was analyzed using a GC (Agilent 6820) with a HP-1(60 m*0.25 mm*1.00μm) capillary column and a FID detector.

## 3. Results and Discussion

### *3.1. The Characterization of Hβ-Cn and Pd/Hβ-Cn Bifunctional Catalysts*

*3.1.1. XRD.* The XRD characteristic diffraction peaks of all Hβ and Hβ-Cn shown in Figure 1 are exclusively attributed to β topological structure, indicating that the skeleton structure of all Hβ-Cn was not damaged. The relative crystallinity of Hβ-Cn samples increased somewhat after the treatment of citric acid, which was caused by the partial non-skeleton species eluted from the molecular sieve.

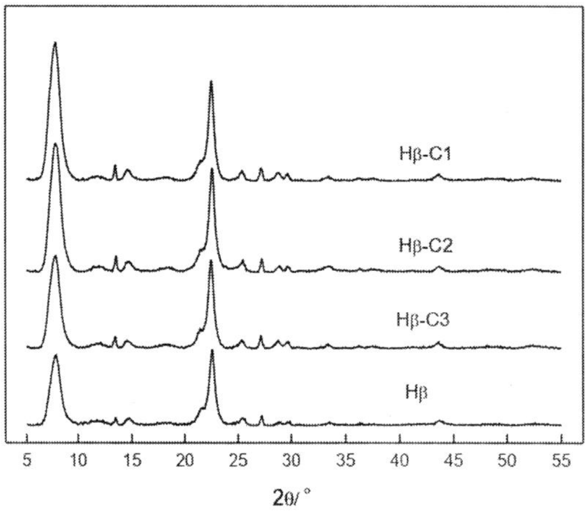


**Figure 1.** XRD powder patterns of Hβ and Hβ-Cn samples

*3.1.2. SEM.* In order to investigate the effect of citric acid dealumination treatment on the morphology of Hβ molecular sieves, Hβ and Hβ-Cn samples were characterized by SEM, and the results were shown in Figure 2.

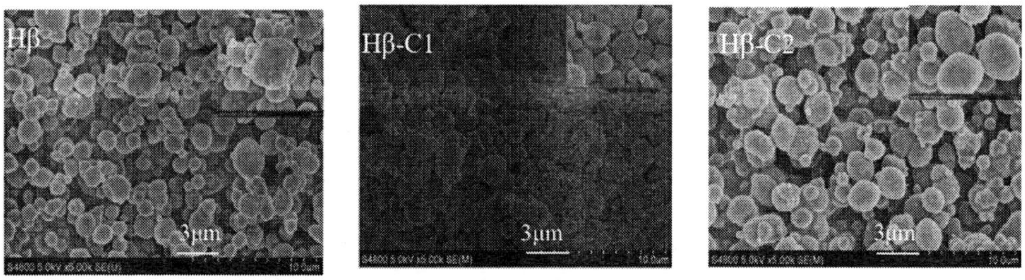


**Figure 2.** SEM images of Hβ and Hβ-Cn samples

It can be seen from Figure 2 that with the change of the concentration of citric acid treatment, there was not a significant difference of the morphology and structure between Hβ and Hβ-Cn samples. All the samples present elliptic spherical aggregates with the typical morphology of β molecular sieve. The size of the aggregates is 1-1.5μm and their diameter is about 1μm.

*3.1.3. $N_2$ physical adsorption.* To investigate the effect of citric acid dealumination treatment on the specific surface area and pore volume of Hβ molecular sieves, Hβ and Hβ-Cn samples were characterized by $N_2$ physical adsorption. $N_2$ adsorption-desorption isotherms of all samples are shown in Figure 3 and the adsorption data are listed in Table 1.

As shown in Figure 3, the dsorption-desorption isotherms of all Hβ-Cn samples are corresponding to a type I isotherm, which indicating that Hβ molecular sieves modified by citric acid dealumination treatment are still microporous materials. A hysteresis loop appears at a relatively high partial pressure, which indicating the existence of mesoporous structure in Hβ-Cn samples.

The BET surface area and pore volume of Hβ-C1 sample shown in Table 1 increases slightly, which is because the dealumination treatment with low concentration of citric acid did not cause obvious damage to the skeleton structure of Hβ molecular sieve, but washed out some non-skeleton species adsorbed on the outer surface of the molecular sieve. With the increase of citric acid concentration, BET surface area, micropore surface area and micropore volume of Hβ-C3 samples were all reduced to a certain extent, but the decrease was small. This mainly due to that the interaction between citric acid and skeleton aluminum species leads to chelate dealumination.

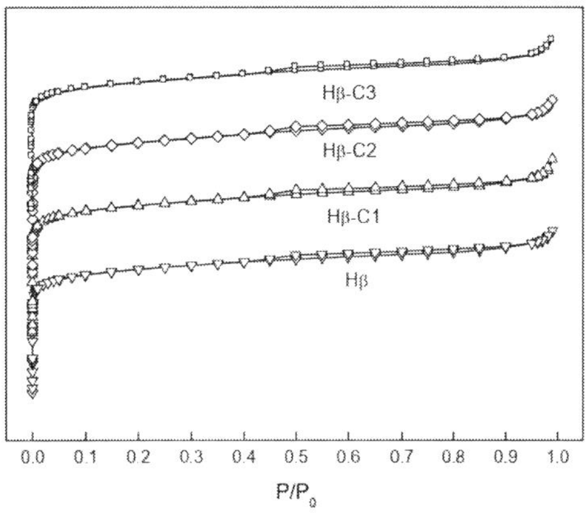


**Figure 3.** $N_2$ adsorption-desorption isotherm of Hβ and Hβ-Cn samples

**Table 1.** $N_2$ adsorption data of Hβ and Hβ-Cn samples

| Sample | Surface area($m^2/g$) | | | Pore volume ($cm^3/g$) | | |
|---|---|---|---|---|---|---|
| | BET | Micropore | External | Total | Micropore | Mesopore |
| Hβ | 648 | 545 | 103 | 0.341 | 0.216 | 0.125 |
| Hβ-C1 | 661 | 548 | 113 | 0.352 | 0.218 | 0.134 |
| Hβ-C2 | 650 | 541 | 109 | 0.343 | 0.214 | 0.129 |
| Hβ-C3 | 638 | 535 | 103 | 0.338 | 0.212 | 0.126 |

*3.1.4. $NH_3$-TPD.* In order to investigate the effect of citric acid dealumination modifition with different concentration on the acid strength and acid amount of Hβ molecular sieve, Hβ-Cn samples were characterized by $NH_3$-TPD shown in Figure 4 and Table 2.

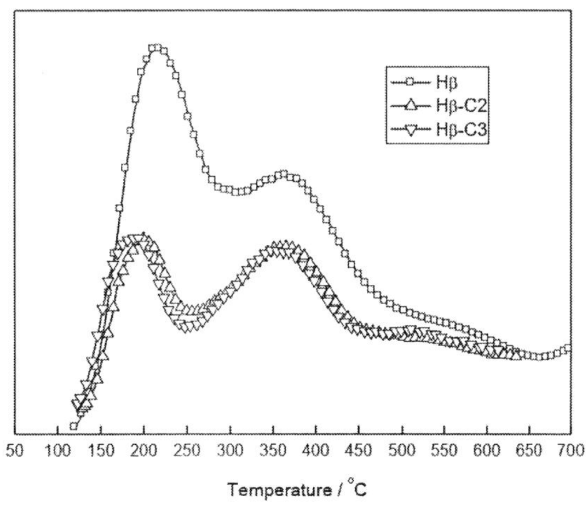


**Figure 4.** $NH_3$-TPD profiles of Hβ and Hβ-Cn samples

**Table 2.** Acid datas of Hβ and Hβ-Cn samples

| Sample | Acid amount (mmol/g) | | | T/°C | |
|---|---|---|---|---|---|
| | Weak | Strong | Total | LT | HT |
| Hβ | 0.29 | 0.56 | 0.85 | 218 | 365 |
| Hβ-C2 | 0.11 | 0.42 | 0.53 | 218 | 363 |
| Hβ-C3 | 0.12 | 0.43 | 0.55 | 210 | 363 |

As can be seen from Figure 4 and Table 2, the amount of weak acid, strong acid and total acid of Hβ-Cn samples treated with 1.0 mol/L and 1.5 mol/L citric acid all decreased significantly, but the strength of strong acid did not change significantly, indicating that the citric acid dealumination treatment partially removed the skeleton aluminum and non-skeleton aluminum species of Hβ.

*3.1.5. $H_2$ chemical absorption.* Pd/Hβ and Pd/Hβ-Cn bifunctional catalysts were characterized by $H_2$ chemical adsorption analysis to investigate the effect of citric acid dealumination on Pd dispersion shown in Table 3.

**Table 3.** Metal Pd dispersion and sites of Pd/Hβ and Pd/Hβ-Cn catalysts

| Catalysts | Dispersion of Pd (%) | $C_{Pd}$ (μmol/g) |
|---|---|---|
| Pd/ Hβ | 41.6 | 19.6 |
| Pd/ Hβ-C1 | 34.8 | 16.4 |
| Pd/ Hβ-C2 | 33.4 | 15.8 |
| Pd/ Hβ-C3 | 31.1 | 14.7 |

Pd dispersion on Pd/Hβ-Cn bifunctional catalysts shown in Table 3 that were lower than that on Pd/Hβ before dealumination, and inversely proportional to the concentration of citric acid, which was consistent with the change of total acid amount and surface area of bifunctional catalysts. This indicates that the acid sites are helpful to the dispersion of metal particles on surface of Hβ.

### *3.2. Effect of Citric Acid Dealumination on Catalytic Hydroisomerization Performance of n-Hexadecane over Pd/Hβ-Cn Bifunctional Catalysts*

The isomerization (main reaction) and cracking reaction (side reaction) of n-alkanes occurs mainly at the acidic sites of bifunctional catalyst. The pore structure, acidic density and acid strength of Hβ molecular sieve all influence the selectivity of isomerization reactions and the distribution of reaction products [6]. The hydroisomerization of n-hexadecane (n-$C_{16}H_{34}$) over Pd/Hβ and Pd/Hβ-Cn bifunctional catalysts was carried out to investigate the effect of citric acid dealumination on catalytic performance, and the results were shown in Figure 5~6.

It can be seen from Figure 5 that the hydroisomerization reaction temperature over Pd/Hβ-Cn slightly raises under the condition of the same conversion of *n*-$C_{16}H_{34}$. This is because the acid site density and Pd dispersion decreased after dealumination. The conversion of *n*-$C_{16}H_{34}$ over all of bifunctional catalysts can reached more than 90% at the reaction temperature of 245°C. It can be seen from Figure 6 that the amount of multi-branched *iso*-$C_{16}H_{34}$ in hydroisomerization products over Pd/Hβ and Pd/Hβ-Cn at *n*-$C_{16}H_{34}$ conversion of 50%, especially Pd/Hβ-C2 (60.51%)and Pd/Hβ-C3 (57.84%) was much higher than that over Pd/Hβ (36.99%).The partial skeleton aluminum species with strong acid sites are removed by citric acid, which effectively inhibited the further cracking of the muti-branched isomerized products.

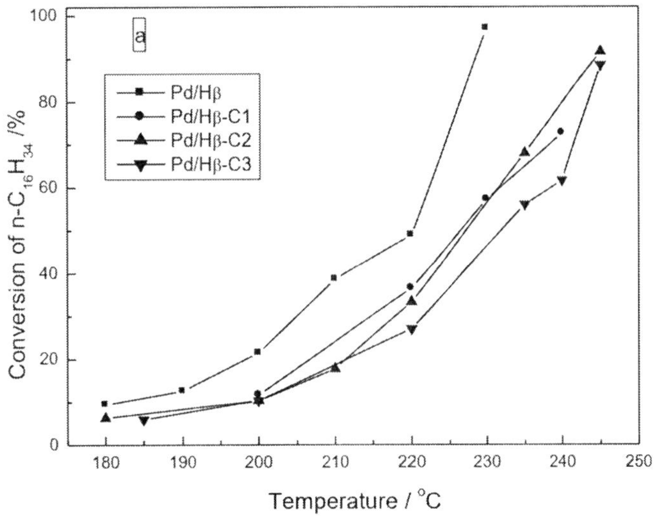


**Figure 5.** Hydroisomerization conversion of *n*-$C_{16}H_{34}$ over Pd/Hβ and Pd/Hβ-Cn catalysts

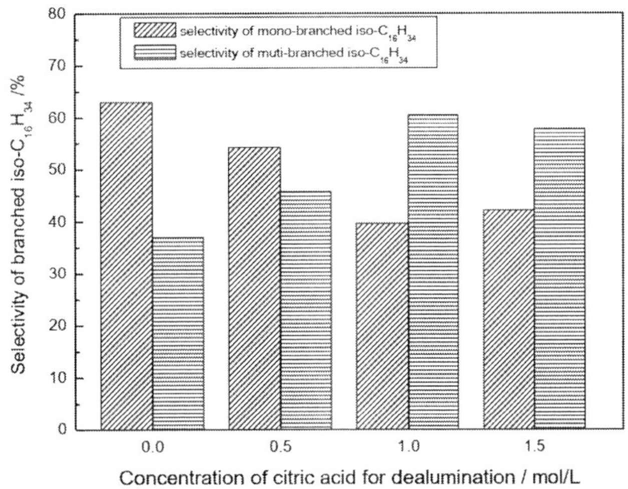


**Figure 6.** Selectivity of mono-branched *iso*-$C_{16}H_{34}$ and multi-branched *iso*-$C_{16}H_{34}$

## 4. Conclusion

In summary, Hβ molecular sieves ($SiO_2/Al_2O_3$ molar ratios of 40) synthesized by hydrothermal method were dealuminated by citric acid solutions with concentrations of 0.5mol/L, 1.0mol/L and 1.5mol/L respectively. The Hβ and modified Hβ-Cn loading 0.5 wt% palladium bifunctional catalysts (Pd/Hβ-Cn) were prepared by incipient wetness impregnation. The skeleton structure of all Hβ-Cn was not damaged and their relative crystallinity increased somewhat after the treatment of citric acid. The effect of citric acid ealumination on the morphology, structure and pore properties of Hβ was very little, but the amount of weak acid, strong acid and total acid of Hβ-Cn samples, and Pd dispersion for Pd/Hβ-Cn decreased significantly. The hydroisomerization reaction temperature over Pd/Hβ-Cn slightly rises under the condition of the same conversion of *n*-$C_{16}H_{34}$. The amount of multi-branched *iso*-$C_{16}H_{34}$ over Pd/Hβ-Cn was much higher than that over Pd/Hβ. The remove of partial skeleton aluminum species with strong acid sites effectively inhibited the further cracking of the muti-branched isomerized products.

**Acknowledgements**
This work was financially supported by National Natural Science Foundation of China (No. 21676074), Project of Heilongjiang Science and technology (Grant No. ZY17A06) and Science Foundation of Heilongjiang Academy of Sciences (Grant No. YY2017SH01& ZNBZ2018SH01).

**References**
[1] Gerhard K,Luis F R 2017 *Progress in Energy and Combustion Science* 58 *36*
[2] William P, James M, Juan R, et al 2017 *Fuel* 209 442
[3] Liu S, Ren J, Zhang H,et al 2016 *Journal of Catalysis* 335 11
[4] Zhang J W, Wu H M, Zhao A J 2017 *J Porous Mater* 24 437
[5] Chavarria J,Ramirez J,Gonzalez H,et al 2004 *Catalysis Today* 235
[6] Liu S, Ren J, Zhang H, et al. 2016 *Journal of Catalysis* 335 11

# The Impact Compressive Properties of 7075-T651 Aluminium Alloy over a Wide Range of Strain Rates and Temperatures

**Xin Guo and Li Song**

Mechanics and Materials Science Research Center, Ningbo University, Ningbo, 315211, China. Email: songli@nbu.edu.cn

**Abstract.** The mechanical properties of 7075-T651 Aluminium Alloy was studied under dynamic compressive loading over a wide range of strain rates from $600s^{-1}$ to $16000s^{-1}$ and the temperature range from 25°C to 400°C. A classic split Hopkinson pressure bar (SHPB) and a radiant-heating furnace are used for the experiment and measurement. Some improved test and data processing technique are used to get more accurate test result. It is found that the compressive stress-strain response is insensitive to the applied strain rate but depends on experimental temperature. The experiment results provide a good basis to build a more valid and accurate material models for 7075-T651 Aluminium Alloy.

## 1. Introduction

7075-T651 aluminium alloy has been extensively used in commercial and military air carriers. Its mechanical properties over a wide range of strain rates and temperatures is of interest. A considerable amount of work has been done toward understanding and modelling its mechanical behaviour under low strain rates and various temperatures [1-4]. However, up to now, there has been less work concerning the effects of high strain rate and temperature and little work has been done related to its mechanical behaviour under a ultra-high strain rate and high temperature. The purpose of this paper is to experimentally study and characterize the behaviour of a 7075-T651 aluminium alloy during dynamic compression over a wide range of strain rates and temperatures.

## 2. Experimental

The material used in this study is the 7075-T666 aluminum alloy. Its chemical composition is shown in Table 1. The cylindrical specimens were prepared from the plates with the compression axis parallel to the rolling direction. For the test of strain rate below 6000s-1, specimens were 6.0 mm in diameter and 4.0 mm in height and for higher strain rate experiments, a smaller specimen with a diameter of 3.0 mm and a height of 2.0 mm was used.

**Table 1.** Chemical composition of aluminum alloy 7075-T651 (WT. %)

| Ti | Si | Mn | Mg | Fe | Cr | Zn | Cu | Others |
|---|---|---|---|---|---|---|---|---|
| 0.2 | 0.4 | 0.3 | 2.1-2.9 | 0.5 | 0.18-0.28 | 5.1-6.1 | 1.2-2.0 | 0.15 |

Dynamic compression tests were conducted using a split Hopkinson pressure bar (SHPB) [1] (Fig.1). The diameter of the striker, incident bar and transmission bar is 14.5 mm. The length of incident bar and transmission bar is 1000mm and the lengths of the strikers are 40mm, 30mm and 15mm. A geometrically optimized thin copper disc (pulse shaper) is used for ensuring the stress equilibrium within the specimen during the test [5] (Fig.2). A wide band strain amplifier (frequency response from DC to 2MHz) and a Tektronix 3014B Digital Oscilloscope are used and it guarantees almost no additional signal loss during the test. The tri-wave data processing method which calculates the stress, strain rate and strain using incident wave signal, reflect wave signal and transmission wave signal was used to obtain a more accurate stress–strain curve comparing to the classic two-wave method[6]. When using tri-wave method, the time shifts from gage I and gage II to the bar-specimen contact surfaces need to be calibrated carefully.

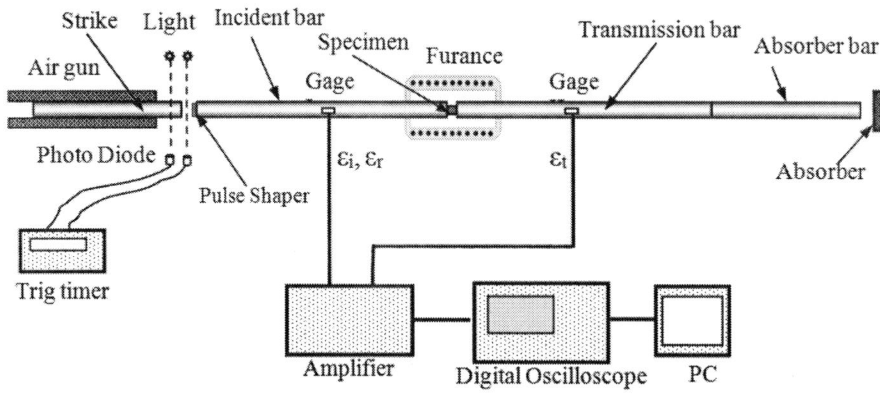


**Figure 1.** Schematic presentation of the compression split Hopkinson bar apparatus.

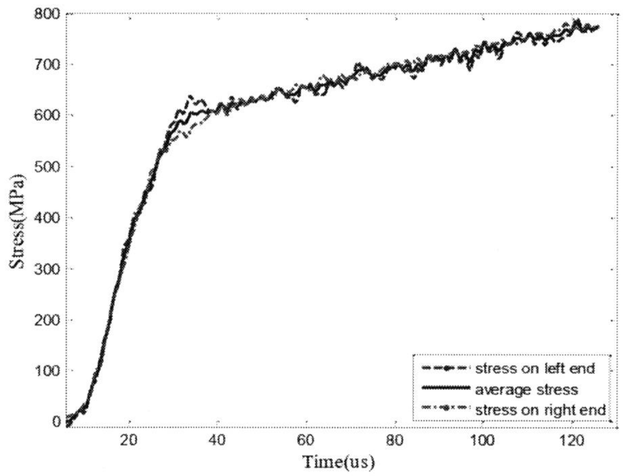


**Figure 2.** Stress equilibrium during a test.

A chamber radiant-heating furnace is used to provide heating temperature from 25°C to 400°C. The specimen and part of the pressure bars in the furnace were heated together and it ensures the perfect contact between the specimen and the pressure bars. Since the highest temperature in the experiment is 400°C, the effect of temperature on the propagation of stress wave in the rods can still be ignored. The temperatures of the specimen were monitored using a thermocouple (TC-2) attached on the transmission bar near the contact end with the specimen and a carefully calibration is made (Fig.3). Though directly attaching the thermocouple (TC-3) on specimen is the best way to measure the temperature, it may introduce some interference to the mechanical response of the specimen especially when a small specimen is tested. The thermocouple (TC-1) can also be just put in the chamber, but it is found that a temperature gradient exists within the chamber and a maximum 50°C measurement error may be introduced.

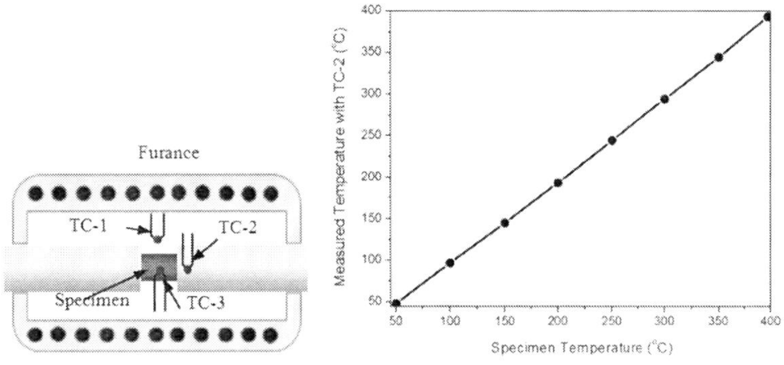


(a) (b)

**Figure 3.** Temperature measurement (a) and calibration (b)

## 3. Experimental Results

The experiments under the true strain rates from $600s^{-1}$ to $16000s^{-1}$ and temperature range from 25°Cto 400°C was carried out. In order to eliminate signal noise, a 500 kHz digital low pass filter is used on the recorded data. The tri-wave method is used for measurement data processing. After the voltage-strain relation and time shift calibration are completed, no manual intervention is needed in the data processing process. Figure 4 shows a set of typical test signals and experimental results. Theoretically, based on the tested true-stress–true-strain curves, the material flow stress, work hardening rate and maximum stress at high strain rate conditions can be derived. However, the elastic response of a material cannot be measured reliably using a SHPB apparatus [1] and it is difficult to determine the exact yield stress as a qusi static experiment. For this reason, the material flow stress based on the 5% plastic strain is introduced [7].

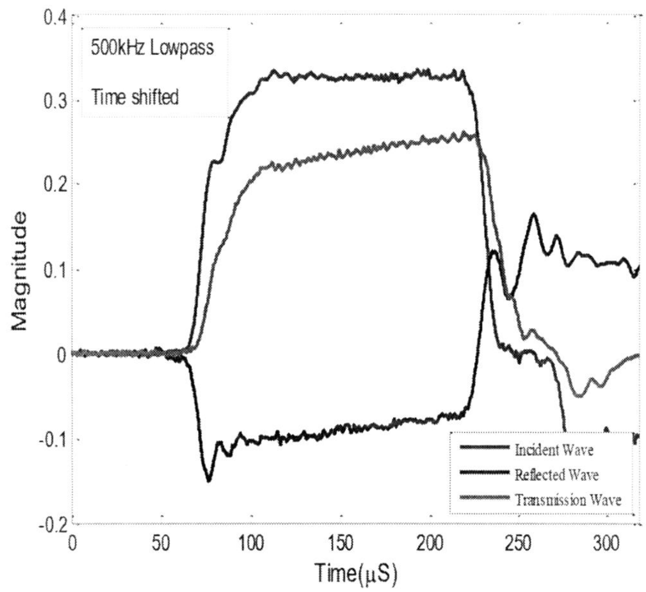


(a)

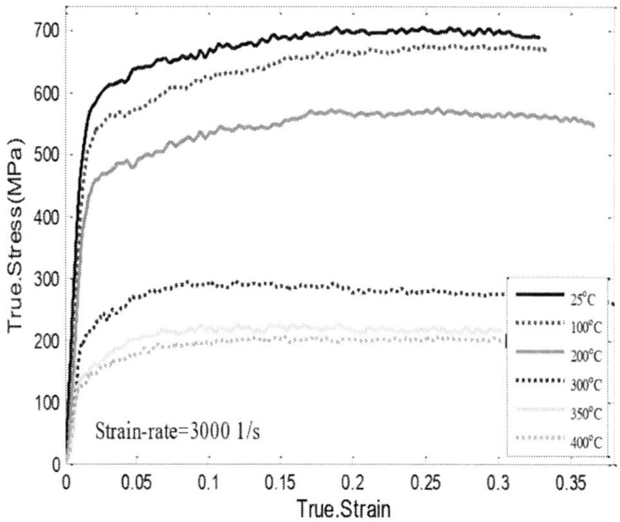

(b)

**Figure 4.** Typical experiment signal (a) and test results (b).

### *3.1. Strain Rate and Temperature Dependence of Flow Stress*

The flow true stress at 5% true strain as a function of strain rate and temperature is shown in Fig.5 (a) and (b). It can be seen (Fig.5 (a)) that the 7075-T651 aluminum alloy shows a weak strain rate sensitivity. The flow stress shows slightly positive strain rate dependence only at the room temperature. When the temperature is higher than 100°C and lower 350°C, the flow stress is almost independence of strain rate. The flow stress even shows a negative strain rate dependence when the temperature is higher than 350°C and It can be explained by strain rate thermal softening due to Adiabatic temperature rise when the strain rate is very high.

As shown in Fig.5 (b), the flow stress declines rapidly with increasing temperature and Drops sharply between 200°C and about 320°C. When the temperature is above 320°C, the drop rate of the flow stress becomes much slower. It seems that the temperature dependence of the flow stress is independent of strain rate.

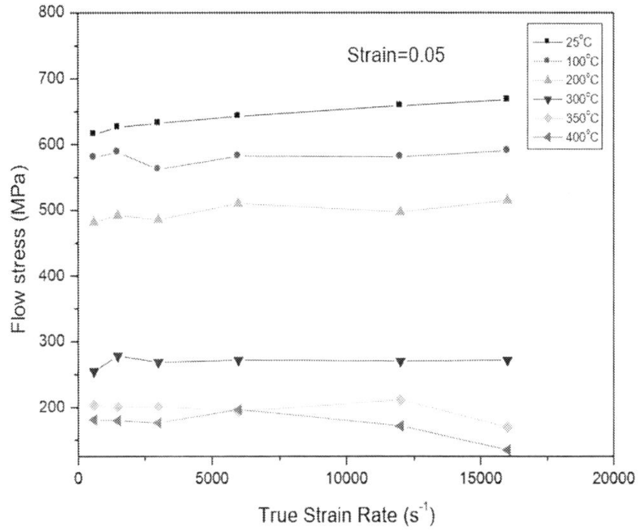


(a)

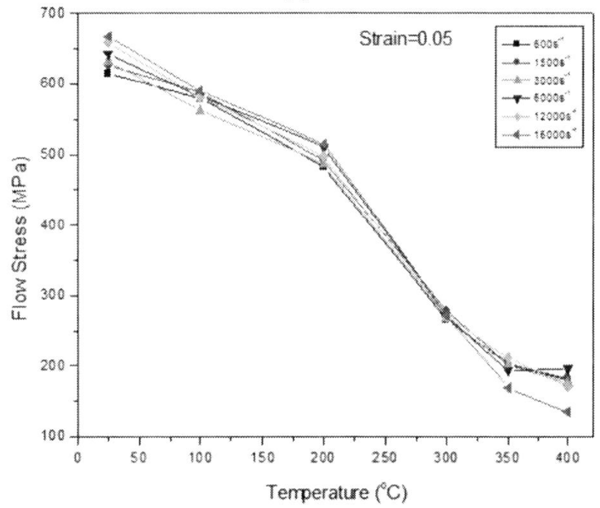


(b)

**Figure 5.** The dependency of flow stress on (a) strain rate and (b) temperature for the 7075-T651 aluminum alloy.

### *3.2. Work Hardening Rate and Plastic Instability*

The stress-strain curves are plotted in Fig.6 for different temperature and strain rate. It can be seen that the work hardening rates of the alluminum alloy only slightly changed with the temperature change (Fig.6(a)) and the work hardening rates have little change for different strain rates (Fig.6(b)).

The dependence of dynamic plastic instability point that the slope of the stress-strain curve becomes negative on specific strain rate and temperature is not mentioned in literatures because the maximum strain during this experiment usually was not enough to show this phenomenon. It is shown that this dynamic plastic destabilizing behavior clearly affected by temperature and strain rate. In Fig.6 (a), the instability strain decreases with increased temperature when temperature is below 300°C but when temperature is above 350°C, the plastic instability did not appear. In Fig.6 (b), the plastic instability strain decreases with strain rates are between $16000s^{-1}$ and $6000s^{-1}$. For lower strain rate, since the experimental maximum strain is too small, no conclusion can be reached. In general, the plastic deformation of the alloy can be divided into three regions: the work hardening region, the platform region and the plastic instability region.

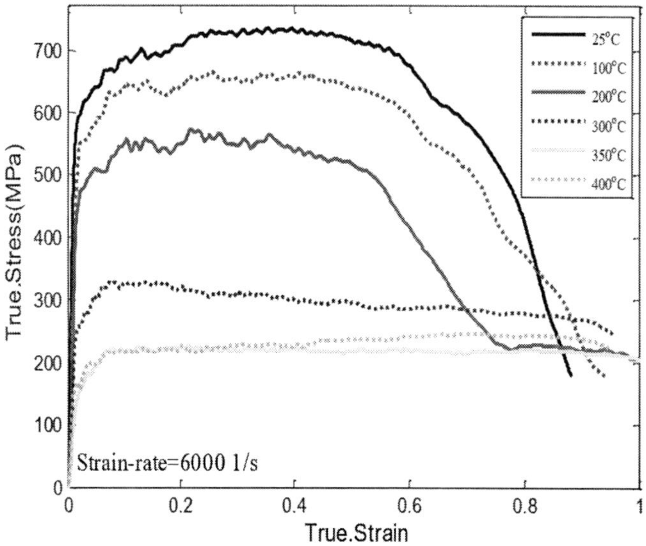


(a)

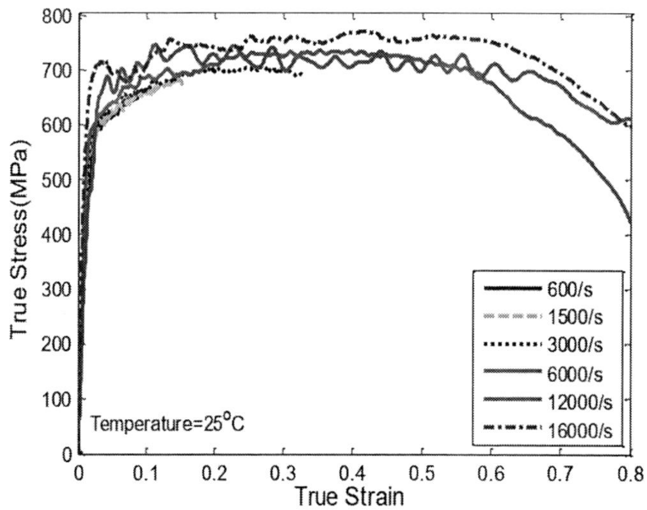


(b)

**Figure 6.** The true stress-strain curves for (a) different temperature and (b) different strain rate

## 4. Conclusions

The mechanical properties of 7075-T651 Aluminum Alloy were studied under dynamic compressive loading over a wide range of strain rates and temperature. Based on this study, following conclusions can be made:

- The flow stress of 7075-T651 aluminum alloy almost is constant with increasing strain rate at higher temperature. The flow stress only slightly increases with increasing strain rate at room temperature.
- The flow stress decreases with increasing temperature and the maximum flow stress decline rate appears in the temperature region between 200°C and 320°C.When temperatures are above 320°C,

the decrease of flow stress with increasing temperature significantly slowdown. The overall relation between flow stress and temperature cannot be simply represented with a power function.

- The work hardening rate is only slightly affected by temperature and strain rate. The dynamic plastic instability strain decreases with increased temperature and strain rate for the existing experimental data and more experiment is needed.

**Acknowledgments**
This work was financially supported by NSF of China, Project No. 11272162.

**References**

[1] Chen W W, Song B. Split Hopkinson (Kolsky) Bar, Design Testing and Applications [J]. 2015.
[2] Lee, Woei Shyan, et al. "The strain rate and temperature dependence of the dynamic impact properties of 7075 aluminum alloy." Journal of Materials Processing Tech 100.1(2000):116-122.
[3] Jin, Zhe, et al. "Mechanical Response of AA7075 Aluminum Alloy over a Wide Range of Temperatures and Strain Rates." Materials Science Forum 331-337(2000):527-532.
[4] Senthil, K., et al. "Study of the Constitutive Behavior of 7075-T651 Aluminum Alloy." International Journal of Impact Engineering 108(2017):171-190.
[5] Li, Song, and Shisheng, H. "Two-wave and three-wave method in SHPB data processing." Explosion & Shock Waves (2005).
[6] Oosterkamp, L. Djapic, A. Ivankovic, and G. Venizelos. "High strain rate properties of selected aluminium alloys." Materials Science & Engineering A 278.1(2000):225-235.
[7] Abotula, Sandeep. Dynamic Constitutive Behavior of Aluminum Alloys: Experimental & Numerical Studies. Experimental and Applied Mechanics, Volume 6. Springer New York, 2011:65-66.

# Heat Transfer Characteristics of Twin Circular Jet Impingement Cooling on the Strip Surface after Rolling

**Jianhui Shi*, Liping Bian, Ruquan Liang and Song Xiao**

School of Mechanical and Vehicle Engineering, Linyi University, Linyi 276000, China
*Corresponding author
Email:sjh051208@126.com

**Abstract.** The heat transfer characteristics of the strip surface due to the twin circular jet impingement were numerically investigated by a three-dimensional finite element program. Systematic parametric research is conducted by varying the jet velocity of nozzle exit of 7.5 m/s, nozzle to the steel plate surface distance (H=6.7, 11.7, 16.7), and the nozzle to nozzle centre spacing (W=10, 15). The general trend of the distribution of the local Nusselt number and the average Nusselt number variation on the impingement surface of strip was obtained under the various parameters. The results show that, the heat transfer capacity of the strip surface perpendicular to the centre of the two jets decreases with the increase of the nozzle to nozzle centre spacing. The average Nusselt number increases by about 2%-20% with the increase of nozzle to the steel plate surface distance from 6.7 to 11.7 for the nozzle to the steel plate surface distance of 100mm and 200mm, while the average Nusselt number decreases by about 11%-25% with the increase of nozzle to the steel plate surface distance from 11.7 to 16.7.

## 1. Introduction

Jet impingement cooling is an extremely effective cooling method to achieve high heat transfer rates. In the steel making industry, water-cooling technologies are widy used for controlling the steel plate temperature. For the steel plate and strip, heat transfer of jet impinging is the key process of the ultra-fast cooling after rolling and it affects the cooling rate and uniformity of them. The study on twin and multi jet impingement cooling is very important because controlling the cooling rate and uniformity can improve the microstructure and mechanical properties of the steel plate and strip.

Xu el al. [1] carried out a numerical investigation on two-dimensional heat transfer behavior in the impingement zone under single circular water jet. Dou el al. [2] experimentally investigated the heat transfer characteristics of a circular water jet impingement on high-temperature steel plate. They found that increasing the surface roughness can provided better heat transfer in the stagnation region. The experimental and numerical study on the transient heat-transfer characteristics of circular air-jet impingement on a flat plate have been carried out by Guo el al. [3]. The study was performed for the inner diameter of circular nozzle of 6 mm and the Reynolds number of 14,000 -53,000. Shi et al. [4] conducted a numerical investigation on the heat transfer symmetry of a single group oblique slot jet impingement for high temperature strip after hot rolling. They reported that the functional relationship between the systematic parameters and the discrepancy of the average Nusselt number or the coordinate values of the stagnation point. Wang et al. [5] investigated the effect of the initial surface temperature, water temperature, and jet velocity on the heat transfer characteristics during the cooling process of a water jet impinging on a hot plate surface. They observed that the wetting delay time was extended by the superheat increase and the sub-cooling and jet velocity decrease.

Frosell et al. [6] investigated the dynamics of the impingement region of a circular turbulent jet for the nozzle diameter of 6.4mm, the distance from the nozzle to the impingement surface of 2.5-8.5 and the Reynolds number of 71,000 -142,000. Their results showed that the size and the movement of the impingement region are observed to increase linearly with the standoff distance of the nozzle, while the size of the impingement region is independent of the fluid velocity at the nozzle. Brian et al. [7] experimentally studied the effects of volumetric quality (0-0.9) on the hydraulic jump radius, local Nusselt number and the pressure at the stagnation point for air-assistant circular water jet impinging. Their results showed that the dimensionless hydraulic jump radius increased with volumetric quality, attained a maximum value at around 0.8 of the volumetric quality, and then decreased. A few studies on the circular jet impingement cooling of the circular cylinder have been presented by Lee [8], Wang [9], Amirhosein [10] and Dushyant [11].

The object of the above studies is single circular jet impinging cooling, while a few studies on two or multi jet impingement cooling were carried out. Abdel [12] conducted experimental and numerical studies on the two dimensional impinging twin circular air jet flow with no-cross flow. The parametric was carried out for Reynolds number of 95,000-224,000, nozzle to plate spacing (h/d) of 3-12, nozzle to nozzle centerline spacing (l/d) of 3, 5 and 8, and jet angle of 0-20°. It was observed that the spreading of jet decreases by increasing nozzle to plate spacing, the intensity of re-circulation zone between two jets decreases by increasing of h/d and jet angle, and the increase of turbulence kinetic energy occurs within high gradient velocity. Greco et al. [13] experimentally study the behavior of impinging single and twin circular synthetic jets in phase opposition. Dushyant et al. [14] carried out an experimental and numerical investigation on flow and heat transfer characteristics of double circular air jet impingement cooling of a heated circular cylinder. They reported that the average Nusselt number over the cylinder surface increased with the increase in non-dimensional distance between double the jets and the confinement at the bottom of the cylinder increases heat transfer.

Christopher [15]2016 carried out a numerical investigation on convective heat transfer of multiple impinging circular jets for a horizontal circular cylinder. They reported that movement away from the cylinder axis, increased axial spacing and lateral offset all lead to degradation of heat transfer from the cylinder for jet-to-cylinder ratios of the jet-to-cylinder diameter ratio great than or equal to 0.23.Satyanand [16]2018 investigated the effectiveness distribution measurements for a row of heated circular jets impinging on a cylindrical convex surface at different inclinations.

## 2. Numerical Procedures

### 2.1. Governing Equations

Reynolds averaged equations can be written using mass Eq. (1), momentum Eq. (2) and energy Eq. (3) for the steady incompressible flow.

$$\frac{\partial u_i}{\partial x_i} = 0 \tag{1}$$

$$\rho u_j \frac{\partial u_i}{\partial x_j} = \frac{\partial P}{\partial x_i} + \frac{\partial}{\partial x_j}\left[\mu\left(\frac{\partial u_i}{\partial x_j} + \frac{\partial u_j}{\partial x_i}\right) - \rho\overline{u'_i u'_j}\right] \tag{2}$$

$$\rho u_j \frac{\partial T}{\partial x_j} = \frac{\partial}{\partial x_j}\left[\frac{\mu}{\Pr}\frac{\partial T}{\partial x_j} - \rho\overline{T'u'_j}\right] \tag{3}$$

Where, $P$ is the mean pressure, $T$ is the mean temperature, $T'$ is the fluctuating temperature, $u_i$ and $u_j$ are the mean velocity, $u_i'$ and $u_j'$ are the fluctuating velocity, $x_i$ and $x_j$ are the coordinate direction, $\rho$, $\mu$, and $Pr$ are the fluid density, the dynamic viscosity, and the Prandtl number, respectively.

### *2.2. Boundary Conditions and Initialization*
Fig. 1 shows the three-dimensional simplified model of the circular jet impinging on the hot rolling strip. The shaded region in the bottom of the model is the section of the strip, and the other cylindrical areas (ABCD) above the shaded region are the fluid calculation regions. The initial medium of fluid calculation regions is static air and the medium of jet is incompressible cooling water. The inlet jet velocity ($V_0$) of the circular jet is uniform. The diameter of the fluid inlet is d and r is the radius. The distances of the circular jet outlet to the strip surface is H (H= h /r), and h is the actual distance of the circular jet outlet to the strip surface. The distance between the centre of two circles jet is W (W= w /r), and w is the actual distance between the centre lines of two circular jet flows. The initial temperature of the cooling water is maintained at 300 K. The velocity component of the wall is zero. The normal pressure gradient of the outlet boundary is zero. The initial temperature of strip is evenly distributed (1123 K). The surface of the strip is specified as isothermal wall.

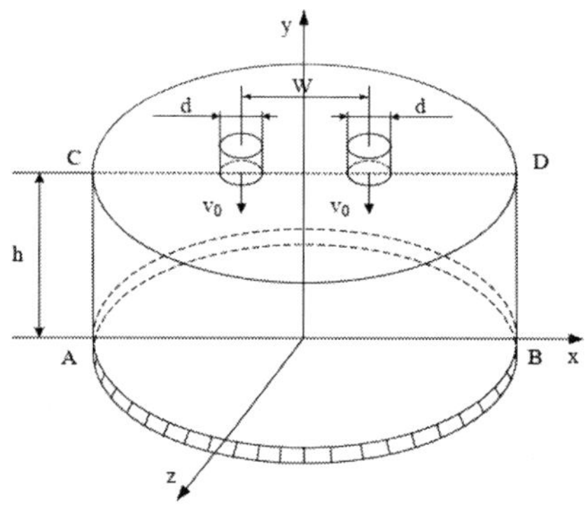


**Figure1.** The three-dimensional simplified model of the twin circular impinging jet

## 3. Simulation results and Analysis
Numerical investigations were carried out, and the parameters are the jet inlet velocity (V=7.5m/s), the nozzle to the steel plate surface distance (H=6.7, 11.7, 16.7), and the nozzle to nozzle centre spacing (W=10, 15) and the diameter of the fluid inlet (d=6mm). The Nusselt numbers of the X-axis and Z-axis on the steel plate surface were obtained under the various parameters. The X-axis is parallel to the centre of the two jets, and the Y-axis is perpendicular to the centre of the two jets.

### *3.1. Local Nusselt Number Distribution Along the Strip Surface*
In order to study the heat transfer features of the twin circular jet impingement cooling processes, the distribution of the local Nusselt number on the strip surface of the strip for X-axis and Z-axis is plotted in Fig. 2 to Fig. 3. These plots are shown for a length of the heat transfer surface of 200mm from the symmetry plane (x = 0 or z=0).

In Fig.2, the overall distribution trend of the local Nusselt number on strip heat transfer surface is that the valley value of the Nusselt number appears at the symmetry point for X-axis, and two peak values of the Nusselt number form near the stagnation point. The valley value of the Nusselt number decreases with the increase of the nozzle to nozzle centre spacing from 10 to 15, and the two peaks near the stagnation point also change from steep to gentle. While the range of the heat transfer surface of the strip gradually increased with the increase of the nozzle to nozzle centre spacing.

In Fig.3, the overall distribution trend of the local Nusselt number for Z-axis is that the multiple peak values of the Nusselt number appear at strip heat transfer surface. When the non-dimensional domain of the two nozzles centre is 10, the general trend of the valley value near stagnation point is closed to the X-axis. As the two jets are relatively close, the main heat transfer region still appears at the middle area. The Nusselt number of the heat transfer surface for X-axis became more uniform with the increase of the non-dimensional domain of the two nozzles centre, and multiple small peaks form along the heat transfer surface, and the peak values decreased. It indicates that the heat transfer capacity of the strip surface perpendicular to the centre of the two jets decreases with the increase of the nozzle to nozzle centre spacing.

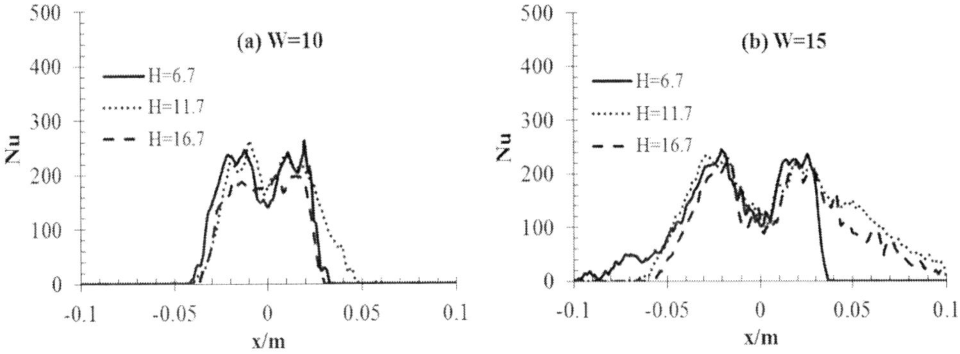


**Figure 2.** Local Nusselt number distribution along the strip surface with jet velocity of 7.5m/s for x

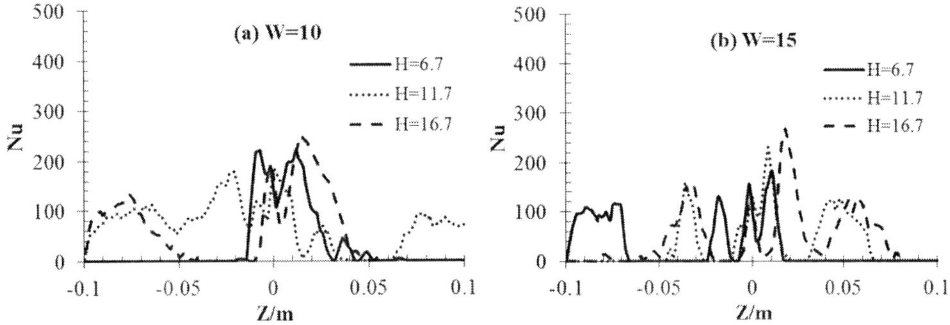


**Figure 3.** Local Nusselt number distribution along the strip surface with jet velocity of 7.5m/s for z

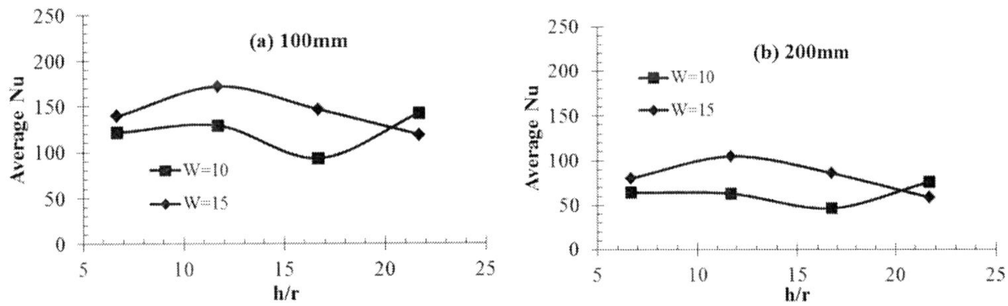


**Figure 4.** Average Nusselt number variation along the strip surface of 7.5m/s for x axis

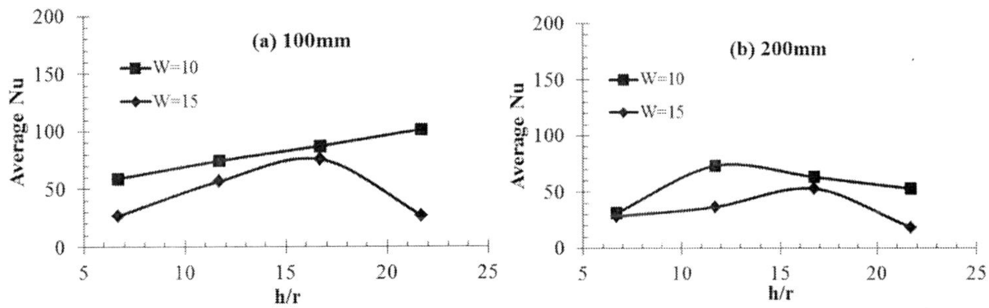


**Figure 5.** Average Nusselt number variation along the strip surface of 7.5m/s for z axis

### *3.2. Average Nusselt Number Variation along the Strip Surface*

Fig. 4 and Fig. 5 show the average Nusselt number variation of the heat transfer surface with the effect of various parameters for X-axis and Z-axis. The analysis length of strip surface is 100mm (-50mm to 50mm) in Fig. 4 (a), and it is 200mm (-100m to 100m) in Fig. 4 (b). It is noticed that the average Nusselt number of the strip surface for X-axis increases with the increase of the nozzle to the steel plate surface distance for H of 6.7 to 11.7. In addition to some irregularities of the average Nusselt number, the general trend is that the average Nusselt number increases by about 6%-15% with the increase of H for the nozzle to the steel plate surface distance of 100mm, and it increases by about 2%-20% for the nozzle to the steel plate surface distance of 200mm. while the average Nusselt number of the strip surface for X-axis decreases with the increase of the nozzle to the steel plate surface distance for H of 11.7 to 16.7. The average Nusselt number decreases by about 11%-22% with the increase of H for the nozzle to the steel plate surface distance of 100mm, and it decreases by about 15%-25% for the nozzle to the steel plate surface distance of 200mm.

## 4. Conclusions

The heat transfer characteristics of the twin circle jet impingement for the parameter of nozzle to the steel plate surface distance and the nozzle to nozzle centerline spacing is investigated numerically using the ANSYS Fluent. The ratio of the the steel plate surface distance is varied from 6.7 to 16.7, and the nozzle to nozzle centerline spacing is forced from 10 to 15. The distribution of the local Nusselt number on the strip surface for X-axis and Z-axis were obtained. The average Nusselt number variation of the heat transfer surface with the effect of various parameters for the analysis length of 100mm to200mm was investigated. The valley value of the Nusselt number decreases with the increase of the nozzle to nozzle centre spacing from 10 to 15, and the two peaks near the stagnation point also change from steep to gentle. The average Nusselt number increases by about 2%-20% with the increase of H from 6.7 to 11.7 for the nozzle to the steel plate surface distance of 100mm and 200mm. while the average Nusselt number decreases by about 11%-25% with the increase of H from 11.7 to 16.7.

## Acknowledgments

This work is supported by the Shandong Provincial Natural Science Foundation of China (Grant NO. ZR2018PEE004).

## References

[1] Xu F C, Gadala M S, Heat transfer behavior in the impingement zone under circular water jet 2006, *Int. J. Heat Mass Transf.* **49** 3785-3799.

[2] Dou R F, Wen Z, Zhou G, Liu X L and Feng X H, Experimental study on heat-transfer characteristics of circular water jet impinging on high-temperature stainless steel plate 2014, *App. Therm. Eng.* **62** 738-746.

[3] Guo Q, Wen Z and Dou R F, Experimental and numerical study on the transient heat-transfer characteristics of circular air-jet impingement on a flat plate 2017, *Int. J. Heat Mass Transf.* **104** 1177-1188.

[4] Shi J H, Yuan G, Jiang L Y, Li Z L, Zhao K and Wang G D, Heat transfer symmetry of the strip surface due to a group oblique slot jet impingement after hot rolling 2015, *Steel Res. Int.* **86** 1548-1557.

[5] Wang B X, Lin D, Xie Q, Wang Z D and Wang G D, Heat transfer characteristics during jet impingement on a high-temperature plate surface 2016, *App. Therm. Eng.* **100** 902–910.

[6] Frosell T, Fripp M and Gutmark E, Dynamics of the impingement region of a circular turbulent jet 2018, *Exp. Therm. Fluid Sci* **91** 399-409.

[7] Brian K F, Tamira D F, Aspen W G and Kyosung C, Experimental study of the hydrodynamic and heat transfer of air-assistant circular water jet impinging a flat circular disk 2017, *Int. J. Heat Mass Transf.* **106** 804–809.

[8] Lee D H, Chung Y S and Kim D S, Turbulent flow and heat transfer measurements on a curved surface with a fully developed round impinging jet 1997, *Int. J. Heat Fluid Fl.* **18** 160-169.

[9] Wang X L, Motala D, Lu T J, Song S J and Kim T, Heat transfer of a circular impinging jet on a circular cylinder in crossflow 2014, *Int. J. Therm. Sci.* **78** 1-8.

[10] Amirhosein H, Mehran R Z, Heat transfer and flow characteristics of impinging jet on a concave surface at small nozzle to surface distances 2018, *App. Therm. Eng.* **138** 534-541.

[11] Dushyant S, Premachandran B and Sangeeta K, Circular air jet impingement cooling of a circular cylinder with flow confinement 2015, *Int. J. Heat Mass Transf.* **91** 969–989.

[12] Abdel F A, Numerical and experimental study of turbulent impinging twin-jet flow 2007, *Exp. Therm. Fluid Sci.* **31** 1061–1072.

[13] Greco C S, Castrillo G, Crispo C M, Astarita T Cardone G, Investigation of impinging single and twin circular synthetic jets flow field 2016, *Exp. Therm. Fluid Sci.* **74** 354–367.

[14] Dushyant S, Premachandran B and Kohli S, Double circular air jet impingement cooling of a heated circular cylinder 2017, *Int. J. Heat Mass Transf.* **109** 619–646.

[15] Christopher C, Anthony G S, Forced convective heat transfer on a horizontal circular cylinder due to multiple impinging circular jets 2016, *App. Therm. Eng.* **105** 290-303.

[16] Satyanand A, Rajendra P V, Effectiveness distribution measurements for a row of heated circular jets impinging on a cylindrical convex surface at different inclinations 2017, *Int. J. Heat Fluid Fl.* **69** 210-223.

IOP Conf. Series: Materials Science and Engineering 472 (2019) 012043 doi:10.1088/1757-899X/472/1/012043

# Study on Recrystallization Organization of Workpiece about Processing Hole Using Friction Heat

**Xiaocun Xu[1,2], Xin Li[1] and Yu Zhao[1]**

[1] School of Mechanical and Power Engineering, Harbin University of Science & Technology, Harbin, Heilongjiang Province 150080, P.R. China
[2] Email: 1120549159@qq.com

**Abstract.** The new technogical process is a kind of green mechanical processing without chip in this paper, the process can form relative deep hole in thin panel by friction heat between the drilling tool and workpiece, and it not only saves material, but also saves man-hours. In the paper, simulated analysis about the influence of different rotate speed and feed speed on the temperature field of forming hole of workpiece, and researched on the influence of processing parameters to recrystallization percent distribution and grain size of the around hole; Finally studied experimentally on the microstructure distribution by FEI Sirion and contrasted analysis the grain size around the hole of different location of test and simulation. The research results show that the simulation results agree well with the actual test results, and the appropriate process parameters are found to form hole under the conditions of this paper. The research results are useful for further optimization of the process in order to produce deep hole workpiece with ideal recryatallization organization.

## 1. Introduction

The current machining process goal is to have the machining process with high efficiency, low consumption and high quality. In the modern manufacturing, hole is a very important part of spare parts. In the traditional processing of thin-wall hole machining, hole depth and thickness are equal, and not directly to machining thread of follow-up process. It is a major task to how to reduce resource consumption and reduce the environment pollution for the current manufacturing [1]. In this paper, the new technological process to form deep hole in relative thin panel by friction heat not only saves material, but also saves man-hours, and the workpiece has a good piece of assembly [2].

The process is a kind of green mechanical processing without chip. The process is similar to the extrusion process [3]. Due to the hole wall is got through the hot deformation of a material, so the metal flow line distribution is more reasonable, and the hole has good mechanical properties. Numerical simulation is a kind of effective methods to research sheet metal stamping and extrusion [4,5]. Currently,the main research contents are for drill friction heat production model, the influence of process parameters on the drilling and the blade geometry design [6,7], and the drill tool [8].

The important problem is studying on workpiece quality for the processing. The Performance of workpirce depends on the workpiece microstructure, and the hole is formed by friction heat, so a key problem of forming hole is about microstructure of forming area of the hole. The paper focused on the recrystallization organization of hole forming area.

## 2. Research Influence of Process Parameters on the Microstructure

### *2.1. Simulation Conditions*

The test workpiece is 20 steel whose thickness is 4mm in this paper. The drilling tool materials is YG6X cemented carbide. The key parameters in simulation: environment temperature is 20°C, deformation temperature is 900~1200°C, drilling tool rotate speed is 1500~3500rpm, feed speed is 0.2~2mm/s, original grain size is 50μm, the friction factor is 0.2.

At first the two geometric models about workpiece and drilling tool are drawn with UG, and convert to STL format into the finite element simulation software Deform. Figure1 shows mesh work.

The rotation speed and feed speed of drill tool are important influence factors on the hole forming process and structure property of workpiece, therefore, the following analysis is made in this paper in order to get the optimum technological parameters in the process of machining deep hole on sheet.

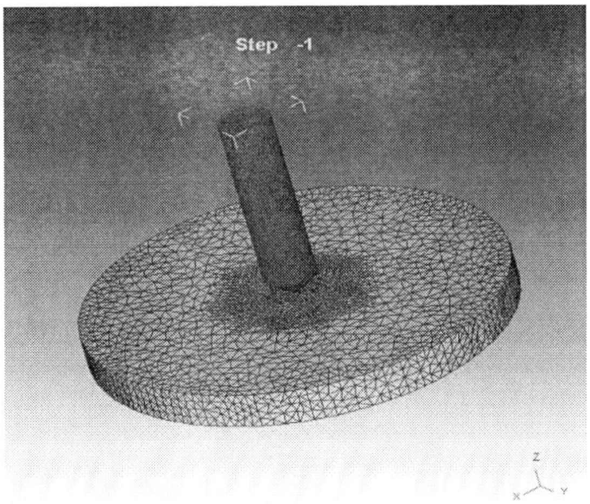


**Figure 1.** Meshed schematic of two objects

### *2.2. Influence of Rotating Speed on the Microstructure*

In the process of processing deep hole in sheet, different rotating speed will produce different frictional heat when the feeding speed of drilling tool is constant. Deformation temperature is important to influence workpiece forming effect and grain size and grain distribution after deformation, the study is useful because the microstructure of the hole decides performance of workpiece.

Figure 2 shows sheet temperature field distribution with different rotational speed when the feed rate of drilling tool is 0.6 mm/s.

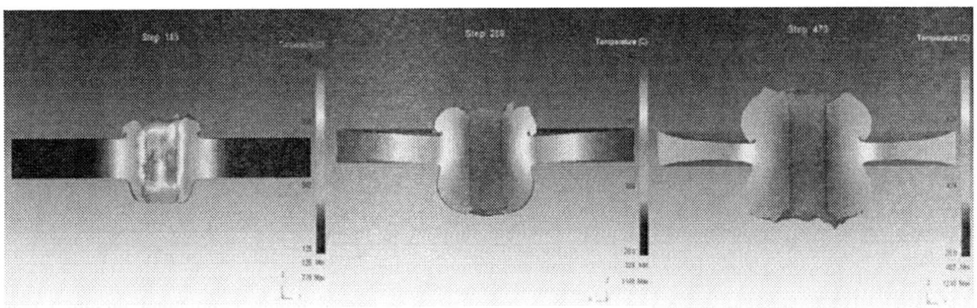

(a) n=1500 rpm (b) n=2500 rpm (c) n=3500 rpm

**Figure 2.** Distribution of temperature field with different speed cases

Figure 2 shows that deformation zone temperature of sheet also increase gradually with the increase of drilling tool rotation speed when the tool feed rate is constant. The highest temperature of deformation area corresponding to three kinds of drilling rate are 778°C, 1140°C and 1230°C respectively. The highest temperature of figure 2(a) and figure 2(c) are beyond the scope of reasonable hot working temperature. Figure 2 shows that shaping effect is not the same under different conditions, the length of the hole wall deformation area also increased with the increase of the maximum temperature deformation area, the reason is that metal plasticity increases with the increase of temperature, the movement of metal atoms under the precondition of the same feed speed, and plastic enhance , metal break after longer time, the length of the hole wall deformation area increases.

Figure 3 is percentage distribution of dynamic recrystallization of deformation area under different rotational speed when feed rate of drilling tool is 0.6 mm/s .

Different deformation zone temperature range agrees with the percentage of dynamic recrystallization and distribution according to figure 2 and figure 3. When rotational speed of drillng tool is 1500rpm, and the temperature ranges is between 125°C and 778°C, a lot of deformation area temperature can't reach appropriate recrystallization temperature of 20 steel, therefore, the recrystallization happened in the wall of hole area is less, the percentage of dyna125°C and 778°C, mic recrystallization is about 0.05. When rotational speed of drillng tool is 2500rpm or 3500 rpm, the simulation results show that both dynamic recrystallization percent peak of deformation area is 1, but distribution area is different, compared to dynamic recrystallization of regional distribution of the former is more reasonable than latter. The highest temperature of latter reached 1230°C which is beyond to the reasonable processing temperature range, and the grains of forming area become big, then the workpiece makes some defects such as superheat, overburning, and decarburization.

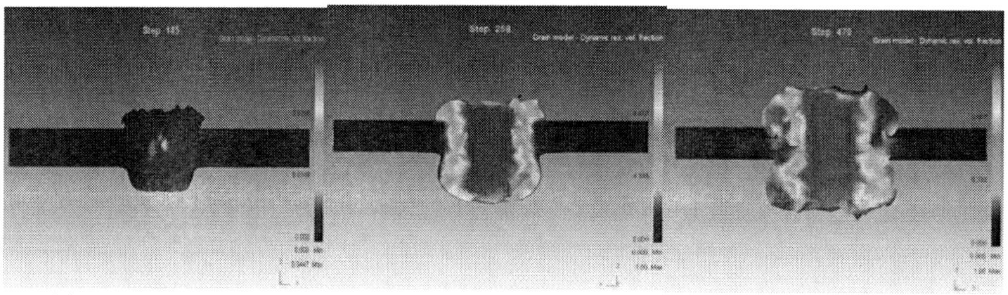

(a) n=1500rpm (b) n=2500rpm (c) n=3500rpm

**Figure 3.** Distribution of dynamic recrystallization with different speed

Drill speed of too high or too low will reduces the deformation of organizational performance of forming hole area. When the drilling tool rate is too low, the deformation temperature of forming hole area can not reach the minimum temperature of recrystallization temperature, and not for dynamic softening, then microstructure not improve; when drilling rate is too high, the temperature of the deformation zone temperature exceeds the reasonable range, grains of workpiece forming area become big, and workpiece performance deteriorate. So its reasonable speed of the drilling tool is about 2500rpm in this paper.

### 2.3. *Influence of Feed Speed on the Microstructure*

In the process to form deep hole in sheet, the feed speed is an important influence factor also for the final shape effect and the performance of important parameters of the product. The feed speed is not only related to the final deformation zone temperature, but also limits the time to happen dynamic recrystallization in workpiece material. It is very necessary to study the effects of the change of the feed speed.

Figure 4 is sheet temperature field distribution with different feed rate when the rotational speed of drilling tool is 2500rpm. Feed speed can also make very different deformation zone temperature shown as figure 4.

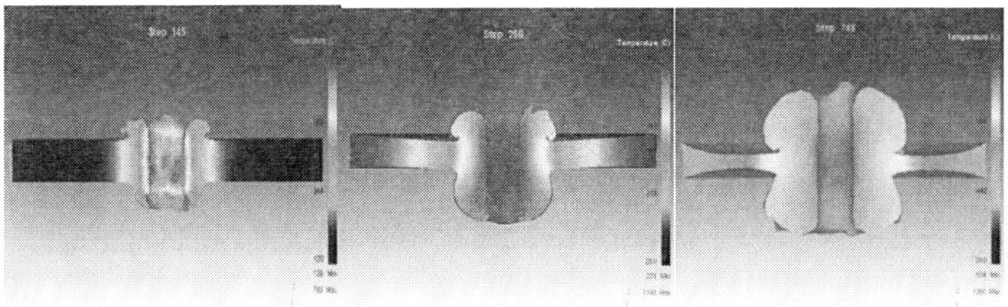

(a) f=2mm/s (b) f=0.6mm/s (c) f=0.2mm/s

**Figure 4.** Distribution of temperature field with different feed speed

Under the condition of the same rotate speed, feed speed of drilling tool is more slow and more long contact time between the drilling tool and the sheet, and the friction heat is more, though time of heat exchange from the sheet to outside air in the process is longer also, the exchange heat is much less than frictional heat , then lead to higher temperature of deformation zone. In addition, because the exchange heat time is long, temperature distribution in the workpiece is relatively even; the contact time of drilling tool and sheet decreases, temperature of workpiece is reduced, and the around temperature of workpiece is lower becauce of short exchange heat time and high temperature of the centre of forming area.

Figure 5 is percentage distribution of dynamic recrystallization of deformation area with different feed rate of drilling tool when rotational speed is constant.

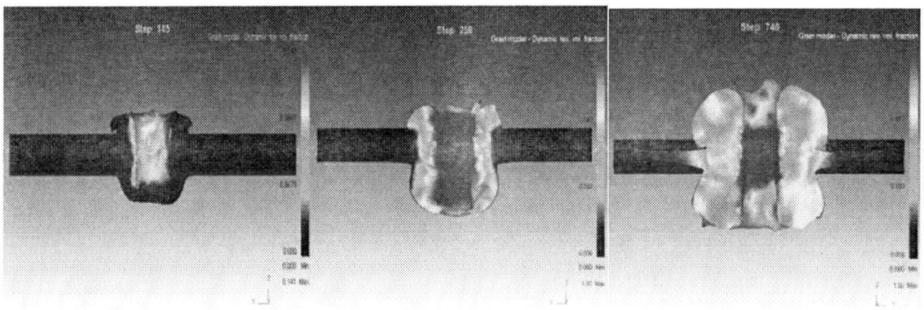

(a) f=2mm/s (b) f=0.6mm/s (c) f=0.2mm/s

**Figure 5.** Layout of dynamic recrystallization with different feed speed

Different feed speeds can cause a lot of difference of temperature of the deformation area, and the deformation of dynamic recrystallization fraction and distribution are obviously different shown as figure 5, so it is similar to the impact of rotate speed of drilling tool to temperature of forming area. In addition, the deformation speed has another influence on the recrystallization distribution of workpiece material, when the feed speed is too fast, the inside of the material grain failed to fully dynamic recrystallization occurs, the recrystallization can't overcome work hardening in a timely manner, which causes high hardness and brittleness material of deformation area, and adverse effect especially for subsequent thread processing. Sometime the workpiece material may break more early, and the depth of forming hole is not ideal.

In a word, reasonable process parameters can achieve ideal appearance size and improve organizational performance. In the paper, relative reasonable process parameters are 2500rpm of rote speed and 0.6mm/s of feed speed by finite element software Deform-3D.

## 3. Simulation Results and Experimental Analysis

Figure 6 is profile of temperature field of cutting plane. The temperature variation range after workpiece deformation is between 223°C and 1140°C shown as figure 6 according to the analysis of cloud picture, the temperature of wall of hole is highest in the picture, which is due to a lot of heat generated in the friction directly when the drilling tool touches sheet. Due to the effect of thermal diffusion of the workpiece itself, the heat of workpiece center diffuses to out the deformation area and results in the decrease of the temperature gradient distribution, uniform, the temperature of the temperature of panel edges decreases to around 220°C which is beyond to the lowest temperature to occur recrystallization.

Figure 7 is percentage distribution of dynamic recrystallization of the symmetry plane. The comparative analysis between figure 6 and figure 7 can conclude that the dynamic recrystallization percentage distribution corresponds with the results of the temperature field distribution. The highest temperature is in the center of whole wall where the deformation is largest in all deformation area, so the dynamic recrystallization of the area is sufficient to happen, and the microstructure is also refined. The temperature of edge of whole wall decreases, but it is still higher than the temperature of recrystallization temperature, it similar to occur the part of the dynamic recrystallization. The temperature of area outside of deformation section is low so that the deformation quantity is little or no deformation, therefore, dynamic recrystallization can't occur.

The grain size and distribution are the direct factors influencing the material performance. Some typical regional feature points in symmetry plane after the simulation on figure 8 are marked and the grains size at these feature points are measured and compared with the experiment results.

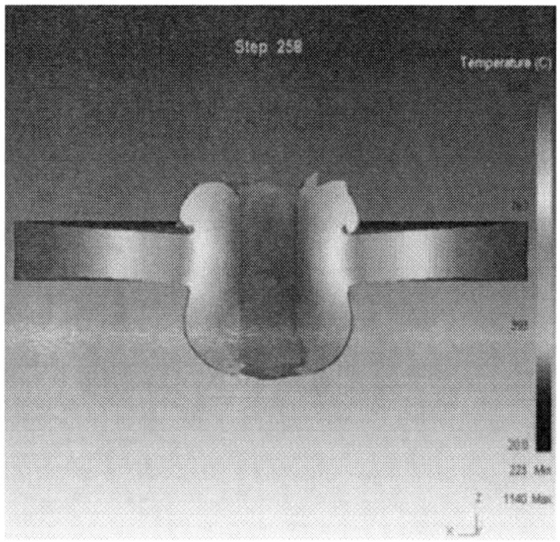


**Figure 6.** Temperature field of symmetry plane

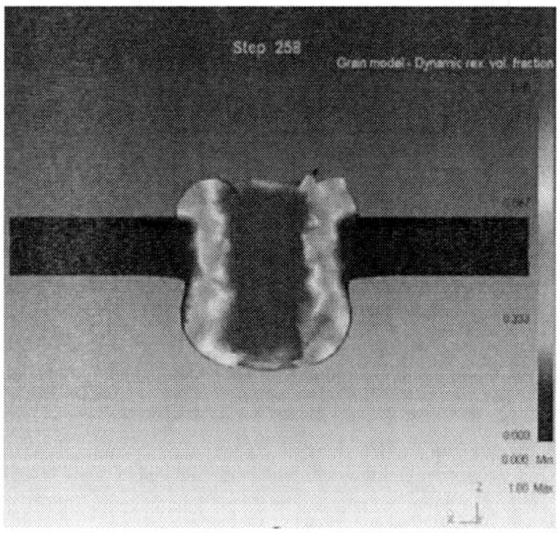


**Figure 7.** Distribution of dynamic recrystallization

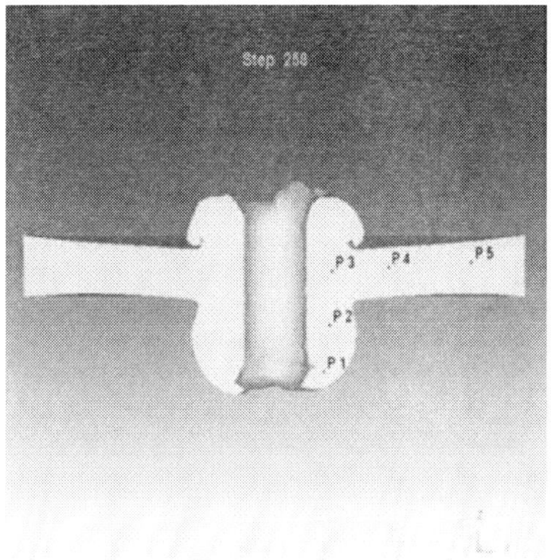


**Figure 8.** The study points

The test workpiece material: 20 steel, sheet thickness is 4mm. The rotation speed and feed speed of drilling tool are 2500 rpm and 0.6 mm/s respectively. The sizes of grains of representative area in the cross section deformation area are measured by FEI Sirion scanning electron microscope, and the distribution of grain size and changes may be observed.

Figure 9 shows distribution diagram of microstructure of some representative area on the cutting plane of test workpiece by FEI sirion scanning electron microscope. In figure 9, grains of deformation area are smaller than the original grains, which show producing dynamic recrystallization. The grain sizes of points P1, P2 are between 12μm and 16μm. The grain size of points P3 is about 25μm. The grain size of points P4 is about 34μm. The grain size of points P5 is about 50μm as same as the original grain size because of almost no recrystallization P5 point on the edge of the billet. Therefore, the deformation degree of different parts is different, percentage of dynamic recrystallization is different also, and the grain size is small where recrystallization is full of regional, on the contrary, dynamic recrystallization in the distal of components can't occur, so the grain size is as big as original grain size.

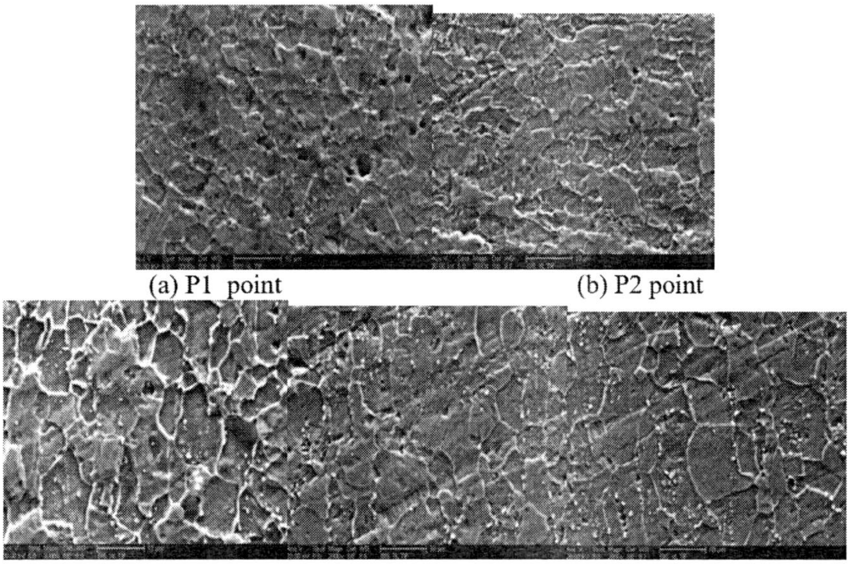

(a) P1 point (b) P2 point

(c) P3 point (d) P4 point e) P5 point

**Figure 9.** Distribution diagram of microstructure patterns on cutting plane

All point of the simulation of the grain size data compared with the estimated value of each point after the experiment shown as figure 10. There is a certain deviation shown as figure10, because the boundary conditions in the simulation process by simulation software are not the same as the actual operation, and although the process of heat conduction and thermal diffusion may be considered, recrystallization mathematical model can not calculate very accurately the deformation in the process of grain size of dynamic recrystallization and grain growth. The error of two groups of data is not big. The research results can prove the validity of the simulation.

Figure 11 shows actual cutting plane figure. Two image contrast can be seen that the simulation results are in conformity with the actual dimension, the simulation effect is feasible.

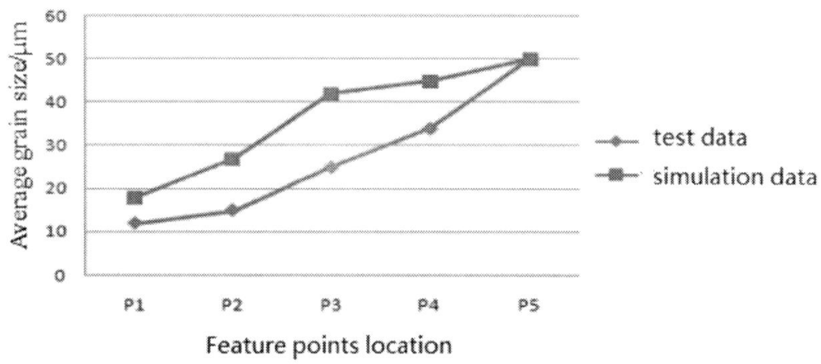


**Figure 10**. Data comparison between simulation and experimental

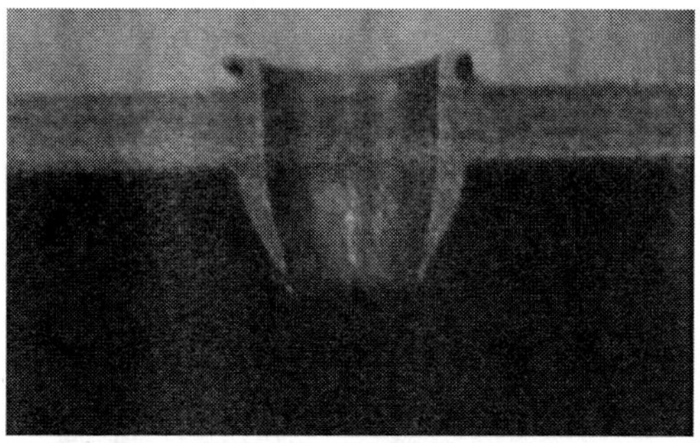

**Figure11**. The workpiece section

## 4. Conclusions

When the other conditions are constant, the rotation speed of drilling tool is too small to get enough high deformation zone temperature and make dynamic recrystallization forming deep hole in sheet, and forming effect is low. When drill speed is high, the deformation zone temperature is enough high to cause overheating phenomenon and get the coarse grains; the feed speed of drilling tool is too fast to get enough time to make material grain to happen adequately dynamic recrystallization forming deep hole in sheet, so the recrystallization cannot overcome work hardening in a timely manner; feed speed is too slow, the contact time between drilling tool and the material deformation area is too long also, then to generate enough heat also to influence grain refinement.

Based on the research of this paper, the suitable condition of process parameters to get recrystallization organization is: rotation speed is 2500 rpm, feed speed is 0.6 mm/s.

## References

[1] Chao C and Jianlong H 2010 Machinery Design & Manufacture, no.9 pp123-5
[2] Yuhang L 2011 Nei Jiang Science & Technology, no.6 pp91-2
[3] Jiandong D, Dayong L and Feng L 2013 J. Hrbust, no.3 pp119-25
[4] Xiaojing L, Bo L, Xiaoxiao C, Qiangrong P and Yansang Z 2015 J. Hrbust, no.1 pp75-9

[5] Xiaojing L, Ran Y, Zhangchao F and Zhe W 2018 J. Hrbust, no.2 pp-129-33
[6] Di Z and Xiaobin.P 2004 Machinery, no.9 pp49-50
[7] Jian W and Ziyue W 2014 Tool Engineering, no.11 pp3-7
[8] Xiaocun X and Shichao W 2017 Cemented Carbide vol.34, no. 3, pp192-6

# Effect of Inorganic Components on Properties of Ceramizable Phenolic Resin Matrix Composites

**Dongpei He, Minxian Shi*,Yuqing Yang and Zhixiong Huang**

School of Materials Science and Engineering, Wuhan University of Technology, Wuhan 430070,China.Email:minxianshi@whut.edu.cn

**Abstract.** In order to meet the thermal protection needs of aerospace, a composite material with low cost and good thermal protection effect was prepared. Boron phenolic resin is used as the matrix, and high silica glass fiber cloth is used as the reinforcing filler,the aluminosilicate mineral powder such as glass frit, feldspar powder, kaolin, talcum powder is a ceramic filler, and ceramic phenolic resin matrix composites were prepared by moulding method.The effects of different mineral powders on the properties of composites were investigated.The results show that with the increase of the amount of inorganic filler, the thermal conductivity of the material increases gradually, the high temperature weight loss rate is first increased and then decreased, and the minimum is reached when the filler content is 30%; The results of XRD analysis show that the inorganic filler in the material undergoes ceramization reaction at 1000°C to form a new material cristobalite, which effectively blocks the penetration of oxygen and heat into the material. The ceramic layer formed at high temperature makes the material flow at high temperature and high speed. The scouring resistance and oxidation ablation performance are improved, and it is expected to become a new type of heat-proof material.

## 1. Introduction

When spacecraft such as returning satellites, spacecraft and deep space detectors re-enter the atmosphere, the surface can reach temperatures above 1000°Cunder aerodynamic heating [1], therefore, the spacecraft must use a thermal protection system to ensure the internal instruments work properly. The thermal protection method can be divided into endothermic method, radiation method and ablation method according to its protection mechanism [2]. Ablative heat protection is one of the most widely used thermal protection systems, the so-called ablation is the consumption of solid surface materials caused by thermochemical and mechanical processes under the action of heat flow when missiles and aircraft re-enter the atmosphere. Ablative heat protection can be adjusted by its own mass exchange and heat exchange. Under certain pressure and enthalpy, the ablation process can reach a quasi-steady state, that is, the thickness of the affected material is nearly constant [3]. As a typical representative of ablative materials, phenolic resin-based composite materials have been widely used in the thermal protection system of deep space detectors such as the Moon and Mars due to their high heat-proof efficiency, reliable operation, and strong ability to adapt to the heat flow environment [4], early AVCOAT-5026 has been used as a representative of this type of ablation material in the "Apollo" spacecraft [5].

With the rapid development of the aerospace industry, the traditional phenolic resin-based composite materials can no longer meet the requirements of use, and the phenolic resin matrix must be modified.The ceramizable polymer matrix composite is a new type of fireproof and heat insulating material.Professor Cheng Yibing of Monash University in Australia invented a ceramizable polymer-based composite material. The fireproof material was prepared by adding mica powder, feldspar powder and inorganic clay to silicone rubber and silicon-based resin and successfully applied to fireproof cables [6], and commercialized production in Australia.Based on the silicon-based ceramics technology, the special functional materials laboratory of Wuhan University of Technology has taken the lead in the domestic research on the ceramization theory of carbon-based polymers. Polymer matrix composites modified by ceramic powders can be ceramized at high temperature to form ceramic phase [7], which improves the ablation resistance and mechanical properties of the materials.The ceramization of carbon-based resin provides a simple and effective way to modify the phenolic resin. The process is simple, the raw materials are cheap and easy to obtain, and the modification effect is obvious.In this paper, low-cost aluminosilicate mineral powders such as feldspar powder, kaolin and talcum powder were used as ceramic fillers to modify phenolic resin, and ceramsite phenolic resin composites were prepared. The effects of different filler contents on the properties of the materials were studied.

## 2. Experimental

### *2.1. Preparation*

Dissolving boron phenolic resin and absolute ethanol in a mass ratio of 1:1 to prepare a solution; Weigh a certain amount of ultra-fine kaolin, glass frit, feldspar powder and talc powder into the phenolic resin solution and mix them evenly into a dipping solution.The dipping solution was evenly scraped onto the high silica fiberglass cloth of a certain quality to make a prepreg cloth. After drying, the preg cloth was cut and stacked. The prepreg cloth was put into a mold with a demoulding cloth and placed on a hot press at 120°C for 2Mpa for 1h. Then the preg-curing pressure was 5Mpa for 1h, 180°C for 2h, and 200°C for 1h, and then ceramic phenolic resin composites were obtained.

### *2.2. Performance Testing*

Thermogravimetric analysis: Synchronous thermal analyzer model STA449c/3/G produced by NETZSCH, Germany. Test conditions: in a nitrogen atmosphere, the airflow rate is 15 ml/min, the heating rate is 10°C/min, and the temperature range is RT-1200.X-ray diffraction analysis: Bruker's D8ADVANCE X-ray diffractometer, Cu-Kα as the target source, working voltage 40KV, working current 40mA, sweeping angle range: 5~80°,scanning speed: 1°/min.Thermal conductivity test: According to the national standard GB/T 10297-1998, QTM-500 thermal conductivity tester manufactured by Kyoto Electronics Industry Co.Ltd., sample size: (100±1) mm×(100±1) mm×10mm, test temperature is room temperature, heating current: $I^2$=0.625.Model ULTRAPLUS-43-13 field emission scanning electron microscope produced by German Zeiss optical instrument.

## 3. Results and Discussion

### *3.1. Effect of Filler on Thermal Conductivity of Materials*

Fig. 1 shows the relationship between the thermal conductivity of the material and the change of inorganic filler. It can be seen from the diagram that the thermal conductivity of the material increases with the increase of filler addition. Firstly, because the inorganic filler with highthermal conductivity phase replaces the position of the original resin, the local thermal conductivity of the matrix is higher than that of the surrounding points. As the content of inorganic filler continues to increase, these local high thermal conductivity points will form a thermal conductivity path [8], leading to the increase of the overall thermal conductivity of the material with the increase of the inorganic filler. Gradually increasing. Secondly, the addition of inorganic fillers reduces the porosity of the material to a certain extent, and improves the density of the material. Therefore, the reduction of porosity is also an important factor in the increase of thermal conductivity of the material.

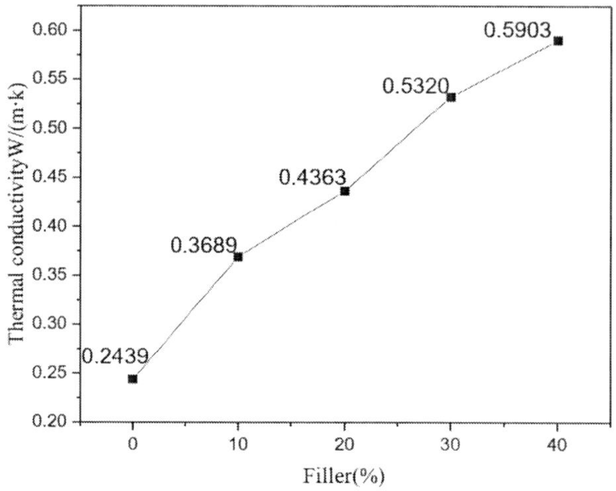


**Figure 1.** Effect of inorganic filler on thermal conductivity of materials

*3.2. Thermal Stability Analysis of Ceramizable Phenolic Resin Matrix Composites*

Figure 2 shows the TG (a) and DTG (b) curves of ceramizable phenolic resin matrix composites. It can be seen from Figure.2(a) that the initial decomposition temperature of the ceramsite based phenolic resin composite after the addition of the filler does not change much, and the thermal stability increases at 300-800°C, the residual carbon ratio of the ceramizable phenolic resin composite at 1200°C increased significantly from 76.47% to 83.74%.According to the DTG curve in Figure.2(b), the thermogravimetric curves of ceramizable phenolic resin matrix composites with different content of fillers are divided into room temperature to 300°C, 300 to 800°C, 800°C to 1200°C three stages by 300°C and 800°C.From room temperature to 300°C, the weight loss rate is small at this stage, mainly due to the evaporation of water and the further reaction of unreacted small molecules; in the stage of 300-800°C, the quality has changed significantly. This is mainly due to the obvious pyrolysis of the phenolic resin. With the molecular chain breaking and rearrangement of the phenolic resin, various small molecular gases such as CO, $H_2O$ and $CH_4$ are produced [9].It can be seen from Fig. 2(b) that the phenolic resin composite material after adding the filler has little change before the addition of the filler, indicating that the filler does not participate in or affect the curing reaction of the phenolic resin, and does not substantially increase the cracking peak temperature of the material. At 800-1200 °C, the mass loss rate is significantly reduced. The thermal weight loss at this stage is mainly due to the further cyclization of the benzene ring, which is caused by the cleavage into a structurally complete glassy hydrocarbon atom combined into hydrogen evolution [10].

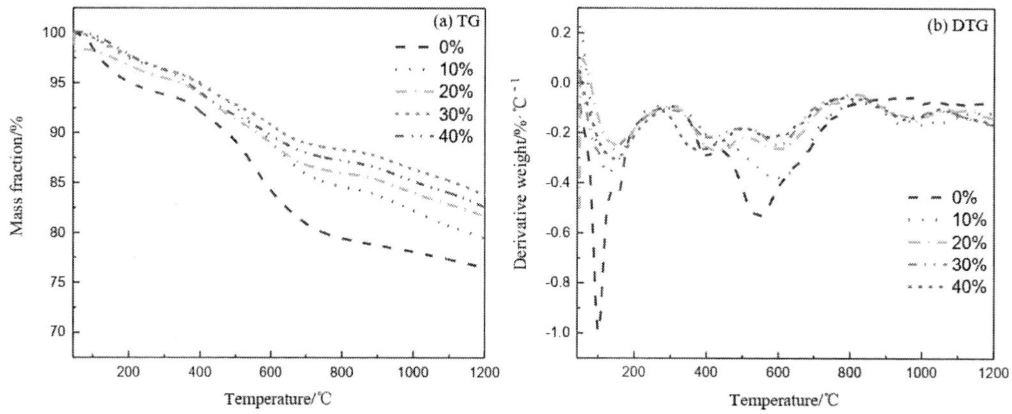


**Figure 2.** TG (a) and DTG (b) curves of ceramizable phenolic resin matrix composites

*3.3. Analysis of the Microstructure of the Material Section*

Figure.3 shows the cross-sectional morphology of a 20% filler phenolic resin matrix composite after treatment at different temperatures. When the heat treatment temperature is 600°C, the composite fibers are debonded with the matrix, some of the fibers are broken, and there are holes left by fiber pulling out and grooves left by fiber shedding. When the heat treatment temperature is 800°C, part of the filler melts at high temperature, filling part of the voids and voids makes the porosity of the material slightly lower. However, due to the high viscosity of the melted liquid at 800°C and the difficulty of flowing, more voids can be observed in the cross section of the material. When the heat treatment temperature is 1000°C, a large number of liquid phases can be observed in the filler melting. The liquid phase flow fills the voids and voids in the material continuously. The melted liquid phase even coats the fiber and matrix, which further reduces the porosity of the material and makes the internal structure of the material more compact.

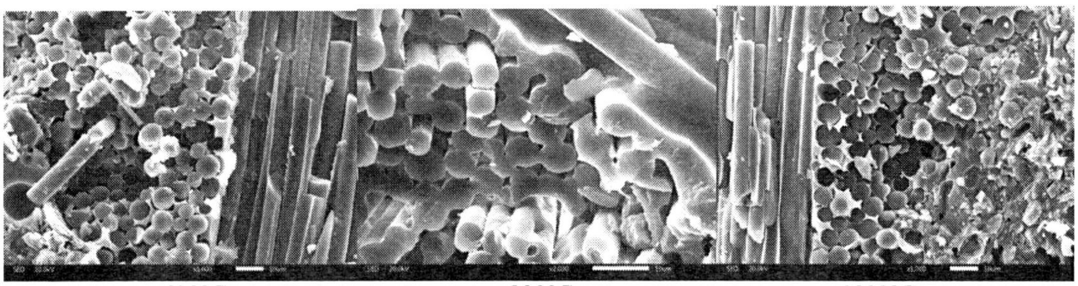


**Figure 3.** Sectional morphology of 20% filler phenolic resin matrix composite after different temperature treatment

*3.4. Phase analysis of Ceramizable Phenolic Resin Matrix Composites*

Figure.4 is an XRD pattern of a ceramizable phenolic resin based composite at various temperatures. It can be seen from the figure that the XRD diffraction peaks of the materials after treatment at 600°C and 800°C are substantially coincident, and there is no significant change, indicating that the cracking products are substantially the same, and no reaction occurs between the inorganic fillers. The characteristic peaks of Na ($AlSi_3O_8$) were weakened and the diffraction peaks of high-temperature cristobalite appeared when the heat treatment temperature was 1000°C, which indicated that ceramization reaction took place between fillers, Na ($AlSi_3O_8$) was consumed and a new high-temperature cristobalite was formed.

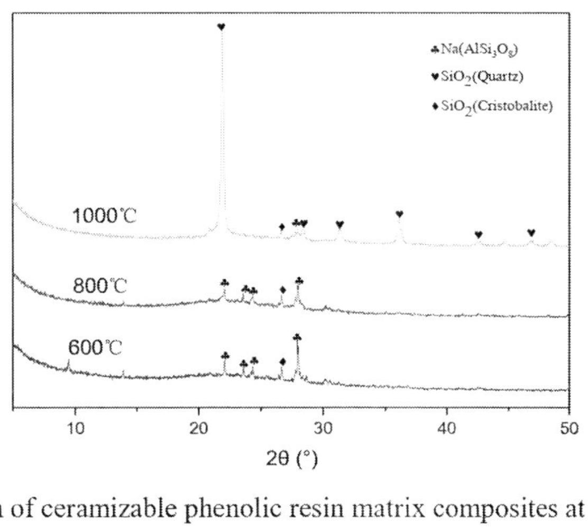


**Figure 4.** XRD pattern of ceramizable phenolic resin matrix composites at different temperatures

## 4. Conclusion

With the increase of inorganic fillers, the thermal stability of the material was gradually improved at 1200°C, and the thermal weight loss decreased from 23.53% to 16.26%.

With the increase of heat treatment temperature, inorganic filler melts gradually to produce alarge number of liquid phase, filling up the pores produced by pyrolysis, reducing the porosity of the material; at the same time, there is an obvious ceramization reaction between fillers, $Na(AlSi_3O_8)$ phase is gradually depleted, reaction to produce a new material high temperature cristobalite.

## Reference

[1] Gao Shouqi, Wei Huazhen,and Li Dayong, et al. Review of ablative materials [J]. New Chemical Materials, 2009, 37(2): 19-21.
[2] Hu Liangquan. Status evelopment of Lightweight/Insulation Functional Materials [J]. Functional Materials Information, 2010(2): 19-23.
[3] Ouyang Shuiwu and Zheng Guoming. Ablation Theory and Experimental Results of High Silicon Oxygen Reinforced Plastics [J]. Journal of Astronautics.1981, (1): 14-22
[4] Fan Zhenxiang, Cheng Haifeng and Zhang Changrui, et al. Progress in thermal protection materials [J].Materials Report, 2005, 19 (1): 13-16.
[5] Schaefer J W.Experimental and Analytical Evaluation of the Apollo Thermal Protection System under Simulated ReentryConditions[R].NASA-CR-65817, 1965.
[6] Mansouri J,Burford R P and Cheng Y B.Pyrolysis Behaviour of Silicone-based Ceramifying Composites[J].Materials Sci-ence and Engineering A,2006,425:7-14.
[7] Huang Zhixiong, Qin Yan,and Rao Zhilong, et al. Ceramicized high carbon poly-mer matrix composite material and preparation method thereof: CN, 103058632A[P].2013-04-24.
[8] Ding jie.Preparation and performance of heat-resistant and ablation-resist inorganic particles modified carbon/phenolic composites [D], 2016.
[9] KA Trick , TE SalibaMechanisms of the pyrolysis of phenolic resin in a carbon/phenolic composite[J]. Carbon 1995, 33 (11):1509-15.
[10] J Wang , H Jiang , N Jiang.Study on the pyrolysis of phenol-formaldehyde (PF) resin and modified PF resin [J].Thermochimica Acta, 2009, 496 (1):136-142.

IOP Conf. Series: Materials Science and Engineering 472 (2019) 012045 doi:10.1088/1757-899X/472/1/012045

# The Effect of Bauschinger Related Parameters on Material Property of Dh780 and Springback Behavior of Longitudinal Beam

**Wenyu Ma[1,2,3], Jianwei Yang[1,2,3], Xuebin Zheng[1,2,3], Baochuan Wang[1,2,3], Chunguang Li[1,2,3], Yongqiang Zhang[1,2,3]**

1. Research Institute of Technology of Shougang Group Co., Ltd.
2. Beijing Key Laboratory of Green Recycling Process for Iron & Steel Production Technology
3. Beijing Engineering Research Center of Energy Steel

Address:No. 69, Yang Zhuang Street, Shijingshan District, Beijing, P. R. China P.C. 100043. Email:wymaustb@163.com

**Abstract.** The application of high strength steel on automobile becomes more and more popular. So it is necessary to study the formability. In this paper, the effect of Bauschinger was studied. The material employed in this study is DH780. The influence of Bauschinger related parameters on material property was analyzed by changing the parameter values. Then, the effect on the springback behavior was studied after the finite element analysis. The part used was a longitudinal beam used in the automobiles. The related Bauschinger related parameters were Transient Softening Rate, Stagnation Ratio, Young's Reduction Factor and Young's Reduction Rate. The results were achieved after the analysis. The transient softening rate K shows a significant effect on the stress-strain curves. The increase of transient softening rate K leads to the decrease of reversed stress and the increase of springback.

## 1. Introduction

The springback of high strength steel gained great attention in recent years [1]. Fei et al. [2] analyzed the springback behavior using experiment and simulation method. The v-free bending equipment was employed. The microstructure was also observed and studied. Ma et al. [3] studied the influence of process parameters on springback of high strength steel in terms of a stamped part and analyzed the relationship between springback and other process parameters in stamping high strength steel. Sumikawa et al. [4] studied the springback behavior using finite element analysis, experiments and mathematical analysis. The springback prediction accuracy resulted from finite element software can be improved by considering the nonlinear elastoplastic behavior and others. Lim et al. [5] researched the time dependence of springback of high strength steel. Several materials were employed in this study. The Young's moduli show significant effect on the time-dependent springback.

At the same time, it is necessary to study the influence of the Bauschinger effect on the springback of high strength steel. To enrich this research area, the effect of Bauschinger related parameters on material property of DH780 and springback behavior was analyzed.

## 2. Design of Test

The parameters related to Bauschinger effect are transient softening rate, stagnation ratio, young's reduction factor, young's reduction rate and so on. And in this research, the three important factors are studied. They are Transient Softening Rate, Stagnation Ratio, Young's Reduction Factor and Young's

Reduction Rate. The effect of these parameters on the stress-strain curves was analyzed. The range of transient softening rate K is from 0.005 to 0.09; the range of stagnation ratio s is from 0 to 1; and the range of young's reduction factor Er is from 0 to 0.5. The basic values of these four parameters, namely transient softening rate, stagnation ratio, young's reduction factor and young's reduction rate are 0.014, 0.9, 0.13 and 40 respectively.

Test 1 is designed for the effect of transient softening rate K on material properties of DH780 and springback behavior of longitudinal beam.

**Table 1.** The values of Bauschinger related parameters are designed for test 1 in terms of the effect of transient softening rate K.

| No. | K | s | $E_r$ | X |
|---|---|---|---|---|
| 1 | 0.005 | 0.9 | 0.13 | 40 |
| 2 | 0.01 | 0.9 | 0.13 | 40 |
| 3 | 0.014 | 0.9 | 0.13 | 40 |
| 4 | 0.05 | 0.9 | 0.13 | 40 |
| 5 | 0.09 | 0.9 | 0.13 | 40 |

Test 2 is designed for the effect of stagnation ratio s on material properties of DH780 and springback behavior of longitudinal beam.

**Table 2.** The values of Bauschinger related parameters are designed for test 2 in terms of the effect of stagnation ratio K.

| No. | K | s | $E_r$ | X |
|---|---|---|---|---|
| 1 | 0.014 | 0.1 | 0.13 | 40 |
| 2 | 0.014 | 0.3 | 0.13 | 40 |
| 3 | 0.014 | 0.5 | 0.13 | 40 |
| 4 | 0.014 | 0.7 | 0.13 | 40 |
| 5 | 0.014 | 0.9 | 0.13 | 40 |
| 6 | 0.014 | 1.0 | 0.13 | 40 |

Test 3 is designed for the effect of Young's reduction factor Er on material properties of DH780 and springback behavior of longitudinal beam.

**Table 3.** The values of Bauschinger related parameters are designed for test 3 in terms of the effect of Young's reduction factor Er.

| No. | K | s | $E_r$ | X |
|---|---|---|---|---|
| 1 | 0.014 | 0.9 | 0.001 | 40 |
| 2 | 0.014 | 0.9 | 0.05 | 40 |
| 3 | 0.014 | 0.9 | 0.1 | 40 |
| 4 | 0.014 | 0.9 | 0.13 | 40 |
| 5 | 0.014 | 0.9 | 0.2 | 40 |
| 6 | 0.014 | 0.9 | 0.3 | 40 |
| 7 | 0.014 | 0.9 | 0.4 | 40 |
| 8 | 0.014 | 0.9 | 0.5 | 40 |

## 3. FE Model

There are several parts in the FE model of a longitudinal beam, including punch, die, blank and so on. Figure 1 (a) shows the blank figure (gray in color) and die (red in color) and Fig. 1 (b) shows the formed part after springback process.

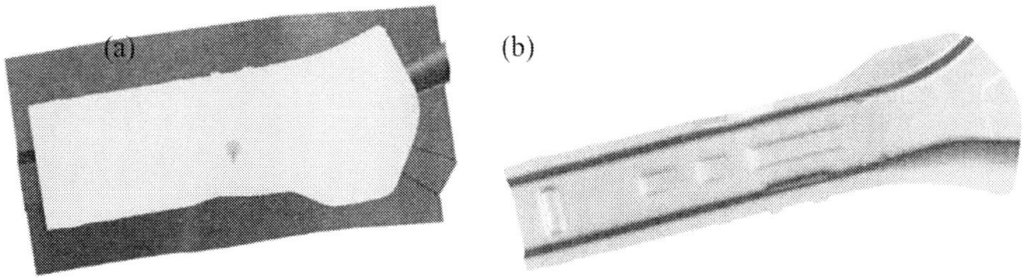


**Figure 1.** The blank figure (gray) and die (red) (a) and the part after springback (b).

The material used in this model is DH780, and the basic properties of DH780 are given in the following. the yield stress is 518 MPa; n is 0.16; Rm is 809.6MPa and $A_G$ is 16.9%. The stress-strain relationship concerning the Bauschinger effect is shown in the following figure (Fig. 2).

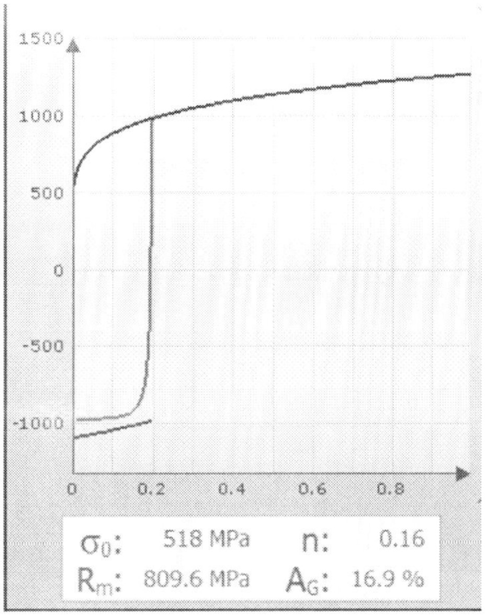


**Figure 2.** The stress-strain relationship concerning Bauschinger effect of DH780

In this paper, the springback behavior is mainly focused and analyzed. And after the forming process, the springback occurs. The effect of Bauschinger related parameters on springback is reflected after changing the value of parameter.

## 4. Results and Discussion

*4.1. Effect of Bauschinger Related Parameters on Material Property of DH780*

1) The effect of transient softening rate K on the stress-strain relationship.

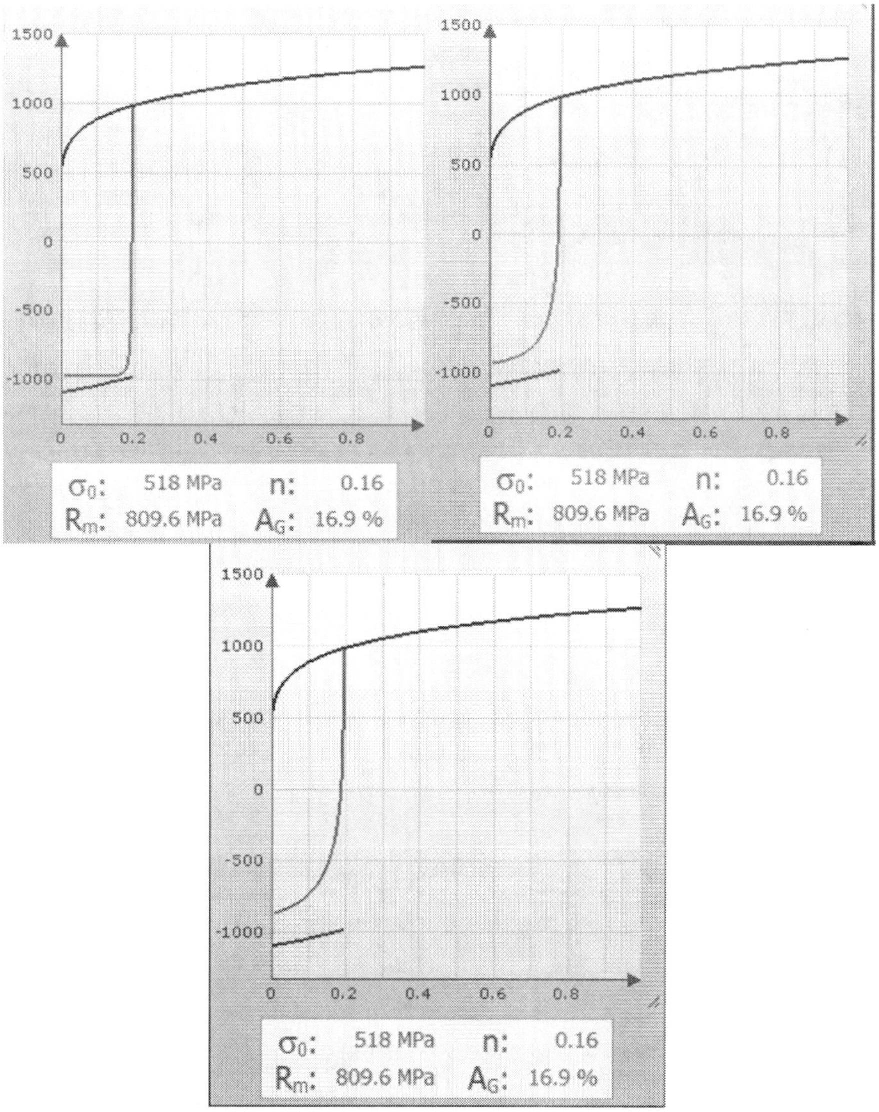


**Figure 3.** The effect of transient softening rate K on the stress-strain relationship with different values of (a) 0.005, (b) 0.05 and (c) 0.09.

It can be seen from the above figures that with the increase of the value of transient softening rate K, the reverse loading curve of stress-strain relationship becomes smoother and the elastic modulus becomes lower, and the reversed stress decreases significantly.

2) The effect of stagnation ratio s on the stress-strain relationship.

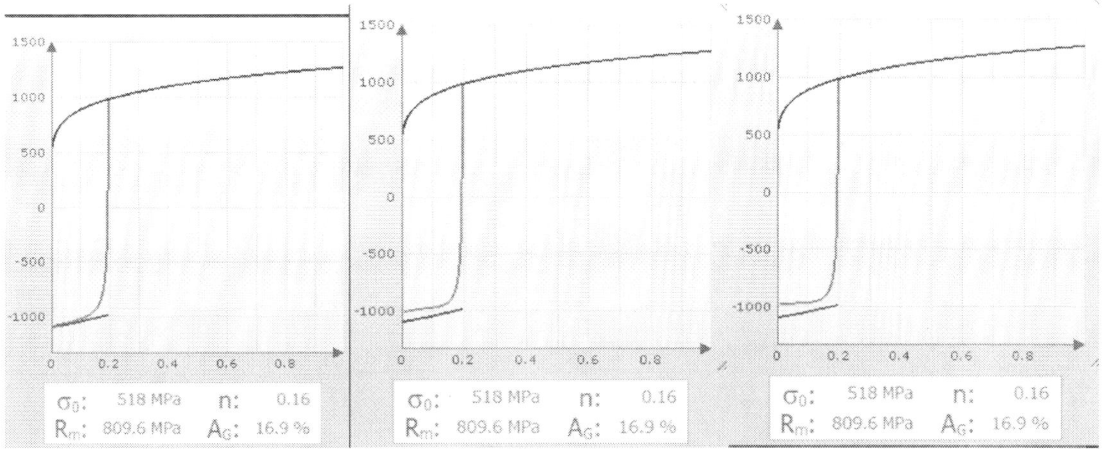


**Figure 4.** The effect of stagnation ratio s on the stress-strain relationship with different values of (a) 0.1, (b) 0.5 and (c) 1.0.

The results show that with increasing the stagnation ratio s value, the stress-strain curve of reverse loading part becomes flat, and the absolute value of reversed stress decreases.

3) The effect of Young's reduction factor Er on the stress-strain relationship.

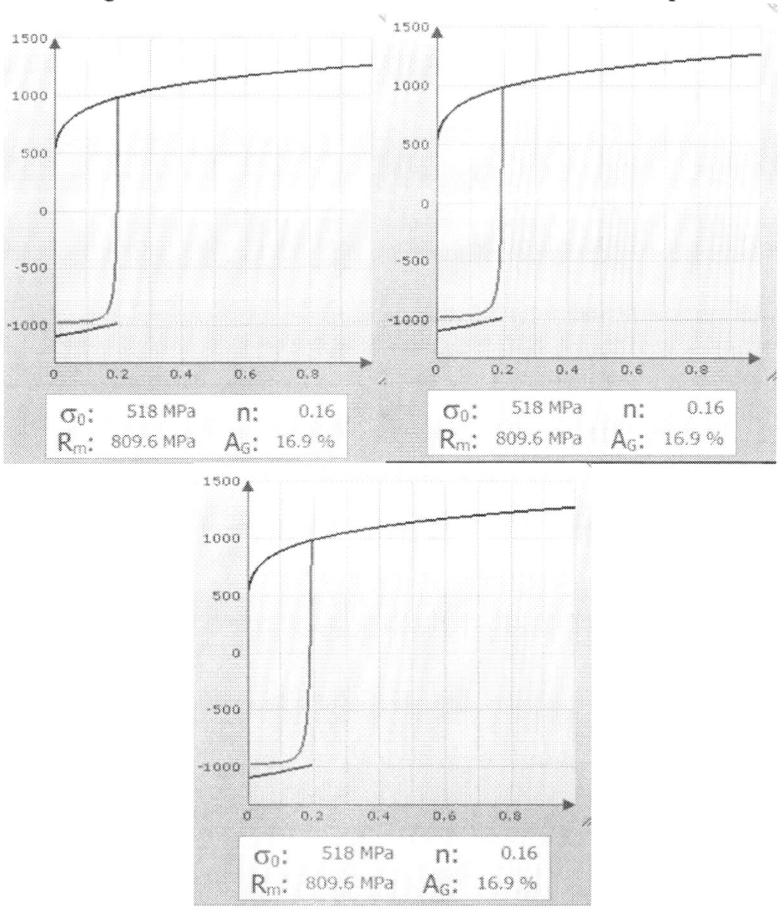


**Figure 5.** The effect of Young's reduction factor Er on the stress-strain relationship with different values of (a) 0.001, (b) 0.2 and (c) 0.5.

It can be seen from the above figure that the effect of Young's reduction factor Er on the stress-strain relationship is insignificant.

*4.2. Effect of Bauschinger Related Parameters on Springback Behavior of Longitudinal Beam*

1) The effect of transient softening rate K on the springback and thickness distribution of the formed longitudinal beam part.

| | |
|---|---|
| 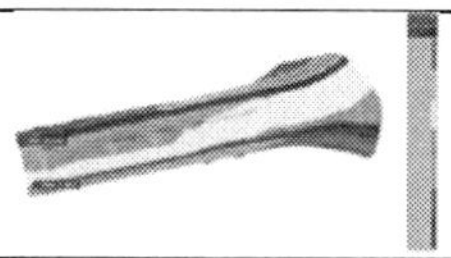 | 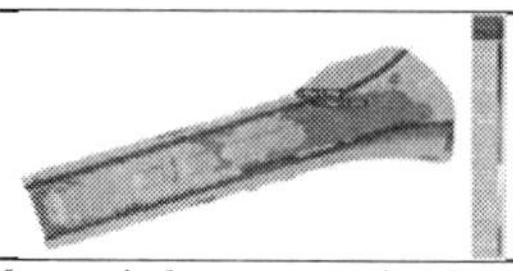 |
| (a) k=0.005, the maximum springback value is 2.667, the minimum springback value is -8.642; the maximum thickness is 2.294, the minimum thickness is 1.736. | |
| 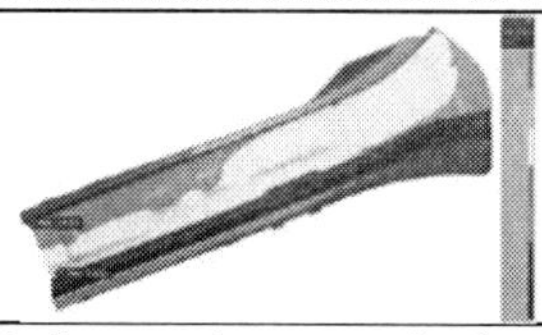 | 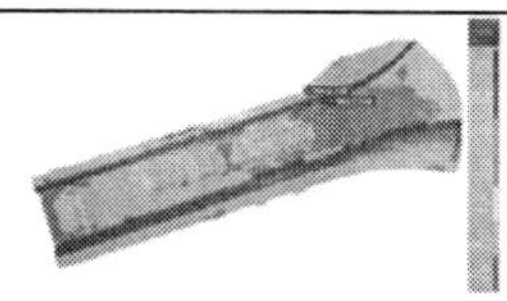 |
| (b) k=0.05, the maximum springback value is 3.812, the minimum springback value is -14.29; the maximum thickness is 2.266, the minimum thickness is 1.728. | |
| 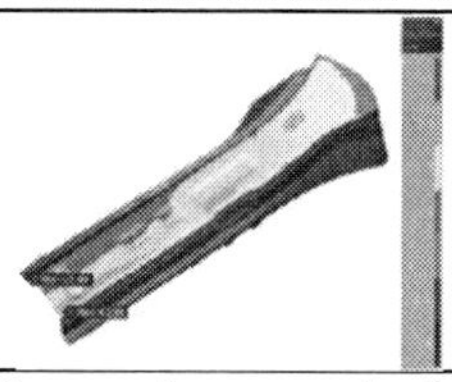 | 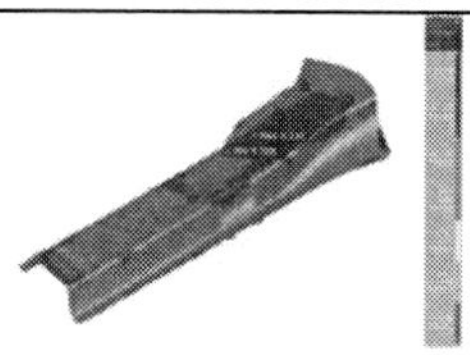 |
| (c) k=0.09, the maximum springback value is 4.239, the minimum springback value is -17.89; the maximum thickness is 2.250, the minimum thickness is 1.728. | |

**Figure 6.** The results for the effect of transient softening rate K on the springback and thickness distribution of the formed longitudinal beam part.

According to the above figures, with the increase of transient softening rate K, the minimum springback value decreases, that is to say, the absolute value of minimum springback increases. Because with increasing transient softening rate K, the transient softening effect increases, the yield stress after the tension decreases, which leads to a large springback value. Thus, the absolute value of springback raised.

2) The effect of stagnation ratio s on the springback and thickness distribution of the formed longitudinal beam part.

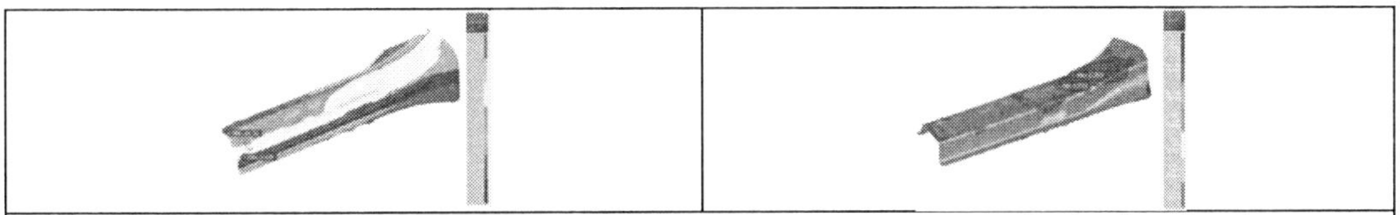

(a) s=0.1,the maximum springback value is 3.160, the minimum springback value is -11.52; the maximum thickness is 2.293, the minimum thickness is 1.735.

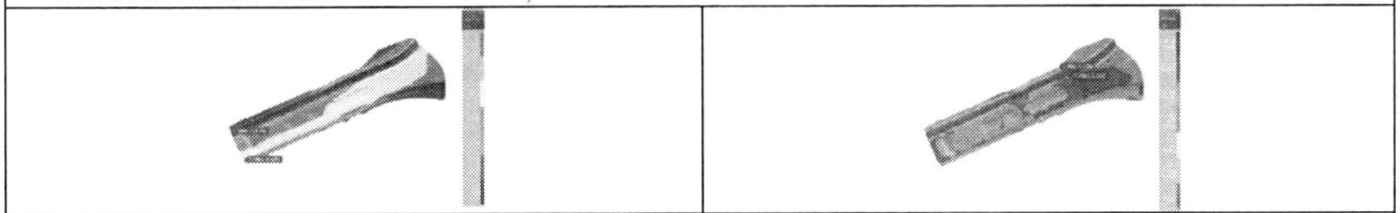

(c) s=0.5, the maximum springback value is 3.139, the minimum springback value is -11.06; the maximum thickness is 2.257, the minimum thickness is 1.734.

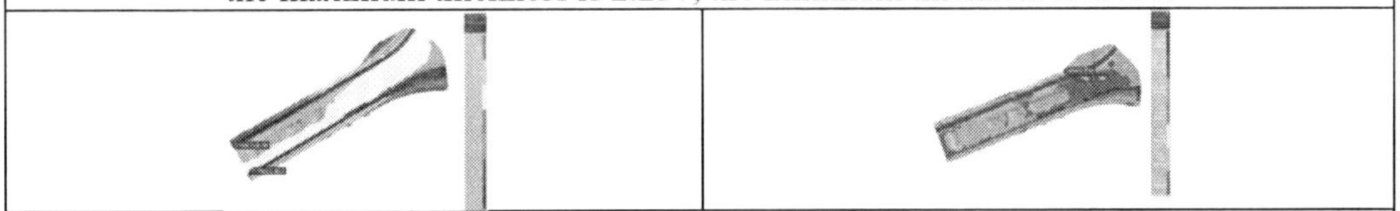

(f) s=1.0, the maximum springback value is 2.823, the minimum springback value is -10.58; the maximum thickness is 2.286, the minimum thickness is 1.733.

**Figure 7.** The results for the effect of the stagnation ratio s on the springback and thickness distribution of the formed longitudinal beam part.

The results above illustrate that with increasing the stagnation ratio s, both the maximum and minimum absolute springback values decrease.

3) The effect of Young's reduction factor Er on the springback and thickness distribution of the formed longitudinal beam part.

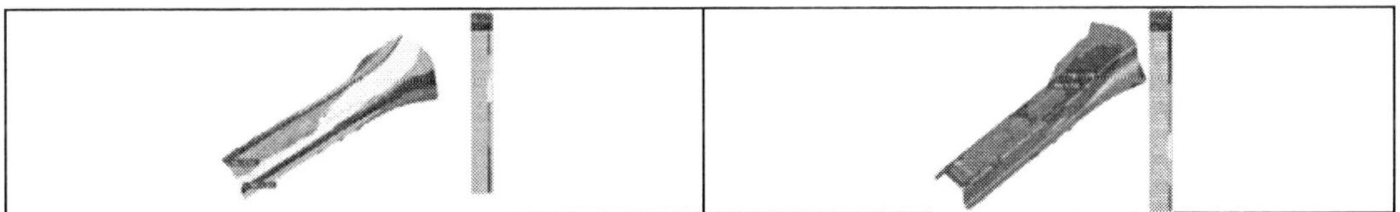

(a) $E_r$=0.001, the maximum springback value is 2.424, the minimum springback value is -9.322; the maximum thickness is 2.298, the minimum thickness is 1.733.

(e) $E_r$=0.2, the maximum springback value is 3.252, the minimum springback value is -11.32; the maximum thickness is 2.286, the minimum thickness is 1.734.

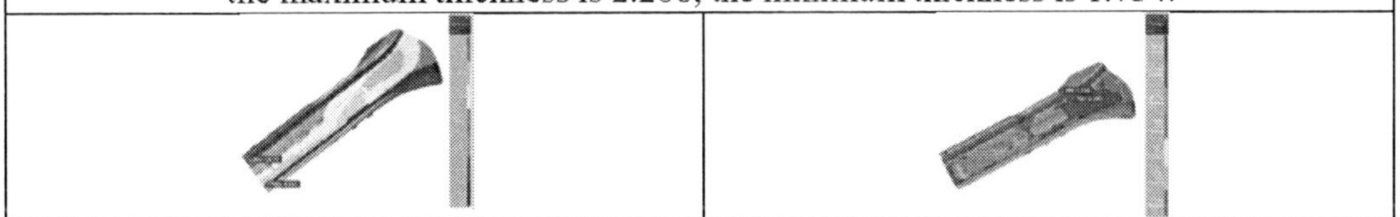

(h) $E_r$=0.5, the maximum springback value is 4.943, the minimum springback value is -16.72; the maximum thickness is 2.281, the minimum thickness is 1.736.

**Figure 8.** The results for the effect of Young's reduction factor Er on the springback and thickness distribution of the formed longitudinal beam part.

What can be seen from the above results is that with the increase of Young's reduction factor Er, both the maximum and minimum absolute values of springback increase. However, it affects the thickness weakly. The maximum thickness value decreases but not obviously.

## 5. Conclusions

1) With the increase of the value of transient softening rate K, the reverse loading curve of stress-strain relationship becomes smoother and the elastic modulus becomes lower, and the reversed stress decreases significantly.

2) With increasing the stagnation ratio s value, the stress-strain curve of reverse loading part becomes flat, and the absolute value of reversed stress decreases.

3) The effect of Young's reduction factor Er on the stress-strain relationship is insignificant.

4) With the increase of transient softening rate K, the minimum springback value decreases, that is to say, the absolute value of minimum springback increases. Because with increasing transient softening rate K, the transient softening effect increases, the yield stress after the tension decreases, which leads to a large springback value. Thus, the absolute value of springback increases.

5) With increasing the stagnation ratio s, both the maximum and minimum absolute springback values decrease.

6) With the increase of Young's reduction factor Er, both the maximum and minimum absolute values of springback increase. However, it affects the thickness weakly. The maximum thickness value decreases but not obviously.

## References

[1] B. Chongthairungruang, et al., Experimental and numerical investigation of springback effect for advanced high strength dual phase steel. Materials and Design, 2012. 39: p. 318-328.
[2] Dongye Fei, P.H., Experimental and numerical studies of springback in air
[3] V-bending process for cold rolled TRIP steels. Nuclear Engineering and Design, 2006. 236: p. 1847–1851.
[4] Ma, W. et al, Study on the cold stamping of oil sump tank for the thickness and springback. IOP Conference Series: Materials Science and Engineering. 381: P.955-959.
[5] S. Sumikawa, A. Ishiwatari and J. Hiramoto, Improvement of springback prediction accuracy by considering nonlinear elastoplastic behavior after stress reversal. 2017. p. 46-53.
[6] H. Lim, M.G.L.J., Time-dependent springback of advanced high strength steels. International Journal of Plasticity, 2012. 29: p. 42-59.

# Experimental Study of Hydrothermal Spallation Drilling

**Mingbo Wang and Ning Wang**

School of Petroleum Engineering, China University of Petroleum, No.66 Changjiangxi Road, Huangdao District, Qingdao City, Shandong Province, China.
Email: wmbyjy@126.com

**Abstract.** Geothermal energy has gained more and more attention from all around the globe for its cleanness and efficiency. In the meantime, wells are drilling more and more deeper to develop oil and gas resources. Low rate of penetration is the common problem faced by geothermal wells and deep oil wells. To speed up the development of geothermal energy and petroleum resources in deep formation, a hydrothermal drilling method is proposed here in this paper and the drilling mechanism along with an in-lab experimental setup have been analyzed.

## 1. Introduction

Conventional drilling method uses mechanical drill bits to crush and mill the rock. As the formation depth increases, both the rock strength and hardness are also increased which in turn cause the low rate of penetration. There is little progress right now in mechanical engineering method to deal with low drilling rate.

For some specific rocks, when the heat fluxes on the order of 0.5-10 MW/m$^2$ are applied, the rocks tend to fail and spalls will be generated and ejected away from the hotspot [1-2]. The spallation process has been used for decades in the mining industry for drilling shallow blast holes using supersonic flame jets. As the well depth increases, the transport of spalls or cuttings will be difficult, sometimes impossible. In conventional drilling method, drilling fluids are used to transport cuttings away from the bit surface, into the annulus and finally to the surface for further disposal. Combining the thermal spallation method with conventional cutting transport method we then have the hydrothermal spallation method.

The previous hydrothermal spallation drilling experimental studies are specifically reviewed here to assess the technologies and techniques used to produce and characterize hydrothermal flames and optimization of the technology.

## 2. Conditions in a Deep Borehole Environment

The downhole environment encountered in drilling deep boreholes (>3km) is much different than atmospheric conditions in shallow boreholes where thermal spallation drilling has been traditionally used. Deep boreholes usually employ a drilling fluid to aid in the drilling process. In conventional rotary drilling, the drilling mud can perform the following functions: 1) it cools and lubricates the drill bit, 2) it pushes cuttings away from the bit surface and transports them to the surface, and 3) it maintain borehole stability by generating hydrostatic pressure along the whole borehole. The hydrostatic pressure exerted by the mud column not only maintains the wellbore stability, but also deal with well control problems caused by the high formation pressure. Conversely, too high of a hydrostatic pressure can cause drilling fluid to seep into the formation, a serious condition known as lost circulation.

From the discussion above, it is clear that any deep thermal spallation drilling system will require a liquid drilling fluid with sufficient density to deal with well control problems and cutting transportation. Conventional thermal spallation systems are not designed to operate under such conditions. While drilling a test hole in a granite formation in Conway, Browning noted that his open-hole spallation drilling system could tolerate some influx of fluids from the formation, but the rate of penetration slowed markedly when water bearing zones were encountered in the rock. The maximum depth reached during drilling was only 304.8 m. Wells deeper than 3,048 m will require the drilling equipment to operate in a fluid-filled hole. In addition to aqueous conditions, much higher pressures will be encountered.

## 3. The Need for Hydrothermal Flame Research

If jet flames are to be used in a deep borehole environment to generate the heat fluxes needed to induce thermal spallation, then the technology to make flames in a high pressure, high density environment is needed. Figure 1 shows that if the water-based drilling fluid is used, the bottom hole pressure will be above the critical pressure of water. Therefore, the ability to produce stable, hydrothermal flames in supercritical water will be needed.

According to the definition, hydrothermal flame is produced in water over 22.1 Mpa and at 374 °C. At early stage, hydrothermal flame study focused on the high-temperature and high-pressure thermodynamic properties of a variety of water and gases' binary systems [3-5]. At conditions over the critical water pressure and temperature, light gases will mix with supercritical water to form homogeneous mixture. This kind of mixture can be ignited to provide the heat needed.

Even though they were first demonstrated in Franck's laboratory more than 2 decades ago, the research of hydrothermal flame remains preliminary. There's still much to do to conduct in-depth analysis of such a technology.

## 4. Properties of Supercritical Water Mixtures

The thermophysical and transport properties of water near and above 374 °C and 22.1 Mpa differ greatly from its properties at ambient conditions. At supercritical conditions, water has densities that are intermediate between liquid and gaseous water, the density and other properties can be tuned rapidly by changing its pressure or temperature. Figure 1 is an example of water properties at the pressure of 25 Mpa. The viscosity, dielectric strength, surface tension and other physical properties of the supercritical combustion environment can be greatly varied by adjusting the pressure, temperature, and composition of the system. For example, the viscosity at 400 °C is about one-tenth of that of liquid water at ambient temperatures, leading to higher diffusion coefficients.

Rather small changes in temperature or pressure near the critical point can cause dramatic changes in the solvation character of water-causing it to transition from a very polar, hydrogen bonding solvent at subcritical conditions that easily dissociates and dissolves ionic salts to a nonpolar solvent at supercritical conditions that solubilizes gases and light hydrocarbons and provides an attractive environment for treating heterogeneous wastes by oxidation.

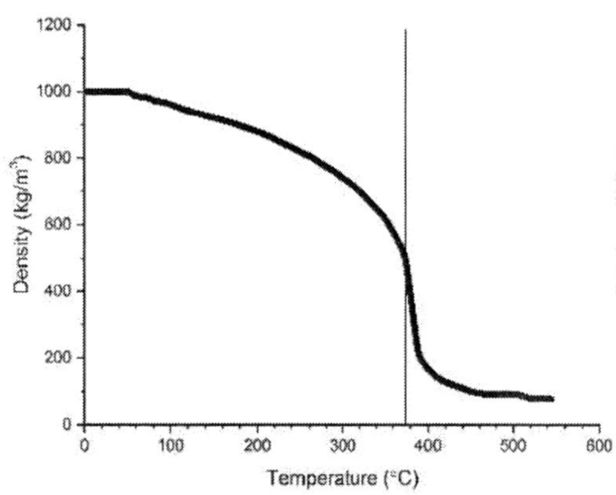


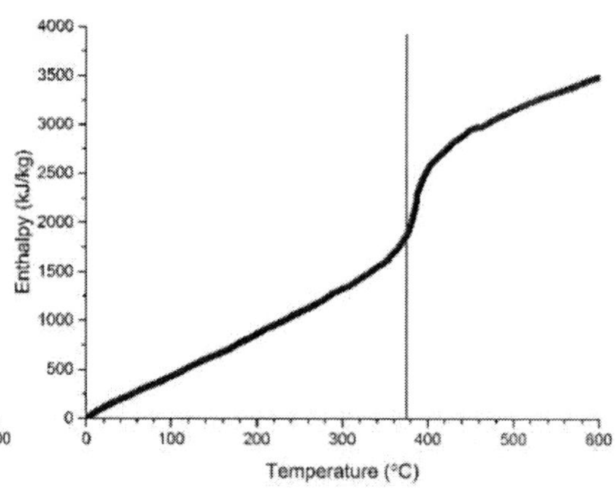

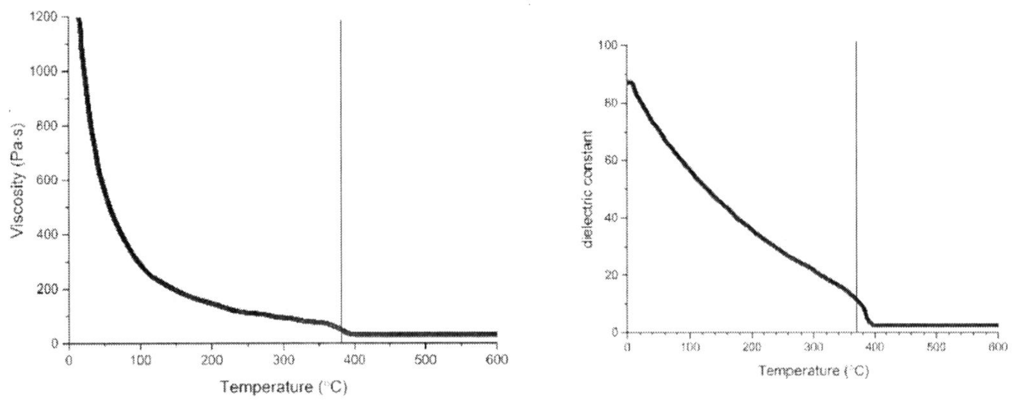


**Figure 1.** Water properties at P = 25 Mpa[6].

More specifically, at ambient condition, the dielectric constant of water at a pressure of 25 Mpa is 80. If you increase the temperature to 400 °C, the constant can decrease to less than 2. These effects, correlated with the decrease in water density, reduces water's hydrogen bonding power, making supercritical water behave like a non-polar solvent such as hexane. Above 375 °C, $N_2$, $O_2$, $H_2$, $CO_2$ as well as benzene, ethane and methane are easy to be mixed with water completely. Under supercritical conditions, only a single phase exists, the interfacial mass transfer resistances among multiphase flow do not exist anymore. Ionic salt is very low in solubility in the supercritical water. It precipitates out of supercritical water mixtures quickly.

Supercritical water oxidation exploits the complete miscibility of $O_2$ and non-polar organic compounds in supercritical water to destroy toxic or non-biodegradable organic waste. It is a moderate temperature process that is usually carried out at temperatures between 450-650 °C and at supercritical pressures of 23 Mpa or higher. At these conditions, oxidation reactions take place in a homogenous, single-phase environment. To facilitate isothermal kinetic studies or to prevent thermal runaway in industrial applications, the concentration of reactive species is low. In order to achieve high reaction rates, much longer residence times are needed. A range of model compounds and waste streams, including municipal sludges and chemical and biological warfare agents have been studied [7].

Despite its attributes, supercritical water oxidation has two technical drawbacks: corrosion and plugging of the reactors and process equipment due to precipitating solids. Corrosion is caused by $Cl^-$ and $F^-$ often present in wastewater streams. These ions, combined with supercritical water and oxygen, can lead to severe corrosion under the condition of high pressure and temperature. To construct the equipment, nickel content alloys are preferred. Corrosion behavior in supercritical water oxidation systems has been well documented [8]. Equipment fouling and plugging is caused by the precipitation of salt particles in the wastewater as it transitions from the sub- to supercritical conditions [9].

Another disadvantage of supercritical water oxidation is the limited operating temperature and working pressure. Even using high-temperature nickel alloys, the temperature of the reactor wall under stress must usually be kept below 650 °C to 700 °C to avoid creep and failure. This limits the temperature at which the reaction can be carried out. Since the rate of the decomposition reactions follow Arrhenius temperature dependence, the reaction temperature must be kept quasi-isothermal to prevent thermal runaway.The lower kinetic rate results in much longer reaction times, requiring large reaction vessels to process a given volume of waste water [10].

## 5. Pressure Vessel and Combustion Chamber Design

A hydrothermal spallation drilling testing tool requires equipment which can withstand high pressures, safely operate hydrothermal flames and at the same time facilitate drilling operation. This section summarizes the designs of a combustion chamber and of a pressure vessel.

### *5.1. Pressure Vessel*

The ship geometry includes its length and diameter as well as its technical and operational specifications (e.g., its materials, the largest operating pressure and temperature). Since the diameter of wells beyond 2500 m is on an average of 0.19-0.30 m, this sets the upper limit of the vessel diameter.

During the spallation drilling, to obtain the necessary thermal stress in rock mass, the thermal constraint of rock sample is required. The rock surrounding the hot zone must be at a lower temperature to avoid the release of thermal stress. Previous literature concluded that the surface heated directly should be less than 10% of the sample's overall surface. Therefore, a small size diameter rock sample can be used to easily achieve this requirement.

The specifications of the vessel designed by Panagiotis S [11] are listed in the following:

- Normal working pressure ranges from 25 and 40 Mpa. The maximum internal pressure is 65 Mpa for safety reasons.
- In the process of normal operation, the highest wall temperature reaches 500 °C.
- 2 windows in the main vessel body and 2 small windows on the upper flange create optical access.
- The vessel is 400 mm and 140 mm in inner length and the diameter, respectively.
- Silver coated stainless-steel rings having a rhomboidal cross section serve as sealings.
- The nickel alloy with high performance serves as the vessel material.

Additional related specifications of the pressure vessel include:

- Via the upper vessel part, the fuel stream is injected in the axial direction, with the highest temperature of 450 °C.
- The oxygen flow is fed radially through the same flange at the same maximum temperature as the fuel flow (450 °C).
- The cooling water stream (CW1) is injected radially in the radial direction at 20 °C.
- The water for the cooling mantle (CW2) is fed in the radial direction, on its upper side and flows downwards.

### *5.2. Combustion Chamber Design*

The hydrothermal spallation drilling tool's combustion chamber should serve as a combustor while forming the flame jets to serve as tools for drilling.

Panagiotis S. proposed a new design of the combustion chamber (Figure 2) which has the following characteristics:

- Via a center hole in the fuel injection nozzle, an igniter is inserted in the chamber.
- In the similar angle as the oxygen flow, the fuel is injected. As a result, a recirculation area is created within the area with inserted igniter. In addition, this injection angle can minimize the impact of the fuel injection on combustion chamber wall.
- The combustion chamber must have a sufficiently large size to accommodate the higher heat load of combustion.
- The nozzle of fuel injection and combustion chamber must be easily adapted.

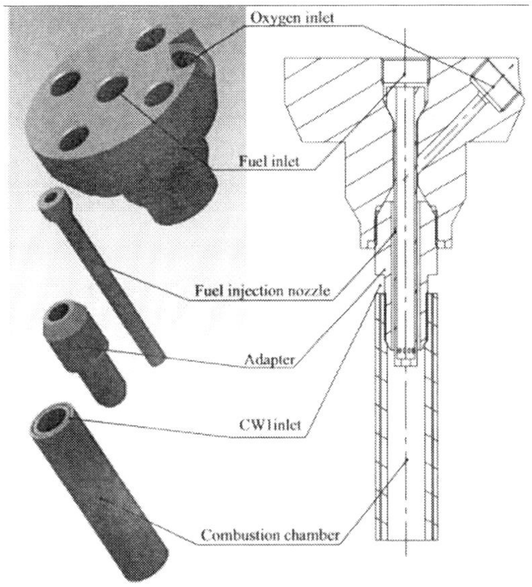


**Figure 2.** Combustion chamber in the high pressure vessel [11]

## 6. Ignition Experiments

For a smooth and reliable underground working condition, there must be a way to ignite the flammable hydrothermal mixture. Under the condition of spontaneous combustion, a typical borehole cannot be achieved at a reactant temperature below 100 to 250 °C. In the meantime, Subsurface heating of the reactants is hard to achieve since it is required for very high electricity. Therefore, forced ignition is still the only way to operate a hydrothermal flame within 2500 meters, and its study is essential for implementing spallation drilling technology. Since some metals are less likely to be oxidized by supercritical water than ceramics, a coiled wire was selected as an alternative to the ceramic igniter. The coil is composed of NiCr 60/15 and is temperature-dependent on its resistance and can withstand high temperatures. The wire thickness is 0.4 mm, while the outer diameter of the coil is 2.5 mm, with an inner diameter of nearly 1.7mm. At 20 °C, its electrical resistance and total length are 7.5 Ω and 30 mm, respectively.

## 7. Heat Flux Sensors and Their Calibration

Drill equipment design highlights the formation of flame jet and the optimization of heat transfer by hydrothermal flame impact. The behavior of hydrothermal flame as a free jet in liquid water differs from the conventional flame's employed by Rauenzahn and Wilkinson [12-16]. The condition near the hydrothermal flame has a great demand on the heat flow sensor especially on the chemical compatibility and temperature robustness. The temperature for maximum operating of the sensor regulates the point the most approaching to the flame that can be measured for heat flow. In the meantime, Rauenzahn reported that during spallation the most obvious temperature rising on the rock surface is nearly 500 °C. Since these sensors are higher than the surface of the peeling rock in thermal conductivity and are able to be cooled further, their maximum temperature for design is established at 700 °C. Likewise, in hydrothermal spallation drilling, heat flux values between 2000 $kW/m^2$ and 500 $kW/m^2$ are expected. Accordingly, the sensor design shall ensure that the maximum expected heat flow does not exceed its maximum operating temperature. Heat flow is a physical quantity that can only be measured indirectly. The simplest heat flux measurement follows temperature differences at two very close points within the material. Godefroy et al. constructed thin film sensors abiding by this principle. Similar sensors were presented from Holmberg and Diller [17] and Mabel and Huxtable [18], and each produces a voltage directly proportional to the incoming heat flow.

## 8. Design of nozzles

During the nozzle design water entrainment has to be controlled and this can be achieved by reducing the fluid density near the flame jet. Panagiotis S, et al [12] proposed a conceptual design of the nozzle. The cooling water is transferred from the flame jet through injection to an angle related to the combustion chamber axis. In addition, the outlet diameter of the central jet and circular jet controls the jet velocity and ratio.

Furthermore, the angle of water injection impacts the volume between impact plate and curtain, determining the amount of heat required to reduce the water density close to the flame. Energy required to reduce surrounding fluid's density will increase, and the efficiency will be lower with the increase of the volume. As a result, this volume will be reduced by a small curtain angle, whereas water inflow will increase because of the oblique impact. Such impact can accurately extinguish the flame at the point of impact. Clearly, a design balancing above two counteracting effects should be found.

## 9. Impingement Experiments

The water hot flame jet in liquid water bath has a strong characteristic of water. This phenomenon extinguishes the flame jet, affecting the efficiency of the technology. The thermal fluid split drilling rig should be able to control but not necessarily minimized the band rate. The dry crack drilling requires no more than the rock sample's brittle plastic transition temperature. The major strength of transcritical fluid is that it has a high convective heat transfer coefficient near its pseudo-critical point. Therefore, the combination of medium impact temperature and high heat transfer coefficient seems to be the best method to operate the drilling tools for hydrothermal spallation.

For the achievement of the above goals, expertise of three scientific disciplines is needed. First, the impact of water entrainment on flame impact and its limit should be analyzed. Second, underwater welding technology faces similar challenges. To protect the impact area of the gas jet and to provide an appropriate environment for the welding process, the water curtain is employed. However, the impact of the water curtain will still cause flow into the stagnation point of the flame. Therefore, the study of the oblique impact of underwater water jet can help reduce the flow in this direction, which is necessary for the design of drilling tools. Three scientific components of the drilling tool design process is shown in Figure 3.

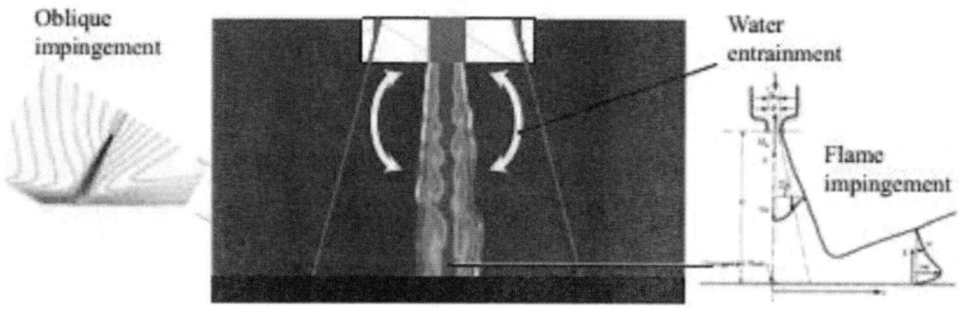


**Figure 3.** Hydrothermal flame jet in a water bath [11]

## 10. Conclusions

To recover geothermal energy and develop oil and gas resources in deep formation, hydrothermal spallation drilling method is proposed in this paper. A review work has been done on the following parts: the borehole condition, the mechanism of hydrothermal spallation drilling method, the equipment setups and the hydrothermal flame jet in a water bath.

## Acknowledgement

This work was financially supported by "the Fundamental Research Funds for the Central Universities" (No.16CX05019A) and China Scholarship Council (No. 201806455009).

**References**

[1] Hu X D, Song X Z, Li G S, Shen Z H, Cui L, Lv Z H, Shi Y and Chi P 2016 Numerical analysisof bottomhole velocity and pressure distributions of high pressure high temperature hydrothermal jet IADC/SPE Asia Pacific Drilling Technology Conference (Singapore: Society of Petroleum Engineers) pp 22–24

[2] Xu D, Wang S Huang C Tang X and Guo Y 2014 Transpiring wall reactor in supercritical water oxidation Chemical Engineering Research and Design 92 2626–2639

[3] Schuler M J, Rothenfluh T, Rohr P R V 2015 Stagnation flow heat transfer of confined, impinging hot water jets under supercritical pressures The Journal of Supercritical Fluids 99 51–60

[4] Wang H, Wang W, Bi Q and Wang L 2015 Experimental study of heat transfer and flow resistance of supercritical pressure water in a SCWR sub-channel Journal of Supercritical Fluids 100 15–25

[5] Meier T, Brkic D, Schuler M J, Kant M and Rohr P R V 2015 The potential of hydrothermal flames to induce spallation gneiss Proceedings of World Geothermal Congress (Melbourne) pp 19–25

[6] Wagner W and Pruss A 2002 The IAPWS formulation 1995 for the thermodynamic properties of ordinary water substance for seneral and scientific use Journal of Physical and Chemical Reference Data 31

[7] Stathopoulos P, Meier T and Rohr P R V 2014 Hydrothermal flame impingement experiments. Combustion chamber design and impingement temperature profiles Journal of Supercritical Fluids 89 48–57

[8] Schuler M J, Rothenfluh T, Stathopoulos P Brkic D, Meier T and Rohr P R V 2015 Simulatingsupercritical water jets with a variable turbulent Prandtl number Chemical Engineering & Technology 37 1896–1902

[9] Franck E U 1993 Combustion and diffusion flames in supercritical aqueous fluids at high pressures High Pressure Chemistry, Biochemistry and Materials Science 401 247–263

[10] Marshall W L and Franck E U 1981 Ion product of water substance, 0–1000°C, 1–10,000 bars –new international formulation and its background Journal of Physical and Chemical Reference Data 10 295–304

[11] Stathopoulos P 2013 Hydrothermal spallation drilling experiments in a novel high pressure pilot plant PhD Thesis ETH Zurich

[12] Stathopoulos P, Rothenfluh T, Schuler M J, Brkic D and Rohr P R V 2012 Assisted ignition of hydrothermal flames in a hydrothermal spallation drilling pilot plant Thirty-seventh Workshop on Geothermal Reservoir Engineering (California: International Geothermal Association)

[13] Rothenfluh T, Schuler M J and Rohr P R V 2011 Penetration length studies of supercritical water jets submerged in a subcritical water environment using a novel optical Schlieren method Journal of Supercritical Fluids 57 175–182

[14] Rothenfluh T 2013 Heat transfer phenomena of supercritical water jets in hydrothermal spallation drilling Nursing Standard 12 31–36

[15] Rauenzahn R M 1986 Analysis of rock mechanics and gas dynamics of flame-jet thermal spallation drilling PhD Thesis Massachusetts Institute of Technology

[16] Wilkinson M A 1989 Computational modeling of the gas-phase transport phenomena and experimental investigation of surface temperatures during flame-jet thermal spallation drilling PhD Thesis Massachusetts Institute of Technology

[17] Holmberg D G and Diller T E 1995 High-frequency heat flux sensor calibration and modelling Journal of Fluids Engineering 117 659–664

[18] Mabel S R, Huxtable S, Gifford A and Driller T E 2005 Design and calibration of a novel high temperature heat flux gage ASME Conference Proceedings (San Francisco) 4 983–988

IOP Conf. Series: Materials Science and Engineering 472 (2019) 012047 doi:10.1088/1757-899X/472/1/012047

# Preparation and Properties of Anti-Insulation Integrated Phenolic Resin Composites

**Duan Ge[1], Minxian Shi[1],Yalin Yao[2], Shican Jiang[1], Dongpei He[1] and Zhixiong Huang[1]**

[1]School of Materials Science and Engineering, Wuhan University of Technology, NO. 122 Luoshi Road, Wuhan 430070, China.
[2]Beijing FRP Research and Design Institute Co. Ltd, No. 1 Xinglong South Street, Yanqing District, Beijing 102101, China.
Email: minxianshi@whut.edu.cn

**Abstract.** This paper mainly studies a kind of anti-insulation integrated light phenolic resin composite material among many thermal protective materials. The study found that the addition of ceramic microspheres can significantly reduce the density of the sheet, improve the thermal insulation properties of the material, and also improve the other properties of the material. Studies have shown that the residual rate increases with increasing volume fraction, and the ablation performance is optimal at a percentage of 30%.

## 1. Introduction

In recent years, research on the addition of hollow microspheres into the system in the field of integrated anti-insulation has attracted much attention. Qin Yan et al [1] used phenolic resin as the matrix, and prepared low-density anti-insulation composite materials with good thermal insulation properties and excellent ablation performance under high temperature conditions by adding functional filler microbeads and porcelain auxiliaries. The porous fiber porous matrix is used as the matrix, and the phenolic resin is filled in the inner hole of the porous fiber skeleton, which fully utilizes the advantages of the skeleton and the resin. Bahramian et al. [2] prepared a porous fiber-based phenolic resin ablative composite material. Chen Yongxin [3] and other phenolic foams with hollow glass microspheres were prepared by compression molding. When the mass fraction of hollow glass microspheres was 10% of the resin mass, compared with pure phenolic foam, it was not pretreated and pretreated. The hollow glass microsphere-reinforced phenolic foam increased the residual carbon rate at 880 °C by 5.4% and 6.4%, respectively. The untreated and pretreated hollow glass microspheres enhanced the phenolic foam compressive strength by 282.9% and 328.6%, respectively, compared with the pure phenolic foam. . The surface insulation of the fuel tank and solid rocket of the US Space Shuttle [4] uses a sprayable composite foam coating of hollow glass beads as an excellent thermal insulation material.

Adding hollow microspheres can ensure the strength requirements of materials to a certain extent while reducing weight and preventing heat insulation. Gao and his team members [5] filled glass microspheres into silicone rubber foam, effectively reducing the average pore size and increasing the hardness and tensile strength of the material. Wouterson et al. created different microstructures by using three different types of microspheres and by volume fraction of 0 to 50% of the microspheres. The results show that the bending strength decreases with increasing filler content, and the bending strength behavior is not affected. The effect of the component microspheres. Huang et al [6] found that the flexural strength and fracture toughness of microsphere-filled phenolic composite foam modified by glutaraldehyde and siloxane decreased less than that of materials without added microspheres.

## 2. Experimental

Firstly, the boron phenolic resin solid is pulverized into a powder and then dissolved in absolute ethanol. The mass ratio of the boron phenolic resin to the absolute ethanol is 1:1, and then placed in an oven at 76 ° C, and stirred once every 20 minutes until the solid boron phenolic resin is completely dissolved in ethanol. Next, the ceramic microspheres were surface treated with a silane coupling agent KH-550.

According to the number of layers and the area of each layer, the quality of the fiber cloth is weighed. The mass of the boron phenolic resin after dissolution was 3.5 times that of the fiber cloth. The percentage of ceramic microspheres added was determined to be 0%, 10%, 20%, 30%, 40 % depending on the quality of the resin. The uniformly mixed resin was coated on the fiber cloth. Allow the air to stand until the prepreg is not sticky, cut, overlap.

Two thickness plates (4mm and 10mm thick) were produced by compression molding. The number of layers of prepreg was 20 and 40, respectively. The content of ceramic microspheres was 0%, 10%, 20%, 30%, 40%. Plate. The 20-layer and 40-layer prepreg are respectively laminated together. The prepreg of the 4 mm plate has a size of 9 mm*14 mm, and the prepreg of the 10 mm thick plate has a size of 9 mm*9 mm.

## 3. Results and Discussion

### *3.1. Density*

The density test was carried out according to GB/T 1463-2005. Since the ceramic microspheres are added, and the inside of the microspheres is hollow, its density is lower than that of the resin, so that a lightweight effect can be obtained, and a composite sheet having a low density relative to the original sheet material can be obtained. Density is measured by weighing, calculating the volume, and determining the density. The density of composite sheets with volume fractions of 0, 10%, 20%, 30%, and 40% ceramic microspheres is shown in Figure 1.

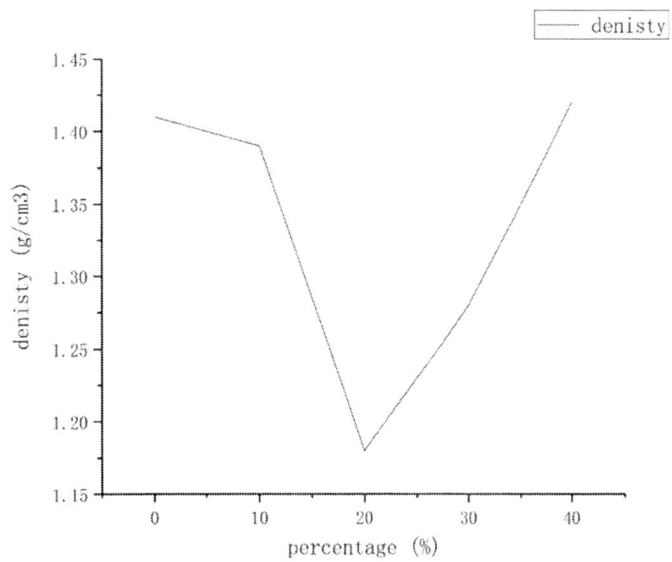


**Figure 1.** Panel density of each compound

It can be seen from the above that the ceramic microsphere content of 10%, 20%, 30% of the sheet density is less than 0%. When the ceramic microsphere content reaches 40%, the composite sheet density is higher than 0%. Studies have shown that the addition of ceramic microspheres can

Effectively achieves a light weight effect, a reduced specific gravity, and an optimum content of about 20%.

*3.2. Mechanical Properties Test (Bending Strength Test)*

In this experiment, 0, 10%, 20%, 30%, and 40% of the sheets were subjected to heat treatment at three temperatures (600 ° C, 800 ° C, and 1000 ° C) except for normal temperature, and a three-point bending test was performed. For the bending test, the universal material testing machine RGM-4100 was used in accordance with the GB/T 1449-2005 standard. The flexural modulus is calculated according to the equation. The average value is required for each set.

$$E=\frac{FS^3}{4\Delta WB^3} \tag{1}$$

(Where F is the maximum force, S is the support span, W is the width, B is the thickness of the specimen, and Δ is the instantaneous displacement according to the instantaneous load.)

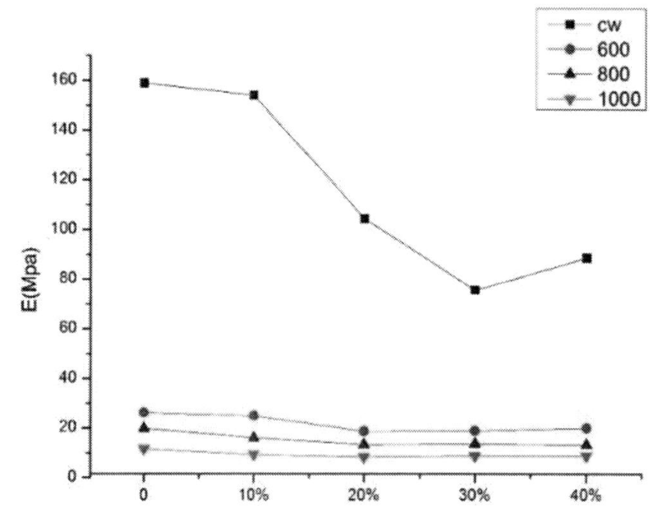


**Figure 2.** Bending strength after ablation of each formulation sheet

It can be intuitively seen from the Figure 3.2 that as the heat treatment temperature increases, the bending strength of the sample decreases remarkably. And as the volume fraction of ceramic microspheres increases, the curve basically shows a downward trend, and all the samples fail at the center of the support span.

*3.3. Section Topography Scanning*

After the bending test is completed, the section is scanned by SEM, and the JSM-IT300 scanning electron microscope is used to measure the observation section, which is helpful for analyzing the reason of the strength. The sample is in the form of a block and is less than 2 cm * 2 cm in size.Observing the cross section by scanning electron micrograph, we can see the brittle fracture of the resin after curing and the breakage of the fiber. Some spherical voids can also be seen in Figure 3, which is caused by the debonding of the microspheres during the bending process. The interaction between the microsphere and the resin is lower than the strength at which the microsphere itself is destroyed. The sample has several components: hollow ceramic microspheres, a phenolic resin matrix, an interface between the microspheres and the matrix, an interface between the resin system and the fibers, and internal voids, and the internal voids are introduced by the molding process, which is unavoidable. When subjected to a load, local phenomena include cracks at the beginning and through the internal void, cracking of the interface, cracking of the microspheres and the resin itself. The crack begins with the internal void formed during the solidification process. When the crack begins, it will propagate to the stage where the lowest energy is needed, the phase that provides the least obstacle to the propagation of the crack front. Due to the introduction of ceramic microspheres, more interfaces were introduced. Although the coupling agent improved the interface, the existence of the interface gave the path of crack propagation. As shown in Figure 3, the cracks along the resin and microspheres can be seen. Interface extension. Therefore, the crack is more likely to propagate along the interface, which also leads to the debonding of the resin, the ceramic microspheres, and the fiber, so that the layering is easy, and the effect of the resin transfer load is greatly reduced, which is also easy to cause stress concentration at the defect. It is also shown in Figure 4 that when the microsphere volume fraction is small, the resin system and fiber interface are much better than the microspheres. On the other hand, the more ceramic microspheres, the more hollow portions are introduced, and the bending strength of the gas inside the ceramic microspheres is necessarily lower than the bending strength of the composite material. Therefore, these aspects lead to the fact that the more the filler content at the four temperatures, the lower the bending strength is substantially. At 40% of the formulation, the flexural strength of the composite sheet slightly rebounds. This may be because 40% of the volume fraction is larger, the content of the microspheres is much, and it is easy to agglomerate, as shown in Figure 4(d), the ceramic microspheres. The agglomeration and fragmentation of many microspheres can be seen on the section of the plate with 40% of the integral number. The microspheres are compressed and destroyed, that is, crushed during the molding process, so that part of the hollow gas in the molding is discharged outside the plate, reducing The inner hollow portion has a slight increase in strength, but cracks and voids are introduced, the interface is deteriorated, and the strength is lower than the original state.

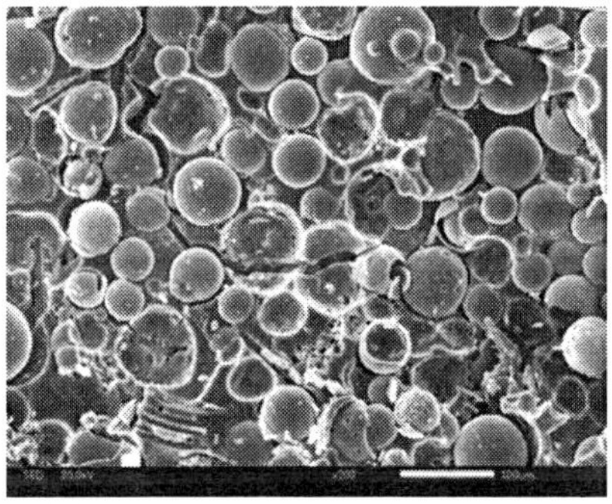

**Figure 3.** Example of section crack extension

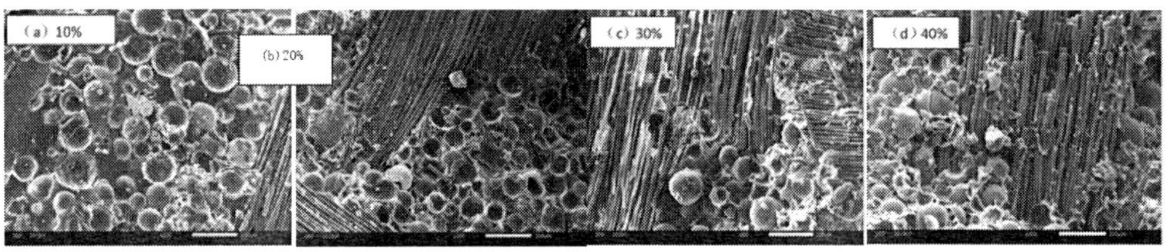


**Figure 4.** Sectional morphology of four formula sheets after bending failure

At the same time, it can be clearly seen that as the heat treatment temperature increases, the bending strength of the plates of each group has been greatly reduced. This may be because the ablation at high temperature causes the resin to decompose and the fibers are damaged. Cracks and voids appear to increase. Figure 4 also shows that the resin of various formulations is severely pulverized at 600 °C, the surface of the fiber becomes rough, and the microspheres are not greatly affected. In the composite material, the main bearing role is the fiber, and the effect of the resin transmitting stress after the heat treatment is not exerted, so the bending strength is greatly lost. However, compared with 0%, that is, the material without filler, the bending strength of the ceramic microspheres 20%, 30%, 40% is less, and the bending strength loss of the 30% sample is the smallest.

Observe the cross-sectional morphology of the heat treated at 600 °C in 30% and 10% of Figure 5(a) (e). It can be seen that the decomposition of the resin is better than 10% in the case of an increase in the content of the microspheres (30%). This may be attributed to the addition of ceramic microspheres, the introduction of a hollow structure, the heat is not easy to transport, the thermal insulation properties of the material are improved, and the heat loss of the internal fiber and resin system is reduced. Observe the heat-treated sample, as shown in Figure 5, to maintain the original shape. Studies have shown that the addition of ceramic microspheres can improve the heat resistance of the sample to a certain extent, and has a certain ability to withstand bending.

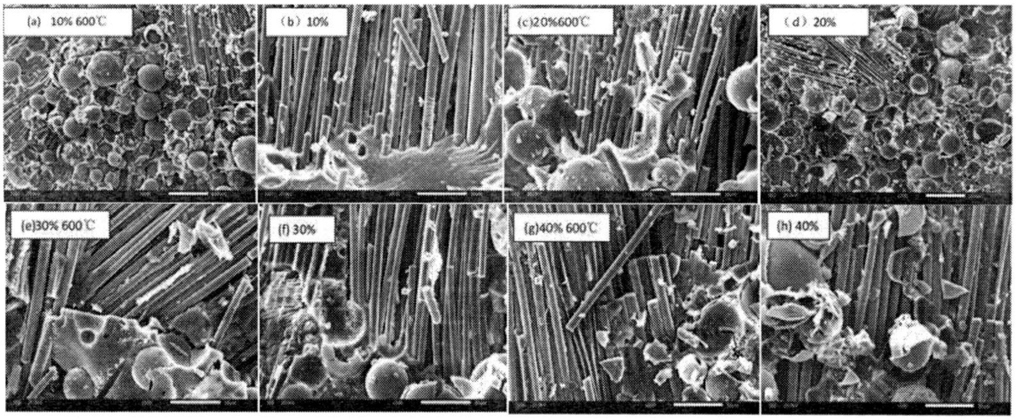


**Figure 5.** Curved failure section before and after heat treatment of four formula plates

### 3.4. *Thermal Insulation Performance Test*

According to the national standard GB/T10297-2015, the QTM-500 thermal conductivity tester is used for testing. The plate size is 5cm*15cm. The experimental tests tested the thermal conductivity of five sheets of 0, 10%, 20%, 30%, and 40%, as shown in Figure 6.

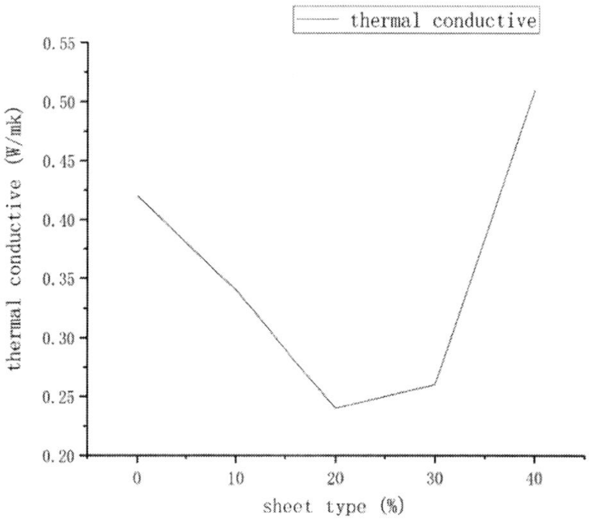


**Figure 6.** Thermal conductivity of five kinds of plates

It can be seen from the above thermal conductivity that the volume fraction from 0 to 20% decreases with the increase of the content of ceramic microspheres. This is because the heat transfer is mainly solid-phase heat conduction, the thermal conductivity of the solid is greater than that of the gas, and the ceramic microspheres are added. Since the interior is hollow, the content of the matrix in the unit volume of the material is reduced, and the solid content is gradually reduced. The increased spherical pores block the solid-phase heat transfer path, so the introduction of the hollow structure can effectively reduce the thermal conductivity of the composite. The thermal insulation performance of the ceramic microspheres with a content of 20% can easily achieve a better effect, and the microspheres continue to increase and easily break, which in turn affects performance. The thermal conductivity increases after the content of ceramic microspheres is more than 20%, which may be due to the excessive content of ceramic microspheres and the extrusion cracking during compression molding.

## 4. Conclusion

The anti-insulation integrated material produced can meet the above and more performance to play a role. Increasing the microspheres can reduce the density, heat insulation and slow the ablation. The boron phenolic resin has good ablation resistance, and the quartz fiber has sufficient stability, but the higher the microsphere content, the better. Experimental research shows that the ceramic microspheres the overall performance of the board with 20% and 30% volume fraction is better.

## References

[1] Qin Yan,Rao Zhilong,Liu Huijuan,Huang Zhixiong.Study on ablative thermal insulation properties of porcelain phenolic composites[J].Fiber Reinforced Plastics / Composites,2012(S1):52-55.

[2] Bahramian A R, Kokabi M, Famili M H N, et al. Ablation and thermal degradation behaviour of a composite based on resol type phenolic resin: process modeling and experimental [J]. Polymer, 2006, 47(10): 3661-3673.

[3] Chen Yongxin, YAO Zhengjun, ZHOU Jintang, et al. Compressive properties and thermal stability of hollow glass microspheres for reinforcing phenolic foams[J].Journal of Composite Materials,2014,31(4):873-879.

[4] LiyingZhang, J.Ma. Effect of coupling agent on mechanical properties of hollow carbon microsphere/phenolic resin syntactic foam [J]. Composites Science and Technology, 2010, 70(8): 1265-1271.

[5] Gao, Jie; Wang, Jibin; Xu, Haiyan; Wu, Chifei.Preparation and properties of hollow glass bead filled silicone rubber foams with low thermal conductivity [J]. Materials and Design, 2013, 46(2):491-496.

[6] Chi Huang, Zhixiong Huang, Xuesong Lv, Guangwu Zhang, Qiong Wang, Bo Wang.Surface modification of hollow glass microsphere with different coupling agents for potential applications in phenolic syntactic foams[J]. Journal of Applied Polymer Science, 2017.

# Synthesis of $Cu_2O$/biochar composites and their antibacterial activity against *Escherichia coli*

**Bujun Yang[1], Haiqian Wang[1], Qing Yang[1], Qikai Zhang[1], Lanmei Zou[1], Min Wu[2], Hongbin Zhang[2], Qiusheng Song[1] and Shaoming Yu[1]***

[1] School of Chemistry and Chemical Engineering, Hefei University of Technology, Hefei 230009, Anhui, PR China
[2] School of Food and Biological Engineering, Hefei University of Technology, Hefei 230009, Anhui, PR China
*Corresponding author. E-mail address: shmyu@hfut.edu.cn

**Abstract.** In this study, distilers' grains were used to prepare biochar, which is renewable and ecofriendly, then cuprous oxides nanoparticles were loaded to the biochar to improve the properties of $Cu_2O$. Their structure and morphology were characterized by X-ray diffraction (XRD), field-emission scanning electron microscope (FE-SEM), and brunauer-emmett-teller (BET) specific surface area. The results showed that the $Cu_2O$ nanoparticles were loaded onto the biochar successfully and the biochar-based $Cu_2O$ nanoparticles ($Cu_2O$/biochar composites) have larger specific surface area than pure $Cu_2O$ nanoparticles. Their antibacterial activity against *Escherichia coli (E. coli)* was examined by tablet colony counting method and optical density (OD) method. The results showed that the $Cu_2O$/biochar composites were more effective and had superior antibacterial performance than pure $Cu_2O$ nanoparticles which indicated that $Cu_2O$/biochar composites have good application prospect in antibacterial field.

## 1. Introduction

Nowadays, inorganic antibacterial agents, especially nanosized materials, have been paid much attention for the increasing demand for environmental sanitation [1~4]. One of the inorganic antibacterial agents is cuprous oxide and it has been used as an antibacterial agent for a long time [5]. However, $Cu_2O$ particles, especially those with nanoscale structure, are easy to aggregate [6,7] , which may weaken its ability in antibacterial activity [8]. *Hrenovic et al.* [9] investigated the antibacterial activity of $Cu_2O$ supported onto natural clinoptilolite in the secondary effluent water and the composites showed excellent antibacterial activity. *Abbasi et al.* [10] reported the preparation of $Cu_2O$ nanoparticles upon silk fibers using the ultrasound technique and they founded that the deposition of $Cu_2O$ thin films on flexible polymer surfaces might be a new path for the fabrication of functional materials for different applications. Therefore, immobilizing $Cu_2O$ onto support materials has been proved to be an effective strategy to improve the performance of $Cu_2O$.

Biochar refers to carbon-based dusty residues obtained from biomass pyrolysis. It has been demonstrated that biochar can reduce bioavailability of some heavy metals and that it has a high adsorption capacity to persistent organic pollutants [11]. Moreover, synthesizing biochar-based nano-composites are a new composite material and combine the advantages of biochar with nano-materials. The resulting composites usually exhibit great improvement in functional groups, pore properties, surface active sites, catalytic degradation ability and easy to separation [12]. In this study, distillers' grains were carbonized by concentrated sulfuric acid (96.5%) to get biochar. Then, a simple deposition method was applied to put $Cu_2O$ onto the biochar. For this is the first time to prepare $Cu_2O$/biochar composites, they were characterizated and the antibacterial properties of them were investigated by observing their antibacterial activities to E. coli through tablet colony counting method and optical density (OD) method. As expected, the $Cu_2O$/biochar composites displayed significantly enhanced antibacterial activity.

## 2. Materials and Methods

### 2.1. Preparation of $Cu_2O$ nanoparticles, biochar and $Cu_2O$/biochar composites

PEG2000, $CuSO_4 \cdot 5H_2O$, NaOH, ascorbic acid, concentrated sulfuric acid, tryptone, yeast extract powder, NaCl, and agar were purchased from Sinopharm Chemical Reagent Co., Ltd. Distillers' grains were obtained from Fuyang (Anhui Province, China). Before experiments, distillers' grains were dried in an oven for 24 h at 105 °C and grinded into a particle.

0.1 g PEG2000 was dissolved in 20 mL of 0.5 mol/L copper sulfate solution under magnetic stirring in 250 mL round-bottom flask. Then, 40 mL of 0.5 mol/L sodium hydroxide solution was added to the above solution under magnetic stirring for 30 min. Thereafter, 50 mL of 0.1 mol/L ascorbic acid was added dropwise and stirred for 30 min. The whole reaction process was in room temperature. After that, the solution was centrifuged, washed and dried to get $Cu_2O$ nanoparticles.

To prepare biochar, 20 g distillers' grains were put into 120 mL concentrated sulfuric acid under magnetic stirring for 2 h in 250 mL Stand-up flask at 60 °C, then washed and dried at 80 °C in the dryer.

The synthesis of $Cu_2O$/biochar composites was similar to $Cu_2O$ nanoparticles. Firstly, 2.145 g of biochar and 0.1 g PEG2000 were put into 20 mL of 0.5 mol/L copper sulfate solution under magnetic stirring in 250 mL round-bottom flask for 12 h to get more impregnation of $Cu^{2+}$ on the biochar. Secondly, 40 mL of a certain concentration of NaOH solution was added to the above solution under magnetic stirring for 30 min. lastly, 50 mL of a certain concentration of ascorbic acid was added dropwise and stirred for 30 min. The whole reaction process was also at room temperature. Then, the solution was centrifuged, washed and dried at 80 °C to get $Cu_2O$/biochar composites.

### 2.2. Characterization Methods

The crystal structures and composition of the synthesized particles were determined by X-ray diffraction (XRD, D/MAX2500V) with a Cu Kα radiation at 40 kV and 100 mA at a scanning rate of 2°/min from 10° to 80° (2θ). The microstructures and morphologies of the particles were determined by field-emission scanning electron microscopy (FE-SEM, SU-8020) at 5.0 kV. The specific surface areas (SBET) of the samples were determined by an automated surface area and pore size analyzer (Autosorb-1, NOVA2200e) based on nitrogen adsorption. The effects of the particles on the growth rates of *E. coli* were determined by UV-visible spectrophotometer (UV-Vis, Perkin–Elmer Lambda 25).

### 2.3. Antibacterial Activity Experiments

Tablet colony counting method [13] was employed to investigate the antimicrobial ability of $Cu_2O$ nanoparticles and $Cu_2O$/biochar composites. *E. coli* was selected as test strain. All the materials were sterilized in an autoclave (121 °C, 20 min) before experiments. Firstly, 1 mL *E. coli* was precultured in 50 mL Luria –Bertani (LB) fluid medium ( 5 g tryptone, 5 g NaCl, 2.5 g yeast extract powder, 500 mL $H_2O$ ) at 37 °C in 250 r/min oscillator for 12 h. Secondly, put a certain amount of $Cu_2O$ nanoparticles and $Cu_2O$/biochar composites into conical flask containing 25 mL solid LB medium ( 5 g tryptone, 5 g NaCl, 2.5 g extract powder, 6 g agar, 500 mL $H_2O$ ). Bind up and then sterilize in an autoclave (121 °C, 20 min). Thirdly, put the above solid LB medium mixed with samples into culture dish cooling to room temperature on super clean bench. Then, using SS-Spreader to spread 200μL bacterial suspension which was diluted from the above bacterial suspension to $10^8$ CFU/mL uniformly. After that, they were incubated at 37 °C for 24 h. Antibacterial activity was evaluated by the number of formed colonies on solid LB medium.

## 3. Results and Discussion

### *3.1. Characterization of Materials*

*3.1.1. XRD.* The XRD patterns of biochar, $Cu_2O$ nanoparticles and $Cu_2O$/biochar composites are shown in Fig.1. Fig.1 (a) shows a broad diffraction peak with 2θ ranged in 20°-30°, attributable to the characteristic peak of C(002). This result indicates that the biochar is composed of amorphous carbon with a low degree of graphitization [14]. The diffraction peaks of Fig.1(b) and Fig.1(c) were indexed as the (110), (111), (200), (220), (311) and (222) reflections, corresponding to the standard structure of cuprous oxide (JCPDS No. 05-0067) which indicates that $Cu_2O$ nanoparticles were successfully loaded onto the biochar. What's more, no significant peaks of impurities were detected, indicating the high purity of the obtained products. The higher (111) diffraction intensity of $Cu_2O$ nanoparticles suggests that the surfaces of the $Cu_2O$ nanoparticles have {111} crystal planes [15], which indicates the products are favorable for inactivating bacteria [16].

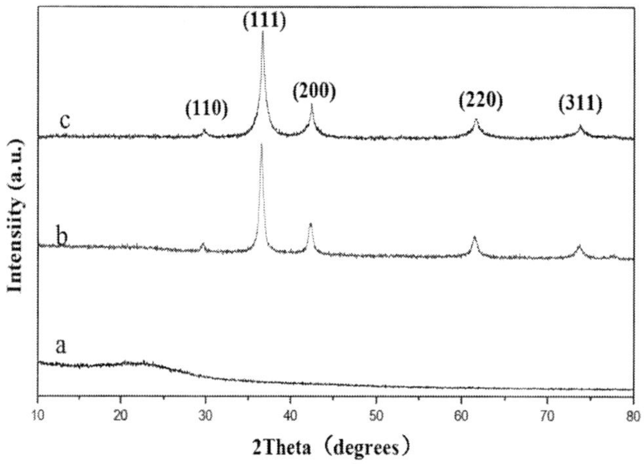


**Figure 1.** X-ray diffraction patterns of: (a) biochar, (b) $Cu_2O$/biochar composites, (c) $Cu_2O$ nanoparticles.

*3.1.2. FE-SEM.* The FE-SEM images of biochar, $Cu_2O$ nanoparticles and $Cu_2O$/biochar composites are shown in Fig.2. Fig.2(a) shows the FE-SEM image of biochar and reveals that it has rough surface which is in favor of loading $Cu_2O$ [17]. Fig.2(b) shows the FE-SEM image of $Cu_2O$ nanoparticles, the partical size of $Cu_2O$ nanoparticles are close to 200 nm, as we can see from the image, it has serious aggregate problem. Fig.2(c) shows the FE-SEM image of $Cu_2O$/biochar composites. In contrast, there was no obvious aggregation of $Cu_2O$ on the surface of composites, the existence of biochar make $Cu_2O$ nanoparticles avoid serious aggregation and disperse homogeneously.

**Figure 2.** FE-SEM images of (a) biochar, (b) $Cu_2O$ nanoparticles, (c) $Cu_2O$/biochar.

*3.1.3. BET specific surface area.* Table 1 presents the BET specific surface area ($S_{BET}$) data of pure $Cu_2O$ nanoparticles, biochar and $Cu_2O$/biochar composites. Compared with pure $Cu_2O$ nanoparticles catalyst and biochar, the $S_{BET}$ of $Cu_2O$/biochar composites show much higher values. It may be due to that $Cu_2O$/biochar composites have less aggregate problem than pure $Cu_2O$ nanoparticles as we can see from the FE-SEM.

**Table 1.** BET specific surface area of pure $Cu_2O$ nanoparticles, biochar and $Cu_2O$/biochar composites.

| Material | pure $Cu_2O$ nanoparticles | biochar | $Cu_2O$/biochar composites |
|---|---|---|---|
| $S_{BET}$ ($m^2/g$) | 7.764 | 11.607 | 24.634 |

*3.2. Comparison of Antibacterial Activity in Killing E. coli between Pure $Cu_2O$ Nanoparticles and $Cu_2O$/biochar Composites*

Before the comparisons of antibacterial activity in killing *E. coli* between pure $Cu_2O$ nanoparticles and $Cu_2O$/biochar composites, a series of experiments has been down to see if the biochar has antibacterial activity. The results are negative. One of the experiments adopted tablet colony counting method. The results were shown in Fig.3. As we can see from it, the biochar has no antibacterial activity.

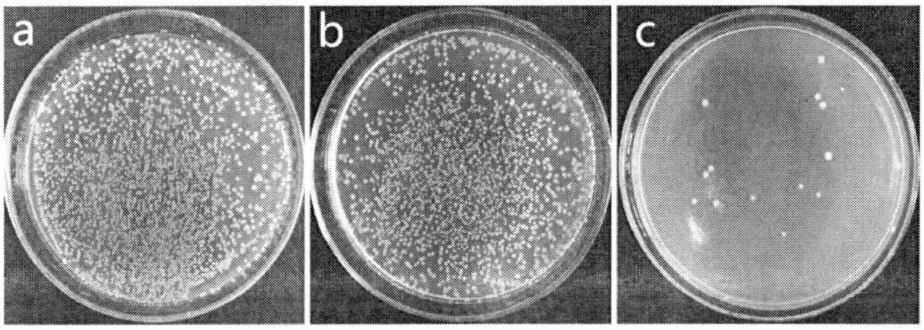


**Figure 3.** The comparison of antibacterial activity between biochar and pure $Cu_2O$ Nps. (a: blank control, b: biochar, c: $Cu_2O$ nanoparticles )

*3.2.1. MIC.* To quantitatively evaluate the antibacterial effects of the $Cu_2O$ nanoparticles and $Cu_2O$/biochar composites, their minimal inhibitory concentrations (MIC) to *E. coli* were detected with agar dilution method [12]. $Cu_2O$ nanoparticles and $Cu_2O$/biochar composites were dispersed into agar medium. After solidification, *E. coli* was inoculated on the above agar medium. Then, the growth of bacteria was observed after cultivated for 24 h in a constant temperature incubator at 37 °C. The lowest concentration, at which no visible bacterial colony could be observed, was considered to be the MIC of the antibacterial materials. The results were shown in the table 2. As we can see from table 2, the MIC of $Cu_2O$/biochar composites is lower than pure $Cu_2O$ nanoparticles, which indicates that $Cu_2O$/biochar composites have better antibacterial activity than pure $Cu_2O$ nanoparticles.

**Table 2.** MICs of the pure $Cu_2O$ nanoparticles and $Cu_2O$/biochar composites.

| Synthetic samples | pure $Cu_2O$ nanoparticles | $Cu_2O$/biochar composites |
|---|---|---|
| MIC(μg/mL) | 56 | 40 |

*3.2.2. The effects of particles on the growth rates of E. Coli.* Prepare three sterilized 50 mL LB fluid medium, add 10 mg $Cu_2O$ nanoparticles and 25 mg $Cu_2O$/biochar composites to two of them respectively, the rest as a blank control. Under UV light irradiation for 30 min in super clean workbench, then access 1mL bacteria which cultivated for 12 h to each liquid medium. The inoculated *E. coli* cells were grown further in a constant temperature incubator at 37 °C for various incubation times. Take out 300μL *E. coli* cells from them and diluted with 2700μL sterile water. Then with sterile water as a reference, the *E. coli* densities were determined by the optical density (OD) at 600 nm. With time as the abscissa, OD value as the ordinate, draw the growth curve [14].

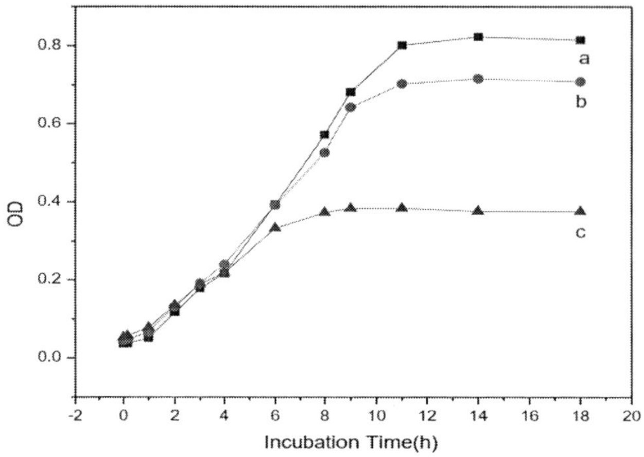


**Figure 4.** The comparison of antibacterial activity between biochar and pure $Cu_2O$ Nps. (a: blank control, b: biochar, c: $Cu_2O$ nanoparticles )

Fig.4 shows that $Cu_2O$ / biochar composites have apparent better antibacterial activity than pure $Cu_2O$ nanoparticles. It may be due to that the cuprous oxide in composite materials has less serious aggregate problem (as we can see from the SEM) and higher specific surface area (as we can see from the table 1). The figure also shows that the biochor has no antibacterial activity. At present, many experimental results have showed that the antibacterial mechanism of cuprous oxide nanoparticles is a contact killing through which bacteria initially suffers severe damage to the cell envelope. Then further damage ensues by an independent pathway of each cuprous oxide nanoparticle [18]. Thus, a higher specific surface area contributes to the improvement of antimicrobial activity [8], which accounts for that $Cu_2O$/biochar composites have apparent better antibacterial activity.

*3.2.3. Antibacterial validity.* In order to explore the antibacterial validity of $Cu_2O$/biochar composites, 20 mg $Cu_2O$ nanoparticles, 50 mg $Cu_2O$/biochar composites were put into 100 mL physiological saline respectively for two months. After two months, the liquid and the solid were seperated. 100 mL saline was added into the solid and marked with b, c respectively, b was solid of $Cu_2O$ nanoparticles and c was solid of $Cu_2O$/biochar composites. a was blank control. After sterilization, 200μL *E. coli* which was cultivated for 12h was added to a, b and c. Put them in 250 r/min oscillator at 37 °C for 10 min and then coat them to the prepared solid LB medium. Antibacterial activity was evaluated after being cultivated in constant temperature incubator for 24 h. The results show in Fig.5. As we can see from the Fig.6, $Cu_2O$ /biochar composites have better antibacterial activity than $Cu_2O$ nanoparticles after two months. That is to say, $Cu_2O$ /biochar composites have longer antibacterial validity.

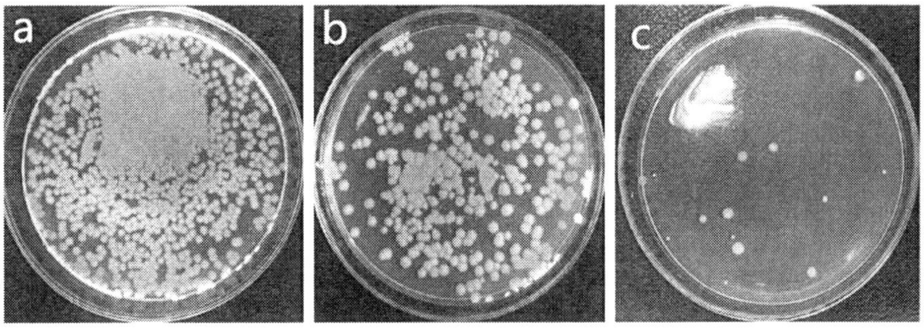


**Figure 5.** Antibacterial validity of pure $Cu_2O$ nanoparticles and $Cu_2O$ /biochar composites. (a: blank control, b: $Cu_2O$ nanoparticles, c: $Cu_2O$ /biochar composites )

## 4. Conclusions

In summary, a simple deposition method was adopted to synthesize $Cu_2O$ nanoparticles. Distillers' grains and concentrated sulfuric acid were used to prepare the biochar. Then chemical precipitation was adopted to load $Cu_2O$ nanoparticles onto the biochar to improve the properties of $Cu_2O$. The samples were characterized by XRD, FE-SEM and BET. The characterizations showed that the particle size of pure $Cu_2O$ nanoparticles is about 200nm with {111} crystal planes, $Cu_2O$ nanoparticles are loaded onto the biochar successfully and $Cu_2O$ /biochar composites have larger specific surface area and lower aggregation than pure $Cu_2O$ nanoparticles. Besides, their antibacterial activity against *E. Coli* was compared by MIC, the growth rates of *E. coli* and antibacterial validity. The results showed that the $Cu_2O$/biochar composites were more effective and had superior antibacterial performance than pure $Cu_2O$ nanoparticles. Thus, biochar is a good carrier for $Cu_2O$ nanoparticles and $Cu_2O$/biochar composites have good application prospect in antibacterial field.

**Acknowledgements**
Financial support from the National Natural Science Foundation of China (No. 21277039, 21671052 and 81573399) is acknowledged.

**References**

[1] Manna L, Milliron D, Meisel A, Scher E and Alivisatos A 2003 *Nature Mater* **2** 382
[2] Knez M, Nielsch K and Niinistö L 2007 *Adv. Mater.* **19** 3425
[3] Chen X, Wang Y, Zhang Y, Chen Z and Liu Y 2014 *Anal. Chem.* **86** 4278
[4] He Y, Li J and Liu Y 2015 *Anal. Chem.* **87** 9777
[5] Hassan I, Sathasivam S, Islam H, Nair S and Carmalt C 2017 *RSC Adv.* **7** 551
[6] Chen J, Zhou P, Li J and Li S 2007 *Carbohyd. Polym.* **67** 623
[7] Liu X, Li Z, Zhang Q and Lei F 2012 *Mater. Res. Bull.* **47** 2631
[8] Adams L, Lyon D and Alvarez P 2006 *Water Res.* **40** 3527
[9] Hrenovic J, Milenkovic J, Daneu N, Kepcija R and Rajic N 2012 *Chemosphere* **88** 1103
[10] Abbasi A, Noori N, Azadbakht A and Bafarani M 2016 *J. Iran. Chem. Soc.* **13** 1273
[11] Maroušek J, Vochozka M, Plachý J and Zak J 2017 *Clean Technol. Environ. Policy* **19** 311
[12] Tan X, Liu Y, Gu Y, Xu Y and Zeng G 2016 *Bioresource Technol* **212** 318
[13] Zhao D, Lu Y, Wang Z, Zeng X, Liu S, and Wang T 2016 *International Journal of Refractory Metals & Hard Materials* **54** 417
[14] Zhou Y, Niu S and Li J 2016 *Energy Conversion & Management* **114** 188
[15] Lee Y, Kim S, Park S, Park H and Huh Y 2011 *Materials Letters* **65** 818
[16] Ren J, Wang W, Sun S, Zhang L and Wang L 2011 *Industrial & Engineering Chemistry Research* **50** 10366
[17] Han Y, Boateng A, Qi P, Lima I and Chang J 2013 *Journal of Environmental Management* **118** 196
[18] Meghana S, Kabra S, Chakraborty S and Padmavathy N 2015 *RSC Advances* **5** 12293

IOP Conf. Series: Materials Science and Engineering 472 (2019) 012049 doi:10.1088/1757-899X/472/1/012049

# Functional Requirements Analysis and Practical Design of Plug-and-Play Energy Interconnect Terminals

**Aiqiang Pan**

EPRI, Shanghai Municipal Electric Power Company, 171 Handan Road, Shanghai, China.

Email: dsy_panaq@126.com

**Abstract.** The energy Internet is a new form of energy supply and a practice of "Internet + energy". Similar to the information internet, plug-and-play is one of the most important features of the Energy Internet, and also is one of the most important purposes of the construction of Energy Internet. Many attempts have been made to achieve this purpose. Power system is the most suitable natural carrier to construct Energy Internet, but the physical laws of power flow and information flow are different, so plug-and-play is hard to achieve and it is the primary problem to solve. This paper analyses the specific functional requirements of plug-and-play, and presents a practical design of plug-and-play energy interconnect terminals, which can realize the plug-and-play of energy terminals. The thought put forward in this paper will contribute to the construction of Energy Internet and promote the effective utilization of distributed energy.

## 1. Introduction

The rapid development of the Internet has not only profoundly changed people's lives, but also changed many traditional industries. The Energy Internet is the product of the integration of energy and the Internet. As a new development direction and important subject of energy industry after smart grid, Energy Internet, as an effective technology to solve the problem of distributed energy consumption, has become a new research hotspot in academia and industry.

Energy Internet was first proposed by American economist Rifkin. Research for Energy Internet has been carried out in recent years, e.g. FREEDM project in US [1] and E-Energy project in Germany [2].

Energy Internet is also a hot research topic in China. The Institute of science and technology for development of China regards the Energy Internet as one of the emerging industries of national strategy. China's State Grid cooperation and Tsinghua University both set up an Energy Internet Research Institute. In April 2016, the State Council issued the "Energy Internet innovation action roadmap". In recent years, China is gradually carrying out power market reform, which means that the development of Energy Internet will be prospective.

Currently, the definition, characteristics and implementation methods of Energy Internet are not unified. The final conclusion about the concrete morphology of Energy Internet is not yet formed. But considering the convenience of transmission, utilization and conversion of electric energy, power grid will play a main role in future Energy Internet. And distribution network will also develop from a pure electric power distribution network to a new type of power distribution network with the function of multi-energy (e.g. electricity, gas, cooling, heating, etc.) conversion, substitution and utilization. With the clean substitution at supply side and the electric energy substitution at demand side, electricity will be the main form of energy production and consumption.

So Energy Internet can be considered as a new power network composed of distributed energy acquisition devices, distributed energy storage devices and various types of loads, which integrates advanced power electronics technology, information technology and intelligent management technology to realize energy exchange and sharing network with bidirectional energy flow.

Plug-and-play is an important feature of the information Internet. Each individual is not only the consumer of information, but also the producer of information. Plug-and-play of heterogeneous devices is realized. Plug-and-play is also one of the basic requirements of the function of the Energy Internet, and is also one of the main purposes of the development of the Energy Internet.

Energy Internet needs to break the monopoly, decentralize, so that different participants in the position of equal trading. It requires plug-and-play of energy generation and consumption and a balance of energy from time to time, which requires that the Energy Internet should be a peer-to-peer, flat and bi-directional flow of energy sharing network, generating devices, energy storage devices and loads can be "plug-and-play".

## 2. Existing Solutions

Plug-and-play of energy equipment is more complex than Plug-and-Play of information, involving energy and information synergies. Referring to Internet devices, the concepts of “energy router”, “energy switch”, “energy hub”, and “energy network card” are proposed in [3], to solve the plug-and-play problem, which are widely recognized as basic components of Energy Internet.

However, the physical characteristics of power flow and Internet information flow are completely different.

Internet is a computer technology that supports information flow. It refers to a network connection between different computers to establish a common language, a bunch of communication protocols, so as to achieve communication between computers. For a particular information exchange process, a specific receiver can be specified and their delivery path can be scheduled. It can buffer, preserve or delay delivery. Electricity is injected into the grid by generating sets and consumed power loads that must be balanced at all times without any cushioning, savings or delays.

If we want to feature the Internet in the power system, we need to store, buffer, and dispatch electricity as easily as information. To achieve this goal, we need to make breakthroughs in two key technologies, power electronics and energy storage technology.

Power electronic technology is the foundation of power electronic transformer, which is considered as the key equipment in Energy Internet. It can achieve energy dispatching and flexible transformation of voltage and frequency. It has obvious technical advantages in improving power quality and system stability, realizing flexible transmission mode and real-time control of power flow. However, the current power electronics technology is mainly used in the case of low voltage and low power, and is doubtful at reliability and efficiency.

Energy storage technology is another critical technology, which improves the quality and the economic efficiency of distributed power generation unit, and ensures the stability of system. Energy storage has been widely used in distributed energy sources but still has obvious bottleneck at present. Although kinds of energy storage technologies are developing, the related technologies and performance are immature with high cost. The developing of key materials, process of manufacture and efficiency of energy conversion are also facing with challenges.

Completely copying the Internet model requires fundamental changes in the overall development of power grids, power electronics and energy storage technologies. But its feasibility, economy and effectiveness are doubtful.

Completely copying of the Internet mode for Energy Internet requires fundamental changes of power grid and the fully developments of power electronics and energy storage technology. But its feasibility, economy and efficiency are doubtful.

It is worth exploring how to realize plug-and-play function required by Energy Internet under the existing power system framework.

## 3. Functional Requirements Analysis

Plug-and-play of energy equipment is more complex than Plug-and-Play of information, which involves the coordination of energy and information. In order to realize plug-and-play, the security and stability of the energy network should be guaranteed automatically. In addition, the following contents should be studied: (1) Standard interface and protocol for plug-and-play of energy and information; (2) Self-organizing management of distributed devices, including automatic perception and recognition of devices, generation and splicing of device models, and automatic communication. (3) Autonomous control of the equipment itself to reduce adverse effects on the system

Therefore, plug-and-play interconnected terminals should have the following functional characteristics: unified physical information interfaces, unified power quality requirements, accurate and rapid protection and control means, as well as distributed measurement and trading means. In addition, the energy terminal should also facilitate remote monitoring, control and operation and maintenance, improving the efficiency of operation and maintenance.

### *3.1. Unified Physical & Information Interfaces*

Whether it is photovoltaic power generation, wind power generation, or energy storage devices, electric vehicles connected to the beat grid, all need different types of power conversion, the external interface is divided into strong electricity connected electrical interface and weak electricity connected communication interface. Due to the lack of unified specifications and standards, equipment manufacturers now define their own product interface forms, standards are not unified, a wide range of engineering applications are complex.

### *3.2. Unified Power Quality Requirements*

Energy terminal is the gateway of energy exchange, and its quality needs to meet the unified requirements. Power-centric Energy Internet, energy quality requirements are the requirements of power quality. It is a basic requirement for energy interconnected terminals to establish a unified power quality requirement system, monitor the power quality situation at all times and make corresponding adjustments according to different power quality conditions.

Specific quality indicators can include voltage quality and current quality. Voltage quality includes voltage frequency, voltage harmonics, voltage deviation, three-phase unbalance, voltage fluctuation and flicker. Current quality includes current frequency, harmonic current, negative sequence current, impulse current, DC component and so on. In addition, it also contains EMC indicators for specific needs. Energy interconnect terminals should be able to monitor the parameters of these voltage waveforms and current waveforms, and in the case of poor indicators, can identify the causes and sources.

### *3.3. Protection and Automation Control*

As the interface between terminal energy and Energy Internet, energy interconnected terminals should have comprehensive protection and control functions, which are mainly embodied in the requirements of islanding protection, component protection, system protection, fault protection, restoration of grid-connected function and so on. It should meet the requirements of reliability, selectivity, sensitivity and speediness. In addition, in some cases, the interconnected terminals should have power control function, voltage control function, frequency response function and so on, which can realize the function of virtual generator.

### *3.4. Distributed Metering And Trading*

The interconnected terminal shall have the function of electric energy metering, and its configuration and accuracy shall meet the requirements of relevant national technical standards and regulations.

The interconnected terminal shall be able to communicate with bus measurement systems and plug-and-play energy interconnect terminals, obtain information, performance logical computing, verifying and solving energy online trading by units of generation, transmission, storage, consumption, and achieving the best trading results through flexible pricing policy.

## 4. A Practical Design

We proposed a design of energy interconnect system which can satisfactorily compatible with the existing power grid structure. The proposed design abandons energy routers and power electronic transformers which are depended on the existing power grid and linear elements. And the functions of frequency conversion, energy storage and energy active dispatching are not considered.

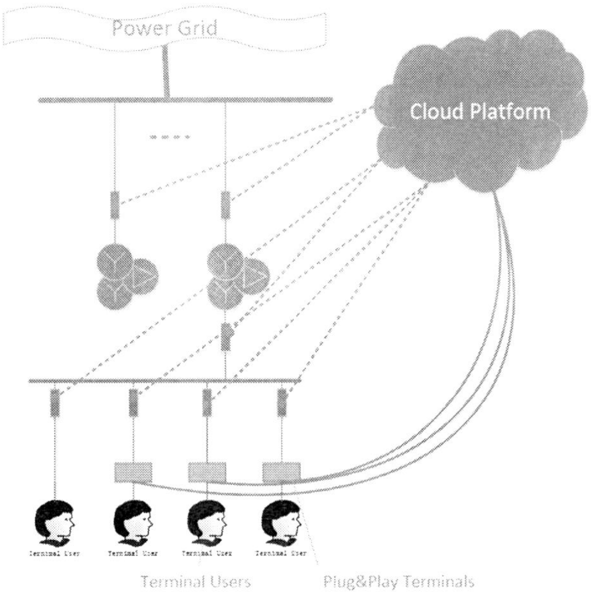


**Figure 1.** The architecture of plug-and-play energy interconnect system.

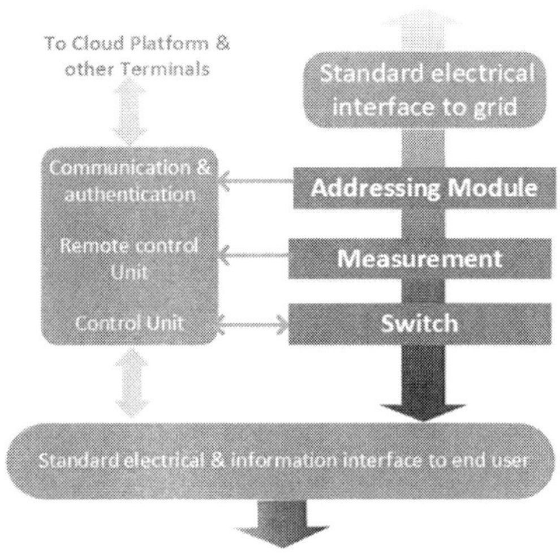


**Figure 2.** The architecture of plug-and-play energy interconnect terminal

The architecture of plug-and-play energy interconnects system shown as Fig.1. Based on the existing power system measurement system, a plug-and-play energy interconnect terminal is connected in series at the front end of the end user.

The plug-and-play energy interconnect terminal is the energy and information interface of power grid and energy terminal devices (such as distributed energy generator, electrical equipment and energy storage devices). Meanwhile, energy interconnect terminal is the combination of energy flow and information flow and connected in series.

The plug-and-play energy interconnect terminal includes:

- Standard electrical interface: the electrical interface can achieve the universal interface for various electrical sources with standard voltage and frequency.
- Standard communication interface: through standard interface and protocol, energy terminal devices can transmit state and instruction to the terminal.
- Measuring module: by monitoring the voltage and current signals flowing through the standard power interface, the real-time metering of electricity and the monitoring of power quality parameters are obtained. Verifying and solving energy online trading by units of generation, transmission, storage, consumption, and achieving the best trading results through flexible pricing policy.
- Switch: performing a current cut-off to achieve the isolation of equipment and power grid.
- Addressing module: getting the location of the equipment access point in the power grid by Power Line Carrier or other communication methods.
- Control unit: achieving protection and control function through logically calculation of the information from measurement modules of terminal, power systems or other terminals. According to the power flow information of the terminal accessing location, giving the instructions of terminal accessing, terminal breaking, and storage charged or discharged, etc.
- Communication and authentication module: performing the communication with power system and other terminals and achieving the corresponding security authentication and identity authentication of the terminal.
- Energy online trading: verifying and solving energy online trading by units of generation, transmission, storage, consumption, and achieving the best trading results through flexible pricing policy.

The design of these components can achieve the function of plug-and-play for power generation, consumption and storage equipment. With its characteristics such as less investment, economy, convenience, efficiency and practicality, the proposed design can be well compatible with current power supply system, improving the effective utilization of distributed energy and promoting the construction and development of Energy Internet.

## 5. Conclusion

Energy Internet is a wide-area system integrating variety of new energy technologies, information technologies, and new energy trading models. Its development is not accomplished overnight.

Judging from the objective law of the development of things, first of all, we should carry out researches on various innovation-enabled technologies within the framework of existing power system. And then, we should integrate relevant technologies to conduct comprehensive demonstration applications within a smaller region, and gradually expand them to a larger scope.

This paper analysed the plug-and-play needs in the Energy Internet and the key elements of its implementation. The plug-and-play energy interconnect terminal proposed in this paper utilizes the measurement information of the existing power system. In the framework of the existing power system, the plug-and-play effect is achieved. Based on the existing power system, the Energy Internet can be realized step by step. The analysis and design provided in this paper will be helpful to the construction of the Energy Internet.

## References

[1] Huang A Q, Crow M L, Heydt G T, Zheng J P, Dale S J. The Future Renewable Electric Energy Delivery and Management (FREEDM) System: The Energy Internet. Proceedings of the IEEE, 2011, 99(1):133 - 148.

[2] Vermesan O, Blystad L C, Zafalon R, Moscatelli A, Kriegel K, Mock R, et al. Internet of energy-connecting energy anywhere anytime. Springer Berlin Heidelberg, 2011: 33-48.

[3] Geidl M, Klokl B, Koeppel G, Andersson G, Frohlich K. Energy hubs for the futures. IEEE Power & Energy Magazine, 2007, 5(1): 24-30.

# Study on Static Recrystallization Behavior of 38MnVTi Non-Quenched and Tempered Steel

**Luwei Dai , Yuanfang Chen*, Hui Zeng , Ding Xie and Tao Zhang**

(College of Materials Science and Engineering, Chongqing University of Technology, Chongqing 400054, China)

Luwei Dai (1993-), female, Guang'an, Sichuan, master degree, mainly engaged in digital mold manufacturing and special processing research.

E-mail:574123364@qq.com.

**Abstract.** The static recrystallization behavior of 38MnVTi non-quenched and tempered steel within the interval of passes was investigated by double-pass compression experiments using Gleeble-1500 thermal simulation machine. The true strain-stress curves of the steel were obtained, and the effects of deformation temperature, strain rate and pre-strain on the static recrystallization behaviours were studied. Single-pass compression experiments for 38MnVTi non-quenched and tempered steel were finished with the same conditions to the double-pass compression experiments. The static recrystallization dynamics model and the static recrystallization grain size model can be established. The predicted results of the model accord well with experimental ones, which indicate the proposed model, can give an accurate prediction of the static softening behaviour for 38MnVTi non-quenched and tempered steel.

## 1. Introduction

38MnVTi non-quenched and tempered steel can achieve good comprehensive mechanical properties and uniform microstructure after forging or rolling, so it is widely used as automobile steering knuckle and semi-axle material [1-3]. During the hot forging and hot rolling, it needs to control the process parameters to control the final microstructure of the product, so as to control the final property of the product. Therefore, in order to predict and control the microstructure of non-quenched and tempered steel 38MnVTi more accurately and optimize the hot working process, it is necessary to study the static recrystallization law of 38MnVTi non-quenched and tempered steel in holding time after thermal deformation [4].

## 2. Experimental Materials and Methods

The experimental material is 38MnVTi non-quenched and tempered steel bar, and the chemical composition is shown in Table 1. In order to meet theexperimental conditions, the bar is processed into a Φ10mm × 12mm cylindrical sample. The true stress-strain curves can be obtained by double-pass compression experiments on Gleeble-1500 thermal simulator.

**Table1.** Chemical composition of 38MnVTi steel (wt%)

| Element | C | Si | Mn | V | Ti | P | S |
|---|---|---|---|---|---|---|---|
| Mass fraction(%) | 0.36 | 0.3 | 1.2 | 0.1 | 0.024 | 0.026 | 0.033 |

The thermal deformation process is as follows: the samples are rapidly heated to 1200 °C at a rate of 10 °C / s, and kept warm for 5 min, and then cooled to the respective deformation temperatures at a rate of 5 °C / s after complete austenitizing. Next, kept warm for 30s in order to eliminate the temperature gradient in the sample. Later carry out the first pass compression with the true strain of 0.15.After a different interval time, the second compression was made with the same true strain.The process flow chart is shown in Figure 1. Three different process parameters shown in Table 2. After hot compression, the samples were longitudinally cut, polished, polished, and etched (The etched was the mixed solution of picric acid and a small amount of SDBS). Then grain size was observed on an optical microscope.

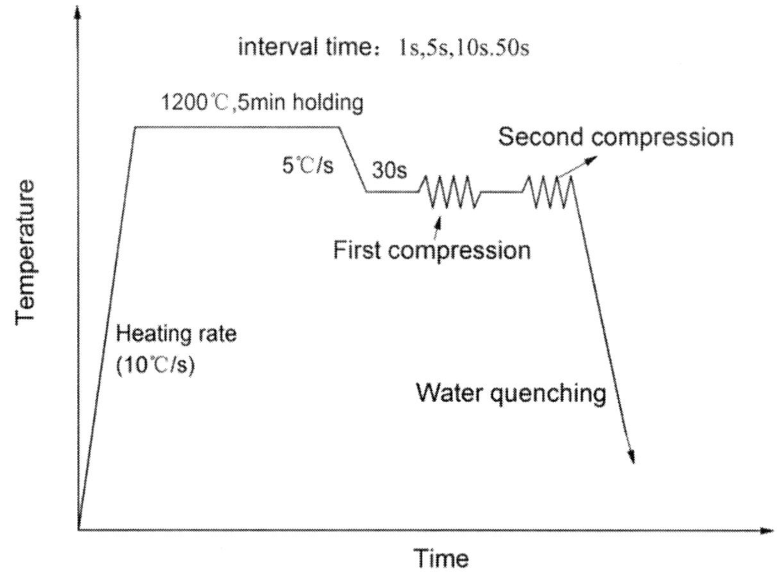


**Figure 1.** Experimental procedure for the double-pass hot compression experiments

**Table 2.** Process parameters for the double-pass hot compression experiments

| Process No. | Deformation temperature(°C) | The first true stain | Stain rate($s^{-1}$) | Interval time(s) |
|---|---|---|---|---|
| 1 | 950,1000,1100,1150 | 0.15 | 0.1 | 1,5,10,50 |
| 2 | 1150 | 0.15 | 0.1,0.5,1 | 1,5,10,50 |
| 3 | 1150 | 0.07,0.1,0.15 | 0.1 | 1,5,10,50 |

## 3. Experimental Results and Discussion

### *3.1. True Stress - Strain Curve of Static Recrystallization*

Figure 2 shows the true stress-strain curve, which obtained by static recrystallization experiment of 38MnVTi non-quenched and tempered steel at different interval times, deformation temperatures, deformation degrees, and strain rates. The change of rheological curve reflects the softening behavior of the material. As Fig.2 (a) shows, when the heat distortion temperature, deformation degree and strain rate remain constant, the deformation resistance of 38MnVTi non-quenched and tempered steel decreases with the increase of the interval time. As shown in Fig. 2 (b), when the interval time, deformation degree and strain rate remain constant, the flow stress decreases significantly with the increase of the deformation temperature. It can be seen from Fig. 2 (c) that when other conditions are certain, the higher the deformation degree is, the lower yield stress of the second pass is, and the yield stress of the first pass is not changed. This is because the higher the degree of pre-strain, the greater the dislocation density, the greater the dislocation. In the second pass deformation, it will quickly reduce the degree of hardening work. From Fig. 2 (d), it can be inferred that when the deformation degree, deformation temperature and interval time remain constant, the flow stress increases with the increase of strain rate. This is because in the thermal deformation process, to improve the deformation temperature or reduce the strain rate will make the material dislocation density decreased, and the degree of work hardening also reduced, finally the flow stress of metal materials reduces during thermal deformation process.

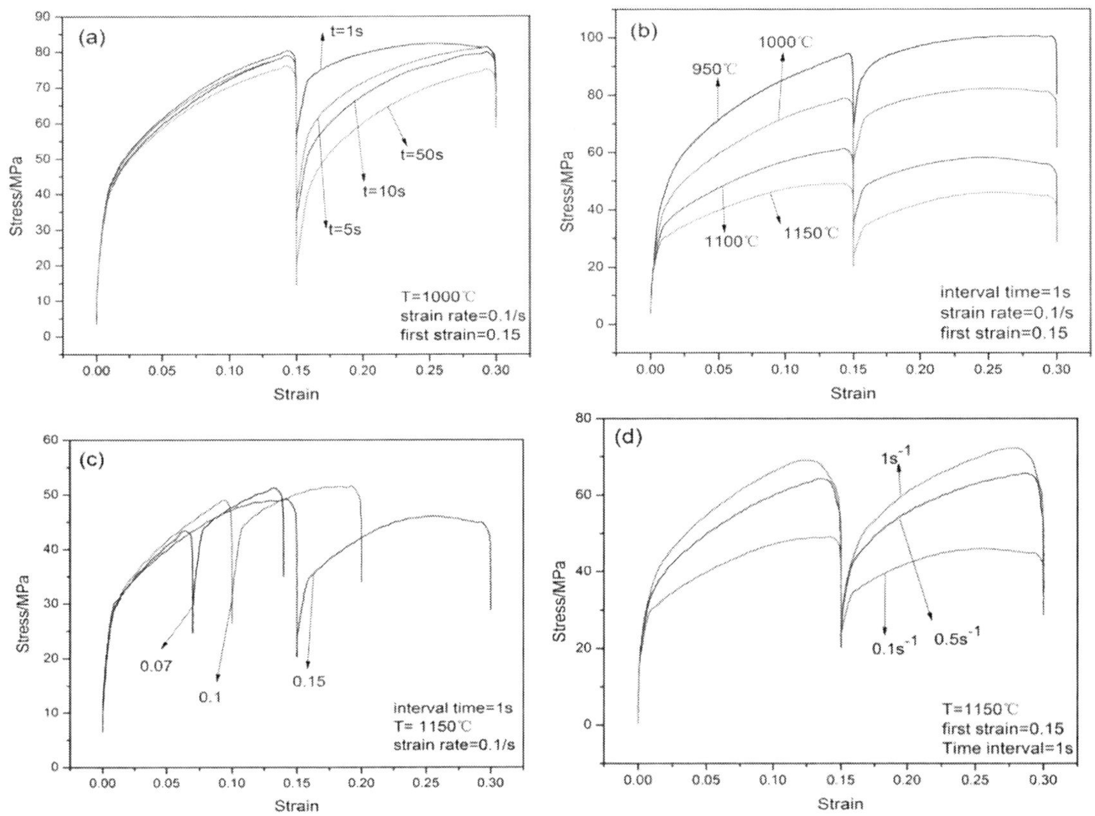


**Figure 2.** Stress-strain curves of double-pass experiment under different hot deformation conditions

### *3.2. Determination of Static Recrystallization Volume Fraction*

In this paper, 0.2% stress compensation method was used to measure the static recrystallization volume fraction $X_{SRX}$ [5]. The static recrystallization volume fraction is calculated by the following equation:

$$X_{SRX} = \frac{\sigma_m - \sigma_2}{\sigma_m - \sigma_1} \tag{1}$$

Where: $\sigma_m$ is the yield stress before unloading for the first pass; σ1 and σ2 are the values of the yield stress deviation of 0.2% at the beginning of the first and second pass compression respectively.

### *3.3. Effect of Deformation Conditions on Static Recrystallization Volume Fraction*

Fig.3 (a) shows the effect of deformation temperature on the static recrystallization volume fraction of 38MnVTi non-quenched and tempered steel. It's in consistent with the Avrami formula [6].When other process parameters are constant, the static recrystallization volume fraction increases with the increasing deformation temperature, and the completion time of static recrystallization is shortened. This is because the static softening process is a thermal excitation process, under other constant conditions, with the temperature rose, on the one hand, the degree of recovery increases. On the other hand, the easier the dislocation climb and the grain boundary migration achieved, the smaller the degree of recrystallization critical deformation got.

Fig.3 (b) is the effect of the pre-strain on the static recrystallization volume fraction of 38MnVTi non-quenched and tempered steel at 1150 °C and strain rate of $0.1s^{-1}$. It can be seen that when the temperature and strain rate remain constant, the static recrystallization volume fraction increases with the increasing pre-strain. This is because the work hardening plays a leading role while the degree of thermal deformation below the critical strain. With the degree of deformation increasing, the material dislocation density increases rapidly and the driving force of static recrystallization increases. So, static recrystallization has basically replaced static recovery, and new recrystallization grains lead to a rapid static recrystallization by dislocation motion, which makes the static recrystallization volume fraction become larger [7].

Fig.3 (c) shows the effect of strain rate on static recrystallization volume fraction of 38MnVTi non-quenched and tempered steel at 1150 °C and pre-strain of 0.15. This is because in the same conditions, as the strain rate increases, the dislocation density rate increases, deformation storage energy increases, static recrystallization driving force becomes large, and the static recrystallization is promoted. It can be found that the static recrystallization fraction of 38MnVTi non-quenched and tempered steel is not sensitive to the strain rate.

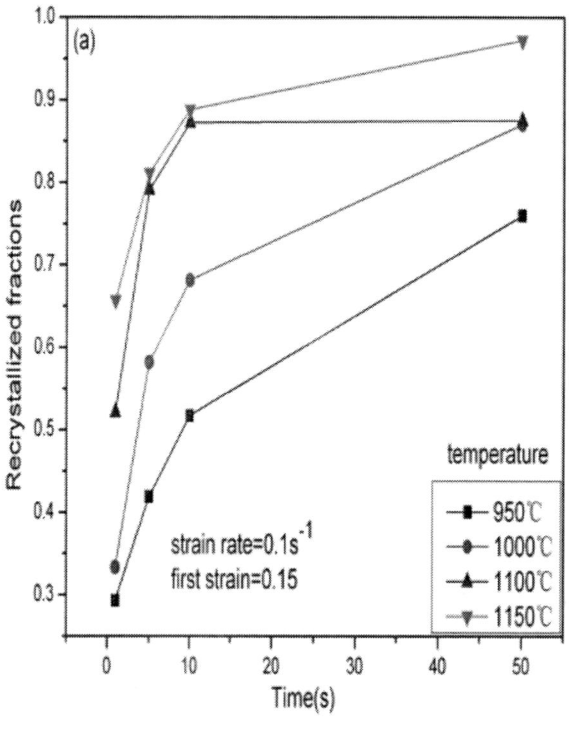

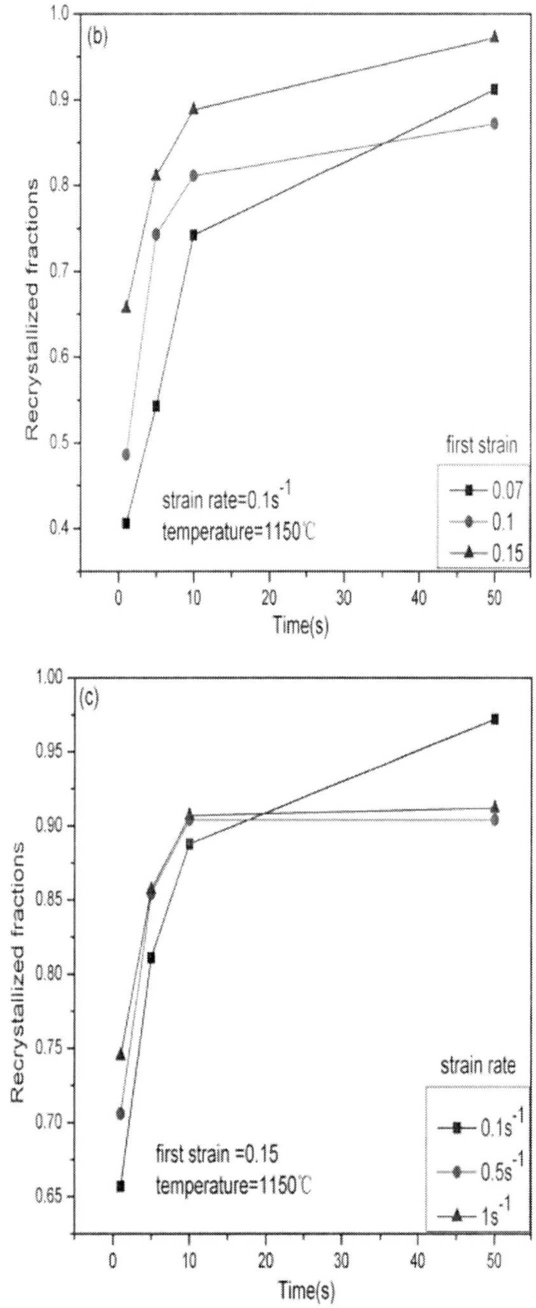


**Figure 3.** Effects of deformation condition on the static recrystallization volume fractions

### *3.4. Establishment of Static Recrystallization Kinetics Model*

A large number of studies have shown that the static recrystallization kinetics equation is generally expressed by the Avrami equation [8]:

$$X_s = 1 - \exp\left[-0.693\left(t / t_{0.5}\right)^n\right] \quad (2)$$

Where: $X_S$ is the static recrystallization volume fraction; t is the interval time; $t_{0.5}$ is the time of the static recrystallization volume fraction to 50%; n is the material constant. Taking its double logarithm on both sides of equation (2), and then simplified:

$$\ln\left(\ln\left(1/\left(1-X_s\right)\right)\right)=\ln 0.693+n\ln t-n\ln t_{0.5} \tag{3}$$

Combined with the above experimental data, it can be calculated that the average value of n is 0.299 by linear regression. Substituting n into equation (2), it can be inferred $t_{0.5}$ under different deformation conditions. $t_{0.5}$ can be expressed as follows:

$$t_{0.5}=Ad_0{}^{s}\varepsilon^{m_1}\dot{\varepsilon}^{m_2}\exp(Q_{srx}/RT) \tag{4}$$

Where: $d_0$ is initial grain size; $\varepsilon$ is strain; $\dot{\varepsilon}$ is strain rate; $Q_{srx}$ is static recrystallization activation energy, kJ/mol; R is molar gas constant (R=8.314J/mol); T is thermodynamic temperature; A, s, m1, m2 are material constants. Considering that the initial austenite grain size is coarse, the initial grain size is taken as 110μm, s = 1.Taking its logarithm on both sides of equation(4) .It can be calculated that $m_1$ is -2.58, $m_2$ is -0.23 and $Q_{srx}$ is 250.27kJ/mol by linear regression. Substituting the above data into the equation, it can be calculated A is $6.62\times10^{-15}$.

Based on the above analysis, the static recrystallization kinetic model of 38MnVTi non-quenched and tempered steel is obtained as follows:

$$X_s=1-\exp[-0.693(t/t_{0.5})^n] \tag{5}$$

$$t_{0.5}=6.62\times10^{-15}d_0\varepsilon^{-2.58}\dot{\varepsilon}^{-0.23}\exp(250270/RT) \tag{6}$$

### *3.5. Validation of Static Recrystallization Kinetics Model*

From the figure4 above, it is obvious that the predicted results of this model accord well with the observed data, which indicates that the static recrystallization kinetics model can predict the static recrystallization behavior of 38MnVTi non-quenched and tempered steel accurately, and it can provide the basic theory for hot deformation process.

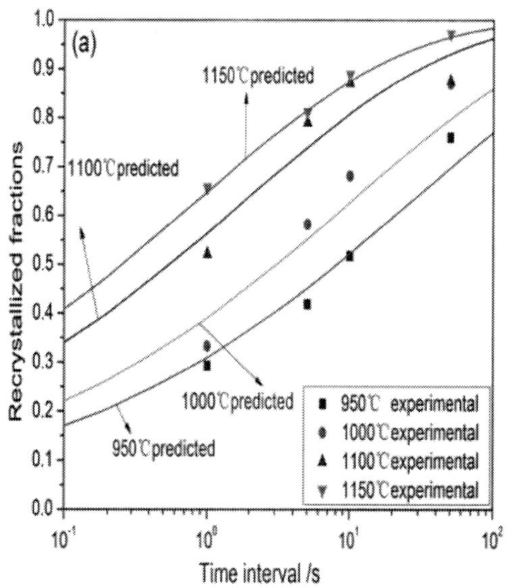

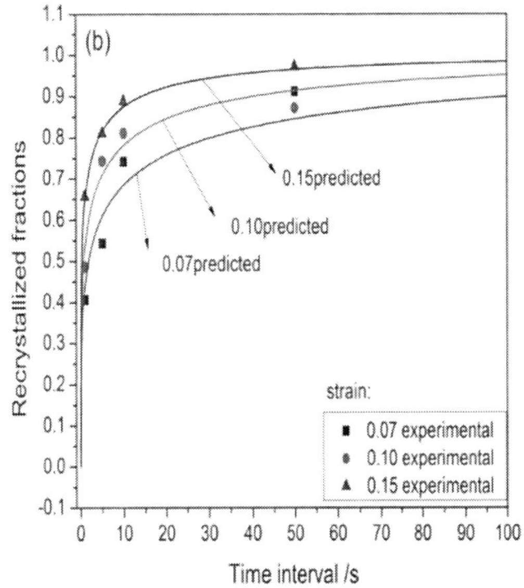


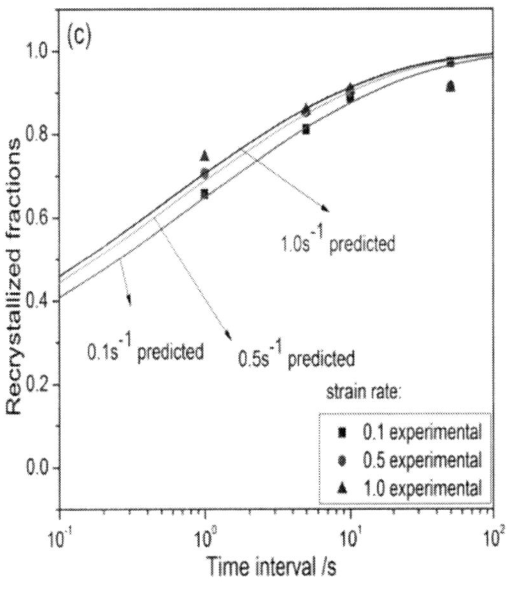


**Figure 4.** Comparison between the experimental static recrystallization volume fractions and the predicted result

*3.6. Establishment of Static Recrystallization Grain Size Model*

After cutting the samples of the double-pass heat compression test along the axial direction, taking them into the mixed in the solution of picric acid and a small amount of SDBS more than 40 minutes until the austenite grain boundary appears in large area, and the average grain size is measured by the intercept method.

As shown in Fig.5, the average grain size of the static recrystallization is directly proportional to the deformation temperature. When the other process conditions are constant, the average grain sizes at deformation temperatures of 950 ° C, 1000 ° C, 1100 ° C and 1150 ° C are 40.2μm, 53.2μm, 61.3μm and 74.2μm, respectively. This is because the dislocation storage which provides energy for the static recrystallized nucleus is reduced by increasing the temperature and the nucleation rate of the static recrystallization decreases accordingly. The nucleated part grows rapidly with the increase of temperature, and annexed the small grains which are not enough to grow up, which makes coarse grain.

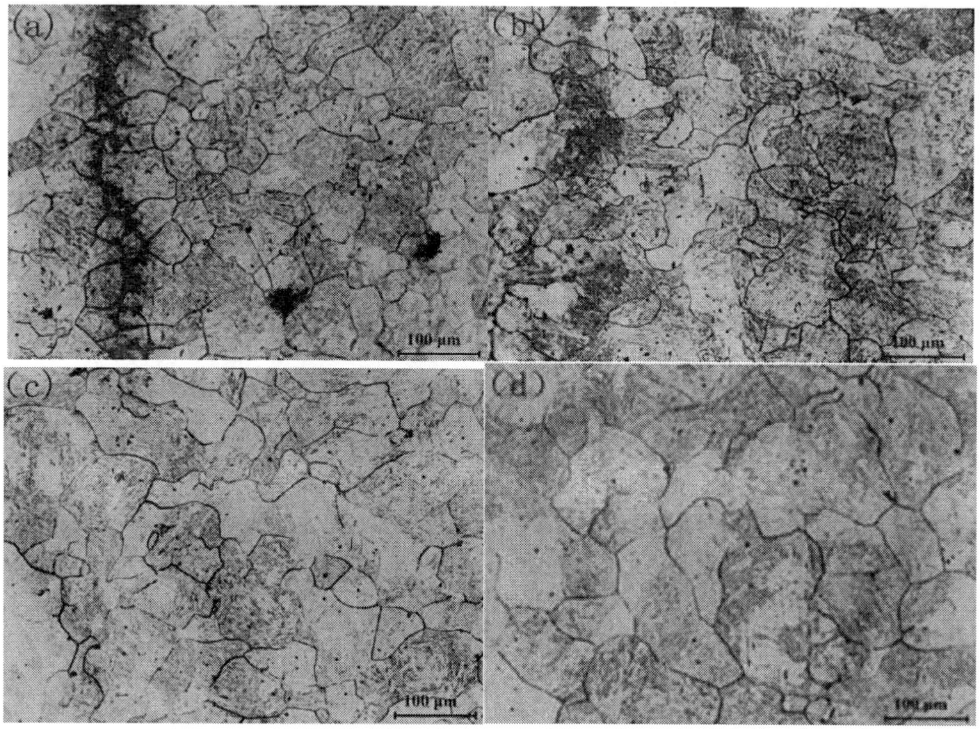


(a)950°C,strain rate:0.1s$^{-1}$,strain:0.15 (b) 1000°C,strain rate:0.1s$^{-1}$,strain:0.15
(c) 1100°C,strain rate:0.1s$^{-1}$,strain:0.15 (d) 1150°C,strain rate: 0.1s$^{-1}$,strain:0.15

**Figure 5.** microstructure of different process parameters

In the static recrystallization process, the grain size model is as follows:

$$D_{SRX} = A_1 d_0 \varepsilon^{b_1} \dot{\varepsilon}^{b_2} \exp(-Q_{DSRX} / RT) \tag{7}$$

Where: $D_{SRX}$ is the grain size of the static recrystallization; $d_0$ is initial grain size; $\varepsilon$ is strain; $\dot{\varepsilon}$ is strain rate; R is a molar gas constant; T is thermodynamic temperature; $A_1$, $Q_{DSRX}$, $b_1$, $b_2$ are material constants. Taking logarithm on both sides of equation(7). Through the same methods of linear regression, it can be inferred that Q = 39719.21J / mol, $b_1$= -0.408, $b_2$=-0.2, $A_1$=5.17. Therefore, the static recrystallization grain size model of 38MnVTi non-quenched and tempered steel is as follows:

$$D_{SRX} = 5.17 d_0 \varepsilon^{-0.41} \dot{\varepsilon}^{-0.2} \exp(-39719 / RT) \tag{8}$$

The experimental values of average grain size of static recrystallization are in good agreement with predicted values. Therefore, the grain size model can be used well for the grain size of 38MnVTi non-quenched and tempered steel during the static recrystallization process, and it can provide theoretical guidance for making and optimizing production process of 38MnVTi non-quenched and tempered steel.

## 4. Conclusion

(1) 38MnVTi non-quenched and tempered steel was researched with double-pass compression experiments on Gleeble-1500. It can be concluded that deformation temperature, strain rate and pre-strain parameters have effects on the static recrystallization behavior of 38MnVTi non-quenched and tempered steel . The higher the deformation temperature, the greater the strain rate and the greater the degree of pre-strain, the greater the static recrystallization volume fraction. The effect of pre-strain and deformation temperature on static recrystallization is greater than strain rate.

(2) The linear recrystallization dynamics model and grain size model were established by linear regression as follows:

$$X_s = 1 - \exp[-0.693(t / t_{0.5})^n]$$

$$t_{0.5} = 6.62 \times 10^{-15} d_0 \varepsilon^{-2.58} \dot{\varepsilon}^{-0.23} \exp(250270 / RT)$$

$$D_{SRX} = 5.17_1 \mathrm{d}_0 \varepsilon^{-0.41} \dot{\varepsilon}^{-0.2} \exp(-39719 / RT)$$

The predicted results of the model accord well with experimental data, which can provide guidance and theoretical basis for the thermal deformation process.

**Reference**

[1] DEARDO A J. Microalloyed strip steels for the 21st century [J]. *Materials Science Forum*, 1998, 284/286:15-26.

[2] GONZALEZ-BAQUET I, KASPAR R, RICHTER J, et al. Microalloying and a new post forging treatment of medium carbonsteels[J].*Materials Science Forum*,1998,284/286:411-418.

[3] NAYLOR D J. Microalloyed forging steels [J].*Materials Science Forum*, 1998, 284/286:83-94.

[4] LIN Yongcheng, CHEN Mingsong, ZHONG Jue. Static recrystallization behaviors of deformed 42CrMo steel [J]. *Journal of Central South University (Science and Technology)*, 2009(2)411-416.

[5] JIANG Bo, YANG Zhong, WEN Cheng, et al. Static recrystallization behaviors of non-quenched steel C38N2 for crankshafts [J]. *Journal of University of Science and Technology*, 2014(8):1039-1044.

[6] Gu S D, Zhang C, Zhang L W,et al. Characteristics of metadynamic recrystallization of Nimonic 80A superalloy[J]. *Journal of Materials Research*, 2015, 30(4):538 -546.

[7] HE Wenwu, SUN Shuli, LIU Jiansheng, et al. Static recrystallization microstructure and model of Mn18Cr18N retaining rings steel [J]. *Materials Science and Technology*, 2014, 06:17-22.

[8] Bai D Q, Yue S, Jonas J J. Static recrystallization of Nb-B steels under continuous cooling conditions [J]. *ISIJ International*. 1996, 36(8):1084-1093.

# Simulation Study on Noise Reduction Performance of Thickness of Sound Absorber Films in Industrial Gas Turbine Muffler

**Jiangang Yi**[1,2,*] **and Siwei Ju** [2]

[1] Hubei Key Laboratory of Industrial Fume & Dust Pollution Control, Jianghan University, Wuhan 430056, China
[2] School of Electromechanical and Architectural Engineering, Jianghan University, Wuhan 430056, China
469883553@qq.com

**Abstract.** The simulated model of industrial gas turbine muffle is established to study the influence of sound absorber films for noise reduction performance. By the analysis of the Finite Element Model, the results show when the thickness of sound absorbing films increases to a certain extent, the flow pressure increases, the flow space decreases, and the flow velocity in the muffler increases, which reduces the distortion parameters and enhances the flow uniformity.

## 1. Introduction
Because of the variation of air density and pressure in industrial gas turbine pipes, it is easy to form larger air noise, which has seriously reduced the efficiency of industrial production and produced obvious noise pollution [1]. In general, it is necessary to install mufflers in industrial gas turbine pipes for noise reduction and environmental protection [2-3]. However, the noise reduction effect of mufflers with different structures used in pipelines is different. Therefore, it is of great significance to carry out the simulation study on the structure parameters and noise reduction performance of industrial gas turbine pipe muffler for understanding the distribution of flow field in the pipe, which is the base of adopting the pertinent structure optimization work.

## 2. Structural Parameters and Finite Element Modeling of Muffle
This paper mainly studies the influence of the thickness of sound absorber films on the flow field in the industrial gas turbine pipe muffler. For this purpose, Fluent is used to simulate the flow field variation under different structural parameters [4-5]. Table 1 is the muffler parameters. The finite element model of flow field is established by ICEM with unstructured grids. The established flow field finite element model of the sheet resistive muffler is shown in Fig. 1.

**Table 1.** The muffler parameters

| Parameters | 1 | 2 | 3 | 4 |
|---|---|---|---|---|
| thickness (*mm*) | 300 | 400 | 500 | 600 |
| angle (°) | 180 | 150 | 120 | 90 |
| number | 4 | 5 | 6 | 7 |

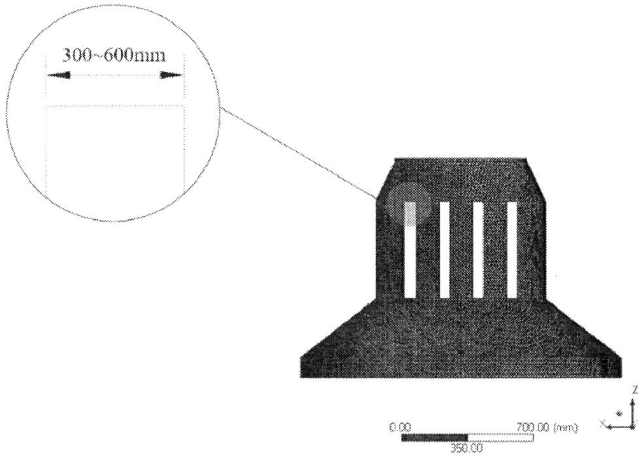


**Figure 1.** Muffle Model

## 3. Simulation Study of Sound Absorbing Films on Flow Field

The finite element model of sound absorbing films with different thickness absorber is very complicated. Therefore, the finite element model is divided by unstructured grids to simplify the calculation. The pressure distributions of sound absorbing films with different thickness are shown in Fig. 2, the velocity distributions are shown in Fig. 3, and the turbulence distributions are shown in Fig. 4. The pressure losses are shown in Table 2.

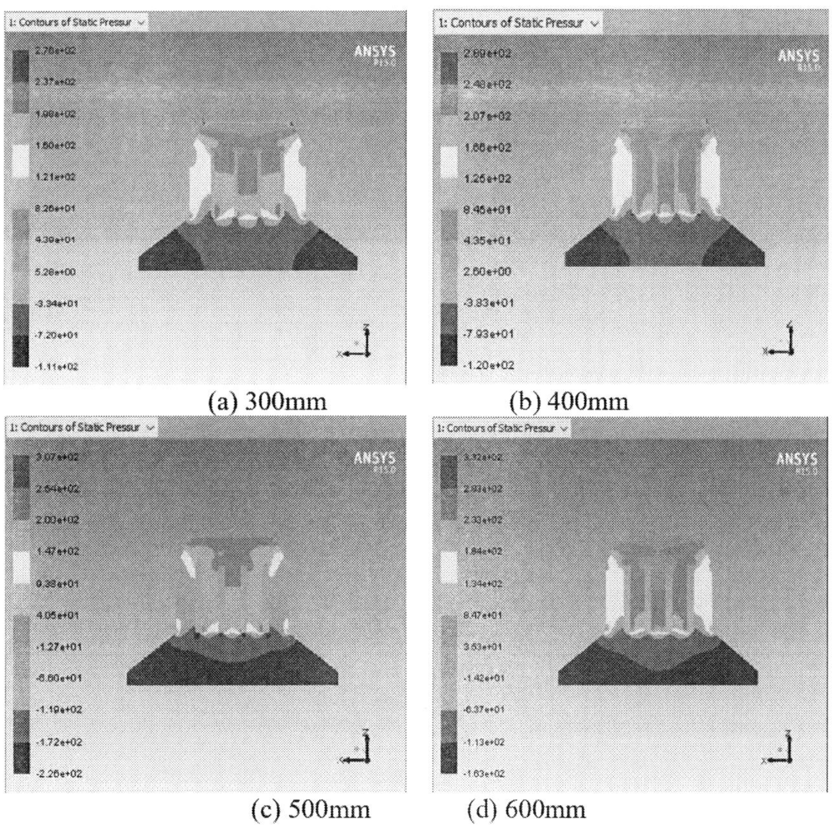


**Figure 2.** Pressure Distribution of Sound Absorbing Films with Different Thickness

IOP Conf. Series: Materials Science and Engineering 472 (2019) 012051 doi:10.1088/1757-899X/472/1/012051

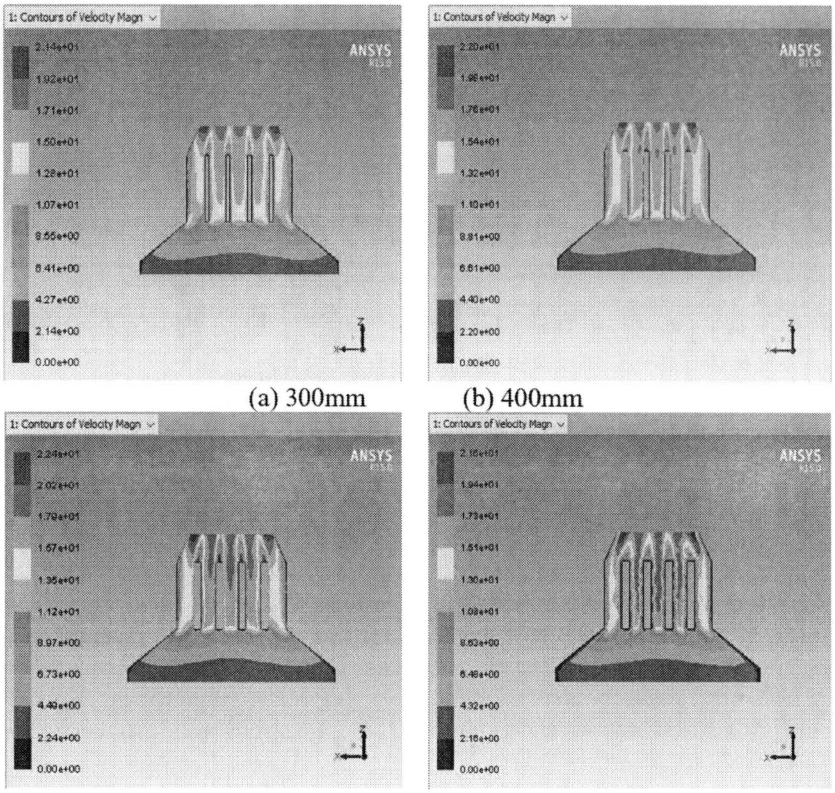


(a) 300mm (b) 400mm

(c) 500mm (d) 600mm

**Figure 3.** Velocity Distribution of Sound Absorbing Films with Different Thickness

Fig. 3 shows that the velocity of the flow field is increased gradually with the inlet and outlet, and the velocity of the inner space of the muffler is increased obviously. The velocity of the inner space of the muffler is increased with the thickness of the sound absorbing films. Therefore, the increase of the pressure of the flow field and the decrease of the flow space will lead to the increase of the velocity in the flow space.

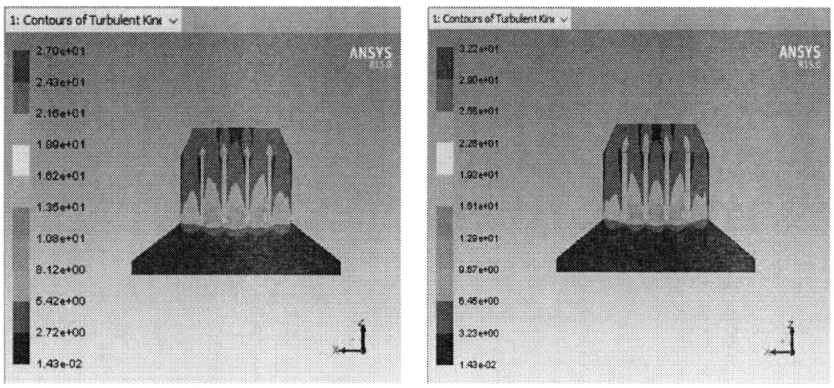


(a) 300mm (b) 400mm

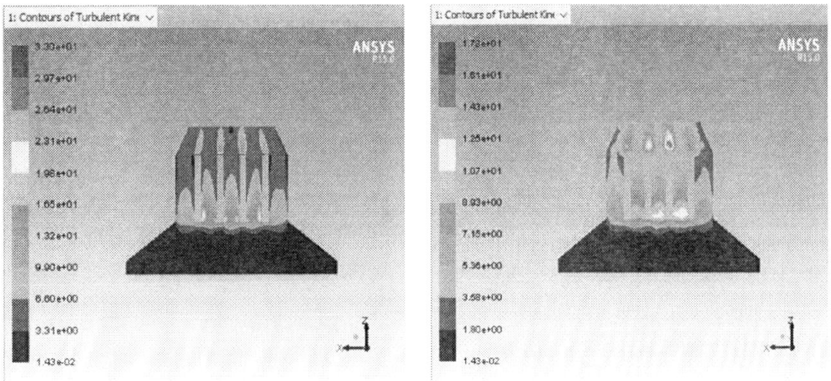


(c) 500mm (d) 600mm

**Figure 4.** Turbulence Distribution of Sound Absorbing Films with Different Thickness

Fig. 4 shows that the turbulence intensity at the inlet and outlet of the muffler is higher than that in other regions on the whole. The turbulence intensity and the area of the turbulence region at the inlet and outlet of the muffler increase with the thickness of the sound absorbing films, while the turbulent intensity at the inlet and the area near the inlet of the muffler tube with 600 mm thick sound absorbing films is decreased. Therefore, the overall turbulence intensity and the area with higher turbulence intensity increase with the increase of the thickness of sound absorbing films, but the overall turbulence intensity will decrease when the thickness of sound absorbing films increases to a certain extent.

**Table 2.** Pressure Loss of Sound Absorbing Films with Different Thickness

| **Thickness(*mm*)** | **Loss(*Pa*)** | **Distortion** |
|---|---|---|
| 300 | 253 | 2.61 |
| 400 | 266 | 2.64 |
| 500 | 284 | 2.64 |
| 600 | 310 | 2.51 |

Table 2 shows that the pressure loss of sound absorbing tube with 300 mm thickness of the absorbing films is up to work. The variation rule is that the pressure loss increases significantly with the increase of the thickness of the absorbing films. The distortion parameters of flow field in tubes with different thickness films do not fit the work standard. The variation rule is that the distortion parameters of flow field increase with the increase of the thickness of the absorbing films at first, but decrease with the increase of the thickness of the absorbing films when they are up to 600mm. Combining with the analysis results of velocity and turbulence distribution, it can be concluded when the thickness of sound absorbing films increases to a certain extent, the flow pressure increases, the flow space decreases, and the flow velocity in the muffler increases, which reduces the distortion parameters and enhances the flow uniformity.

## 4. Conclusion

Through the simulation analysis of the structure parameters of industrial gas turbine muffler, the influence of the thickness, quantity and diversion angle of sound absorbing films on the flow field of the pipeline is studied. It is found that the quantity and thickness of sound absorbing films have great influence on the noise reduction performance of the pipeline, and the noise reduction effect increases with the number and thickness of sound absorbing films. Moreover, the flow field of the pipe decreases with the increase of the number and thickness of the sound absorbing films. On the other hand, the pressure loss and flow uniformity decrease with the increase of the number and thickness of sound absorbing materials. This provides a reference for further optimization of industrial gas turbine muffler structure.

## References

[1] Fan C, Tian Y, Wang Z Q, et al. Structural parameter effect of porous material on sound absorption performance of double-resonance material [J]. 2017, 213(1):012028.

[2] E.A.IZZHEUROV. Calculation of resonant sound absorption parameters for performance evaluation of metal rubber material [J].Science in China (Series E: Technological Sciences), 2009, 52(12):3587-3591.

[3] Han B K, Yu-Xi A N, Bao H Q, et al. Optimization Analysis for Sound Absorption Performance of Porous Metal Materials[J]. Journal of Shandong University of Science & Technology, 2009.

[4] Wang T K, Cai J, Zhang C, et al. Experimental study on effect of porous materials on sound absorption [J]. Journal of Henan University of Urban Construction, 2018.

[5] Garai M, Pompoli F. A simple empirical model of polyester fibre materials for acoustical applications [J]. Applied Acoustics, 2005, 66(12):1383-1398.

# Structural Parameters Analysis of Sound Absorption Films in Industrial Gas Turbine Muffler

**Jiangang Yi**[1,2,*] **and Siwei Ju** [2]

[1] Hubei Key Laboratory of Industrial Fume & Dust Pollution Control, Jianghan University, Wuhan 430056, China
[2] School of Electromechanical and Architectural Engineering, Jianghan University, Wuhan 430056, China
469883553@qq.com

**Abstract.** Aerodynamic noise is the main noise source produced by industrial gas turbines. It brings harm to production and life. Muffler is the key component to suppress aerodynamic noise. In this paper, the influence of the structure parameters such as the diversion angle, films number and porosity of the absorber for the noise reduction effect is studied, which provides the basis for the structure design and optimization of the muffler.

## 1. Introduction
Industrial gas turbine is an advanced and complex set of power machinery and equipment [1]. It is high-power and high-performance power machine that rely on high-temperature gas to drive turbo machinery continuously with air as the medium [2]. However, due to the air flow in the gas turbine pipeline, it produces eddy current, impact, or sudden change of pressure, resulting in air disturbance and a large number of noises [3]. Mufflers are used in air pipes of industrial gas turbine to absorb sound. As the most important part of the muffle, the sound absorption films determine the efficiency of noise reduction of the muffle. Therefore, it is of great significance to study the structural parameters of the sound absorption films.

## 2. Influence of Diversion Angle
The structure parameters analysis of the sound absorption films involves gas flow, acoustics, materials and other subjects, which is a complex problem. The design of mufflers is usually guided by one-dimensional plane wave theory. However, when the frequency is high, the results of one-dimensional plane waves are not accurate and are not applicable. Compared with it, the results of the three-dimensional numerical method are more accurate, the analysis efficiency is higher, and the cost is lower [4-5].

In order to understand the flow field and acoustic characteristics of sound absorption films in industrial gas turbine muffler, the intake pipe flow field of a gas turbine in an environmental protection equipment company was taken as the research object. The flow field and acoustic field of the intake pipe were simulated and analyzed by computational mechanics (CFD) and acoustic finite element method (Acoustics-FEM). Fig. 1 shows that the velocity of the flow field increases gradually with the inlet to outlet, especially in the inner space of the muffler. The overall velocity of the flow field decreases with the decrease of the diversion angle of the muffler. The high-speed flow area at the outlet increases gradually with the decrease of the diversion angle of the muffler. Therefore, the reduction of the diversion angle of the anechoic plate will lead to a small overall decrease of the flow velocity and increase the size of the high-speed air flow area at the outlet. The turbulence intensity at the inlet and outlet of the muffler is higher than that in other areas. Except that the turbulence intensity of the duct with 150 degree guide angle muffler is higher than that of the assembly site muffler, the turbulence intensity of the duct with other guide angles decreases with the decrease of the guide angle of the muffler. The intensity turbulence region increases with the decrease of the diversion angle of the muffler, so the decrease of the diversion angle will reduce the overall turbulence intensity of the flow field and increase the turbulence intensity of the muffler inlet and outlet.

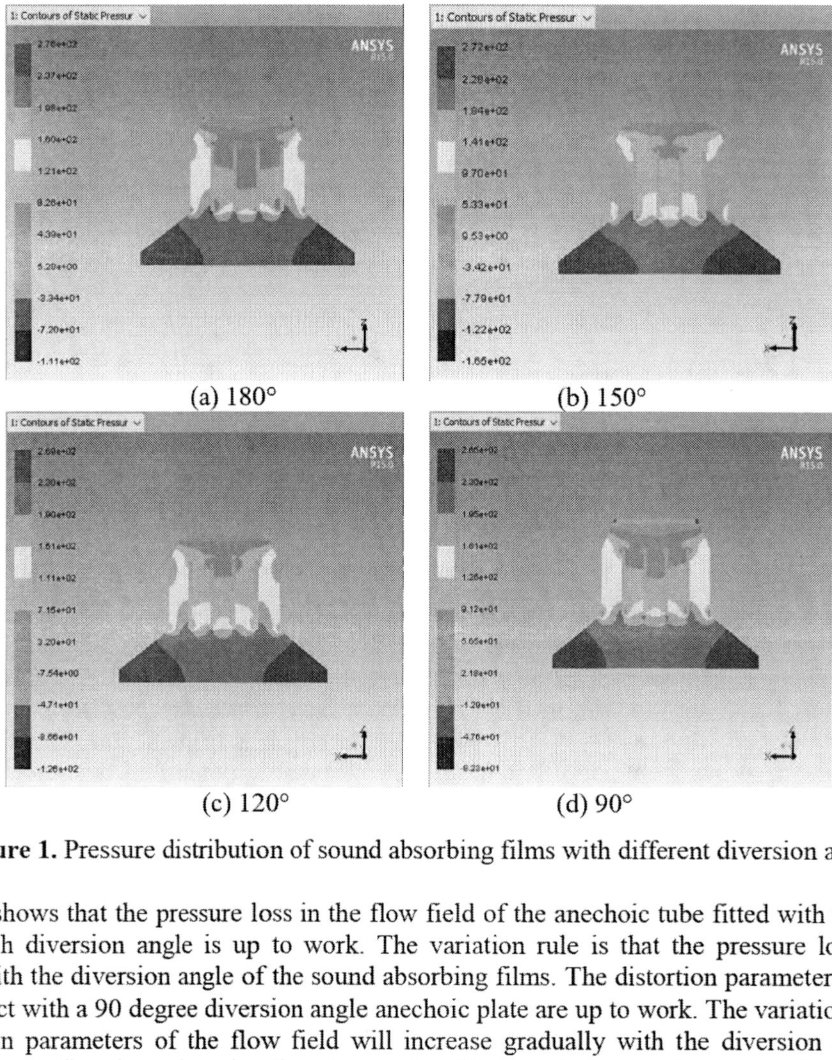


(a) 180° (b) 150°

(c) 120° (d) 90°

**Figure 1.** Pressure distribution of sound absorbing films with different diversion angle

Table 1 shows that the pressure loss in the flow field of the anechoic tube fitted with the anechoic plate of each diversion angle is up to work. The variation rule is that the pressure loss increases gradually with the diversion angle of the sound absorbing films. The distortion parameters of the flow field in a duct with a 90 degree diversion angle anechoic plate are up to work. The variation rule is that the distortion parameters of the flow field will increase gradually with the diversion angle of the anechoic plate at first, but when the diversion angle of the anechoic plate is close to the plane angle, the distortion parameters of the flow field hardly change. Therefore, reducing the diversion angle of the sound absorbing films can reduce the flow resistance of the anechoic tube and improve the flow uniformity.

**Table 1.** Pressure loss of sound absorbing films with different diversion angle

| **Diversion angle (°)** | **Loss(*Pa*)** | **Distortion** |
|---|---|---|
| 180 | 253 | 2.61 |
| 150 | 249 | 2.61 |
| 120 | 247 | 2.52 |
| 90 | 242 | 2.43 |

## 3. Influence of Sound Absorbing Films Number

Except that the overall velocity of the flow with five pieces of sound absorbing films is less than that with four pieces of sound absorbing materials, the overall velocity of the flow field in the anechoic tube with other numbers of sound absorbing films increases with the increase of the number of sound absorbing films, especially near the center of the muffler. Therefore, increasing the number of sound absorbing films can increase the overall flow velocity in the pipe, especially in the area near the center of the muffler.

On the whole, the turbulence intensity at the inlet and outlet of the muffler is higher than that in other areas. Except that the overall turbulence intensity of the muffler with 7 pieces of sound absorbing films is higher than that of the muffler with 6 pieces of sound absorbing films, the overall turbulence intensity of the muffler with other pieces decreases with the increase number of the sound absorbing films. Therefore, increasing the number of sound absorbing films can reduce the overall turbulence intensity of the flow field. However, when the number of sound absorbing films increases to a certain amount, it will increase the overall turbulence intensity of the flow field.

**Table 2.** Pressure loss of sound absorbing films with different number of films

| Number | Loss($Pa$) | Distortion |
|---|---|---|
| 4 | 253 | 2.61 |
| 5 | 265 | 2.50 |
| 6 | 280 | 2.60 |
| 7 | 290 | 2.58 |

Table 2 shows that the pressure loss of the anechoic tube assembled with different amount of sound absorbing films does not meet the requirements. The variation rule is that the pressure loss increases significantly with the increase of the number of sound absorbing films. The flow field distortion parameters of the pipes with 5 sound absorbing films can meet the requirements. Therefore, reducing the number of sound absorbing films to 5 pieces can effectively decrease the pressure loss of the flow field.

## 4. Influence of Sound Absorbing Films Porosity

When the sound wave propagates to the pores of the sound absorbing films, it will cause the small vibration of the fiber in the films. Because of the inhomogeneous distribution of the pores in the sound absorbing films and the friction and viscous resistance of the air, it will make the sound energy incident into the sound absorbing material into thermal energy and be absorbed. If the thickness of sound absorbing films is not enough, it will lead to incomplete absorption of sound energy. Although the noise reduction performance of muffler will increase with the thickness of sound absorbing films, when the thickness of the sound absorbing films increases to a certain value, the increase of noise reduction performance will gradually decrease. Considering the cost and the influence on the characteristics of pipeline flow field, if the thickness of sound absorbing films continues to increase, it will cause unnecessary waste and reduce the flow field characteristics of the pipeline.

| | A | B | C |
|---|---|---|---|
| 1 | Frequency | Real | Imaginary |
| 2 | 20 | 7.00E-13 | 9.89E-12 |
| 3 | 40 | 9.90E-13 | 1.98E-11 |
| 4 | 60 | 1.21E-12 | 2.97E-11 |
| 5 | 80 | 1.40E-12 | 3.96E-11 |
| 6 | 100 | 1.57E-12 | 4.94E-11 |
| 7 | 120 | 1.72E-12 | 5.93E-11 |
| 8 | 140 | 1.85E-12 | 6.92E-11 |
| 9 | 160 | 1.98E-12 | 7.91E-11 |
| 10 | 180 | 2.10E-12 | 8.90E-11 |
| 11 | 200 | 2.21E-12 | 9.89E-11 |
| 12 | 220 | 2.32E-12 | 1.09E-10 |
| 13 | 240 | 2.43E-12 | 1.19E-10 |
| 14 | 260 | 2.53E-12 | 1.29E-10 |
| 15 | 280 | 2.62E-12 | 1.38E-10 |
| 16 | 300 | 2.71E-12 | 1.48E-10 |
| 17 | 320 | 2.80E-12 | 1.58E-10 |
| 18 | 340 | 2.89E-12 | 1.68E-10 |
| 19 | 360 | 2.97E-12 | 1.78E-10 |
| 20 | 380 | 3.05E-12 | 1.88E-10 |
| 21 | 400 | 3.13E-12 | 1.98E-10 |
| 22 | 420 | 3.21E-12 | 2.08E-10 |
| 23 | 440 | 3.28E-12 | 2.18E-10 |
| 24 | 460 | 3.36E-12 | 2.27E-10 |
| 25 | 480 | 3.43E-12 | 2.37E-10 |
| 26 | 500 | 3.50E-12 | 2.47E-10 |
| 27 | 520 | 3.57E-12 | 2.57E-10 |
| 28 | 540 | 3.64E-12 | 2.67E-10 |
| 29 | 560 | 3.71E-12 | 2.77E-10 |
| 30 | 580 | 3.77E-12 | 2.87E-10 |
| 31 | 600 | 3.84E-12 | 2.97E-10 |
| 32 | 620 | 3.90E-12 | 3.07E-10 |
| 33 | 640 | 3.96E-12 | 3.16E-10 |
| 34 | 660 | 4.02E-12 | 3.26E-10 |
| 35 | 680 | 4.08E-12 | 3.36E-10 |
| 36 | 700 | 4.14E-12 | 3.46E-10 |
| 37 | 720 | 4.20E-12 | 3.56E-10 |
| 38 | 740 | 4.26E-12 | 3.66E-10 |
| 39 | 760 | 4.32E-12 | 3.76E-10 |
| 40 | 780 | 4.37E-12 | 3.86E-10 |
| 41 | 800 | 4.43E-12 | 3.96E-10 |
| 42 | 820 | 4.48E-12 | 4.05E-10 |
| 43 | 840 | 4.54E-12 | 4.15E-10 |
| 44 | 860 | 4.59E-12 | 4.25E-10 |
| 45 | 880 | 4.65E-12 | 4.35E-10 |
| 46 | 900 | 4.70E-12 | 4.45E-10 |
| 47 | 920 | 4.75E-12 | 4.55E-10 |
| 48 | 940 | 4.80E-12 | 4.65E-10 |
| 49 | 960 | 4.85E-12 | 4.75E-10 |
| 50 | 980 | 4.90E-12 | 4.85E-10 |
| 51 | 1000 | 4.95E-12 | 4.94E-10 |

**Figure 2.** Pressure distribution of sound absorbing films with different diversion angle

An acoustic finite element model of pipes was built up by ICEM. The simulation results are shown in Fig.2. It can be found that, as the porosity of the sound absorbing material increases, the transmission loss value of the pipeline increases gradually and moves to the low frequency area gradually. The noise attenuation of the pipeline assembled with various porosity in the range of 20-60Hz and 180-300Hz is less than 50dB, which does not meet the requirement. The effect of porosity of sound absorbing material on transmission loss of pipeline increases with the increase of frequency, especially in the range of 20-180Hz, the transmission loss curve of pipeline is almost unchanged. Therefore, the increase of porosity of the sound absorbing films can only slightly increase the muffler volume in the medium frequency range. The flow resistance of sound-absorbing materials has a great influence on the valley value and its adjacent range of the transmission loss curve of the pipeline, and the anechoic value increases with the increase of the flow resistance, while the flow resistance has a little influence on the peak value of the transmission loss and its adjacent frequency range.

## 5. Conclusion

The simulation results show that the diversion angle has little effect on the sound absorption, and only the anechoic Valley and its nearby frequency value decrease slightly with the decrease of the diversion angle. With the increase of the number of sound absorbing materials, the effect of noise reduction is significantly increased. On the other hand, the pressure loss and flow uniformity decrease with the increase of the number of sound absorbing materials and porosity.

## References

[1] Fan C, Tian Y, Wang Z Q, et al. Structural parameter effect of porous material on sound absorption performance of double-resonance material [J]. 2017, 213(1):012028.
[2] Fang Z G, Gao Y X, Fang C. Effect of Structure Modification on the Microwave Absorbing Performance of Carbon Foams[J]. Advanced Materials Research, 2011, 197-198:536-539.
[3] Toyoda M, Sakagami K, Okano M, et al. Improved sound absorption performance of three-dimensional MPP space sound absorbers by filling with porous materials [J]. Applied Acoustics, 2017, 116:311-316.

[4] Lee Y Y, Lee E W M, Ng C F. Sound absorption of a finite flexible micro-perforated panel backed by an air cavity[J]. Journal of Sound & Vibration, 2005, 287(1):227-243.
[5] Qian Y J, Cui K, Liu S M, et al. Numerical study of the acoustic properties of micro-perforated panels with tapered hole [J]. Noise Control Engineering Journal, 2014, 62(3): págs. 152-159.

# The Investigation on Adhesive Contact Problem of Gradient Piezoelectric Coating under Insulating Punch

**Weixing Huang and Tiejun Liu***

Department of Mechanics, College of Science, Inner Mongolia university of Technology, Hohhot 010051, China.
Email: liutiejun@imut.edu.cn

**Abstract.** This paper considers the two-dimensional adhesive contact problem for a rigid insulating cylindrical punch acting on a gradient piezoelectric coating bonded to half-space.It is assumed that the material parameters of functionally gradient piezoelectric materials vary exponentially along the thickness direction.The adhesion between the punch and the contact surface of the coating is described according to the adhesion Model given by Maugis. By using the Fourier integral transform, the adhesive contact problem of gradient piezoelectric coatings is reduced to singular integral equations. the numerical method is applied to solve the singular integral equations with Cauthy kernel.The numerical results show that both non-homogeneous parameters of gradient piezoelectric material and adhesion effect have significiant impact on the adhesive contact behavior of gradient piezoelectric coating.

## 1. Introduction

With the development of science and technology, functionally graded materials (FGM) [1], and functionally graded piezoelectric materials (FGPM) [2] have been received extensive attention from researchers. By using Fourier integral transform, Guler et al [3]. proved that appropriate gradient of shear modulus can obviously change the surface contact force distribution. By using transfer matrix method and Fourier integral transform technology, Ke et al [4]. analyzed the sliding frictional contact problem of a layered half-plane made of functionally graded piezoelectric materials (FGPMs) in the plane strain state. Liu et al [5]. analyzed the effects of the conductivity of the indenter on the distribution of the contact stress and the electric charge. When the microscopic problems are considered, the effect of adhesion has a great impact. Johnson et al [6] proposed the JKR model and measured a larger contact area than Hertz. Chen et al [7] [8] established an adhesion contact model for gradient materials with material parameters vary as power function and obtained a series of closed analytical solutions. Jin et al [9] solved the axisymmetric adhesive contact problem between rigid spheres and functionally graded materials half-space. To the best of author's knowledge, there was no literatures on adhesive contact problem for functionally graded piezoelectric materials.

In this paper, the two-dimensional adhesive contact problem of a functionally gradient piezoelectric coating indented by a rigid insulating cylindrical punch is considered.

## 2. Formulation of the Problem and Solution

As shown in figure 1, the cylindrical insulating punch is pressed into the half-space coated by FGPM under the action of the line load. Hertz contact area $(-a,a)$, and the adhesive area $(-c,c)$ are formed on the surface of piezoelectric coating. It is assumed that the material parameters of the functional gradient piezoelectric coating change exponentially along the thickness direction as follows:

$$\{c_{kl}(z), e_{kl}(z), \varepsilon_{kk}(z)\} = \{c_{kl0}, e_{kl0}, \varepsilon_{kk0}\} e^{\beta z} \tag{1}$$

where, $c_{kl}(z)$, $e_{kl}(z)$, $\varepsilon_{kl}(z)$ is respectively the elastic constant, piezoelectric constant, dielectric constant, $c_{kl0}$, $e_{kl0}$, $\varepsilon_{kl0}$ is the material parameter of homogeneous half space, $\beta$ is the material gradient parameter of FGPM coating.

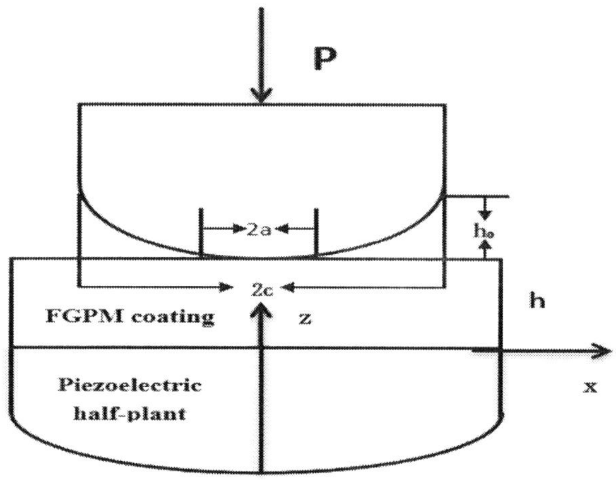


**Figure 1.** The rigid insulating cylindrical indenter acting on the functional gradient piezoelectric coated half-space

According to the linear piezoelectric constitutive relationships, static equilibrium equation and Maxwell's equation, we have

$$\begin{aligned} & c110\frac{\partial u_{x1}^2}{\partial x^2} + c440\frac{\partial u_{x1}^2}{\partial z^2} + (c130 + c440)\frac{\partial u_{z1}^2}{\partial z \partial x} + (e130 + e150)\frac{\partial \phi_1^2}{\partial z \partial x} \\ & + \beta\left[c440\left(\frac{\partial u_{x1}}{\partial x} + \frac{\partial u_{z1}}{\partial z}\right) + e150\frac{\partial \phi}{\partial x}\right] = 0 \end{aligned} \tag{2a}$$

$$\begin{aligned} & c440\frac{\partial u_{z1}^2}{\partial x^2} + c330\frac{\partial u_{z1}^2}{\partial z^2} + (c130 + c440)\frac{\partial u_{x1}^2}{\partial z \partial x} + e150\frac{\partial \phi_1^2}{\partial x^2} + e330\frac{\partial \phi_1^2}{\partial z^2} \\ & + \beta\left[c130\frac{\partial u_{x1}}{\partial x} + c330\frac{\partial u_{z1}}{\partial z} + e330\frac{\partial \phi_1}{\partial z}\right] = 0 \end{aligned} \tag{2b}$$

$$\begin{aligned} & e150\frac{\partial u_{z1}^2}{\partial x^2} + e330\frac{\partial u_{z1}^2}{\partial z^2} + (e150 + e310)\frac{\partial u_{x1}^2}{\partial z \partial x} - \varepsilon 110\frac{\partial \phi_1^2}{\partial x^2} - \varepsilon 330\frac{\partial \phi_1^2}{\partial z^2} \\ & + \beta\left[e310\frac{\partial u_{x1}}{\partial x} + e330\frac{\partial u_{z1}}{\partial z} - \varepsilon 330\frac{\partial \phi_1}{\partial z}\right] = 0 \end{aligned} \tag{2c}$$

According to the Maugis' theory, the normal stress includes two parts: Hertz contact stress $\sigma_1$ and adhesive stress $\sigma_0$:

$$p(t) = \sigma_1(t) - \sigma_0, |t| \le c \tag{3}$$

where the values of normal stress are positive. It is noted that the positive value represents the direction of stress along the coordinate system direction and negative value represents the direction of stress opposites to the coordinate system direction.

IOP Conf. Series: Materials Science and Engineering 472 (2019) 012053 doi:10.1088/1757-899X/472/1/012053

Because the contact area is far less than the punch radius, the shape of the punch can be expressed as:

$$u_{z1}(x) = \delta_0 - \frac{x^2}{2R}, \qquad \frac{\partial u_{z1}(x)}{\partial x} = -\frac{x}{R} \tag{4}$$

where, $\delta_0$ is the maximum indentation depth.

Considering the asymptotic behavior of $M_{21}(s,h)$, one may prove

$$\lim_{s\to\pm\infty} sM_{21}(s,h) = \pm M_{21}^{\infty}$$

Then, extracting the normal displacement component, can be expressed as

$$u_{z1}(x,h) = \frac{-M_{21}^{\infty}}{\pi}\int_{-a}^{a} \ln|x-t|\,p(t)dt + \frac{1}{\pi}\int_{-a}^{a} Q(x,t)p(t)dt \tag{5}$$

where $$Q(x,t) = \int_0^{\infty}[M_{21}(s,h) - \frac{M_{21}^{\infty}}{s}]\times\cos[s(x-t)]ds$$

The derivative of equation (4) with respect to x, we can obtain

$$\frac{\partial u_{z1}(x,h)}{\partial x} = \frac{M_{21}^{\infty}}{\pi}\int_{-a}^{a}\frac{\sigma_1}{t-x}dt - \frac{M_{21}^{\infty}}{\pi}\int_{-c}^{c}\frac{\sigma_0}{t-x}dt + \frac{1}{\pi}\int_{-a}^{a} Q_1(x,t)\sigma_1 dt - \frac{1}{\pi}\int_{-c}^{c} Q_1(x,t)\sigma_0 dt \tag{6}$$

where $$Q_1(x,t) = -\int_0^{\infty} s[M_{21}(s,h) - \frac{M_{21}^{\infty}}{s}]\sin[s(x-t)]ds$$

According to the adhesive contact theory given by Maugis [10], the displacement in the adhesive contact area is as follows:

$$\begin{aligned} u_{z1}(\mathrm{a},0) - u_{z1}(c,0) = {} & \frac{-M_{21}^{\infty}}{\pi}\int_{-a}^{a}\ln\left|\frac{a-t}{c-t}\right|\sigma_1 dt + \frac{M_{21}^{\infty}}{\pi}\int_{-c}^{c}\ln\left|\frac{a-t}{c-t}\right|\sigma_0 dt \\ & + \frac{1}{\pi}\int_{-a}^{a}(Q(a,t)-Q(c,t))\sigma_1 dt - \frac{1}{\pi}\int_{-c}^{c}(Q(a,t)-Q(c,t))\sigma_0 dt \end{aligned} \tag{7}$$

According to the equilibrium equation, the contact pressure satisfies:

$$\int_{-a}^{a}\sigma_1 dt - \int_{-c}^{c}\sigma_0 dt = P \tag{8}$$

By introducing the following variables:

$$t = a\eta,\ x = a\varsigma,\ -a \le (t,x) \le a,\ -1 \le (\eta,\varsigma) \le 1,\ c = ma \tag{9}$$

the contact stress can be expressed as:

$$\sigma_1(\eta) = f(\eta)\sqrt{1-\eta^2} \tag{10}$$

Then, equation (5), (6) and (7) are reduced to

$$-\frac{a\varsigma_r}{R} = \sum_{l=1}^{N}\frac{(1-\eta_l^2)f(\eta_l)}{(N+1)}\left[\frac{M_{21}^{\infty}}{\eta_l - \varsigma_r} + aQ_1(\varsigma_r,\eta_l)\right] - \frac{\sigma_0 M_{21}^{\infty}}{\pi}\ln(\frac{m-\varsigma_r}{m+\varsigma_r}) - \frac{ma\sigma_0}{\pi}\int_{-1}^{1} Q_1(a\varsigma_r, a\eta_l)d\eta \tag{11}$$

$$\frac{(m^2-1)a^2}{2R}-h_0=\sum_{l=1}^{N}\frac{(1-\eta_l^2)f(\eta_l)}{(N+1)}\left\{a\left[Q(a,a\eta_l)-Q(\mathrm{m}a,a\eta_l)\right]-aM_{21}^{\infty}\ln\left|\frac{1-\eta_l}{m-\eta_l}\right|\right\}$$
$$+\frac{ma\sigma_0 M_{21}^{\infty}}{\pi}\int_{-1}^{1}\ln\left|\frac{1-\eta_l}{m-\eta_l}\right|d\eta_l-\frac{ma\sigma_0}{\pi}\int_{-1}^{1}[Q(a,a\eta_l)-Q(ma,a\eta_l)]d\eta_l \tag{12}$$

$$\sum_{l=1}^{N}\frac{(1-\eta_l^2)f(\eta_l)}{(N+1)}(\pi a)=P+2ma\sigma_0 \tag{13}$$

where $\eta_l=\cos[l\pi/(N+1)]$, $\varsigma_r=\cos[\pi(2r-1)/2(N+1)]$, $r=[1,N+1]$

After given the Hertz contact area $a$ and the ratio of the adhesive area and contact area $m$, the stress distribution $f(\eta_l)$ can be obtained from equation (11). If $m$ and $f(\eta_l)$ satisfied equation (12). We can get the contact stress $\sigma_1(\eta)$ and applied force $P$. Otherwise. Another value of $m$ is chose to solve equation (11) until $m$ and $f(\eta_l)$ satisfied equation (12).

## 3. Numerical Results and Discussion

In this paper, the adhesive contact problem of functional gradient piezoelectric coating is analyzed by making use of the commercially available pzt-4 piezoelectric material [11]. It is assumed that the thickness of the gradient coating $h$ equals to $0.01m$, the radius of the cylindrical punch $R$ is $0.08m$. Figure 2 show the effects of adhesion parameter $\sigma_0$ on the relation between the contact area $a$ and the applied linear loads $P$(a), the ratio of the adhesive area and contact area $m$(b), and the distribution of the contact stress $p(x)$(c). It can be seen from the figure 2a that the contact radius increases with the increase of the adhesion $\sigma_0$ for the same applied force $P$. As the adhesive parameter $\sigma_0$ gradually decreases, the applied force $P$ decreases when $P<0$. Figure 2b show that $m$ decreases and gradually becomes flat with the increasing of the contact radius. For the same Hertz contact radius, the $m$ value decreases as the adhesive force $\sigma_0$ increases. Figure 2c shows that the maximum contact stress gradually increases with the increase of adhesive force $\sigma_0$ when external loads $P=1000\,N/m$. Figure 3 show the effects of gradient index $\beta$h on Hertz contact area $a$(a), the ratio of the adhesive area and contact area $m$(b) and the distribution of the contact stress $p(x)$(c) when the adhesion stress $\sigma_0=5\times10^6$. It can be seen from the figure 3a that the Hertz contact radius decreases as the gradient index increases when the applied force $P$ are the same. Figure 3b show that with the increase of contact radius, the m value decreases and gradually becomes flat. when Hertz contact radius is the same, the m value decreases with the gradient index increases. Figure 3c shows that the maximum contact stress gradually increases with the increase of gradient index when applied force $P=1000\,N/m$.

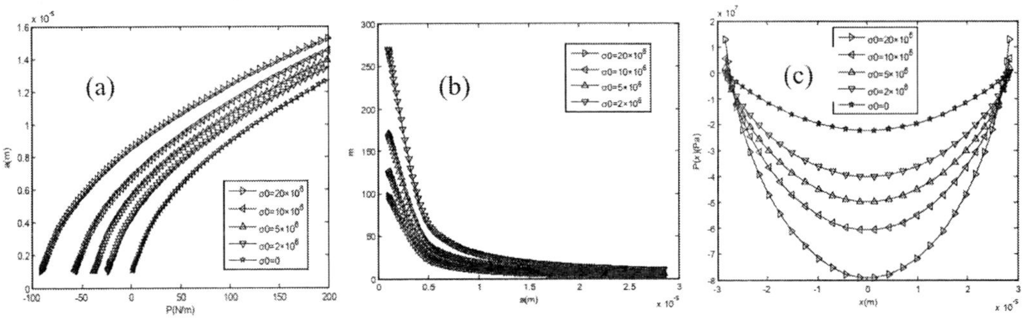


**Figure 2.** When $\beta$h=0.2, the effect of adhesive stress $\sigma_0$ on the contact radius a (a), length ratio of adhesive area to contact area m (b) and stress $p(x)$ (c)

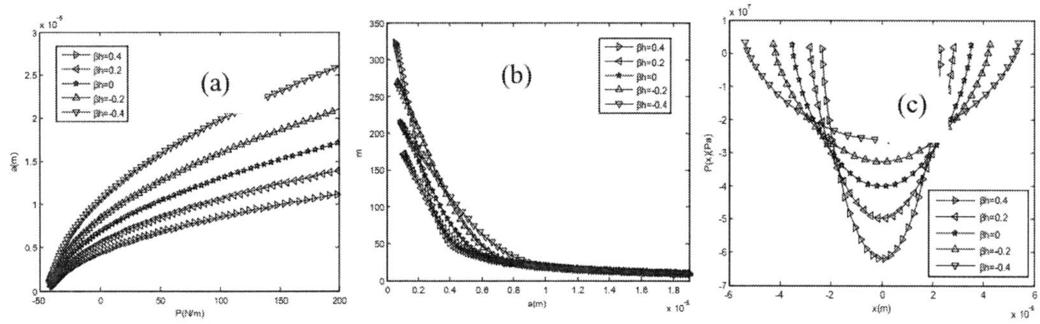

**Figure 3**. When $\sigma_0$=$5\times10^6$, the influence of gradient index ( $\beta$h ) on the contact radius (a) m value (b) and stress (c)

## 4. Conclusion

The gradient index of FGPM coating and the adhesion stress has obvious influence on the relation between applied force and the Hertz contact area, the adhesive contact area and the distribution of the contact stress. The adhesive contact behavior can be changed by adjusted the gradient index of FGPM coating.

## Acknowledgments

The support from the National Natural Science Foundation of China under Grant Nos. 11662011 and 11002065 are gratefully acknowledged. The support from study abroad project of Inner Mongolia is also acknowledged.

## References

[1] Delfosse D. Fundamentals of Functionally Graded Materials[J]. *Materials Today*, 1998, **1**(4):18.
[2] Ke L L, Yang J, Kitipornchai S, et al. Electro-mechanical frictionless contact behavior of a functionally graded piezoelectric layered half-plane under a rigid punch[J]. *International Journal of Solids & Structures*, 2008, **45**(11):3313-3333.
[3] Guler M A, Erdogan F. Contact mechanics of graded coatings[J]. *International Journal of Solids & Structures*, 2004, **41**(14):3865-3889.
[4] Ke L L, Wang Y S, Yang J, et al. Sliding frictional contact analysis of functionally graded piezoelectric layered half-plane[J]. *Acta Mechanica*, 2010, **209**(3-4):249-268.
[5] Liu T J, Zhang C. Axisymmetric conducting indenter on a functionally graded piezoelectric coating[J]. *International Journal of Mechanical Sciences*, 2016, **115-116**:34-44.
[6] Johnson K L, Sridhar I. Adhesion between a spherical indenter and an elastic solid with a compliant elastic coating[J]. *Journal of Physics D Applied Physics*, 2001, **34**(5):683.
[7] Chen S, Yan C, Soh A. Adhesive behavior of two-dimensional power-law graded materials[J]. *International Journal of Solids & Structures*, 2009, **46**(18–19):3398-3404.
[8] Chen S H, Yan C, Zhang P, et al. Mechanics of adhesive contact on a power-law graded elastic half-space[J]. *Mech Phys Solids*,2009,**57**:1437-1448.
[9] Jin F, Guo X, Gao H. Adhesive contact on power-law graded elastic solids: The JKR–DMT transition using a double-Hertz model[J]. *Journal of the Mechanics & Physics of Solids*, 2013, **61**(12):2473-2492.
[10] Maugis D, Barquins M. Adhesive contact of sectionally smooth-ended punches on elastic half-spaces: theory and experiment[J]. *Journal of Physics D Applied Physics*, 1983, **16**(10):1843.
[11] Liu T J, Zhang C, Wang Y S. Analysis of axisymmetric indentation of functionally graded piezoelectric coating or substrate systems under an insulator indenter[J]. *Journal of Intelligent Material Systems & Structures*, 2016, **28**(1):23-24.

# The Quality Detection Technology and Characterization Marking of Knife Edge

**Zhonghua Su***, **Taizhong Huang, Lichang Fan, Haipeng Mi, Xiong Deng and Xiaodong Deng**

Chongqing Academy of Metrology and Quality Inspection, Chongqing 401123, China
*Email: Su_water@163.com

**Abstract.** The quality and detection of kitchen knives at home and abroad were introduced in this paper. The knife edge toughness index was proposed to measure the comprehensive ability of the blade to resist chipping and curling, and the testing method and testing equipment for the cutter and the chopper were developed. Besides, aiming at the sharpness, durability and toughness which are the key indicators of the knife edge quality, the corresponding detection and identification methods were given.

## 1. The Industry and Testing Situation of The Kitchen Knife

The kitchen knife product is one of the key products in the daily hardware industry. It is an indispensable kitchen tool in people's life. The production process includes more than forty production processes such as material selection, cutting, forging, quenching heat treatment, grinding and polishing, balance point positioning, tool holder preparation, cutting, and opening. Each step requires extremely strict quality control [1]. Compared with foreign kitchen knife production technology, domestic manufacturers have yet to improve in terms of design, surface treatment, processing precision, heat treatment technology and raw material use. As far as kitchen knives are concerned, the most widely recognized kitchen knives in the world are German and Japanese kitchen knives. The German knives are characterized by precision and durability. And the Japanese knives are characterized by lightness and sharpness. There are thousands of kitchen knife production enterprises in China, and the output of products has increased year by year. Most of them are private enterprises, mainly distributed in coastal provinces such as Guangdong and Zhejiang, and Longshui Town in Dazu, Chongqing. Due to the large scale and advanced equipment, the market share of Guangdong and Zhejiang kitchen knives has increased year by year, and some well-known kitchen knives have emerged, such as Yangjiang 18, Zhang Xiaoquan, and Qiao Fu [2], but the quality of kitchen knives is still not as good as Europe, America and Japan.

Standardization is the key to improving the quality of Chinese kitchen knives. The standards for the quality and testing methods of kitchen knives at home and abroad are mainly DIN EN ISO 8442-5-2005 "Materials and products with food contact and cutlery. Part 5: knives sharpness and blade maintenance test", QB/T 1924-1993 "Chopper", DB50/T 261-2014 "Chopper" [3-4]. And the corresponding testing items include material (component), surface defects, blade thickness, resistance corrosion performance, tool strength, handle firmness, hardness, sharpness, durability, drop resistance, resistance to deformation of non-metallic handles in hot water, sanitary requirements, surface roughness, metallographic structure, quenching width, and steel, etc [5-8]. However, the blade toughness which consumers are most concerned about is only clearly required by the foreign tool associations, and this index is not clearly required in the relevant standards at home and abroad, let alone in the classification of grades. The blade toughness index is easy to understand for non-professionals and can directly reflect the quality and durability of cutting tools. Therefore, establishing the evaluation index in the knife standard and characterizing the knife quality are beneficial to the control of the tool quality by the kitchen knife manufacturer, and it is easy for consumers to understand and consume. In this paper, the testing method, testing equipment and quality characterization of knife edge toughness were systematically studied.

## 2. The Toughness Detection Technology of Kitchen Knife

Common failure modes of household kitchen knives are chipping, curling and blunting. The failure problem of blunting can be solved by re-grinding, however, the other two kinds of failure problems, such as chipping and curling, are difficult to solve by grinding them again. Therefore, the ability of the blade to resist the chipping and the curling is the key index to evaluate the quality of the kitchen knife. The chipping and the curling edge are mainly affected by the hardness of the blade edge. When the hardness is high, the chipping is easy, and while the hardness is low, which is easy to roll. The scientific value of the hardness value is not conclusive, because the reasonable hardness value also needs to consider the shape and the material of the knife edge and the heat treatment process. In order to provide guidance for the scientific production of the kitchen knife production enterprise and make the purchaser of kitchen knives consume clearly, the toughness of the kitchen knife edge is proposed and studied in this paper. The toughness index of the knife edge is mainly to characterize the ability of the knife edge to resist the chipping and curling when it meets a hard object during use, and it is also a key indicator for evaluating the quality of the kitchen knife. But it is necessary to study the scientific and reasonable testing methods and equipment before making a reasonable evaluation of the index. Since the kitchen knife usually includes a cutter and a machete, this paper will separately explain the detection method and testing equipment for the toughness of the cutter and the chopper.

### *2.1. The Toughness Detection Method of Cutting Knife Blade*

In daily life, the main reason of cutting and curling edge is to cut meat food containing bone. Therefore, the method of measuring the toughness of cutting edge will be used to cut the specimen containing hard material. In order to ensure the consistency of the test conditions, the pencils with different hardness of the same brand were selected and the hardness range was 9B-9H.

A soft pencil was selected at the beginning of the experiment, and then put into the pen tray, immediately afterwards the pencil was cut vertically at a speed of Vmm/s at the position 10 mm from the end of the pencil. After repeating the operation N times, a magnifying glass was used to magnify the cutting edge of the pencil by X times to measure the size of the chipping and curling in the image. Chipping and curling edges are effective defects only when the size of the blade direction or perpendicular to the edge direction is $\geq$ L micron. If there is no effective defect in the knife edge, then choose a pencil with a larger hardness value to repeat the previous experiment, and the test is stopped until a $\geq$ L micron chipping or curling occurs on the edge of the knife. The toughness of the knife edge is the pencil hardness value used before the effective defect is generated. Figure 1 shows the magnification of the cutter before and after the toughness test.

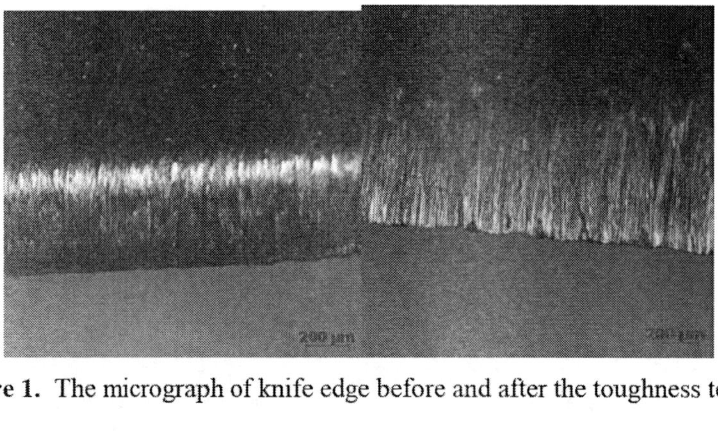


**Figure 1.** The micrograph of knife edge before and after the toughness testing

### *2.2. The Toughness Testing Equipment of the Cutting Knife Edge*

The cutter edge toughness detecting device comprises a base , a sample pusher , a magnifying glass , a bracket , a cutter vertical cutting drive , a cutter holding device , a pencil cutting board and a cutting board fine adjustment device , corresponding to the numbers 1 to 10 in Figure 2. The base in figure 2 is mainly used to support the whole equipment to ensure the smoothness of the equipment placement; the sample pusher is fixedly mounted on the base for pushing the pencil sample, and when the pencil is cut, the pusher automatically pushes the pencil forward by 10mm for the next cutting; the magnifying glass is fixedly mounted on the base for magnifying and observing the size of the chipping and curling in the cutting position of the blade; the bracket is vertically fixed on the base, and is mainly used for supporting the cutter vertical cutting drive and the cutter clamping device; the fixed height of the cutter vertical cutting drive could be adjusted, and the cutter clamping device is fixed on the drive shaft at the lower end of the drive, besides, the cutter clamping device and the cutter move vertically up and down with the drive shaft, and the moving speed or the maximum vertical shear force can be adjusted as needed; the cutter clamping device is mainly used to fix the cutter, and the cutting board fine adjustment device is set because the thickness of the cutter sample is inconsistent and the cutter edge is not necessarily aligned with the blade protection seam on the cutting board after the cutter is fixed; the lower end of the cutting block is fixed on the upper surface of the cutting board fine adjustment device, and the surface of the cutting board is crisscrossed with a sample slot and a blade protection joint, the sample slot is used for placing a pencil sample, and the blade protection joint is used for protecting the blade edge, thereby to avoid unexpected damage to the blade.

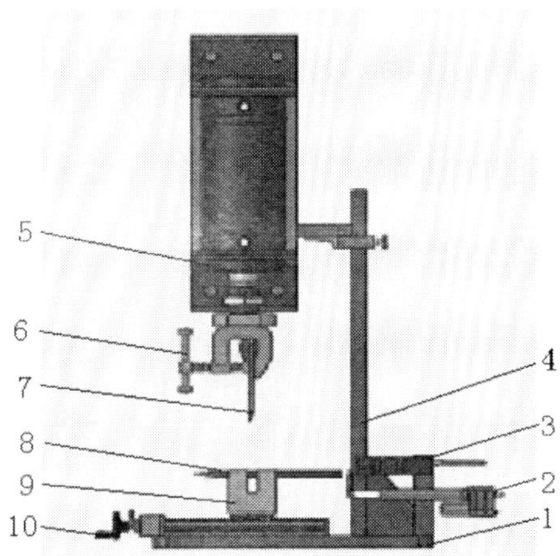


**Figure 2.** Cutter blade toughness testing equipment

### *2.3. The toughness Detection Method of the Hacking Knife Edge*

The main reason for the chopper's chipping and curling is the cutting of thick and hard bones. Therefore, when testing the toughness of the knife edge, it is necessary to select a sample whose hardness value is close to the bone. In order to ensure the consistency of the test conditions, a pencil of different hardness of a certain brand was selected before the experiment, and the hardness range of which includes 9B-9H. At the beginning of the experiment, a pencil was chosen with a hardness lower than that of the bone, then pencil was placed in the pen tray, and cut at a speed of Vmm/s at a position 10 mm from the end of the pencil. After repeating the operation N times, a magnifying glass was used to magnify the cutting edge of the pencil by X times to measure the size of the chipping and curling in the image. Chipping and curling edges are effective defects when the size of the blade direction or perpendicular to the edge direction is ≥ L micron. If there is no effective defect in the knife edge, then choose a pencil with a larger hardness value to repeat the previous experiment, and the test is stopped until a ≥ L micron chipping or curling occurs on the edge of the knife. The toughness of the knife edge is the pencil hardness value used before the effective defect is generated.

### *2.4. The Toughness Testing Equipment of the Chopper Knife Blage*

Chopper knife toughness testing equipment includes base, cutting board fine adjustment device, pencil cutting board, chopper clamping device, bracket, cutting drive, connecting rod, rocker arm, rocker arm limiting device, for details, see numbers 1 to 3 and 6 to 11 in figure 3. The base in the figure is mainly used to support the whole equipment to ensure the smoothness of the equipment placement; the cutting board fine adjustment device is mainly used for adjusting the position of the pencil cutting board fixed on the surface, and ensuring that the cutting edge does not directly contact the cutting board after the chopper cuts the pencil, but enters the protection seam of the cutting edge, thereby avoiding unintended damage of the cutting edge; the cutting board is used to place the pencil sample, the upper surface is crisscrossed with the sample slot and the blade protection joint; the chopper clamping device is mainly used to fix the machete; the bracket is vertically fixed on the base, and is mainly used for supporting the cutting drive and the chopper clamping device; the cutting drive provides driving force for the cutting motion of the chopper, and the driving force is transmitted to the rocker arm through the connecting rod, and the rocking arm drives the clamping device and the chopper to cut, thereby realizing the movement mode of simulating the artificial cutting of the bone. To avoid damage to the blade due to excessive cutting of the chopper, the rocker limit device is specially designed to limit the upper and lower limits of the chopper. In addition, a sample pusher and a cutting edge magnifier are also provided in the apparatus.

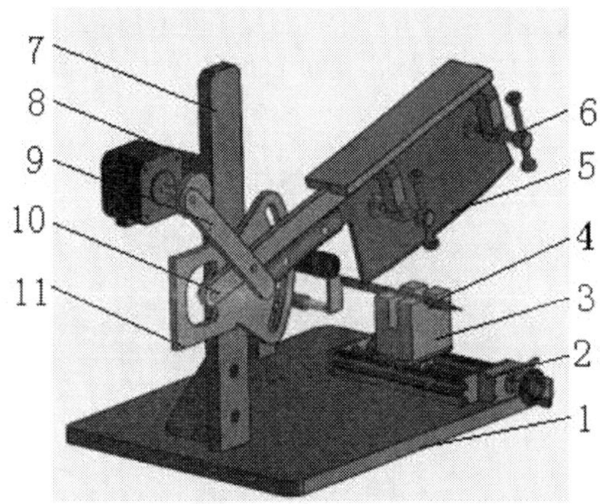


**Figure 3.** Chopper blade toughness testing equipment.

## 3. The Characterization and Identification Method of Knife Edge Quality

Good steel used on the blade is the key method to improve the quality of the knife, which fully shows that the quality of the knife edge is very important to the whole knife. Therefore, we should focus on evaluating the quality of a kitchen knife when we evaluate the quality of the knife. This paper will expound the methods of characterizing and identification for the sharpness, durability and toughness which are the key elements of the knife edge quality.

The sharpness and durability of the knife mainly depend on the geometric angle and material properties of the blade. The properties of the material include material size, compactness, wear resistance, hardness, elasticity and forming texture. In this paper, the method used to characterize the sharpness and durability of the kitchen knife is DB50/T 261-2014 "Chopper". And the sharpness value is the sum of the cutting depths of the first three cutting weeks, and the durability value is the sum of the cutting depths of 30 weeks during the test process. The toughness of the kitchen knife mainly depends on the geometric angle, hardness, elasticity, grain size and compactness of the cutting edge. The characterization method uses the previously proposed blade toughness detection method.

In the case where the sharpness, durability and toughness characterization methods are determined, this paper will present a clear identification method for identifying the edge quality. The kitchen knife manufacturing company usually adopts mechanization in the production process, and the sharpness, durability and toughness of the kitchen knife are relatively consistent. Therefore, it is only necessary to sample and characterize the continuous batch product before characterizing the knife edge quality. If the test results are as follows, sharpness is 60mm, durability is 260mm, and toughness is 3H, then the mechanical printing mark F60N260J3H is printed near the trademark position of the kitchen knife, and the quality of the knife edge can be understood by the consumer through the mark.

## 4. Conclusions

In view of the technical requirements, testing methods and testing equipments of domestic kitchen knives without anti-fracture and curling edge, this paper proposed a blade toughness testing project and developed corresponding testing techniques, including the testing method and equipment of the cutting tool and the detection method and equipment of the chopper.

Based on the criticality of sharpness, durability and toughness on the quality of the knife edge, the methods of detecting and characterizing the key indicators were given, which would increase the market driving force for the quality improvement of the kitchen knife product.

## References

[1] Tang Jialu, Wang Tao and Wang Tuo. The investigation on the production process of the kitchen knife in Shilibao, Shandong Province [J]. Journal of Shandong University of Art and Design, 2009, 58: 68-69.
[2] Li Zilin. Research of the laser deposition process and performance of the TiC composites layer on the kitchen knife blade [D]. Lanzhou: Lanzhou University of Technology, 2016.
[3] Zhou Liansheng, Liu Yuexia, Ye Na, et al. A study of the sharpness testing methods of edge products [J]. Southern Metals, 2010, 177: 51-53.
[4] Zhang Mao, Qiu Yaling, Xiao Xiaohua, et al. A rapid test method for tool durability [J]. Modern Machinery, 1997, 37: 40-41.
[5] Jing Xuwen and Tang Xinru. Research on magnetic technology of improving cutter lifespan [J]. Mechanical Design & Manufacture, 2002, 5: 81-82.
[6] Ye Weichang. The methods for increase in duration of wear-resistance of cutter [J]. Journal of Huaiyin Industrial College, 1994, 3: 58-64.
[7] Zhou Jinquan. Several ways to improve tool durability [J]. Journal of Mechanical Engineer, 1999, 8: 37-38.
[8] Zheng Zhijun, Ping Yibo, Shi Lin, et al. Welding crack analysis on dissimilar steel of certain kitchen knife [J]. Hot Working Technology, 2015, 6(11): 239-242.

IOP Conf. Series: Materials Science and Engineering 472 (2019) 012055 doi:10.1088/1757-899X/472/1/012055

# Preparation of Catalyst Supported by CuO-CoO-MnO/$SiO_2$ Nanocomposite Aerogel and its Catalytic Performance in the Synthesis of Diphenyl Carbonate

**Yueqing Zhao[1], Xizhe Zhao[2], Mingxi Zhang[1] and Qianyi Jia[1]**

[1] College of Materials Science and Engineering, North China University of Science and Technology, 21 Bohai Road, Caofeidian Xincheng, Tangshan 063210, P. R. China.
Email: zhaoyueqing123@163.com

[2] Department of Chemistry, College of Staten Island, City University of New York, 2800 Victory Boulevard, Staten Island, New York 10314, USA

**Abstract.** A novel catalyst supported by nano-porous CuO-CoO-MnO/$SiO_2$ composite aerogel was prepared by impregnation method. It was used in the catalytic synthesis of diphenyl carbonate (DPC). The effects of temperature, pressure and reaction time on the yield of DPC were investigated. The results show that Phenol and DPC can smoothly pass through the nano-pores of the composite aerogel during the catalytic synthesis of DPC. When the temperature, pressure and reaction time of DPC synthesis are 110 °C, 6.0 MPa and 6 h, respectively, the yield of DPC is 27.14 mass%, which is much higher than that of catalysts supported by other porous carriers.

## 1. Introduction

Aerogel is a porous, non-crystalline solid material with large specific surface area, high porosity, low density and better chemical property. Therefore, it has been used as catalyst carrier, gas absorbent and filter, heat insulator, sound deadening, drug adsorption and release, etc [1-5]. Nano-porous composite aerogel with transition metal oxides can be used as catalyst or catalyst carrier in catalysis [6-8]. In the past decade, some catalyst carriers, such as porous ceramics, 3A molecular sieve, HZSM molecular sieve, activated carbon, $SiO_2$ xerogel, $SiO_2$ aerogel, CuO/$SiO_2$ composite xerogel, CuO/$SiO_2$ composite aerogel, CuO-CoO-MnO/$SiO_2$ composite xerogel and the like have been used in the synthesis of diphenyl carbonate (DPC). DPC can be used as the precursor in the synthesis of polycarbonate (PC) by polymerization, PC is an engineering thermoplastic material and it has been widely used in mechanical parts and elements, electrical insulators of various electrical and electronic products and so on [9-12]. However, DPC yields are usually less than 10.0%. This is because of lack of active transition metal-oxides used as co-catalysts or because of small specific surface area and low porosity in above-mentioned catalyst carriers. The properties of catalysts supported by composite aerogels can be further optimized through the combination of structural constituents and redox constituents [13-15]. CuO-CoO-MnO/$SiO_2$ composite aerogel is a nano-porous material with multiple active transition metal oxides, a specific surface area of 664.4-695.8 $m^2/g$ and a pore size distribution of 2-16 nm [15-16]. Catalyst supported by the composite aerogel has been used in the environment-friendly synthesis of DPC, the yield of DPC is much higher than that of catalysts supported by traditional catalyst carriers [18-20].

In this study, a novel catalyst used for the catalytic synthesis of DPC was prepared by impregnation method. The catalytic mechanism of the supported catalyst in the synthesis of DPC was explored. The effects of temperature, pressure and reaction time on the yield of DPC was investigated. The catalytic performances of the catalysts in the synthesis of DPC were investigated.

## 2. Experimental

### *2.1. Preparation of Supported Catalyst*

Catalyst supported by nano-porous CuO-CoO-MnO/$SiO_2$ composite aerogels was prepared by impregnation method. The composite aerogel was firstly ground into fine particles in a mortar and then the fine particles were sieved in standard sieves. The sieved fine particles of the composite aerogel were put into a vacuum drier and then dried at 110 °C for 2 h so as to remove the residual moisture in the pores of the catalyst carrier. $PdCl_2$ is used as the main catalyst and $Co(OAc)_2$ is used as the co-catalyst in the heterogeneous catalytic system. They were dissolved in 5 mL aqueous ammonia, the concentrations of $PdCl_2$ and $Co(OAc)_2$ in the solution were 0.034 mol/L and 0.24 mol/L, respectively, and then the impregnation solution was obtained. The fine particles of the nano-porous composite aerogel with a mass of 1 g were firstly dispersed in the prepared impregnation solution, then stirred at room temperature for 2.0 h, aged the mixture for 12 h, dried in a vacuum dryer at 60 °C for 3 h, and then calcined in a muffle furnace at 550 °C for 7 h so as to activate the supported catalyst, finally the supported catalyst was obtained.

### *2.2. Application in the Synthesis of DPC*

The synthesis of DPC was carried out in an autoclave reactor, and the autoclave was equipped with a magnetic stirrer, the reactor was pre-filled with 50 mL $CH_2Cl_2$ (dichloromethane) used as a solvent, 1 g supported catalyst, 0.440 4 g $C_6H_4(OH)_2$ (hydroquinone or $H_2BQ$) used as an organic co-catalyst, 7.528 8 g phenol used as a reactant, 3 g 3A molecular sieve with size of 10-230 μm used as a drying agent, 0.967 4 g $(C_4H_9)_4BrN$ (tetrabutylammonium bromide or TBAB) used as a stabilizer and a surfactant. After sealing the autoclave, CO was injected into the reactor so as to replace the air and prevent the oxidation of phenol. When the pressure and the temperature were 0.5 MPa and 40 °C, respectively, $O_2$ and more CO were injected into the autoclave reactor until the total pressure was 6.0 MPa. When the temperature was 80 °C, the reaction of oxidative carbonylation started in the autoclave reactor. After 6 h, the temperature was 110 °C, the heating apparatus was shut off, and the autoclave was cooled down to room temperature. Gases in the autoclave reactor were exhausted slowly, and then the pressure was decreased gradually. The mixture of the solvent, liquid products and residual phenol was taken out of the autoclave reactor for analysis.

## 3. Results and Discussion

### *3.1. Geometric Parameters Calculation and Molecular Configuration Simulation*

The geometric parameters of phenol and DPC were calculated, and the molecular configurations of phenol and DPC were simulated by using Material Studio 7.0 software so as to prepare suitable catalyst carrier and optimize the structure of catalyst in the synthesis of DPC. Table 1 shows the geometric parameters of a single molecule of phenol. Table 2 shows the geometric parameters of a single molecule of DPC. Figure 1 shows the steric configurations of a single molecule of phenol and a single molecule of DPC.

**Table 1.** Geometric parameters of a single molecule of phenol

| Band angle | | Band length | | Molecule size | |
|---|---|---|---|---|---|
| Key atom | $\theta$/(°) | Key atom | $\gamma$/nm | Direction | Size/nm |
| C(2)-C(1)-O | 122.310 | C(1)-C(2) | 0.138 7 | *X* | 0.618 8 |
| C(3)-C(1)-O | 117.611 | C(1)-O | 0.135 7 | *Y* | 0.464 2 |
| C(1)-O-H | 108.261 | H-O | 0.098 2 | | |

**Table 2.** Geometric parameters of a single molecule of DPC

| Band angle | | Band length | | Molecule size | |
|---|---|---|---|---|---|
| Key atom | $\theta$/(°) | Key atom | $\gamma$/nm | Direction | Size/nm |
| C(4)-C(1)-O(8) | 123.826 | C(4)-C(1) | 0.138 4 | *X* | 1.054 9 |
| C(5)-C(1)-O(8) | 134.495 | C(1)-C(5) | 0.138 4 | *Y* | 0.709 0 |
| C(1)-O(8)-C(2) | 121.328 | C(1)-O(8) | 0.138 1 | *Z* | 0.452 6 |
| O(8)-C(2)-O(9) | 127.285 | O(8)-C(2) | 0.134 4 | | |
| O(9)-C(2)-O(10) | 121.682 | C(2)-O(9) | 0.119 6 | | |
| O(8)-C(2)-O(10) | 110.866 | C(2)-O(10) | 0.133 9 | | |
| O(10)-C(3)-C(7) | 116.428 | C(3)-O(10) | 0.137 4 | | |
| O(10)-C(3)-C(6) | 122.240 | C(3)-C(7) | 0.138 4 | | |
| C(4)-C(1)-C(5) | 121.665 | C(3)-C(6) | 0.138 7 | | |
| C(6)-C(3)-C(7) | 121.071 | | | | |

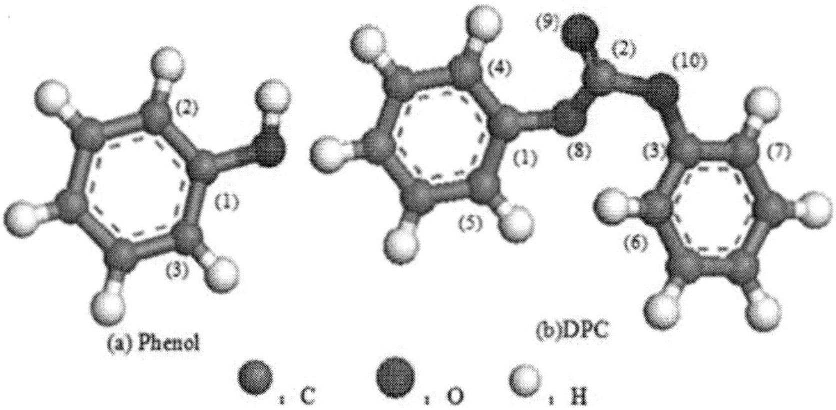


**Figure 1.** Steric configurations of a single molecule of phenol and a single molecule of DPC

The geometric dimensions of phenol are *X*=0.618 8 nm and *Y*=0.464 2 nm, and the geometric dimensions of DPC are *X*=1.054 9 nm, *Y*=0.709 0 nm and *Z*=0.452 6 nm (Table 1 and Table 2), whereas the average pore size of the nano-porous composite aerogel is 7.68 nm [16, 20]. Therefore, phenol and DPC can smoothly pass through the nanopores of the composite aerogel and then DPC can gradually disperse in the mixture of solvent and residual phenol during the heterogeneous catalytic synthesis of DPC because the average pore size of the composite aerogel is obviously bigger than that of phenol and DPC.

CuO-CoO-MnO/$SiO_2$ composite aerogel is a nano-porous material with a specific surface area of 664.4–695.8 $m^2 \cdot g^{-1}$, and the mole fraction of active transition metals in the nano-porous composite aerogel is 0.71 mol%-13.77 mol% [16]. The structure and composition can not only promote the loading of $PdCl_2$ and $Co(OAc)_2$, but also make full use of the catalysis of active transition metal oxides as co-catalysts.

### *3.2. Effect of Technological Parameters on the Yield of DPC*

*3.2.1. Effect of temperature on the yield of DPC.* Effect of temperature in the autoclave reactor on the yield of DPC was investigated in the temperature range of 80-130 °C. Figure 2 shows the effect of temperature in the reactor on the yield of DPC. According to chemical reaction kinetics, the reaction rate increases with the increase of temperature in the autoclave reactor, and then the yields of DPC increase with the increase of temperature in the temperature range of 80-110 °C. On the other hand, the yields of DPC decrease in the temperature range of 110-130 °C, because the reaction of oxidative carbonylation for DPC synthesis is an exothermic reaction [20], higher temperature in the autoclave reactor is not favorable to the synthesis of DPC. Therefore, the optimum temperature of the catalytic synthesis of DPC is 110 °C.

*3.2.2. Effect of Pressure on the Yield of DPC.* Effect of pressure in the autoclave reactor on the yield of DPC was investigated at a constant temperature of 110 °C. Figure 3 shows the effect of pressure in the reactor on the yield of DPC. The reaction of oxidative carbonylation for the catalytic synthesis of DPC is shown by equation (5).

$$2PhOH+CO+1/2O_2 \xrightarrow[T,\ P]{Cat} PhO(CO)OPh+H_2O \quad (1)$$

According to the equation, because the number of molecules decreases after the reaction, the increase of pressure in the reactor is favorable to the increase of resultants. In addition, according to Henry's law or $x=kP$, the increase of pressure is also favorable to the increase of the solubility of CO or $O_2$ in solvent or $CH_2Cl_2$, and the reaction probability between reactants and catalysts or co-catalysts increase. Accordingly, the reaction rate increases with the increase of pressure, the yields of DPC increase with the increase of pressure in the autoclave reactor, and the yield of DPC increases to maximum at 6.0 MPa (Figure 3).

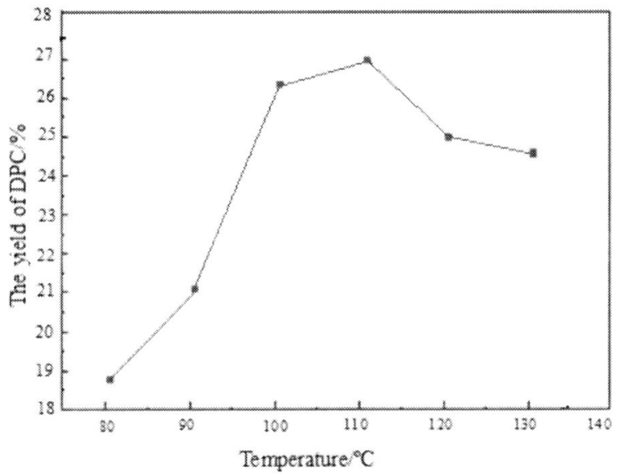


**Figure 2.** Effect of temperature on the yield of DPC.

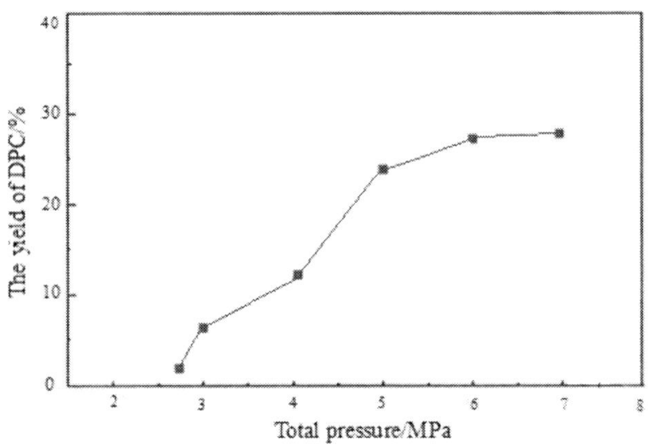


**Figure 3.** Effect of pressure on the yield of DPC

### *3.2.3. Effect of Reaction Time on the Yield of DPC*
When the pressure in autoclave reactor was 6.0 MPa, the partial pressures of CO and $O_2$ were 5.0 MPa and 1.0 MPa, respectively, the effect of reaction time on DPC yield was investigated at 110 °C. Figure 4 shows the effect of reaction time in the reactor on DPC yield. The yields of DPC obviously increase with the increase of reaction time in 6 h, but the yields of DPC increase slightly after 6 h and the yields of DPC is almost constant, because oxidative carbonylation almost reaches chemical equilibrium after 10 h. Therefore, the optimum reaction time of DPC synthesis is 6 h (Figure 4).

### *3.3. Evaluation of Catalytic Performances*
Figure 5 shows the gas chromatograms of chemical substances. The abscissa of the graph shows peaks appearance time of $CH_2Cl_2$, phenol and DPC, and the ordinate of the graph indicates the energy release intensity after burning of $CH_2Cl_2$, phenol and DPC. The left peak in the graph at 0.490 min corresponds to $CH_2Cl_2$, and it is used as the solvent in the catalytic system. The middle peak in the graph at 1.725 min indicates the phenol, and it is used as a reactant. The right peak in the graph at 6.255 min represents the target product or DPC (Figure 5).

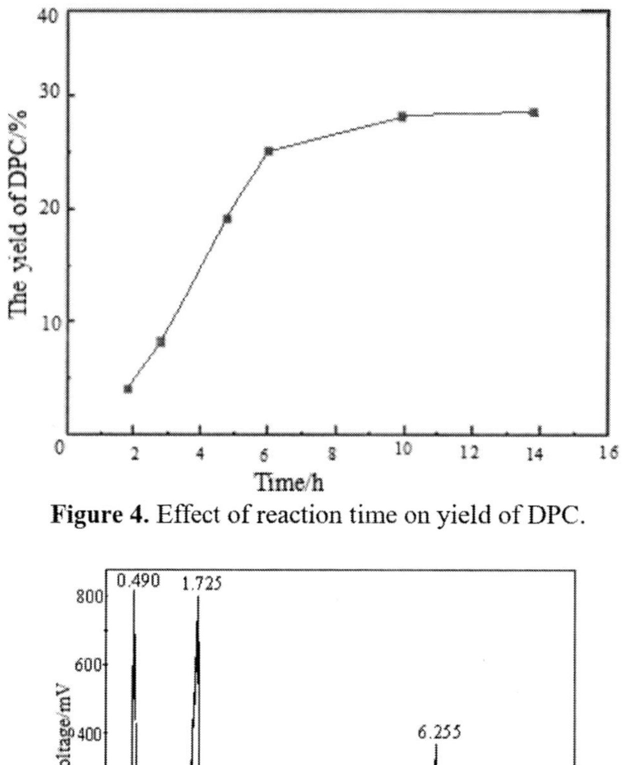

**Figure 4.** Effect of reaction time on yield of DPC.

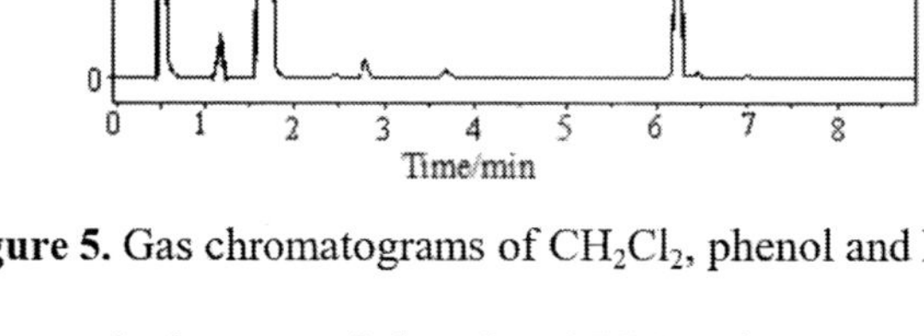

**Figure 5.** Gas chromatograms of $CH_2Cl_2$, phenol and DPC

In order to measure the standard curves of phenol and DPC, using $CH_2Cl_2$ as the solvent, five kinds of standard solutions with different concentrations were prepared by dilution method, and their compositions were similar to those of the reaction liquids. Fig. 6 shows the standard curve of phenol. Fig. 7 shows the standard curve of phenol and DPC. The linear equations were obtained by least square regression analysis. The linear regression equation of phenol is shown by equation (6), and the linear regression equation of phenol and DPC is shown by equation (7).

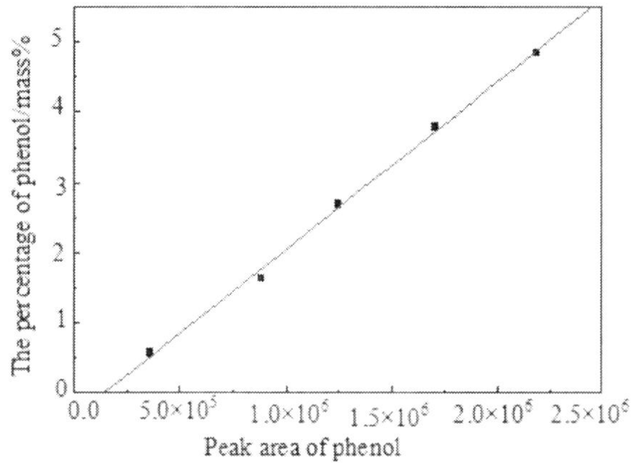


**Figure 6.** The standard curve of phenol

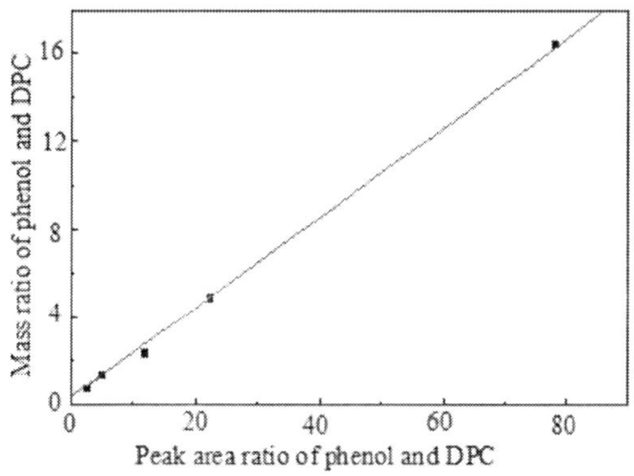


**Figure 7.** The standard curve of phenol and DPC.

$$w = -0.2876 + 3.07 \times 10^{-6} A \tag{2}$$

Where $A$ is the peak area of phenol, $w$ is the percentage of phenol, and 0.99883 is a correlation coefficient.

$$y = 0.79176 + 1.324x \tag{3}$$

Where $x$ is the peak area ratio of phenol and DPC, $y$ is the mass ratio of phenol and DPC, and 0.99816 is a correlation coefficient.

The analysis results show that the yield of DPC catalyzed by the supported catalyst is 27.14 mass%. The comparison of catalytic performances of catalysts supported by different carriers in the synthesis of DPC is shown in Figure 8. The yield of DPC catalyzed by the novel supported catalyst is much higher than that of catalysts supported by other carriers, such as porous ceramics, 3A molecular sieve, activated carbon, HZSM molecular sieve, $SiO_2$ xerogel, $SiO_2$ aerogel, $CuO/SiO_2$ composite xerogel, $CuO/SiO_2$ composite aerogel and $CuO$-$CoO$-$MnO/SiO_2$ composite xerogel (Figure 8).

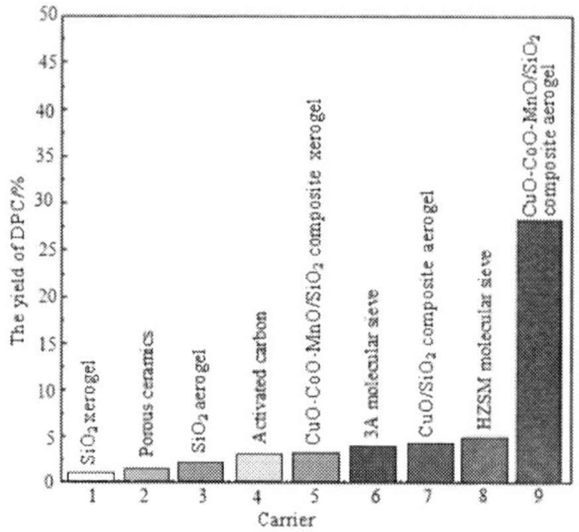


**Figure 8.** Comparison of catalytic performances of catalysts supported by different carriers in the synthesis of DPC

## 4. Conclusions

Phenol and DPC can smoothly pass through the nano-porous CuO-CoO-MnO/$SiO_2$ composite aerogel during the catalytic synthesis of DPC because the average pore size of the composite aerogel is obviously bigger than that of phenol and DPC.

When the temperature, pressure and reaction time of DPC synthesis are 110 °C, 6.0 MPa and 6 h, respectively, the yield of DPC is 27.14 mass%, which is much higher than that of catalysts supported by other porous catalyst carriers .

## Acknowledgements

The authors wish to thank the financial support of Department of Human Resources and Social Security of Hebei Province, China (No. CL201615). The authors also wish to thank Professor Hai-lei ZHAO from University of Science and Technology Beijing for their support and assistance.

## References

[1] Fricke J and Tillotson T 1997 Thin Solid films 297 212-218
[2] Rolison D R and Dunn B 2001 J. Mater. Chem. 11 963-980
[3] Jing X, Kai X and Yimin C 2000 J. Mol. Catal. A-Chem. 14 71-73
[4] Chiara P, Liliana O and Michela S 2000 J. Catal. 14 286-292
[5] Dong G, Gao Y and Chen S 1999 Catal. Lett. 58 37-41
[6] Muller C A, Maciejewski M and Mallat T 1999 J. Catal. 184 280-293
[7] Ernst B, Libs S and Chaumette P 1999 Appl. Catal. A-Gen. 186 145-168
[8] Wei X and Xinqiang Z 2005 Catal. Commun. 6 431-436
[9] Hongia G, Hongping C, Yinghua L, Yulai. Rui, Jingde L and Zhanda F 2008 Chin. J. Chem. Eng. 16 223-227
[10] Vavasori A and Toniolo L 1999 J. Mol. Catal. A-Chem. 139 109-119
[11] Moiseev I I, Vargaftik M N and Chernysheva T V 1996 J. Mol. Catal. A-Chem. 108 77-85
[12] Song H Y, Park E D and Lee J S 2000 J. Mol. Catal. A-Chem. 154 243-250
[13] Yimin C, Jing X and Kai X 200 Mater. Eng. 5 43-46
[14] Jing X, Kai X and Yimin C 2002 New Chem. Mater. 30 32-34
[15] Mohana n J L and Brock S L 2003 Chem. Mater. 15 2567-76
[16] Yueqing Z, Hailei Z, Yinghua L, Qianyi J and Bobo Z 2010 T. Nonferr. Metal Soc. 20 1463-69
[17] Yueqing Z, Yinghua L, Xizhe Z and Qianyi J 2011 Prog. Nat. Sci.-Mater. 21 1-6
[18] Yueqing Z, Hailei Z, Yinghua L, Qianyi J and Yi S 2008 J. Chin. Ceram. Soc. 36 658-662
[19] Yueqing Z, Xizhe Z, Mingxi Z and Qianyi J 2018 Micropor. Mesopor. Mater. 261 220-226
[20] Hongxia G, Hongping C, Yinghua L and Yulai R 2008 Chin. J. Chem. Eng. 16 223-227

# Development of a Solid-Phase Microextraction-Surface-Enhanced Raman Scattering Approach to Detect Ferbam

**Shan Wang, Jinhui Zhao, Guoqiang Wu, Weijun Leng and Chunrong Wang***

School of Food Science and Technology, Jiangxi Agricultural University, Nanchang, 330045, China

Email: crwang07@126.com

**Abstract.** By coupling solid-phase microextraction (SPME) with surface-enhanced Raman scattering (SERS), a facile method was developed for the determination of ferbam, a pesticide often used in fruits and vegetables. It is a potential carcinogen. The SPME substrate was made by free settling of rod-like ZnO nanomaterials on a aluminum sheet. The extraction procedure for ferbam was performed in a simple solution, and then the ferbam was determined using SERS after uniformly dropping a SERS-active substrate (gold nanoparticles, AuNPs) onto the surface of the SPME substrate. In this method, qualitative measurement in extraction precedure takes only 5 minutes. The lowest detectable concentration of ferbam is 0.01 ppm. The newly developed method is fast, convenient and sensitive in the determination of ferbam in solution.

## 1. Introduction

Surface-enhanced Raman scattering (SERS) is a highly sensitive technique, based on the gaint electromagnetic enhancement induced by nanotechnology. SERS has been widely explored to detect organic chemicals. For example, the weak Raman molecular "fingerprint" can be tremendously enhanced through the placement of the analyte on gold nanostructures [1-4]. On the other hand, solid-phase microextraction (SPME) [5-8] is proved to be an effective method for gathering analyte and often used to detect organic compounds coupled with gas chromatography [9] and mass spectrometry [10-12]. In our study, we developed a new approach combining SPME with SERS to perform a simple, sensitive, and easy operation for the determination of ferbam. In the preparation of the extraction substrate, rod-like ZnO nanomaterials were deposited on an aluminum support sheet for SPME [13]. The ferbam was absorbed and concentrated by the ZnO, and the content of ferbam was then measured using SERS. Compared to other ferbam detecting methods such as GC and HLCP the SPME-SERS provided a simple, rapid, and sensitive approach for the determination of ferbam in solution.

## 2. Experimental Sections

### 2.1. Chemicals and Instruments

Chmicals such as Zn $(CH_3COO)_2{\bullet}2H_2O$ (99.0%), hydrogen tetrachloroaurate hydrate (99.999%) and sodium chloride (99.5%) were of analytical grade, and procured from Sigma Aldrich (St. Louis, MO). Hydrochloric acid (34 to 37.5%), acetone (99.9%), ethanol (100%), ammonia (25.0 to 28.0%), and methanol (99.9%) were purchased from Fisher Scientific (Fair Lawn, NJ).

### *2.2. Preparation of Gold Nanoparticles*

Au nanoparticles were prepared through the reduction of by trisodium citrate as described by Frens.[14] Briefly, 30 mL of $HAuCl_4$ aqueous solution (0.01%, w/w) was heated at 100 °C under vigorous stirring. While boiling, 0.6 mL of trisodium citrate aqueous solution (1%, w/w) was added very quickly. The solution turned wine red in colour within 2 min. After 15 min of reaction, the heating was stopped, but the stirring was maintained until the solution cooled to room temperature. The product was stored at 4 °C.

### *2.3. Preparation of Raman Substrate*

The detailed experimental process is described as follows. Firstly, an aluminum foil (with dimensions of $10\times10\times0.1mm^3$) was pretreated by dilute hydrochloric acid and acetone assisted by ultrasonic to remove surface metal ions and surface oil contaminant, and was then used as the working substrate for the ZnO assemlage. The solution of 0.1 mol $L^{-1}$ Zn $(CH_3COO)_2$ with pH value of 11 was prepared. Then the solution of 0.3 mol $L^{-1}$ Zn $(CH_3COO)_2$ was prepared and sprayed onto the aluminium foil which was heated at 280 °C, and the spraying course was lasted for 30 min. This heated amluminium foil was cooled down and was submerge in the bottle containing 0.1 mol $L^{-1}$ Zn $(CH_3COO)_2$ solution. The bottle was sealed and heated to 90 °C for 4 h. After the bottle was cooled down, the aluminum foil covered with the synthesized crystal film (ZnO/Al) was taken out, washed with deionized water several times, and dried under vacuum at 80 °C for 4h.

### *2.4. SPME-SERS Measurement of Ferbam*

In the SPME-SERS of ferbam, as described in Scheme 1, the as-prepared substrate preconcentrated the analyte in solution by the direct immersion solid-phase microextraction method for 10 min at ambient temperature. The substrate was dropped with 10 μl gold sol and allowed to air-dry, and Raman signals were collected.

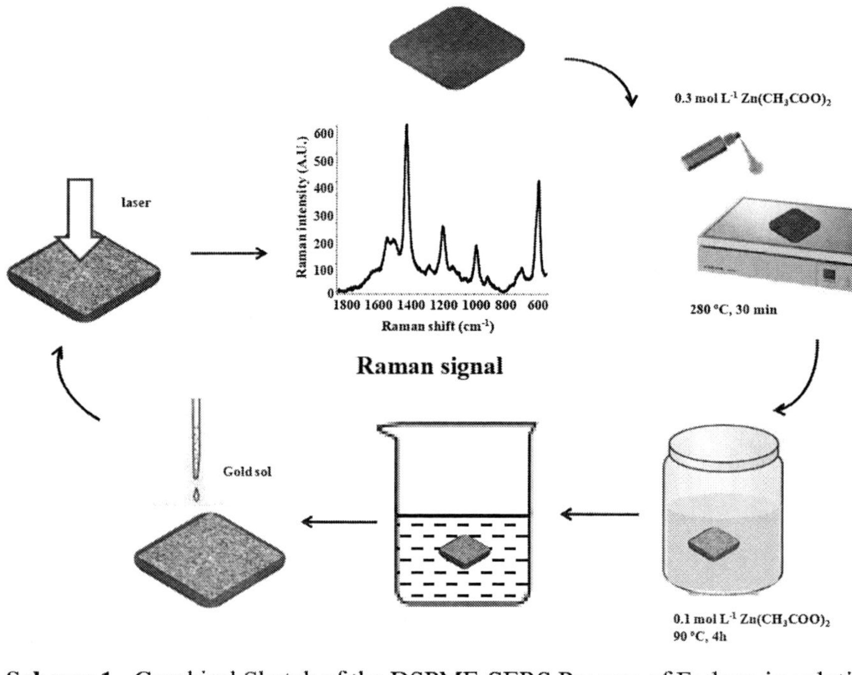


**Scheme 1.** Graphical Sketch of the DSPME-SERS Process of Ferbam in solution

### *2.5. Characterization Techniques and SERS Detection*

Scanning electron microscopy (SEM) images were obtained from a FEI Magellan 400 microscope (FEI, Oregon, USA). A DXR Raman microscope (Thermo Fisher Scientific, Madison, Wis., USA) with a 785 nm laser and a 50× confocal microscope objective was used in this study. OMNIC™

software version 9.7 was used to control the Raman instrument. SERS spectral data were analysed using the TQ Analyst software (version 9.7) from Thermo Scientific.

## 3. Results and Discussion

### *3.1. Preparation and Characterization of Nano ZnO*

A simple and practicable one-step reaction procedure was used to prepare nano ZnO for SPME. As shown in Figure 1a, the surface contains lots of rod-like ZnO nanparticles and form a rough substrate. This substrate is convenient for SERS measurements since the SERS-active substrate (AuNPs) could be well-dispersed on it. Figure 1b reported the typical Raman spectrum for ferbam (50 ppm). In fact, we can clearly observed the ferbam's characteristic intensity increase of the $\rho(CH_3)$, $\delta(CH_3)$ and $\nu(CN)$ at 1517.65 $cm^{-1}$, $\delta(CH_3)$ and $\nu(CN)$ at 1380.62 $cm^{-1}$, and $\rho(CH_3)$ and $\nu(CN)$1149.48 $cm^{-1}$, respectively. The highest peak of ferbam was observed at 1380 $cm^{-1}$ which was associated with CN stretching. As a result, it was chosen as the main characteristic peak for the following experiments.

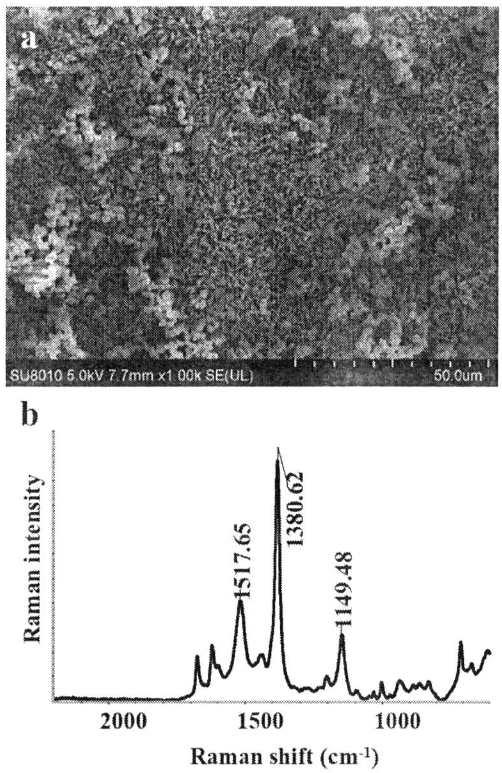


**Figure 1.** (a) SEM image of ZnO/Al substates and (b) typical Raman spectrum of Ferbam.

### *3.2. Analytical Performance of the SPME-SERS for Ferbam*

No strong background signal was detected using SERS substrate (Figure 2). This is very beneficial for SERS detection because all detected peaks are from the analyte, which increases the sensitivity and makes the analysis easier and clearer. The effect of the extraction time was the most important parameter for the analytical Method. SERS signals of ferbam appeared after 5 min of extraction, which means that this method can greatly reduce the detection time, making this method of great significance for fast detection. The SERS signal intensity of the ferbam increased when the extraction time was increased to 60 min and then remained constant or slightly decreased, even when a longer extraction time was selected.

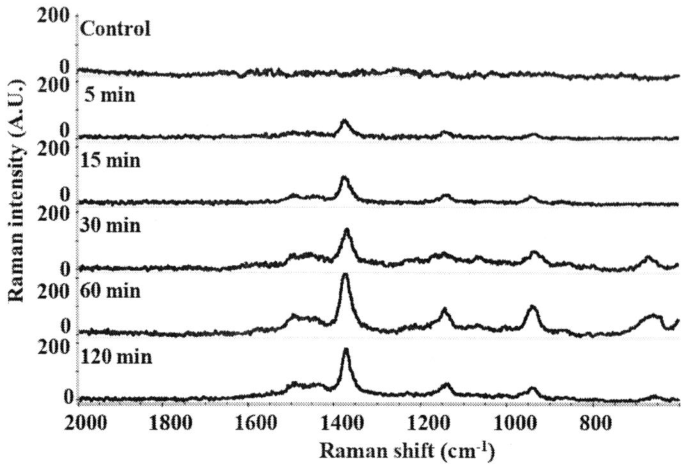


**Figure 2.** Raman spectra of ferbam (1 ppm) with different extraction time

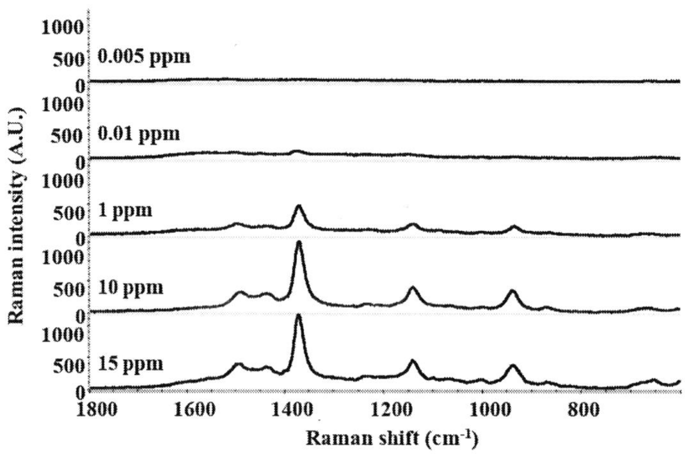


**Figure 3.** Raman spectra with different concentration

The lowest detectable concentration is one of the most important detection index for a new analyzing method. According to Figure 3, the lowest detectable concentration of ferbam was 0.01 ppm, which indicated that we were able to detect ferbam as low as 0.01 ppm for ferbam. The lowest detectable concentration at 0.01 ppm in a solution sample is comparable to the magnetic solid phase extraction with gas chromatography-flame photometric detection (GC-FPD) and high-performance liquid chromatography with diode array detection (HPLC-DAD).[ 15, 16]

## 4. Conclusion

In summary, we developed a simple and sensitive method combing SPME with SERS to analyze ferbam, a widely used in fruit. Rod-like ZnO nanomaterials were synthesized and used as the substrate for SPME. Coupled with SERS, the method was successfully applied in the investigation of ferbam in solution. As a result of the significant enrichment superiority of the nano ZnO toward ferbam and the characteristics of the ferbam SERS spectra, ferbam could be detected in solution at the lowest detectable concentration of 0.01 ppm, and the results were closely consistent with other reported methods. The characteristics of simpler operation, shorter analysis time, and less sample consumption make it a promising approach for the fast analysis.

**Acknowledgments**
This work was supported by the Jiangxi provincial key research and development plan (20161BBG70074) and Science and technology research project of Jiangxi Provincial Education Department (GJJ170281).

**References**
[1] Qin X, Xiaoyu G, Li X, Ye Y, Yiping W, Ying W and Haifeng Y 2016 Sens. Actuators B Chem. 241 1008-13
[2] Juhui K, Chankil L and Jaebum C 2015 J. Hazard. Mater. 285 11-7
[3] Chunying L, Yiqun H, Keqiang L, Barbara A R and Yuxia F 2016 Food Control 65 99-105
[4] Seth L F, Alexandra B F, Francis L L, Soon H O, Elizabeth A D and Jeremy D D 2017 Analyst 142 4456-67
[5] Camilla F S, Keyller B B and Clebio S N 2017 Analyst 143 141-9
[6] Chaojun H, Xiangmeng G, Yumin Y, Yanfu X, Fengyang J and Wenbo G 2016 Anal. Methods 8 4727-35
[7] Chin-Hsiung W, Hung S, Jo-Han C, Min-Zong H, Hsing-Jung L and Jentaie S 2018 Anal.Chim. Acta 1021 60-8
[8] Luís B, Geoffroy D, Tuukka R, Jevgeni P, Kari H, Lydia H, Enna H, Matti J, Markku K and Marja L R 2018 ATMOS MEAS TECH 11 1-17
[9] Mohammad K, Forouzan A, Hamid Z, Omid S and Ezzatollah N 2013 B. Environ. Contam. Tox 90 291-5
[10] Changhua X, Guosheng C, Zhenhai X, Yuxia F, Xichang W and Yuan L 2016 Trac-trend Anal. Chem. 80 12-29
[11] Soheila S, Hassan S and Yaghoub A 2012 J. Chromatrgr. A. 1219 61-5
[12] Corinne C, Olivier D and Martine P 2002 Spectrochim. Acta B 57 173-87
[13] Zhuo D, Xuexu C, Yiru W, Enhua F, Zhigang Z and Xi C 2015 Anal. Chem. 87 633-40
[14] G. Frens 1973 Nature 241 20-2
[15] Peng W, Mai L, Donghui L, Jing Z, Xueke L, Fang W, Zhiqiang Z and Peng W 2018 J. Chromatogr. A 1535 9-16
[16] Chunxia W, Liu N, Qiuhua W, Chun W and Wang Z 2010 Anal. Chim. Acta 679 56

# Development and Performance Evaluation of a New Compound Humidity Control Building Material

**Yuehua Shen**[1,2]

[1] Department of Chemistry, Fudan University, Shanghai, P.R.China
[2] Shanghai Hengyuan Macromolecular Materials Co., Ltd., Rm. 304B, Huanan Building, No.1988, Dongfang Rd., Shanghai, P.R.China
Email: sandra@shhyco.com

**Abstract.** A power-free "precise" humidity control building material is made of porous inorganic minerals as carriers to be coated with the liquid of the humidity control composition, which is used to regulate humidity in indoor environment. The selected mineral substrates are made of inorganic materials such as fibers, cement, quartz sand and expanded vermiculite. Porous vermiculite is the main adsorption carrier. In order to increase the speed of moisture absorption and desorption, mineral substrates with thickness of 5mm and weight of 7kg/m2 are selected. According to the actual conditions of museums all over the country, the design requirement of suitable humidity-regulating board should involve the following points: it does not need to depend on the maintenance of power system, and can presuppose "intelligent" humidity-regulating board with "precise humidity". It has the characteristics of large humidity-regulating capacity, rapid moisture absorption and desorption, high cost performance and convenient construction.

## 1. Introduction

According to the regulation of indoor environmental humidity, there are two kinds of active and passive energy consumption. "Active" regulation, also known as "active" method, uses energy-consuming power systems to humidify and dehumidify; passive regulation, also known as non-dynamic humidity regulation, regulates humidity in relatively closed territory by material's own moisture adsorption and desorption capacity, which is an ecological regulation mode [1].

Japan has developed a kind of non-power humidity-regulating building materials for indoor use very early, such as NHML humidity-regulating board produced by JIC Company of Japan, which is a special calcium silicate crystallization by hydrothermal reaction of lime, silicate raw materials and water, and then added some auxiliary materials such as reinforcing fibers to the crystallization, finally compacted and formed after drying. This kind of humidity-regulating building material requires strict waterproof and airproof treatment of the space, and needs "maintenance" before use so that the indoor humidity can be adjusted to the target humidity range by constant temperature and humidity power system. The humidity in the space would be in a stable state only when the corresponding "balance" is achieved.

The important application area of humidity-regulating building materials is the heritage protection industry, forming an appropriate space for the protection of valuable cultural heritage materials. It not only has a stable humidity range, but also has a special function space for the protection of cultural relics such as fire prevention, inferior (anti-aging), mildew prevention, insect prevention and so on.

For the collection industry in China, the temperature and humidity control of the warehouse has always been a problem. Especially in some economically backward areas, the warehouse does not have a constant temperature and humidity system installed, even if there is, because of the cost problem, it can not achieve 24-hour power system to maintain the warehouse temperature and humidity, forcing some warehouses to become "concentrated destruction of cultural relics" place. Therefore, the development and application of high-efficiency and energy-saving composite humidity-regulating building materials has important application value and social significance for energy saving and improving the quality of goods preservation.

## 2. Experiment

### *2.1 Raw Material*

The porous mineral substrates selected here are made of inorganic materials such as fibers, cement, quartz sand and expanded vermiculite. The fibre reinforced calcium silicate plate products are manufactured by belt slurry conveying drum pressing, high temperature pressure curing, drying and grinding.

The mineral substrates contain expanded vermiculite, which is a group of hydrated trioctahedral iron-aluminium-magnesium flake or mica silicate minerals, consisting of lattice layered minerals such as chlorite, hydromica and clay minerals [2]. Vermiculite has higher specific surface area, negative surface charge, higher surface porosity, and more developed pores. Vermiculite is chemically inert, non-toxic and does not release toxic gases during combustion. So it is very suitable for building materials.

In order to increase the speed of moisture absorption and desorption, mineral substrates choose "thin plate" instead of "thick plate" to increase the surface area of adsorption. The thickness of JIC in Japan is 12 mm and 20 mm, while the thickness of mineral substrates selected in this study is 5 mm and the weight is 7 kg/m2.

In addition to the mineral substrate itself has a certain moisture absorption and desorption performance, it will also be coated with the composite solution of precise humidity, forming a new type of composite humidity control building materials.

The composition solution is made up of at least one mixture of polyols, macromolecules and water. Since RH40~60% is the basic requirement of the cultural relics warehouse, this research work is to coat the composite solution of targeted humidity range on the mineral substrates, not only can the "precise humidity" be entrusted to the ordinary mineral substrates with certain humidity-regulating function, but also greatly improve the moisture absorption and desorption capacity of the original substrates.

### *2.2 Preparation of Compound Humidity Control Building Materials*

1)Select 100*100mm's mineral substrate. In order to eliminate the error of water content of mineral substrates, the samples were dried in oven at 120°C for 2 hours.

2)The RH 55% solution was prepared. After all dissolution, it was coated on the mineral substrates according to 4g, 6g, 8g, 10g and 12g quantitatively. Two sides were coated in turn. Because the viscosity of the hot liquid is low, it can quickly penetrate into the porous structure of vermiculite on the mineral substrate.

3) According to the standard JC/T 2002-2009 "Test Method for Moisture Absorption and Desorption Performance of Building Materials"[3], the mineral substrates with different coatings will be prepared. After 24-hour balanced in RH 75%, then 24-hour balanced in RH 50%, the weight change will be recorded.

4) Choose the right amount of coating.

The specific data are shown in Table 1.

**Table 1.** Moisture absorption and desorption properties of different coating amounts on 100*100mm substrate

| Coating amount(g) | Weight of coated sheet (g) | Weight of balanced in RH75% (g) | Moisture absorption in RH75% (g) | Weight of balanced in RH50% (g) | Moisture desorption in RH50% (g) |
|---|---|---|---|---|---|
| 4g | 72.04 | 73.27 | 1.23 | 71.95 | 1.32 |
| 6g | 73.76 | 75.61 | 1.85 | 72.94 | 2.67 |
| 8g | 76.19 | 78.71 | 2.52 | 75.89 | 2.82 |
| 10g | 77.89 | 81.02 | 3.13 | 77.53 | 3.49 |
| 12g | 79.54 | 83.4 | 3.86 | 78.15 | 5.25 |

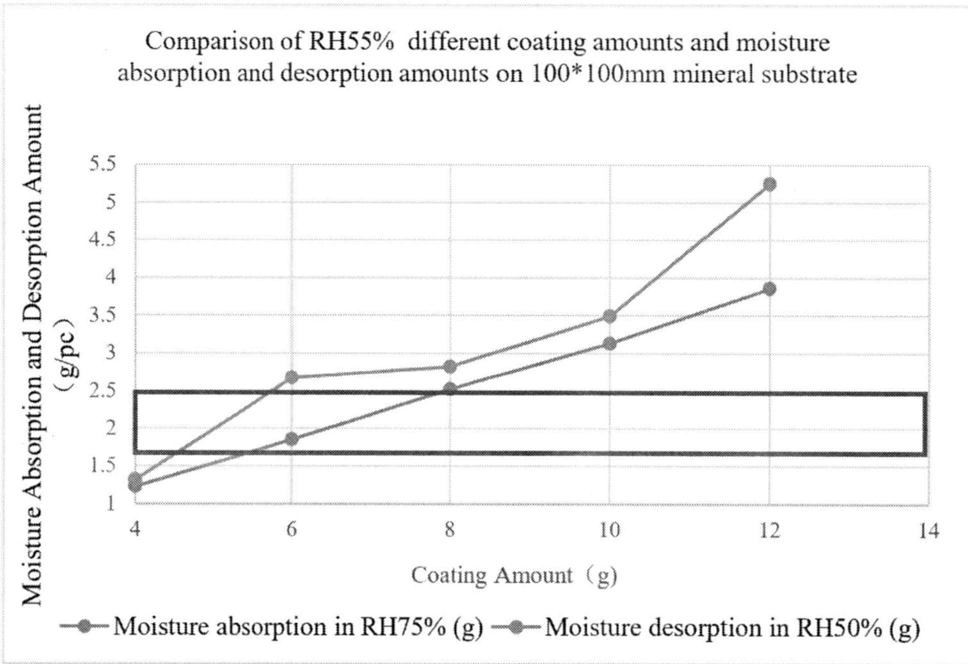


**Figure 1.** Comparison of RH55% different coating amounts and moisture absorption and desorption amounts on 100*100mm mineral substrates

Combined with Table 1 and Figure 1, we can see:

1) Because the coating solution is the formulation of RH 55%, the moisture absorption increment after moisture absorption equilibrium under RH 75% condition is less than the total moisture release after moisture release equilibrium from RH 75% to RH 50%.

2) With the increase of coating amount, the moisture absorption and moisture release amount also increased.

3) Considering that the market demand is a suitable cost-effective product, and the cost is also the key consideration factor of this research work. So the improvement target is 2-3 times of the existing standard, and the corresponding coating quantity of the points in the red box of Fig. 1 can be used.

## 3. Performance Evaluation

Usually, the evaluation of "humidity-regulating materials" includes two parts: humidity-regulating capacity (moisture absorption and desorption capacity) and humidity-regulating speed (moisture absorption and desorption speed). "Precise humidity" is added in this research as a new evaluation index.

Comparing the samples of JIC from Japan, the samples of this study (named "humigic") and the national standard moisture absorption and desorption standards, we will further evaluate the performance of humidity control board.

Taking 250*250mm size as test sample:

Sample of "humigic": Humidity control plate coated with RH 50% composite solution does not need maintenance;

Sample of JIC from Japan should be maintained in RH50% for 24 hours.

The specific comparison is shown in Table 2.

**Table 2.** Comparative data of JIC, humigic and GB

| Item | JIC | Humigic | GB |
|---|---|---|---|
| Thickness(mm) | 12 | 5 | ≥5 |
| Size(mm) | 251*251 | 250*250 | 250*250 |
| Area($m^2$) | 0.063 | 0.0625 | 0.0625 |
| RH50% Maintenance | Need | No need | Need |
| Weight of balanced in RH50%(g) | 457.7 | 487.5 | N/A |
| Moisture Absorption after 3h (g) | 3.29 | 6.67 | 1.56 |
| Moisture Absorption after 6h (g) | 4.95 | 9.33 | 2.19 |
| Moisture Absorption after 12h (g) | 7.08 | 15.79 | 3.13 |
| Moisture Desorption after 3h (g) | 3.08 | 6.4 | 1.41 |
| Moisture Desorption after 6h (g) | 4.02 | 8.22 | 1.75 |
| Moisture Desorption after 12h (g) | 4.94 | 12.6 | 2.19 |
| Moisture absorption per unit area after 3h($g/m^2$) | 52.2 | 106.7 | 25 |
| Moisture Desorption per unit area after 3h($g/m^2$) | 48.9 | 102.4 | 22.5 |
| Desorption/Absorption after 3h(%) | 93.70% | 95.97% | 90% |
| Moisture absorption per unit area after 6h($g/m^2$) | 78.6 | 149.3 | 35 |
| Moisture Desorption per unit area after 6h($g/m^2$) | 63.8 | 131.5 | 28 |
| Desorption/Absorption after 6h(%) | 81.10% | 88.10% | 80% |
| Moisture absorption per unit area after 12h($g/m^2$) | 112.4 | 252.7 | 50 |
| Moisture Desorption per unit area after 12h($g/m^2$) | 78.4 | 201.6 | 35 |
| Desorption/Absorption after 12h(%) | 69.80% | 80% | 70% |

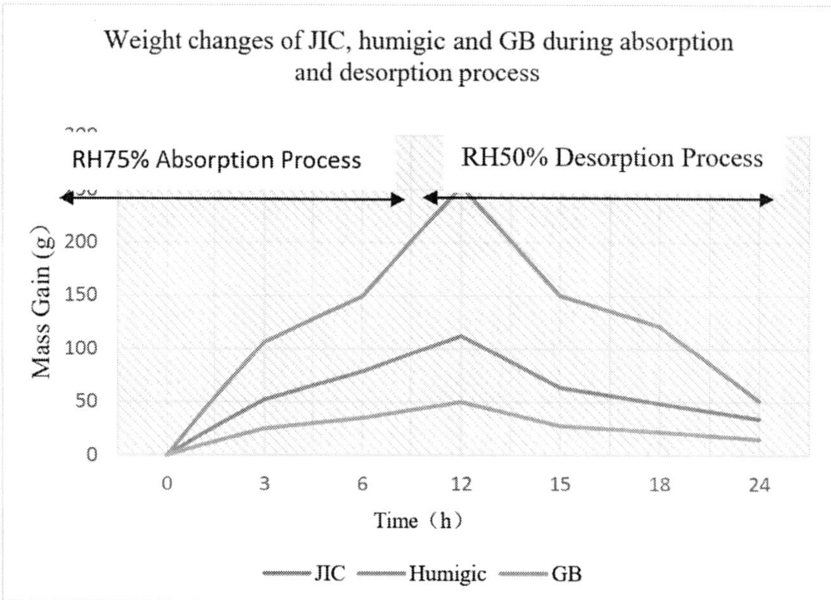


**Figure 2.** Weight changes of JIC, humigic and GB during absorption and desorption process

Combining Table 2 and Figure 2, we can see that the Humigic in this study is superior to JIC and national standard data in terms of moisture absorption and desorption capacity and speed.

In addition, “Humigic” does not need to be maintained. It can be called "intelligent" humidity control building materials. Compared with ordinary humidity control building materials, it is more efficient and convenient to use.

In order to meet the application requirements of building materials, besides the "humidity regulating performance", the "humigic" also needs to meet other relevant standards, including non-flammability, formaldehyde content, TVOC content, safety, screw holding force and so on. Therefore, after testing and confirmation, the enterprise standard (Q31/0230000039C006-2018) has been formulated to make "humigic" humidity-controlling building materials become standardized products to the market.

## 4. Conclusion

A new type of compound “precise” humidity-controlling building material is made of porous inorganic minerals as carriers to be coated with the liquid of "precise RH formula" composition in the thermal solution state. When the solution is cooled to form supersaturated state, the crystals will evenly form in the porous structure of inorganic minerals, thus increasing the humidity capacity and having "precise humidity-regulating" property.

It can be seen from the standard [4] and related test methods that: (1) There is no concept of "humidity control accuracy" in the standard; (2) The standard level of humidity control capacity and speed of moisture absorption and desorption is low, especially in most commonly used range RH40-60%, so that energy efficiency and application fields of humidity-controlling building materials are limited.

The important application and development of this research is to develop a humidity-regulating board which is generally applicable in the domestic cultural warehouse. It needs to meet the following characteristics:

1) Preset "precise humidity", no need maintenance before use.

2) Installation and construction are simple and convenient to replace.

3) The humidity control capacity is large, at least several times higher than the existing standard. Only 8 hours of power system each day to maintain 24 hours of humidity.

4) Cost of both material and construction is 2~3 times lower than that of Japan JIC

**Reference**

[1] Wu Yi. Study on properties and evaluation methods of humidity-regulating materials [D]. Hunan: College of Civil Engineering, Hunan University,
[2] Wang Jing Liang. Application and Prospect of vermiculite [J]. comprehensive utilization of mineral resources. 1996, 3:18
[3] JC/T 2002-2009. Test method for moisture absorption and desorption properties of building materials[S]. Beijing: China Federation of building materials, 2009:
[4] JC/T 2082-2011. Humidity control function indoor building decoration material [S]. Beijing: China Building Materials Federation, 2011:

# Research on Optimization Design of Local Damping Vibration Reduction

**Simi Tang**

Naval Research Academy, 100161, Beijing
Email: oldcandy@163.com

**Abstract.** Damping technology is widely used in navigation and aerospace. It is an important research how to achieve weight reduction and efficiency enhancement with damping layer. Reasonable local damping design is an effective method. Based on modal analysis, Ansys software is used as the simulation platform to design a local damping laying based on modal vibration mode. In order to verify the designed performance of local damping weight reduction, a beam with fixed support at both ends was analyzed, and the damping design of local damping and global damping was carried out. The simulation results show that the designed local damping can achieve good damping effect under the condition of reducing the weight of the damping layer, and it has reference value for engineering applications.

## 1. Introduction

Damping material is the most commonly used technique in modern vibration and noise reduction engineering, including free damping structure and constrained damping structure. Free damping is applied directly to the surface of the vibrating body, while constrained damping is after coating the damping layer, and then attached a layer of metal structure to control the deformation of the free surface. However, the two principles are the same, that is, when subjected to strain; the damping structure consumes vibration energy by bending deformation or shear deformation. Damping materials are the most commonly used viscoelastic materials, which have been studied by a large number of scholars [1-6].

In order to achieve the best vibration damping performance, it is often chosen to dampen the entire vibrating body. However, in engineering applications, especially aerospace equipment, there are strict limits on the weight of the structure. Therefore, the weight of the damping layer must be strictly limited. In this paper, a local damping design method is proposed. In the different damping coating stages, the maximum deformation of modes in different states are analyzed. The damping performance of undamped beam, fully constrained damped beam and local constrained damping are compared and analyzed.

## 2. Constitutive Equation of Damping Structure

The free damping structure is to directly adhere a layer of material with large damping to the machine parts or structural parts that need to be subjected to vibration and noise reduction treatment. This additional material is called free damping layer, and the structural part is called basic elastic layer. When the structure vibrates, the damping layer vibrates with the structural member, the damping material undergoes alternating tensile and compressive stresses and strains, and the strain lags behind the stress, so that the vibration energy of the structure is lost, then the purpose of vibration reduction is achieved. The longitudinal section of the free-damping composite structure is shown in Figure 1(a).

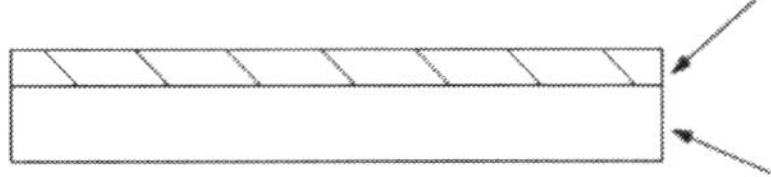


(a) free damping

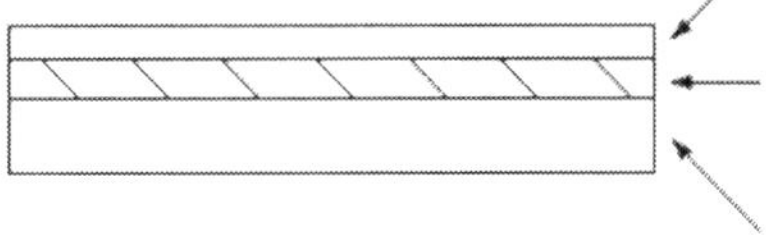


(b) constrained damping

**Figure 1.** The sketch of vibration suppression with damping

The constrained damping structure is to stick a layer of damping material on the basic elastic layer, and then firmly adhere a layer of elastic material on the upper part of the damping layer, which is called a constraining layer, and its longitudinal section is as shown in Fig. 1(b). When the basic elastic layer is subjected to bending vibration to elongate the damping layer, the elongation of the constraining layer is much smaller than the elongation of the damping layer. Conversely, when the damping layer is compressed, the constraining layer in turn prevents the damping layer from compressing, so that the covered elastic layer becomes the constraining layer. Since the elongation and compression of the damping layer are hindered by the constraining layer, the upper and lower surfaces of the damping layer respectively generate different deformations of compression and stretching, so that the damping layer is subjected to shear stress and strain, thereby the vibration energy of the structure is dissipated.

The constraining layer, the viscoelastic layer and the elastic layer are denoted by Nos. 1, 2, and 3 respectively, and their thicknesses are $h_i$ (i = 1, 2, 3), they are represented by *c*, *v*, and *s* respectively.

### 2.1. Elastic Layer Constitutive Relation($-\frac{h_1}{2} \le z \le \frac{h_1}{2}$)

A point is take in the elastic layer, the displacement *z* is the point from the elastic layer, the initial displacement of the surface in the elastic layer is *u* and *v*, and the phase shift is *W*. According to the thin shell theory, the point is in the *x* and *y* directions respectively. The positive strain is:

$$\varepsilon_{sx} = \frac{\partial u}{\partial x} - z\frac{\partial^2 W}{\partial x^2}, \qquad \varepsilon_{sy} = \frac{\partial v}{\partial y} - z\frac{\partial^2 W}{\partial y^2} \tag{1}$$

The shear strain at this point is:

$$\gamma_{sxy} = \frac{\partial u}{\partial y} + \frac{\partial v}{\partial x} - 2z\frac{\partial^2 W}{\partial x \partial y} \tag{2}$$

### 2.2. Constraint Layer Constitutive Relation($\frac{h_1}{2} + h_2 \le z \le \frac{h_1}{2} + h_2 + h_3$)

A point is take in the elastic layer, the displacement *z* is the point from the elastic layer, the initial displacement of the surface in the elastic layer is *u* and *v*, and the phase shift is *W*. According to the thin shell theory, the point is in the *x* and *y* directions respectively. The positive strain is:

$$\varepsilon_{cx} = \frac{\partial u}{\partial x} - z\frac{\partial^2 W}{\partial x^2} + h_2\frac{\partial \alpha}{\partial x}, \qquad \varepsilon_{cy} = \frac{\partial v}{\partial y} - z\frac{\partial^2 W}{\partial y^2} + h_2\frac{\partial \beta}{\partial y} \tag{3}$$

The shear strain at this point is:

$$\gamma_{cxy} = \frac{\partial u}{\partial y} + \frac{\partial v}{\partial x} - 2z\frac{\partial^2 W}{\partial x \partial y} + h_2\frac{\partial \alpha}{\partial x} + h_2\frac{\partial \beta}{\partial y} \tag{4}$$

### 2.3. Constraint Layer Constitutive Relation($\frac{h_1}{2} \le z \le \frac{h_1}{2} + h_2$)

A point is take in the elastic layer, the displacement $z$ is the point from the elastic layer, the initial displacement of the surface in the elastic layer is $u$ and $v$, and the phase shift is $W$. According to the thin shell theory, the point is in the $x$ and $y$ directions respectively. The positive strain is:

$$\varepsilon_{vx} = \frac{\partial u}{\partial x} - z\frac{\partial^2 W}{\partial x^2} + (z - h_1/2)\frac{\partial \alpha}{\partial x}, \quad \varepsilon_{vy} = \frac{\partial v}{\partial y} - z\frac{\partial^2 W}{\partial y^2} + (z - h_1/2)\frac{\partial \beta}{\partial y} \tag{5}$$

The shear strain at this point is:

$$\gamma_{vxy} = \frac{\partial u}{\partial y} + \frac{\partial v}{\partial x} - 2z\frac{\partial^2 W}{\partial x \partial y} + (z - h_1/2)\frac{\partial \alpha}{\partial x} + (z - h_1/2)\frac{\partial \beta}{\partial y} \tag{6}$$

## 3. Local Damping Design Method

In practical applications, the vibration of the structure will be required to be controlled within a certain range, but it does not mean that the damping material must be completely coated on the structural surface. If the largest part of the strain of the structural member can be found, the local application of the damping material at this location may also cause the vibration of the structural member to meet the index requirements. Which parts of the structural member are subjected to damping treatment is the core of the local damping design. The basic design process is as follows:

- Determine the design goal of the vibration reduction and give the target value of the vibration reduction effect;
- Through the modal analysis, find the first-order modal deformation, analyze the position of the maximum deformation;
- Appropriate damping materials and constraining layer materials is selected, and local damping parameters such as thickness and length is set initially;
- Perform modal analysis on the construction with first-order local damping, the first-order modal deformation is obtained again, the maximum deformation position is analyzed;
- The vibration damping effect is analyzed;
- If the damping effect has basically reached the target, the first-order modal deformation is analyzed again to study the application of the local damping material;
- The vibration reduction effect is analyzed again;
- If the design goal is reached, the design is completed;
- If the design goal is not reached, return to step (2) and adjust the damping parameters.

The flow chart of the local damping design program is shown in Figure 2.

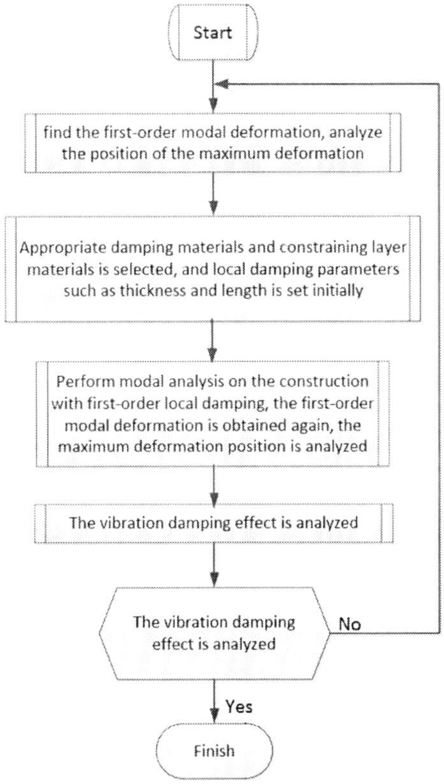


**Figure 2.** The local damping optimization program flow chart

## 4. Beam FEM Example

In order to verify the optimized design method of the previous section, a local constrained damping design is applied to a beam. Firstly, it was analyzed by ANSYS software. The beam was installed with fixed support at both ends. The beam length was 5 meters, width was 0.05 meters, and thickness was 0.005 meters. The beam is stainless steel. The modal frequencies of the first few stages and the corresponding vibration modes are shown in Fig. 3.

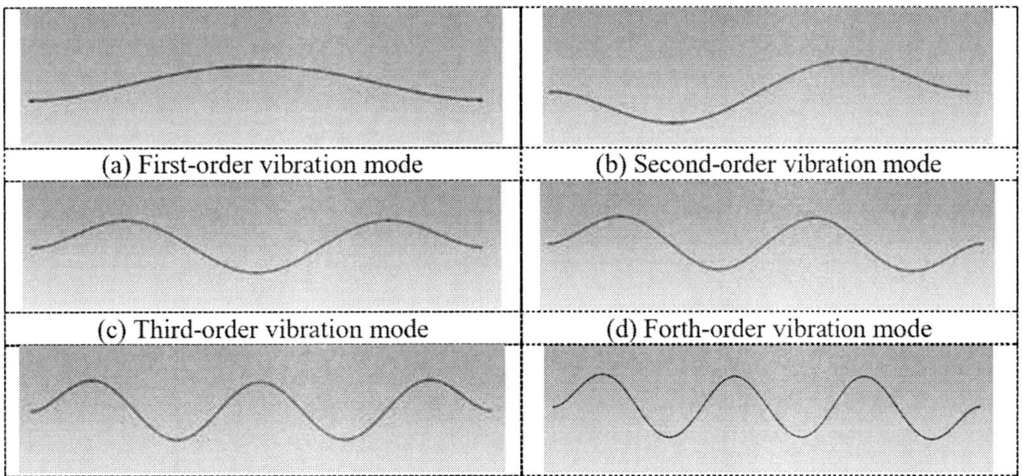

(a) First-order vibration mode
(b) Second-order vibration mode
(c) Third-order vibration mode
(d) Forth-order vibration mode
(e) Fifth-order vibration mode
(f) Sixth-order vibration mode

**Figure 3.** Modal analysis of fixed beam at both ends.

At different modal frequency, the maximum response of structural vibration is in different places. Now the damping design of the beam is carried out separately. The following schemes are available:

### *4.1. Global Constrained Damping Design*

A layer of damping material is applied to the entire beam, and then the aluminum alloy beam is used to restrain the damping material above the damping material, so that the freely damped material layer becomes the constrained damping layer. When the structural layer is subjected to bending vibration and the damping layer is compressed or elongated, the deformation of the constraining layer is much smaller than the deformation of the damping layer, so that shear strain and shear stress are generated inside the damping layer, as shown in figure 4.

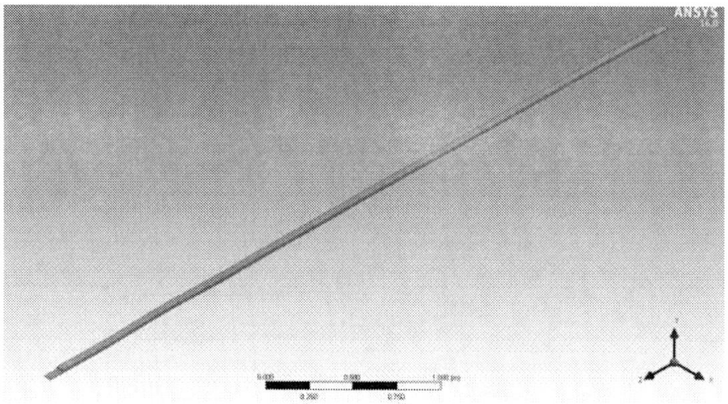


**Figure 4.** Constrained Damping Beam

### *4.2. Locally Constrained Damping Design*

It can be seen from the modal analysis that the violent regions of the modal deformation are distributed at different positions at different modal frequencies. The damping energy is caused by the deformation of the structure. In the place where the structural deformation is relatively small, even if the constrained damping layer is applied, it will not have much effect. If the local damping is performed in a region with a large deformation, it is possible to obtain a better damping effect under the premise of increasing the weight. The local damping layer should cover the area with large vibration mode as much as possible, as shown in Figure 5.

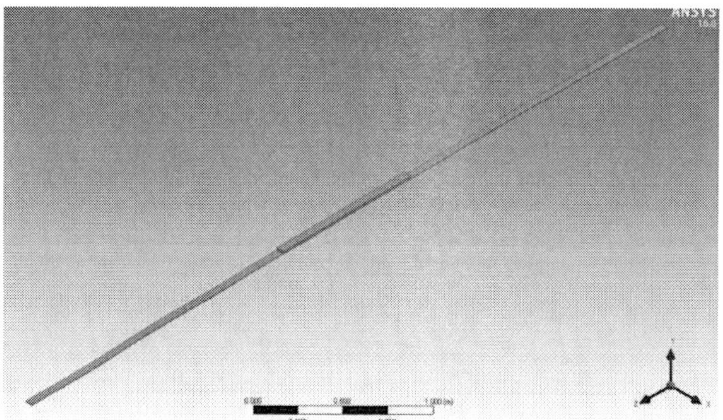


**Figure 5.** Local Constraint Damping

The harmonic response analysis is carried out on the beam, the local constrained damping beam and all the constrained damping beams. A force load is applied at the central point of the beam. The frequency range is 0~200Hz. The frequency response curve of displacement and velocity at the central point of the above three beams are Calculated. The results are as shown in Figure 6 and 7.

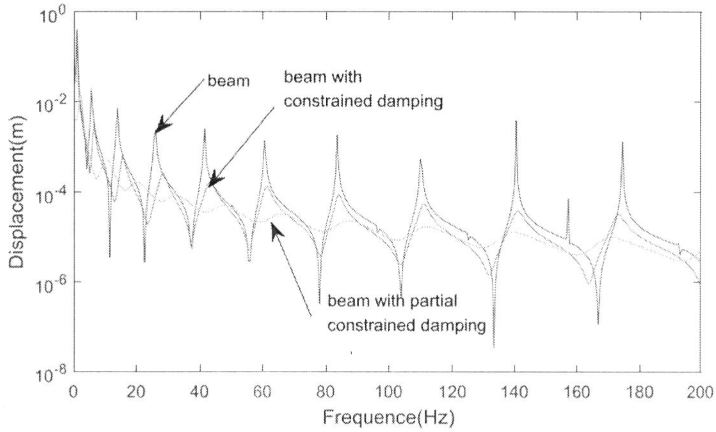


**Figure 6.** Comparison of the central point amplitudes of the three beams

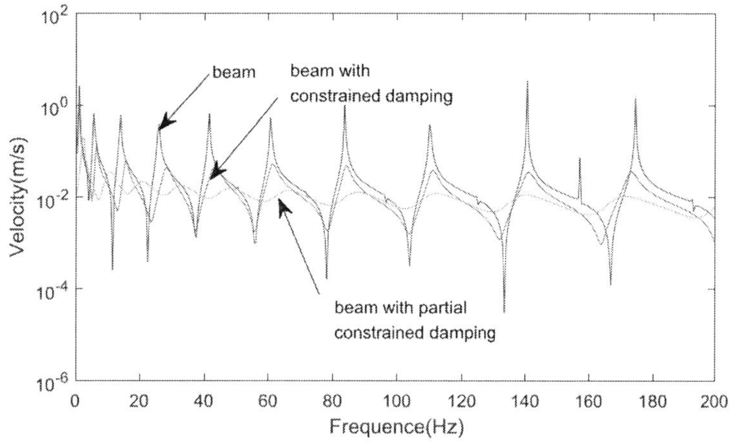


**Figure 7.** Comparison of central point velocity of three beams

## 5. Conclusion

In this paper, Ansys software is used as the simulation platform. Based on the results of modal analysis, the constrained damping layer is laid on the region with large mode deformation to reduce the additional weight. It can be seen from the numerical simulation results that the local damping can greatly reduce the area of the conventional global damping layer and excellent performance of vibration reduction effect when the weight of the constrained damping layer is reduced by 80%. This method has a good significance for the design of structural vibration control.

## References

[1] Soroka W W. Note on the relations between viscous and structural damping coefficients [J]. Journal of the Aeronautical Sciences, 2012, 16(7): 30-45.
[2] V Pradeep, N. Ganesan. Vibration behavior of ACLD treated beams under thermal environment [J]. Journal of Sound and Vibration, 2006, 292(3-5): 1036-1045.
[3] Mohammadi F,Sedaghati R. Linear and nonlinear vibration analysis of sandwich cylindrical shell with constrained viscoelastic core layer [J] . International Journal of Mechanical Sciences, 2012, 54(1): 156-171.

[4] Zheng C. Cai G,Pau S H,et al. Minimizing vibration response of cylindrical shells through layout optimization of passive constrained layer damping treatments[J]. Journal of Sound and Vibration, 2005, 279(3-5): 739-756.
[5] Lepoittevin G,Kress G. Optimization of segmented constrained layer damping with mathematical programming using strain energy analysis and modal data[J]. Materials and Design, 2010, 31(1): 14-24.
[6] Ninoslav Truhar. An efficient algorithm for damper optimization for linear vibrating systems using Lyapunov equation [J]. Journal of Computational and Applied Mathematics, 2004, 172: 169-182.

IOP Conf. Series: Materials Science and Engineering 472 (2019) 012059 doi:10.1088/1757-899X/472/1/012059

# Effect of Hydrogen Nutrition Denitrifying Bacteria on Nanoscale Zero-Valent Iron Settlement

**Qing Xia[1], Ruixue Liu[1], Chenzi Xu[1] and An Yi[1*]**

[1]Innovative Team of Monitoring and Precaution for Cropland Environment, Institute of Agro-environmental Protection, Ministry of Agriculture, Tianjin, 300191, China

*Corresponding author:Simon8601@126.com

**Abstract.** Nanoscale zero-valent iron (nZVI) particles has an extremely small size and can be suspended in groundwater for long periods of time to remove contaminants. However, more and more experimental results show that the migration performance of nZVI particles in groundwater is far lower than expected, and it is difficult to meet the actual needs of in-situ remediation. In order to study the effect of hydrogenotrophic denitrifying bacteria on Nano-iron, the sedimentation test was used as the main research method to discuss the demonstration. The experimental results showed that within the pure water system and in the culture solution system, the sedimentation rate of Nano-iron was decreased after adding bacteria. As the amount of bacteria in the system increases, the sedimentation rate of Nano-iron tends to decrease. When the amount of bacteria reaches a certain amount, it will have a saturating effect on the sedimentation effect of Nano-iron. XRD spectra showed that $Fe_3O_4$ and $Fe_2O_3$ were the main iron oxides in the initial oxidation, while $Fe_2O_3$, $Fe_3O_4$ and $Fe_2O_3$ were obtained in the final time of sedimentation. Amorphous δ-FeOOH (corrosion product of zero-valent iron in pure water) replaces $Fe_2O_3$ and $Fe_3O_4$ as major corrosion products after the addition of bacterial reaction.

## 1. Introduction

The need for safe water has forced developing countries to develop a water treatment method that is innovative and economically valuable, including improved Nano-iron drug delivery and in situ remediation of groundwater [1]. Therefore, the huge potential of Nano-iron in the treatment of groundwater pollution has aroused widespread concern. In order to effectively carry out in-situ groundwater remediation, Nano-iron needs to have a low sedimentation rate and agglomeration level, otherwise the migration ability of Nano-iron in the ground will be greatly limited [2].

The ability of Nano-iron to migrate and settle in a groundwater environment depended on many factors [3], such as particle size, pH, ionic strength, soil composition, groundwater flow rate, etc. Agglomeration was a major factor causing a decrease in the migration properties of Nano-iron. Due to magnetic attraction and van der Waals force, Nano-iron was easily agglomerated, and the particle size and surface potential of the agglomerate made it easy to settle in the water environment, thus seriously affecting its migration ability [4]. In addition, the oxidation products formed by anaerobic corrosion of Nano-iron in water will also greatly reduce its migration properties through adsorption, sedimentation [5].

The existing research focused on the effects of hydrological parameters and water quality conditions on the mobility of Nano-iron, but the influence of indigenous microorganisms in groundwater cannot be ignored. It is of great significance to study the influence of microorganisms on the sedimentation performance of Nano-iron and its mechanism in groundwater [6]. Therefore, this study focused on the sedimentation of Nano-iron under the action of hydrogen autotrophic denitrifying bacteria and attempted to explore the mechanism.

## 2. Materials and Methods

### 2.1. Cultivation of Microorganisms

Alcaligenes eutrophus was purchased from China Center of Industrial Culture Collection (Beijing, China). Medium for the bacterial growth was prepared with 52 mg·$L^{-1}$ of $ZnCl_2$, 190 mg·$L^{-1}$ of $CoCl_2 \cdot 6H_2O$, 100 mg·$L^{-1}$ of $MnSO_4 \cdot 7H_2O$, 24 mg·$L^{-1}$ of $NICl_2 \cdot 6H_2O$, 29 mg·$L^{-1}$ of $CuCl_2 \cdot 2H_2O$, 36 mg·$L^{-1}$ of $Na_2MoO_4 \cdot 2H_2O$ and 30 mg·$L^{-1}$ of $H_3BO_3$, respectively, as described by Shin and Cha [7]. The culture tank contained 5 L concentrated medium and was seeded with 50ml cell suspension which contained 20 mg·$L^{-1}$ of A. eutrophus. The culture tank was purged with argon gas for 20 min following inoculation to remove dissolved oxygen, and incubated in an illumination incubator (250-D, Jiangsu) at 30 °C. The gas tank was filled with 5.0 g commercial iron powder (300 meshes) and 50ml HCl(1M) to produce $H_2$ gas, and then the top cap was screwed in position. According to the consumption of nitrate, the culture solution is updated in time until the absorbance of the bacterial solution reaches 6005 or higher.

### 2.2. Preparation of NZVI particles

The preparation of NZVI used in this experiment was carried out under a flow of Ar gas. In addition, all solvents were also saturated with Ar gas. Based on previous work [8], NZVI particles in this study were synthesized by reduction of ferrous iron solution by $KBH_4$ in ethanol–water solution. After the reaction, the iron nanoparticles were repeatedly washed by deionized water and anhydrous ethanol and then dried under the protection of Ar gas. The average particle size of NZVI was approximately 30 nm.

### 2.3. Batch Experiments

*2.3.1 Feasibility test* Dilute 8 mL of the bacterial solution into the colorimetric tube, and introduced 10 ml of Nano-iron suspension (unmodified, average particle size 30 nm) by vacuum line, and dilute to 25 mL with deionized water. After adjusting the pH of the reaction system to 7.0 with HCl, it was sealed and stored, and the whole process was protected with nitrogen. Based on the preparation method of the above-mentioned iron bacteria synergy system, and appropriately adjusted according to the test requirements (7 ml culture solution was added in the batch test containing the culture solution). Four kinds of reaction systems (Table 1) such as "Nano-iron + bacteria + culture solution", "Nano-iron + bacteria + water", "Nano-iron + culture solution", and "Nano-iron + water" were prepared. Three batches of each batch were repeated to reduce the test error and ensure the accuracy of the test.

**Table 1.** Batch test on the effect of hydrogen autotrophic denitrifying bacteria

| Number | NZVI content (mL ) | Bacteria ( mL ) | Culture medium ( mL ) | $H_2O$ ( mL ) |
|---|---|---|---|---|
| A1 | 10 | 8 | 15 | 0 |
| A2 | 10 | 8 | 0 | 15 |
| A3 | 10 | 0 | 25 | 0 |
| A4 | 10 | 0 | 0 | 25 |

*2.3.2 Effect of iron bacteria content ratio* In the case of the coexistence of Nano-iron and hydrogen autotrophic bacteria, the experiment in the previous section pointed out those hydrogen autotrophic bacteria will interfere with the sedimentation rate of Nano-iron, and the effect of bacteria on Nano-iron may be multi-faceted. We focused on whether the effect of bacteria on Nano-iron in different proportions of iron bacteria (Table 2) will have different results, so as to observe the sedimentation law of Nano-iron under different bacterial concentrations.

**Table 2.** Batch test on the effect of hydrogen autotrophic denitrifying bacteria under different dosages

| Number | NZVI content (mL ) | bacteria (mL) | $H_2O$ ( mL ) |
|---|---|---|---|
| B1 | 10 | 0 | 15 |
| B2 | 10 | 4 | 11 |
| B3 | 10 | 8 | 7 |
| B4 | 10 | 12 | 2 |

## 3. Results and Discussion

### *3.1. Feasibility Test*

As shown in Fig. 1, in the pure water system and the culture solution system, the sedimentation rate of Nano-iron decreased after the addition of bacteria. In the pure water system, when sterilized, the concentration of Nano-iron was reduced by 70% within 40 minutes, and after the addition of bacteria, the concentration of Nano-iron was only 60% in the same time; In the culture system, when sterilized, the concentration of Nano-iron decreased by 65% within 40 minutes, and after the addition of bacteria, the concentration of Nano-iron decreased by only 58% in the same time. At the same time, the results of the t-test showed that there was a significant difference between the sedimentation data of the sterile group and the additive group ($p<0.05$). This indicated that the presence of bacteria can significantly inhibit the sedimentation rate of Nano-iron in water (pure water system or culture system).

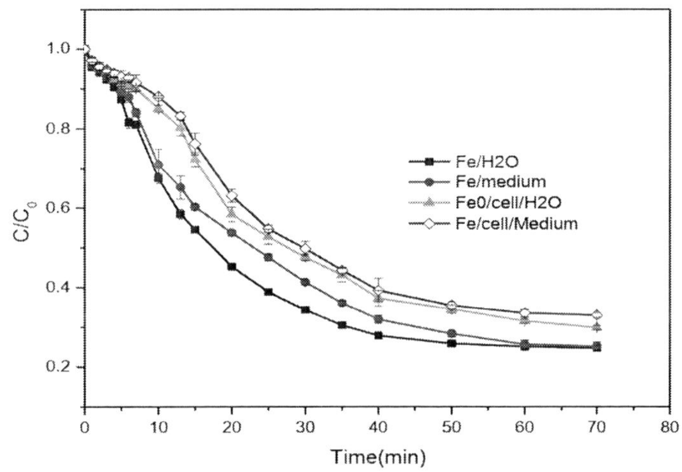


**Figure 1.** Nano iron concentration curve under different treatments

On the other hand, the survival of bacteria depended on the supply of nutrients. Therefore, we have also conducted corresponding research on the interference effect of the culture solution. The results showed that the presence of the culture solution had a significant effect on the sedimentation of the Nano-iron during sterility, and the presence of the culture solution significantly inhibited the sedimentation of the Nano-iron in 40 minutes. However, after the addition of bacteria, the interference effect of the culture solution was reduced. In the bacteria system, the difference of sedimentation rate of Nano-iron in the pure water system and the culture solution system was reduced from 0.057 $min^{-1}$ to 0.039 $min^{-1}$.

In addition, in the 40 minutes of the test of each group of Nano-iron sedimentation rate showed a slow decline trend. However, there is a phenomenon here that is worth noting. In the water group, the sedimentation rate of Nano-iron was greatly different in the first 40 minutes between the pure water system and the culture solution system, but the difference between the two groups were almost completely eliminated by 70 minutes. It showed that the culture solution influence the sedimentation dynamics instead of affecting the final degree of settlement.

### *3.2. Effect of Iron Bacteria Content Ratio*

As shown in Fig. 2, as the amount of bacteria in the system increased, the sedimentation rate of Nano-iron tended to decrease. In a pure water system, the concentration of Nano-iron was reduced by 70% within 40 minutes when sterile. In the case of 4 ml of bacteria, the concentration of Nano-iron decreased by 66% in 40 minutes. The results of t-test showed that there was no significant difference between the sedimentation data of the none bactericidal and the 4 mL addition group ($p>0.05$). This means that when the amount of bacteria in the system was small, the concentration change of Nano-iron with time was not much different from the blank test, that is, it has no effect on the sedimentation rate of Nano-iron. In the presence of a small amount of bacteria, the effect on the settlement of Nano-iron was not significant. The reason may be that Nano-iron has a toxic effect on bacteria to a certain extent. When the amount of bacteria was too small, the activity of bacteria in the system was affected by Nano-iron. In the past decade, some scholars [9-11] have studied the effects of Nano-iron particles on organisms and the environment, and there are some indications of the toxic effects of Nano-iron.

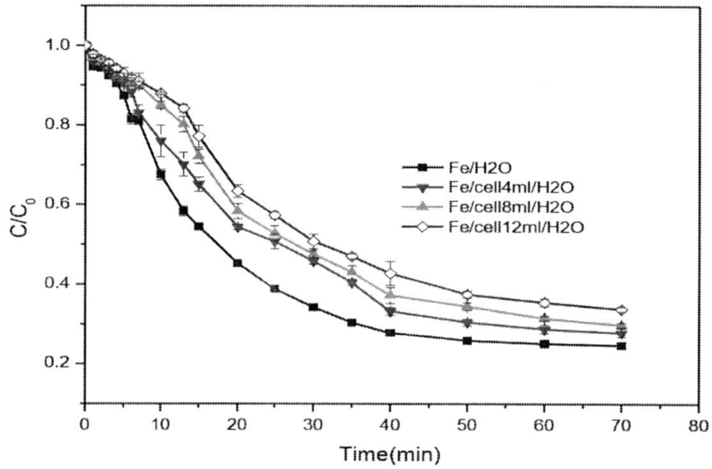


**Figure 2.** Concentration change of Nano-iron in different time periods under different bacterial amount

However, after adding 8 mL of bacteria, the concentration of Nano-iron decreased by 58% within the same time. At the same time, the results of the t-test showed that there was a significant difference ($p<0.05$) between the nanoscale iron deposition data of the sterile group and the 8 mL and 12 mL addition groups. It showed that the presence of bacteria can significantly inhibit the sedimentation rate of the Nano-iron in the water system (pure water system).

Under the condition of 8ml and 12ml of bacteria, the concentration ratio of Nano-iron at the same time tends to be consistent. According to SPSS data analysis, the data of the two groups of batches test results is $p>0.05$, and the difference between the two groups is not significant. It means that when the amount of bacteria reaches a certain amount, it will have a saturating effect on the sedimentation effect of Nano-iron. Even if the amount of bacteria is increased, the sedimentation rate of Nano-iron cannot be continuously increased.

### *3.3. Analysis of Agglomeration Behavior of Nano-Iron*

In the pure water system, the state of presence of Nano-iron in water at the initial time is a full spherical shape, and the Nano-iron particles are arranged in clusters, and the overall surface morphology is uniform and densely arranged (Figure 3a). Over time, Nano-iron will agglomerate, which was quite different from the initial morphology of Nano-iron (Fig. 3b). After the addition of bacteria, the Nano-iron dispersed in the pure water in a disperse state. From the TEM image, it can be seen that the Nano-iron is full spherical; the Nano-iron particles have no adhesion clusters, and the sedimentation rate of the Nano-iron changes. Compared with the aseptic state, the degree of oxidation of the surface of the Nano-iron was reduced, which was characterized by relatively less gray-white oxide on the surface of the Nano-iron in the transmission electron micrograph (Fig. 3d), and the degree of agglomeration was lighter.

We used XRD technology to analyze the oxidation products of Nano-iron. The results showed that in the pure water system, when it was aseptic (Fig. 4a), the absorption peak at 36° characterized the existence of Fe0, and the absorption peak at 43.5°, characterized the existence of $Fe_2O_3$. The absorption peak at 43° and 62° characterized the presence of $Fe_3O_4$. After adding bacteria (Fig. 4b), the absorption peak at 43° and 62° characterized the existence of $Fe_3O_4$, the absorption peak at 36°, characterized the existence of Fe0, the absorption peak at 43.5°characterized the existence of $Fe_2O_3$ and FeOOH.

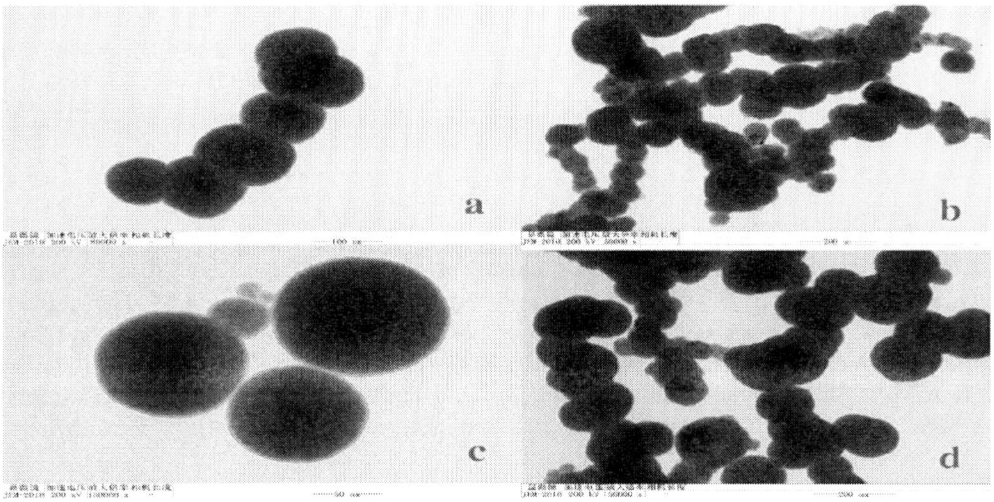


**Figure 3.** Transmission electron micrograph of Nano-iron

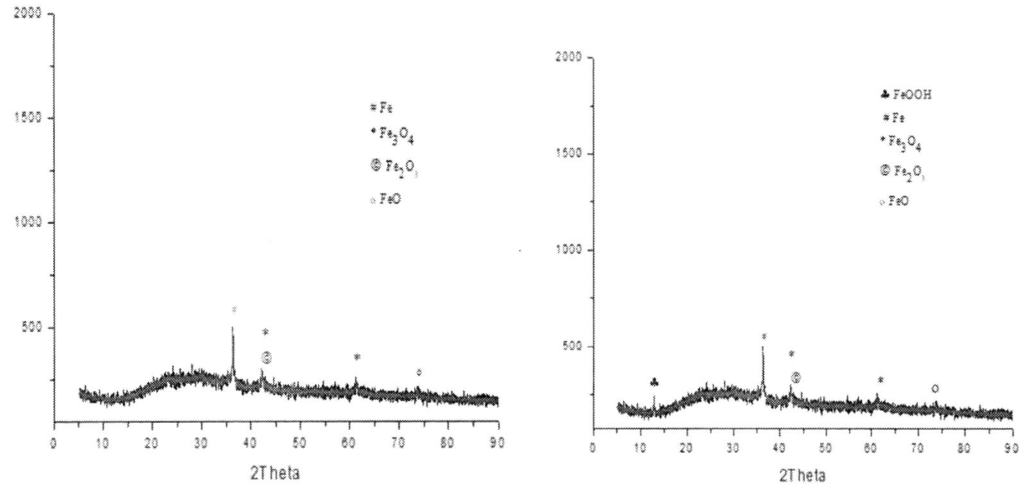


(a) (b)

**Figure 4.** XRD characterization of Nano-iron

## 4. Conclusions

Based on the results obtained, the following conclusions can be drawn:

1. Within the pure water system and in the culture solution system, the sedimentation rate of Nano-iron was decreased after adding bacteria. As the amount of bacteria in the system increases, the sedimentation rate of Nano-iron tends to decrease. At the same time, the results of the t-test showed that there was a significant difference ($p<0.05$) between the nanoscale iron deposition data of the sterile group and the 8 mL and 12 mL addition groups. It showed that the presence of bacteria can significantly inhibit the sedimentation rate of the Nano-iron in the water system (pure water system).

2. When the amount of bacteria reaches a certain amount, it will have a saturating effect on the sedimentation effect of Nano-iron. Even if the amount of bacteria is increased, the sedimentation rate of Nano-iron cannot be continuously increased.

3. XRD spectra showed that $Fe_3O_4$ and $Fe_2O_3$ were the main iron oxides in the initial oxidation, while $Fe_2O_3$, $Fe_3O_4$ and $Fe_2O_3$ were obtained in the final time of sedimentation. Amorphous δ-FeOOH (corrosion product of zero-valent iron in pure water) replaces $Fe_2O_3$ and $Fe_3O_4$ as major corrosion products after the addition of bacterial reaction.

## References

[1] Bystrzejewska-Piotrowska Gm Golimowski J, Urban PL. Nanoparticles: Their potential toxicity, waste and environmental management. Waste Management, 2009, 29:2587–2595.
[2] Kanel SR, Nepal D, Manning B, Choi H. Transport of surface-modified iron nanoparticle in porous media and application to arsenic (III) remediation. J Nanopart Res, 2007, 9 (5):725-735.
[3] Wei C, Li X. Surface coating with Ca (OH)(2) for improvement of the transport of nanoscale zero-valent iron (nZVI) in porous media. Water Sci Technol, 2013, 68(10):2287-2293.
[4] Phenrat T, Saleh N, Sirk K, Tilton RD, Lowry GV. Aggregation and sedimentation of aqueous nanoscale zerovalent iron dispersions. Environ Sci Technol, 2007, 41:284-290.
[5] Fajardo C, Saccà ML, Martinez-Gomariz M, Costa G, Nande M, Martin M. Transcriptional and proteomic stress responses of a soil bacterium Bacillus cereus to nanosized zero-valent iron (nZVI) particles. Chemosphere, 2013, 93(6):1077-1083
[6] Yi A, Li T, Jin Z, et al. Nitrate degradation and kinetic analysis of the denitrification system composed of iron nanoparticles and hydrogenotrophic bacteria [J]. Desalination, 2010, 252(1):71-74.
[7] Shin K, Cha DK. Microbial reduction of nitrate in the presence of nanoscale zero-valent iron. Chemosphere, 2008, 72:257–262.
[8] W. Wang, Z.H. Jin, T.L. Li, H. Zhang, S. Gao, Chemosphere 65 (2006) 1396–1404.
[9] El-Temsah YS, Joner EJ. Impact of Fe and Ag nanoparticles on seed germination and differences in bioavailability during exposure in aqueous suspension and soil. Environ toxicol, 2012, 27(1):42-49
[10] Hjorth R, Hansen SF, Jacobs M, Tickner J, Ellenbecker M. The applicability of chemical alternatives assessment for engineered nanomaterials. Integr Environ Assess Manag, 2017, 13(1):177-187
[11] Ševců A, Eltemsah YS, Joner EJ, Černík M. Oxidative stress induced in microorganisms by zero-valent iron nanoparticles. Microbes Environ, 2011, 26 (4) :271

IOP Conf. Series: Materials Science and Engineering 472 (2019) 012060 doi:10.1088/1757-899X/472/1/012060

# Field Electron Emission from the Tip of a Graphite Fiber

**Jin Zhang[1], Ergang Chen[1], Rongli Wang[2], Yiming Xiao[1], Fanquan Kong[1], and Hai Wang[2]**

[1]Department of Physics, Yunnan University, Kunming 650091, P.R.China
[2] School of Physics, Kunming University, Kunming 650214, P. R. China
Email: zhangjin96@aliyun.com

**Abstract.** Usually carbon wires cannot be used for obtaining remarkable field electron emission because the difficulty of sharpening its tip with a sufficient small radius of curvature to fit the electrical field. However, a large number of small fragments are formed on tips of graphite fibers after a discharge treatment with 10 kV DC high voltage on graphite fibers. These fragments observed by SEM and HRFEM are spheres, flakes and needle-like nanostructures with the size around 50 nm. In this case, each of these spheres and flakes can be an emission center in field emissions and some detectable emission currents can be obtained with a low applied voltage. Experimental results show that a single graphite fiber of 20μm diameter can generate near 0.2 mA emission current with 4 kV applied voltage and the stability have been measured to be less than 10 %. It is expected that graphite fibers as a novel cold cathode material will provide a wide range of applications such as in flat panel displays, high-power and high-brightness vacuum electronic devices, and vacuum microelectronic devices.

## 1. Introduction

For all kinds of vacuum electron apparatus and instruments, obtaining free electrons is a basic condition. Thermal emission is one of the most conventional methods to generate free electrons. These electrons escape by heating from the surface of emitters. Although the way is simple, heating emitters need consuming energy with luminescence of thermal emitters. It usually affects properties of apparatus or instruments. Field emission (FE), a kind of electron emission without heating and luminescence, extracts electrons from the surface of emitters into vacuum under the influence of an applied electric field. Field emission is therefore a better way of acquiring free electrons. There have been moves to use cold cathodes to replace all hot cathodes and more importantly to explore new applications of cold cathodes in areas such as large-area electronics, where hot cathodes are unsuitable [1].

Cold cathodes utilize the local field enhancement at the top of each emitter to lower the threshold voltage enabling field emission. However, the fabrication facility required and the fabrication processes of field emission cathodes are relatively expensive and complicated. In seeking appropriate material and technique, field emission from various carbon structures has attracted attention of many investigators in the field [2-5]. Since the discovery of carbon nanotubes [6], much attention has been paid to explore the use of this ideal one-dimensional nanomaterial as cold cathodes [7-9]. Most of studies have been carried out to characterize the field emission properties of carbon nanotubes. It was reported that carbon nanotubes have excellent property of field emission [10-11]. Relative measurements showed that carbon nanotubes as emitters can yield current density of $4A/cm^2$ [12-13].

We report here on our field emission experiments performed on tips of graphite fibers as electron field emitters. After graphite fibers discharge, it can be observed that many corpuscles with the size range of nm-μm appeared at tips of graphite fibers. These tips can emit stable intensive current which has a wide range of potential applications in display techniques.

## 2. Experimental Details and Results

In field electron emission of graphite fiber, it is the most important thing to prepare the tip of graphite fiber as an emitter (cold cathode). The tip fabrication and measurements carried out in a vacuum coating machine with a glass jar (see Fig.1). Details of preparing process as following: cut an industrial graphite fiber with about 5 mm long and 20 μm diameter, and then adhere one end of this fiber to an angle (<300) top of an acute angle triangle of aluminum piece electrode, and leave another end for a tip to be treated. Then the aluminum piece with a graphite fiber was nipped at the cathode on a vacuum experimental table. There was a plane metal anode with fluorescence material placed around 3cm against the tip. After installing, vacuum glass jar was covered.

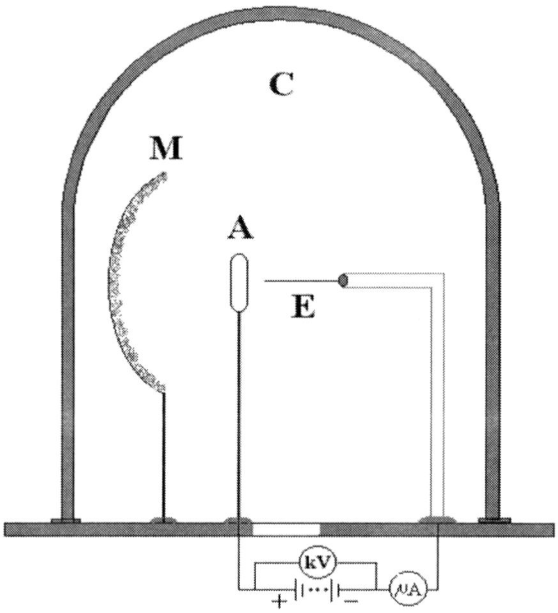


**Figure 1.** The experimental Instrument of Graphite fiber emission. Graphite fiber E was used as an emitter, Symbol A denotes an anode, the voltage is supplied to A and E. Symbol C and M stand for vacuum glass jar and luminescent screen, respectively.

Adjusting the outer power of jar made the anode voltage gradually rise from zero, at that time, because the tip of a graphite fiber was not treated, it didn't emit even raising the voltage to several thousand volts yet. When the voltage up to ten thousand volts, the emission current increased suddenly from zero to around 1mA, and the tip began to glow and become red. If the voltage was kept at ten thousand volts, the graphite fiber with 5 mm length might be burned out completely and the emission current became unstable, finally became zero. However, if we reduced the anode voltage below to 5 kV when the current just began to change suddenly, not only the tip never became red, but also steady current of several hundred μA could be obtained. The treated tip of graphite fiber can emit much more intense current increasing linearly with the increase of the anode voltage, which was not similar to the beginning process of raising the anode voltage for the untreated tip. The results of measurements were showed at the I-V curve in Fig.2(a), and the morphology of the treated tip with spheres and needle-like nanostructures were also confirmed by high resolution field electron microscope (HRFEM), shown in Fig.2(b). The field emission characters of the graphite fiber would not change as long as the voltage below 5 kV. Experimental results show that a single graphite fiber of 20 μm diameter can generate near 0.2 mA emission current with 4 kV applied voltage and the stability have been measured to be less than 10%.

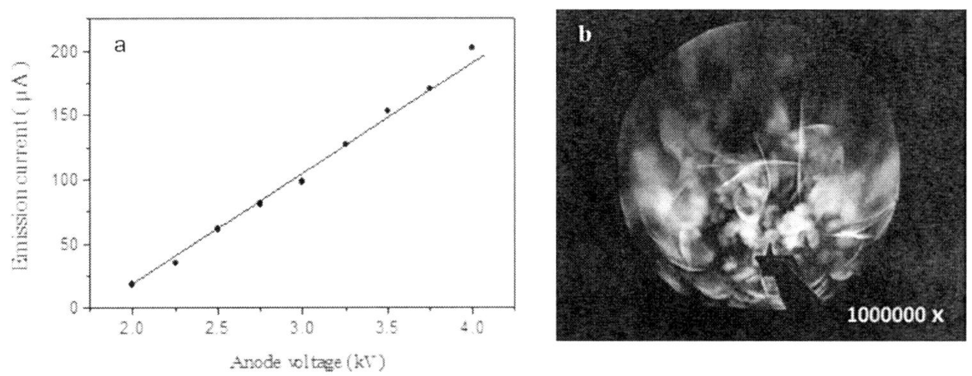


**Figure 2.** I-V curve (a) of field electron emission from the treated tip of a graphite fiber, and the FEM image (b) of field electron emission from the tip of a graphite fiber.

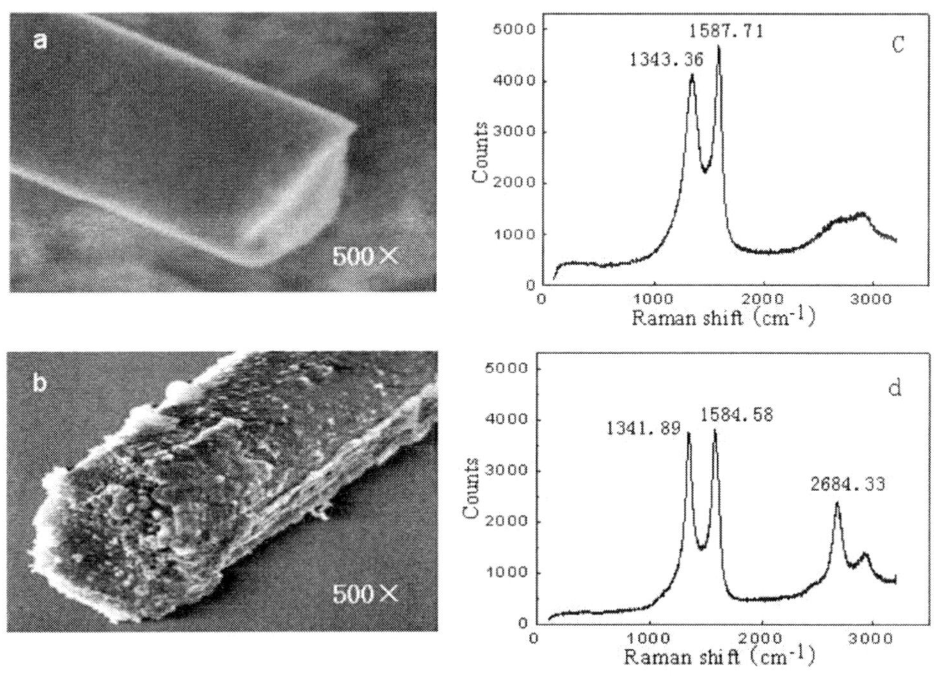


**Figure 3.** SEM images and Raman spectra of a graphite fiber tip before (a, c) and after (b, d) a discharge treatment.

To understand the emission mechanism of graphite fibers, tips of graphite fibers had observed by scanning electron microscope (SEM) before and after discharge of 10 kV voltages. SEM images of graphite fibers show in Fig.3. It is seen that the untreated tip was clean and only appeared natural rupture trail, but after discharge treatment it formed a compound tip with kinds of fragments between several nm and several μm, shapes as balls, pieces, and nanotubes grown on carbon fibers [14]. The bright dots in Fig.3 reveal sites of emitting electrons from graphite fibers.

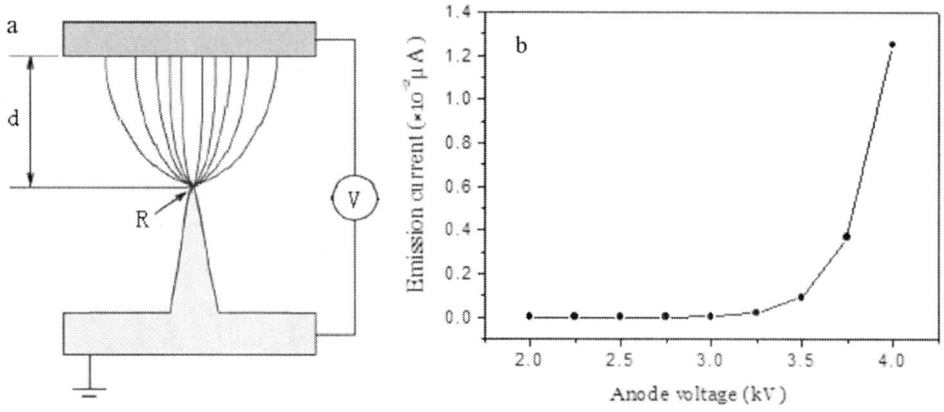


**Figure 4.** Schematic diagram (a) and I-V curve (b) of field electron emission from the tip of a graphite fiber.

**3. Discussion**

The principle of field emission is based on the application of a very high electric field to extract electrons from emitters. The current generated by the process of field emission is given by the Fowler–Nordheim (FN) equation [14]:

$$I = (aA\beta^2E^2/\phi)\exp(-b\phi^{3/2}/\beta E) \tag{1}$$

where A is the emitting area J is the Fowler–Nordheim current density. Properties of field emission electron are characterized by field emission current density J, the emission current on the unit surface of emitter (cathode), described by the Fowler–Nordheim (FN) law as:

$$J = CE^2 \times e^{-D/E} \tag{2}$$

where D = 6.8×107 $\Phi^{3/2}$, E is the electric field at the tip apex, Φ is the work function, and C is a constant. For carbon Φ=4.7 eV and C=6.2×$10^{-7}$. Most influence for these parameters is both the work function of cathode material Φ and field intensity E at the top of cathodes.

From the above formula, it is important to choose a material with smaller Φ value in field emission. For common materials, their work functions Φ are about 2-6 eV. Thus, we choose the value of E at $10^7$-$10^8$ V/cm, a very intense electronic field, to gain enough J. At the electrostatic field calculation, it is hardly to yield the electron field with parallel plate electrode. But if an electrode (anode) is a plate electrode, and another one (cathode) is a tip with small ball-like shape and its curvature radius R, shown in Fig. 4 the electron field around the surface of the tip can be approximately calculated by the following formula:

$$E = V/(NR) = V/d \tag{3}$$

where V is the voltage at two electrodes, and N is shape coefficient. For simplicity, we let N value equal to 2. Hence, the above formula is true due to d = 2R.

According to the formula (3), if the distance d of metal wire as a cathode is very small, such as d = $10^{-4}$cm, when the voltage V equals to several thousand volt, we may get electron intensity needed. Therefore, when the tips of tungsten wire as a cathode is utilized for field emission, the radius R of the tungsten wire etched is $10^{-4}$ cm. While the voltage between two electrodes is in the range of several kV to several 10 kV, we can acquire expecting emission current density such as J=$10^4$-$10^5$A/cm$^2$ according to the formula (2).

However, keeping the tip of $10^{-4}$ cm endure the bombardment from $10^4$-$10^5$A/cm$^2$ current and remainder air be not breakdown, only a few of metal wires with refractory such as tungsten suitable for field emission, and vacuum must be up to $10^{-7}$-$10^{-8}$ Pa for normal work. Even this, it is really available field emission current only 1μA.

Field electron emission from graphite fibers obeys Fowler-Nordhein law, formula (2), but the I-V curve (see Fig.2) from our experiment is linear. It can be explained that the measured current I is practically the total of current emitted by each small corpuscle on the tip according to Fowler-Nordhein law. If we estimate an average value on the size and number of these fragments on the tip, make all corpuscles share the emission current I, at the same time calculate the electronic field E at each corpuscle in their average diameter, the result of calculation is in good agreement with Fowler-Nordhein law.

In fact, we can reasonably estimate from Fig.3 that the tip consists of 1000 small balls and flakes with the size around 50nm. Each of these spheres and flakes can be an emission center in field emissions and some detectable emission currents can be obtained with a low applied voltage. From formula (1) and (2), we could get

$$i = 9.74 \times 10^{-7} V^2 \exp(-7.28 \times 10^4 / V). \quad (4)$$

When V equal to from 2 kV to 4 kV, we drew the i-V curve for each small corpuscle of the tip of a graphite fiber, shown in Fig.6.

In our experiments, the total of the field emission current from a graphite fiber can up to several hundred μA, and the departure angle of the total electron emission very large, beyond $180^0$. So field emission electrons from graphite fibers can emit to a large area. The merits make it possible to apply in large-area electronics. As a novel cold cathode material, graphite fibers will provide a wide range of applications such as flat panel displays with high-brightness, vacuum electronic devices, and vacuum microelectronic apparatus, new-type fluorescent sources, and high voltage rectifier. Field electron emission from graphite fibers may finally lead to a very significant change in display techniques.

## 4. Conclusions

Field emission of graphite fibers has the following merits: First, the total of the field emission current can up to several hundred μA, but that of tungsten as a conventional emitter filament only around 1 μA. Second, the requirement of vacuum is low. It means the tip of graphite fiber can endure the bombardment of remainder air. Third, the fragments distributed on the whole surface of the tip, which are as a result of the discharge treatment of graphite fiber, so the departure angle of the total emission current is very large, beyond 1800. Thus, electrons can emit to a large area. These merits make it possible to apply in broad plane display devices, new type fluorescent sources, high voltage rectifier, and vacuum microelectronic apparatus.

## Acknowledgments

This work was supported by the National Natural Science Foundation of China (Grant Nos. U1402273, 11364045, and 11664044), Department of Science and Technology of Yunnan Province (Grant No. 2016FC001), and by Yunnan University (2016MS14).

## References

[1] N.S. Xu, S. Ejaz Huq, "Novel cold cathode materials and applications", Materials Science and Engineering R, Vol.48 (2-5), pp.47–189, 2005.
[2] M. J. Colgan, M. J. Brett, "Field emission from carbon and silicon films with pillar microstructure", Thin Solid Films, Vol. 389, pp.1-4, 2001.
[3] A. T. Sowers, B. L. Ward, S. L. English, R. J. Nemanich, "Measurement of field emission from nitrogen-doped diamond films", Diamond and Related Materials, Vol.9(9-10), pp.1569-1573, 2000. J.
[4] C. F. Chen, C. L. Tsai and C. L. Lin, "Study of carbon nanoemitters using CO2–CH4 gas mixtures in triode-type field emission arrays", Diamond and Related Materials, Vol.11( 3-6), pp.788-792, 2002.

[5] N.A.Fox, M.J.Youh, W.N.Wang, J.W.Steeds, H-F.Cheng, and I-N. Lin, "Properties of electron field emitters prepared by selected area deposition of CVD diamond carbon films", Diamond and Related Materials, Vol. 9(3-6), pp.1263-1269, 2000.
[6] S. Iijima, "Helical microtubules of graphitic carbon", Nature, Vol.354, pp.56-58, 1991.
[7] Jean-Marc Bonard, Frederic Maier, Thomas Stockli, et al., "Field emission properties of multiwalled carbon nanotubes", Ultramicroscopy, Vol.73 (1-4), pp.7-15, 1998.
[8] Do-Hyung Kim, Hyeong-Rag Lee, Man-Woo Lee, et al., "Effect of the in situ Cs treatment on field emission of a multiwalled carbon nanotube", Chemical Physics Letters, Vol.355(1-2), pp.53-58, 2002.
[9] Young-Rae Cho, Jin Ho Lee, Yoon-Ho Song, et al. Photolithography-based carbon nanotubes patterning for field emission displays. Materials Science and Engineering B, Vol.79 (2), pp.128-132, 2001.
[10] J. Zhang, T. Feng, W. Yu, et al., "Enhancement of field emission from hydrogen plasma processed carbon nanotubes", Diamond and Related Materials, Vol.13, pp. 54-59, 2004.
[11] Jun Chen, X. Zhou, S. Z. Deng and N. S. Xu. "The application of carbon nanotubes in high-efficiency low power consumption field-emission luminescent tube", Ultramicroscopy, Vol. 95, pp.153-156, 2003.
[12] Hee Jin Jeong, Ha Kyu Choi, Gil Yong Kim, et al., "Fabrication of efficient field emitters with thin multiwalled carbon nanotubes using spray method", Carbon, Vol.44, pp.2689-2693, 2006.
[13] Chao Hsun Lin, Hui Lin Chang, Ming Her Tsai, et al., "Growth mechanism and properties of the large area well-aligned carbon nano-structures deposited by microwave plasma electron cyclotron resonance chemical vapor deposition. Diamond and Related Materials", Vol.11 (3-6), pp. 922-926, 2002.
[14] L. Zeng, W. Wang, D. Lei, et al., "High-field electron emission of carbon nanotubes grown on carbon fibers", Physica B 403, pp2662-2665, 2008

IOP Conf. Series: Materials Science and Engineering 472 (2019) 012061 doi:10.1088/1757-899X/472/1/012061

# Application Research in Accident Classification Based on Fuzzy Evaluation System

**Qiang Ren and Genyin Cheng**

College of Gradate, North China Institute of Science and Technology, Yanjiao, Beijing, China, 065201.

E-mail:979615680@qq.com

**Abstract**. In view of the fuzzy boundary problem in the criterion of accident grade classification, the number of deaths, the number of injuries and the direct economic losses of accidents are taken as evaluation elements respectively, and the judgment matrix of accident grade is constructed. The influence weight of indicators is calculated by AHP and the consistency of judgment matrix is checked. Taking some typical production accidents as an example, the membership function is constructed according to the existing accident indexes. In this paper, an accident grade classification method based on fuzzy evaluation system is put forward, and the accident evaluation matrix is obtained, and the accident grade is determined according to the principle of maximum membership degree. To a certain extent, it can provide scientific reference and reference basis for the qualitative analysis of accident grade.

## 1. Introduction

The determination of the accident level is divided into different scopes according to different classification principles. The consequences of an accident may belong to multiple categories at the same time. In order to ensure the scientific and fairness of the accident investigation results, it should be accurately determined according to the scientific and rational division principle. So it is necessary to use the fuzzy evaluation system for in-depth research.

Relevant scholars carried out detailed analysis and discussion on the classification of road traffic affairs. The fuzzy mathematics theory was used to comprehensively evaluate the multiple indicators of road traffic accidents. In the selection of evaluation indicators, the classification criteria of indirect economic loss accidents were introduced [1]. Feng-qin Li applied the fuzzy hierarchy theory to the gas outburst hazard analysis, comprehensively evaluated the established element sets, analyzed various indicators in detail, constructed the judgment matrix and relative weights, identified the most dangerous indicators and determined their outstanding dangers [2].

In view of the frequent occurrence of various types of accidents in China, comprehensive quantitative statistics and strict qualitative analysis are carried out on the consequences of accidents. Based on the known casualties and economic losses caused by accidents, the fuzzy evaluation method is not clear about the unclear existence. The problem is given to the system's argumentation analysis, and then the objective evaluation and supplement of the nature of the accident is judged, which has certain reference significance for the fairness of the accident investigation result.

## 2. Overview of Fuzzy Evaluation

### *2.1. Evaluation Method Basis*

Under the principle of the same division, the number of casualties caused by a single accident and the direct economic losses may belong to different divisions, which directly affect the fairness of the accident investigation results. The factors of the accident level should be comprehensively measured and utilized. The fuzzy definition of mathematics analyzes the different indicators of accidents, which can make up for the irrationality of the classification of the severity of accidents by a single index [3]. In this paper, fuzzy system theory is used to classify several representative production safety accident levels in the industry, and the membership function of fuzzy accident level is established, which gives a fuzzy evaluation method that can be used to define the accident level. Compared with other existing methods, this paper gives the method has obvious reference and complementary.

### 2.2. *Influencing Factors Weight*

Weight is the quantitative evaluation of the relative importance of each factor in fuzzy decision making. In order to eliminate subjective bias, the classical method of solving multi-objective decision-making problem--AHP is adopted, which ensures the objectivity of weight coefficient better [4]. Accuracy is the mathematical programming of decision makers for a complex decision-making process. It essentially determines the weighting effect of each indicator on the ultimate goal, and determines the final degree of influence of an element on the event itself [5]. The factors that determine the accident level when assessing the result of the accident level are: (1) the number of deaths; (2) the number of injured; (3) direct economic losses, according to the weight of the relevant experts on the factors affecting the accident, construct the judgment matrix $P$, namely:

$$P = \begin{bmatrix} 1 & 4 & 9 \\ 1/4 & 1 & 5 \\ 1/9 & 1/5 & 1 \end{bmatrix} \tag{1}$$

Using the MATLAB software's characteristic function [v d]=*eig(P)*, the maximum eigenvalue max=3.071 is obtained, and the corresponding characteristic vector is $w'$=[0.947 0.309 0.081]T. The $w'$ is normalized to obtain w=[$w_1$,$w_2$,$w_3$]= [0.708 0.231 0.061], the matrix $w$ at the time is the weight value of each indicator of the accident [6].Then judge the matrix consistency index test, it need to calculate the consistency index value CI=($\lambda_{max}$-n)/(n-1)=0.0355, and find out that the average third-order consistency index RI=0.52, then CR=CI/RI = 0.0355/0.52 = 0.068 < 0.1, the judgment matrix consistency test is passed.

## 3. Establish Fuzzy Evaluation Matrix

According to the relevant regulations of China's "*Production Safety Accident Report and Investigation and Handling Regulations*" [7], production safety accidents are classified according to the number of casualties or direct economic losses as shown in Table 1.

**Table 1.** Divide accident rating criteria based on the number of casualties or direct economic losses

| Accident indicator | Accident level | | | |
|---|---|---|---|---|
| | general accident | bigger accident | major accident | extraordinary accident |
| number of deaths/$R_1$ (Unit: person) | $R_1$<3 | 3≤$R_1$<10 | 10≤$R_1$<30 | $R_1$≥30 |
| number of injured/$R_2$ (Unit: person) | $R_2$<10 | 10≤$R_2$<50 | 50≤$R_2$<100 | $R_2$≥100 |
| direct economic loss/$R_3$ (Unit: 10 million yuan) | $R_3$<0.1 | 0.1≤$R_3$<0.5 | 0.5≤$R_3$<1 | $R_3$≥1 |

Establishing indicator set $U$= {$u_1$, $u_2$, $u_3$}= {number of deaths, number of injured, direct economic loss}, accident level set V={$v_1$, $v_2$, $v_3$, $v_4$}={general accident,bigger accident, major accident, extraordinary accident}, the i-factor evaluation of the *i-th*(*i*= 1, 2,...3) indicators $R_i$ is $R_i$= {$r_{i1}$, $r_{i2}$, ... $r_{i5}$}, which is regarded as a fuzzy subset on $V$, $r_{ij}$ (i = 1, 2, ..., 3; j=1, 2, ..., 4) indicates that the *i-th* index $u_{ij}$ is rated as the membership degree of the *j-th* level $v_i$, that is, the membership probability [8].

The judgment matrix formed by the indicator set and the comment set can be expressed as:

$$R = \begin{bmatrix} r_{11} & r_{12} & r_{13} & r_{14} \\ r_{21} & r_{22} & r_{23} & r_{24} \\ r_{31} & r_{32} & r_{33} & r_{34} \end{bmatrix} \tag{2}$$

In the formula, $r_{ij} = U_R\{u_i, v_j\}$ $(0 < r_{ij} \leq 1)$

Fuzzy evaluation $B$ is a fuzzy subset on $V$, which can be written as:

$B = w \bullet R$, that is:

$$B = \begin{bmatrix} w_1 & w_2 & w_3 \end{bmatrix} \bullet \begin{bmatrix} r_{11} & r_{12} & r_{13} & r_{14} \\ r_{21} & r_{22} & r_{23} & r_{24} \\ r_{31} & r_{32} & r_{33} & r_{34} \end{bmatrix} \tag{3}$$

In order to avoid the data difference caused by subjectivity, the relevant data experience is used to establish the membership function of the four levels of indicators related to the severity of the accident. The accidents are classified into general accident, bigger accident, major accident, extraordinary accident according to the number of deaths,. Its membership function is represented by $D_1(x), D_2(x), D_3(x), D_4(x)$, respectively, and is divided by trapezoidal fuzzy sets.

$$D_1(x) = \begin{cases} 1 & 0 \leq x < 3 \\ (5-x)/2 & 3 \leq x < 5 \\ 0 & \text{others} \end{cases} \tag{4}$$

$$D_2(x) = \begin{cases} x/2 & 3 \leq x < 5 \\ 1 & 5 \leq x < 10 \\ (20-x)/10 & 10 \leq x < 20 \\ 0 & \text{others} \end{cases} \tag{5}$$

$$D_3(x) = \begin{cases} x/10 & 10 \leq x < 20 \\ 1 & 20 \leq x < 25 \\ (30-x)/5 & 25 \leq x < 30 \\ 0 & \text{others} \end{cases} \tag{6}$$

$$D_4(x) = \begin{cases} x/5 & 25 \leq x < 30 \\ 1 & x \geq 30 \\ 0 & \text{others} \end{cases} \tag{7}$$

Similarly, in line with the number of injured people caused by production safety accidents, the accidents are classified into general accident,bigger accident, major accident, extraordinary accident.

The membership functions are$E_1(x), E_2(x), E_3(x), E_4(x)$respectively. (x) indicates that trapezoidal fuzzy set partitioning is still used.

$$E_1(x) = \begin{cases} 1 & 0 \leq x < 10 \\ (30-x)/20 & 10 \leq x < 30 \\ 0 & \text{others} \end{cases} \tag{8}$$

$$E_2(x) = \begin{cases} x/20 & 10 \leq x < 30 \\ 1 & 30 \leq x < 50 \\ (70-x)/20 & 50 \leq x < 70 \\ 0 & \text{others} \end{cases} \tag{9}$$

$$E_3(x) = \begin{cases} x/20 & 50 \leq x < 70 \\ 1 & 70 \leq x < 90 \\ (100-x)/20 & 90 \leq x < 100 \\ 0 & \text{others} \end{cases} \tag{10}$$

$$E_4(x) = \begin{cases} x/10 & 90 \leq x < 100 \\ 1 & x \geq 100 \\ 0 & \text{others} \end{cases} \tag{11}$$

According to the size of the direct economic loss caused by the accident, the accident is divided into general accident, bigger accident, major accident, extraordinary accident. The membership function is represented by $F_1(x), F_2(x), F_3(x), F_4(x)$respectively, still using trapezoidal fuzzy set.

IOP Conf. Series: Materials Science and Engineering 472 (2019) 012061 doi:10.1088/1757-899X/472/1/012061

$$F_1(x)=\begin{cases}1 & 0\le x<0.1\\ (0.3-x)/0.2 & 0.1\le x<0.3\\ 0 & \text{others}\end{cases} \quad (12) \qquad F_2(x)=\begin{cases}x/0.2 & 0.1\le x<0.3\\ 1 & 0.3\le x<0.5\\ (0.7-x)/0.2 & 0.5\le x<0.7\\ 0 & \text{others}\end{cases} \quad (13)$$

$$F_3(x)=\begin{cases}x/0.2 & 0.5\le x<0.7\\ 1 & 0.7\le x<0.9\\ (1-x)/0.1 & 0.9\le x<1\\ 0 & \text{others}\end{cases} \quad (14) \qquad F_4(x)=\begin{cases}x/0.1 & 0.9\le x<1\\ 1 & x\ge 1\\ 0 & \text{others}\end{cases} \quad (15)$$

According to the given accident statistics, the values of each factor are calculated separately and the specific element values of the evaluation matrix $R$ are obtained. After the weight calculation, the accident level can be determined according to the principle of maximum membership degree [9-11].

## 4. Example Application

Now select industry-specific production safety accidents as research objects,the construction accident was caused by the collapse of the "11.24" Power Plant construction platform of Fengcheng of Jiangxi province, the chemical accident of the "6.5" liquefied gas tank explosion accident of Petrochemical Co., Ltd at Jinyu of Shandong province, the road traffic accidents with Zhejiang province"7.23"Yong-wen line railway train rear-end accident as well as taking the "9.5" gas explosion accident in Yongdingzhuang Mine as an example (Table 2).

**Table 2.** Statistics of accident data

| Accident | number of deaths (Unit: person) | number of injured (Unit: person) | direct economic loss (Unit: person) |
|---|---|---|---|
| the "11.24" construction platform of Jiangxi Fengcheng Power Plant | 73 | 2 | 1.0197 |
| the "6.5" liquefied gas tank explosion accident of Petrochemical Co., Ltd | 10 | 9 | 0.4468 |
| Zhejiang"7.23"Yong-wen line railway train rear-end accident | 40 | 172 | 1.9371 |
| the "9.5" gas explosion accident in Yongdingzhuang Mine | 31 | 16 | 0.01 |

The number of deaths, injuries, and direct economic losses in the accidents in Table 2 are taken into equations (4)-(7), (8)-(11), and (12)-(15), respectively. Judgment matrix of the collapse of the city power plant is:

$$R_1=\begin{bmatrix}0&0&0&1\\1&0&0&0\\0&0&0&1\end{bmatrix}$$

Judgment matrix of the "6.5" liquefied gas tank explosion accident of Petrochemical Co. Ltd at Jinyu of Shandong province is:

$$R_2 = \begin{bmatrix} 0 & 1 & 1 & 0 \\ 1 & 0 & 0 & 0 \\ 0 & 0 & 1 & 0 \end{bmatrix}$$

Judgment matrix of Zhejiang"7.23"Yong-wen line railway train rear-end accident is:

$$R_3 = \begin{bmatrix} 0 & 0 & 0 & 1 \\ 0.7 & 0 & 0 & 1 \\ 0 & 1 & 0 & 0 \end{bmatrix}$$

Judgment matrix of the "9.5" gas explosion accident in Yongdingzhuang Mine is:

$$R_4 = \begin{bmatrix} 0 & 0 & 0 & 1 \\ 0.7 & 0.8 & 0 & 0 \\ 1 & 0 & 0 & 0 \end{bmatrix}$$

The evaluation matrix of each accident can be calculated according to (3) of formula: B1= [0.231 0 0 0.769], $B2$= [0.231 0.708 0.769 0], $B3$= [0.462 0.061 0 0.939], $B4$= [0.223 0.185 0 0.708]. In accords to the principle of maximum membership degree, the four accidents can be graded as: extraordinary accident, extraordinary accident, major accident, and extraordinary accident. The evaluation results are consistent with the official conclusions, indicating that the fuzzy evaluation system determines the accident level has a high accuracy.

## 5. Conclusion

Based on the number of casualties and direct economic losses caused by production safety accidents, this paper uses AHP to establish a judgment matrix, and obtains the membership function according to the accident classification criteria to obtain the severity level of the accident. Combined with the full text, the following conclusions are drawn:

(1) According to the comprehensive consideration of the weight of the factors affecting the accident by experts in the field, it can be known that the number of deaths caused by the accident, the number of injured persons and the judgment weight corresponding to the direct economic loss are $w$=[0.708 0.231 0.061], the evaluation matrix of each accident: $B_1$=[0.231 0 0 0.769],$B_2$=[0.231 0.708 0.769 0],$B_3$=[0.462 0.061 0 0.939],$B_4$=[0.223 0.185 0 0.708], determined according to the principle of maximum membership degree The level of consequences of an accident, the law provides to some extent relevant experience for the qualitative analysis of other types of accidents.

(2) The fuzzy evaluation analyzes the weight of the factors that meet the consistency test from the microscopic point of view, and constructs the membership function for number of deaths, number of injured, direct economic loss according to the *Regulations on Production Safety Accident Report and Investigation and Treatment*. Then, through the weight calculation and the principle of maximum membership degree, the accident level is judged, which provides reference and reference for the final accident classification.

(3) Based on the number of casualties and economic losses, assist in determining the severity of the accident, make a certain contribution to determine the level of the accident, effectively reduce the subjectivity of human decision-making, and have the feasibility of the result of the accident division.

**Acknowledgement:**
I am very grateful to my tutor and classmates for their support and counseling for my thesis.

## References

[1] Yu LIN.Analysis of Grades of Road Traffic Accidents——Fuzzy Evaluation Model [J].*Journal of Anhui university of science and technology (Natural Science)*, 2005(03):pp 33-36.
[2] Fengqin Li, Kunyun Tian. Evaluation of Fuzzy Hierarchy Theory in the Analysis of Gas Outburst

Hazard [J].*Coal technology*, 2015, 34(11): pp 149-151.

[3] Zuzeng Peng, Yuyu Sun. *Fuzzy mathematics and applications* [M]. Wuhan: Wuhan University Press, 2002.

[4] Yang Xu, Yan Zhou, Xin Sun, Yue Yu. Comprehensive Evaluation of Mine Safety Based on Fuzzy Analytic Hierarchy Process [J]. *Chinese journal of safety science*, 2009, 19(05): pp 147-152+179.

[5] Jun Zhang, Jianpeng Wang, Wenguang Yang. Study on fuzzy comprehensive prediction model of caving height in fully mechanized mining face[J].*Journal of china university of mining & technology*,2014,43(03):pp 426-431.

[6] Wei Wang, Hongxi Xu, Yu Lu, Rui Li.A Study on Urban Road Traffic Accident Early Warning Based on Fuzzy Comprehensive Evaluation Model[J].*Modern transportation technology*,2016,13(06):pp 85-89.

[7] State Administration of Work Safety, *Regulations on production safety accident report and investigation and treatment* [Z].2007, 4,20.

[8] Xiaoxin Liu. Application of Fuzzy Comprehensive Evaluation Method in Coal Mine Safety Management Evaluation [J].*Shaanxi coal*, 2015,34(02):pp 91-93+106.

[9] Quan-zhong Jing, Xiu-hui Jiang, Jian-song Yang and Yanfeng Zhou. Research on Index System of Coal Mine Safety Production Capacity Based on Analytic Hierarchy Process [J]. *Chinese safety science journal*, 2006(09): pp 74-79+145.

[10] Xiao-hong Zhu, Yu-shi Lu and Dehong Zhou.Application of Fuzzy Evaluation Mathematical Model in Road Traffic Safety Evaluation [J].*Safety and environmental engineering*, 2006(03):pp 102-104.

[11] Lina Yang, Baoju Hun, Chao Zhang. Application of Fuzzy Comprehensive Evaluation Method in the Evaluation of Mine Gas Accident Hazard Sources [J].*Hebei coal*, 2007(02):pp 22-24.

IOP Conf. Series: Materials Science and Engineering 472 (2019) 012062 doi:10.1088/1757-899X/472/1/012062

# Study on Aerodynamic Characteristics of Flapping Wings Motion Mode

**Cui Wang[1], Hongde Wang[2], Jiangbo Sun[1] and Qiang Liu[2*]**

[1]Jilin Vocational and Technical Institute of Communications, School of Mechanical Engineering, Changchun, Peoples R China

[2]School of Mechanical and Aerospace Engineering, Jilin University, Changchun, China2

Email: liuqiang2012@jlu.edu.cn, Qiang Liu

**Abstract.** Flapping Wing Micro Aerial Vehicle (FWMAV) is a new type of MAV simulating the flight of insects or birds. FWMAV can generate lift and thrust simultaneously to achieve the function of lifting, forward flying or hovering by only flapping wings. It also can change flight mode flexibility by adjusting flapping parameters, which makes it have nice maneuverability and high flight efficiency and more suitable for military mission. So it has great development and research value. In this paper, insect is selected as biomimetic research object, using the computational fluid dynamics method to make a numerical study on aerodynamic characteristics of flapping wings motion mode.

## 1. Introduction

At present, most of the numerical studies on flapping wing aerodynamic characteristics of bionic flapping-wing micro air vehicles are carried out under a single and specific "8" flapping trajectory [1][2][3]. However, the "8" shaped flutter of different insects in nature is different from that of flying insects. Therefore, it is of great significance to further analyze the differences of aerodynamic characteristics between different types of flapping trajectories and different states of the same type, and to provide a solid theoretical basis for the design of flapping-wing micro air vehicles with complex motions. In this chapter, the X-Flow software based on lattice Boltzmann method is used to calculate the different symmetrical "8" flapping trajectories of rigid model wings by increasing the degree of freedom of motion in the absence of free inflow. The differences of aerodynamic characteristics are analyzed, and the influence of different flapping wing modes on aerodynamic forces is explored [4][5].

## 2. .The Influence of Horizontal Symmetry "8" Shaped Flapping Wing on its Aerodynamic Characteristics

The kinematic functions of the horizontal "8" flapping wing mode are as follows:

Flutter angle around OX axis (First down and back up):

$$\psi(t) = -0.5\Phi cos(2\pi ft) \tag{1}$$

The twist angle corresponding to the OZ axis in one cycle:

$$\alpha(t)=\begin{cases}(\pi-\alpha_d)-\theta((t+0.125T)-\frac{t_r}{2\pi}sin(\frac{2\pi(t+0.125T)}{t_r})) & 0\le t<0.125T\\ \alpha_d & 0.125T\le t<0.375T\\ \alpha_d+\theta((t-0.375T)-\frac{t_r}{2\pi}sin(\frac{2\pi(t-0.375T)}{t_r})) & 0.375T\le t<0.625T\\ \pi-\alpha_d & 0.625T\le t<0.875T\\ (\pi-\alpha_d)-\theta((t-0.875T)-\frac{t_r}{2\pi}sin(\frac{2\pi(t-0.875T)}{t_r})) & 0.875T\le t<T\end{cases} \quad (2)$$

In the above equation, Φ is the flutter amplitude, take Φ=120°,f for flutter frequency, take f=10Hz, $\alpha_d$is the middle angle of the lower swing, and $\alpha_d$=40°, $t_r$ is reverse time and take $t_r$=0.25T, $\theta=(\pi-2\alpha_d)/t_r$.

The variation of flutter angle and twist angle with time in one motion cycle is shown in Figure 1. It can be seen from the diagram that the torsion angle is fixed at the middle of the attack. At the end of the flutter up, the model wings begin to flip and the twist angle decreases until the first quarter of the swoop, and the twist angle decreases to a fixed value. The flapping rule of downward flapping is similar to that at the end. Under this motion function, the projection of the wing tip trajectory on the XY plane is shown in Figure 2:

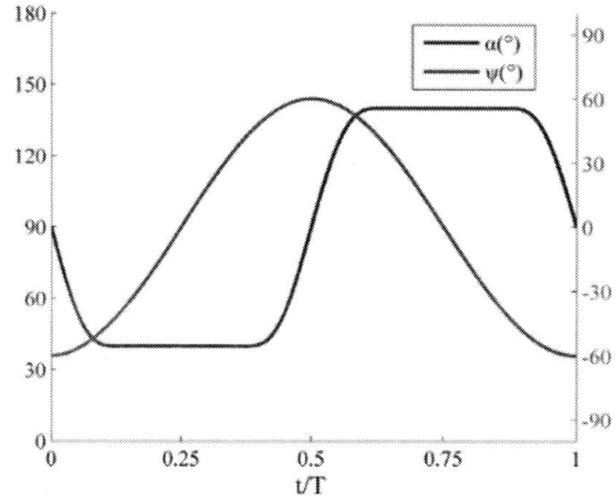


**Figure 1**. Angle of flutter and twist of horizontal symmetry "8"

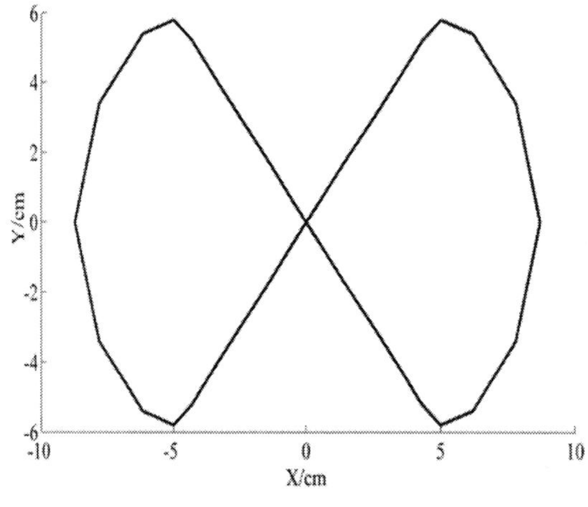


**Figure 2**. Projection of wingtip trajectory in XY plane

The numerical results of lift coefficient and drag coefficient varying with time in a motion cycle are shown in Figure 3 below:

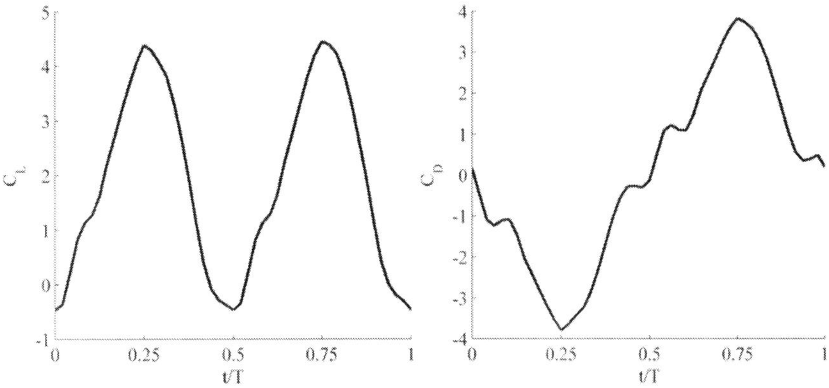


**Figure 3.** $C_L$ and $C_D$ of model flapping in horizontal "8" flutter mode (Negative resistance is thrust.)

In order to explore more intuitively the change of aerodynamic force in the campaign cycle under the flutter mode, the pressure cloud chart of the extension section of the model wing 1/2 at different times is shown in Table 1:

**Table 1.** pressure cloud chart of 1/2 horizontal section "8" shape at different times

| | | | |
|---|---|---|---|
| 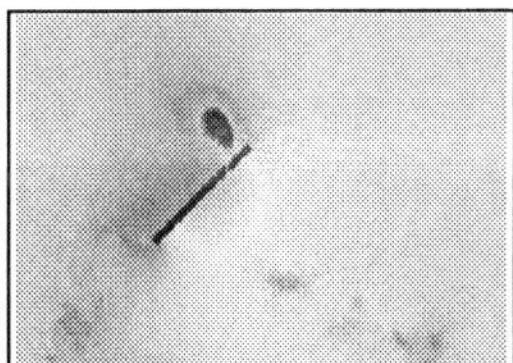 | 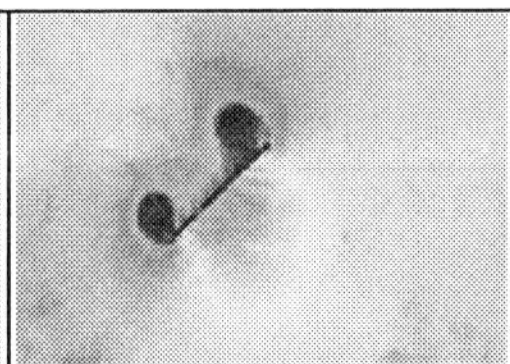 | 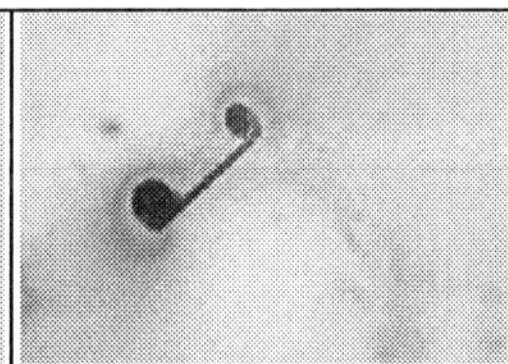 | 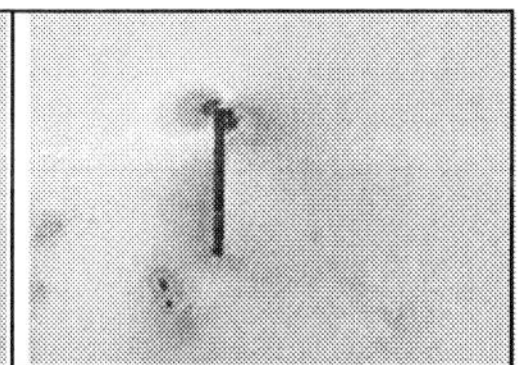 |
| (a)1/8time | (b)2/8time | (c)3/8time | (d)4/8time |
| 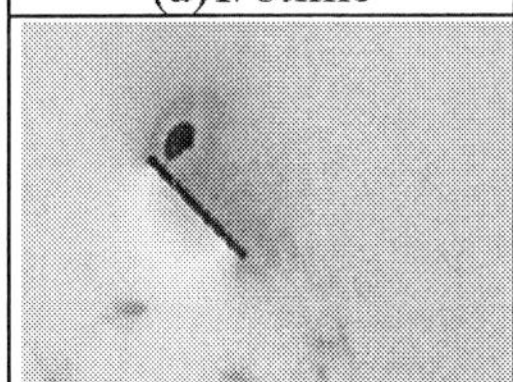 | 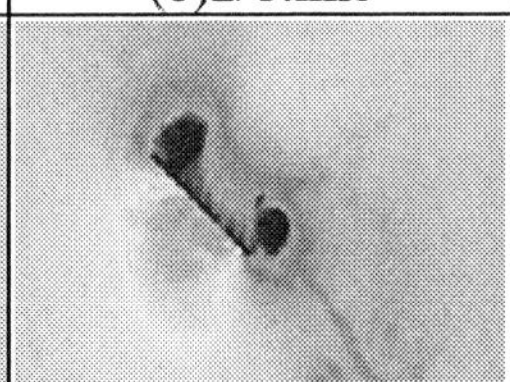 | 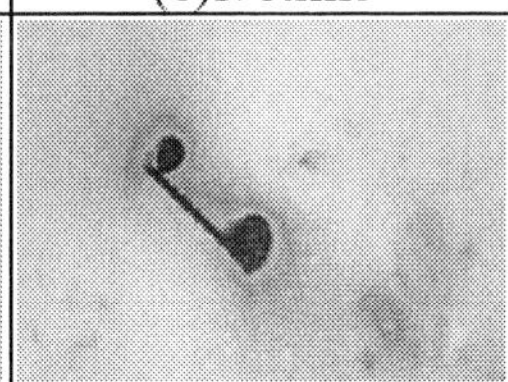 | 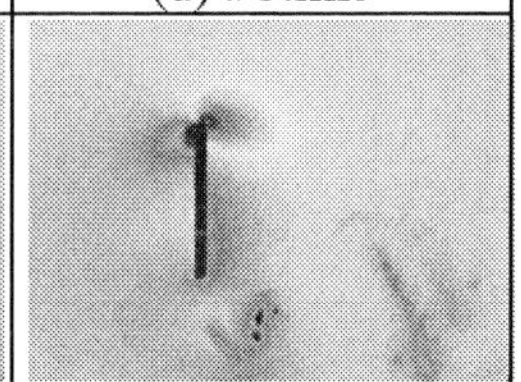 |
| (e)5/8time | (f)6/8time | (g)7/8time | (h)8/8time |

Based on the cross section pressure cloud chart and aerodynamic coefficient map at different time points, we can draw the following conclusions: During the upstroke and down stroke of the horizontal symmetrical "8" flapping wing in the next movement cycle, the pressure at the lower end of the wing is always greater than that at the upper end, and the lift force is generated. This is due to the existence of twisting motion, when fluttering down the direction of the movement of the wings is downward and backward and When fluttering upward the direction of the movement of the wings is forward and downward. At the same time, due to the torsion movement, the effective area of the model wing with air contact is reduced. The peak value of lift coefficient is less than that of linear flapping wing. However, due to the symmetrical motion trajectory of the downward and upward flutters, the thrust produced by the downward dive is equal to the resistance produced by the upward flutter, so that the average thrust coefficient is 0 in a movement cycle. Therefore, this type of flutter is suitable for takeoff flight or hovering flight.

IOP Conf. Series: Materials Science and Engineering 472 (2019) 012062 doi:10.1088/1757-899X/472/1/012062

## 3. The Effect Of Vertical Symmetry "8" Shaped Flapping Wing on its Aerodynamic Characteristics

The kinematic functions of the horizontal "8" flapping wing mode are as follows:

Flutter angle around OX axis (First down and back up):

$$\psi(t) = -0.5\Phi cos(2\pi f t) \tag{3}$$

A twist angle around a OZ axis in a cycle:

$$\alpha(t) = \begin{cases} \alpha_d \, sin(\frac{\pi t}{t_r}) & 0 \le t < 0.2T \\ \alpha_d & 0.2T \le t < 0.3T \\ \alpha_d \, cos(\frac{\pi(t-0.3T)}{t_r}) & 0.3T \le t < 0.7T \\ -\alpha_d & 0.7T \le t < 0.8T \\ -\alpha_d \, cos(\frac{\pi(t-0.8T)}{t_r}) & 0.8T \le t < T \end{cases} \tag{4}$$

In the above equation, $\Phi$ is the flutter amplitude, take $\Phi$=120°, f for flutter frequency, take f=10Hz, $\alpha_d$is the middle angle of the lower swing, and $\alpha_d$=45°, $t_r$ is reverse time and take $t_r$=0.4T.

The variation of flutter angle and twist angle with time in one motion cycle is shown in Figure 4. It can be seen from the diagram that the variation law of the torsion angle of the motion mode is similar that of the horizontal symmetrical "8" flapping wing. In the middle of the downward flutter and upward flutter, the twist angle is set. At the end of the flutter up, the model wings begin to turn over, and the twist angle decreases until the middle of the flutter, and the twist angle decreases to a fixed value. The flapping rule of downward flapping is similar that at the end. Under this motion function, the projection of the wing tip trajectory on the XY plane is shown in Figure 5:

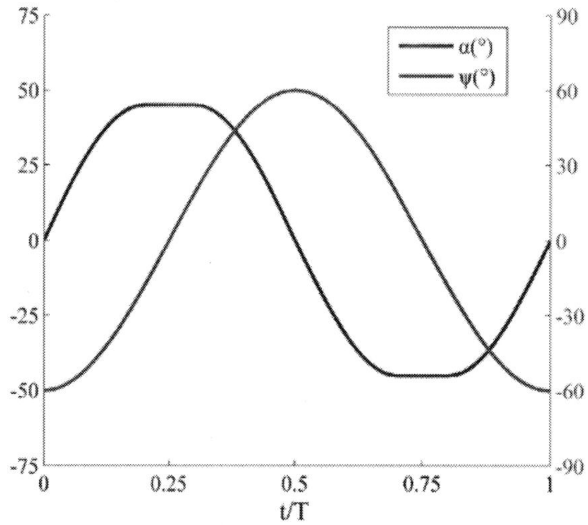


**Figure 4.** angle of flutter and twist of vertical symmetry "8"

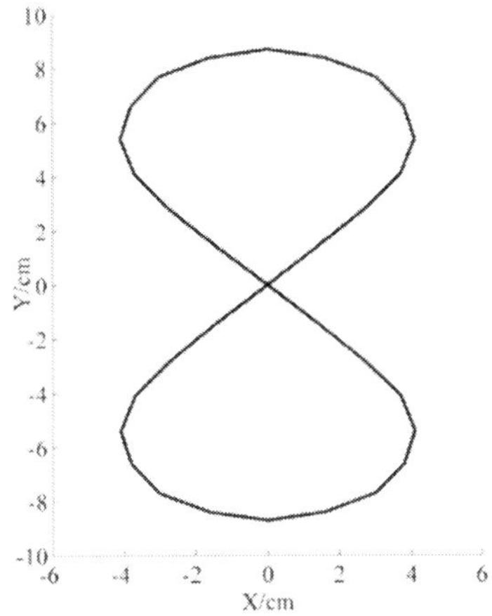


**Figure 5.** Projection of wingtip trajectory in XY plane

The numerical results of the time-dependent variations of lift and drag coefficients during a period of motion are shown in Figure 6 below:

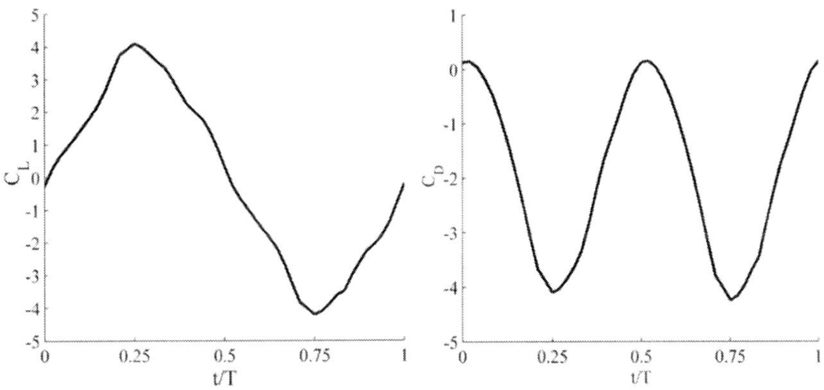


**Figure 6.** $C_L$ and $C_D$ of the vertical "8" flutter model wings. (Negative resistance is thrust)

For a more intuitive study of the aerodynamic changes in the next cycle of the flutter, the pressure the cloud chart at the 1/2 extension of the model wing at different times are shown in Table 2 below.

**Table 2.** pressure cloud chart of 1/2 vertical section of "8" shape at different times

| | | | |
|---|---|---|---|
| 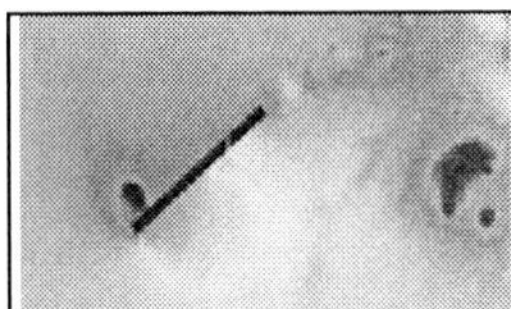 | 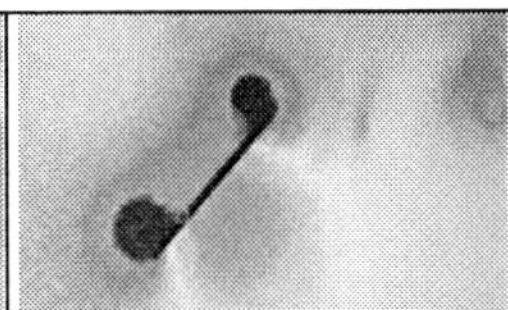 | 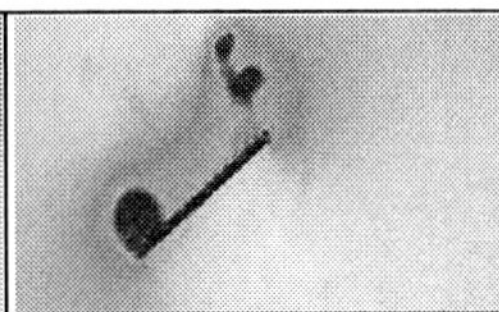 | 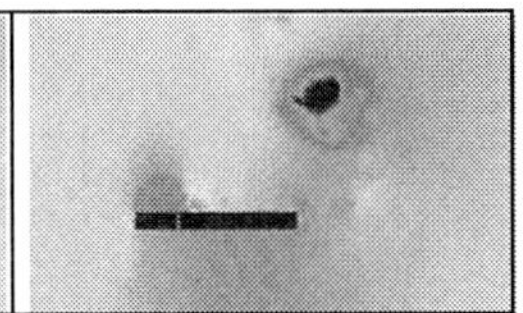 |
| (a)1/8time | (b)2/8time | (c)3/8time | (d)4/8time |
| 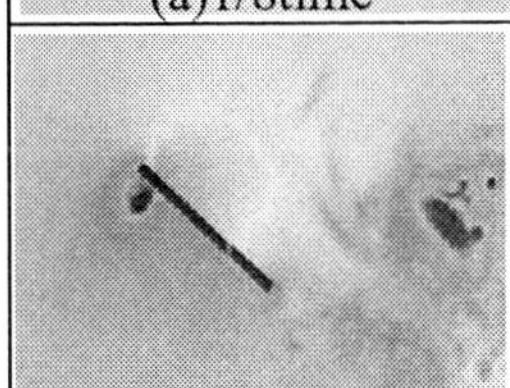 | 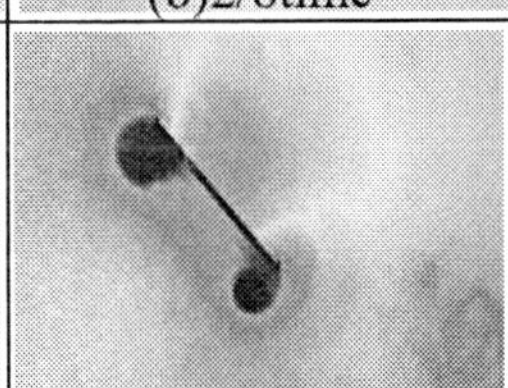 | 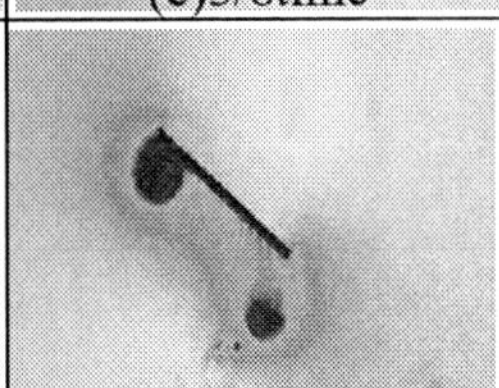 | 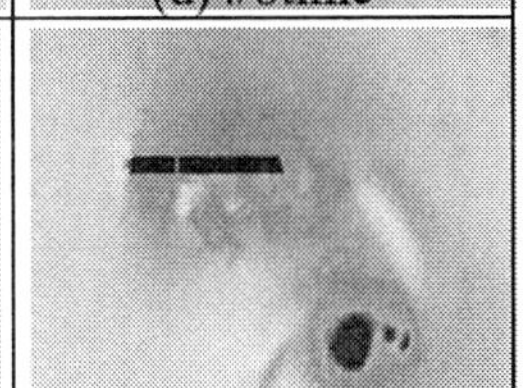 |
| (e)5/8time | (f)6/8time | (g)7/8time | (h)8/8time |

Based on the cross section pressure cloud chart and aerodynamic coefficient map at different time points, we can draw the following conclusions: During the upstroke and downstroke of the vertical symmetrical "8" flapping wing in the next movement cycle, the pressure at the rear end of the wing is always greater than that of the front section, and there is negative resistance (i.e. thrust). This is due to the existence of twisting motion, when fluttering down the direction of the movement of the wings is downward and backward and when fluttering upward the direction of the movement of the wings is upward and backward. However, due to the symmetry of the downward and upward motions, the downward motions produce the same lift and the upward motions produce the same negative lift, so that the average lift coefficient is 0. Therefore, the flutter mode is suitable for the diving acceleration flight state in a short time.

## 4. Conclusion

In this paper, the following conclusions can be drawn from the numerical study of computational fluid dynamics. The symmetrical "8" shape flutter can produce larger periodic lift than the pure linear flutter. The horizontal symmetry "8" shaped flapping wing is suitable for take-off or hovering flight. The vertical symmetry "8" shaped flapping wing is suitable for diving acceleration flight in a short time. The periodic average aerodynamic force produced by the symmetrical "8" zigzag motion (two degrees of freedom) with flutter and torsion is related to the inclination of the "8" zigzag motion. The inclination angle increases, the average lift decreases, and the average thrust increases. Therefore, by adjusting the size of $\alpha_m$ and $\alpha_n$ properly, different aerodynamic effects can be produced.

## Acknowledgments

This work was supported by China Postdoctoral Science Foundation(No. 2014M551177, No. 2015T80300), Fundamental Research Funds for the Central Universities [JCKY-QKJC11], Department of Science and Technology of Jilin Province(20180101321JC) and Education Department of Jilin Province ( JJKH20170287KJ)

## References

[1] Zhu Jianyang,2016,Numerical study of flapping trajectories on aerodynamic characteristics of flapping wing, Engineering mechanics, Vol.33 No.1 246-251.
[2] Zhang Hongmei and Yang Wenqing, 2016 ,Aerodynamic characteristics of flapping wing bionic "0" and "8" shaped flutter modes, Progress in Aeronautical Engineering,Vol.7 No.1 44-50.
[3] Zhang Yongli, Zhao Chuangxin and Xu Liangjin, 2006, Influence of flapping wing trajectory on aerodynamic force, Science Bulletin, Vol.51 No.6 634-640.
[4] Xu Ding, Chen Gang and Wang Xian,2013, Direct Numerical Simulation of Channel Turbulence Based on Multi-GPU Lattice Boltzmann Method, Applied mathematics and mechanics, Vol 34 No.9 956-963.
[5] Chen S and Doolen G D, 1998, Lattice Boltmann method for fluid flows, Annual review of fluid mechanics, Vol 30 No.1 329-364.

# Research on Test Method for Low-Temperature Performance Consistency of Power Battery Pack

**Hongwei Wang[1], Jun Liu[2], Qiang Ma[1], Yanling Fu[1], Ziqiang Tao[1], Haiqing Xiao[1] and Hua Bai[1]**

[1]Institute of Industrial and Consumer Product Safety, Chinese Academy of Inspection and Quarantine, No. 11, Ronghua South Road, Yizhuang, Daxing, Beijing, 100176, China
[2]Beijing Petrochemical Engineering Co., Ltd., Beijing 100107, China
Email: wanghongwei-1978@163.com

**Abstract.** Under low temperature condition, the inconsistency that the internal resistance of the battery due to the influence of processing and manufacturing processes can't be effectively tested at constant temperature, therefore, this paper proposes a method for test of the electric performance consistence of the lithium-ion batteries at low temperature, this method can test the electric performance consistency of the lithium-ion batteries at low temperature, effectively solve the problem of cells that have good electric performance consistency under normal temperature condition, but have considerably different electric performance under low temperature condition.

## 1. Introduction

This project proposes a method for test of the electric performance consistence of the lithium-ion batteries at low temperature, this method can test the electric performance consistency of the lithium-ion batteries at low temperature, effectively solve the problem of cells that have good electric performance consistency under normal temperature condition, but have considerably different electric performance under low temperature condition, and can test a large quantities of lithium-ion cells to screen out the acceptable lithium-ion cells to form appropriate battery pack, to ensure the battery pack have a better consistency under low temperature environment, improve the quality of lithium-ion battery pack and save the costs of consumers for battery replacement, and has a certain economic and social significance.

## 2. Method and Procedures

The test steps are as follows:

- Adopt the voltage value reduction method to select the lithium-ion cells with higher self-discharge.

  The specific steps are as follows: (1) conduct chemical storage and aging storage of more lithium-ion cells for a certain period of time in turn, and conduct 10 cycles of charging and discharging activation of lithium-ion cells, and then charge to fully charged state; (2) measure the voltage value V1of each lithium-ion cell, store them for 7 days at room temperature (20°C±2°C) and measure the battery voltage value V2again, and then calculate the voltage drop value $\Delta V = V1 - V2$ before and after the storage of each lithium-ion cell; (3) if the $\Delta V \geq$set voltage drop threshold, the lithium-ion cell has higher self-discharge and is unacceptable lithium-ion cell, if the $\Delta V <$ set voltage drop threshold, the lithium-ion cell has lower self-discharge value and is acceptable lithium-ion cell B1. The set voltage drop threshold can be determined according to the battery type and use requirements.

- For the lithium-ion cell B1, select out the lithium-ion cells of the same capacity difference criterion after processing through capacity check.
  The specific steps are as follows: (1) measure the actual battery capacity C1 of each lithium-ion cell B1; (2) grade the lithium-ion cells B1according to the capacity differences: use 10% of the nominal capacity of lithium-ion cell as the capacity difference criterion$\Delta$C1 for grading, i.e. to select the cell Bc1 by (actual battery capacity C1- nominal capacity) / nominal capacity x100%= ±10%.
- Charge and discharge the lithium-ion cells Bc1 at low temperature before grading again the lithium-ion cells BC1.
  The specific steps are as follows: (1) put the lithium-ion cells BC1 in a place under low temperature condition (low temperature range: -40°C - 0°C) for 4 hours, conduct 10 charge-discharge cycles and then charge them to fully charged state; (2) measure the actual battery capacity BC2 of each lithium-ion cell BC1; (3) grade the lithium-ion cells BC1 according to the low-temperature capacity difference criterion $\Delta$C2: use 20% of nominal capacity of cell as the low temperature difference criterion $\Delta$C2 for grading (i.e. (actual battery capacity C2 - nominal capacity) / nominal capacity x 100%= ±20%) to select by conducting capacity grading again to obtain more cells BC2 meeting the above capacity difference criterion; (4) grade and select, and then charge all the cells BC2 to the same state of charge.
- Select more lithium-ion cells BC2 to form appropriate battery pack.

## 3. Experimental Procedure

Test sample. The model and the parameters of cell in this paper as follows:

- The model: domestic battery, rated voltage 3.7 V, rated capacity 11 Ah, internal resistance < 6 mΩ, weight < 320 g.
- Dimensions: 133 mm×66 mm×18 mm.
- The composing of the battery: the cathode material is $LiMn_2O_4$, the anode material is graphite, the electrolyte is $LiPF_6$, EC and DMC, and battery separator is celgard 2325.

Test Instrument: thermostat box, model SPHH-101; Integrated battery tester, MACCOR battery performance test systems; Data acquisition system, model MV2000.

## 4. Results and Discussion

The examples for application of this method are as follows:

Conduct 10 charge-discharge cycles of 10 lithium-ion cells under room temperature (20°C±2°C) condition by using the constant-current and constant-voltage charge system and constant-current discharge system, i.e. discharge the batteries to be tested at 3500 mA under the test temperature condition until the battery voltage reaches 2.7 V, put them in a place for 1 h, and conduct constant current charge at 3500 mA until the battery voltage reaches 4.2 V, conduct constant-current charge and then stop charging as the charging current drops to 350 mA, so repeat 10 charge-discharge cycles, and measure the voltage value $V_1$of each lithium-ion cell and store them at room temperature (20°C±2°C) for 7 days, then measure the voltage drop value $\Delta V= V_1- V_2$ of each cell before and after storage, and measure and calculate the results as shown in Table 1.

**Table 1.** The measure and calculate result

| Sample No. | voltage $V_1$/V | voltage $V_2$/V | voltage drop value ΔV |
|---|---|---|---|
| 1 | 4.197 | 4.175 | 0.022 |
| 2 | 4.197 | 4.191 | 0.006 |
| 3 | 4.196 | 4.187 | 0.009 |
| 4 | 4.199 | 4.196 | 0.003 |
| 5 | 4.198 | 4.194 | 0.004 |
| 6 | 4.198 | 4.192 | 0.006 |
| 7 | 4.196 | 4.191 | 0.005 |
| 8 | 4.193 | 4.173 | 0.020 |
| 9 | 4.197 | 4.185 | 0.012 |
| 10 | 4.197 | 4.191 | 0.006 |

The voltage drop threshold set for this experiment is 0.20V, we can see from Table 1 that, for the lithium-ion cells No. 1 and No. 8, the ΔV≥ set voltage drop threshold, i.e. ΔV≥0.020V, they are lithium-ion cells with higher self-discharge (unacceptable lithium-ion cells), and the rest of cells are acceptable cells $B_1$ (battery No.: 2,3,4,5,6,7,9,10).

Measure the actual battery capacity $C_1$ of each lithium-ion cell $B_1$, and calculate their corresponding capacity differences, as shown in Table 2.

**Table 2.** The calculate results

| Sample No. | actual battery capacity $C_1$/mAh | $\Delta C_1$ /% | capacity differences /mAh |
|---|---|---|---|
| 2 | 10683.87 | -0.03 | 6.84 |
| 3 | 10675.01 | -0.03 | 6.75 |
| 4 | 10659.94 | -0.03 | 6.60 |
| 5 | 10646.68 | -0.03 | 6.47 |
| 6 | 10352.08 | -0.06 | 3.52 |
| 7 | 10364.25 | -0.06 | 3.64 |
| 9 | 10352.88 | -0.06 | 3.53 |
| 10 | 10338.21 | -0.06 | 3.38 |

Table 2 shows that, the capacity differences of lithium-ion cells are in the range of capacity difference criterion $\Delta C_1$ ($\Delta C_1$= (actual battery capacity $C_1$ -nominal capacity)/nominal capacity × 100%), therefore all lithium-ion cells listed in Table 2 are batteries $B_{C1}$ after grading of capacity differences (battery No. 2, 3, 4, 5, 6, 7, 9, 10), and charged to 100% state of charge.

Put the cells $B_{C1}$ in a place under low temperature condition (-30°C ±2°C) for 4 hours, conduct 10 charge-discharge cycles, and calculate the cycle life $L_{10}$ after $10^{th}$charge-discharge cycle (i.e. the discharge capacity at $10^{th}$charge-discharge cycle / discharge capacity at $1^{st}$charge-discharge cycle x 100%), as shown in Table 3.

**Table 3.** The results

| Sample No. | cycle life / % |
|---|---|
| 2 | 95.38 |
| 3 | 94.86 |
| 4 | 95.97 |
| 5 | 95.57 |
| 6 | 93.11 |
| 7 | 96.21 |
| 9 | 93.98 |
| 10 | 95.66 |

The cycle life criterion of lithium-ion cells set for this experiment is 95%, i.e. the lithium-ion cells with cycle life $L_{10}$>95% are the batteries selected under low temperature environment. The cycle lithium-ion cells with cycle life $L_{10}$>95% selected according to Table 3, are the batteries No. 2, 4, 5, 7, 10, they are the acceptable lithium-ion cells$B_{C2}$selected under low temperature environment.

In the lithium-ion cells $B_{C2}$, select the batteries no.5 and No.10 to connect in series into a battery pack 1. Figure 1 and Figure 2 are the charge capacity curve and discharge capacity curve, respectively, when conducting 5 charge-discharge cycles of battery pack 1 and pack 2 (packs obtained by using ordinary method to conduct voltage difference, internal resistance difference and capacity difference screening) at 0.2C rate under condition of -30°C±2°C.We can see from the comparison in Figure 1 and 2 that the battery packs obtained by using the method of this invention have low decay of charge and discharge capacity, relatively high discharge capacity and good consistency of packs under low temperature.

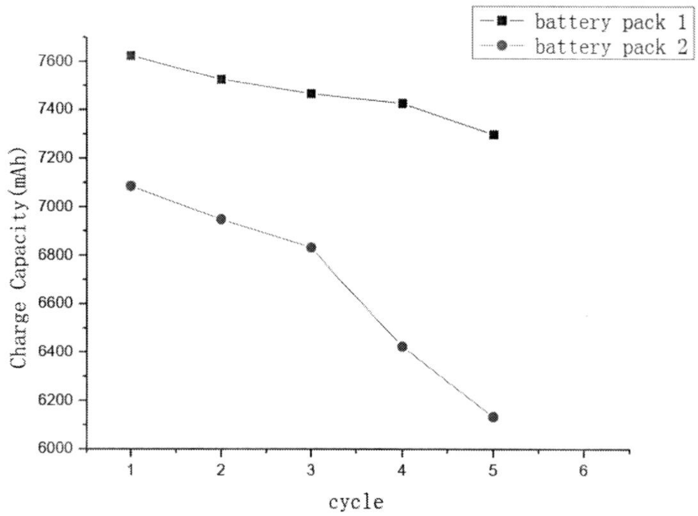


**Figure 1.** Charge Capacity Curve.

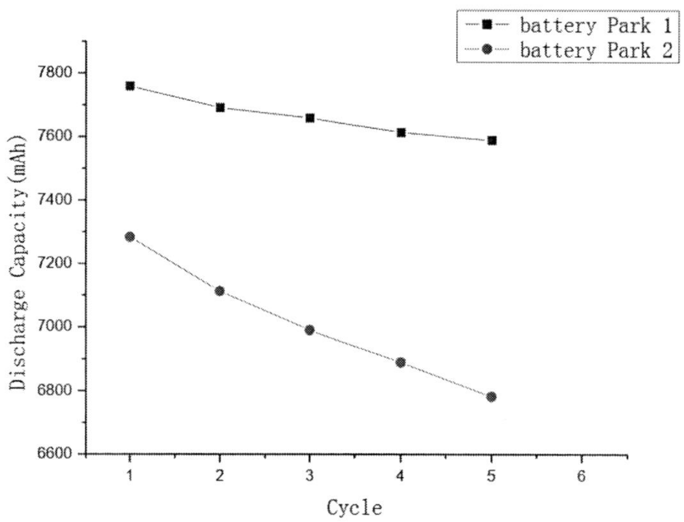


**Figure 2.** Discharge Capacity Curve.

## 5. Conclusions

It is thus clear that, this project provides a test method for consistency of electric performance of the lithium-ion battery compared with the existing technologies, which can test the consistency of electric performance of the lithium-ion battery under low-temperature condition in a fast and convenient way, effectively solve the disadvantage of various cells that have good electric performance consistency under normal temperature condition, but have considerably different electric performance under low temperature condition, and can be promoted to be applied in testing large quantities of batteries and screening to group batteries. It can improve the quality of lithium-ion battery pack and save the costs of consumers for battery replacement, and has certain economic and social significance.

## Acknowledgments

The authors would like to acknowledge the financial support provided for this research work by Fund: Fundamental scientific research projects of Chinese Academy of Inspection and Quarantine (CAIQ) 2017JK024, National key R & D program of China (2017YFF0210703).

## References

[1] Hu Xin-guo. Power battery technology and application 2009. Chemical Industry Press
[2] Lin Cheng-tao, Li Teng and Tian Guan-yun. The life test of Li-ion power battery for electric vehicle 2010. Battery Bimonthly, 40, p 23
[3] GB/Z18333.1-2001 Lithium-ion batteries for electric road vehicles 2001
[4] QCT 743-2006 Lithium-ion Batteries for electric vehicles. National Development and Reform Commission 2006
[5] Wei Hong-bing, Wang Cai-juan and Zhao Yong. Investigation of impact test of Li-ion cells 2011. Chinese battery industry, 16, p 209

# Glutaraldehyde/Polyvinyl Alcohol Crosslinked Nanosphere Modified Corn Stalk Reinforced Polypropylene Composite

**Shanshan Liu, Heyi Ge, Yu Zou and Juan Chen***

School of Materials Science and Engineering, University of Jinan, Jinan 250022, People's Republic of China

*corresponding author: mse_chenj@ujn.edu.cn

**Abstract.** In this paper, polypropylene (PP) based wood-plastic composite (WPC) was prepared from corn straw fiber (CSF), and a new green way was found for the treatment of agricultural waste. In order to improve the interfacial compatibility between CSF and PP matrix, the glutaraldehyde (GA) and polyvinyl alcohol (PVA) crosslinked modified CSF was used. GA and PVA were crosslinked to form polymer particles on the surface of CSF. The WPC was characterized by SEM, FTIR, XRD and electronic universal testing machine. When the loading of modified CSF was 30 wt%, the flexural strength of WPC was up to 62 MPa, which was 12.3% higher than that of WPC reinforced by untreated CSF. Meanwhile, the composite reinforced by modified CSF had lower water absorption compared with that of the untreated CSF composite.

## 1. Introduction

Wood-plastic composite (WPC) is a green, high-performance material processed by different molding processes using wood powder and thermoplastic resin. The waste corn stalk is an ideal reinforce of WPC, which has good tensile strength, tensile modulus, hardness, flexural strength and flexural modulus. Compared with wood fiber, corn stalk fiber has better dimensional stability, processing properties, good environmental performance and low cost [1].

Interfacial compatibility between natural fibers and plastics is an important topic for scholars. Plant fibers contain a large number of polar groups, while plastic surface has non-polar groups. As a result, the compatibility between the two materials is poor, influencing the mechanical properties of the composites. Some methods have been adopted to improve the interfacial compatibility between plant fibers and plastics, including physical methods and chemical methods [2].

Polyvinyl alcohol (PVA) is a hydrolysis product of polyvinyl acetate, a polymer obtained by polymerization of a monomer without polymerization of a monomer [3]. It has excellent solvent resistance, biocompatibility, non-toxicity and good film-forming properties, and can form a smooth, strong tear-resistant film [4]. Glutaraldehyde (GA, $C_5H_8O_2$), is a transparent oily liquid with a pungent odor. It is soluble in hot water and can be used as a processing aid for food industry, wood preservatives, pharmaceuticals and synthetic raw materials for polymers. Chiellini et al. confirmed that PVA could cross-link with GA and GA could react with the hydroxyl functional group of PVA polymer chain to form an acetal bond [5].

Researchers have conducted a lot of research on the poor combination between natural fibers and thermoplastic substrates. However, the method for improve the interface compatibility by modifying CSF using GA and PVA has not been reported so far. In this paper, GA and PVA were modified on the CSF by a crosslinking reaction and the modified CSF was extruded with PP. The modified CSF and the composite were named as GA/PVA-CSF and GA/PVA-CSF-PP, respectively. GA/PVA-CSF had better compatibility with PP matrix than the untreated CSF (UT-CSF). GA/PVA-CSF-PP had higher interface bonding strength and mechanical properties such as flexural strength and flexural modulus.

## 2. Experiment

### *2.1. Materials*

Corn stalks were collected directly from farmland. PP was supplied by Lanyu Plastic Raw Materials Co., Ltd., Dongguan. $MgCl_2 \cdot 6H_2O$, PVA, GA and liquid were purchased from Aladdin Reagent Company, Shanghai. All chemical reagents were analytical reagent (AR).

### *2.2. Preparation of GA/PVA-CSF*

After corn stalks were dried at 105 °C for 12 h, the inner core was stripped out and the skin was crushed by two steps of coarse grinding and fine grinding. The standard test sieve was used to sieve the CSF. The particle size was about 80-100 mesh. 295 g deionized water, 5 g PVA and 50 mL liquid paraffin were added to the three-neck flask, and the system was stirred at 95 °C for 1 h. 5 g CSF was added into the flask and stirred for 1 h. 3 mL GA and 4.5 g $MgCl_2 \cdot 6H_2O$ were added in the system and stirred at 95 °C for another 3 h. The product was filtered, washed with absolute ethanol and deionized water until the washing solution was clear and dried at 85 °C for 24 h to get GA/PVA-CSF.

### *2.3. Preparation of Composites*

CSF and PP coupling agent were uniformly mixed (the mass ratio of CSF to PP was given in Table1), which were slowly added to the hopper of the twin-screw extruder at a feed rate of 5 r/min and an extrusion rate of 35 r/min. The temperature of the set twin-screw extruder was 160°C, 165°C, 170°C and 175°C. The hopper temperature of the micro injection molding machine was 205°C. The mold temperature was 40°C and the injection pressure was 5 MPa.

**Table 1.** Constituents of composites and abbreviations for respective samples

| Samples | GA/PVA-CSF Content (wt%) | Samples | UT-CSF Content (wt%) |
|---|---|---|---|
| 10GA/PVA-CSF-PP | 10 | 10UT-CSF-PP | 10 |
| 20GA/PVA-CSF-PP | 20 | 20UT-CSF-PP | 20 |
| 30GA/PVA-CSF-PP | 30 | 30UT-CSF-PP | 30 |
| 40GA/PVA-CSF-PP | 40 | 40UT-CSF-PP | 40 |

## 3. Results and Discussion

Figure 1a is a view of the surface morphology of the UT-CSF. The UT-CSF has a smooth surface. After modified with GA/PVA, as can be seen from Figure 1b, many white polymer particles are formed on the surface of the CSF, which are the product of the cross-linking. It can be seen from Figure 1c that the spherical particles have a diameter of about 100 nm.

Figure 1d is the image of 40UT-CSF-PP section. It can be seen that the CSF is not tightly bonded to the PP matrix, and there is a gap between them. This is because the UT-CSF has a smooth surface, and it is difficult to be compatible with PP. It can be seen from Figure 1e,f that the GA/PVA-CSF is closely combined with PP, and there is no void between the two parts. The reason is that the surface of CSF has produced a lot of polymer particles after cross-linking modification, which increases the roughness of the CSF surface. Therefore, the modification of CSF can strengthen the physical bond between CSF and PP matrix.

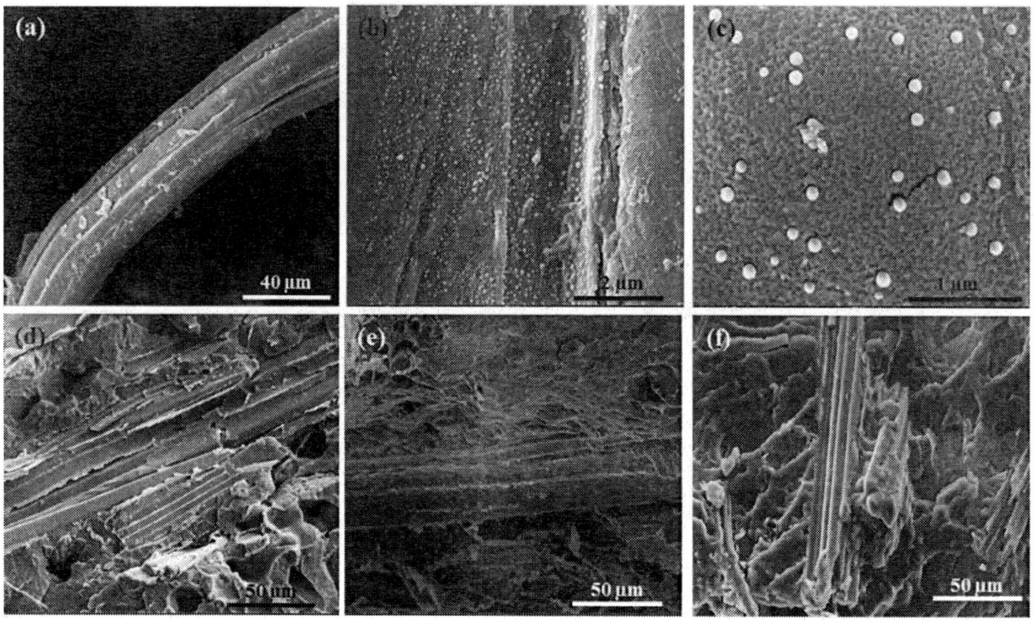


**Figure 1.** SEM of the samples: (a) UT-CSF, (b) GA/PVA-CSF, (c) magnified GA/PVA-CSF, (d) 40UT-CSF-PP, (e) 40GA/PVA-CSF-PP and (f) 30GA/PVA-CSF-PP.

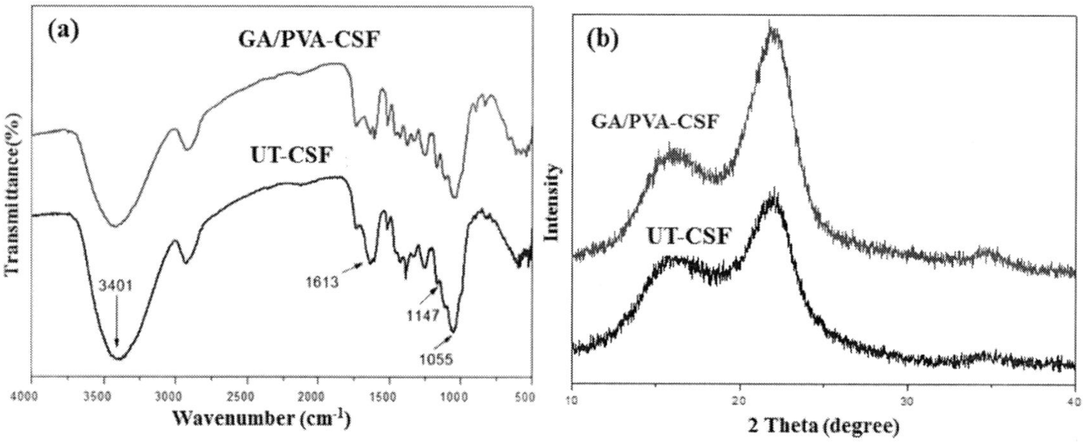


**Figure 2.** FTIR of CSF (a) and XRD of CSF (b)

The FTIR of the two kinds of CSF is shown in Figure 2a. The band at 3500~3300 $cm^{-1}$ is correspond to the stretching vibration of cellulose hydroxyl group (-OH), and the band at 1600~1900 $cm^{-1}$ is the stretching vibration peak of C=O bond in cellulose. The band at 1100~1300 $cm^{-1}$ is the stretching vibration peak of C-O [6-8]. It can be seen from Figure 2a that the single peak at 1631$cm^{-1}$ has been splited into two independent peaks around 1600 $cm^{-1}$, and the peak at 1147$cm^{-1}$ is increased on the FTIR spectrum of GA/PVA-CSF compared to UT-CSF, indicating that the polymer particles have formed by the cross-linking reaction of GA and PVA on the surface of CSF.

The XRD spectrum of the two kinds of CSF is shown in Figure 2b. Obviously, there are two main peaks at 16° and 22°, reflecting the (100) and (002) planes of hemicellulose and lignin. The peak at 16° represents the amorphous phase of hemicellulose and lignin, and the peak at 22° is closely related to the crystalline cellulose phase of CSF. The crystallinity index (CrI) calculated from equation (1) is a parameter used to measure the content of crystalline cellulose in cellulosic materials [9].

$$CrI\ (\%)=\frac{I_{002}-I_{am}}{I_{am}}\times 100\ \% \qquad (1)$$

where $I_{002}$ is the peak intensity at a 2θ angle close to 22 and $I_{am}$ is the amorphous counter reading at a 2θ angle of around 18. Compared with UT-CSF, the CrI of GA/PVA-CSF increases from 33.6% to 35%, which is mainly due to the damage the amorphous substance at high temperature and the formation of polymer particles on the surface of CSF.

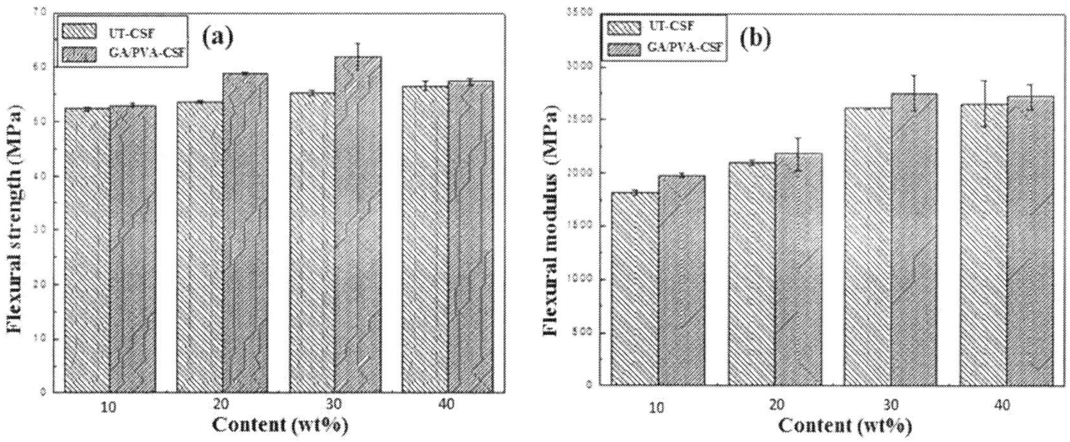


**Figure 3.** Flexural strength (a) and flexural modulus (b) of the composites.

Figure 3 is a graph of flexural strength and flexural modulus of WPC. For the two kinds of CSF, both the flexural strength and the flexural modulus increase when the CSF loading is less than 30 wt%. It can be seen from Figure 3a that the flexural strength of GA/PVA-CSF-PP is higher than that of UT-CSF-PP because the surface of the modified CSF generates polymer particles, which makes the surface of CSF rougher. The contact area between the CSF and the PP becomes larger, and the interface bonding strength is increased, so that the flexural strength is enhanced. The 30GA/PVA-CSF-PP has the highest flexural strength of 62 MPa. When the loading of GA/PVA-CSF is more than 30 wt%, the flexural strength of the composite is reduced. Meanwhile, it can be seen from Figure 3b that the flexural modulus of GA/PVA-CSF-PP is higher than that of UT-CSF-PP, indicating that the ability of GA/PVA-CSF-PP to resist flexural deformation within the elastic limit is further improved. It is verified that the interfacial bonding strength between the modified CSF and the PP matrix is enhanced.

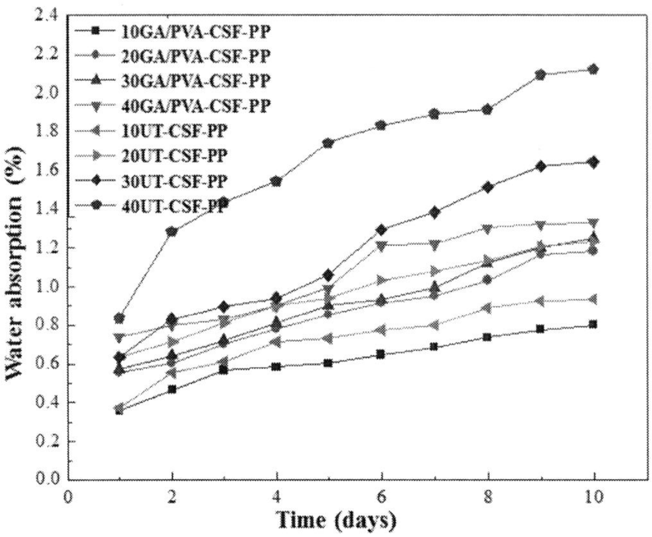


**Figure 4.** Water absorption rate of the prepared composites.

The interface compatibility of WPC is important factor affecting the water absorption of WPC. If the water absorption rate is too high, the defects such as damage of interface, fracture deformation, and mildew will occur. Therefore, it is of great significance to reduce the water absorption of WPC.

Figure 4 shows the water absorption of the prepared WPCs. It can be seen that after 10 days, the water absorption rate of 10GA/PVA-CSF-PP is the lowest; the highest water absorption of 40UT-CSF-PP is 2.12%. The water absorption of GA/PVA-CSF-PP is lower than that of UT-CSF-PP because the PP matrix is more tightly bound to CSF in GA/PVA-CSF-PP. The more the percentage of CSF is added, the higher the water absorption rate. This is because there is a large amount of hydroxyl groups in the cellulose, which makes CSF polar, and CSF has the same polarity as water, so the water absorption increases.

## 4. Conclusion

This study provides a new method to modify CSF to improve the interface compatibility of WPC. The microstructures of GA/PVA-CSF and GA/PVA-CSF-PP were characterized. The flexural properties and the water absorption of WPC were studied. Through the cross-linking reaction, a large number of nanoparticles were formed on the surface of the CSF. The presence of these particles could increase the roughness of the surface of the CSF, thereby increasing the occlusal force area between CSF and PP. The interfacial compatibility of the composite was improved. Moreover, it was found by flexural performance test that the flexural strength and the flexural modulus of GA/PVA-CSF-PP were higher than those of UT-CSF-PP. The water absorption performance test results showed that the water absorption of GA/PVA-CSF-PP was less than that of UT-CSF-PP. Therefore, GA/PVA modified CSF can effectively improve the compatibility and the strength of CSF-PP composite. This study provides an effective solution to the interface compatibility problem of WPC.

## Acknowledgments

This work is supported by National Key R&D Program of China (Grant No.2016YFC0701005).

## References

[1] Krishnaiah P, Ratnam C, Manickam S 2017. Ultrason. Sonochem. **34** 729
[2] Kumar N, Mireja S, Khandelwal V, Arun B and Manik G 2017 Compos. Part B-Eng **109** 277

[3] Alves P, Carvalho R, Moraes I, Luciano C, Bittante A and Sobral P 2011 *Food Hydrocolloid.* **25** 1751
[4] Alfayyadh A, Lotfy S, Basfar A and Khalil M 2017 *Prog. Nat. Sci-Mater.* **27** 316
[5] Mansur H, Sadahira C, Souza A Mansur A 2008 *Mat. Sci. Eng. R* **28** 539
[6] Mattos B, Misso A, Cademartori P, Lima E, Magalhães W and Gatto D 2014 *Constr. Build. Mater.* **61** 60
[7] Ismail H, Edyhan M and Wirjosentono B 2002 *Polym. Testing* **21** 139
[8] Stana-Kleinschek K, Strnad S and Ribitsch V 1999 *Polym. Eng. Sci.* **39** 1412
[9] Motaung T and Mtibe A 2015 *Matei. Sci. & Appl.* **6** 1022

# Effect of the Cross-Linking Modification on Corn Stalk Fiber Reinforced Polypropylene Composites

**Shanshan Liu, Ke Li, Yu Zou and Juan Chen***

School of Materials Science and Engineering, University of Jinan, Jinan 250022, People's Republic of China

*Corresponding author: mse_chenj@ujn.edu.cn

**Abstract.** In this study, corn stalk fiber (CSF) was used as a reinforcer to prepare wood plastic composite (WPC). The primitive CSF was subjected to alkali treatment in a low alkali solution to expose the cellulose fiber. Alkali-treated CSF (AT-CSF) was cross-linking modified with chitosan and glutaraldehyde. Then, the untreated CSF (UT-CSF) and the cross-linked modified CSF (CL-CSF) were added to polypropylene (PP) with different contents (10 wt%, 20 wt%, 30 wt%, 40 wt%), respectively. The morphology, mechanical, thermal and water absorption behaviours of WPC were investigated in detail. The results indicated that the crosslinking modification is an efficient method to improve the mechanical properties of WPC. Compared with UT-CSF, the crystallinity index (CrI) of CL-CSF was increased from 33.8% to 39.5%. The cross-linking modification promoted an increase in flexural strength of 19.2% and an increase in flexural modulus of 8.3% compared to the composite reinforced by UT-CSF. The water absorption of the composite reinforced by CL-CSF was lower than the composite reinforced by UT-CSF.

## 1. Introduction

In recent years, natural fibers as reinforcements in wood plastic composites (WPC) have attracted the attention of many researchers. Low density, high specific tensile strength, high specific stiffness, and high aspect ratio are the main advantages of natural fibers [1], which make it possible for natural fibers to replace glass fiber [2].

Chitosan is the deacetylated form of chitin that is extracted from the shells of crustaceans. It has a unique advantage in some respects because of its biocompatibility, biodegradability, hydrophilicity, nontoxicity and nonantigenicity [3]. In addition, there is a tremendous potential for chitosan to develop as an excellent bio-coupling agent with its presence of rich hydroxyl and amino groups. Cross-linking modification of chitosan has wide application in the field of preparation of adsorbents and food packaging films [4].

Glutaraldehyde (GA) is a well-studied commercial adhesive, which has the efficient ability to react with the -$NH_2$ groups of chitosan [5]. Chitosan can be converted into a stable network structure with the help of crosslinking modification [6]. Hydrophilic amino groups are consumed resulting in the increase of hydrophobicity of the cross-linked product.

In this paper, the chemical crosslinking technique was supposed to increase the compatibility between the corn stalk fiber (CSF) and polypropylene (PP) matrix. CSF was modified via the crosslinking of chitosan with GA. The composite reinforced by the modified CSF was obtained by injection molding. The loading of CSF was increased from 0 to 40 wt%. Moreover, the mechanical properties and the water stability of the composite were investigated in detail.

## 2. Experiment

### *2.1. Treatment of CSF*

Corn stalks were harvested on a local farm in Shan Dong Province, China. The stalks were crushed and screened using 80 mesh sieves to get powder. The dried powder was soaked in an alkali solution (3 wt%) at room temperature for 2 h. Then, washed several times with distilled water to pH= 7 and dried at 105°C for 24 h to get alkali-treated CSF (AT-CSF).

1 g chitosan was dissolved into 60 mL acetic acid solution (2 wt%) with stirring at room temperature for 30 min. 10 g AT-CSF were distributed in the chitosan solution with the help of mechanical stirring. Simultaneously, 60 mL liquid paraffin and 3 mL span 80 were added to the above system. The system was stirred at 40 ° C for 30 min to get emulsion.

GA (5 wt %) was added into the above emulsion and the emulsion was adjusted to pH = 9-11. The system was stirred at 60 °C for 120 min and the obtained powder was washed with distilled water and absolute ethanol until the cleaner was transparent. The powder was dried to obtain cross-linked modified corn stalk fiber (CL-CSF).

### *2.2. Preparation of Composite*

The formulations of the composite studied in this paper are given in table 1. The CSF used in this experiment was dried before mixing and the components were homogenized. The homogenized mixture was melt mixed by using a twin screw extruder (SJZS-10B with average screw diameter of 20 mm and average L/D ratio of 40). The hopper of the twin-screw extruder at a feed rate was 5 r/min and extrusion rate was 35 r/min. The temperatures of the four regions of twin-screw extruder were: 160°C, 165 °C, 170 °C and 175 °C. The hopper temperature of the micro injection molding machine was 205 °C. The mold temperature was 40 °C and the injection pressure was 5 MPa. The composite combined CSF and PP was named as CSF-PP.

**Table 1.** Constituents of composite and abbreviations for respective samples

| Samples | Fiber Content (wt%) | Samples | Fiber Content (wt%) |
|---|---|---|---|
| 10UT*-CSF-PP | 10 | 10CL-CSF-PP | 10 |
| 20UT-CSF-PP | 20 | 20CL-CSF-PP | 20 |
| 30UT-CSF-PP | 30 | 30CL-CSF-PP | 30 |
| 40UT-CSF-PP | 40 | 40CL-CSF-PP | 40 |

*UT: untreated

### *2.3. Characterization*

The Fourier transform infrared spectrophotometer (FTIR) spectra and the scanning electron microscope (SEM) were used to characterize fibers and composites. The flexural test of the WPC was carried out according to the ISO standards 527-1 using an advanced universal testing machine (Shanghai Shenli Testing Machine Company, China). The impact strength of all samples was measured on an impact type test machine (Chengde Jinjian Testing Machine Company, China) in accordance with ISO standards 179-1. The water absorption test was conducted in accordance with ASTM: D570-98 (2010).27.

The crystallinity of CSF was analysed with an X-ray diffractometer (D8 ADVANCE, Bruker, Germany). The crystallinity index (CrI) of CSF was calculated according to peak height method (equation (1)) [7].

$$CrI\ (\%)=\frac{I_{002}-I_{am}}{I_{am}}\times 100\ \% \tag{1}$$

where $I_{002}$ is the peak intensity at a 2θ angle close to 22° and $I_{am}$ is the amorphous counter reading at a 2θ angle of around 18°.

## 3. Results and Discussion

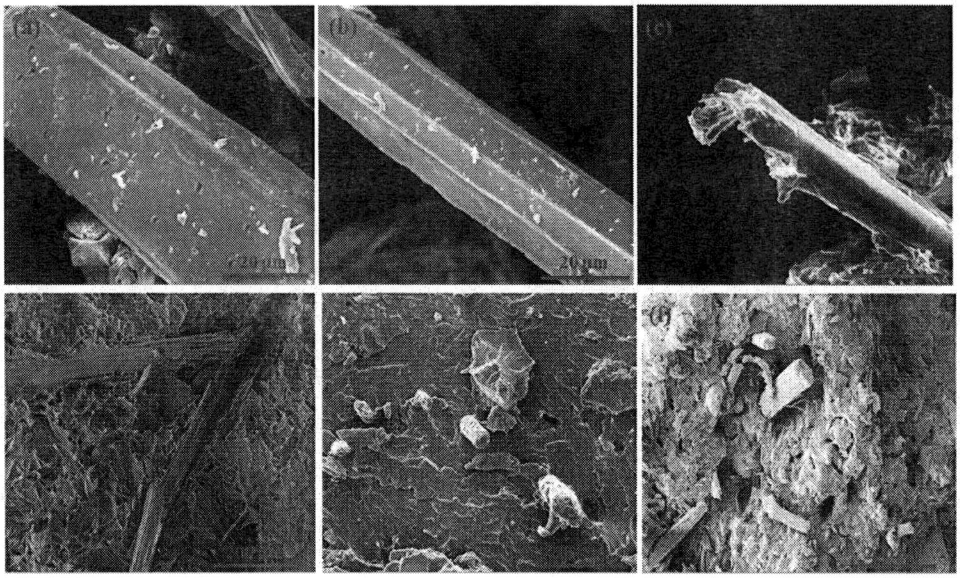

**Figure 1.** SEM images of samples: (a) UT-CSF, (b) AT-CSF, (c) CL-CSF, (d) 20UT-CSF-PP, (e) 20CL-CSF-PP and (f) 40CL-CSF-PP

The different morphology of the CSF is shown in figure 1a-c. It is clear that the surface of UT-CSF contains impurities such as waxy and gummy materials. After the alkali treatment, the pectin, waxy and other impurities on the surface of CSF were removed and the cleaner and rougher surface can be observed in figure 1b. Besides, the alkali treatment made the structure of AT-CSF loosened, which is conductive for the crosslinking modification of CSF. Figure 1c shows that the surface of the CL-CSF is covered with transparent film, which is a product of the crosslinking of chitosan with glutaraldehyde. It can be clearly observed that the polymer film shows a wrinkled and distorted appearance on the surface of the CL-CSF, indicating that the polymer film has good toughness.

Figure 1d-e shows the fracture surface of the composites reinforced by UT-CSFs and CL-CSFs. Figure 1d shows the poor interfacial bonding of PP matrix to UT-CSF. Obviously, there are fiber pullouts and fiber breakage in the fracture surface, which indicates the poor interfacial adhesion between UT-CSF and PP matrix. The CL-CSF and PP matrix are in close contact with each other from the observation of figure 1e-f. The right hand twisted fibers can be observed, which are attributed to the stress transfer between the CSF and PP matrix. The excellent wetting of CL-CSF in PP matrix is presented in figure 1e-f. The cross-linking modification results in the better interfacial adhesion of the composite, which will facilitate mechanical properties and water resistence.

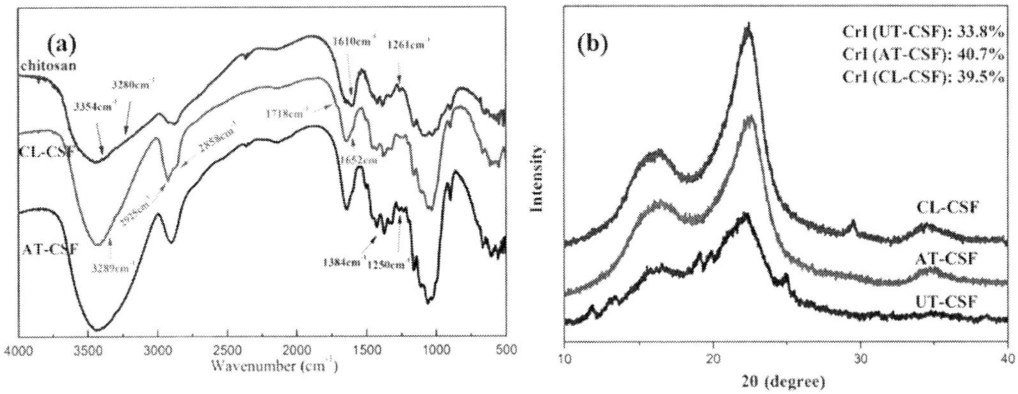


**Figure 2.** FTIR spectra of samples (a) and XRD patterns of samples (b)

The FTIR spectra of chitosan, CL-CSF and AT-CSF are shown in figure 2a. The broad peaks of chitosan located around 3354 $cm^{-1}$ and 3280 $cm^{-1}$ correspond to the stretching vibrations of N-H and O-H respectively [8]. However, some changes have occurred in the FTIR of CL-CSF. The two independent peak of CL-CSF at 3354 $cm^{-1}$ and 3280 $cm^{-1}$ have been converted to single peak at 3289 $cm^{-1}$ could be attributed to the consumption of amino groups on the surface of chitosan [9]. Compared with chitosan, the FTIR of CL-CSF presents C=N stretching vibration at 1652 $cm^{-1}$, indicating that chemical reaction occurred during the cross-linking.

In figure 2b, XRD of three CSF shows two main peaks at 16° and 22°, which are assigned to (1 0 0) and (0 0 2) reflection planes [7]. The peak at $2\theta=16°$ refers to the amorphous phase of hemicellulose and lignin. The sharp peak with high intensity at $2\theta=22°$, which is closely related to the crystalline cellulose phase of CSF. Fig. 2b clearly indicated that the alkali treatment and the crosslinking modification affect crystalline phase of CSF. The crystallinity index (CrI) is an important parameter for interpreting changes in the cellulose structure of plant fibers [8]. The CrI of AT-CSF increased to 40.7%, which was mainly related to the removal of amorphous material of CSF. Compared to AT-CSF, the CrI of CL-CSF is reduced to 39.5%. The decrease mainly caused by the crosslinking product attached to the surface of the fiber. Therefore, the cross-linked product covered by CSF surface is a network structure [9].

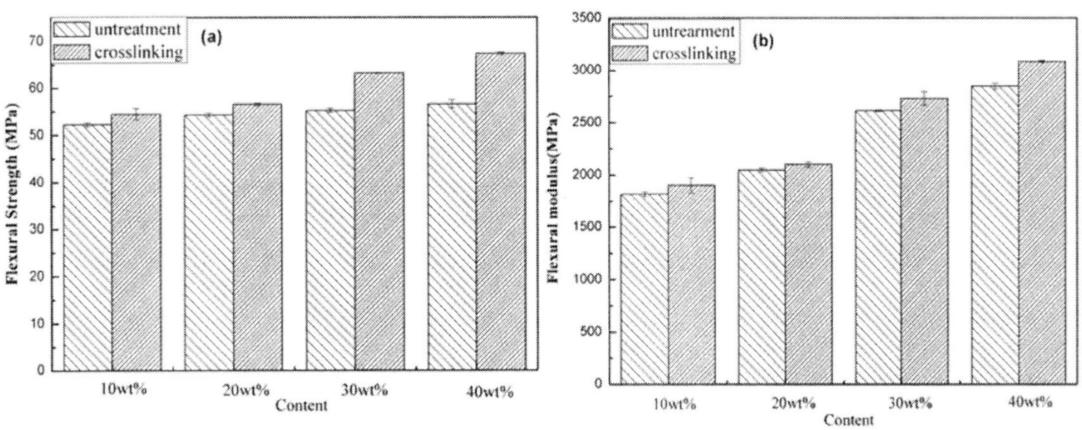


**Figure 3.** Flexural properties of composites reinforced by UT-CSF and CL-CSF: (a) bending strength and (b) flexural modulus

The flexural strength and modulus of composites reinforced by UT-CSF and CL-CSF have been presented in figure 3. It can be seen that both the flexural strength and the flexural modulus of the composites increase as the content of CSF increases. The loading of CL-CSF (10 wt%, 20 wt%, 30 wt%, 40 wt%) promotes an increase in flexural strength of 4.2%, 4.0%, 14.5% and 19.2%, an increase in flexural modulus of 4.6%, 2.5%, 4.5% and 8.3% compared to those of UT-CSF composite.

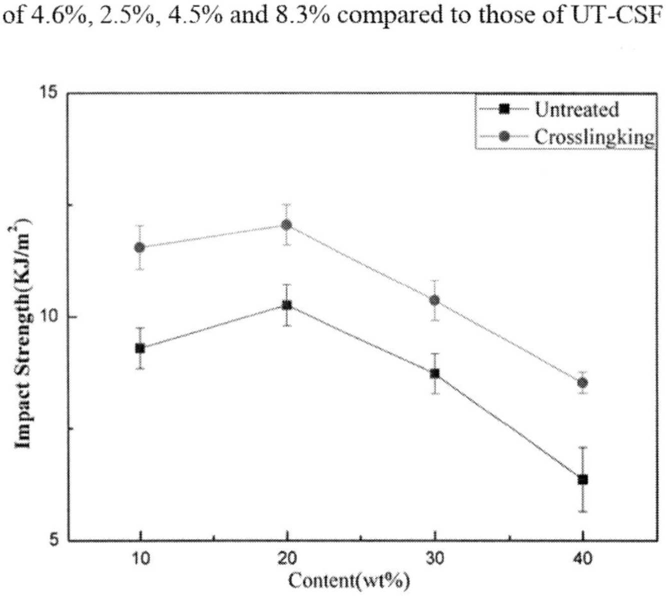


**Figure 4.** Impact strength of composites

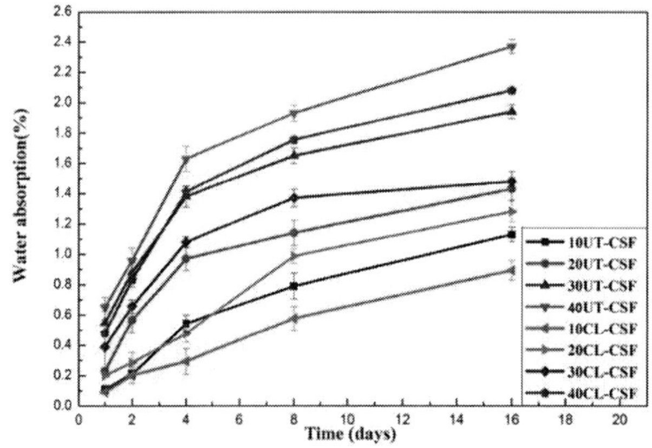


**Figure 5.** Water absorption curves of composites

The impact strength of the composites is shown in figure 4. The impact strength of the two composites increases first and then decreases with the increase of CSF content. When the CSF content is 20 wt%, the impact strength is the highest. Compared to UT-CSF-PP, the impact strength of CL-CSF-PP with different CSF content is increased 24.2%, 17.5%, 18.7%, and 34%, respectively. This is because the polymer film layer on the CL-CSF can improve the wettability between the CSF and the PP matrix.

Water absorption is of great significance for the application of WPC. The water absorption of composites over time is shown in figure 5. Obviously, CL-CSF-PP has lower water absorption than UT-CSF-PP. The water absorption of 40CL-CSF-PP (2.08 %) was lower than that of 40UT-CSF-PP (2.37 %) at 16 days. The lower water absorption implied that the cross-linking modification improved the interfacial compatibility between the CSF and PP.

## 4. Conclusion

In this study, the PP composites reinforced by UT-CSF and CL-CSF were fabricated by an injection molding machine. CL-CSF significantly enhanced the mechanical properties of the composite. Compared with UT-CSF, the CrI of CL-CSF was increased from 33.8% to 39.5%. The addition of CL-CSF promoted an increase in flexural strength of 4.2%, 4.0%, 14.5% and 19.2% and an increase in flexural modulus of 4.6%, 2.5%, 4.5% and 8.3% compared to UT-CSF composite. The water absorption of CL-CSF-PP was lower than that of UT-CSF-PP. The results indicate that the crosslinking modification is a promising method to improve the mechanical properties of WPC.

## Acknowledgments

This work is supported by National Key R&D Program of China (Grant No.2016YFC0701005).

## References

[1] Madsen B, Thygesen A and Lilholt H 2009 Compo. Sci. Technol. 69 1057
[2] Wambua P, Ivens J, and Verpoest I 2003 Compos. Sci. Technol. 63 1259
[3] Cheung H, Lau K, Lu T and Hui D 2007 Compos. B 38 291
[4] Kobayashi T and Furukawa I 1995 J. Antibact Antifung Agents 23 343
[5] Migneault I, Dartiguenave C, Bertrand M and Waldron K 2004 Biotechniques 37 790
[6] Ji X, Li B, Yuan B and Guo M 2017 Carbohydr. Polym. 176 273
[7] Motaung T E and Mtibe A 2015 Matei. Sci. & Appl. 6 1022-1032
[8] Li B, Shan C, Zhou Q, Fang Y, Wang Y, Xu F, Han L, Ibrahim M, Guo L, Xie G and Sun G 2013 Mar. Drugs 11 1534
[9] Kildeeva N, Perminov P, Vladimirov L Novikov W and Mikhailov S 2009 Russ. J. Bioorg. Chem. 35 360

# The Combustion Characteristics of Coal Blends under Low Oxygen Atmosphere Using the Thermogravimetric Analyzer

**Liangliang Liu[a], Zhiqing Wang[a,*] and Yukun Qin[a,b]**

[a] National Engineering Lab for Coal-fired Pollutants Emission Reduction, Shandong Provincial Key Lab of Energy Carbon Reduction and Resource Utilization, Shandong University, Jinan, China, 250061.

[b] School of Power Engineering, Harbin Institute of Technology, Harbin, China, 150001

Email[*]: jackywzq@sdu.edu.cn

**Abstract.** Experiments on coal blends combustion were carried out using simultaneous TGA dynamic runs under low oxygen atmosphere. With the increasing of bituminous coal, the initial combustion temperature increases, however there is no obvious change in the burnout temperature. During coal blends combustion, lignite is dominant in the ignition stage, while bituminous coal has more influence on the burnout stage. The combustion characteristic index (S) indicates that adding bituminous coal will reduce the combustion characteristic of coal. The activation energy (E) is calculated by Coats–Redfern method. The results indicate that with the addition of bituminous coal, the activation increases from 12.58 to 38.4 KJ/mol.

## 1. Introduction

China is the world's largest consumer of coal, and it is difficult to change the dominant position of coal power in the short term. However, low energy efficiency, and high pollutant emissions restrict the coal-based energy production and consumption [1]. Dynamic coal blending is a clean coal technology suitable for China's national conditions, which is conducive to improving the utilization rate of coal thermal energy and reducing the emission of pollutants from coal burning [2].

Domestic and foreign scholars have conducted a series of researches on the combustion characteristics and the release of NOx of mixed coal. Tianyang Chi [3] investigated the characteristic parameters (i.e., relative ignition temperature, maximum ignition points, oscillation frequency, and fluctuation ratio and combustion dynamic energy) with the flame monitoring system. Anxin Weng [4] surveyed the burn-out performance of mixed coal can be improved with the graded combustion. And when the proportion of bituminous coal was 25%, the NOx emission was the lowest. Fernando Rubiera [5] researched the combustion behavior of coal blends in TGA.

Coal pyrolysis is the initial stage of gasification, liquefaction and combustion, which is very important for its transformation. Pyrolysis will vary depending on the conditions (coal type, final temperature, particle size, atmosphere, heating rate, pressure) used in the pyrolysis process [6-10]. However, most of the current investigations are focusing on pyrolysis under inert conditions and combustion with oxygen-rich conditions [11-12]. According to efficient low-nitrogen combustion technology [13] proposed by our research group, coal will be first pyrolyzed under the low-oxygen tail gas, and then burned out in the furnace. Therefore, it is necessary to study the pyrolysis characteristics of coal under the atmosphere of low oxygen.

In this research, the pyrolysis characteristics of lignite, bituminous coal and there blends under 8%$O_2$ atmosphere were studied using TGA analysis. In addition, a reasonable evaluation of coal pyrolysis and kinetic analysis were used to calculate the activation energy of the process.

## 2. Experimental

### *2.1. Materials*

The lignite and bituminous coal were ground to 80-100 mesh sizes. According to mass ratio, the coal samples were mixed evenly. The coal ratio and serial number of mixed coal were shown in table 1. Ultimate and proximate analysis of coal samples were shown in Table 2.

**Table 1.** The serial number of coals

| Samples | S1 | S2 | S3 | S4 | S5 |
|---|---|---|---|---|---|
| Lignite, | 100% | 70% | 50% | 30% | 0% |
| Bituminous | 0% | 30% | 50% | 70% | 100% |

**Table 2.** Proximate and ultimate analysis of the samples

| | Ultimate analysis | | | | | Proximate analysis | | | |
|---|---|---|---|---|---|---|---|---|---|
| | $C_{ad}$,% | $H_{ad}$,% | $N_{ad}$,% | $S_{ad}$,% | $O_{ad}$,% | $M_{ad}$,% | $A_{ad}$,% | $V_{ad}$,% | $FC_{ad}$,% |
| **S1** | 69.41 | 2.27 | 0.92 | 0.37 | 27.03 | 30.6 | 6.46 | 58.2 | 34.92 |
| **S2** | 69.11 | 2.89 | 0.95 | 0.47 | 26.58 | 22.47 | 7.8 | 51.93 | 39.38 |
| **S3** | 70.44 | 3.3 | 0.99 | 0.51 | 24.76 | 17.39 | 8.6 | 48.64 | 44.14 |
| **S4** | 70.78 | 3.65 | 1.02 | 0.54 | 24.01 | 12.91 | 9.27 | 44.12 | 45.40 |
| **S5** | 70.64 | 4.32 | 1.04 | 0.7 | 23.3 | 4.23 | 10.38 | 37.43 | 52.55 |

### *2.2. Experimental Method*

Pyrolysis behavior of two coals and the blends were studied with 50 ml/min flow in a simultaneous thermal analyzer (model TGA/SDTA851e). Samples were submitted to dynamic runs carried out up to 900°C at the heating rate ($\beta$=dT/dt) of 15°C/min.

## 3. Results and Discussion

### *3.1. Thermal Analysis of Samples*

Fig.1a and b shows the weight loss and the instantaneous weight loss rate along with different oxygen concentrations. The separate combustion characteristics of lignite and bituminous coal were greatly different, which indicated that coal rank had great influence on combustion characteristics.

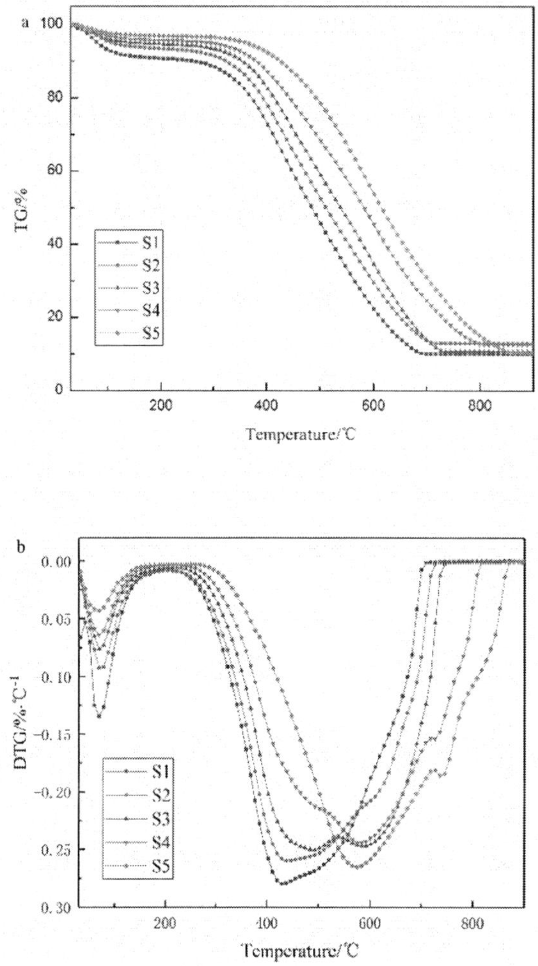


**Figure 1.** (a) The weight loss. (b) the instantaneous weight loss rate.

In this paper, the ignition temperature (Ti) was defined by the TG-DTG tangent method [14]. The burnout temperature (Tb) was detected as temperature of 98% of the final weight loss. DTGmax was maximum weight loss rate, Tmax was temperature of maximum weight loss rate and DTGmean was the average weight loss rate. In addition, the combustion characteristic index (S) is also determined using the below mentioned equations to evaluate the combustion performance of the samples and the higher S the better the combustion reactivity.

$$S = \frac{\left(\frac{dw}{dt}\right)_{max}\left(\frac{dw}{dt}\right)_{mean}}{T_i^2 T_b} \quad (1)$$

These characteristic parameters of TG-DTG curves were shown in Table3, which showed more evident differences between the samples. As shown in Table3, with bituminous coal ratio increasing, Ti increased from 330°C to 439°C. The change of ignition temperature may be caused by the volatile content of the coals. However, Tb had similar temperatures at about 880°C except the S1, which indicates that the bituminous coal has great effect on the burnout temperature. Tmax rose from 430°C to 583°C with the bituminous coal ratio increasing. In addition, the combustion characteristic index (S) had a downward trend with the bituminous coal ratio increasing.

**Table 3.** The characteristic parameters of TG-DTG curves

| Samples | $T_{max}$ | (dw/dt)max | (dw/dt)mean | Ti | Tb | S |
|---|---|---|---|---|---|---|
| | °C | $(\%\cdot°C^{-1})$ | $(\%\cdot°C^{-1})$ | °C | °C | $(\%^2\cdot°C^{-5}\cdot10^{-10})$ |
| S1 | 430 | 0.2798 | 0.1024 | 330 | 810 | 3.25 |
| S2 | 448 | 0.2597 | 0.0989 | 341 | 884 | 2.5 |
| S3 | 493 | 0.2506 | 0.1014 | 363 | 879 | 2.19 |
| S4 | 583 | 0.2417 | 0.0992 | 391 | 879 | 1.78 |
| S5 | 583 | 0.2644 | 0.1016 | 439 | 884 | 1.58 |

*3.2. Kinetic Analysis*

Generalizing, the rate of macromolecular compound can be generally described by:

$$\frac{d\alpha}{dt} = kf(\alpha) \tag{2}$$

where α is the conversion degree; t is time; k the temperature-dependent constant. α and f(α) is defined as:

$$\alpha = \frac{m_0 - m_t}{m_0 - m_\infty} \tag{3}$$

$$f(\alpha) = (1-\alpha)^n \tag{4}$$

where $m_0$ is the mass of initial sample; $m_t$ is the mass of t time sample; $m_\infty$ is the mass of experiment ended sample; n is the reaction series.

According to the Arrhenius:

$$k = A\exp(-E/RT) \tag{5}$$

where A is the frequency factor; E is the activation energy; R is the ideal gas constant and $R = 8.3145 J/(mol\cdot°C)$.

The above rate expression can be transformed into a non-isothermal rate expression describing reaction rate as a function of temperature at a constant $\beta(\beta = dT/dt)$:

$$\frac{d\alpha}{dT} = A\beta^{-1}\exp(-E/RT)f(\alpha) \tag{6}$$

Coats-Redfern method [15] was used to process the experimental data and obtain the inverse of coal kinetic parameters. First, the data after processing is fitted with linear relevancy. The n value corresponding to the line closest to 1 was the reaction series, and then the n value of the line is used. The activation energy E was calculated.

The activation energy (E) and correlation coefficients ($R^2$) for samples are shown in Table4. With the bituminous coal ratio increasing, the activation energy increases from 12.58 kJ/mol to 38.4 kJ/mol.

**Table 4.** E and $R^2$ of the samples

| Samples | S1 | S2 | S3 | S4 | S5 |
|---|---|---|---|---|---|
| E(KJ/mol) | 12.58 | 23.42 | 30.83 | 35.07 | 38.4 |
| $R^2$ | 0.998 | 0.997 | 0.995 | 0.995 | 0.999 |

## 4. Conclusions

In this study, the combustion characteristics and kinetics of coal blends were investigated by TGA.

With the increase of bituminous coal, there are other obvious changes in the TG-DTG curves. The lignite had the lower ignition temperature ($T_i$) and burnout temperature ($T_b$). As the proportion of bituminous coal increases, the initial combustion temperature increases (from 330°C to 439°C), however there is no obvious change in the burnout temperature at about 880°C. In the process of coal blends combustion, lignite plays a leading role in the ignition stage, while bituminous coal has more influence on the burnout stage. The combustion characteristic index (S) (from $3.25\times10^{-10}$ down to $1.58\times10^{-10}$) indicates that higher proportion of bituminous coal and the worse combustion of coal blend.

In addition, the activation energy (E) was calculated by Coats–Redfern method. The results indicated that with the addition of bituminous coal, the activation energy goes higher, which increases from 12.58 KJ/mol to 23.43 KJ/mol, 30.83 KJ/mol, 35.07 KJ/mol and 38.4 KJ/mol, respectively.

**Acknowledgments**
This work was sponsored by the National Key Research and Development Program of China (Contract No. 2017YFB0602002).

**References**

[1] Na Zhang, Noam Lior, et al. The energy situation and its sustainable development strategy in China. Energy. 36(2011) 3639-3649.
[2] Guangkui Liu, Zhengqi Li, et al. Effect of the anthracite ratio of blended coals on the combustion and NOx emission characteristics of a retrofitted down-fired 660-MWe utility boiler. Applied Energy. 95 (2012) 196-201.
[3] Tianyang Chi, Hongjian Zhang, et al. Investigations into the ignition behaviors of pulverized coals and coal blends in a drop tube furnace using flame monitoring techniques. Fuel. 89 (2010) 743–751.
[4] Anxin Weng, Hao Zhou, et al. Experimental research of the NOx generation and burnout characteristics during the combustion of blended coals. Journal of Engineering for Thermal Energy and Power. 19(2004) 242-246.
[5] Rubiera F, Arenilla A, et al. Modification of combustion behavior and NO emissions by coal blending. Fuel Processing Technology. 77(2002) 111-117.
[6] Haiping Yang, Hanping Chen, et al. Influence of pressure on coal pyrolysis and char gasification. Energy Fuels. 21 (2007) 3165-3170.
[7] Cuiguang Yang, Songgeng Li, et al. Pyrolysis behavior of large coal particles in a lab-scale bubbling fluidized bed. Energy Fuels. 27 (2012) 126-132.
[8] Marcela Safarova, Jaroslav Kusy, et al. Pyrolysis of brown coal under different process conditions. Fuel. 84 (2005) 2280-2285.
[9] Tao Wei, Yanrong Zou, et al. Study of the influence of pressure on enhanced gaseous hydrocarbon yield under high pressure-high temperature coal pyrolysis. Fuel 89 (2010) 3590-3597.
[10] Junhong Wang, Juan Du, et al. Study on the structure and pyrolysis characteristics of Chinese western coals. Fuel Processing Technology. 91 (2010) 430-433.
[11] A. Arenillas, F. Rubiera, J.J. Pis. Simultaneous thermogravimetric-mass spectrometric study on the pyrolysis behavior of different rank coals. Journal of Analytical and Applied Pyrolysis. 50 (1999) 31–46.
[12] Zhijun Zhou, Xin Hu, et al. Oxy-fuel combustion characteristics and kinetic parameters of lignite coal from thermo-gravimetric data. Thermochimica Acta. 553 (2013) 54-59.
[13] Zhiqiang Wang, Liangliang Liu, et al. A kind of layer combustion furnace using tail recirculation and superfine pulverized coal reburning and its combustion technology. China, ZL 201510746618.7.
[14] Qingzhao Li, Changsui Zhao, et al. Comparison of pulverized coal combustion in air and in O2/CO2 mixtures by thermo-gravimetric analysis, Journal of Analytical and Applied Pyrolysis. 85 (2009) 521-528.
[15] Liangliang Liu, Zhiqiang Wang, et al. Research on the release of gases during the bituminous coal combustion under low oxygen atmosphere by TG-FTIR. Journal of the Energy Institute. 91 (2018) 323-330.

IOP Conf. Series: Materials Science and Engineering 472 (2019) 012067 doi:10.1088/1757-899X/472/1/012067

# Investigation on Weeds in Vegetable Fields in the Northern Suburbs of Meizhou City, South China

**Qihe Yang, Caixia Zhang , Xiaojuan Yin and Hesheng Yang**

Jiaying University, Meizhou, Guangdong, P. R. China, 514015

**Abstract**. There were a total of 81 weed species in the northern suburbs of Meizhou, belonging to 34 families and 71 genera, and most of them were dicotyledons. Asteraceae and Gramineae were the dominant families. The genera were mainly distributed in the tropical zone, accounting for 71.83% of the total species, with a strong tropical. The short-lived weeds in the first and second year accounts for 53.09% of the total species, and there were more annual weeds. From the light intensity adaptability point of view, these vegetable weeds were dominated by light-demanding and shade-tolerant plants. From the standpoint of water adaptability, most of them were mesophytes. Weeds with a high degree of damage to vegetable crops were usually herbs, meeting the R countermeasures, having strong ecological adaptability, strong reproduction ability, and high competitiveness, and can spread rapidly in the wild. In addition, a total of 19 alien invasive species were found in this weed community, with the largest number in the Asteraceae family and most originating from tropical America. Among them, *Alternanthera philoxeroides, Wedelia trilobata*, etc. can form a single superior community, causing great harm. These weeds have strong resistance to herbicides, insecticides, domestic sewage, etc., so how to control them is still a difficult problem.

Weeds refer to all herbs or subshrubs that are unfavorable to human activities or harmful to production sites. They are a class of highly evolved plant groups under the pressure of both human and natural selection (Li et al., 2013; Fang, 2013). There are over 300,000 plant species named globally, among which about 8000 species are considered weeds (Liang and Qiang, 2014). More than 36,000 plants were found in the records of books and periodicals in China, among them, there are about 1200 weed species belonging to 119 families. Compared with all kinds of crops planted by humans, weeds have strong resistance to stress (Fang, 2013; Fu, 2014). These characteristics make them easy to compete with crops for water, nutrients, light and living space. They cause serious harm to crops, reduce their yield and quality, and finally bring great economic losses to humans. In addition, the pollen or seeds of some weeds, such as *Ambrosia artemisiifolia*, contain toxins that can poison humans, animals and endanger their health. As a result, the integrated control of weeds has become the focus of scientific research. Vegetable field weeds are always important factors to improve the yield and quality of vegetable crops; they are often entangled with vegetables, which lead to the difficulty of manual extraction. In addition, some weeds are also the intermediate hosts of diseases and pests of vegetable crops such as whitefly (*Trialeurodes vaporariorum*), aphid, and spot hidden fly (*Liriomyza maculatahas*), which cause serious harm to various vegetable crops planted in vegetable fields (Fu, 2014) .In this paper, the species composition, floristic origin, life type, light intensity adaptation and water adaptation ecological type of vegetable field weeds in the northern suburb of Meizhou were studied by field survey, literature review and statistical analysis. This paper summarizes the characteristics of weed communities and damage in vegetable field in northern suburb of Meizhou City, and provides scientific basis for the comprehensive control of weeds in vegetable fields in South China.

## 1. The General Situation of the Studied Area and Survey Methods

The vegetable fields under investigation is located in the northern suburb of Meizhou City, North-eastern Guangdong Province, South China (116°8'17"E, 24°20'17"N). The climate type is subtropical monsoon, where enjoys a warm and humid subtropical monsoon climate. The average elevation is about 92.6m, temperature is 20.6~21.4°C, annual sunshine duration is 1714.6~2010.5 h, rainy days are about 150 d, annual rainfall is 1483.4~1798.4 mm, 75% of which concentrated in months from Apr. to Sept., and the frost-free period is about 309 d. The rainy season is mainly from April to September. The soil is mostly red loam, its pH value of red loam soil is 4.5-6.8, which is slightly acidic.

## 2. Results

### *2.1. Species Composition of Weeds*

Through the investigation in the vegetable fields, 81 weeds were found, belonging to 71 genera and 34 families. Among the 81 weeds, there were only 3 species of ferns, belonging to 3 genera and 3 families, accounting for 3.70% of total species, 4.22% of total genera, and 8.82% of the total families respectively. They were *Cercis chinensis*, *Dryopteris chinensis* and *Pteris vittata* respectively. There were 13 species of monocotyledons belonging to 12 genera and 4 families, which account for 16.05% of the total species, 16.90% of the total genera and 11.76% of the total families, which include *Miscanthus floridulus*, *Digitaria sanguinalis*, *Eleusine indica*, *Leptochloa chinensis*, *Kyllinga brevifolia*, *Paspalum thunbergii*, *Setaria viridis*, *Commelina communis*, etc. Dicotyledons, 65 Species, 56 genera, 27 families, which account for 80.25% of the total species, 78.87% of genera, 79.41% of families, such as *Alternanthera philoxeroides*, *Gonostegia hirta*, *Phyllanthus urinaria*, *Anemone hupehensis*, *Pouzolzia zeylanica*, *Conyza canadensis*, *Bidens pilosa*, *Lysimachia alfredii*, *Crassocephalum crepidioides*, *Siegesbeckia orientalis*, *Spilanthes paniculata*, *Achyranthes aspera*, *Amaranthus lividus*, *Euphorbia hypericifolia*, *Abelmoschus moschatus*, *Wedelia trilobata*, *Ageratum conyzoides*, *Artemisia argyi*, *Portulaca oleracea*, *Amaranthus spinosus*, *Ipomoea cairica*, etc. The majority of weeds in the vegetable fields are dicotyledons, followed by moncotyledons, and very few are ferns, indicating that dicotyledons are the dominant group of weeds in the vegetable fields.

There were abundant weeds in the vegetable fields, and the number of species in each family varied greatly. Among the 34 weed families, 19 families contain only one weed . They were Primulaceae, Plantaginaceae, Tiliaceae, Pteridaceae, Cucurbitaceae, Thelypteridaceae, Malvaceae, Crassulaceae, Onagraceae, Verbenaceae, Portulacaceae, Ranunculaceae, Equisetaceae, Vitaceae, Thymelaeaceae, Guttiferae, Araceae, Thelypteridaceae and Commelinaceae; 12 families containing 2~3 species: Labiatae, Leguminosae, Polygonaceae, Umbelliferae, Cyperaceae, Scrophulariaceae, Onvolvulaceae, Oxalidaceae (2 species); Rubiaceae, Solanaceae, Cruciferae, Urticaceae (3) ; 3 families containing 4~9 weed species, including Euphorbiaceae (6), Gramineae (9) and Amaranthaceae (6), and the most dominant family containing more than 10 species was the Compositae (14 ).

The dominant families of weeds in the vegetable fields are defined as the family containing over 5 weed species. There were 4 families, which are Euphorbiaceae, Gramineae, Compositae and Amaranaceae, which account for 11.76% of the total families, including 35 species and 30 genera of weeds, which account for 42.25% of the total genera and 43.21% of total species respectively, which showed that the weeds of the aforementioned 4 families were the main weeds in the vegetable fields. Among these dominant families, Compositae and Gramineae were the first and second dominant family, containing 23 species belonging to 23 genera, which accounted for 32.39% of the total genera and 28.40% of total species, and was also a widely distributed weed in the world. The individual number of Compositae and Gramineae were usually the most in the relatively arid regions, but in the area near water, those of Cyperidae and Amaranaceae tended to be more than those of the two families mentioned above.

### 2.2. Floristic Composition of Weeds

According to distribution types of the genus of spermatophyte and pteridophyte propose by Wu (1991) and Lu (2004), 71 genera of these weeds in the vegetable fields were analyzed (Table 1). There were 51 genera in the tropical distribution type, 34 genera in the pantropical distribution type, which accounted for the largest proportion in the weeds and reached 47.89% of the total genera. Among them, the genus of angiosperms included *Setaria, Cynodon, Cenchrus, Leptochloa, Paspalum, Eleusine, Kyllinga, Commelina, Conyza, Euphorbia, Clerodendrum, Ludwigia, Hedyotis, Ipomoea, Borreria, Corchorus* , *Centella, Spilanthes, Pilea, Eclipta, Alternanthera, Portulaca, Achyranthes, Wedelia, Pharbitis, Celosia, Hydrocotyle, Acalypha, Siegesbeckia, Phyllanthus, Mimosa, Pouzolzia* and the genus of pteridophyte included *Pteris, Cyclosorus.* There were all 4 genera in old world tropical distribution type, tropical Asia to tropical Oceania type, tropical Asia (Indo-Malaysia) types, which accounted for 5.63% of the total genera respectively. The old world tropical distribution type included *Zehneria, Abelmoschus, Cayratia* and *Emilia*; the tropical Asia to tropical Oceania distribution type included *Lophatherum, Epimeredi, Wikstroemia* and *Mazus*; the tropical Asia distribution type included *Typhonium, Pueraria, Ixeris* and *Gonostegia*. There were 3 genera in tropical Asia and tropical America distribution type, which accounted for 4.23% of the total genera and included *Ageratum, Scoparia* and *Parthenium*. But there were only 2 genera in tropical Asia to tropical Africa distribution type, namely *Miscanthus* and *Crassocephalum*, which accounted for 2.82% of the total genera.

In addition to the tropical type, the most common type is the world distribution. There were 15 genera, accounting for 21.13% of the total genera, including *Digitaria, Cyperus, Plantago, Bidens, Rorippa, Hypericum, Polygonum, Galinsoga, Solanum, Salvia, Cardamine, Amaranthus, Anemone, Lysimachia, Oxalis*, all of which belonged to angiosperms. The third was the north temperate distribution type, which included 5 genera and accounted for 7.04%, which was *Equisetum* of pteridophyta, and *Artemisia, Seriphidium, Capsella, Aster* and *Sedum* of angiosperm. This weed community was rich in species and complex in geographical composition, and dominated by tropical distribution type which included 51 genera of the total genera and accounted for 71.83% of the total genera, so the flora of the vegetable fields had a strong tropical characteristics.

**Table 1.** Distribution types of weeds in vegetable fields

| Distribution type | Number of genera | Percentage% |
|---|---|---|
| 1.World /Cosmopolitan | 15 | 21.13 |
| 2.Pantropical distribution/Pantropic | 34 | 47.89 |
| 3.Tropical Asia and tropical America discontinuous distribution/ Tropical Asia and Tropical America disjunct | 3 | 4.23 |
| 4.Old World Tropic | 4 | 5.63 |
| 5.Tropical Asia to tropical Oceania distribution/Tropical Asia and Tropical Australasia | 4 | 5.63 |
| 6.Tropical Asia to tropical Africa/Tropical Asia toTropical Africa | 2 | 2.82 |
| 7.Tropical Asia (Indo-Malaysia) | 4 | 5.63 |
| 8. North Temperate | 5 | 7.04 |
| Total | 71 | 100 |

### 2.3. Life form Composition of Weeds

For plants, their life forms are the types of vegetation which can be reflected by their appearance because of their long-term adaptation to the comprehensive environmental conditions (Yang et al., 2015; Huai et al., 2003). The weed life forms in these vegetable fields are classified into three types: annual, biennial and perennial. The statistical results are shown in Fig. 1.

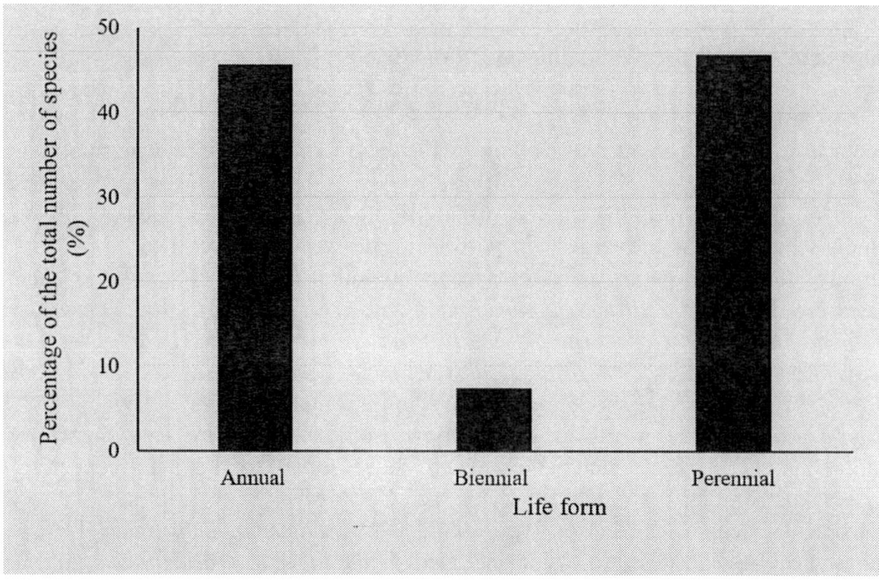


**Figure 1.** Lifeform composition of weeds in the vegetable fields

There were 37 annual weeds belonging to 15 families, accounting for 45.68% of the total species, including *Commelina communis*, *Phyllanthus urinaria*, *Digitaria sanguinalis*, *Conyza canadensis*, *Bidens pilosa*, *Polygonum hydropiper*, *Solanum photeinocarpum*, *Crassocephalum crepidioides*, *Siegesbeckia orientalis*, *Spilanthes paniculata*, *Amaranthus lividus*, *Euphorbia hypericifolia*, *Setaria viridis*, *Eleusine indica*, *Ageratum conyzoides*, *Portulaca oleracea*, *Amaranthus spinosus*, *Corchorus aestuans*, *Pilea microphylla*, *Polygonum aviculare*, *Eclipta prostrata*, *Acalypha australis*, *Mazus japonicus*, *Cenchrus echinatus*, *Parthenium hysterophorus*, *Euphorbia hirta*, *Euphorbia thymifolia*, *Leptochloa chinensis*, *Borreria stricta*, *Celosia argentea*, *Solanum nigrum* var. *pauciflorum*, *Pharbitis nil*, *Galinsoga parviflora*, *Hedyotis diffusa*, *Cardamine hirsuta*, *Hypericum japonicum* and *Emilia sonchifolia*. There were only 6 species of biennial weeds belonging to 4 families, accounting for 7.41% of the total species number of weeds. They were *Abelmoschus moschatus*, *Salvia plebeia*, *Capsella bursa-pastoris*, *Rorippa indica*, *Plantago depressa* and *Epimeredi indica* respectively.

There were 38 perennial weeds belonging to 25 families, accounting for 46.91%of total species. They were *Alternanthera philoxeroides*, *Miscanthus floridulus*, *Gonostegia hirta*, *Anemone hupehensis*, *Pouzolzia zeylanica*, *Setaria viridis*, *Clerodendrum kwangtungense*, *Aster tataricus*, *Lysimachia alfredii*, *Kyllinga brevifolia*, *Chyranthes aspera*, *Alternanthera sessilis*, *Paspalum thunbergii*, *Wedelia trilobata*, *Artemisia argyi*, *Cyperus rotundus*, *Ipomoea cairica*, *Oxalis corniculata*, *Pteris vittata*, *Solanum torvum*, *Scoparia dulcis*, *Lophatherum gracile*, *Centella asiatica*, *Hydrocotyle sibthorpioides*, *Ludwigia octovalvis*, *Oxalis corymbosa*, *Ixeris denticulata*, *Wikstroemia indica*, *Mimosa pudica*, *Equisetum ramosissimum*, *Cyclosorus parasiticus*, *Typhonium divaricatum*, *Sedum emarginatum*, *Euphorbia esula*, *Zehneria indica*, *Cayratia japonica*, *Hedyotis auricularia* and *Pueraria lobata*.

Among these weeds in the vegetable fields, the annual and biannual species accounted for 53.09% of total species, which was slightly more than the perennial species. Their seed germination rate is high, seedlings grow fast, young period of plants is short and mature quickly, so they can be successfully domesticated and constructed (Ma, 2013; Guo and Wang, 2009)' such as *Conyza canadensis*, *Eleusine indica* and *Digitaria sanguinalis*. The proportion of annual plants increases with the increase of human activity, which is closely related to the high intensity and high frequency of human disturbance in vegetable fields. The perennial weeds usually have strong asexual reproduction ability, causing vegetable soil to flip, and the fragmentation of nutrients does not affect the survival of the species.

### *2.4. Light Intensity Adaptation Ecological Types of Weeds*

According to the different requirements for light intensity, they are divided into three categories: light-demanding plants, shade plants and shade tolerant plants (Qiang, 2010). The statistical results of ecological types of adaptation to light intensity of these weed communities are shown in Table 2. In the vegetable fields, the shade tolerant plants were the most, there were 47 species belonging to 26 families, accounting for 58.02% of the total species in the vegetable fields. They were *Anemone hupehensis*, *Pouzolzia zeylanica*, *Clerodendrum kwangtungense*, *Aster tataricus*, *Eclipta prostrata*, *Kyllinga brevifolia*, *Achyranthes aspera*, *Alternanthera philoxeroides*, *A. sessilis*, *Pteris vittata*, *Solanum torvum*, *S. nigrum*, *S. nigrum* var. *pauciflorum*, *Lophatherum gracile*, *Centella asiatica*, *Hydrocotyle sibthorpioides*, *Wikstroemia indica*, *Mimosa pudica*, *Equisetum ramosissimum*, *Sedum emarginatum*, *Euphorbia esula*, *Zehneria indica*, *Cayratia japonica*, *Hedyotis auricularia*, *Capsella bursa-pastoris*, *Rorippa indica*, *Commelina communis*, *Phyllanthus urinaria*, *Bidens pilosa*, *Polygonum hydropiper*, *P. aviculare*, *Crassocephalum crepidioides*, *Siegesbeckia orientalis*, *Spilanthes paniculata*, *Amaranthus lividus*, *Pilea microphylla*, *Acalypha australis*, *Mazus japonicus*, *Galinsoga parviflora*, *Hedyotis diffusa*, *Cardamine hirsuta*, *Hypericum japonicum*, *Euphorbia thymifolia*, *Gonostegia hirta*, *Oxalis corniculata*, *O. corymbosa*, *Lysimachia alfredii*. There were 32 light-demanding weeds belonging to 15 families, accounted for 39.51% of the total species, including *Miscanthus floridulus*, *Wedelia trilobata*, *Cynodon dactylon*, *Paspalum thunbergii*, *Artemisia argyi*, *Cyperus rotundus*, *Ipomoea cairica*, *Scoparia dulcis*, *Ludwigia octovalvis*, *Ixeris denticulata*, *Pueraria lobata*, *Abelmoschus moschatus*, *Salvia plebeia*, *Plantago depressa*, *Epimeredi indica*, *Digitaria sanguinalis*, *Conyza canadensis*, *Euphorbia hypericifolia*, *Setaria viridis*, *Eleusine indica*, *Ageratum conyzoides*, *Portulaca oleracea*, *Amaranthus spinosus*, *Corchorus aestuans*, *Cenchrus echinatus*, *Parthenium hysterophorus*, *Euphorbia hirta*, *Leptochloa chinensis*, *Borreria stricta*, *Celosia argentea*, *Pharbitis nil* and *Emilia sonchifolia*. There were only 2 species of relatively typical shade-demanding plants belonging to 2 families, accounting for 2.47%, which were *Cyclosorus parasiticus* and *Typhonium divaricatum*.

**Table 2.** Analysis of light intensity adaptation ecological type composition of weeds in vegetable fields

| Light intensity adaptation ecological type | Light-demanding plant | Shade plant | Shade tolerant plant |
|---|---|---|---|
| Species number | 32 | 2 | 47 |
| Percentage of total species/% | 39.51 | 2.47 | 58.02 |

The weed community in the vegetable fields was dominated by light-demanding plants, followed by shade tolerant plants, and the typical shade plants were rare. This is because shade plants require wetter habitats and can grow well under lower light intensity, while most parts of the vegetable fields can not provide suitable conditions for their good growth and development.

### *2.5. Water Adaptation Ecological Types of Weeds*

According to the water condition of habitat, plants can be divided into aquatic plants and terrestrial plants, and the terrestrial plants can be divided into three types: xerophyte, hygrophyte and mesophyte (Sun et al., 2016). As the weeds in the vegetable fields could be classified into three types: xerophyte, hygrophyte and mesophyte. The statistical results are shown in Table 4. In this weed community, there were 9 hygrophytes belonging to 7 families, which accounted for 11.11% of the total species and included *Commelina communis*, *Polygonum hydropiper*, *Pouzolzia zeylanica*, *Ludwigia octovalvis*, *Centella asiatica*, *Centipede brevifolia*, *Hydrocotyle sibthorpioides*, and *Pilea microphylla* and *Equisetum ramosissimum*. The other 72 weeds all were mesophytes belonging to 29 families, which accounted for 88.89%, such as *Miscanthus floridulus*, *Cynodon dactylon*, *Digitaria sanguinalis*, *Setaria viridis*, *Paspalum thunbergii*, *Lophatherum gracile*, *Cenchrus echinatus*, *Leptochloa chinensis*, *Eleusine indica*, *Kyllinga brevifolia*, *Cyperus rotundus*,

*Typhonium divaricatum*, *Commelina communis*, *Lysimachia alfredii*, *Plantago depressa* and *Salvia plebeia*.

The weed community in the vegetable fields was obviously dominated by mesophytes, which was nearly 90% of the total species, and accords with the water supply situation under most vegetable habitats, mesophytes grow more luxuriantly than hygrophytes.

**Table 3.** Water-adapted ecotype composition analysis of weeds in vegetable fields

| Water adaptation ecological type | Xerophyte | Hygrophyte | Mesophyte |
|---|---|---|---|
| Species number | 0 | 9 | 72 |
| Percentage of total species/% | 0 | 11.11 | 88.89 |

*2.6. Harm Degree of Weeds to Vegetables*

By competing with crops for water, nutrients and light energy, weeds, thus affect the yield and quality of crops and cause economic losses (Guo and Wang, 2009). Through field investigation and statistics on weeds in the vegetable fields, according to the frequency and quantity of weeds, the harm degree of weeds to vegetable crops was divided into four grades: very serious, serious, moderate and slight ( Fig. 2).

In this weed community, there were very seriously harmful 11 weeds belonging to 7 families, which accounted for 13.58% of the total species. They were *Miscanthus floridulus*, *Commelina communis*, *Cynodon dactylon*, *Digitaria sanguinalis*, *Alternanthera philoxeroides*, *Gonostegia hirta*, *Phyllanthus urinaria*, *Anemone hupehensis*, *Pouzolzia zeylanica*, *Conyza canadensis*, and *Bidens pilosa*. These weeds are herbaceous, widely distributed and adaptable. They were composed of annual and perennial weeds. Most of the annual weeds had short life cycle, fast growth, seed propagation, long flowering and fruiting period, and strong fecundity, such as *Digitaria sanguinalis*. Most perennial weeds had strong asexual reproduction, rapid diffusion and strong viability, such as *Cynodon dactylon*.

There were 27 seriously harmful weeds belonging to 14 families, which accounted for 33.33% of total species, including *Kyllinga brevifolia*, *Setaria viridis*, *Paspalum thunbergii*, *Eleusine indica*, *Cyperus rotundus*, *Pteris vittata*, *Polygonum hydropiper*, *Solanum nigrum* var. *pauciflorum*, *Clerodendrum kwangtungense*, *Aster tataricus*, *Lysimachia alfredii*, *Crassocephalum crepidioides*, *Siegesbeckia orientalis*, *Spilanthes paniculata*, *Achyranthes aspera*, *Alternanthera sessilis*, *Amaranthus lividus*, *Euphorbia hypericifolia*, *Abelmoschus moschatus*, *Wedelia trilobata*, *Ageratum conyzoides*, *Artemisia argyi*, *Portulaca oleracea*, *Amaranthus spinosus*, *Ipomoea cairica*, *Oxalis corniculata* and *Cenchrus echinatus*. These weeds usually grow quickly, has strong fecundity and strong adaptability, so they widely distribute and are common weeds in orchard and vegetable fields.

28 moderate harmful weeds belonging to 20 families accounted for 34.57% of total species. There were *Lophatherum gracile*, *Leptochloa chinensis*, *Cyclosorus parasiticus*, *Equisetum ramosissimum*, *Solanum torvum*, *Salvia plebeia*, *Corchorus aestuans*, *Scoparia dulcis*, *Centella asiatica*, *Hydrocotyle sibthorpioides*, *Pilea microphylla*, *Polygonum aviculare*, *Eclipta prostrata*, *Ludwigia octovalvis*, *Oxalis corymbosa*, *Ixeris denticulata*, *Acalypha australis*, *Capsella bursa-pastoris*, *Wikstroemia indica*, *Rorippa indica*, *Mazus japonicus*, *Mimosa pudica*, *Parthenium hysterophorus*, *Plantago depressa*, *Euphorbia hirta*, *E. thymifolia*, *Borreria stricta*, and *Celosia argentea*. These weeds usually require relatively stringent habitat. They prefer warm and wet environment, and loose and fertile soil, reproduces mainly by seeds, while two ferns reproduce by spores.

15 slightly harmful weeds belonging to 13 families accounted for 18.52%. They were *Typhonium divaricatum*, *Solanum nigrum*, *Epimeredi indica*, *Sedum emarginatum*, *Euphorbia esula*, *Zehneria indica*, *Pharbitis nil*, *Galinsoga parviflora*, *Cayratia japonica*, *Hedyotis diffusa*, *H. auricularia*, *Cardamine hirsuta*, *Hypericum japonicum*, *Pueraria lobata* and *Emilia sonchifolia*.

They were relatively rare in the fields and mainly propagated by seeds in the wild.

Among these weeds, there were 19 invasive alien species belonging to 11 families, 17 species of which were native to the tropics and 12 species were from South America, which were *Alternanthera philoxeroides*, *Conyza canadensis*, *Eleusine indica*, *Cyperus rotundus*, *Crassocephalum crepidioides*, *Abelmoschus moschatus*, *Wedelia trilobata*, *Ageratum conyzoides*, *Amaranthus spinosus*, *Ipomoea cairica*, *Cenchrus echinatus*, *Solanum torvum* , *Scoparia dulcis*, *Pilea microphylla*, *Oxalis corymbosa* , *Mimosa pudica*, *Parthenium hysterophorus*, *Celosia argentea* and *Galinsoga parviflora*.

In addition, among the 19 alien weeds, which of them brought great harm to the crops in the vegetable fields were *Alternanthera philoxeroides*, *Conyza canadensis*, *Crassocephalum crepidioides*, *Eleusine indica*, *Abelmoschus moschatus*, *Wedelia trilobata*, *Ageratum conyzoides*, *Ipomoea cairica*, *Amaranthus spinosus* and *Cyperus rotundus*. The 10 weeds accounted for 52.63% of the total species, which showed that over half of invasive weeds had a great impact on the growth and development of crops. Among them, the harm characteristic of *Alternanthera philoxeroides* is that it can rapidly increase its coverage and form a single superior community (Guo and Wang, 2009) in a short period of time by rapid cloning and branching. *Conyza canadensis* is one of the most widely distributed and invasive species. By producing a large number of achenes and spreading with the wind, it spreads very quickly, and can secrete allelochemicals to inhibit the growth of neighboring plants (Wei et al., 2017; Wu et al., 2014), thus causing serious harm. *Wedelia trilobata* and *Ipomoea cairica* are widely distributed. They can release allelochemicals to the soil by means of stem and leaf leaching, residue decay, root exudation, etc., which inhibit the growth of surrounding plants, crowd out or even kill native species, and form a single superior community (Wei et al., 2017; Wu et al., 2014). The biodiversity of the population diffusion area of these weeds is obviously reduced.

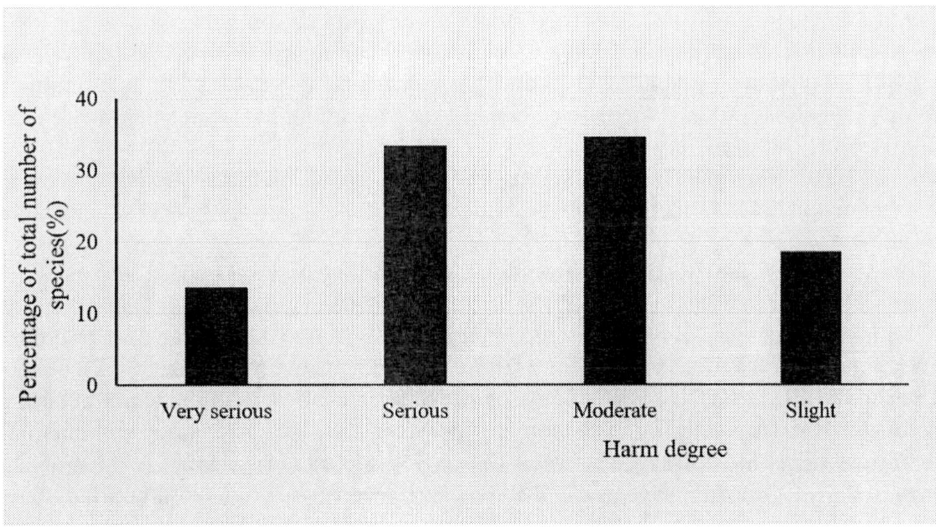


**Figure 2.** Percentage of various harm degree weeds in the vegetable fields

## 3. Discussion and Conclusion

The composition of vegetation in the vegetable fields is usually dominated by herbs and subshrubs due to the frequent artificial removal. The species of woody plants are rare and can only stay in the seedling stage. The results showed that there were 81 weeds belonging to 71 genera and 34 families in the vegetable fields in the northern suburb of Meizhou. Among them, Compositae and Gramineae weeds are widely distributed weeds in the world, and they are also the dominant group of weeds in the vegetable fields, which is inextricably related to the resistance, ecological adaptability, growth potential, and fruitfulness of Compositae and Gramineae weeds (Li et al., 2014;

Xie et al., 2018). The weeds in the vegetable fields have the typical characteristics of continuous cropping and the dicotyledonous weeds account for the majority of the weed species.

The floristic composition of weeds in the vegetable fields was mainly tropical distribution, which shows that the flora of weeds has a strong tropical character, but it has a weaker tropical character than the *Syzygium odoratum* community in Shenzhen Dapeng Peninsula (Zhang et al., 2007) which is also located in the subtropical zone. Moreover, the investigation was conducted only on weeds, and other plants in the area were not investigated and recorded. The floristic composition obtained by statistical analysis could not fully represent the floristic composition of the overall vegetation in this area.

The weeds in the vegetable fields were rich in annual and biennial short-lived weeds, which is consistent with the R-strategy of survival and reproduction of short-lived plants in high-intensity and high-frequency disturbed habitats. Because of their biological characteristics of growth and reproduction, they can be successfully domesticated to establish a population. In the vegetable fields, water and fertilizer conditions are more suitable. Although artificial removal can remove larger individual seedlings and adult plants, but at the same time under the disturbance of continuous tillage, it is beneficial to the germination of weed seeds, so the high intensity and frequency of human interference also increases the proportion of annual plants. Perennials, however, usually have strong ability of asexual reproduction, rotation of vegetable soil and fragmentation of nutrient body will not affect the survival of these weeds, and even contribute to the spread and diffusion of their propagates (Huai et al., 2003; Saito et al., 2010; Scutte, 2014).

The weed community in these vegetable fields was dominated by light-demanding plants and shade tolerant plants. This is due to that light-demanding weeds require sunlight for growth, the vegetable fields usually does not have tall erect woody plant, so it is very suitable for their growth. The tolerance of weeds to light intensity is more extensive, and they can grow well in open space or under the shelter of high crops. While the shade-demanding weeds are fond of shade but not tolerant to strong light irradiation, and they only grow under other higher crops (Sun et al., 2016). In addition, the majority of the weed communities are mesophytes, which are due to those mesophytes have loose requirements for moisture or air humidity than hygrophytes and can be widely distributed in vegetable fields, while hygrophytes need to grow well under humid or high air humidity conditions. Among the 81 weeds in this investigation, 38 species belonging to 21 families caused serious and even very serious harm to crops, most of which were the weeds of Compositae, Gramineae and Amaranaceae. The serous and very serious harmful weeds have no great strict requirement for habitat, so they can be widely distributed and also survive tenaciously in bad environment.

Compared with the weed communities in Hainan Asparagus Orchard (Li et al., 2014) and Qinzhou Pitaya Orchard (Xie et al., 2018) in Guangxi Zhuang Autonomous Region in South China, the common characteristics of the 3 weed communities are the Compositae as the dominant family. *Eleusine indica* also caused great harm to Hainan Asparagus Orchard, Qinzhou Pitaya Orchard and the vegetable fields in the north suburb of Meizhou, which is a gramineous annual plant with developed roots, strong ability of competing for water and fertilizer, rapid reproduction and large amount of seed yield. However, the weeds in the vegetable fields, Hainan Asparagus Orchard and Qinzhou Pitaya Orchard, have great differences between the dominant families and the harm degree of weed species. On the one hand, they may be related to the local specific environment, including soil, water, light, etc. On the other hand, it may be that because of the different planting crops, different growth environment is required, which leads to the difference of human activities in the local area, resulting in the difference of weed species and malignant weed species in different regions (Nagase et al., 2013; Chen et al., 2014). There were 19 alien invasive weeds belonging to 11 families in these vegetable fields. The most common invasive species are the Compositae, such as *Alternanthera philoxeroides*, *Conyza canadensis*, *Eleusine indica*, *Wedelia trilobata*, *Ageratum conyzoides* and *Ipomoea cairica*. Among them, the growth of neighboring plants can be inhibited by allelopathy, such as *Alternanthera philoxeroides*, *Ipomoea cairica*, and *Wedelia trilobata*, which can obviously reduce the biodiversity in the area of population diffusion and cause serious harm to the invading area.

In the vegetable fields, the frequency and amount of application of herbicides, pesticides, domestic sewage and fertilizers could not be small. So these weeds should have strong tolerance and adaptability to these substances. It is essential for these weeds to be uprooted or cleared before fruit or seed ripens to reduce the harm of weeds. The individual number of Compositae and Gramineae are usually the most in the relatively arid regions, but in the area near water, those of Cyperidae and Amaranaceae tend to be more than the two aforementioned families. It is still a technical problem to develop a reasonable physical or chemical method to control weeds in vegetable fields to improve the yield and quality of vegetables.

**Acknowledgement**
This work was financially supported by Key Project of Science and Technology Innovation of Colleges and Universities in Guangdong Province (cxzd1132) and Special Funds of Talent Introduction of Colleges and Universities in Guangdong Province.

**References**
[1] Chen J, Lou T Q, Zhao P L, He J Z 2014 Current situation and countermeasures for control of harmful plants in Yiwu. *Journal of Zhejiang Forestry Science and Technology* 34(4):93-96.
[2] Cirujeda A, Aibar J, Zaragoza C 2011 Remarkable changes of weed species in Spanish cereal fields from 1976 to 2007. *Agronomy for Sustainable Development* 31(4):675-688.
[3] Fang Y S 2013 Analysis of biological characteristics of weeds. *Modern Agricultural Science and Technology*, (7):170-174.
[4] Fu C M, Wang Y T, Zhang F P, Wang Y, Wei Y C 2014 Investigation on diseases,pests and weeds of vegetables in sunlight greenhouse in Jiuquan City. *Gansu Agricultural Science & Technology* (9):26-29.
[5] Guo L J, Wang T 2009 Impact of invasion of exotic plant *Alternanthera philoxeroides* on interspecies association and stability of native plant community. *Chinese Journal of Eco-Agriculture* 17(5):851-856.
[6] Huai H Y, Gao H M, Zhang B, Wu X X, Sun G R 2003 Population Characteristics of *Alternanthera philoxeroides* during its invasion to *Ophiopogon japonicus* turf. *Grassland & Turf* (4):36–38.
[7] Li F H, Sun G J, Ji M, Chu Y F, Wu Y F, Zhang H Y, Han M, Yuan F, Qiu SF, Wang H Q 2013 Analysis of weed species diversity in Jiangsu tea garden. *Acta Agriculturae Jiangxi*, 25(5): 69-71+74.
[8] Li X X, Huang Q Q, Yi K X, Fan Z W, Shen Y D, Chen H T 2014 Weeds in asparagus plantations in Hainan and their control techniques. *Chinese Journal of Tropical Agriculture* 34(7):73-79.
[9] Liang D Y, Qiang S 2014 *List of main crop weeds in China*. (Beijing: China Agricultural Science and Technology Press).
[10] Lu S G 2004 An introduction to the pteridophyte fauna of China *Advances in Plant Science (Vol. 6)*. ed. Li C S (Beijing: higher Education Press), p 29-42.
[11] Ma J S 2013 *A list of invasive plants in China* (Beijing: Higher Education Press)
[12] Nagase A, Dunnett N, Choi M S 2013 Investigation of weed phenology in an establishing semi-extensive green roof *Ecological Engineering*, 58(10):156-164.
[13] Qiang S 2010 Current status and development strategy for weed science in China *Plant Protection*, 36(4):1-5.
[14] Saito K, Azoma K, Rodenburg J 2010 Plant characteristics associated with weed competitiveness of rice under upland and lowland conditions in West Africa.*Field Crops Research* 2010, 116(3):308-317.
[15] Schutte B J et al 2014 and investigation to enhance understanding of the stimulation of weed seedling emergence by soil disturbance *Weed Research* 54(1):1–12.
[16] Sun Y D, Liu J X, Xiong H B, Feng Y H, Li T 2016 Effect of gap light factors on understory plants species distribution in different forest types. *2016 International Conference on Economy, Management and Education Technology*, p 6.

[17] Wei C, Xu YY, Zhuang S Y, et al. Allelopathic effects of water extract and decay solution on seed germination of rice[J].Jiangsu Agricultural Sciences, 2017,45(09),67-70
[18] Wu Y, Si C C, Sun J F, Wang C Y, Du D L 2014 Allelopathic effects of two chemicals in *Wedelia trilobata* root exudates [J]. *Acta Agriculturae Boreali-Occidentalis Sinica*, 23(10):170-178.
[19] Wu Z Y 1991 the areal-types of Chinese genera of seed plants *Acta Botanica Yunnanica*, (supplement Ⅳ):1-139.
[20] Xie H Y, Zhou J Q, Lao J X,Xie S F, Pan W C 2018 Investigation on weeds and their control measures of pitaya orchard in hilly land of Qinzhou City *Modern Agricultural Science and Technology* (1):129-130.
[21] Yang Q H, Pan S F, Lai W N, Yang H S, Kuang W 2015 Investigation on the community characteristics and species diversity of Qiaoxi Geomantic Forest in Meizhou *Guihaia* (6): 833-841.
[22] Zhang Y X, Chen H F, Qin X S, Zhang R J, Xing F W 2007 Characteristics and species diversity of *Syzygium odoratum* community in Dapeng Pninsula,Shenzhen *Guihaia* (4):596-603.

# Comprehensive Evaluation of Geological Environment Carrying Capacity for Lintong District, Xi'an

**Renwei Li and Nianqin Wang**

College of Geology and Environment, Xi'an University of Science and Technology, Xi' an 710054, China.
Email: 13299178276@163.com

**Abstract.** Geological environment carrying capacity is one of the standards that judges the human activities and geological environment whether or not compatible. Carrying out geological environment carrying capacity evaluation and adjusting human activity intensity is the key to achieve sustainable development of regional economy. Based on the variation coefficient method, the paper used the comprehensive evaluation model to evaluate the geological environment carrying capacity of Lintong district, Xi'an. The results show that the carrying capacity of 4 sub-districts excellent, 7 sub-districts is good, 8 sub-districts is medium, and 4 sub-districts is poor. The results can provide references for the development of study area, and achieve the goals of making full use of land resources, regulating the geological environment changing trend and reducing disasters.

## 1. Introduction

Geological environment carrying capacity (GECC) is derived from environmental carrying capacity and is an important part of it [1]. GECC reflects the dialectical relationship between human activities and geological environment, and establishes the link between human activities and geological environment [2]. In 1992, William et al. proposed to use the ecological footprint method to evaluate the GECC by comparing the size of ecological footprint on the level of supply and demand. In 2003, Wackemagel et al. improved further the ecological footprint method [3, 4]. In 2004, Garry et al. evaluated the GECC of New Zealand based on ecological footprint method [5]. In 2008, Yu zhishan et al. improved the concept, characteristics and connotation of the GECC by taking the urban GECC evaluation as an example [6]. In 2014, Ren guangyuan et al. studied the GECC of Ningxia Yanhuang economic zone by 3S technology [7].

Generally, the domestic and foreign scholars have achieved fruitful results on geological environment bearing capacity. However, small regions (such as districts and counties) are not involved and the classification of larger regions is not accurate. The paper takes Lintong district as the study area and evaluates the GECC of study area by the coefficient of variation method on the basis of in-depth understanding of geological environment characteristics.

## 2. Description of the Study Area

Lintong district lies in the northeast of Xi'an and covers an area of 915Km$^2$. By the end of 2017, a total population of 710,000 people and 23 sub-districts were involved in study area. The geographical location of the study area is shown in figure1. From south to north, the landforms of study area are bedrock mountainous area, loess hilly, and loess tableland, diluvium plain and alluvial plain, respectively. Metamorphic rocks, quaternary loess, torrential and alluvium are the main rock-soil masses [8]. The main groundwater type is pore water and it decreases gradually from the north to the south. There are more than ten faults in the region, of which the Linton-Chang'an fault is more active. The maximum annual rainfall of the study area was 954.9mm (2003), oppositely, the minimum annual rainfall was 302.3mm (1995) and the average annual rainfall was 587.7mm. With the development of social economy and the rich historical resources in Lintong district, careless development often results in the destruction of the ecological environment. For the unreasonable mining activities, the slope appearance has changed. As a result, there are many geological environment problems in study area.

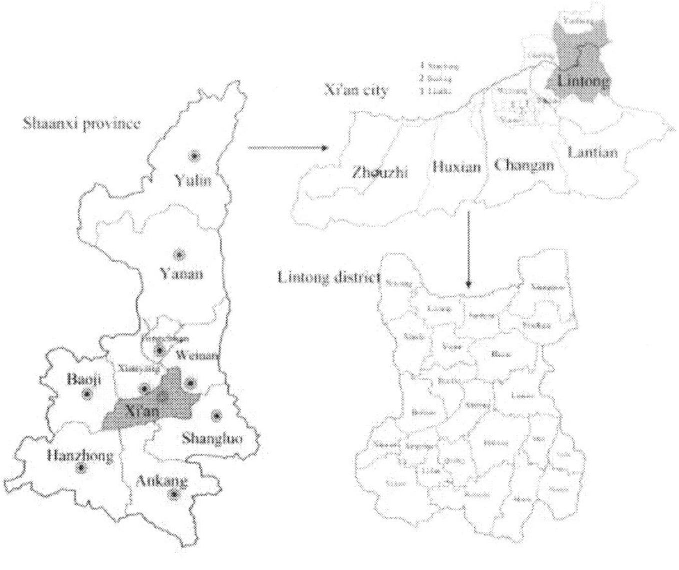


**Figure 1.** The geographical map of study area

## 3. Description of the Study Area

### *3.1. Evaluation Index System*

Referring to administrative division and data collection, the paper takes sub-districts as the evaluation units. As there are no standards to select evaluation indexes [9], the paper takes previous studies and survey data as references. Based on the relationship between bearing carrier and bearing object, 13 evaluation indexes of GECC are finally determined, as shown in figure 2. Evaluation indicators include: altitude, geological hazards, type of rock-soil masses, distances to faults, population density, annual per capita income, industrial types, regional planning, mineral development, slope, annual rainfall, vegetation coverage and available land resources. In order to fully reflect the information contained in the original data and make it easy to calculate, paper takes the bigger and better type assign indexes attribute values. In conclusion, the classification of evaluation indexes is shown in table 1.

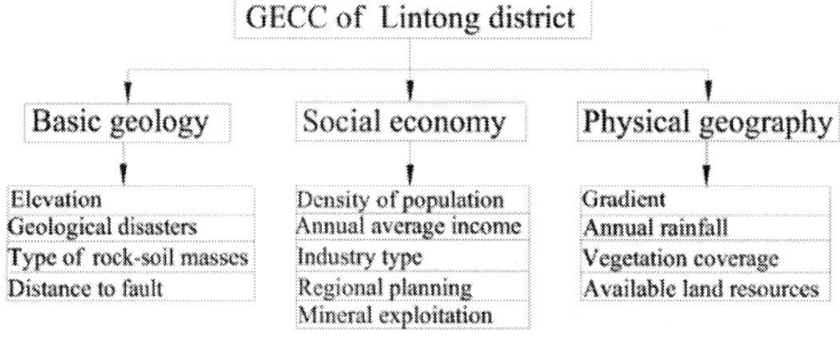


**Figure 2.** The evaluation index system of GECC

**Table 1.** The evaluation index categories of GECC

| Name \ Grades | Excellent(5) | Good(4) | General (3) | Poor(2) | Very poor (1) |
|---|---|---|---|---|---|
| elevation(m) | ≤500 | (500,600] | (600, 700] | (700,800) | ≥800 |
| geological disasters | 2 | 4 | 6 | 8 | ≥10 |
| Type of rock-soil masses | Gravel | Sand | Loess | Silt soil | Silt |
| Distance to fault(Km) | ≥12 | (9, 12) | (6, 9] | (3, 6) | ≤3 |
| Density of population (people/Km$^2$) | ≤300 | (300,600] | (600, 900] | (900,1200) | ≥1200 |
| Annual average income (1,000 dollar) | ≤2 | (2.0, 3.0] | (2.0, 2.5] | (2.5,3.0) | ≥3.0 |
| Industry type | Agriculture——4;Service industry——3.5;Industry——3 | | | | |
| Regional planning | ≥6.0 | (5.0, 6.0) | (4.0, 5.0] | (3.0, 4.0) | ≤3.0 |
| Mineral exploitation | Terrestrial heat——4;Construction material——3.5;Metals——3 | | | | |
| Gradient (°) | ≤3 | (3, 6) | [6, 9) | [9, 12) | ≥12 |
| Annual rainfall (mm) | ≤500 | (500,540] | (540, 580] | (580, 620) | ≥620 |
| Vegetation coverage (%) | ≥30 | (25, 30) | (20, 25] | (15, 20] | ≤15 |
| Available land resources (hectare) | ≥1,400 | (1,400, 1,000] | (1,000, 600] | (600, 200] | ≤200 |

### *3.2. Dimensionless Treatment of Indicators*

In order to reduce the impact on the calculation results, the evaluation indexes are treated dimensionless. The dimensionless calculation processes are as follow:

Positive indicators:
$$Y_{ij} = \frac{X_{ij} - X_{i\min}}{X_{i\max} - X_{i\min}} \quad (1)$$

Negative indicators:
$$Y_{ij} = \frac{X_{i\max} - X_{ij}}{X_{i\max} - X_{i\min}} \quad (2)$$

Where, $Y_{ij}$ is the standardization of the $i^{th}$ class and the $j^{th}$ index. $X_{ij}$ is the actual value of the $i^{th}$ class and the $j^{th}$ index. $X_{i\max}$ is the maximum value of class $i$ index; $X_{i\min}$ is the minimum value of class $i$ index.

### *3.3. Determination of Evaluation Index Weight*

This paper uses the variation coefficient method to calculate the weight of each evaluation index. The variation coefficient method is an objective calculation method, which can avoid the influence caused by subjective factors. This method can use the information contained in each indexes to obtain the weights directly. The calculation process is as follows:

Average:
$$\bar{X}_i = \frac{1}{n}\sum_{j=1}^{n} X_{ij} \quad (3)$$

Standard deviation:
$$S_i = \sqrt{\frac{1}{n-1}\sum_{j=1}^{n}(X_{ij} - \bar{X}_i)} \quad (4)$$

Variable coefficient:
$$Z_i = \frac{S_i}{\bar{X}_i} \quad (5)$$

Weight:
$$W_i = \frac{Z_i}{\sum_{i=1}^{n} Z_i} \quad (6)$$

Where: $i$ is the category of index; $j$ is the specific indicator under each type of index.

### *3.4. The Selection of Evaluation Method*

The evaluation methods of GECC mainly include ecological footprint method, state space method and comprehensive evaluation method. The advantages and disadvantages of each evaluation method are shown in table 2. In this paper, a comprehensive evaluation model was established based on the comprehensive evaluation method. The comprehensive evaluation method is a method to evaluate multiple objects using multiple indexes, also known as the multi-variable comprehensive evaluation method, for example, to evaluate the social development level of different regions and the economic strength of different countries. The comprehensive evaluation model is an effective evaluation model with the advantages of comprehensive analysis, easy calculation and easy to understanding [10]. The comprehensive evaluation model established in this paper can be expressed by the following formula:

**Table 2.** The comparative table of evaluation method

| Evaluation methods | Advantages | Disadvantages |
|---|---|---|
| Ecological footprint | Strong inter-regional contrast | Ignoring the impact of human activity |
| State space | The results are vivid and comprehensive | hard to build models |
| Comprehensive evaluation | Hierarchy is clear and comprehensive | Large data processing |

Comprehensive evaluation method:

$$P = \sum_{i=1}^{n} Y_i \bullet W_i \qquad (7)$$

Where, P is the GECC of the evaluation object. $Y_i$ is the dimensionless value of the first evaluation index; $W_i$ is the weight of the $i^{th}$ evaluation index.

## 4. Evaluation of Geological Environment Carrying Capacity

### *4.1. Values of Geological Environment Carrying Capacity*

The values of sub-districts were obtained by the method mentioned above. The calculation results are shown in table 3.

**Table 3.** The GECC values of sub-districts in study area

| Sub-districts | Values | Sub-districts | Values | Sub-districts | Values | Sub-districts | Values |
|---|---|---|---|---|---|---|---|
| Xuyang | 0.72 | Liyang | 0.75 | Xiquan | 0.50 | Xingzhe | 0.82 |
| Jiaokou | 0.69 | Xiangqiao | 0.62 | Qinling | 0.83 | Daiwang | 0.54 |
| Xinshi | 0.52 | Yujin | 0.49 | Ma'e | 0.48 | Tielu | 0.25 |
| Hezai | 0.68 | Youhuai | 0.83 | Xiekou | 0.51 | Lishan | 0.65 |
| Beitian | 0.87 | Renliu | 0.51 | Renzong | 0.28 | Muzai | 0.20 |
| Xinfeng | 0.66 | Lingkou | 0.52 | Xiaojin | 0.26 | | |

### *4.2. Classification STandards and Zoning Map of GECC*

The values of geological environment carrying capacity were graded by trying standard deviation method, equal interval method, natural breakpoint method, and definition interval method. At last, equal interval method was adopted to divide it into four grades: excellent, good, medium and poor. The classification standard is shown in table 4. According to the classification standards in table 4, MapGIS6.7 was used to obtain the GECC zoning map of Lintong district, as shown in figure 3.

**Table 4.** The standard categories of GECC

| Grades | Excellent | Good | Medium | Poor |
|---|---|---|---|---|
| Intervals | ≥0.8 | (0.55, 0.8) | (0.3,0.55] | ≤0.3 |

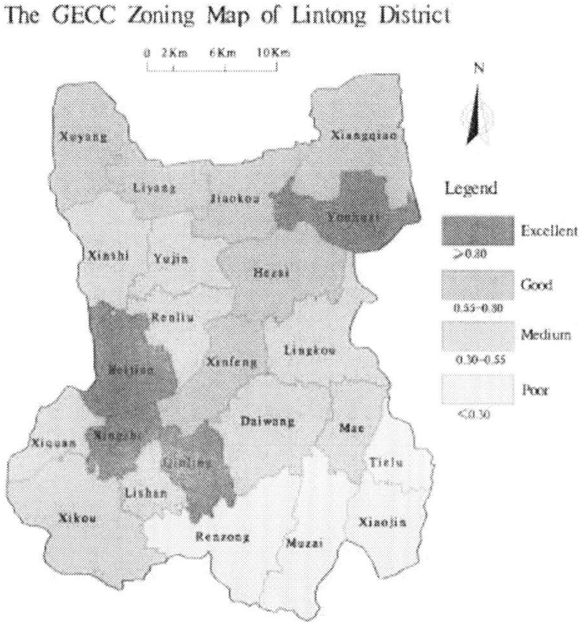


**Figure 3.** The GECC zoning map of Lintong district

*4.3 Test of Evaluation Results*

Due to the particularity of GECC evaluation, there is no effective quantitative method to test the evaluation results. The paper tests the evaluation results from the qualitative aspect. Therefore, the geographical disaster vulnerability zoning map and geomorphology map were used to examine the evaluation results from a qualitative perspective. The geographical hazard vulnerability zoning map and geomorphological map of the study area are shown in fig. 4 and fig. 5.

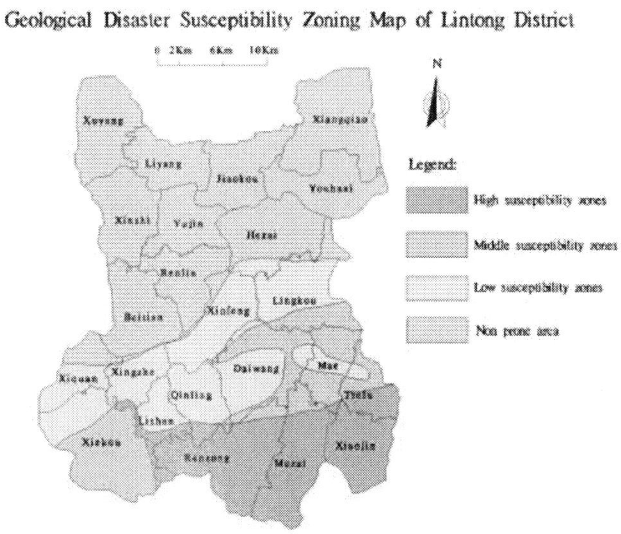


**Figure 4.** Geological disaster susceptibility zoning map

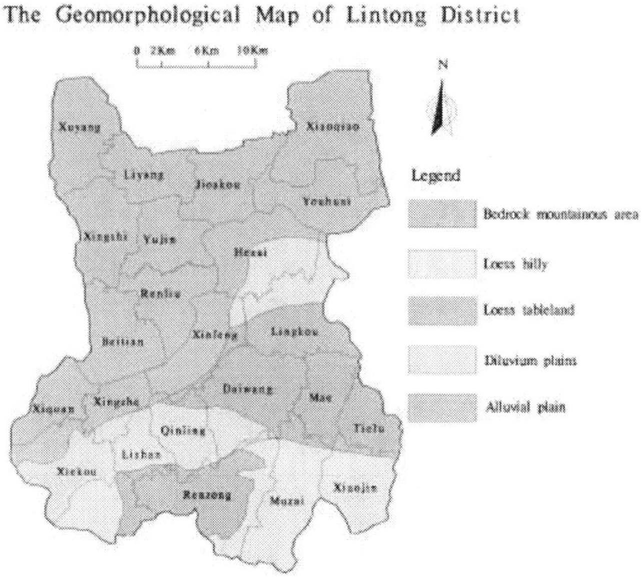


**Figure 5.** The geomorphological map of Lintong district

As shown in figure 4, the bedrock mountainous area and loess hilly area in the south of the study area are medium and high risk areas of geological disasters, the diluvial plain and loess tableland in the central part of the study area are low risk areas of geological disasters, and the alluvial plain in the north of the study area is not prone to geological disasters. Combined with figure 3, it is easy to know that the GECC of the alluvial-diluvial plain is good, and the carrying capacity grade is excellent or good, loess tableland in the middle is medium, bedrock mountainous area and loess hilly region is poor. After testing, the evaluation results are consistent with the real situation and hav*e high accuracy.*

## 5. Conclusions

Based on the variation coefficient method, a comprehensive evaluation model with 13 evaluation indexes was established, which can accurately evaluate the GECC of the study area.

The GECC of study area was evaluated through the comprehensive evaluation model. The carrying capacity was divided into four categories: excellent, good, neutral, medium and poor by the equal interval method and the zoning map of GECC was obtained.

By the evaluation of GECC in study area, the accurate classification of carrying capacity grade was achieved and the results can provide references for the development planning of the research area.

## Acknowledgments

The research was supported by National Natural Science Foundation of China (Grant No. 41572287) and Science & Technology Co-ordination and Innovation Projection of Shaanxi Province (Grant No. 2016KTCL03-19).

## References

[1] Carballo Penela, A., Sebastián Villasante, C., 2008. Applying physical input–output tables of energy to estimate the energy ecological footprint (EEF) of Galicia (NW Spain). *Energy Policy*, 36, 1148-1163.

[2] Clancy, S.A., Worrall, F., Davies, R.J., Gluyas, J.G., 2018. An assessment of the footprint and carrying capacity of oil and gas well sites: The implications for limiting hydrocarbon reserves. *Science of the Total Environment*, 618, 586-594.

[3] Jiang, X., Lu, W.-x., Zhao, H.-q., Yang, Q.-c., Chen, M., 2015. Quantitative evaluation of mining geo-environmental quality in Northeast China: comprehensive index method and support vector machine models. *Environmental Earth Sciences*, 73, 7945-7955.

[4] Kirschbaum, M.U.F. et al., 2012. Comprehensive evaluation of the climate-change implications of shifting land use between forest and grassland: New Zealand as a case study. *Agriculture, Ecosystems & Environment*, 150, 123-138.
[5] Lenzen, M., Murray, S.A., 2001. A modified ecological footprint method and its application to Australia. *Ecological Economics*, 37, 229-255.
[6] Lynch, H.N., Greenberg, G.I., Pollock, M.C., Lewis, A.S., 2014. A comprehensive evaluation of inorganic arsenic in food and considerations for dietary intake analyses. *Science of the Total Environment*, 496, 299-313.
[7] Peng, J., Du, Y., Liu, Y., Hu, X., 2016. How to assess urban development potential in mountain areas? An approach of ecological carrying capacity in the view of coupled human and natural systems. *Ecological Indicators*, 60, 1017-1030.
[8] Pietrapertosa, F., Cosmi, C., Macchiato, M., Salvia, M., Cuomo, V., 2009. Life Cycle Assessment, ExternE and Comprehensive Analysis for an integrated evaluation of the environmental impact of anthropogenic activities. *Renewable and Sustainable Energy Reviews*, 13, 1039-1048.
[9] Wackernagel, M., Monfreda, C., Erb, K.-H., Haberl, H., Schulz, N.B., 2004. Ecological footprint time series of Austria, the Philippines, and South Korea for 1961–1999: comparing the conventional approach to an 'actual land area' approach. *Land Use Policy*, 21, 261-269.
[10] Wang, Z., Wang, Y., Wang, L., Zhang, T., Tang, Z., 2017. Research on the comprehensive evaluation system of eco-geological environmental carrying capacity based on the analytic hierarchy process. *Cluster Computing*.

IOP Conf. Series: Materials Science and Engineering 472 (2019) 012069 doi:10.1088/1757-899X/472/1/012069

# Full Waveform Inversion Based on Wavefield Correlation

**Xue Chen and Helong Yang***

Hebei GEO University, Shijiazhuang 050031, China.
E-mail: 839377060@qq.com

**Abstract.** Low frequencies and initial model play a crucial role in full waveform inversion (FWI). However, in traditional seismic exploration, the frequency range of signals recorded by receivers is limited, and a good initial model for FWI is not available easily, leading to FWI algorithms easily fall into local minima. We propose a wavefield correlation method to improve cycle skipping issue of FWI. Test of 2D frequency domain FWI based on Marmousi model shows that the result of inversion has a better result than conventional method.

## 1. Introduction

Full waveform inversion (FWI) is an optimization procedure of gradually minimizes the waveform difference between modeled data and observed data. The wellknown advantage of FWI is the ability to obtain high resolution and accurate velocity model in complex structure areas, and has recently gained wide interests with specific application in exploration and producing.

As FWI algorithms are a procedure of local optimization, both frequency domain FWI and time domain FWI will be easily trapped in local minima when cycle skipping occurs, leading to the failure of obtaining an accurate velocity model. Therefore, many efforts have been made to solve cycle skipping issue. The most straightforward strategy is the two-step velocity model building [1], that is, first obtain a smooth velocity model with a traditional traveltime tomography, and then FWI is applied on this smooth model to get a more accurate velocity model. Another strategy is the combination of wave equation tomography and full waveform inversion [2-3], but huge computational cost is the drawback of this method. Shin and Cha (2008) proposed Laplace domain FWI algorithms[4], which can recover large scale structures without low frequency data, but this method is sensitive to signal-to-noise ratio and the accuracy of source estimation. Bi and Lin (2014) developed an effective cycle skipping reduction strategy through adaptive data selection for full waveform inversion[5], which assures all input data for FWI is within a half-cycle difference compared with the predicted data by discarding bad traces and muting bad data.

In this paper, we proposed a novel FWI method based on wavefield correlation to overcome cycle skipping issue. We perform conventional FWI method and wavefield correlation FWI method on Marmousi model. Numerical tests show that the new FWI method can obtain a higher resolution result without low frequencies in seismic data, while conventional FWI is trapped in local minima due to cycle skipping issue.

## 2. Full Waveform Inversion Theory

Seismic forward modeling is an important part of FWI. FWI algorithm is an iterative process, and each iteration of inversion should carry out the forward modeling at least twice (for the incident wavefield and the back-propagated residual wavefield), so that the accuracy and costs of the inversion highly depend on the forward modeling. In this paper, we choose the finite difference method based on the two dimensional frequency-domain acoustic wave equation [6]:

$$\frac{\omega^2}{\kappa(x,z)}p(x,z,\omega)+\frac{\partial}{\partial x}\left(\frac{1}{\rho(x,z)}\frac{\partial p(x,z,\omega)}{\partial x}\right)+\frac{\partial}{\partial z}\left(\frac{1}{\rho(x,z)}\frac{\partial p(x,z,\omega)}{\partial z}\right)=-s(x,z,\omega), \quad (1)$$

where $p(x,z,\omega)$ , $\kappa(x,z)$ , and $s(x,z,\omega)$ denote the wavefield, bulk modulus, and source term, respectively, $\omega$ is angle frequency and $\rho(x,z)$ denotes the density.

In the inversion problem, misfit function $C(m)$ is expressed as the difference between real seismic data $d^{obs}$ and simulated seismic data $d^{cal}$:

$$C(m)=\frac{1}{2}[d^{obs}-d^{cal}]^T[d^{obs}-d^{cal}]=\frac{1}{2}\left\|d^{obs}-d^{cal}\right\|_2^2, \quad (2)$$

where 'T' denotes conjugate transpose; $\|\cdot\|_2^2$ denotes $l2$ norm, and $m$ is geological model, such as subsurface velocity model.

## 3. Wavefield Correlation Full Waveform Inversion Method

The process of local optimization is to find the minimum value of the misfit function around the starting model $m_0$, which requires $m_0$ is in the vicinity of true model. If not, or seismic data lacks low frequency components, FWI algorithms will be easily trapped in local minima and yield a bad result. However, low frequency components and good starting model are not always available in the application of exploration and producing. In this paper we propose a novel method to mitigate the effect of cycle skipping issue, named wavefield correlation full waveform inversion. Here we assume the model is elastic and with no attenuation, and the phase don't change along wave propogation:

(1). Compute cross-correlation coefficients of the phase between calculated data and observed data trace by trace before inversion:

$$Cr_i(\tau)=\frac{1}{N}\sum_{t=0}^{N-1}\varphi_i^{obs}(x,t)\varphi_i^{cal}(x,t+\tau), i=1,2,...N, \quad (3)$$

where $Cr$ is cross-correlation coefficients, $\varphi$ denotes phase, $i$ denotes trace number, and $N$ denotes total trace number.

(2). Select the value of $\tau$ when $Cr_i(\tau)$ is maximum. And shift the modeled data as follow:

$$d_i^{\prime cal}(x,t)=d_i^{cal}(x,t+\tau), \quad (4)$$

where $d_i^{\prime cal}$ is the shifted calculated data. In frequency domain, equation (3) can be written as:

$$D_i^{\prime cal}=D_i^{cal}\cdot e^{i\omega\tau}, \quad (5)$$

where $D_i^{cal}$ is frequency domain calculated data, and $D_i^{\prime cal}$ is the shifted frequency domain calculated data.

Thus, the deformation between calculated data and observed data caused by model perturbation is reduced after shifting, which mitigate the probability of cycle skipping.

## 4. Numerical Test

We demonstrate our new method on a modified 2D Marmousi P-wave velocity model. We add a water layer above and don't update the water layer during the nonlinear procedure. Assuming the medel is elastic and with no attenuation. Figure 1(a) and 1(b) are true model and starting model, respectively. The starting velocity model is a smoothed version of true model. We use a Ricker wavelet with peak frequency of 12Hz. The source is located on the surface with the interval 24m. The grid size is 24m and time step is 4ms. All the grid points on the surface acts as receivers. The inversion frequency band is 7~30Hz. Figure 2(a) is the result of conventional FWI method, and figure 2(b) is the result of wavefield correlation FWI method. Figure 3 shows the 1D vertical velocity curves extracted from inversion results (location at receiver No: 241) by different methods. As we start the inversion at a relatively high frequency, conventional FWI method fails to obtain a reliable result due to cycle skipping issue, while our new method obtains a higher resolution result.

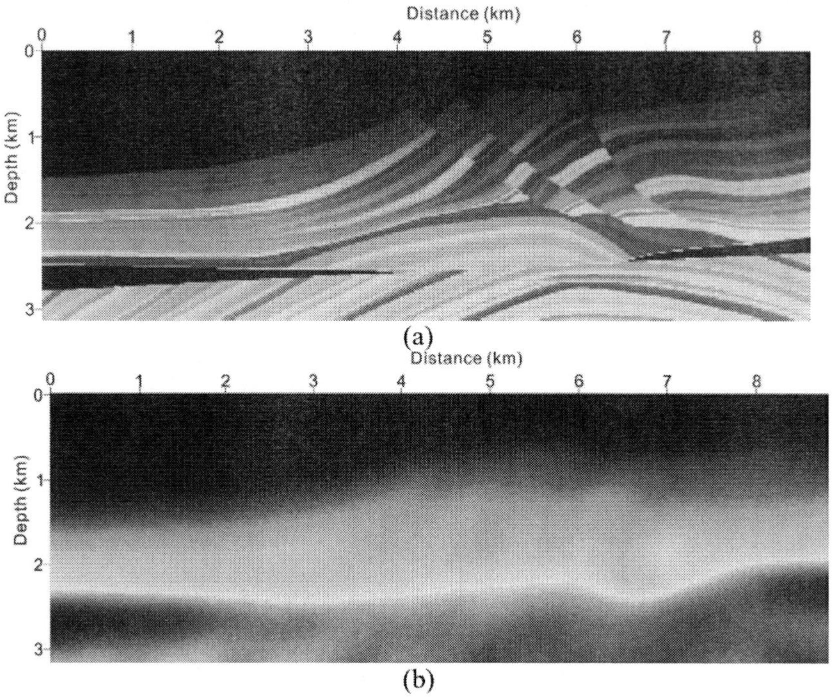


**Figure 1.** FWI model: (a) True velocity model; (b) Starting velocity model for FWI.

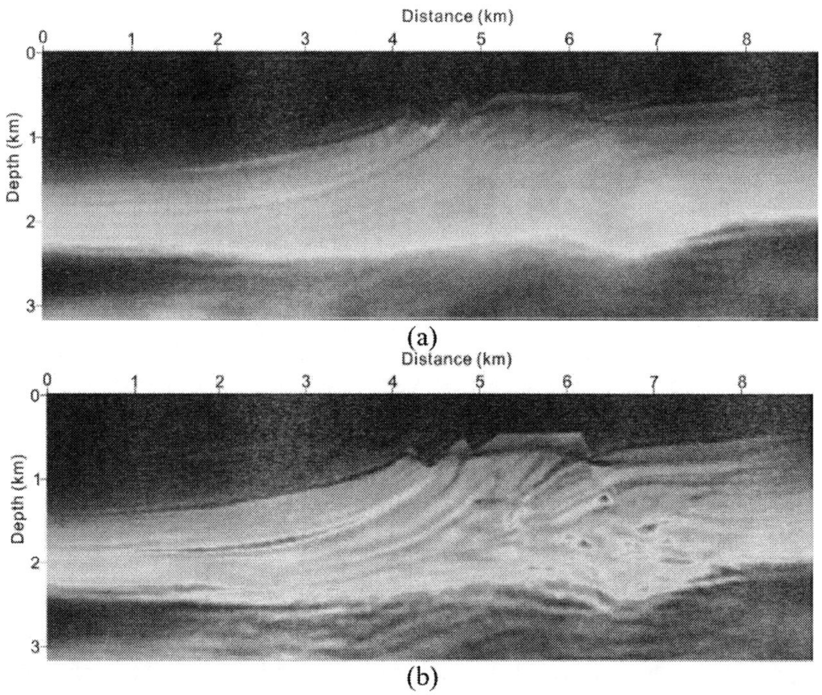


**Figure 2.** Inversion result of FWI: (a) conventional FWI method; (b) wavefield correlation FWI method.

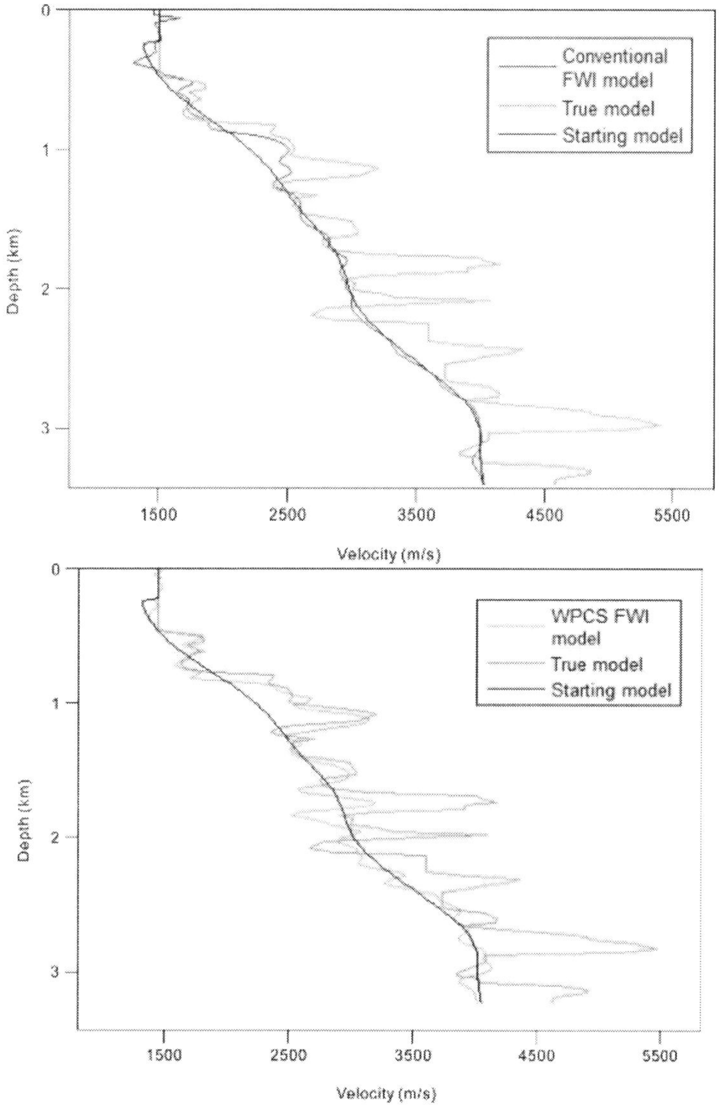


**Figure 3.** 1D vertical velocity curves extracted from inversion models (location at receiver No: 241) by different methods (Black line represents the starting model, green line represents the real model, and red line and blue line represent the conventional FWI model and wavefield correlation FWI model respectively).

## 5. Conclusion

The advantage of FWI is the high resolution of the inversion result by using a large quantity of waveform information. But many weaknesses are still existed in FWI, such as easy to get stuck in local minima and cycle-skipping if the starting model is too far from the actual model or seismic data lacks of low frequencies.

In this paper we propose a wavefield phase correlation shifting FWI method to mitigate the effect of cycle skipping issue on FWI. Seismic wavefield perturbation caused by inaccurate velocity model can be regarded as a 'deformation' deviating from the observed wavefield, and FWI inversion will fall into local minima if the deformation is too large and seismic data without low frequencies. So we can estimate the position of this deformation by correlation algorithm and shift the simulated data to reduce the deformation before inversion. Numerical examples on Marmousi model show that the new method can mitigate the effect of cycle skipping issue and obtain a better result with the absence of low frequencies than conventional FWI method.

**Acknowledgments**
We would like to thank Scientific Research Foundation of Hebei GEO University (BQ 2017026) for the funding support.

**References**

[1] Operto S, Ravaut C, Improta L, Virieux J, Herrero A and Dell'Aversana P 2004 Quantitative imaging of complex structures from dense wide-aperture seismic data by multi-scale traveltime and waveform inversions: a case study *Geophysical Prospecting* **52** 625-51.
[2] Symes W W 2008 Migration velocity analysis and waveform inversion *Geophysical Prospecting* **56** 765–90.
[3] Biondi B and Almomin A 2012 Tomographic full-waveform inversion (TFWI) by combining full waveform inversion with wave-equation migration velocity analysis *82nd Annual International Meeting SEG Expanded Abstracts.*
[4] Shin C and Cha Y H 2008 Waveform inversion in the Laplace domain *Geophysical Journal International* **173** 922-31.
[5] Bi H and Lin T 2014 Effective Cycle Skipping Reduction through Adaptive Data Selection for Full Waveform Inversion *76th EAGE Conference and Exhibition Extended Abstracts.*
[6] Operto S and Virieux J 2007 3D finite-difference frequency-domain modeling of visco-acoustic wave propagation using a massively parallel direct solver: A feasibility study *Geophysics* **72** 195-211.

IOP Conf. Series: Materials Science and Engineering 472 (2019) 012070 doi:10.1088/1757-899X/472/1/012070

# Experience of Protecting World Heritage in Norway: a Case Study of Røros Town

**Chang Liu**

China academy of urban planning and design, Chegongzhuang Dajie, Haidian,Beijing, 10044, China.
Email: txybj@yahoo.de

**Abstract.** As an important public resource, the economic value of heritage has been increasingly concerned by stakeholders in the process of social development.Currently, there is an on-going discussion between professionals, politicians and local owners concerning how and what to preserve the the economic value of heritage. This research takes Røros town, a world heritage in Norway as case study. It attemps to discuss what is the sustainableapproachto protect world heritage from Norway's practise. How it can be adoptedby China.

## 1. Introduction

As an important public resource, the economic value of heritage has been increasingly concerned by stakeholders in the process of social development. According to the theory of conservation economics, heritage is an economic commodity that has an asset life cycle much longer. It's also a "cultural capital" with both cultural attribute and economic attribute. The experience of other countries in the world shows that the heritage is also a non-renewable resource, which makes it a priceless treasure for human. Through the combination of heritage and business, economic benefits could be created in tourism industry. The combination will also promote the development of other derivative industries, like the tertiary industry around the heritage site. In a word, heritage plays an active role in market economy and local economy.

The heritage has direct value and indirect value as its two major economic values. The direct value includes the income mainly from the rent, commercial stores, tourism activities, entertainment industry, etc. The indirect value includes its culture value and aesthetic value, which are unique and have potential economic growth point in the future. This research focus on investigating what is Norway's experience to protect its world heritage site.

IOP Conf. Series: Materials Science and Engineering 472 (2019) 012070 doi:10.1088/1757-899X/472/1/012070

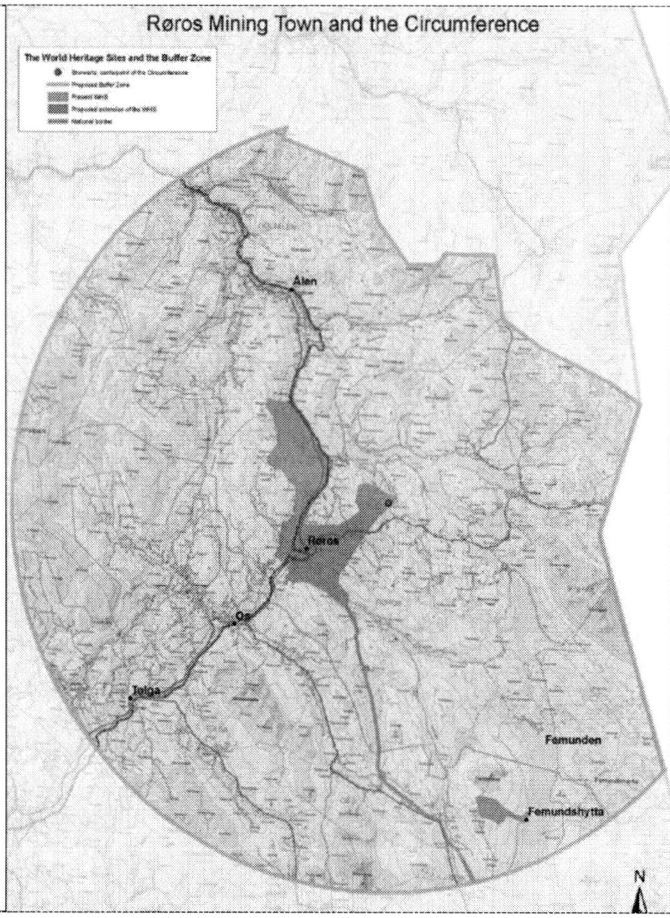


**Figure 1**.The World Heritage sites and the Buffer zone

**Figure 2.** The Town and Cultural Landscapes

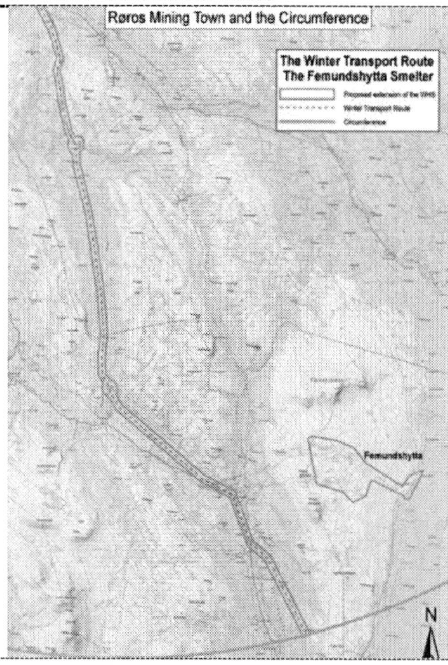


**Figure 3.** The Winter Transport Route

## 2. Historic Background

Røros is an old mining town situated on a mountain plateau in a remote part of Trøndelag County in Norway. After a copper ore was discovered in 1644 a town was established and Røros Copper Works was active for 333 years. In 1977 it closed down after the mining company was declared bankrupt. In 1980 it was inscribed on UNESCOs World Heritage List, and in 2010 extended to include a buffer zone of approximately 480 ha.

The owners of the works were mostly rich citizens of Trondheim, the closest town approximately 100 kilometres away. The king gave the owners rights to a number of privileges. They had the right to use all the ores they found and free access to waterfalls and forest. In addition, they had the right to labour from people in the area, for a reasonable pay. After some time the Copper Works were to become one of the richest mines in Europe.

At the start Røros had a very small population. However, the possibility for paid work was very attractive and the population grew quickly. This was mainly from surrounding areas, but also from bordering areas in Sweden. A few German mining experts also came, introducing technology the Norwegian did not have. Main part of the workers came from simple backgrounds, probably tenants and younger sons of farmers. There was also a small upper-class consisting of Danish-speaking owners of the copper mines. In 1700 the population had grown to around 2000, from almost zero.

This new society developed a unique culture different from the medieval farming system they came from. First of all a working class developed, being wage earners. In addition to this they were smallholders, holding cattle for their own consumption. This was important for surviving through hard times of unemployment or wars. This combination of wage earning and farming has been typical for Røros up to our time.

The people owned their townhouses, and if no one had a claim to land they could take it for production of grass to the cattle. Most households had two to three plots of land around the town. The last household to keep cattle in the outhouse building in town ended in the mid-1980s.

Included in the heritage site is the town, consisting of one and two story wooden houses surrounding the smelter and waterway system? The industrial rural cultural landscape and urban agricultural practise are preserved in addition to the transportation system, including a winter route open from November to May. The totality of this traditional settlement shows how shaping of the cultural landscape and a unique way of life developed.

## 3. Røros Town

The most visible landmarks of Røros today are the church, the slag heap that grew steadily, and the lack of threes after the forest being felled very quickly to be used in the production.The houses were at first mainly consisting of one floor, with a basement. Later two floored became more common. Up until the end of the 19th century the facades towards the streets were dominated by blackened logs. In the back of the house, surrounding a yard, were the cowsheds and stables, in addition to other buildings used for the necessities of running a farm. It is a very dense and efficient use of space. From around the 1880s important changes took place. Swiss-style dominated more and more, and the shape of the houses and roofs changed. There were sporadic discussions about creating a museum and preserve some of the buildings. In 1920 this finally came through. The Protection of Building Act preserved eight houses and building groups.

The inter-war period was a very difficult time for people living in Røros. This meant, among other things, that houses in bad condition was repaired, and not torn down which would probably have been the case under normal circumstances. This is one of the reasons Røros is relatively well preserved. In the 1960s the Swiss style was removed with the intent to bring the style back to 1700 -1800 Røros. Reasonable loans from the governmental Housing Bank and support from the governmental office of Cultural Heritage were given to house owners that wanted to repair and conserve their homes.

This combination of farming and townhouse we find in Røros is unique in Norway. With good financial help the homes, and the facades towards the street was taken care of, preserving a most valuable and charming town. However, the outhouses surrounding the yard suffered. By the 1950s most of them were no longer in use, quickly leading to deterioration. In 1996 "The Outhouse Project" started to rescue this unique combination of farming and townhouse. The project is still going, the goal being to renovate and repair 400 outhouses. Much attention is paid to traditional craftsmanship, and remove as little as possible of the old material.

## 4. Experience of Protecting of Røros town

There is an on-going discussion between professionals, politicians and local owners concerning how and what to preserve. This includes everything from what kind of windows are acceptable, what colour is the original to what kind of new buildings should be allowed. That many of the houses and the ground are in private ownership adds to these disagreements. An intangible aspect is also a part of this. There are for example residence requirements for anyone buying a house inside the town centre, requiring the buyer to live there at least 6 months of the year. This is to avoid Røros becoming a holiday resort.

Being on UNESCOs World Heritage List comes with responsibility. To increase the importance and value of the site the area under preservation was recently extended to include a buffer zone, identical to the old "Circumference". This is the area where the Copper Work was given privileges to water, forest and workers. This gives us an almost complete picture of human adaptation in a remote area with extreme weather condition. It now includes the mining, smelter, transport systems, cultural landscape, technology and a way of life.

Today Røros is a vibrant town, even though the cobber mines closed a long time ago. Tourism, conferences and income from cabin owners are of course an important part of the new economy. This is reflected in the shops, with for example many clothing stores and locally produced handicraft. In the yearly marked, arranged in accordance with tradition in February, local traditional food products are for sale. This includes both Norwegian and South-Sami traditions. It also includes not so traditional items, with products you find in any market or fairground. It is likely to assume that the architectural and spatial environment is a vital part of the success. In addition, Røros has managed to attract new industry, replacing the old copper mine.

The tourism industry development of Røros brings a obvious benefit. On the one hand, tourism development of Røros brings more job opportunities for the residents, and increases their income; they get economic benefits by selling tourism products; they have more opportunities to communicate with outside. On the other hand, tourism development of Røros has both positive and negative impact on the community. For instance, population load of Røros increases constantly, if not strictly controlled, it will surpass the population carrying capacity, and natural environment of heritage site will be destroyed. Tourism development means increasing visitors, and shock to local lifestyle, ideology and heritage concept.

From a larger perspective, it's clear that the market economy also has both positive and negative effects to the heritage protection. It will provide the financial support to the heritage protection projects, and will also help to build up a competitive mechanism for the heritage protection; sometime it could even inspire the citizens to participate in the process of heritage protection.

Cultural tourism is one of the fastest growing tourism sectors. It is often seen as a generator of the resources necessary to preserve and enhance cultural heritage (Chiabai et al., 2013). It is also considered to be the main motivation for tourists and a recreational activity since the early 1980s (Lord, 1999; Kowsurat, 2010; UNWTO, 2015). According to the International Council on Monuments and Sites (ICOMOS, 1976) cultural tourism should exert positive effects on these sites by contributing to the maintenance, protection, and socio-economic benefits of the local population (ICOMOS, 1976). Also, Robinson and Smith (2006) in their book "Cultural Tourism in a Changing World" argue that tourism can be seen as the main way to generate rapid economic development in countries with limited natural resources. On the other hand, they further elaborate that the development of a (cultural) tourism industry includes additional responsibility in terms of interpretation and representation of events and activities (Robinson and Smith, 2006). Indeed, Timothy (2011) stresses the role of interpretation in revealing the importance of places, peoples, and events, as well as the need for considering the community in the interpretive planning process.

However, while tourism may contribute in socio-economic development, it can also lead to irreversible damage and destruction of heritage resources (Orbaşli, 2000; Russo and Borg.Jan, 2002; Dumont et al., 2005; EUA, E. a., 2009; Kowsurat, 2010), as well as negative impacts on the local community in terms of social disruption and increased costs of living (Albert et al., 2012). Therefore, the appropriate management of tourism is needed to create harmony between conservation management and tourism planning in a way that satisfies the goals of both and reduces its negative impacts. So the concept of sustainable tourism emerged to meet the needs of tourists and host communities while simultaneously maintaining the integrity of cultural sites (UNEP and WTO, 2005; Boniface, 2013; UNWTO, 2015) without compromising the ability of future generations to meet their own needs (WCED, W.C., 1987; Swarbrooke, 1999; EUA, E. a., 2009; UNEP and WTO, 2005). This requires a multi-faceted approach aiming at fulfilling the economic, socio-cultural and environmental needs of the targeted communities as well as tourist ambitions and expectations.

In the recently years, the attention that Røros also pays attantion to the intangible cultural heritage prompts the protection and usage of it and then make more of it be shown, to be protected, be restored and be further developed. Many protective measures have been discussed and undergone experimental tests. The experience of protecting world heritage in Norway could be shared by China.

**Acknowledgments**

This research is funded by NSFC "The research on cultural diffusion mechanism of Chinese intangible cultural heritage: the case study on Chinese shadow theatre's diffusion in EU" (41501166); Research project from International Science and Technology Cooperation Program of China, MoST, No. S2016G4553; European Union Commission H2020 Framework Program, No. H2020-SC6-ENG-GLOBALLY-2017.

**References**

[1] Berger Peter 1973The Homeless Mind. Modernization and Consciousness (Vintage Books) p 75
[2] Harrison, Rodney: "Forgetting to remember, remembering to forget: late modern heritage practices, sustainability and the "crises" of accumulation of the past". Routledge, 2013.
[3] Ødegård, Sverre A. and Øverås, Ola H.1983IntroductionFredede hus og anlegg 3, Røros Bergstad (Universitetsforlaget)chapter 1 pp 1–8

# Overview of High-Value Reuse and Grinding at Sub-Zero Temperature of Scrap Rubber Tires

**Qiucheng Wang, Zijun Huang and Zhengqing Liu**

Institute of Vehicle Engineering, Zhejiang University of Technology, Hangzhou 310023, China.
Email: wqc@zjut.edu.cn.

**Abstract.** It is of good application prospects for the low carbon remanufacturing process to product fine reclaimed rubber powder by grinding scrap rubber tire at low temperature. Through review of the recycling technology of scrap tire at low temperature and the high-value reuse of the regenerated rubber powder of domestic and abroad, the advantages and disadvantages of several grinding processes at low temperature are compared, the main existing problems of the cryogenic grinding technology are put forward. In addition, the development trend of the cryogenic grinding technology of waste tire and the high-value reuse of the regenerated rubber powder is prospected. The key point to solve the problem of high-value reuse is to improve the manufacturing process of fine reclaimed rubber powder and the development of equipment.

## 1. Introduction

The number of vehicles in China reached 310 million in 2017.Among them, there are 217 million cars. It is estimated that the number of scrap tires would be 379.8 million in 2018, and the weight would be 14.59 million tons. And it will grow about 5%-6% annually. It has attracted more and more attentions that how to recycle waste tires [1].

There are several ways to recycle waste tires [2], the most valuable way is to crush them into rubber powder and to recycle them in order to become new value product. Among the ways to crush waste tires, they are divided into grinding at room temperature and grinding at cryogenic temperature.

At room temperature, it is not easy to crush the elastomeric rubber tires which is easily deformed by heating and the material whose viscoelastic behavior will lead to plastic deformation in the process of mechanical force. At low temperatures brittle fracture can occur easily (as brittle as glass) if the material is subjected to and then crushed. In order to produce high value fine rubber powder, the cryogenic grinding technology will lead to the recycling process of scrap tires in the future.

## 2. Cryogenic Grinding Technology of Waste Tire

Cryogenic grinding is a process to comminute brittle materials (tires, rubber and other elastic materials) by use of the characteristics of brittleness of materials at low temperatures (such as tires in 193K become very brittle and easy to comminute or separate) [3]. Cryogenic grinding can produce more finely crumb powder than normal grinding, and fine rubber powder is of high value-added [4]. At present, the cryogenic grinding methods of scrap tires (according to the different cold sources) mainly divided into the cryogenic grinding technology with liquid nitrogen, with air expansion refrigeration, with LNG (liquefied natural gas) cold energy refrigeration and so on. The scrap rubber tire blocks are crushed in cold brittleness condition, after they are cooled in refrigeration equipment [5].

IOP Conf. Series: Materials Science and Engineering 472 (2019) 012071 doi:10.1088/1757-899X/472/1/012071

*2.1. Cryogenic Grinding Technology with Liquid Nitrogen*

Cryogenic grinding technology with liquid nitrogen is a kind of technology which uses liquid nitrogen as cooling medium to lower scrap tire below the critical glass state temperature for comminution. As the boiling point of liquid nitrogen is 77K, it is easy to cool down the material below the glass state temperature [6]. In addition, as an inert substance, liquid nitrogen can prevent the material from being oxidized in severe comminution processes.

At present, cryogenic grinding processes with liquid nitrogen for scrap tires are mainly divided into two types: One is to freeze the tire directly to the glass state temperature with liquid nitrogen. Another is to crush the tire into a certain size of rubber block at room temperature and then freeze it with liquid nitrogen below the glass state temperature before grinding. Compared the two processes, the second one is more economical and energy-saving.

*2.2. Sub-Zero Grinding Technology with Air Expansion Refrigeration*

It is proved that the friction heat generated from the grinding can be counteracted when the temperature of the rubber is about 193K [7]. Based on this point of view, in order to improve the economic feasibility of sub-zero grinding of waste tires and reduce the grinding cost, the sub-zero grinding technology using air expansion refrigeration has been developed. The main raw material comes from air which is abundant resources, non-polluted. Besides it is of effective refrigeration at low cost. The basic principle of the sub-zero grinding technology is to compress the air to a certain pressure and produce low temperature air below 153K by gasification and expansion of turbine, and then this low temperature air is used as a refrigerant. The sub-zero grinding process of air expansion refrigeration mainly operate sequentially at environment temperature and at sub-zero temperature.

*2.3. Cryogenic Grinding Technology with LNG (liquefied Natural Gas) Cold Energy Refrigeration*

With the development with LNG industry, the utilization of LNG cold energy has attracted people's attention. Compared with the traditional cryogenic grinding technology, the cryogenic grinding technology with LNG cold energy refrigeration is of lower cost [8]. LNG is a liquid gas at 111K cryogenic temperature under atmospheric pressure, which will release a large amount of cold energy when it gasifies. The released cold energy can be used as a refrigerant when the scrap tires are comminuted at low temperature, so as to prevent the waste of cold energy.

There are two kinds of cryogenic grinding processes using LNG cold energy as refrigerant: one is to produce liquid nitrogen as refrigerant; the other is to directly recover the cold energy of LNG by using nitrogen as refrigerant.

The advantages and disadvantages of the above cryogenic grinding technology are compared in Table 1.

**Table 1.** Comparison of several low temperature grinding methods (Different cold source).

| **Methods** | Advantages | Disadvantages |
|---|---|---|
| Liquid nitrogen refrigeration | *The boiling point is 77K, the refrigeration effect is good, and the precooling time is short., and liquid nitrogen can prevent oxidation as an inert material.* | *The refrigerant consumption is much large and the cost is high.* |
| Air expansion refrigeration | *Making full use of cooling capacity and dry air, rubber does not need to freeze to about 168K of embrittlement temperature and low power consumption.* | *A large amount of refrigeration equipment is needed* |
| LNG cold energy refrigeration | *Cold energy released directly by LNG during gasification reheat process, Less waste caused by cold energy emissions.* | *Large investment and low energy consumption* |

## 3. Current R&D and Future Development Trend

The industrialization of cryogenic grinding technology first came true in 1948, the first company that achieved and utilized liquid nitrogen cryogenic commercial grinding is LNP in America [9].

### *3.1. R&D Status Abroad*

Liquid nitrogen cryogenic grinding has the benefits of having short pre-cooling time, ample material of cold source and so on, which is the prime way to comminute junk tires in developed countries. The American invented this method in advance, which came up with a device that is able to smash plastic and rubber by means of cryogenic grinding with liquid nitrogen in 1960: it pre-cool plastic or rubber with liquid nitrogen at first, and then transport them to grinder by spiral conveyor [10].

According to the patent [11], they put rubber into a grinding equipment below the glass state temperature, and simultaneously add cryogen (such as liquid nitrogen) for cryogenic grinding.

UCC is one of the earliest companies to utilize the technology of cryogenic grinding in the world, whose product line of fine rubber powder has developed from the first generation to the fifth at present [12]. The whole process is sound environmentally. The 5th generation product line adopts the technology of 77k sub-zero treatment. The processes of grinding of junk tires consist of input, crushing, cooling grinding of the whole tire and separating steel wires and fiber from rubber tires, through which the degree of purity of rubber powder can reach up to 99.99%.

It was exploited by Institute of cryogenic physical engineering of the National Academy of Sciences of Ukraine that preparation technology of rubber powder by cryogenic grinding waste tires with liquid nitrogen [13]. The main processes of crushing and grinding are performed at cryogenic temperature, which can manufacture rubber powder with particle size of 0.05-1.25mm according to the needs of the market.

### *3.2. Domestic R&D Status*

The production of waste rubber powder in China started commercial production and application in 1980s. In the early 1990s, it came to an active period. After several decades of development, it has been equipped with certain capacity to manufacture rubber powder.

But the technology of cryogenic grinding in China started relatively late. Due to the high cost, it has not been industrialized in the development. In the 1990s, the first cryogenic grinding with air expansion refrigeration was developed independently in China, which is of the merits of abundant cold source and non-polluted, compared with other cryogenic grinding methods.

China has made a series of improvement as following on the base of cryogenic grinding technology: the cryogenic grinding method with electric pulse was researched by W Wu etc. [14], which makes high voltage electric current pulse in liquid nitrogen for discharging, thus trigger liquid-electric effect to crush waste rubber. Meanwhile, Qingdao green leaf Rubber Co. Ltd and Shenzhen Institute of mechanical and electrical technology jointly exploited LY cryogenic method with liquid nitrogen [13]. The rubber powder with particle size of 75-180μm could be obtained by this method. Zhao B invented a kind of waste tire cryogenic product line. The size of the rubber powder produced under the product line is below 120 mesh, and the purity can reach more than 99.7%[15].The related cryogenic grinding processes also be investigated by the 609th Research Institute of Chinese Aviation Industry [16] [17], Cryogenic Technology Experimental Centre of Chinese Academy of Sciences [18], and Dalian University of Technology.

### *3.3. Prospect of Development Trend*

Compared with developed countries such as Europe and the United States, the normal grinding technology of rubber powder still occupies a dominant position in the domestic production process. However, the proportion of the rubber powder with high added value is still not high. For the sake of efficiently recycling of waste tire, it is preferable to make fine rubber powder by grinding at low temperature currently.

Compared with grinding rubber powder at normal temperature, cryogenic grinding method requires more investment and energy consumption. However, the cryogenic grinding technology with the air expansion refrigeration and LNG cold energy described above has provided a new way to prepare fine rubber powder at low temperature and low cost. At the same time, liquid nitrogen refrigeration has obvious advantages such as good refrigeration effect, preventing rubber oxidation and low power required for grinding. However, the consumption of liquid nitrogen in grinding waste tires with liquid nitrogen refrigeration is still much large. One of the key factors to be solved is how to reduce the consumption of liquid nitrogen or its production cost.

Because of the advantages of fine rubber powder prepared from waste rubber at cryogenic temperature, it is of great significance to the high value application of waste tires. It is an economical and efficient method to produce fine rubber powder combined at room temperature and at cryogenic temperature. Nowadays, environmental protection is paid to more and more attention, the cryogenic grinding method, as a clean and pollution-free process, will deserves to be favored by people.

## 4. High Added-Value Reuse of Fine Reclaimed Rubber Powder

For the high-value reuse of fine recycled rubber powder, one way is used in the production of new tires, in which part of the recycled rubber powder instead of the original rubber powder. Due to the high purity, small particle size and non-pollute of the rubber powder produced by cryogenic grinding, it can be used in some important products to achieve high-value return. For example, it will greatly improve the performance of automobile tires and reduce the production cost if the raw rubber powder is instead of about 20% fine recycled rubber powder obtained by cryogenic grinding [19].

In the 172nd meeting of the rubber society of the American Chemical Society, the latest study released that [20]: the smaller the size of rubber particles, the better the positive effect on physical properties. the rubber powders can be used to manufacture new thermosetting rubber complexes. The application of rubber powder with different numbers to different occasions is shown in Table 2 below:

IOP Conf. Series: Materials Science and Engineering 472 (2019) 012071 doi:10.1088/1757-899X/472/1/012071

**Table 2.** Application of different mesh rubber powder

| **Methods** | Particle size of rubber powder | Application |
|---|---|---|
| Cryogenic grinding | *40~60 mesh (fine rubber powder)* | *Rubber filled products, rubber modified products.* |
| | *60~80 mesh (fine rubber powder)* | *Automobile tires, rubber products, building materials* |
| | *80~200 mesh (micro powder)* | *Rubber products, military products* |

Another high-value reuse for the new energy automotive industry is to explore further extraction of carbon black materials from recycled rubber powder for lithium battery capacitors. A low temperature pyrolysis process made of recycled rubber powder was developed by Delta Energy Company in the United States, which can recover carbon black-based solids from used tires rubber and reduce the production amount of harmful polycyclic aromatic hydrocarbons. Its quality is equivalent to that of carbon black directly produced. The calculation shows that the net energy recovery value of the process is higher than that of the traditional process.

In addition, a new system producing electricity was developed by Israel from used tires, which sells electricity generated by combustion to the State Grid. At the same time the by-products such as steel and sulphur are also obtained. The whole process is safe, clean and pollution-free.

It is put forward to by Z Li of Qingdao University of Science and Technology that the idea of used rubbers can be cracked and high-value reuse of its pyrolysis products can be utilized [21]. It is believed that the microwave pyrolysis technology of used rubber has the advantages of fast cracking speed, high product quality, low energy consumption and no pollution., It can realize automation and mass production.

Chinese government is stepping up to support the new products that are environmental, efficient and energy-saving. The fine rubber powder of high value-added will be the main way of recycling used tires. It will become the development goal of used tire recycling industry that utilizes high quality recycled rubber powder to promote the replacement of synthetic rubber and natural rubber to achieve the goal of saving resources.

## 5. Summary and Prospect

The recycling of scrap tires is of great value. It not only protects the environment, but also makes reasonable use of waste materials. There is higher utilization value for the fine recycled rubber powder produced by cryogenic grinding than coarse powder produced at room temperature. The technology of sub-zero grinding has been studied deeply at home and abroad, however, it has not been industrialized promotion in China.

Although the rubber powder produced by cryogenic grinding of waste tires has the characteristics of fine particle size and low oxidation degree, the cold source is an obstacle to realize freezing scrap tires grinding. Because cold source is directly related to the cost whether it is worthy of cryogenic grinding. Therefore, the efficient refrigeration technology, development of cryogenic grinding equipment and optimization of cryogenic grinding process are the keys of future research in order to improve the economic feasibility of cryogenic grinding, which is also an important prerequisite for promoting cryogenic grinding technology.

## Acknowledgments

The authors would like to acknowledge the funding of Zhejiang Public Welfare Technology Application Research Project [2017C31013] in conducting this research.

**References**

[1] 2018 B. Research report on the dynamics of China's scrap tire market in 2018-2024 years. Wisdom education advisory group
[2] Li X 2013 J. Recyclable Resources and Circular Economy. Experience and reference of waste tire recycling in developed countries 03 40-44
[3] Du L 2006 D. Application of cold energy integration in rubber cryogenic comminution and cold storage technology (Guangzhou: South China University of Technology)
[4] Zheng H, Xu W, Bian A, Fan S, Luo X and Hou J 2009 J. Chemical Industry and Engineering Progress. Research progress in low temperature crushing technology of waste rubber 12 2242-47
[5] Cheng H and Pan C 2002 J. Journal of Southwest University of Science and Technology. Study on cryogenic comminution system 01 42-45
[6] He Y and Liu Y 2001 M. Production and application of rubber powder: new technology of waste rubber recycling (Beijing: China Petrochemical Press) 88-89
[7] Li L 2011 J. Recyclable Resources and Circular Economy. Research progress and trend of low temperature grinding technology for waste rubber powder 10 38-41
[8] Xiong Y, Hua B and Luo D 2007 J. Modern Chemical Industry. The pressure of natural gas pipe network can be used for the refrigeration equipment of waste rubber crushing 1 49-52
[9] Wang C, Wang D, Zhang L and Zhuang S 2012 J. China Rubber. Development of low temperature rubber crushing technology 02 39-43
[10] Zhou C 2005 J. Journal of Beijing Institute of Petrochemical Technology. Cryogenic grinding technology and its application in resource recovery 01 23-26
[11] Waznys PJ,Meckert GW, Cialone AM 2006 P. Comminuting rubber particles to produce rubber powder,involves introducing rubber particles into comminuting device, concurrently introducing inert gas coolant into the comminuting device, and comminuting the rubber particles (US:7108207)
[12] Zhang H 2008 D. Study on cold energy utilization of liquefied natural gas in low temperature comminution of waste tire (Lanzhou: Lanzhou University of Technology)
[13] Kou L, WEI Q 2007 J. Chemial Technology Market. Production methods and status of fine rubber powder 30(8) 25-27
[14] Wu W, Su H and Xu W 1997 J. High Voltage Engineering. Study on pulverizing waste tyres by low temperature pulse discharge 3 87-88
[15] Zhao B 2005 P. Technology and equipment for treating waste tyres (China: 1590060)
[16] Suo T and Li Z 20004 J. Jiangsu Chemical Industry. Recycling technology of waste rubber 31 1-6
[17] Jia Y and Deng Z 2006 P. Method and device for preparing rubber powder by air refrigeration and freezing 1(China:1063546)
[18] Wang P and Liu S 1998 J. Cryogenics. Air turbine refrigeration and low temperature comminution method for producing fine rubber powder from waste rubber 6 1-5
[19] Ramarad S, Khalid M,Ratnam CT, Luqman Chuah A and Rashmi W 2015 J. Waste tire rubber in polymer blends: A review on the evolution, properties and future, Progress in Materials Science 72 100-140
[20] 2011 J. Gxjn. Progress in recycling technology of waste rubber abroad 10804 34-36
[21] Li Z, Liu F 2016 J. Special Purpose Rubber Products. Research progress in high value utilization cracking technology of waste rubber 01 60-64

IOP Conf. Series: Materials Science and Engineering 472 (2019) 012072 doi:10.1088/1757-899X/472/1/012072

# Study on the Determination of Free Formaldehyde in Carbon-Coated Air Purifying Material by Water Extraction Method

**Ruiyan Zhang, Haijiao Zhang, Dan Ma and Xuehong Han**

China Building Material Test & Certification Group Co., Lid. No.1 courtyard of Guanzhuang Dongli, Chaoyang District, Beijing

Email: zry@ctc.com.cn

**Abstract.** The free formaldehyde in carbon-coated air purifying material was determined by water extraction method. The effects of solid-liquid ratio, extraction time and extraction temperature on the content of free formaldehyde in carbon-coated air purifying material were studied. In this paper, orthogonal tests and single factor tests were designed, and the optimal extraction conditions were 1:40 solid-liquid ratio, 90 min extraction time and 40 °C extraction temperature. Under this extraction condition, the relative standard deviation (n=8) of the repeated test is 0.80%, and the recovery rate is 90%-100%, which indicates that the analytical method for determining the free formaldehyde in carbon-coated air purifying material by water extraction method is feasible.

## 1. Introduction

Formaldehyde [1-3] is a colorless, strongly irritating gas and also a protoplasmic poison. Formaldehyde has intense stimulation effect to human body organ. Excessive formaldehyde can cause irritation to the eyes, nasal cavity and respiratory tract, which may cause itching, inflammation, etc., and even cause serious diseases such as tumors and leukemia, because of the degree of harm to the human body is extremely serious. At present, formaldehyde has been recognized as a carcinogenic and teratogenic substance by the World Health Organization. It is a recognized source of allergic reactions, and long-term exposure will lead to genetic mutations.

With the rapid development of the social economy and the improvement of people's living standards, the more interior decoration materials, office equipment and household appliances using, the more formaldehyde probably produced. Therefore, the air purifier has become one of the important tools to remove indoor formaldehyde. Carbon-coated air purifying material is a common filter material in the air purification industry. It is formed by uniformly spreading the broken activated carbon between two non-woven fabrics to form a carbon cloth, and then folded and fixed to combine the effect of the HEPA filter and the activated carbon filter, which can effectively filter PM2.5 particles and gaseous pollutants such as formaldehyde [4]. However, formaldehyde and its derivatives, as the basic chemical materials, are inevitable in the production process of carbon-coated air purifying material. There are safety risks caused by the volatile formaldehyde in carbon-coated air purifying material. The presence of free formaldehyde in the carbon-bonded cloth is easily volatilized and poses a safety risk. Therefore, it is necessary to strictly limit the content of formaldehyde in the carbon-coated air purifying material.

In this paper, the free formaldehyde in carbon-coated air purifying material was determined by water extraction method [6-8]. The optimal extraction conditions were determined by orthogonal tests and single factor tests, which provided a reference for the control and detection of formaldehyde in carbon-coated air purifying materials.

## 2. Experimental

### *2.1. Materials and Instruments*

*2.1.1. Materials.* Ammonium acetate (analytically pure AR), Sinopharm Chemical Reagent Co., Ltd.; glacial acetic acid (analytically pure AR), Sinopharm Chemical Reagent Co., Ltd.; formaldehyde standard solution (10.4mg/ml), Shanghai Anpu Experimental Technology Co., Ltd.; Acetone (analytically pure AR), Sinopharm Chemical Reagent Co., Ltd.; deionized water, 18 MΩ.cm, laboratory-made.

*2.1.2. Nstruments* .Water phase needle filter (0.45μm), Shanghai Anpu Experimental Technology Co., Ltd.; precision electronic balance ME203E/02, METTLER TOLEDO instrument (Shanghai) Co., Ltd.; UV-visible spectrophotometer UV-1780, shimadzu Japan; constant temperature water bath oscillator SHZ-C, Laizhou Yuanmao Instrument Co., Ltd.

### *2.2. Preparation of Standard Curve Working Fluid*

Pipette 1 ml of formaldehyde standard solution into a 100 ml volumetric flask, dilute to the mark with deionized water, shake well, and prepare a formaldehyde intermediate solution with a concentration of 104 mg/L. And the pipette 1 ml of formaldehyde solution from the formaldehyde intermediate solution into a 100 ml volumetric flask, dilute to volume, shake well, and prepare a working solution of 1.04 mg/L. Then pipette formaldehyde working solution 0, 2.0, 4.0, 6.0, 8.0, 10.0 ml to 10 ml colorimetric tubes separately, dilute to the mark with deionized water, add 2 ml acetylacetone solution, shake and heat the mixture was in a boiling water bath for 3 min, taken out and cooled to room temperature. And the absorbance was measured at a wavelength of 413 nm using a 1 cm absorption cell with water as a reference. The absorbance A value measured in the above series of standard solutions is subtracted from A0 value of reagent blank (zero concentration) to obtain the calibrated absorbance y value. The calibrated absorbance y is the ordinate and the formaldehyde concentration x (g) is the abscissa to draw the standard curve. After linear regression, the standard curve equation of concentration and absorbance of standard solution was obtained: $y=0.0200x-1.3810$, $R^2= 0.9991$. Y is the absorbance, and x is the concentration, which are shown in Figure 1.

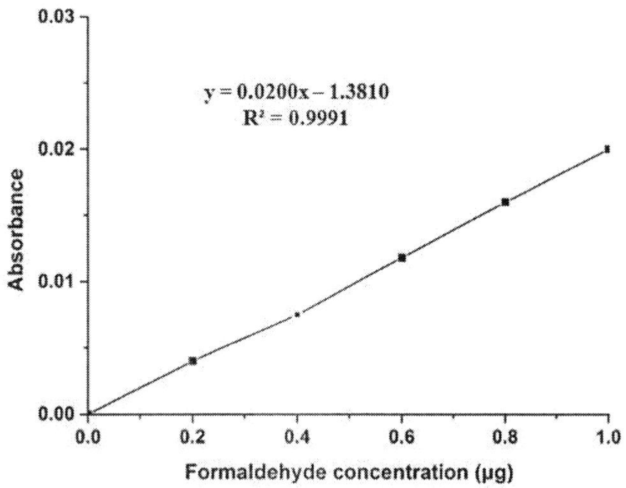


**Figure 1.** Standard curve

### 2.3. Experimental Method

The sample of the carbon-coated air purifying material was cut into a sample to be tested with an area of 10±1 $cm^2$. The samples with a certain quality of cut were weighed as M g and put directly into 250 ml bottle with plug triangle. Add V ml deionized water according to the different solid-liquid ratio, close the lids, and put it into the constant temperature water bath oscillator at the set temperature, oscillate for a certain time, and filter it with filter to another flask for analysis. After cooling to room temperature, remove 10 ml filtrate into 25 ml colorimetric tube, add 2 ml acetone solution and cover the plug. After being heated in boiling water bath for 3 min and cooled in the dark condition, the deionized and 2 ml acetone solution mixed under the same oscillation condition were taken as a blank. The absorbance value was measured at 412 nm and the formaldehyde concentration C (indicated by the formaldehyde standard curve) was calculated. Then the free formaldehyde content is calculated as follows:

$$\text{Free formaldehyde extraction}(\mu g/g)=\frac{C\times V}{10\times M} \quad (1)$$

*C*, formaldehyde concentration value calculated according to the standard curve of formaldehyde, μg;
V, Extraction volume, ml;
10, conversion factor of solution volume of extraction liquid volume removed from the extraction solution, color reaction, ml;
M, the mass of the sample, g.

### 2.4. Orthogonal Experimental Design of Free Formaldehyde Extraction in Carbon-Coated Air Purifying Material [5]

**Table 1.** Factors and levels

| Levels | Factors | | |
|---|---|---|---|
| | *A* | *B* | *C* |
| | Solid-liquid ratio | Extraction time /min | Extraction temperature /°C |
| 1 | 1:10 | 30 | 30 |
| 2 | 1:20 | 60 | 40 |
| 3 | 1:30 | 90 | 50 |
| 4 | 1:40 | 120 | 60 |

In order to further study the effects of solid-liquid ratio, extraction time and extraction temperature on the amount of free formaldehyde extracted from carbon-coated air purifying materials, three factors and four levels test were selected, according to the orthogonal design method on table $L_{16}(4^3)$ to find the most suitable conditions, and the free formaldehyde extraction amount was taken as the investigation index. The factor level arrangement is shown in Table 1.

## 3. Results and Discussion

### 3.1. Orthogonal Test Results

The results of the orthogonal test are shown in Table 2. In the table, $\overline{K_1}$, $\overline{K_2}$, $\overline{K_3}$ and $\overline{K_4}$ are the average values of free formaldehyde extractions obtained from the corresponding levels, and the calculated value larger the level conditions better. It can be seen from Table 2: (A) $\overline{K_4}>\overline{K_2}>\overline{K_3}>\overline{K_1}$, (B) $\overline{K_3}>\overline{K_2}>\overline{K_4}>\overline{K_1}$,(C) $\overline{K_2}>\overline{K_3}>\overline{K_4}>\overline{K_1}$. Therefore, the optimum combination extraction condition of free formaldehyde in the carbon-coated air purifying material is A4B3C2, that is, the solid-liquid ratio is 1:40, the extraction time is 90 min, and the extraction temperature is 40 °C. R is the extreme difference, which indicates the primary and secondary order and the larger the R value, the better the main effect. It can be seen from the variance in Table 2 that R1>R3>R2,which shows that the solid-liquid ratio in the test range has a significant effect on the extraction amount of free formaldehyde in the carbon-coated air purifying materials, and the extraction temperature is significant for the free formaldehyde extraction, while the extraction time has less impact.

**Table 2.** Results of orthogonal experiment

| No. | Factor | | | Free formaldehyde extraction(μg /g) |
|---|---|---|---|---|
| | *A* | *B* | *C* | |
| 1 | 1 | 1 | 1 | 0.263 |
| 2 | 1 | 2 | 2 | 0.728 |
| 3 | 1 | 3 | 3 | 0.429 |
| 4 | 1 | 4 | 4 | 0.536 |
| 5 | 2 | 1 | 2 | 0.321 |
| 6 | 2 | 2 | 1 | 1.190 |
| 7 | 2 | 3 | 4 | 0.930 |
| 8 | 2 | 4 | 3 | 1.184 |
| 9 | 3 | 1 | 3 | 0.520 |
| 10 | 3 | 2 | 4 | 0.874 |
| 11 | 3 | 3 | 1 | 0.897 |
| 12 | 3 | 4 | 2 | 1.108 |
| 13 | 4 | 1 | 4 | 0.421 |
| 14 | 4 | 2 | 3 | 0.878 |
| 15 | 4 | 3 | 2 | 0.924 |
| 16 | 4 | 4 | 1 | 0.412 |
| $\bar{K}_1$ | 0.381 | 0.690 | 0.489 | |
| $\bar{K}_2$ | 0.917 | 0.770 | 1.056 | |
| $\bar{K}_3$ | 0.795 | 0.903 | 0.850 | |
| $\bar{K}_4$ | 0.960 | 0.691 | 0.659 | |
| $R$ | 0.579 | 0.212 | 0.567 | |
| Optimal level | $A_4$ | $B_3$ | $C_2$ | |

*3.2. Single Factor Test*

*3.2.1. Effect of Different Solid-Liquid Ratio on Free Formaldehyde Extraction in Carbon-Coated Air Purifying Material.* The solid-liquid ratio was set to 1:10, 1:20, 1:30, 1:40, 1:50, 1:60 at the optimum level of orthogonal test, which the extraction time was 90 min, and the extraction temperature was at 40 °C. The effect of solid-liquid ratio on the amount of free formaldehyde extracted from the carbon-coated air purifying material was studied, and the optimal leaching ratio was selected. The results are shown in Figure 2.

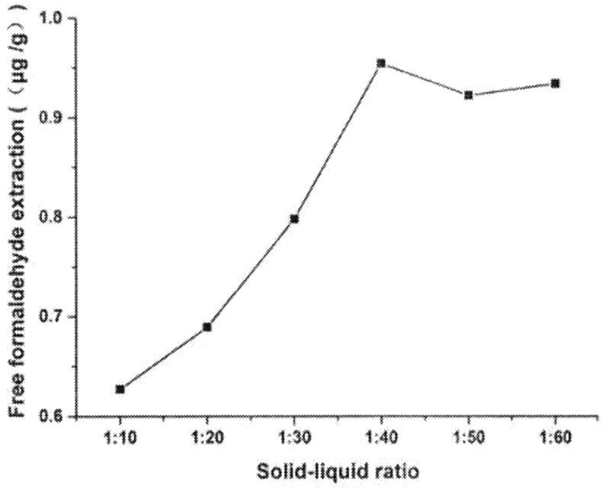


**Figure 2.** Variation of free formaldehyde extraction at different solid-liquid ratios

It can be seen from Figure 2 that the extraction amount of free formaldehyde in the carbon-coated air purifying material increases with the increase of the solid-liquid ratio of the extraction. When the solid-liquid ratio of the extraction does not exceed 1:40, the amount of free formaldehyde extracted increases with the increase of the ratio. When the ratio of leaching to solid-liquid ratio exceeds 1:40, the extraction amount tends to be stable. Therefore, the optimum solid-liquid ratio of the single factor test design is 1:40.

*3.2.2. Effect of different extraction time on the amount of free formaldehyde extracted from carbon-coated air purifying material.* The extraction time was set to 30 min, 60 min, 90 min, 120 min, 150 min, and 180 min at the optimum level of orthogonal test, which the solid-liquid ratio was 1:40, and the extraction temperature was at 40 °C. The effect of the extraction amount of free formaldehyde in the carbon-coated air purifying material was studied, and the optimal extraction time is selected.

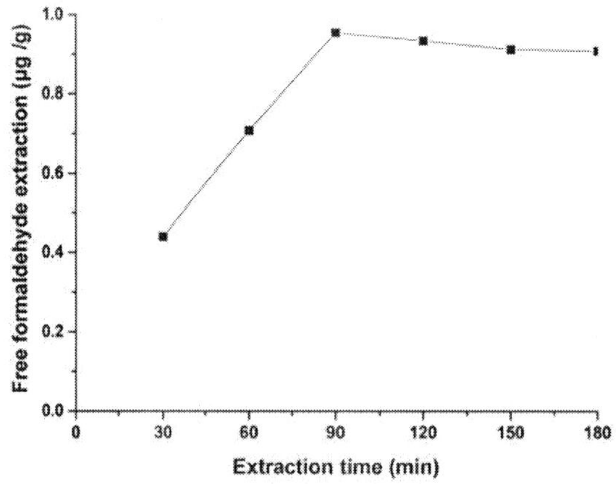


**Figure 3.** Variation of free formaldehyde extraction at different extraction times

It can be seen from Figure 3 that the extraction amount of free formaldehyde in the carbon-coated air purifying material increases with the increase of extraction time. When the extraction time does not exceed 90 min, the amount of free formaldehyde extraction increases with time, and the extraction amount decreases gradually after the time exceeds 90 min. Therefore, the optimal extraction time for the single factor test design is 90 min.

*3.2.3. Effect of different extraction temperatures on the free formaldehyde extraction from carbon-coated air purifying material.*The extraction temperature was set to 30°C, 40°C, 50°C, 60°C, 70°C, 80°C at the optimum level of orthogonal test, which solid-liquid ratio was 1:40, and the extraction time was 90 min. The effect of temperature on the amount of formaldehyde extracted from the air-cleaning material in the carbon cloth is selected to optimize the extraction temperature.

It can be seen from Figure 4 that when the extraction temperature is lower than 40°C, the free formaldehyde extraction in the carbon-coated air purifying material increases with the increase of temperature, and when the extraction temperature exceeds 40°C, the formaldehyde extraction amount decreases as the temperature increases. Therefore, the optimum extraction temperature for the single factor test design is 40°C.

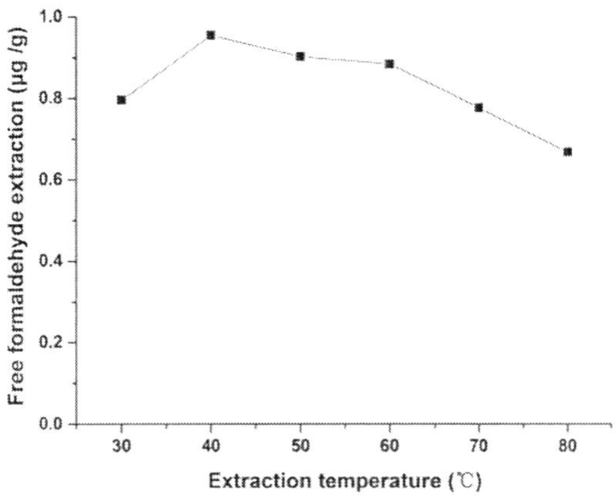


**Figure 4.** Variation of free formaldehyde extraction at different extraction temperatures

*3.3. Verification Test*

*3.3.1 Precision Verification.* According to the results of orthogonal test and single factor test, the optimum conditions for the determination of free formaldehyde in carbon-coated air purifying material by water extraction method were as follows: the solid-liquid ratio was 1:40, the extraction time was 90 min, and the extraction temperature was 40 °C. The best test conditions were tested in parallel for 8 replicates.

**Table 3.** Results of precision of method

| No. | 1 | 2 | 3 | 4 | 5 | 6 | 7 | 8 | Average value | RSD% |
|---|---|---|---|---|---|---|---|---|---|---|
| Free formaldehyde extraction (μg/g) | 0.954 | 0.950 | 0.962 | 0.966 | 0.952 | 0.950 | 0.971 | 0.959 | 0.958 | 0.80 |

*3.3.2 Recovery Verification.* The determination of recovery is used to assess the suitability of the experimental method. The test was carried out by adding a known concentration of formaldehyde solution to the sample solution and then performing the measurement. Using the existing three carbon-coated air purifying material samples, each sample was taken from two erlenmeyer flasks respectively, adding 3.0 g sample, 120 ml of deionized water, and then adding a certain amount of formaldehyde solution of known concentration to one of the samples. And then the condition of extraction was the extraction time of 90 min and the extraction temperature of 40 °C. The free formaldehyde extraction amount of the samples in the two bottles was determined according to the test method, and finally the recovery rate was calculated by the following formula. The results are shown in Table 4.

$$K = \frac{A-B}{C} \times 100\% \qquad (2)$$

K, recovery rate;

A, the formaldehyde extraction after adding a known concentration of formaldehyde solution; B, the free formaldehyde extraction from the sample;

C, the formaldehyde extraction solution known to be added.

**Table 4.** Result of the recovery rate of method

| Sample | $B$(μg /g) | $C$(μg /g) | $A$(μg /g) | Recovery rate $K$(%) |
|---|---|---|---|---|
| 1 | 0.958 | 1.667 | 2.587 | 97.7 |
| 2 | 0.795 | 0.667 | 1.457 | 99.3 |
| 3 | 0.489 | 0.333 | 0.817 | 98.5 |

It has been proved by experiments that the recovery rate of the free formaldehyde content in the carbon-coated air purifying material by water extraction method is between 90% and 100%.

## 4. Conclusion

The orthogonal parameters and single factor experiments show that the optimal parameters for the determination of free formaldehyde in the carbon-coated air purifying material by water extraction method are: solid-liquid ratio 1:40, extraction time 90min, and extraction temperature 40 °C. Through precision experiments and recovery experiments, it is shown that this method is feasible for determining the free formaldehyde content in carbon-coated air purifying material.

## Acknowledgement

This article is founded by the "National Key R&D Program of china". This paper belongs to "The Research on Key Performance and Safety Testing Technology of Household Environmental Purification Products (Project No.: 2017YFF02099)" and "The Fundamental Research on Purification Performance and Safety Testing Technology of Air Purification Materials and core components (Question No.: 2017YFF0209901)".

## References

[1] Wen Jiang. The harm and prevention of formaldehyde [J]. Environmental Education, 2009(3):70-71.
[2] Jiwen Hong. Research progress on formaldehyde pollution of indoor air pollutants and its harmfulness to human body [J]. Chinese Medicine Guide, 2013, 11(7):87-88.
[3] J. Leng, C.-W. Liu, H. J. Hartwell, et al. Evaluation of Inhaled Low Dose Formaldehyde Induced DNA Damage by Liquid Chromatography-Tandem Mass Spectrometry [J]. Toxicology Letters, 2018, 10, 155-156.
[4] General Administration of Quality Supervision, Inspection and Quarantine of the People's Republic of China. GB/T 2912.1-2009, Textiles-Determination of formaldehyde-Part 1 : Free and hydrolyzed formaldehyde (water extraction method ) [S]. Beijing: China Standard Press, 2009.
[5] Zhixi LI. Test Optimization Design and Statistical Analysis [J]. Science Press, 2010.
[6] Fangling Wang, Jianzhong Yang. Formaldehyde detection methods and research progress [J]. Progress in textile technology, 2008, 30 (3):58-62.
[7] Zhongjun Li, Xunzhong Zou, Yu Li, etal. Over view of the process of formaldehyde determination method [J]. Guangdong Trace Element Science, 2014, 12(6):29-30.
[8] Motyka K, Mikuska P. Continuous chemiluminescence de-termination of formaldehyde in air based on Trautz-Schori-gin reaction [J]. Analytica Chimica Acta, 2006, 562 (2):236-244.
[9] Monkawa A, Gessei T, Takimoto Y, etal. Highly sensitive and rapid gas biosensor for formaldehyde based on an en-zymatic cycling system [J]. Sensors and Actuators B: Chemical, 2015 210:241-247.

IOP Conf. Series: Materials Science and Engineering 472 (2019) 012073 doi:10.1088/1757-899X/472/1/012073

# Study on Preventive Measures for Serious Water Inrush Accident in Underground Tunnel Engineering--the Case of Water Inrush in a Tunnel Project

**Yuxue Bao**

Liaoning Geology Engineering Vocational College, Dandong, Liaoning 118008, China

Email:baoyuxue1@126.com

**Abstract.** Based on a case of water inrush accident in a tunnel project, analyzed the accident cause. Combining with practice and production, three measures are put forward to prevent water inrush accidents scientifically in underground shaft and tunnel construction: Before construction The investigation of hydro geological conditions must be done well; Advanced water exploration drilling holes should be designed in engineering design; During construction, The site management should be strengthened,timely correction and guidance . Ensure the project is completed on schedule and safely.

## 1. Introduction

With the growth of China's comprehensive strength, many important underground tunnel construction projects have been successfully completed in recent years, indicates that China has reached a new level in the construction technology and construction management standards of underground tunnels.

The construction of underground tunnels is carried out underground or in mountains. The construction of these projects must have special requirements and technical design [1]. Many years of underground tunnel construction experience tells us that scientific construction design and standardized construction management is an effective guarantee for efficient and safe completion of construction tasks. It requires us to strictly abide by scientific construction methods and standardized construction technology in engineering construction, otherwise we may cause undue losses.

## 2. The Introduction of Accident Cases

In Southwest China, A serious water inrush accident occurred at a key traffic tunnel in the summer of 2018. On the construction site, the underground river was suddenly opened, and a large amount of groundwater quickly gushed out, The river rushed the equipment and personnel of the construction site out of the tunnel. Directly. The accident caused many casualties on the construction site, damaged major production equipment with a value of tens of millions of dollars, the production was forced to stop, Production progress and safety and efficiency of construction personnel have been seriously affected. As shown in Figure 1.

**Figure 1.** Accident Scene

## 3. Formatting the Text

### *3.1. Have Insufficient Investigation Work on Hydro Geological Conditions*

The unexpected nature of the accident indicates that the geological conditions and the hydro geological conditions are insufficient.

The construction area of the project is limestone stratum, the climate of this area belongs to temperate monsoon climate, humid and rainy, and the project area belongs to the typical Karst landform area.

The underground river is very develops in Karst area, and groundwater mainly occurs in Karst and underground rivers. The characteristics of groundwater recharge and migration in karst area are as follows: The instantaneous discharge of groundwater is large and the recharge rate of groundwater is fast.

In the karst area, we must fully investigate the hydro logical and geological conditions in advance and take corresponding design measures according to the relevant geological characteristics and geological conditions in order to ensure the scientific construction of the tunnel[2~3].

The sudden occurrence of this event indicates that the construction may not have been fully hydro geological or engineering geological survey, or the investigation is insufficient. Otherwise, this unexpected accident will not happen.

### *3.2. The Design of the Construction Scheme is not Scientific Enough*

From the point of view of the extent of the accident, large equipment at the scene was rushed out of the tunnel. It shows that such a serious accident has not been considered in the construction of project, which shows that the scientific design scheme obviously has serious defects.

The engineering specifications require that the underground hydro geological conditions and engineering design schemes of different regions should be targeted[2~3].In the construction of tunnels in areas with complex hydro geological conditions, First, we should consider the use of underground drilling with advanced exploration, prevention and drainage in the design of the specific construction plan. Secondly, we should give priority to drilling some advanced water exploration holes, in order to continue the further excavation of the tunnel or shaft under the premise of ensuring the construction safety in the field construction process.

### 3.3. *Construction Site Management and Preventive Measures are not in Place*

From the loss of the site, the construction site management is not in place. It is mainly reflected in:

*3.3.1. Construction site observation and safety supervision is not in place.* Specifications require that geological and safety supervisors on the construction site regularly inspect the site to observe whether there are special geological phenomena and security risks. Generally the sudden gushing of groundwater has certain symptoms, such as the wall of the pit seeping and falling, the color of the groundwater becomes turbid, or the sound of the water can be heard faintly.

*3.3.2. Accident prevention mechanism is not perfect.* Many emergency measures such as avoidance of danger, emergency rescue and disaster relief can not be taken, resulting in greater losses of personnel and property, after the incident.

*3.3.3. The distribution of construction site is unreasonable.* The main manifestations are as follows: (1) the emergency safety chamber inside the tunnel is not standardized; (2) the construction site outside the tunnel is unreasonable, construction materials piled up disorderly. It is precisely because of the above three reasons that led to this serious production accident.

## 4. Prevention Measures for Water Inrush Accidents in Underground Tunnels

Based on the above reasons and combined with practical experience, the prevention of serious water inrush accidents in underground tunnels is mainly considered in the following aspects.

### 4.1. *Standard Construction Procedure, Scheme Design and Investigation of Geological and Hydro geological Conditions Before Construction*

According to the requirements of the construction code [2~5], scientific geological and hydro geological conditions must be investigated strictly to provide accurate geological data for scientific design and construction.

The main contents of the investigation include: (1) the geological conditions in the construction area. It mainly includes lithological and geological conditions and structural geological conditions. (2) The hydro geological and climatic conditions in the construction area. It mainly includes groundwater types, occurrence and recharge conditions, and the correlation between groundwater discharge and seasonal climate. (3) The influence of geological and hydro geological conditions on the construction of the construction area.

### 4.2. *Carry out Scientific Construction Design to Avoid Anticipated Risks*

On the basis of grasping the geological and hydro geological conditions of the construction area, the design should take targeted measures according to the actual situation, mainly from the following aspects:

*4.2.1. Have scientific site selection to avoid geological disasters.* On the premise of obtaining precise geological and hydro geological conditions, the location of tunnel engineering should be designed scientifically to avoid as much as possible areas that may cause major geological disasters. If it can not be avoided, try to get through the lot at the shortest possible distance. We should not ignore construction safety risks or future engineering risks because of construction costs.

*4.2.2. Design advanced borehole for water detection.* In view of the complicated underground hydrology section, in the construction design of the tunnel or shaft, it is necessary to scientifically design the advanced water exploration hole and carry out the design of exploration, prevention and drainage within the construction scope. Eliminate the occurrence of hidden danger of water bursting.

The design of advanced exploration[6], prevention, drainage hole design, can refer to the relevant exploration and construction specifications and hydro geological conditions, as shown in Figure 2.According to the actual geological conditions on the spot, the advanced water exploration boreholes are arranged for some key areas which may have hidden dangers. For example, exploring underground rivers, underground water caverns, etc.

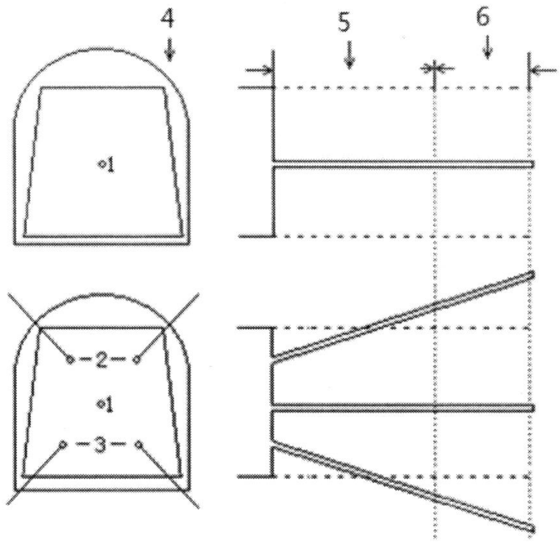


1.Forward borehole drilling 2.Upward borehole drilling 3.Downward borehole drilling
4. Tunnel Cross Section 5. Safe Distance Once drilling 6. Safe Distance

**Figure 2.** Diagram of drilling design for advanced water detection

*4.3. To Standardize Construction Management, Improve Construction Efficiency and Reduce Construction Risk*

Scientific construction management must be well considered in the design of construction. In the production site, the standard management is mainly carried out in the following aspects.

*4.3.1. To strengthen safety supervision on construction site and eliminate hidden dangers.* On the premise of meeting the requirements of construction codes, we should strengthen the guidance of geological safety and require the safety staff to supervise and eliminate the safety traps in time [7~9].

*4.3.2. Geological staff catalogue and grasp the production site geological data in time.* During the construction process, geological workers are required to catalog and obtain the first geological data in time, and according to the geological phenomena in the construction site, the design scheme is compared to verify the scientificity of the design and construction scheme. If the design scheme has major defects, they should be promptly put forward, if the hidden dangers cannot be timely checked, they should be timely contact with decision-makers, proposed to amend and improve the original construction design scheme.

*4.3.3. To standardizing construction management and truly implementing the third party supervision mechanism [9].* According to the regulations, the construction supervision and management system shall be strictly implemented. Truly implement the third party supervision and management mechanism.

## 5. Conclusion

In tunnel construction, it is necessary to design construction scheme scientifically and standardize construction management on the basis of strengthening geological survey in strict accordance with construction specifications. Avoid negligence and deficiency in the design and construction process, and prevent major production accidents and economic losses.

## References

[1] Merchants Chongqing traffic research and Design Institute Co, Ltd. Tunnel Engineering Office,Study For Classification of highway tunnels and concept of Highway Tunnel Group.[EB/OL]https://wenku.baidu.com/view/0a48bf72f46527d3240ce0a9.html?rec_flag=default&sxts=1531566384897
[2] Ministry of Communications of the People's Republic of China,2004.11 Code for design of Road tunnel(JTG D70-2004)[S]
[3] Ministry of Railways of the People's Republic of China,2005.4 Code for design on Tunnel of Railway(TB10003-2005,J449-2005) [S]
[4] Ministry of Railways of the People's Republic of China, 2002.3 Code for Construction on Tunnel of Railway(TB10204-2002,J163-2002) [S]
[5] Ministry of Communications of the People's Republic of China,2009Technical code for construction of highway tunnel(JTG F60-2009) [S]
[6] YuXue Bao, MeiYing Wang, XiaoYun Xiong. Geology of mine [M]. Beijing, Geological Publishing House.2009.7.P54
[7] Sixin Ye.Safety supervision in tunnel construction[J].SiChuan Architecture, 2007,27(2): P228-229
[8] Haifeng Qu, Zhigang Liu, Hehua Zhu, etc. Comprehensive Advanced Geological Prediction Technology in Tunnel Informatization Construction [J]. Journal of Rock Mechanics and Engineering, 2006, 25 (6)
[9] Xiangdong Shen. Application of monitoring and measuring technology in tunnel construction [J]. ShanXi architecture, 2018, 3.P155

# Study on Optimization of New Type of Seepage Well Structure in Sponge City

**Zewei Wu[1], Chaoxian Deng[2], Jianhui Si[1], Kan Zhao[1], Yue Ma[2] and Hanghang Liang[2]**

[1] College of civil engineering, Xi'an University of Technology, Shaanxi Xi'an 710048, China;

[2] Management Committee of Feng Xi new town, Xi xian New District, Shaanxi xi'an 712000, China)

**Abstract.** The seepage well is an indispensable part of the construction of the sponge city. It has a good "elasticity" for the city to adapt to the environmental changes and respond to natural disasters. In the face of the current decline in the groundwater level and the water logging in the city, it is urgent to study the new type of seepage well under the complex geological conditions. In this paper, the advantages and disadvantages of the current seepage well structure are analyzed, and combined with the hydro-geological conditions in the area of Feng Xi new city of xi'an, three new types of seepage wells are put forward, such as reinforced concrete filter pool + mufti glass pipe combined structure seepage well, masonry filter pool + single glass pipe combined structure seepage well and prefabricated reinforced concrete structure seepage well. The advantage, disadvantage and applicability of the well are compared, and case analysis is provided, which provides a basis for the construction of seepage wells in sponge cities.

## 1. Introduction

Under the background of urbanization, the newly built buildings in China are equivalent to half of the total world buildings. With the acceleration of urbanization, the impervious area of the city is gradually increasing, and the city is evolving into a "hard shell city" constructed by reinforced concrete. Rainwater resources can not be effectively utilized, and urban water circulation system and ecosystem are further deteriorated. Urban waterlogging, on the surface, is due to the urban underground drainage system lagging behind urban construction, but the root cause is that construction have changed the surface runoff and increased the burden of underground pipeline network. The problems of water ecology and water resources shortage and water safety faced by cities are mainly attributed to the backwardness of traditional urban engineering pipeline gray drainage infrastructure, flood control planning and drainage engineering planning, and the weak awareness of rational utilization of rainwater resources. Seepage wells, as an important part of sponge cities, can be greatly alleviated. Urban waterlogging, serious water accumulation and a series of other problems; seepage well seepage drainage mode is to use the total amount of catchment control, the rainwater collected in the downward excavation channel through catchment wells, drainage pipes into seepage wells for temporary water storage, while through seepage wells into the underground permeable layer to play a drainage role.

## 2. Research Status of Seepage Well

China's Ministry of Housing and Urban-Rural Construction issued in 2014 "Technical Guidelines for Sponge City Construction - Construction of Rainwater System for Low Impact Development" (Trial).

The purpose is to guide the construction of new urbanization in various areas, promote and apply the low impact development and construction model, and realize the utilization of rainwater resources through the combination of various technologies. . Constantly promote the development of "sponge city" with natural accumulation, natural infiltration and natural purification. Sponge city is defined as a city that is able to adapt to environmental changes like a sponge and has good flexibility in dealing with natural disasters. That is to say, it absorbs water, purifies water, stores water and recycles the stored water when it is raining, so as to realize the sustainable development of the city [1]. The seepage well as a convenient and efficient advanced process has received wide attention.

At present, the structure of seepage wells in China can be divided into masonry structure, reinforced concrete structure and steel tube structure according to different materials, but they all have certain limitations.

### *2.1. Masonry Structure Seepage Well*

Masonry structure is used as a common structural form for early seepage well, and it is constructed by common brick or sector brick. Masonry structure is easy to use local materials, low cost, and does not need formwork and special construction equipment compared with concrete structure, so it is widely promoted. However, it also has many shortcomings. Compared with reinforced concrete structure, the strength of masonry is lower, so the section size of component is larger and the amount of material is more. The masonry is basically manual, with a large amount of labor. The masonry's tensile and shear strength is very low, its seismic resistance is poor, and the compressive strength of bricks and stones can not be fully exerted, so it is limited in use; masonry structure is liable to cause structural damage after settlement, so with the development of new materials and new technology, seepage wells of masonry structure have already been coagulated by reinforcing bars. The structure of soil is replaced by steel tube structure, so its structure needs to be further improved.

### *2.2. Seepage of Reinforced Concrete Structure*

The infiltration well of cast-in-situ reinforced concrete structure has good integrity, stable structure, large runoff capacity, strong adaptability and good durability compared with masonry structure; but the construction of cast-in-situ concrete structure is relatively complex, which consumes more formwork, has larger self-weight and is prone to settlement compared with masonry structure.

### *2.3. Steel Bellows Structure Seepage Well*

Steel bellows can be divided into spiral bellows, annular bellows and corrugated steel sheets. The structure of steel bellows seepage well is mainly used to reduce the water accumulated by the roadbed of expressway, and has been applied to many Expressway sections with remarkable effect. The construction of steel bellows is convenient, and it is generally passed through. Bottom-hole seepage is difficult in urban drainage construction because of its high cost, low permeability rate, difficulty in filling fillers and uncertain opening rate.

## 3. New type of Seepage Well Structure

In order to maximize the function of rainwater storage, infiltration and purification in urban areas, promote the utilization of rainwater resources and ecological environment protection, and focus on solving the problems of water environment, water ecology and waterlogging in urban construction, this paper puts forward a proposal on the basis of traditional seepage well structure design, considering comprehensively the hydrogeological conditions of Shaanxi Province. Rainwater drainage and storage system suitable for Shaanxi area and a more economical and effective seepage well structure form are put forward.

Including: reinforced concrete filter + multiple FRP composite seepage well structure, masonry structure filter + single FRP composite seepage well structure and reinforced concrete prefabricated pipe seepage well structure.

*3.1. Reinforced Concrete Filter Tank + Multiple Glass Fiber Reinforced Plastic Combined Shaft Structure (as Shown in Figure 1).*

The structure of reinforced concrete filter plus multiple FRP composite seepage wells is composed of upper reinforced concrete structure, lower FRP tube and cushion layer at the bottom of concrete. The bearing capacity of reinforced concrete structure is large, but for the area with large flow, the flow confluence area is large, the internal force distribution is complex, The reinforced concrete structure with better integrity can effectively ensure the stability and safety of the structure, and it has good durability, strong adaptability to the environment, and can work normally in extreme environments.

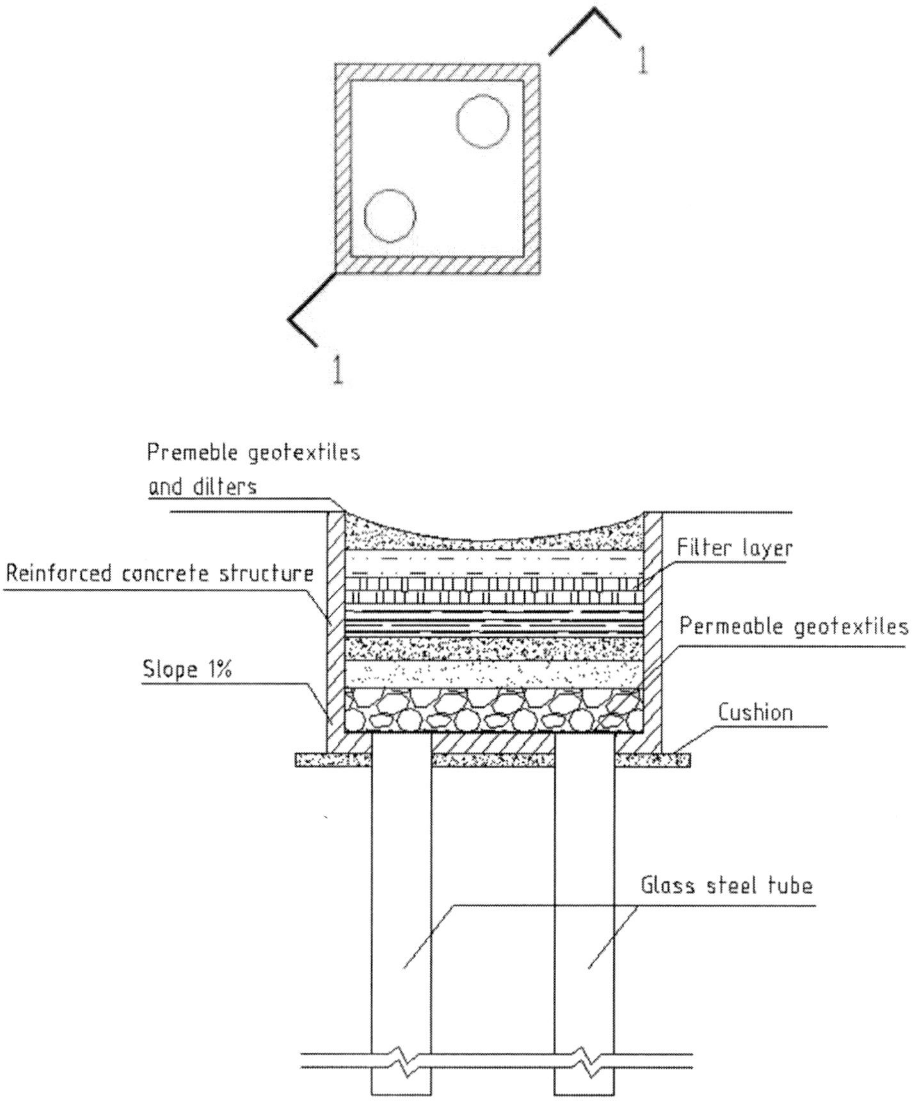


**Figure 1.** Reinforced concrete filter tank plus several glass fibers reinforced plastic combined shaft structures

*3.2. Masonry Structure Filter Material Pool + Single Glass Steel Pipe Combined Seepage Well Structural Form (as Shown in Figure 2).*

The masonry structure filter + single FRP composite seepage well structure is composed of masonry structure filter and single FRP pipe. The FRP pipe is drilled down to the sand layer, and the concrete cushion is set near the upper part of the FRP pipe. The masonry structure is used as the seepage well filter. Compared with the reinforced concrete structure, this structure has light weight and low cost. Therefore, it is economical and applicable in areas with small runoff. However, its structural stability may be affected in a large flow area.

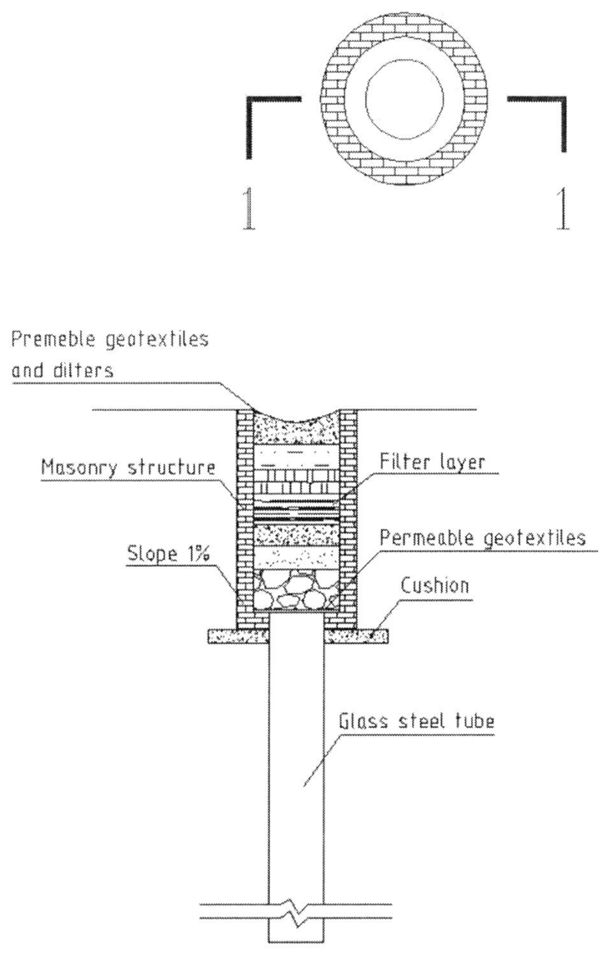


**Figure 2.** The structure of composite shaft with masonry structure filter material pool and single glass steel pipe.

A new material FRP is used as the seepage pipe. FRP has many advantages: smooth inner wall, strong conveying capacity, no scaling, no rust, good frost resistance, good wear resistance, good corrosion resistance and low water resistance. Below zero 20 degrees Celsius, there will be no frost crack after freezing. As the main material of FRP is composed of unsaturated polyester resin and fiberglass, it can effectively resist the corrosion of acid, alkali, salt and other media, and the erosion of untreated domestic sewage, corrosive soil, chemical wastewater and many chemical liquids. In general, it can keep pipelines for a long time safe operation; FRP pipe does not need to be rust-proof, pollution-proof, insulation and other measures and maintenance, its maintenance cost is low. So FRP pipe can work together with masonry structure and concrete structure, at the same time, it can greatly improve the performance of seepage wells.

### *3.3. Prefabricated Reinforced Concrete pipe seepage Well Structural form (as Shown in figure 3)*

The prefabricated reinforced concrete pipe seepage well structure is composed of a series of prefabricated reinforced concrete pipes per unit height. It seeps through the bottom and bottom wall of the pipe. The reinforced concrete pipe will be placed in the underground sand layer, the prefabricated concrete pipe will be overlapped, and at the same time, the hole will be opened under one meter away from the sand layer to improve its penetration speed. The structure form is shown in the figure.

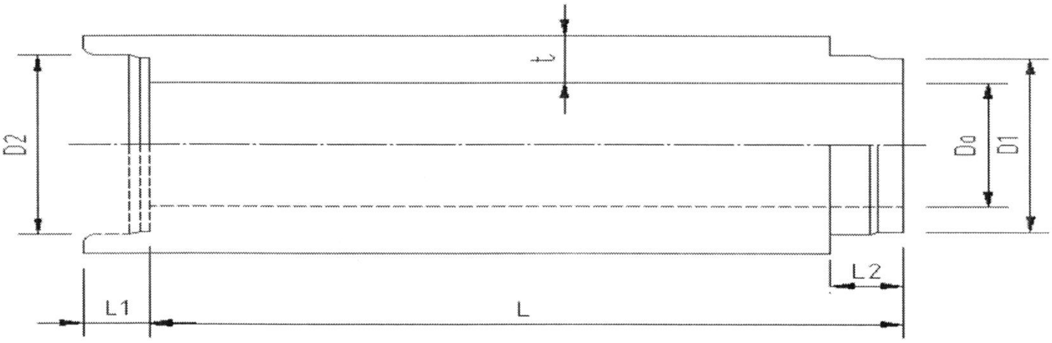


**Figure 3.** schematic diagram of precast reinforced concrete pipe

## 4. Case Analysis

The construction method of seepage well should be combined with the local geological and hydrological conditions, as well as the technical strength, construction equipment, take corresponding measures, formulate detailed construction plan, and strive for safe and effective completion of construction work.

### *4.1. Hydrogeological Survey*

The project is located in Xianyang Vocational and Technical College. Through two field surveys, it is found that the pilot site in this section is low-lying, with a large amount of water and silt. Therefore, two sites were selected as a pilot to study the applicability and rationality of the optimized structure of the well.Fig. 4.1 and4.2 2 are field maps of Pilot 1 and Pilot 2, respectively.

**Figure 4.** pilot 1

**Figure 5.** pilot 2

### *4.2. Seepage Well Construction Plan*
In order to prevent seepage wells from collapse and blockage in the future, glass tube is installed at the lower part of seepage wells, and the depth of caisson is more than 8 meters (the depth reaches the middle sand layer). According to the drainage volume and drainage time in the campus, two wells are planned at the lower part. The upper structure is reinforced concrete sedimentation tank with filter layer or brick sedimentation tank. In order to ensure the normal use of seepage wells in the future, the wellhead located on the bottom plate is covered with steel mesh, and a screen with steel wire spacing of 2 cm is added to the mesh. The matching facilities of climbing ladder well cover are set up at the roof checkpoint to facilitate garbage cleaning and maintenance in the future.

Pilot 1 adopts seepage well structure as shown in Fig 4. The seepage well is composed of rectangular reinforced concrete pool and double FRP pipe. The wall and bottom of the pool are made of C30 concrete, and the reinforcement is double row 14@150. Pilot 2 adopts the seepage well structure shown in Figure5. Which is composed of brick masonry well and single FRP pipe, and two seepage wells adopt FRP circular pipe with wall thickness of 15mm, which opens on the wall of the permeable layer, with the opening rate of 1%-3%. The seepage well needs to extend into the middle sand layer, which is more than 0.5m, and the highest groundwater level is 1m. The wellhead is covered with steel mesh, and a layer of wire spacing 2cm is added on the steel mesh.

### *4.3. Penetration Results*
After water test on seepage well structure, the permeability efficiency of seepage well structure is 2.7 *10-3m/s and 1.24 *10-3m/s respectively (the test results are affected by groundwater level trend and the location of monitoring points). At present, the permeability rate of existing seepage wells is generally 1 *10-5m/s. The comparison of the permeability results of three seepage well structures is shown in the table below, which is not from the table. It is difficult to see that the permeability rate of the new seepage well structure is obviously higher than that of the ordinary seepage well structure, and the permeability efficiency is 100 times higher than that of the traditional seepage well, showing good permeability effect.

## 5. Conclusion
Based on the hydrogeological conditions of Shaanxi Province, through the analysis of the advantages and disadvantages of traditional seepage well structure, this paper puts forward several new seepage well structure forms suitable for Shaanxi Province. The following conclusions are drawn:

(1) Seepage wells, as an effective drainage facility in sponge cities, can greatly alleviate a series of problems such as urban waterlogging and serious water accumulation.

(2) The structure form of reinforced concrete filter + multiple FRP composite seepage wells should be suitable for the larger area of confluence; the structure form of masonry filter + single FRP composite seepage wells should be suitable for the smaller area of confluence; at the same time, they all have the advantages of convenient replacement of filter materials, fast seepage speed, simple construction and geological conditions. The demand is not high;

(3) Three different types of seepage wells are proposed in this paper, but there are many factors affecting the safety and stability of prefabricated reinforced concrete pipe structures. Further research is needed on the diameter, thickness, depth of seepage wells, size and number of openings.

**Acknowledgement**
Supported by Shaanxi Provincial Housing Urban and Rural Construction Science and Technology Project (2017-K16); Fundamental Research Project of Sponge City in Fengxi New Town, Xixian New District (2017610002000088)

**Reference**
[1] Jiancheng Letter No. 275, Ministry of Housing and Urban-Rural Construction, Circular on the Issue of "*Technical Guidelines for Sponge City Construction--Construction of Rainwater System for Low Impact Development (Trial)",* Ministry of Housing and Urban-Rural Construction, Republic of China, 2014:4-37.
[2] Zhuxi, Li Shunqun, Feng Yanfang, Wang Yanyang. *Application of rainwater seepage wells for construction waste in sponge cities* [J].Journal of Guangxi University (Natural Science Edition), 2017, 42 (04): 1415-1421.
[3] Study on well construction technology and recharge test of seepage wells in thick aerated zone by Liu Pengfei, Liu Shaoyu, Wang Zhe, Zhou Xiaoni --- *Taking seepage wells in Hutuo River alluvial fan and Gaocheng section as an example* [J]. Science and technology and engineering, 2016, 16(01): 36-41.
[4] Liao Xiaohang. *Application effect of seepage well technology on Hengda Expressway* [J]. Highway Transportation Science and Technology (Applied Technology Edition), 2015, 11 (02): 36-38.
[5] Wang Chengwei, Li Yuanmeng. *Application of YTHG steel bellows seepage well in Beijing-Shijiazhuang reconstruction and extension project* [J].Highway Transportation Science and Technology (Applied Technology Edition), 2015, 11 (01): 59-62.
[6] Liao Xiaohang. *Application effect of seepage well technology on Hengda Expressway* [J]. Road traffic and safety, 2014, 14 (06): 1-5.
[7] Zhang Qingyu. *Numerical simulation of sidewall structure of concrete permeation* [D]. HeFei University of Technology, 2014.
[8] Wang Si Nan. *Main points of seepage well construction* [J]. Traffic world (transportation, vehicles), 2013 (08): 169-170.
[9] Wang Yun, Wu Wanping, Ruan Yanbin, Xie Songlin, Fu Wei, He Bin. *Influencing factors of unsaturated seepage drainage by regulating and storing seepage wells* [J]. Highway, 2013 (02): 20-24.
[10] Yan Hua. *Application of seepage well technology in highway drainage works* [J]. Traffic standardization, 2012 (06): 54-56.
[11] Li Zhen. *Study on drainage scheme and seepage well construction technology of low embankment expressway passage* [D]. Changsha University of Technology, 2011.
[12] Tong Xin Bo. *The construction technology for the middle and lower excavated channel of the expressway.* [J]. traffic standardization, 2010 (19): 124-126.

# Trivalent Chromium Conversion Coating Formed on Hot Dip Zn-55Al-1.6Si Coated Steel

**Yan Li[1*], Jie Pan[1] and Jiantao Qi[2]**

[1] School of Material Science and Engineering, China University of Petroleum (East China), Qingdao, Shandong 266580, PR China.
[2] College of Chemical Engineering, China University of Petroleum (East China), Qingdao, Shandong 266580, PR China.
*E-mail: yanlee@upc.edu.cn

**Abstract.** Trivalent Chromium Process (TCP) coating on hot dip Zn-55Al-1.6Si coated steel was investigated by XPS, SEM and EDS. The XPS revealed that aluminium oxides enriched the conversion coating surface. SEM and EDS analyses indicated that composition and microstructure of the conversion coating were not uniform on the dendrite and inter-dendrite regions within the Zn-Al alloy coating, and a small amount of spherical precipitation with the same composition as the conversion coating was detected in the zinc-rich inter-dendrite region. The optimal forming parameters of dilute passivator determined by salt spray test and orthogonal test is: 40°C, 50 vol.% and 3 min. In addition, the salt spray corrosion resistance of the optimal conversion coatings can reach 192 h, and the optimal TCP coating can reduce the cathodic corrosion current density in 0.05 M NaCl solution by an order of magnitude due to the electrochemical polarization measurements.

## 1. Introduction

Hot dip coated steel (Zn-55Al-1.6Si) is widely used in hardware decoration, automotive industry and other fields [1-2]. Trivalent chromium passivation process (TCP) can form trivalent chromium Cr (III) chemical conversion coating on the surface of zinc, aluminum and other alloys, thus it has significant potential to enhance the corrosion resistance of hot dip coated steel. Moreover, the trivalent chromium passivation technology will be the focused research in this field due to the environmental injunction and the global emphasis on eliminating hexavalent chromium [3-4].

However, due to the special surface condition of Zn-55Al-1.6Si hot dip coating (containing zinc-rich inter-dendritic region and aluminium-rich dendrite region), the formation process of trivalent chromium chemical conversion coating on its surface is different from that of common various aluminum alloys.

In this study, Cr (III) Alodine 5928 passivator was applied to Zn-55Al-1.6Si hot dip coating surface. Moreover, the optimum technological forming parameters was determined by corrosion results of neutral salt spray tests and the composition and structure characteristics of the thin conversion coating were determined by optical high resolution surface analyses.

## 2. Experimental Methods

Commercialized hot dip Zn-55Al-1.6Si coated steel (hereinafter referred to as Zn55Al) was used as the experimental materials. The hot dip alloy coatings present on both sides, 20-25 μm thick, the specific coating and substrate composition are shown in table 1 respectively. Panels (0.8 mm thick) were mechanically cut into dimensions of 50×50 mm or 15×15 mm for salt spray tests, surface characterization techniques and electrochemical measurements. They were then cleaned followed by immersion in acetone, ethanol and deionized water. Before the trivalent chromate conversion process, chemical pre-treatment were then employed to the cut panels: etching for 0.5 min in 5 wt.% sodium hydroxide solution at 50°C and desmutting for 10 s in 10 vol.% nitric acid (99.9%) at 25°C.

**Table 1.** Coating and matrix composition of hot-dip steel panel (wt.%).

| Elements | C | Mn | Si | P | S | Al | Zn | Fe |
|---|---|---|---|---|---|---|---|---|
| Matrix | / | / | 0.34 | / | / | 52.14 | 37.66 | 1.62 |
| Hot dip coating | 0.035 | 0.220 | 0.008 | 0.012 | 0.008 | 0.049 | / | 99.67 |

To form the conversion coating, the degreased and deoxidized specimen (as described above) was immersed in the RTU Bonderite T-5928 coating bath (pH 3.63) at 40°C for deferent time (2 min, 3 min and 6 min), with very gentle periodic beaker agitation, as needed, to dislodge any gas bubbles ($H_2$) that formed on the metal surface.

A Nava Nano SEM450 microscope associated with an integral EDS detector, operated at an accelerating voltage of 15 kV, were used to investigate surface morphologies and compositions. The chemical states of coating species were determined by X-ray photoelectron spectroscopy (XPS), using an ESCALAB 250Xi system equipped with an Al/K Alpha source. The microfocus monochromator was operated at a spot size of 500 μm, etching depth was about 10 nm, including the dendrite and inter-dendritic regions and the high resolution intensities of the Cr 2p, Zr 3d, O 1s, Al 2p, Zn 2p, etc. photoelectron lines were recorded separately. The background intensity was subtracted from the spectra using the Shirley method. All data were corrected by a linear shift such that the peak maximum of the C 1s binding energy of adventitious carbon corresponded to 284.6 eV.

For the electrochemical measurements, a three-electrode cell was used, with a Solartron 1287 potentiostat system. The cell contained an Ag/AgCl/$Cl^-$ (saturated KCl) electrode, a platinum sheet counter electrode and a working electrode with an exposed area of 1 $cm^2$. Measurements were carried out in naturally aerated or nitrogen deoxidizing 0.05 M NaCl solution at room temperature. The cathodic polarization curves was obtained by scanning the potential from OCP to 0.15 V vs. the OCP, with a scanning rate of 1 mV/s. CView software (version 3.1) was used to fit the experimental data.

The salt fog test was conducted according to ASTM B117 (5 wt.% NaCl). At the end of the test period, the specimens were removed, rinsed thoroughly with ultrapure water, and dried under a stream of $N_2$ gas. After photos were taken by Canon digital camera, any corrosion product was removed by ultrasonication (room temperature) in ammonium acetate acid.

## 3. Results and Discussion

### *3.1. Optimization of Processing Parameters and Neutral Salt Spray Test*

To verify the applicability of the two kinds of passivating agents to the Zn55Al, orthogonal tests with three factors and four levels were designed to determine reasonable technological parameters, and the operating conditions are listed in table 2.The appearance corrosion rating of each group of samples after the 192-h NSS test was assessed based on ISO 10289, and was used as quantitative criteria in the orthogonal test. The optimal experimental parameters are shown also in table 2. The order of the influence of parameters obtained from the orthogonal test result shows that the concentration is the primary influencing factor, and the next two are the temperature and time.

Figure 1 is the surface photos of the panels treated with 9 sets of test parameters under 0 h, 96 h and 192 h salt spray corrosion. After corrosion in a salt fog environment for 96 h, the corroded area exhibited a coating deterioration type of "J" (Crow's feet or star-shaped defects type). The black corrosion zone spreads from the edge of the sample to the center, marked by the red scissors in the figure 1(b). This may be due to the imperfection of the epoxy seal at the specimen edge or the poor TCC formation in the edge regions [5]. In addition, when the corrosion proceeded to 192 h, not only new corrosion areas appeared, but also the original corrosion areas increased in varying degrees, as shown by the red elliptical markers in figure 1(c).

**Table 2.** Preset experimental parameters of orthogonal design and optimal experimental parameters.

| Passivator | pH | Concentration (vol. %) | Bath temperature (°C) | Immersion time(min) |
|---|---|---|---|---|
| Preset parameters | 3.4/3.6/3.8 | 30/40/50 | 30/40/50 | 2/3/6 |
| Optimal parameters | 3.6 | 50 | 40 | 3 |

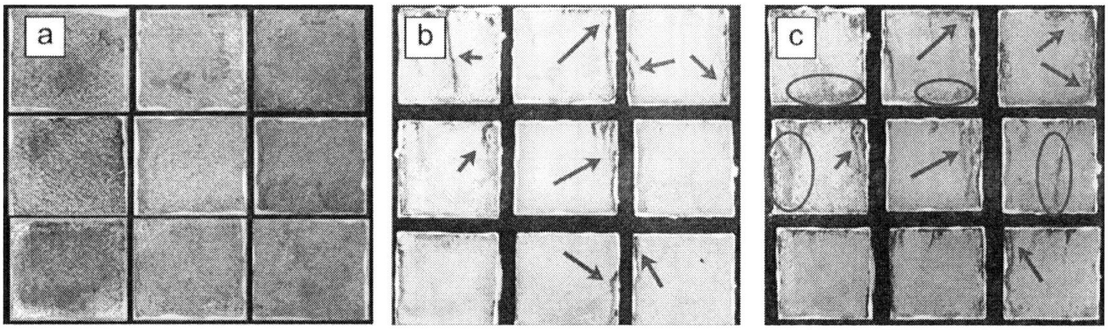


**Figure 1.** Samples after (a) 0 h; (b) 96 h and (c) 192 h salt spray tests treated with combination of different Alodine 5928 passivation parameters in orthogonal test.

### *3.2. Surface Morphology and Chemical Composition of the Pretreated and Conversion Coated Specimens*

Overview XPS spectra for optimal Alodine 5928 treated Zn55Al sample surface is shown here in figure 2. The atomic percentage of the main elements is shown in table 3.

The XPS analysis shows the content of major elements on the surface of the passivated specimen. The intensity ratio of Zr 3d to Cr 2p was fixed at 1.76:1. This proportion is consistent with previous studies [6]. In addition, nearly 8.6 at.% of F element was detected. The concentration of $F^-$ in passivation water sample is about $9.29\times10^3$ ppm by means of Metrohm 883 IC ion chromatograph, which indicates that the passivator has a strong oxidation property [7]. Additionally, the passivation solution also contains about 118.8 ppm $Cl^-$ ions and 72.2 ppm $NO_3^-$ ions, which probably originates from $Cr_2(SO_4)_3$ or $CrCl_3$ in the Alodine 5928 formulation.

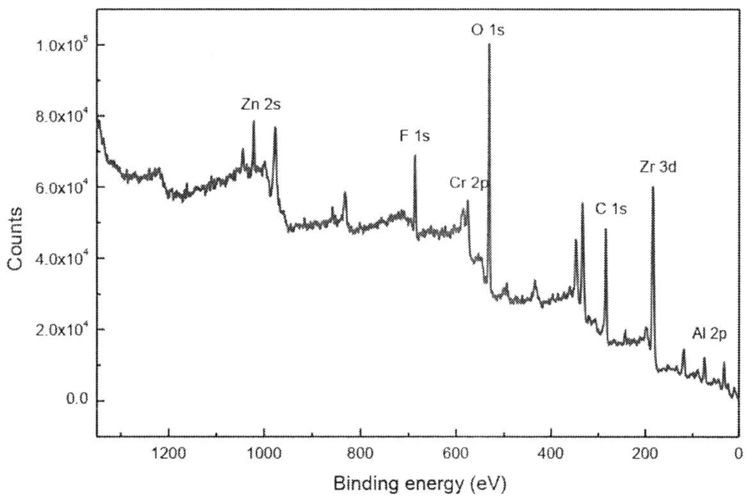


**Figure 2.** Overview XPS spectra for Alodine 5928 treated Zn55Al surface (optimal sample).

**Table 3.** Atomic (%) table of main elements for optimal Alodine 5928 treated Zn55Al by XPS.

| Atomic ( % ) | Zr 3d | F 1s | Cr 2p | O 1s | Zn 2p | Al 2p | C 1s |
|---|---|---|---|---|---|---|---|
| Element | 6.73 | 8.6 | 3.82 | 26.96 | 2.54 | 15.64 | 35.73 |

The examination of the chemical states of coating species was carried out using high-resolution spectra for the O 1s, Al 2p, Zn 2p, Zr 3d and Cr 2p regions. Due to the limitation of space, it is not shown here. The fitting results showed that about 45.11% of the substances of the chemical conversion coating were oxides at a depth of 10 nm, most in the form of alumina. In addition, a small amount of $AlF_3$ and $ZrF_4$ were also detected during the fitting analyze in high-resolution spectra of Al 2p and Zr 3d. A small amount of Cr (VI) ~1% appeared in the formed TCP coating after it was placed in air for a period of time, which may be related to the oxidation of the conversion film with aging; however, the appearance of hexavalent chromium can play a role of "self-healing" for the thin film to a certain extent. Its mechanism and the shelf in the conversion film need further exploration [8].

Zn55Al can be inferred to be passivated by the film formed from the trivalent chromium solution in the following formation process: (i) the zinc and aluminium in the hot-dip coating are oxidized to $Zn^{2+}$ and $Al^{3+}$ ions in the acidic trivalent chromium passivation solution; (ii) with the progress of oxidation and oxygen reduction reaction, a large amount of hydrogen ions are consumed in the reaction between the panel and the solution interface, and the consumed hydrogen ions cannot be replaced, resulting in a rapid rise in the pH and the formation of a thin alkaline layer; (iii) when the gel film formed from the trivalent chromium solution is dried after washing with water, some of the hydroxide is oxidized to oxide, forming a conversion coating composed mainly of oxide blends with extra hydroxide.

Figure 3 reveals a cross-section SEM micromorphology of the optimal Alodine 5928 treated Zn55Al sample and EDS elemental maps of aluminium, oxygen, zirconium, chromium and fluorine. It indicates the preferential precipitation of zinc in zinc-rich inter-dendritic regions due to the cathodic behavior of de-alloying process, leading to the heterogeneous film formation of the conversion film [9]. It is inferred that the preferential dissolution of zinc in the inter-dendritic zinc-rich phase provides a powerful impetus for the film formation process. This results in a slightly larger thickness of the chemical conversion film in the interdendritic region, and obvious spherical precipitation is found, indicating that the redox reaction here is more intense [10-11].

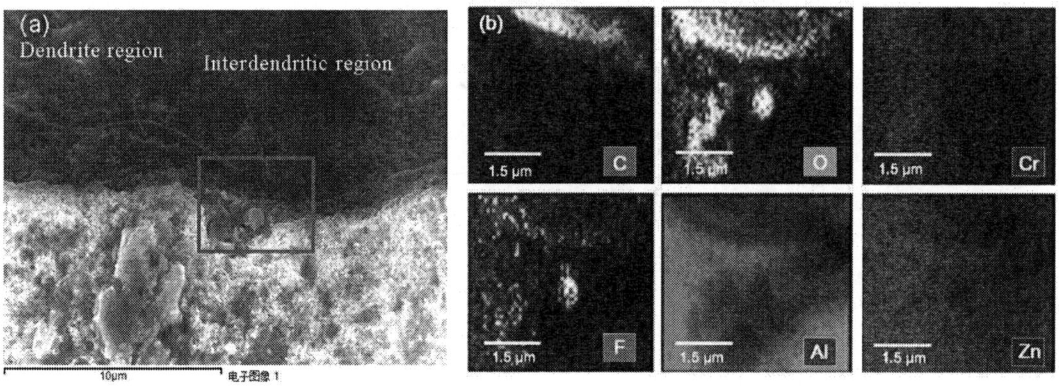


**Figure 3.** Composition and structure of Zn55Al surface with optimal Alodine 5928 treatment: (a) SEM micromorphology and (b) EDS elemental maps.

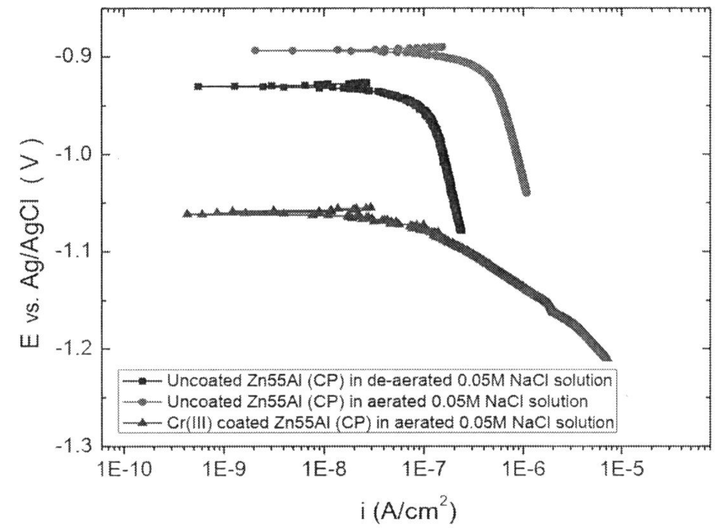


**Figure 4.** Slow-scan cathodic polarization curves for uncoated Zn55Al and Cr (III) coated Zn55Al in aerated and de-aerated 0.05 M NaCl.

**Table 4.** Fitting results of cathodic polarization curves recorded in figure 4.

| Samples | | $E_{corr}$ (V) | $i_{corr}$ (A/cm$^2$) |
|---|---|---|---|
| Uncoated Zn55Al in aerated 0.05M NaCl | CP | -0.8936 | 2.13E-07 |
| Uncoated Zn55Al in de-aerated 0.05M NaCl | CP | -0.9305 | 5.21E-08 |
| Cr(III)-coated Zn55Al in aerated 0.05M NaCl | CP | -1.0598 | 4.81E-08 |

### *3.3. Basic Electrochemical Properties of the TCP Coating*

In figure 4, the cathodic corrosion current density of uncoated Zn55Al at corrosion potential is much smaller in the de-aerated 0.05 M solution, compared with the current density in the aerated solution. The presence of oxygen accelerates the corrosion process by depolarizing the cathodic areas [8]. Since the corrosion of Zn55Al surfaces takes place under cathodic reaction control, it follows that a reduction in the corrosion rate can be brought about by inhibiting the cathodic reaction. The corrosion of Zn55Al bare surface can be retarded by hindering the transport of oxygen to the interface [12]. As shown in figure 4 and table 4, although the TCP coating is very thin, cathodic current at potentials close to $E_{corr}$ measured on Cr(III) treated Zn55Al sample in the aerated 0.05 M NaCl solution is significantly smaller than that on the bare sample, to an extend of an order of magnitude (with 2.13E-07 V and 4.81E-08 V separately). Hence, it is clear that Cr (III) coatings formed an effective barrier layer which hindered the transport of oxygen to the Zn55Al substrate.

## 4 Conclusions

In this paper, the optimal experimental parameters for the formation of chemical conversion coating on Zn-55Al-1.6Si coated surface using Cr (III) Alodine 5928 passivator were obtained as follows: 3 min to formation time, 40°C to passivation temperature and 50 vol. % to dilute concentration. The thin layer of TCP is aluminium-rich, with the presence of oxide and fluoride species, and the less than 1% of the Cr (VI) exists on the surface of the conversion coating. In addition, the neutral salt spray corrosion resistance of this thin film can reach nearly 192 h and TCP coating can inhibit Zn55Al corrosion in dilute NaCl solutions by hindering the transport of oxygen to the substrate metal. The electrochemical polarization results indicate that the cathodic corrosion current of Alodine 5928-treated Zn55Al sample is an order of magnitude smaller than that of untreated sample.

## References

[1] Persson D, Thierry D and Karlsson O 2017 *Corrs. Sci.* **126** 152

IOP Conf. Series: Materials Science and Engineering 472 (2019) 012075 doi:10.1088/1757-899X/472/1/012075

[2] Li Y 2001 *Corrs. Sci.* **43** 1793.
[3] Li Y, Wang H, Hou B, Feng F and Wei X 2001 *Br. Corrs. J.* **36** 56.
[4] Kendig M and Buchheit R 2003 *Corrs.* **59** 379.
[5] Munson C, McFall-Boegeman S, and Swain G 2018 *Electrochim. Acta.* **282** 171.
[6] Qi J, Hashimoto T, Walton J R, Zhou X, Skeldon P and Thompson G 2015 *Surf. Coat. Technol.* **280** 317.
[7] Qi J, Hashimoto T, Walton J, Zhou X, Skeldon P and Thompson G 2016 *J. Electrochem. Soc.* **163**(2) C25.
[8] Zhang X, Bos C, Sloof W, Hovestad A, Terryn H and Wit J 2005 *Surf. Coat. Technol.* **199** 92.
[9] Campestrini P, Westing E and Wit J 2001 *Electrochim. Acta.* **46** 2553.
[10] Munson C and Swain G 2017 *Surf. Coat. Technol.* **315** 150.
[11] Viroulaud R, Swiatowska J, Seyeux A, Zanna S, Tardelli J and Marcus P 2017 *Appl. Surf. Sci.* **423** 927.
[12] Chang Y, Wen N, Chen W, Ger M, Pan G and Yang T 2008 *Corrs. Sci.* **50** 3494.

# Application of Modified Organic-Inorganic Hybrid Zirconium Phosphate Material ZAMPS-PVPA-Salen-Mn(III) in Chemical Fixation of $CO_2$

**Liang Zhou[1,4], Biwei Gong[2], Zhiyong Yang[1], Deju Wei[1] and Wei Lu[3]**

[1] School of Chemical Engineering, Guizhou Institute of Technology, Guiyang 550003, China;
[2] College of Chemistry and Chemical Engineering Southwest University, Chongqing 400715, China;
[3] College of Mechanical Engineering, Guangxi University, Nanning, 530004, Guangxi, China.
[4] E-mail:cosa007cosa007@126.com

**Abstract**. The modified organic-inorganic hybrid zirconium phosphonate materials were used to catalyze the addition reaction of propylene epoxide with carbon dioxide. The catalytic materials of different carbon chains were investigated. The optimal catalytic materials were screened out. At the same time, the reaction time, pressure, temperature and the optimum amount of catalytic materials were also studied. The results show that the catalytic material 1c modified by Salen Mn(III) has good catalytic performance for the carbon dioxide addition reaction, the yield of propylene carbonate are over 90%. In addition, the modified catalytic material 1c is easily separated from the reaction system, and can be recycled at least 5 times, and the catalytic activity is not significantly reduced.

## 1. Introduction

Carbon dioxide, which exists in the three states of gas, liquid and solid, is widely used as a product in various sectors of industry and the national economy. According to statistics, the amount of carbon dioxide emitted by various fossil fuels (such as coal, oil, natural gas) in the world is rapid increased [1]. The waste of carbon dioxide resources has intensified the global warming tendency [2]. The average temperature of the earth is increased by 1.5 to 4.5°C annually [3]. Therefore, the control of carbon dioxide emissions, the recovery, fixation, utilization and recycling of carbon dioxide have become a matter of great concern to all countries in the world [4].

The high and new technology of using $CO_2$ as carbon resource for chemical utilization has become the research focus of the scientific community [5]. The chemical utilization of $CO_2$ is to convert $CO_2$ into bulk basic chemicals, organic fuels or to fix it directly into macromolecule materials. At present, industrialized projects on chemical utilization of carbon dioxide include synthesis of urea, salicylic acid, organic carbonate, inorganic carbonate and so on. The synthesis of cyclic carbonates from $CO_2$ and epoxy compounds is one of the most typical success stories of $CO_2$ resource utilization. During the reaction, without the use of organic solvents, the process is an atomic economic response, conforming to the standards of green chemistry. The product cyclic carbonates are widely used in textiles, printing and dyeing, polymer synthesis and electrochemistry, as well as in pharmaceutical and fine chemical intermediates [6].

A large number of catalysts have been developed to catalyze the coupling of $CO_2$ with epoxy compounds to form cyclic carbonates, such as alkali metal salts [7], metal oxides [8], transition metal compounds [9], Schiff base [10], ion exchange resin [11], functional polymer material [12], quaternary ammonium salt and quaternary phosphonium salt [13-15], gold nanoparticle support resin [16-18] and so on. However, many catalytic systems have disadvantages such as low catalytic stability and low catalytic activity, or requiring harsh conditions such as organic solvents. Both homogeneous [19] and heterogeneous systems have their own advantages and disadvantages. Usually the homogeneous catalytic system has higher catalytic activity than the heterogeneous catalytic system, but the latter is favored because of its simple preparation, easy separation, easy operation, good stability and recyclability.

The organic-inorganic hybrid zirconium phosphate is an excellent carrier material for the heterogeneous catalytic system [20]. One of its outstanding features is the designability of organic group. The organic-inorganic hybrid zirconium phosphate with different organic group can meet a variety of purposes [21]. Meanwhile the organic-inorganic hybrid zirconium phosphate is a kind of multifunctional material with a layered structure, and has high thermal stability, good resistance to acid and alkali [22]. As a peculiar material, it can be modified by various catalytic groups. It has become a new research hotspot in recent years [23]. In this paper, we have extended the research of application of modified organic-inorganic hybrid zirconium phosphate material to the field of activated carbon dioxide reaction, and found that it has a good effect on the catalytic carbon dioxide and propylene epoxide to form cyclic carbonate [24-26].

## 2. Materials and Methods

### *2.1. Chemicals and Instruments*

Propylene epoxide (PO) was supplied by Alfa Aesar, $CO_2$ was technical grade and was received from Guiyang shenjian Air Company. Other commercially available chemicals were laboratory grade reagents from local suppliers. Both the salen ligand and the salen Mn(III) homogeneous catalyst were prepared according to the reference [26]. The yield of propylene carbonate (PC) was determined by gas chromatography (Shimadzu GC-14C).

### *2.2. Preparation of Organic-Inorganic Hybrid Zirconium Phosphate Materials (Scheme 1)*

According to the reference [26], organic-inorganic hybrid zirconium phosphate materials (ZAMPS-PVPA) a-c was successfully prepared. And their compositions and structures have been proved by $^{1}H$ NMR, $^{31}P$ NMR and FT-IR.

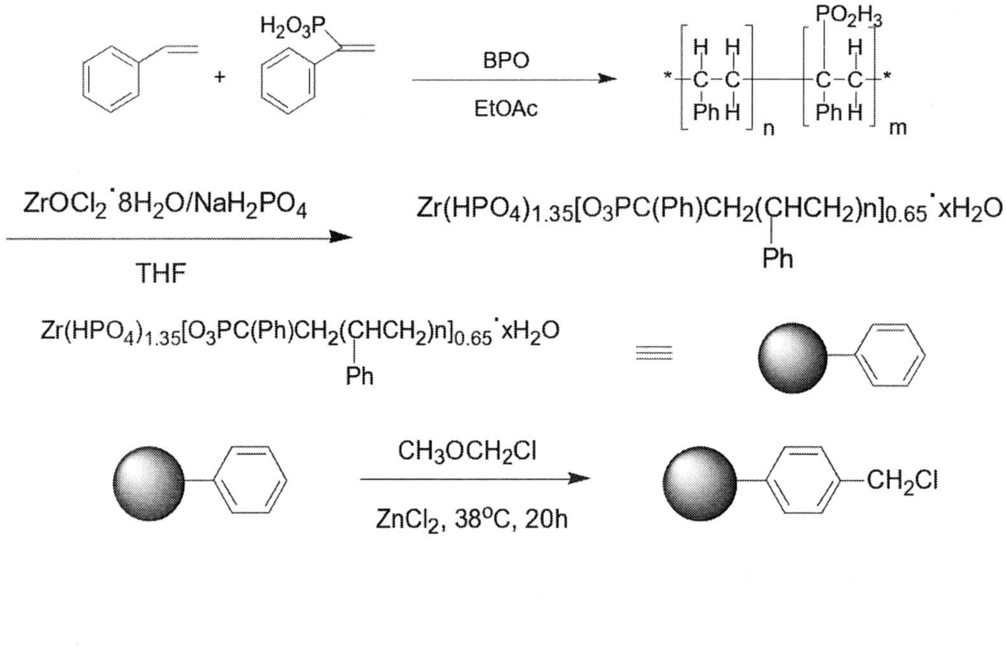

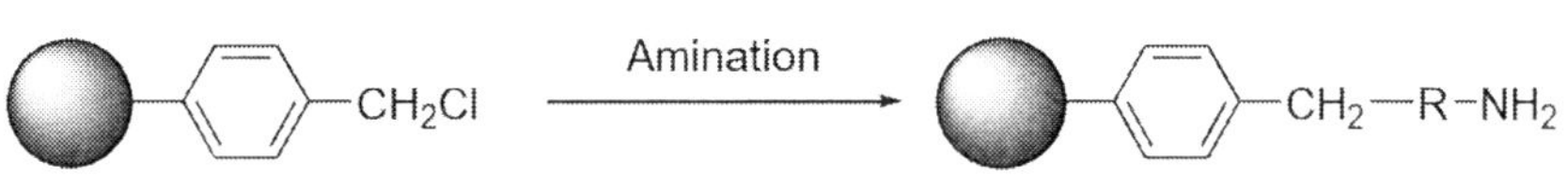


R = NH2-(CH2)2-, NH2-(CH2)3-, NH2-(CH2)4-

**ZASPS-PVPA (a-c)**

**Scheme 1.** Preparation of the materials ZAMPS-PVPA (a-c)

### *2.3. Materials Modified with Salen-Mn(III) as Catalytic Materials 1a-1c (Scheme 2)*

ZAMPS-PVPA a-c (0.5 g) separately, salen Mn(III) (2.5 g), NaOH (0.35 g) and 60 mL of tetrahydrofuran as solvent was added to a 150 mL three-necked flask. The mixture was stirred and heated to reflux for 24 hours. The mixture was filtered and washed with $CH_2Cl_2$ until no Mn ions were detected in the filtrates. The filter cake is dried under vacuum to obtain tan powdery solids catalytic materials as 1a, 1b, 1c.

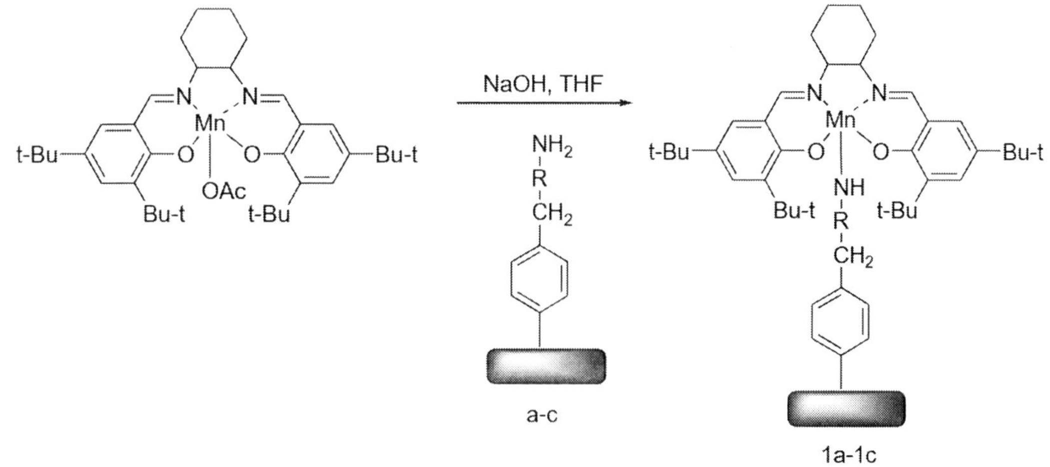


R = NH2-(CH2)2-, NH2-(CH2)3-, NH2-(CH2)4-

**Scheme 2.** The modified route to catalytic materials 1a-1c

### *2.4. The Addition Reaction of Propylene Oxide with Carbon Dioxide*

20.0 mL of propylene oxide, catalytic material and co-catalyst $ZnBr_2$ were mixed in a 50 mL autoclave. After sealing the autoclave, a certain pressure of $CO_2$ was introduced to remove the air in the reaction vessel to prevent oxidation of the reactants by $O_2$ during the reaction. This reaction was carried out at a certain pressure, time and temperature. After completion of the reaction, the autoclave was cooled to remove residual gas. 10.0 mL of ethyl acetate was poured into the reaction vessel. The reaction product was taken out and washed twice with ethyl acetate to remove the catalyst. Finally the solvent in the product was removed by a rotary evaporator to give a colorless liquid.

## 3. Results and Discussion

### *3.1. Catalytic Properties of the Different Catalytic Materials 1a-1c*

This study investigated the catalytic properties of the different modified materials 1a-1c. The results are shown in figure 1. At 130 °C, with $ZnBr_2$ (0.2mmol) as co-catalyst, along with the growth of the carbon chain of catalytic materials from 1a to 1c (0.5 mmol based on Mn in it), after a catalytic reaction for 6h, the yields of PC was increased from 80.3% to 90.2%. It can be seen that as the length of the arm increases, the catalytic effect of the modified materials is significantly improved. Mainly because in the catalytic process, the longer "molecular arm" makes the complex molecules away from the skeleton structure of the organic-inorganic hybrid zirconium phosphate material, and has more spatial freedom, which is beneficial to the deformation of the molecule and the participation of the promoter molecules, and at the same time the activity of the reaction. In general, catalytic material 1c has the best catalytic effect.

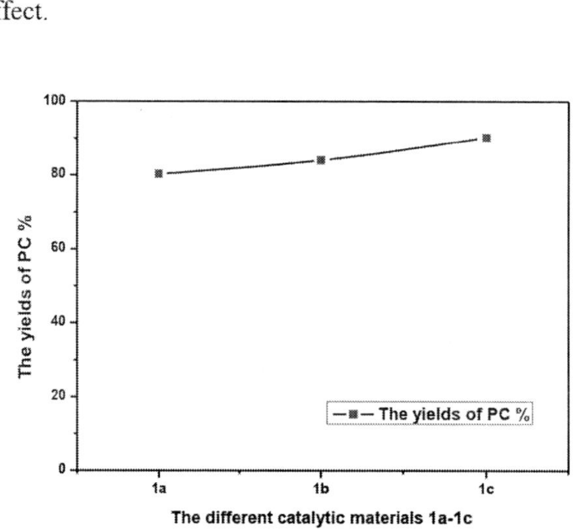


**Figure 1.** Catalytic properties of the different catalytic materials 1a-1c

### 3.2. *Effect of various factors on catalytic performance of catalytic material 1c*

The optimum conditions for the cycloaddition reaction are: $ZnBr_2$ (0.2mmol) as co-catalyst, reaction temperature 130 °C, reaction time 6h, filling with $CO_2$ 5MPa, catalytic material 1c dosage 0.5 mmol based on Mn in it. Under the above conditions the yield of PC is the highest 90.2%.

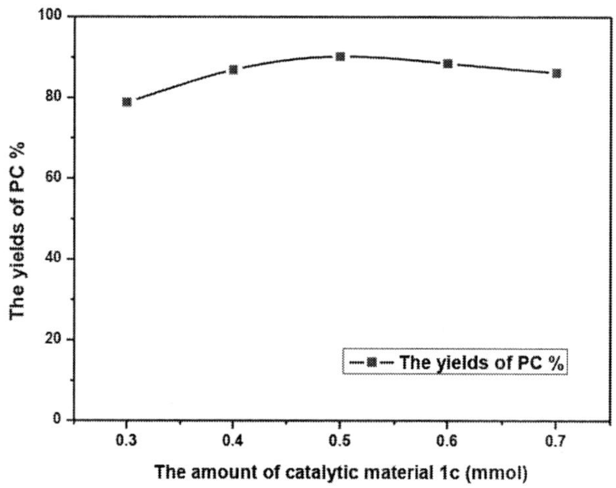

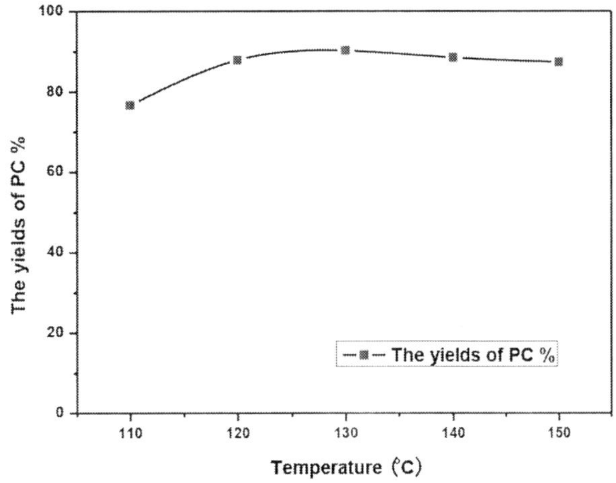


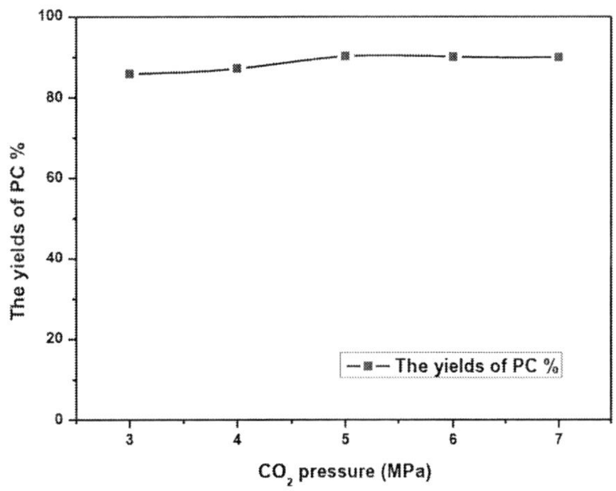


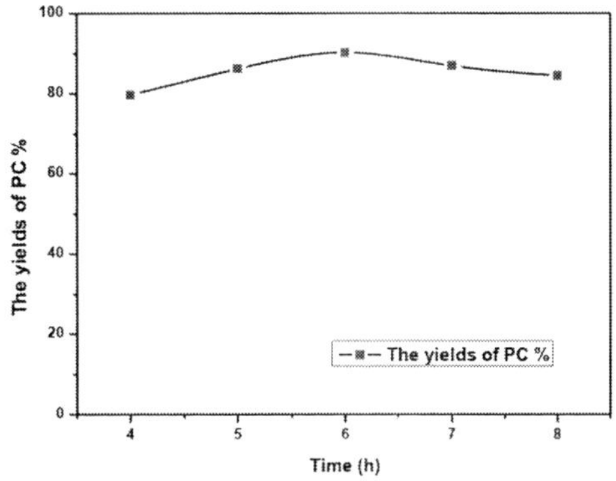


**Figure 2.** (a) Effect of the dosage of catalytic material 1c on the reaction; (b) Effect of reaction temperature on catalytic performance of 1c; (c) Effect of $CO_2$ pressure on catalytic performance of 1c; (d) Effect of reaction time on catalytic performance of 1c

### *3.3. The Recycling of the catalytic material 1c*

In order to detect the stability and recyclability of the catalytic material, we investigated the 1c under the optimal conditions. The catalytic performance of 1c for the catalytic cycloaddition of PO and $CO_2$. It can be seen from figure 3 that the catalytic results of the catalysts 1c was repeated 5 times. It is obvious that after 5 times of use, the catalyst 1c has a decrease in catalytic performance slightly.

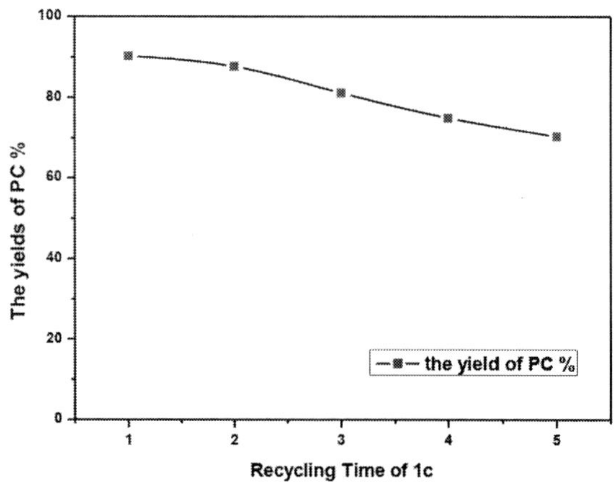


[a] Reaction conditions: catalytic material 1c (0.5 mmol based on Mn in it), co-catalyst $ZnBr_2$ (0.2mmol), PO (20 mmol), temperature 130°C, $CO_2$ pressure 5.0 MPa, reaction time 6h

**Figure 3.** Recycling of 1c in the cycloaddition of PO and $CO_2$ [a]

### *3.4. Proposed Mechanism of the Cycloaddition Reaction*

We believe that the quaternary phosphonium salt and the salen Mn (III) groups on the modified organic-inorganic hybrid zirconium phosphate material can activate propylene oxide in the course of the reaction. So we proposed the acceleration mechanism of the catalytic material and zinc bromide, as shown in the scheme 3.

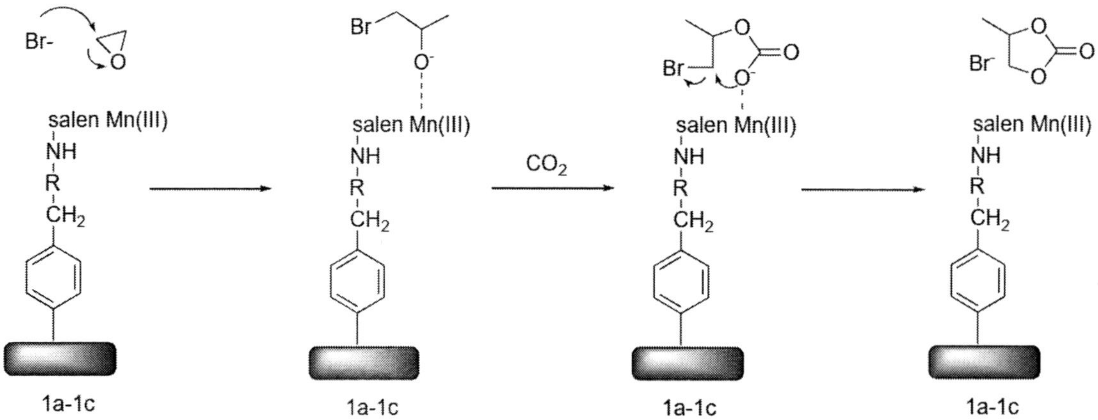


**Scheme 3.** Possible acceleration mechanism on the catalytic material and $ZnBr_2$

## 4. Conclusions

In this study, propylene oxide was used as substrates to carry out the cycloaddition reaction, and the catalytic performances of the catalytic materials were evaluated. The effects of the different modified materials 1a-1c, reaction temperatures, the dosage of catalytic material, $CO_2$ pressure and reaction time were investigated. The results indicate that the cyclic carbonate reaction can be activated by the catalytic material 1c. The yield of propylene carbonate can reach up to 90.2%. Furthermore, the reusability of the catalytic material 1c was investigated in the PO catalytic system. The results showed that the catalytic activity of 1c was not significantly decreased after 5 cycles, and the yield was still higher than 70%. The research in this article is of great significance for the stable catalytic performance of the catalytic material 1c. At the same time, it also opens up a new field for the application of organic phosphonic acid zirconium materials.

**Acknowledgements**
This research was financially supported by the Natural Science Research Project of Guizhou Provincial Education Office in 2015 ([2015] 416) and supported from Guizhou Province Science and Technology Fund in 2016 ([2016]1059).

**References**

[1] Sakakura T, Choi J C, Yasuda H 2007 *Chem. Rev.* **107** 2365
[2] Han X, Poliakoff M 2012 *Chem. Soc. Rev.* **41** 1428
[3] Aresta M, Dibenedetto A, Angelini A 2014 *Chem. Rev.* **114** 1709
[4] He M Y, Sun Y H, Han B X 2013 *Angew. Chem. Int. Ed.* **52** 9620
[5] D'Alessandro D M, Smit B, Long J R 2010 *Angew. Chem. Int. Ed.* **49** 6058
[6] Wang Y, Qin Y, Wang F 2014 *Catal. Sci. Technol.* **4** 3964
[7] Zhao T, Han Y, Sun Y 1999 *Phys. Chem. Chem. Phys.* **1** 3047
[8] Bhanage B M, Fujita S, Ikushima Y 2001 *Appl. Catal. A-Gen.* **219** 259
[9] Sit W N, Ng S M, Kwong K Y, et al 2005 *J Org Chem.* **70** 8583
[10] Shen Y M, Duan W L, Shi M 2004 *J Org Chem.* **69** 3080
[11] Bobbink F D, Vasilyev D, Hulla M, et al 2018 *ACS Catal.* **8** 2589
[12] Yu J I, Choi H J, Selvaraj M 2011 *React Kinet Mech Cat.* **102** 353
[13] Kim H S, Kim J J, Lee S D 2003 *Chem. Eur. J.* **9** 678
[14] Li F, Xia C, Xu L, et al 2003 *Chem. Commun.* 2042
[15] Lu X, Liang B, Zhang Y, et al 2004 *J. Am. Chem. Soc.* **126** 3732
[16] Man M L, Lam K C, Sit W N, et al 2006 *Chem. Eur. J.* **12** 1004
[17] Bai D, Zhang Z, Wang G, et al 2015 *Appl. Organomet. Chem.* **29** 240
[18] Chen D, Luo R, Li M, et al 2017 *Chem. Commun.* **53** 10930
[19] Wang Y, Chen M, Liu J 2014 *Journal of Yunnan University of Nationalities.* **23** 331 (in Chinese)
[20] Hu A, Ngo H L, Lin W 2003 *J. Am. Chem. Soc.* **125** 11490
[21] Costantino U, Curini M, Rosati O 2004 *Curr. Org. Chem.* **8** 591
[22] Amicangelo J C, Leenstra W R 2003 *J. Am. Chem. Soc.* **125** 14698
[23] Huang K, Yu J, Wang G 2004 *Eur. J. Inorg. Chem.* **2004** 2956
[24] Brunet E, Mata M, Juanes O 2004 *Angew. Chem. Int. Ed.* **43** 619
[25] Fry J A, Samanamu C R, Montcharnp J L, et al 2008 *Eur. J. Inorg. Chem.* **2008** 463
[26] Gong B W, Fu X K, Chen J X, et al 2009 *Journal of Catalysis.* **262** 9

IOP Conf. Series: Materials Science and Engineering 472 (2019) 012077 doi:10.1088/1757-899X/472/1/012077

# A Hybrid Energy Storage System Controlled by Two Degree-of-Freedom PID for Wind Power Generations

**Daming Cai, Zongxiao Yang*, Yanfeng Wu and Gensheng Li**

Henan University of Science and Technology, Luoyang 471003, China
*Corresponding author Email: zxyang@haust.edu.cn

**Abstract.** In order to make the unstable three-phase power output of wind turbine can charge the lithium battery pack steadily through the control of topology circuit and controller, a hybrid energy storage system controlled by two degree-of-freedom (DoF) PID is proposed to realize stable charging of lithium battery pack in this paper. The circuit combines the control circuit, the detection circuit and the protection circuit to achieve normal charging of the lithium battery pack. The system model composed of each module is built and simulated by MATLAB/Simulink software. The simulation results show that when the unstable three-phase voltage of wind turbine varies in the range of 0.4V-400V, the hybrid energy storage system keeps the charge current of the lithium battery group within the range of the normal range of the 2.5A-3.5A. And the system can effectively charge the lithium battery pack with the unstable three-phase power output of wind turbine.

## 1. Introduction

Because of the intermittent and unstable characteristics of wind resources, the wind power has random volatility which is difficult to avoid. When the wind turbine charges lithium batteries, the unstable current will inevitably affect the life of lithium battery pack and even cause charging safety accidents. Therefore, it is necessary to use the appropriate adjustment system to improve the charging characteristics of the unstable three-phase power of the wind turbine to the lithium battery pack, which can realize the stable charging of the lithium battery pack.

Based on the energy buffering and closed-loop feedback technology and the piecewise linear method, a capacitor buffering dual-parameter control energy storage strategy for off-grid wind power system was proposed and simulated in MATLAB/Simulink [1]. A Fractional order fuzzy PID controller to control the frequency and power of an hybrid power system was proposed and the controller used the quasi-oppositional harmony search algorithm to improve performance of system [2]. A hybrid operation strategy for wind power systems with battery energy storage systems was proposed [3]. The strategy uses active power tracking for maximum power point tracking and adjusts short-term frequency response based on dynamic discharge control.

An adaptive fuzzy PID controller that allows the wind turbine to always output at maximum power was proposed [4]. However, this controller requires a lot of human experience, and the model accuracy needs to be verified.

Corresponding controllers and control strategies for different hybrid energy storage systems. Their performance has been verified by simulation [5-7].

The hybrid energy storage system of Fala capacitor+ lithium battery pack is used to build a hybrid energy storage system model in MATLAB/Simulink and simulate it in this paper. The simulation results show that the system can effectively realize the stable charging of lithium battery pack by unstable three-phase electricity output from wind turbines.

## 2. Topology Design of Rectifying Circuit and Chopper Circuit

### 2.1. Topology Design of Rectifier Circuit

In order to convert the unstable three-phase power output from the wind turbine to DC output, we use a three-phase bridge rectifier circuit to convert the three-phase power into a pulsating DC. The Q1-Q6 is uncontrollable two-stage tube, as shown in Figure 1. Assuming the three-phase voltage of the wind turbine is $U_2$, the maximum value of the average output voltage of the rectifier circuit at no load is 2.45 $U_2$.

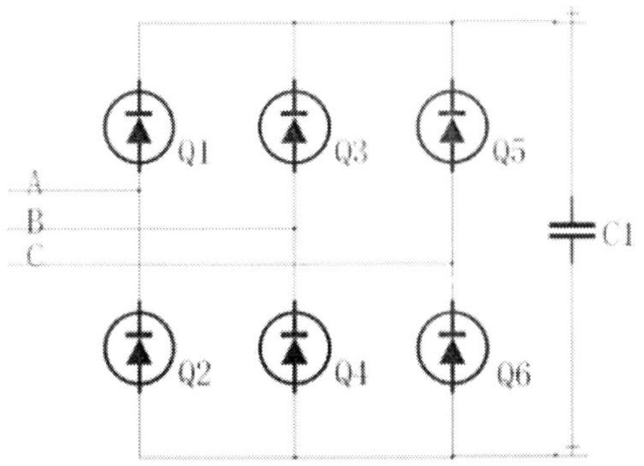


**Figure 1.** Rectifying and filtering circuit.

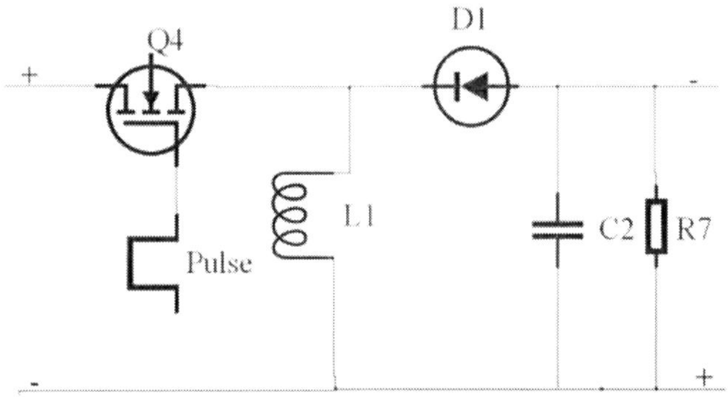


**Figure 2.** Boost-buck chopper circuit.

### 2.2. Topology Design of Chopper Circuit

DC chopper circuit is a commonly used DC-DC conversion circuit, which is divided into three types: boost-buck circuit, boost circuit and buck circuit. The maximum cut-off voltage of the whole lithium battery pack is 25.6V, because this paper aims at eight lithium batteries with a voltage of 3.2V. Due to the instability of the wind speed, the DC voltage of the output voltage of the wind turbine may be higher than 25.6V after the rectifier filter, and may be lower than 25.6V. Therefore, this paper chooses the boost-buck circuit to get the charge voltage of the lithium battery group. As shown in Figure 2.

The chopper circuit controls the conduction time of the MOSFET tube by controlling the duty cycle of the PWM, thus controlling the output voltage $U_{out}$ of the circuit. The value of the output voltage is as shown in Formula 1.

$$U_{out}=\frac{D}{1-D}\cdot U_{in}\ . \tag{1}$$

where $D$ is the duty cycle, $U_{in}$ is the input voltage, and $U_{out}$ is the output voltage.

### 2.3. Design of DC-DC Voltage Controller

Due to the instability of the wind speed, the three-phase power output by the wind turbine is not stable after rectification and filtering, which leads to the fluctuation of the charge current of the lithium battery pack, which will have an adverse effect on the charging time and the cycle life of the lithium battery pack. Therefore, in order to keep the output current constant, this paper adds a DC-DC voltage controller to the circuit. By collecting the output voltage $U_{out}$ of the boost-buck chopper circuit, the controller compares the $U_{out}$ with the preset reference voltage $U_{ref}$, and the controller takes the difference $E_{error}$ of the two as the input, and outputs the control signal through PID control. The control signal is usually represented as a certain duty ratio. The two degree of freedom PID DC-DC voltage controller MATLAB/Simulink model is built, as shown Figure 3. And the control signal output as shown in Formula 2.

$$D=K_p\left(bU_{ref}-U_{out}\right)+K_i\frac{T_s z}{z-1}\left(U_{ref}-U_{out}\right)+K_D\frac{N}{1+NT_s\frac{z}{z-1}}\left(cU_{ref}-U_{out}\right)+FF. \tag{2}$$

where $D$ is the control signal, expressed as a duty cycle. $K_p$ is the proportional gain. $K_i$ is the integral gain. $K_d$ is the differential gain. $N$ is the filter coefficient. $T_s$ is the sample time. $U_{ref}$ is the reference voltage. $U_{out}$ is the measured voltage. *FF* is the feed forward input.

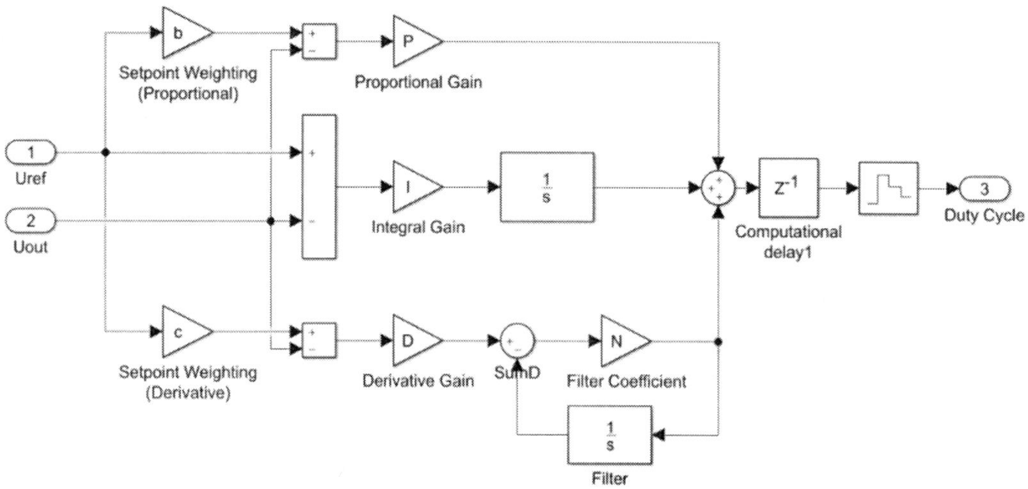


**Figure 3.** Basic principle block diagram of DC-DC voltage controller.

## 3. Design of Hybrid Energy Storage System for Super Capacitor and Lithium Battery Pack

Design of control strategy for hybrid energy storage control system. The power output fluctuation of wind turbines can be controlled by using the flexible charging and discharging capacity of capacitors. The specific control strategy is shown in Figure 4.

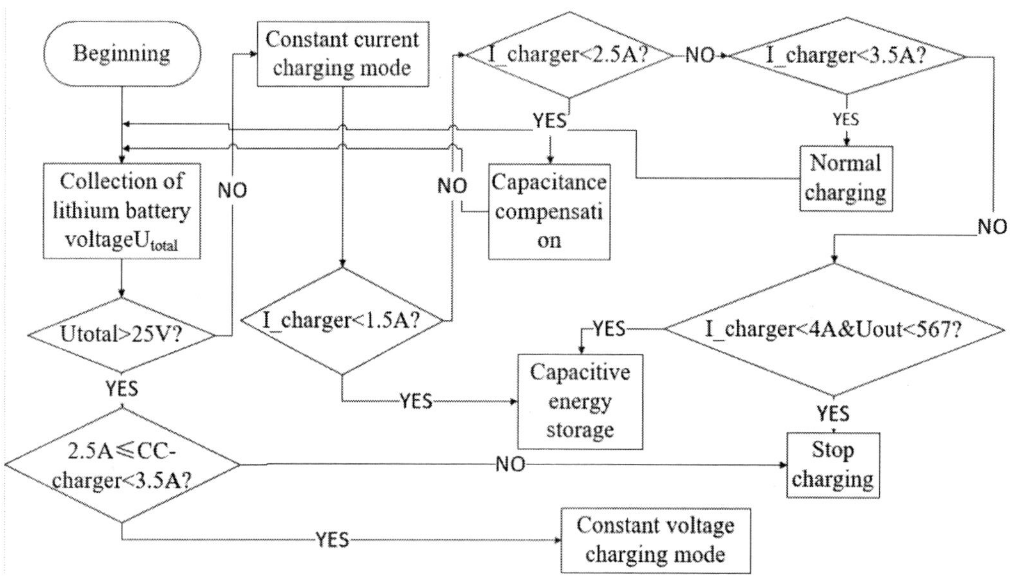


**Figure 4.** Flow chart of control strategy for hybrid energy storage control system.

## 4. Model Simulation and Analysis of Simulation Results

In order to simulate the unstable three-phase power produced by the wind turbine, the Three-Phase Programmable Voltage Source module in MATLAB/Simulink is used to program the three-phase power. The initial voltage of Fala capacitor is set to 0.5V. In the process of simulation, CC-charge shows the output current after the three-phase electricity is rectified and chopped circuit. The I-charge represents the charging current of the lithium battery pack, and U-c indicates the voltage at both ends of the Fala capacitor. The unsteady three-phase power output of the wind turbine simulated by the system is shown in Figure 5.

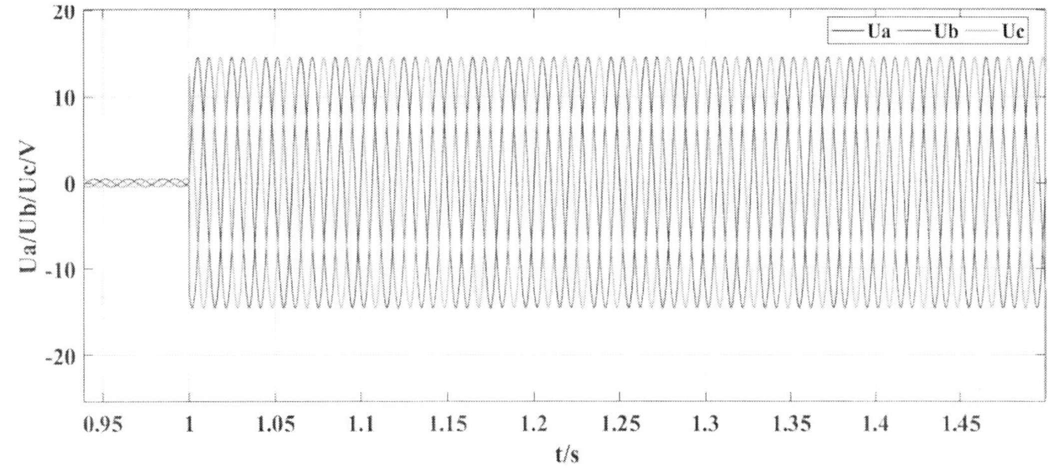


**Figure 5.** Unstable three-phase voltage simulated by system.

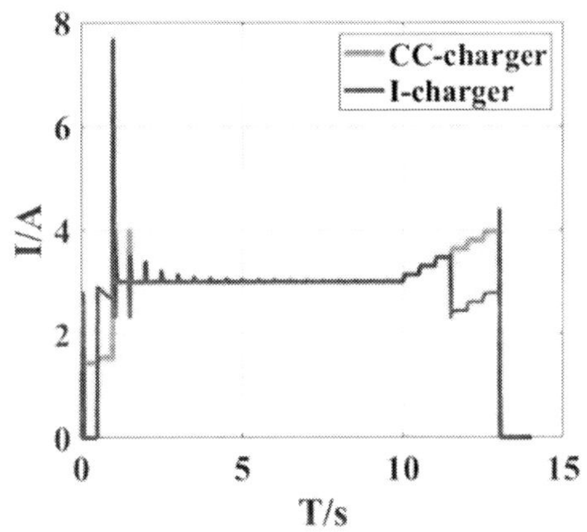


**Figure 6.** The currents changing

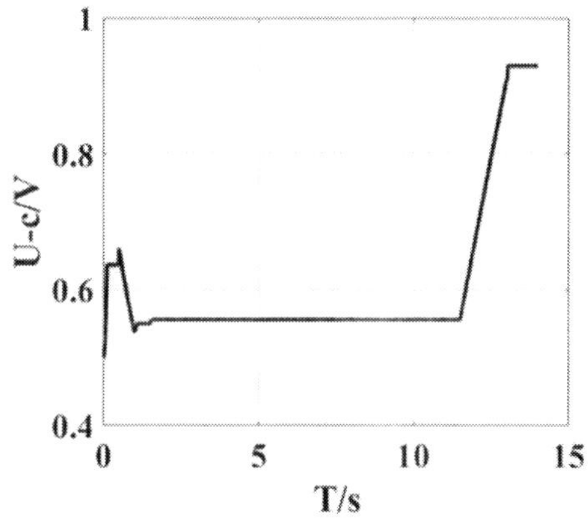


**Figure 7.** Fala capacitance voltage changing

From Figure 6-Figure 7, we can see that when the CC-charge is less than the threshold 1.5A, the charge current I-charge of the lithium battery pack is 0. Fala capacitance is in charge state. When CC-charge is greater than 1.5A and less than 2.5A, the Faraday capacitor is in discharge state. Fala capacitor completes charging compensation for lithium battery pack. When the CC-charge is greater than the threshold 2.5A and less than 3.5A, the circuit of the Fala capacitor is open. I-charge is consistent with CC-charge. From Figure 8-Figure 9, we can see that the SOC and terminal voltages of the lithium ion batteries have increased steadily under the regulation of the hybrid energy storage system. The reason why the voltage curve increases at 1 second is due to a large increase in the phase voltage in one second, and the phase voltage increases from 0.4V to 12.62V at a moment. Because the voltage rise is instantaneous, the DC-DC controller needs time to respond to it.

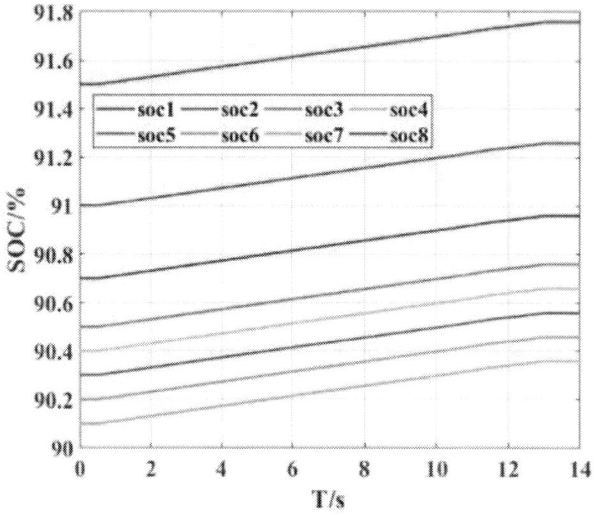


**Figure 8.** SOC changing of lithium batteries.

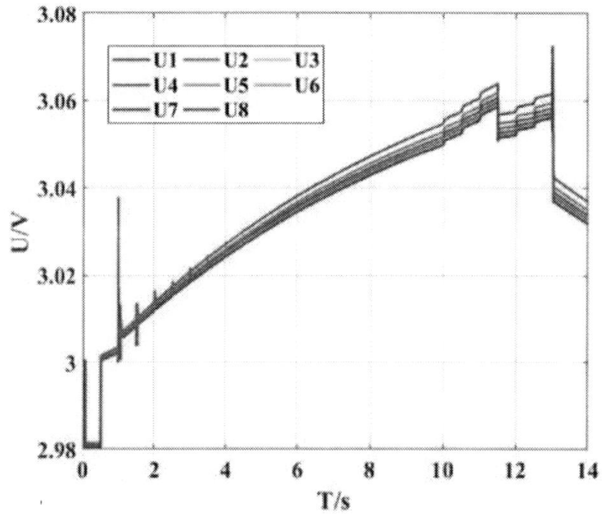


**Figure 9.** Terminal voltage changing of lithium battery.

## 5. Conclusion

Through the above theoretical analysis and MATLAB/Simulink simulation, the wind energy storage system can better realize the stable charging of the unstable three-phase power of the wind turbine to the lithium battery pack. When the system adopts the constant current charging mode and the CC-charger is less than 1.5A, the Fala capacitor is in the charge state and stores the energy, and the charge current of the lithium battery pack is 0.When the CC-charger is greater than or equal to 1.5A and less than 2.5A, the Faraday capacitor is in a discharge state and the lithium battery pack is compensated for charging. When the CC-charger is greater than or equal to 2.5A and less than 3.5A, the system can charge the lithium battery pack normally. When the CC-charger is greater than or equal to 3.5A and less than 4A, the Faraday capacitor is charged and dynamically stored to absorb part of the power. When the CC-charger is greater than or equal to 4A and the output voltage of the rectifier circuit is greater than 567V, the system stops charging the lithium battery pack.

**Acknowledgement**
The financial support from the National Natural Science Foundation of China (NSFC) under the grant No. 70671035 and No. 71071078 is acknowledged gratefully.

**References**
[1] Z.X. Yang, L.X. Hao, B.B. Yang. Capacitor buffer dual parameter controller energy storage strategy for wind turbines. Power grid technology. 38 (2014) 1257-1263.(in Chinese)
[2] T. Mahto, V. Mukherjee. Fractional order fuzzy PID controller for wind energy-based hybrid power system using quasi-oppositional harmony search algorithm [J], IET Generation, Transmission & Distribution. 11(13)(2017)3299-3309.
[3] J.W. Choi, S.Y. Heo, M.K. Kim. Hybrid operation strategy of wind energy storage system for power grid frequency regulation. IET Generation, Transmission & Distribution 10(2016)736-749.
[4] F. Yang, T. Yang, X.H. Yang. Fuzzy PID Control and Simulation of Wind Power Generation Yaw System. Advanced Materials Research.953-954 (2014)353-356.
[5] L. Kouchachvili, W. Yaïci, E. Entchev. Hybrid battery/supercapacitor energy storage system for the electric vehicles. Journal of Power Sources. 374(2018)237-248.
[6] A. Derrouazin, M. Aillerie, N. Mekkakia-Maaza, J.P. Charles. Multi input-output fuzzy logic smart controller for a residential hybrid solar-wind-storage energy system. Energy Conversion and Management. 148(2017)238-250.
[7] B.H. Zhang, H.F. Guo, Y.X. Li. Optimization of Smoothing Control in Hybrid Wind and Battery Energy Storage System. Advanced Materials Research. 852(2014)703-707.
[8] H. Moradi, G. Vossoughi. Robust control of the variable speed wind turbines in the presence of uncertainties. A comparison between $H_\infty$ and PID controllers. Energy 90 (2015)1508-1521, doi:10.1016/j.energy.2015.06.100.

IOP Conf. Series: Materials Science and Engineering 472 (2019) 012078 doi:10.1088/1757-899X/472/1/012078

# Research on the Water Area Extraction in Suzhou Based on Remote Sensing

**Zhilong Xi , Jiang Liu and Dandan Liu**

School of Surveying and Mapping Engineering, Heilongjiang Institute of Technology, Harbin 150050, P. R.China.

Email: 940953087@qq.com

**Abstract.** In order to accurately obtain the information and calculate the area of water, the traditional visual interpretation method was used to extract the information of water in Suzhou with the high-resolution remote sensing image. By on-the-spot sampling measurement verification, the precision of visual interpretation was proved to meet the practical requirements. The research in experimental area indicate that the visual interpretation method is operable. The water area of Suzhou was obtained as 3205 square kilometers, and the area is 994 square kilometers except for the area of Taihu lake and the Yangtze river.

## 1. Introduction

With the rapid development of space information technology, it is possible to acquire remote sensing images with multi temporal, high resolution and accuracy [1]. The extraction of water information has been a basic work in the field of remote sensing information extraction [2], and the 3S technology plays a more and more important role in building base database and monitoring illegal activities such as occupying water area illegally. There are two methods for extracting water information from high-resolution remote sensing images. One is visual interpretation, the other is semi-automatic or automatic information extraction. The accuracy of visual interpretation is relatively high, which depends on subjective judgment of human beings, but the efficiency is low. The kinds of semi-automatic or automatic information extraction methods are various, Li Changyou [3] used ETM to extract the water of Ulansuhai Lake by multi-band combination algorithm and multi-spectral mixture analysis method, but the accuracy of the semi-automated method can't be verified quantitatively, and the band combination can't be excluded only in specific water conditions. Although the automatic or semi-automated extraction method improves the efficiency to a certain extent, it lacks the universality of the application, most of which only adapt to the specific research area. There is no unified standard for the evaluation method of precision for the specific resolution of remote sensing images, and the extracted target is basically water, and it can't be effectively extracted. Therefore, such research does not form a relatively complete and mature theoretical system. In view of the large scope of the researching area and the complex distribution of rivers and lakes, and the need to calculate water area, the accuracy of water area is higher. In order to minimize the influence of errors, the visual-interpretation method is used for extracting information of water in Suzhou city.

## 2. The Basic Situation of Researching Area and Data Source

Suzhou is located in the downstream of Yangtze River and Taihu Lake, and its territory is flat. The river ports and lakes are crisscrossed and scattered, and there are more than 20 000 rivers, 300 lakes in it. Water has played a huge role in flood control, drainage, irrigation, water supply, shipping, ecological landscape and water culture in Suzhou. In order to investigate the total area and distribution of water areas in Suzhou accurately, obtaining high-precision water extraction information, as well as using aerial images of different phase to monitor the change of them and getting information of water-occupied behaviour, we need to use the high-resolution aerial images combined with large-scale topographic maps to finish the survey in same condition.

In this research, the basic images used for interpretation were taken in 2011 and 2013, and all of them are colourful aerial images with resolution of 0.1-0.3 m. Which have been processed by orthophoto correction and band fusion. At the same time, what were regarded as auxiliary information for interpretation included topographic map of Suzhou with 1:1000 scale, topographic map of Jiangsu Province with 1:10000 scale, and electronic map of Suzhou City.

## 3. Extraction of Background Information in Water Area

### *3.1. Technology Roadmap*

The roadmap for extraction of background information in water area is shown as figure 1.

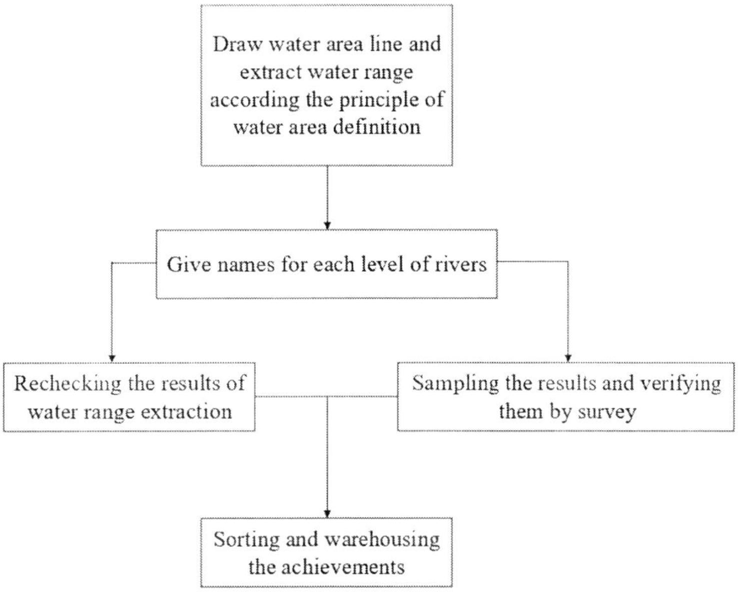


**Figure 1.** The roadmap for extraction of background information in water area

*3.1.1. Definition for water area.* In principle, the definition of water area is based on two basic conditions: hard coastline and no-hard coastline: (1) Hard coastlines include embankment, hard revetment, revetment, retaining walls and roads. Generally, it appears white stripes on aerial images. When there is a rigid coastline or house within 5 meters outside the water surface, the boundary of the water is directly drawn along the edge of the rigid shoreline or the side of the water side of the building, as shown in Figure 2. (2) When there are no artificial structures along the coastal waters, we draw water boundary along the water and land trace lines on the base of waterfront of the water area combined with high-resolution aerial images at different periods, just as shown in Figure 3.

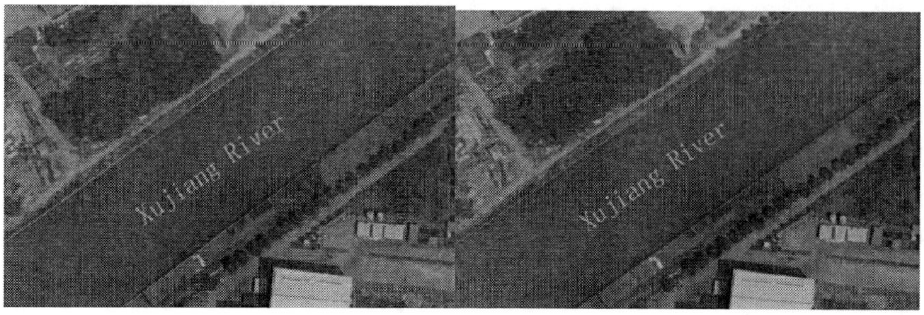


**Figure 2.** Sketch map of waterfront area with hard slope protection, revetment and retaining wall along the coast

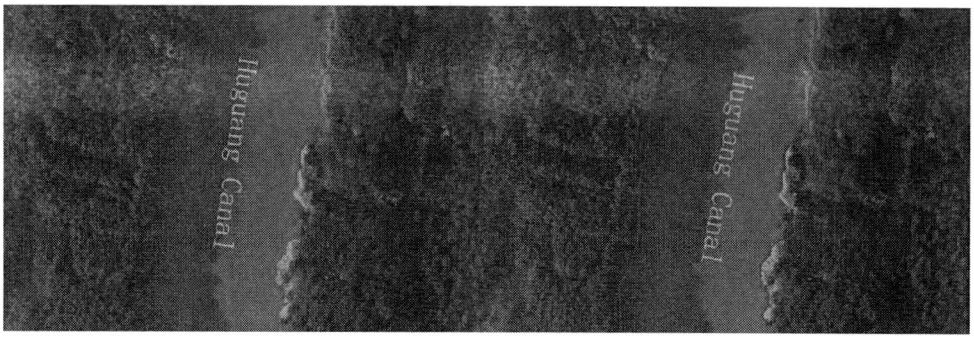


**Figure 3.** Sketch map of the water area of a riverway with no hard coastline along the coast

*3.1.2. Definition of river channel range of entry and exit.* According to the "Suggestions on strengthening management and protection of lakes" and actual situation, the principle of defining the scope of the lake channel is formulated: (1) The water range of lakes and rivers should be bounded by the structural centerline that controls the vertical flow of buildings, when there are sluices and other controlled buildings in the estuaries.(2) If there are no sluices and other controlled buildings in the estuaries but the lake is managed by municipal government, we should trace to 500 meters of the lake for the boundary, as shown in Figure 4.(3) If there are no sluices and other controlled buildings in the estuaries and the lake is not managed by municipal government, or the lake is managed by municipal government and the river is a catchment river, the area of rivers and lakes is directly bounded by the entrance and exit of rivers and lakes.(4) When the length of the river between two cities is less than 1 000 m, the water area of the lake is bounded by half the length of the river.

**Figure 4.** The scope of the entry and exit of the lake without control buildings

*3.1.3. Special situation processing* .(1) When the two rivers run through, the middle line of the river channel should run through the connecting place, that is the length of the two rivers includes the range of the perforation.

(2) When the two rivers run through, if the water of the intersection doesn't belong to each area, the water is incorporated into the rivers with higher or larger width, and the other is disconnected at the intersection.

(3) When the river is connected with other types of water, if the water area of the intersection belongs to an independent water type, when the river area is drawn, both ends of the river are disconnected at the intersection of the intersection.

(4) When the sandbank and beach in the middle of the river and lake is full of large artificial buildings, we should process them by hollow treatment, as shown in Figure 5.

(5) When the sandbars and beaches in the middle of the river and lake don't have the above situation, they will be submerged at high water level, so they belong to the waters.

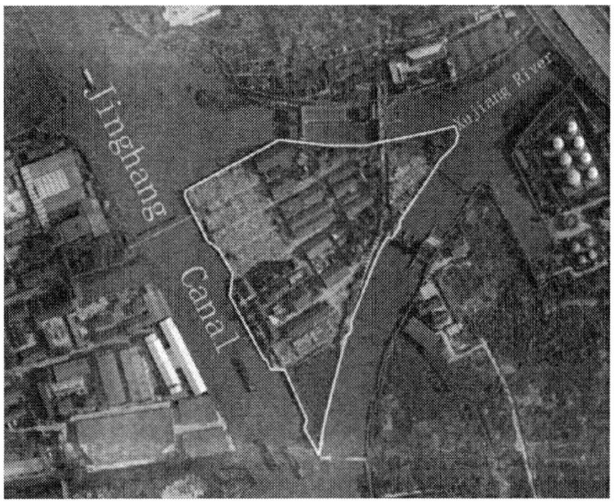


**Figure 5.** A schematic diagram for determination of water area (Blue line represents the water area line, and the yellow line represents the dike line)

*3.2. The Result of Extracting Information of Water Area for Suzhou City in 2013*

Including Taihu Lake and Yangtze River, the whole water area of Suzhou city is about 3 205.005 km2, and if not including Taihu Lake and Yangtze River, the number is about 994.348 km2. Including Taihu Lake and Yangtze River, the area of lakes is the most components of water area, accounting for 68.6% of the total area of the city's water. The second is rivers, the proportion is about 29.9%, and the area of ponds is the smallest, about 1.5%. If it does not contain Taihu Lake and Yangtze River, rivers' area of Suzhou is the largest, which accounts for about 50.5% of the total water area. The second is lakes, whose proportion is about 44.6% and the area of ponds is still the smallest, that is about 4.9%.

## 4. Verification of the Results of Water Extraction by Actual Survey

The RTK technology is used to measure the boundary line of the water area. The error of the calculated point position is ±0.16 m, less than ±0.3 m, which is the limitation of digital mapping technology. Based on the above discussion, we can get the conclusion that the method of water area extraction based on remote sensing is accurate and reliable, and it can meet the actual demand, too.

**Table 1.** Information of water area for sampling

| Name of rivers | Level | Latitude and longitude | |
|---|---|---|---|
| | | Segment A | Segment B |
| Taipu River | Catchment | 120.5586E, 31.0051N | 120.6185E, 30.9984N |
| Jinghang River | Catchment | 120.4793E, 31.3998N | 120.6431E, 31.0028N |
| Xitang River | Municipal | 120.5499E, 31.3825N | 120.5881E, 31.3343N |
| Wusong River | Municipal | 120.8546E, 31.3091N | 120.7177E, 31.1998N |
| Laoqiputang River | Municipal | 121.2099E, 31.6087N | 121.1173E, 31.5924N |
| Nanhengtao-Qigan River | County | 120.6515E, 31.8759N | 120.6880E, 31.8516N |
| Shitou River | County | 121.1139E, 31.5968N | 121.1330E, 31.5783 N |

**Table 2.** Results of sampling and verifying

| Level | Middle Error(m) | Extraction area of internal industry ($m^2$) | Measurement area of external industry ($m^2$) | Relative error of area | Rivers' length of measuring (km) | Actual length of rivers (km) | Sampling ratio |
|---|---|---|---|---|---|---|---|
| Catchment | ±0.11 | 12585383.7 | 12596809.8 | -0.10% | 82.51 | 168.55 | 49.0% |
| Municipal | ±0.18 | 35221021.5 | 35158727.6 | 0.20% | 416.34 | 764.01 | 54.5% |
| Country | ±0.19 | 3899779.2 | 3889801.3 | 0.30% | 74.10 | 172.31 | 43.0% |
| Town | ±0.20 | 1076975.9 | 1080126.6 | -0.30% | 51.11 | 65.49 | 78.0% |
| Total | - | 52783160.3 | 52725465.3 | 0.10% | 624.05 | 1170.36 | 53.3% |

## 5. Conclusions

If the survey situation is complex, the range is large and resolution of images is high, the method of visual interpretation has high efficiency compared with traditional and manual inspection. At the same time, it can also take the accuracy of extraction and actual needs into account. Using remote sensing and GIS technology to establish a more accurate base library of water information, combined with high-resolution remote sensing images, we can get the basic information of the specific area clearly, intuitively and efficiently. The frequency of data renewal will be accelerated, and it can provide decision-making support for the timely investigation and investigation of illegal occupation. The upgrading of information management level is of great significance. The accuracy of visual interpretation is closely related to the experience of the interpreters, but there will inevitably be errors. In order to avoid the errors of human factors, the automatic extraction method is still the focus of future exploration.

**Acknowledgments**

This work was financially supported by the key Lab Open Project Fund of Heilongjiang Ordinary University-the Comprehensive Lab of Space Geographic Information (KJKF-14-04).

## References

[1] Zhang Rui 2007 *Semi-automatic extracting methods of surface features in high-resolution remote sensing images* (Jinan: Shandong University of Science and Technology Press) p 2-4.

[2] Hu Qizhong and Qi Jianyong 2009 *Method for extracting water and water information in remote sensing images* Engineering Investigation **2** p 475-481.

[3] Li Changyou, Yu Ruihong and Wang Li 2004 *Water extraction method and analysis of remote sensing image of Ulansuhai Lake* Journal of Inner Mongolia Agricultural University **25** p 1-4.

[4] He Zhiyong, Zhang Xiaocan, Huang Zhicai, et al. 2004 *A water extraction technology for high-resolution remote sensing images* Journal of Zhejiang University (SCIENCE EDITION) **31** p 701-707.
[5] Ma Hongxu, Guo Shenglian and Zhou Yanlei 2013 *An improved method of water information extraction based on remote sensing images* Water Resources Research **2** p 127-133.

# Analysis of Key Environmental Impact Factors Based on Fuzzy Clustering and Grey T' s Correlation Degree Method

**Gaofei Zhao[1], Yujie Huo[1], Zhaofeng Zhang[1] and Zhaoyuan Peng[2]**

[1]Beijing HAZT Testing Technology Co., Ltd. No. 9 Shun Xing Road, Shunyi District, Beijing
[2]Beijing Suplet Power Co., Ltd.
Email: zhaogaofei@hazt.cn

**Abstract.** This paper expounds the basic principle of fuzzy clustering and grey T' s correlation degree method. It takes a power module used in communication device and worked outfield as an example to analysis the impacts of natural environment factors on product, that based on failure rate. Firstly, fuzzy clustering method is used to classify the environmental factors into four categories. Secondly, grey T' s correlation degree method is used to compare and rank the degree of influence of the four types of factors on the failure rate of power module. Finally, the analysis method and the key environmental factors affect the failure of the power module are given. The analysis of key influencing factors lays a foundation for the analysis of failure causes and mechanism of products, and provides a basis for reliability test design.

## 1. Introduction

In recent years, with the wide application of electronic products, the reliability of electronic products has been increasing, especially those used in outdoor environments. Communication equipment has been distributed throughout the country and a power module must be with them. Our territory is vast and the natural environment is complex and changeable. Therefore, the power supply module of communication equipment is subjected to various harsh natural climatic factors, and the reliability of the product is threatened. We can't intuitively determine which factors have the same or similar effects on product failure and which factors play a major role in the process of product failure because of the complex and variable environmental impact factors. This brings difficulties to the analysis of fault mechanism, the design of acceleration test and the design of product reliability.

Cluster analysis can classify complex environmental factors and classify similar factors into the same category. We choose representative factors to characterize such environmental factors and achieve the purpose of reducing species of the influencing factors [1]. Jia Zhijun used principal component analysis to evaluate the climatic factors of 13 atmospheric corrosion sites in China. He divided the climatic factors of the 13 sites into 6 categories [2]. Zhang Lunwu classified the environment of 8 atmospheric corrosion test stations by fuzzy clustering based on the environmental factors of the atmospheric corrosion test stations in China for more than ten years. The correlation between 17 environmental factors affecting atmospheric corrosion was comprehensively evaluated [3]. But the final result of cluster analysis is to classify environmental factors, which can't reflect the impact of environmental factors on product failure. Through clustering analysis, there are still many kinds of stress after the reduction of species. It is difficult to identify and analysis the key causes of product failure. We need to further analysis and calculate the key factors that affect product failure.

Grey correlation analysis can describe the correlation between various variables effectively and has a good measure for various nonlinear correlation relations [4], [5]. The grey system theory proposes the concept of grey correlation analysis for each subsystem, and intends to seek numerical relationships between subsystems or factors in the system through certain methods. Therefore, grey correlation is an effective means to measure the closeness of two variable factors [6].

Based on the weather factors and failure rate data of the service area of power module within a communication apparatus, this paper uses fuzzy clustering method to cluster the environmental influencing factors. Then the key factors affecting the fault are quantitatively extracted by means of grey correlation analysis, which lays a foundation for the analysis of fault mechanism and reliability test of design.

## 2. Fuzzy Clustering Analysis of Environmental Impact Factors

### *2.1. Fuzzy Cluster Analysis Method*

The results of cluster analysis reflect the potential structure, characteristics and similarity of the raw data [7]. In general, cluster analysis can be divided into two categories. They're absolute clustering and fuzzy clustering. Absolute cluster analysis means that there is only one class identifier for a data object while data objects belong to multiple classes at the same time with different probabilities in fuzzy clustering. The specific processes are as follows.

- List the original data matrix

$$X = \begin{bmatrix} X_{11} & \cdots & X_{1m} \\ \vdots & \ddots & \vdots \\ X_{n1} & \cdots & X_{nm} \end{bmatrix} \quad (1)$$

$X_{ij}$ is the average number of ith environmental factors in the jth time period. N is the number of environmental factors. M is the number of time periods, so the matrix $X=[X_{ij}]_{n\times m}$ is obtained in the formula.

- Transform the value to a range

$$X_{ij}^1 = \frac{x_{ij} - \wedge_{i=1}^{n}(x_{ij})}{\vee_{i=1}^{n}(x_{ij}) - \wedge_{i=1}^{n}(x_{ij})} \qquad (i=1,2,\cdots,n) \quad (2)$$

The denominator is the difference between the maximum and the minimum of the j column. Obviously $X_{ij}^1 \in [0,1]$, so $X^1 = [X_{ij}^1]_{n\times m}$.

- Seeking similarity relation matrix

The similarity relation matrix is obtained by the maximum minimum method as follow.

$$r_{ij} = \frac{\sum_{k=1}^{m}(X_{ik}^1 \wedge X_{jk}^1)}{\sum_{k=1}^{m}(X_{ik}^1 \vee X_{jk}^1)} \quad (3)$$

$r_{ij}$ is the similarity coefficient between the environmental factor I and the environmental factor J. m is the number of environmental factors. $X_{ik}^1$ is the value of the I environmental factor in the K period. $X_{jk}^1$ is the value of the jth environment factor in the kth period. The formula $X_{ik}^1 \wedge X_{jk}^1$ represents the minimum value between the $X_{ik}^1$ and $X_{jk}^1$. $X_{ik}^1 \vee X_{jk}^1$ indicates the maximum value between the two. According to the formula (3), we can get similar matrix $R = [r_{ij}]_{n\times n}$ by processing related data.

- Equivalence relation matrix

Using the transfer closure method to find the equivalent relation matrix

$$t(R) = \cup_{i=1}^{\infty} R^{I} \quad (4)$$

- Drawing dynamic cluster diagram

Draw dynamic cluster diagram and do cluster analysis by selecting appropriate similarity coefficient lambda value and using the equivalence relation matrix.

### 2.2. Data Collection of Environmental Factors

The collection of environmental factor data refers to the study of natural environment and climate factors for the areas in which the subjects are servicing. If a certain type of power module is mainly served in the A area, we conduct data analysis and collation of the climatic factors that may affect the failure rate of A. As shown in Table 1, all climatic factors in the table mean the average of this month.

**Table 1.** Climate Statistics of the Area A in 2017

| 2017 | Temperature | humidity | PM2.5 | radiation intensity | wind speed | Precipitation | sunshine time |
|---|---|---|---|---|---|---|---|
| **January** | 3.32 | 78.13 | 114.29 | 2.81 | 2.58 | 0.25 | 6.8 |
| **February** | 7.64 | 64.39 | 69.86 | 3.69 | 2.64 | 2.03 | 8.59 |
| **March** | 14.13 | 67.26 | 62.68 | 4.80 | 2.61 | 7.87 | 8.45 |
| **April** | 23.70 | 72.33 | 53.27 | 5.86 | 2.90 | 0 | 8.86 |
| **May** | 29.77 | 74.10 | 59.42 | 6.36 | 3.61 | 28.95 | 9.09 |
| **June** | 30.97 | 81.83 | 42.47 | 6.05 | 2.60 | 80.78 | 7.48 |
| **July** | 31.61 | 95.03 | 52.06 | 5.22 | 2.00 | 63.51 | 4.95 |
| **August** | 30.13 | 97.71 | 38.06 | 4.93 | 2.06 | 199.9 | 7.05 |
| **September** | 27.87 | 95.23 | 57.40 | 4.57 | 2.2 | 3.05 | 6.72 |
| **October** | 17.23 | 97.45 | 56.94 | 3.73 | 2.32 | 32.77 | 3.99 |
| **November** | 9.57 | 80.23 | 45.43 | 2.80 | 2.83 | 0.00 | 5.15 |
| **December** | 4.48 | 72.48 | 43.65 | 2.41 | 2.58 | 0.00 | 5.32 |

### 2.3. Cluster Analysis of Environmental Factors

The original data matrix of natural environment factors for area A in 2017 is as follows.

$$X = \begin{bmatrix} 3.32 & 7.64 & 14.13 & 23.7 & 29.77 & 30.97 & 31.61 & 30.13 & 27.87 & 17.23 & 9.57 & 4.48 \\ 78.13 & 64.39 & 67.26 & 72.33 & 74.1 & 81.83 & 95.03 & 97.71 & 95.23 & 97.45 & 80.23 & 72.48 \\ 114.29 & 69.86 & 62.68 & 53.27 & 59.42 & 42.47 & 52.06 & 38.06 & 57.4 & 56.94 & 45.43 & 43.65 \\ 2.81 & 3.69 & 4.8 & 5.86 & 6.36 & 6.05 & 5.22 & 4.93 & 4.57 & 3.73 & 2.8 & 2.41 \\ 2.58 & 2.64 & 2.61 & 2.9 & 3.61 & 2.6 & 2 & 2.06 & 2.2 & 2.32 & 2.83 & 2.58 \\ 0.25 & 2.03 & 7.87 & 0 & 28.95 & 80.78 & 63.51 & 199.9 & 3.05 & 32.77 & 0 & 0 \\ 6.8 & 8.59 & 8.45 & 8.86 & 9.09 & 7.48 & 4.95 & 7.05 & 6.72 & 3.99 & 5.15 & 5.32 \end{bmatrix}$$

Get the following matrix according to the equation (2) to transform the value to a range.

$$X^1_{ij} = \begin{bmatrix} 0.03 & 0.08 & 0.18 & 0.33 & 0.37 & 0.36 & 0.32 & 0.14 & 0.28 & 0.16 & 0.12 & 0.06 \\ 0.68 & 0.92 & 1 & 1 & 1 & 1 & 1 & 0.48 & 1 & 1 & 1 & 1 \\ 1 & 1 & 0.93 & 0.74 & 0.79 & 0.5 & 0.54 & 0.18 & 0.59 & 0.57 & 0.57 & 0.6 \\ 0.02 & 0.02 & 0.03 & 0.08 & 0.04 & 0.04 & 0.03 & 0.01 & 0.03 & 0.01 & 0.03 & 0.03 \\ 0.02 & 0.01 & 0 & 0.04 & 0 & 0 & 0 & 0 & 0 & 0 & 0.04 & 0.04 \\ 0 & 0 & 0.08 & 0 & 0.36 & 0.99 & 0.66 & 1 & 0.01 & 0.32 & 0 & 0 \\ 0.06 & 0.1 & 0.09 & 0.12 & 0.08 & 0.06 & 0.03 & 0.03 & 0.05 & 0.02 & 0.06 & 0.07 \end{bmatrix}$$

Find the similarity matrix according to the equation (3).

$$R = \begin{bmatrix} 1 & 0.22 & 0.3 & 0.57 & 0.06 & 0.32 & 0.59 \\ 0.22 & 1 & 0.36 & 0.04 & 0.01 & 0.55 & 0.07 \\ 0.3 & 0.36 & 1 & 0.05 & 0.02 & 0.21 & 0.1 \\ 0.57 & 0.04 & 0.05 & 1 & 0.34 & 0.05 & 0.52 \\ 0.06 & 0.01 & 0.02 & 0.34 & 1 & 0 & 0.18 \\ 0.32 & 0.55 & 0.21 & 0.05 & 0 & 1 & 0.08 \\ 0.59 & 0.07 & 0.1 & 0.52 & 0.18 & 0.08 & 1 \end{bmatrix}$$

Seek the equivalent relation matrix according to the equation (4).

$$t(R) = \begin{bmatrix} 1 & 0.32 & 0.32 & 0.57 & 0.34 & 0.32 & 0.59 \\ 0.32 & 1 & 0.36 & 0.32 & 0.32 & 0.55 & 0.32 \\ 0.32 & 0.36 & 1 & 0.32 & 0.32 & 0.36 & 0.32 \\ 0.57 & 0.32 & 0.32 & 1 & 0.34 & 0.32 & 0.57 \\ 0.34 & 0.32 & 0.32 & 0.34 & 1 & 0.32 & 0.34 \\ 0.32 & 0.55 & 0.36 & 0.32 & 0.32 & 1 & 0.32 \\ 0.59 & 0.32 & 0.32 & 0.57 & 0.34 & 0.32 & 1 \end{bmatrix}$$

Draw a dynamic cluster diagram as figure 1.

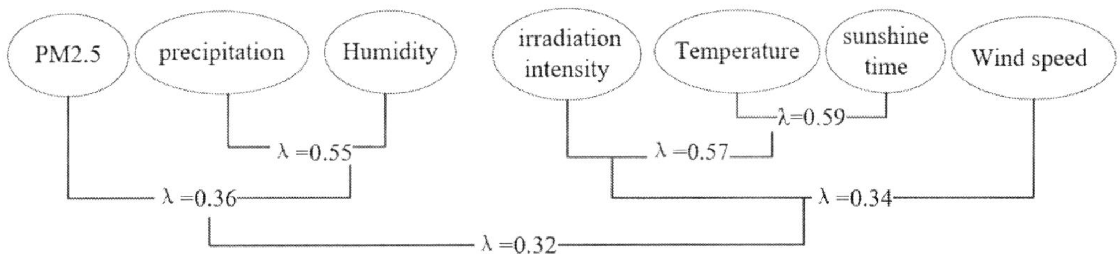


**Figure 1.** Dynamic Cluster Diagram

It is known from the dynamic cluster map that when >0.55, the climate factors can be divided into four categories. The first category includes temperature, irradiation intensity and sunshine time. The second category obtains precipitation and humidity. The third category is PM2.5. The other one is the wind speed.

We choose the easily controlled factors in the test process to characterize the climatic factors as follows.

- Temperature
- Humidity
- PM2.5
- Wind speed

*2.4. Conclusion*

The environmental factors that affect the failure rate of the power module in the A area can be characterized by four indexes. They are temperature, humidity, PM2.5 and wind speed.

## 3. Analysis of Key Factors Based on Failure Rate

### *3.1. Grey T' s Correlation Degree Method*

T' s correlation degree method is a kind of grey correlation degree methods, and revised Deng's correlation degree. Because of its symmetry, uniqueness, comparability and normalization, it is suitable for analysis of influence factors [8]. It can analysis the environmental impact factors of power failure, extract the key factors effectively, and provide theoretical basis for the design of power module, failure analysis and environmental test design.

*3.1.1. Principle of Grey T' s Correlation Degree Method.* The grey T' s correlation degree is calculated according to the degree of relative change of the time series curve of the factor. For discrete time series, the degree of approximation of the relative trend of the two curves is determined by the size of the increment after normalization the original variable during the two time series at every period of Δt=Δt(k)-Δt(k-1) (k=2,3,…,n)[9]. The correlation degree of two time series is defined as the weighted average of the correlation coefficient between each period of Δt(k) and the weight is Δt(k).

*3.1.2. Model of Grey T' s Correlation Degree Method*

- Standardizing

The process of standardization makes up for the lack of comparability of grey correlation degree, making it possible to analysis the influencing factors [10].

$$D_0 = \frac{1}{n-1}\sum_{k=2}^{n}|X_0(t_k) - X_0(t_{k-1})| \tag{5}$$

$$D_j = \frac{1}{n-1}\sum_{k=2}^{n}|X_j(t_k) - X_j(t_{k-1})| \tag{6}$$

$$y_0 = \{\frac{X_0(t_k)}{D_0}, k = 1,2,3, \dots, n\} \tag{7}$$

$$y_j = \{\frac{X_j(t_k)}{D_j}, k = 1,2,3, \dots, n\} \tag{8}$$

- Seeking Incremental sequence

$$\Delta y_0 = \{\Delta y_0(t_k) = y_0(t_k) - y_0(t_{k-1}), k = 2,3, \dots, n\} \tag{9}$$

$$\Delta y_j = \{\Delta y_j(t_k) = y_j(t_k) - y_j(t_{k-1}), k = 2,3, \dots, n\} \tag{10}$$

- Calculation of correlation coefficient at each time period

$$\varepsilon(t_k) = \mathrm{sgn}(\Delta y_0(t_k) \cdot \Delta y_j(t_k)) \cdot \frac{\min(|\Delta y_1(t_k)|, |\Delta y_j(t_k)|}{\max(|\Delta y_1(t_k)|, |\Delta y_j(t_k)|} \tag{11}$$

$$\varepsilon(t_k) = 0, \text{when } \Delta y_0(t_k) \cdot \Delta y_j(t_k) = 0$$

In the upper equation， the positive and negative of correlation coefficient is determined by $\varepsilon(t_k)$. $\varepsilon(t_k)$>0 when $\Delta y_0(t_k) \cdot \Delta y_j(t_k)$>0 means that $X_0$ and $X_1$ change from $t_{k-1}$ to $t_{k-1}$ in the same direction, that is, positive correlation. $\varepsilon(t_k)$<0 when $\Delta y_0(t_k) \cdot \Delta y_j(t_k)$<0 means that $X_0$ and $X_1$ change from $t_{k-1}$ to $t_{k-1}$ in different directions, that is, negative correlation. $\varepsilon(t_k)$=0 when $\Delta y_0(t_k) \cdot \Delta y_j(t_k)$=0 means that $X_0$ and $X_1$ are irrelevant during $\Delta(t_k)$[11].

*3.1.3. Definition of Grey T' s correlation degree.* The correlation degree between the original time series $X_1$and $X_2$is defined as r(X1, X2).

$$r(X1, X2) = \frac{1}{b-a}\sum_{k=2}^{n} \Delta(t_k) \cdot \varepsilon(t_k) \tag{12}$$

### *3.2. Analysis of Key Factors*

Through fuzzy clustering analysis, complex and changeable natural environment and climate factors are classified into different categories, and the categories of influencing factors are reduced. However, we can't clearly understand the impact of each type of natural environment and climate on the failure rate of the product. For the communication power module, temperature, humidity, wind and PM2.5 are the 4 dimensions that affect the failure rate. Based on the T correlation degree method, the correlation degree of the 4 dimensions to the power failure rate can be obtained.

Taking the power module of a certain communication equipment as an example, the quantitative calculation of the influence degree of natural environment factors based on the failure rate is carried out. The failure rate and environmental climate statistics are shown in Table 2.

**Table 2.** Failure Rate of a Certain Communication Power Module and Corresponding Environmental and Climatic Data

| **Time** | Failure rate($10^{-6}$) | Temperature | Humidity | **PM2.5** | Wind speed |
|---|---|---|---|---|---|
| **January** | 3.50 | 3.32 | 78.13 | 114.29 | 2.58 |
| **February** | 2.50 | 7.64 | 64.39 | 69.86 | 2.64 |
| **March** | 3.00 | 14.13 | 67.26 | 62.68 | 2.61 |
| **April** | 3.50 | 23.70 | 72.33 | 53.27 | 2.90 |
| **May** | 4.00 | 29.77 | 74.10 | 59.42 | 3.61 |
| **June** | 5.50 | 30.97 | 81.83 | 42.47 | 2.60 |
| **July** | 8.50 | 31.61 | 95.03 | 52.06 | 2.00 |
| **August** | 9.00 | 30.13 | 97.71 | 38.06 | 2.06 |
| **September** | 8.00 | 27.87 | 95.23 | 57.40 | 2.2 |
| **October** | 8.50 | 17.23 | 97.45 | 56.94 | 2.32 |
| **November** | 4.50 | 9.57 | 80.23 | 45.43 | 2.83 |
| **December** | 3.00 | 4.48 | 72.48 | 43.65 | 2.58 |

- Standardizing

According to equation (5) to (8), the result is shown in Table 3.

**Table 3.** Standardized Sequences

| **Time** | Failure rate ($10^{-6}$) | Temperature | Humidity | **PM2.5** | Wind speed |
|---|---|---|---|---|---|
| **January** | 0.59 | 0.66 | 11.21 | 8.93 | 7.50 |
| **February** | 1.00 | 1.52 | 9.24 | 5.46 | 7.67 |
| **March** | 1.89 | 2.81 | 9.65 | 4.9 | 7.60 |
| **April** | 3.68 | 4.71 | 10.38 | 4.16 | 8.43 |
| **May** | 4.86 | 5.92 | 10.63 | 4.64 | 10.5 |
| **June** | 5.41 | 6.16 | 11.74 | 3.32 | 7.56 |
| **July** | 6.07 | 6.28 | 13.63 | 4.07 | 5.81 |
| **August** | 6.14 | 5.99 | 14.02 | 2.97 | 6.00 |
| **September** | 5.10 | 5.54 | 13.66 | 4.48 | 6.40 |
| **October** | 3.54 | 3.42 | 13.98 | 4.45 | 6.75 |
| **November** | 1.82 | 1.90 | 11.51 | 3.55 | 8.24 |
| **December** | 0.68 | 0.89 | 10.40 | 3.41 | 7.50 |

- Calculate increment sequence

The increment sequence is calculated according to equation (9) and (10), and the result is shown in Table 4.

**Table 4.** The Increment Sequence

| Period | Failure rate ($10^{-6}$) | Temperature | Humidity | PM2.5 | Wind speed |
|---|---|---|---|---|---|
| **1~2** | 0.41 | 0.86 | -1.97 | 0.17 | 1.29 |
| **2~3** | 0.89 | 1.29 | 0.41 | -0.08 | 1.63 |
| **3~4** | 1.79 | 1.9 | 0.73 | 0.83 | 1.56 |
| **4~5** | 1.18 | 1.21 | 0.25 | 2.07 | 0.74 |
| **5~6** | 0.54 | 0.24 | 1.11 | -2.94 | -0.46 |
| **6~7** | 0.66 | 0.13 | 1.89 | -1.74 | -1.22 |
| **7~8** | 0.07 | -0.3 | 0.38 | 0.19 | -0.43 |
| **8~9** | -1.04 | -0.45 | -0.36 | 0.39 | -0.53 |
| **9~10** | -1.56 | -2.12 | 0.32 | 0.36 | -1.24 |
| **10~11** | -1.72 | -1.52 | -2.47 | 1.48 | -1.37 |
| **11~12** | -1.13 | -1.01 | -1.11 | -0.73 | -0.57 |

- Calculate correlation degree

According to equation (11) and (12), the correlation coefficient and correlation degree of each period are calculated, as shown in Table 5.

**Table 5.** Correlation Coefficient and Correlation Degree in Each Period

| Period | Failure rate($10^{-6}$) | Temperature | Humidity | PM2.5 | Wind speed |
|---|---|---|---|---|---|
| **1~2** | 1 | 0.48 | -0.21 | -0.12 | 0.42 |
| **2~3** | 1 | 0.69 | 0.46 | -0.63 | -0.09 |
| **3~4** | 1 | 0.94 | 0.41 | -0.41 | 0.47 |
| **4~5** | 1 | 0.98 | 0.21 | 0.41 | 0.57 |
| **5~6** | 1 | 0.44 | 0.49 | -0.41 | -0.18 |
| **6~7** | 1 | 0.19 | 0.35 | 0.88 | -0.38 |
| **7~8** | 1 | -0.24 | 0.18 | -0.06 | 0.38 |
| **8~9** | 1 | 0.43 | 0.34 | -0.69 | -0.38 |
| **9~10** | 1 | 0.74 | -0.2 | 0.02 | -0.23 |
| **10~11** | 1 | 0.88 | 0.7 | 0.52 | -0.86 |
| **11~12** | 1 | 0.89 | 0.98 | 0.12 | 0.65 |
| T' s correlation | 1 | 0.58 | 0.34 | -0.03 | 0.03 |

*3.3. Results*

The calculation results show the correlation degree between temperature, humidity, PM2.5 or wind speed and failure rate

- The grey T' s correlation degree of temperature and failure rate is 0.58.
- The grey T' s correlation degree of humidity and fault rate is 0.34.
- The grey T' s correlation degree of PM2.5 and fault rate is -0.03.
- The grey T' s correlation degree of wind speed and failure rate is 0.03.

Obviously, the correlation between temperature, humidity and failure rate is relatively large. However, the correlation between PM2.5 or wind speed and fault rate is very small.

## 4. Conclusions

In this paper, a method of reducing the dimension of environmental factors that affect product failure based on fuzzy clustering is proposed, and the grey T' s correlation degree analysis method is used to extract the key environmental factors that affect the product failure. Taking the power module of a certain type of communication equipment as an example, collecting and investigating related data and analyzing the main environmental factors that affect its failure are temperature and humidity.

According to 12 refs, 52% of the products caused by environment, 21% of which are temperature and 10% of humidity, are the main influencing factors [12]. This is in accordance with the conclusions of this paper, which illustrates the feasibility of the method described in this paper, thus laying a foundation for the analysis of the mechanism of product failure, the design of reliability test and the improvement of the reliability of the product.

**References**

[1] S Z Song, G Y Wang, S Y Wang. Progress in Data Processing of Natural Environmental Corrosion of Materials in China *J. Journal of corrosion and protection of China*, 2003, **23** (1) pp 56-64.
[2] Z J Jia, Q l Zhao, C F Dong, et al. The Principal Component Analysis of The Climate Factors of 13 Atmospheric Corrosion Sites in China *J. China Journal of corrosion and protection*, 2009, **29** (5) pp 388-393.
[3] L W Zhang, X H Wang, X H Feng, et al. Fuzzy Clustering Analysis of Atmospheric Corrosion Environment and Environmental Factors *J. Chinese Journal of corrosion and protection*, 2003, **23** (2) pp 112-115.
[4] J L Deng. The Basic Method of Grey System *M.* Huazhong University of Science and Technology Press, 1987.
[5] M X Cao. Grey Relational Analysis Model and Its Application *D.* Nanjing University of Aeronautics & Astronautics, 2007.
[6] J H Wu, X J Wang, Y Zhang. The Choice of Copula Function in Correlation Analysis *J. Statistical Research*, 2014, **31** (10) pp 99-107.
[7] X Bai. Similarity Measurement and Application in Cluster Analysis *D.* Beijing. JiaoTong University, 2012.
[8] Y G Sun, Y G Dang. Grey Grey T' s correlation degree Improved *J. System Engineering Theory and Practice*, 2008, **28** (4) pp 135-139.
[9] M Tian, S F Liu, Z K Bu. Review of Grey Relational Algorithm Models *J. Statistics and Decision Making*, 2008, 2008 (1) pp 24-27.
[10] W X Tang. Grey T' s correlation degree and Its Calculation Method *J. Mathematical Statistics and Management*, 1995 (1) pp 34-37.
[11] Z G Mei. Grey Absolute Correlation Degree and Its Calculation Method *J. System Engineering*, 1992, **10** (5).
[12] Y T Gao, Y Xu, J Chang. Multi Environmental Test of Sight and Theory of Thermal Performance of Infrared Optical Materials *M.* Tsinghua University press, 2017.

IOP Conf. Series: Materials Science and Engineering 472 (2019) 012080 doi:10.1088/1757-899X/472/1/012080

# Numerical Simulation and Experimental Study of Three-Dimensional Density-Stratified Flow

**Ming Yang, Ming Wang, Fangxiu Zhang and Ruixun Lai**

Sediment Research Centre, Yellow River Institute of Hydraulic Research, YRCC, MWR, No. 45 Shunhe Rd. Zhengzhou, China

**Abstract.** The numerical relationship between vertical turbulence P/S $\sigma_{tz}$ turbulence and stratification feature parameters is established by analyzing the flux expression of turbulent stress/ flux algebraic mode. Corresponding density change rate parameters reflecting changes in temperature and sediment concentration are introduced to correct the vertical diffusion coefficient of turbulent. A three-dimensional turbulence model with multi-phase coupling of water flow, temperature and sediment is established on this basis. The established standard Johnson water channel model is adopted for experimental study on the density-stratified flow model of temperature and sediment concentration and the experimental results are compared with the calculation results of three-dimensional density-stratified flow mathematical model.

## 1. Introduction

Density-stratified flow is a common natural phenomenon. For example, after a reservoir is built, due to the wide water surface and large water body in the dam area, the reservoir shows a special water-temperature stratified structure in a state of water temperature-stratified flow; the turbid density current moving downstream along the riverbed due to fine sediment mixed in water sinking in the bottom of the reservoir is also a sediment density- stratified flow.

Turbulence is a relatively complex physical phenomenon. If there is simultaneous influence of temperature, sediment and other density stratification, the mixing mechanism of turbulence will be more complicated and the simulation will be very difficult.

In this study, under the consideration of effects of temperature and sediment stratification on the turbulent diffusion coefficient, the numerical relationship among vertical turbulence P/S, turbulence and stratification feature parameters is established by analyzing the flux expression of turbulent stress/ flux algebraic mode, and corresponding density change rate parameters reflecting changes in temperature and sediment concentration are introduced to correct the vertical diffusion coefficient of turbulent. On this basis, a three-dimensional turbulence model with multi-phase coupling of water flow, temperature and sediment is established, and the established standard Johnson water channel model is adopted for experimental study on the density-stratified flow model of temperature and sediment concentration. The experimental results are compared with the calculation results of three-dimensional density-stratified flow mathematical model.

## 2. Governing Equation

Continuity equation:

$$\frac{\partial \rho}{\partial t}+\frac{\partial \rho u_j}{\partial x_j}=0 \tag{1}$$

Momentum equation

$$\frac{\partial u_i}{\partial t} + u_j \frac{\partial ui}{\partial xi_{x_j}} = \Gamma_\phi \frac{\partial^2 u_i}{\partial x_j^2} - \frac{\partial p}{\partial x_i} + f_i[1 - \beta(T_0)T_0 + \beta(T)T + \eta(T,S)S] \tag{2}$$

Temperature diffusion equation:

$$\frac{\partial T}{\partial t} + \frac{\partial (u_j T)}{\partial x_j} = \frac{\partial}{\partial x_j}\left[\Gamma_T \frac{\partial T}{\partial x_j}\right] + \frac{J}{\rho C_p} \tag{3}$$

Sediment motion equation:

$$\frac{\partial S_l}{\partial t} + \frac{\partial u_j S_l}{\partial x_j} - \omega_L \frac{\partial S_L}{\partial z} = \frac{\partial}{\partial x_j}\left(\Gamma_S \frac{\partial S_l}{\partial x_j}\right) \tag{4}$$

Riverbed deformation equation:

$$\gamma' \frac{\partial z_{bl}}{\partial t} = \omega_l \left(S_{bl} - S_{bl}^*\right) \tag{5}$$

Where $f_i$ is mass force; P is pressure ( $N/m^2$ ); $\Gamma_\phi$ is turbulent diffusion coefficient ( $kg/m/s$ ). It is necessary to introduce a turbulence model to solve it. $T_0$ is reference temperature. The reference temperature is taken to be ambient water temperature. $\rho_0$ is water density of clean water at temperature of $T_0$. $\beta(T)$、 $\eta(T,S)$、 $\rho_0$ are the followings respectively:

$$\beta\ (T) = \overline{\beta}(T)/\rho_0, \eta(T,S) = \overline{\eta}(T,S)/\rho_0, \rho_0 = a_0 + \overline{\beta}(T_0)\cdot T_0$$

Where a0=999.842594kg/m3. The β(T) is equivalent to the negative value of the thermal expansion coefficient, and β has a nonlinear relationship with T.

$\Gamma_T = K_j + \frac{v_t}{P_n}$ is turbulent heat diffusion coefficient. $K_j = \lambda/\rho_0 C_p$ . $C_p$ is specific heat at constant pressure and $\lambda$ is fluid heat transfer coefficient.

$J$ is the net heat absorbed by water (W/m$^2$), which mainly includes radiation, evaporation and conduction. The heat flux entering water body through water surface is:

$$J = \varphi_{sn} + \varphi_{an} - \varphi_{br} - \varphi_c - \varphi_e$$

Where $S_l$ is group sediment concentration ( $kg/m^3$ ); $l$ is sediment particle size group number; $\omega_l$ is sediment sinking speed ( $m/s$ ) and $\Gamma_S$ is sediment diffusion coefficient. $z_b$ is bed elevation ( $m$ ); $\gamma'$ is the dry density of sludge near the riverbed surface ( $kg/m^3$ ), and $S_{bl}$ and $S_{bl}^*$ are the sediment concentration and sedimentation force near the riverbed surface ( $kg/m^3$ ).

$k \sim \varepsilon$ correction equation:

Turbulent diffusion coefficient in water flow equation is $\Gamma = v + \frac{v_t}{\sigma_{ti}}$ .

$$\begin{cases} \dfrac{\partial(\rho k)}{\partial t} + \dfrac{\partial(\rho u_j k)}{\partial x_j} = \dfrac{\partial}{\partial x_j}\left(\Gamma_k \dfrac{\partial k}{\partial x_j}\right) + S_k \\ \dfrac{\partial(\rho \varepsilon)}{\partial t} + \dfrac{\partial(\rho u_j \varepsilon)}{\partial x_j} = \dfrac{\partial}{\partial x_j}\left(\Gamma_\varepsilon \dfrac{\partial \varepsilon}{\partial x_j}\right) + S_\varepsilon \end{cases} \tag{6}$$

Where

$$\Gamma_k = \mu + \frac{\mu_t}{\sigma_k}; \Gamma_\varepsilon = \mu + \frac{\mu_t}{\sigma_\varepsilon}; S_k = G - \rho\varepsilon; \ S_\varepsilon = c_{1\varepsilon}\frac{\varepsilon}{k}G - c_{2\varepsilon}\rho\frac{\varepsilon^2}{k}; G = \mu_t \frac{\partial u_i}{\partial x_j}\left(\frac{\partial u_i}{\partial x_j} + \frac{\partial u_j}{\partial x_i}\right)$$

Where $\sigma_k$ and $\sigma_\varepsilon$ are Prandtl numbers corresponding to the turbulent kinetic energy $k$ and turbulent dissipation rate $\varepsilon$, and they are 1.0 and 1.3; $c_\mu$, $c_{1\varepsilon}$ and $c_{2\varepsilon}$ are the empirical constants and they are 0.09, 1.44 and 1.92; $G$ is the generation item of turbulent kinetic energy due to average velocity gradient.

The expansion equation of corresponding $G$ is

$$G = \mu_t \frac{\partial u_i}{\partial x_j}\left(\frac{\partial u_i}{\partial x_j} + \frac{\partial u_j}{\partial x_i}\right) = \mu_t \left\{\left(\frac{\partial u_i}{\partial x_j}\right)^2 + \frac{\partial u_i}{\partial x_j}\frac{\partial u_j}{\partial x_i}\right\} \tag{7}$$

Where Turbulent Prandtl (Ratio of viscosity coefficient to diffusion coefficient, the same below) turbulent pulsation) is $\mu_t = c_\mu \rho k^2 / \varepsilon$ , the turbulent kinetic energy $k$ and turbulent dissipation rate $\varepsilon$ are obtained by solving the standard $k \sim \varepsilon$ equation.

Considering that the horizontal diffusion is less affected by density gradient, while the vertical diffusion is greatly affected by it, the horizontal turbulent P/S $\sigma_{tz}$ is taken to be a constant (=0.9), while the vertical turbulence P/S $\sigma_{tz}$ is considered to be related to the turbulence and stratification feature parameters based on the analysis of the flux expression of turbulent stress/ flux algebraic mode. The following formula is obtained through simulation calculation of measured data of the thermal stratified flow:

$$\sigma_{tz} = 0.54308 + 0.00075g\frac{k^2}{\varepsilon^2}(\beta\frac{\partial T}{\partial z} + \eta\frac{\partial S}{\partial z}) \tag{8}$$

Where $\beta$ and $\eta$ are density change rates corresponding to changes in temperature and sediment concentration respectively.

$$\beta = -\frac{1}{\rho_0}\frac{1}{V}\frac{dV}{dT} \tag{9}$$

$$\eta = \frac{1}{\rho_0}\left(1 - \frac{\rho_0}{\rho_s}S\right) \tag{10}$$

Where $v$ is the viscosity coefficient of water flow movement ($m^2/s$).

## 3. Boundary Conditions

The boundary conditions of the solution area include inlet boundary, outlet boundary and wall boundary.

(1) Upstream inlet boundary. The flow rate, temperature, and sediment concentration are given as essential boundary conditions.

(2) Downstream outlet boundary. Control is performed based on free pressure outflow.

(3) Free surface boundary.

The VOF method is adopted to calculate the free surface and the symmetrical boundary conditions are adopted as the hydrodynamic variables of the free surface, that is, the normal gradient of horizontal velocities $u$ , $v$, $k$ and $\varepsilon$ is 0, and the vertical velocity of them is $w = 0$.

In the case of surface heat exchange, the definition of the temperature boundary is treated separately for different situations.

(4) Lower surface boundary and side boundary. In this study, the wall boundary is treated by the convection wall function method.

## 4. Numerical Method

The SIMPLE-C method is adopted to solve the governing equations. The specific process is as follows:

1) Give the initial predicated value of the pressure field $p^*$ .

2) Obtain $u^*$, $v^*$ and through the momentum discrete equation. Firstly, the coefficients $A_p$, $A_i$ and the source term $S_\phi$ are solved; secondly, $A_P/\alpha_p$ is solved and the source term is $S_{\phi u} + \phi_P^*(1-\alpha_p)A_P/\alpha_p$ ; finally, $u^*$, $v^*$ and $w^*$ are obtained through the momentum discrete equation, and they are the solution after the sub-relaxation.

3) The pressure correction equation is adopted to solve the pressure correction value $p'$. Based on $u^*$, $v^*$ and $w^*$, the interface control coefficients $a_E$, $a_W$, $a_N$, $a_S$, $a_T$ and $a_B$ and source term $S_\phi$ are obtained; finally the pressure correction equation is introduced and $p'$ is obtained.

4) The corrected pressure and flow rate are calculated. $u'$, $v'$ and $w'$ are obtained, and then the corrected pressure flow rate is obtained based on $p = p^* + p'$, $u = u^* + u'$, $v = v^* + v'$ and $w = w^* + w'$.

5) The corrected $p$, $u$, $v$ and $w$ are taken as new initial values to calculate the coefficients and source terms of the momentum discrete equation. Go back to the second step and repeat the whole process until a convergent solution is obtained.

## 5. Model Application Calculation

### *5.1. Basic Situation*

The built model is applied to the Johnson water channel model to conduct an experimental study on the density-stratified flow of temperature and sediment concentration.

The Johnson water channel model layout is shown in Figure 1. The model is 24.39m long and the depth gradually changes from 0.3m at the inlet to 0.91m at the dam site. According to the shape change of the section, the model reservoir can be divided into two sections. The first section is a section with length of 6.1 m, horizontal bottom and width varying linearly from 0.3 m to 0.91 m. The second section is a section with a unchanged width of 0.91m and bottom elevation lower than the length of 18.29 m by 0.61 m.

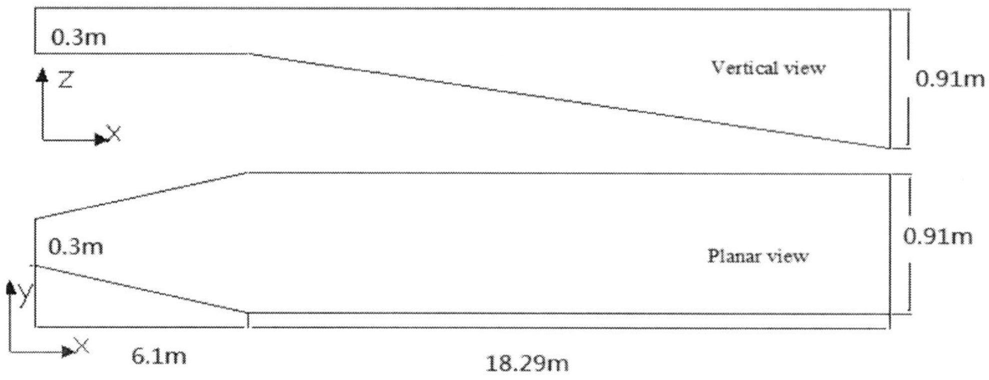


**Figure 1.** Standard Johnson water channel schematic diagram

*5.2. Calculating Conditions:*

(1) Water temperature density-stratified flow test:

The reservoir was filled with water at 21.44 °C first, and then the cold water at 16.67 °C was introduced into the reservoir from the porthole 0.15 m high at inlet near the bottom. The outlet of the dam is located at the site 0.15 m above the bottom. The hole height is 0.15 m, and the inflow velocity is 0.014 m /s.

(2) Sediment density-stratified flow test:

The inlet sediment particle size is d50=0.006mm and the sediment concentrations are set to $38kg/m^3$ and $100kg/m^3$;

*5.3. Calculating Results*

Figure 2 shows the temperature stratified flow condition. The longitudinal flow velocity distribution of mid-perpendicular line on the cross section 11.43m away from the inlet at t = 11 min in model calculation is compared with the actual measured distribution.

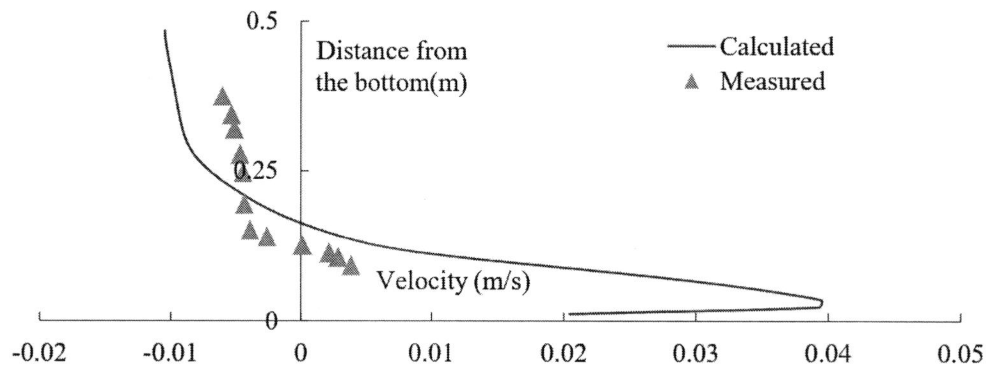


**Figure 2.** Longitudinal flow velocity distribution of mid-perpendicular line 11.43m away from the inlet at t = 11 min

According to the flow direction of longitudinal flow velocity, water body of the section can be divided into two parts: the subsurface flow layer and the upper water body. The flow velocity of the bottom cold water layer is large, the thickness of the subsurface flow layer is about 0.12 m, and the reverse vortex flow appears in the upper water body of the subsurface flow with small flow velocity.

The comparison of prediction results of the model and the measured results shows that submerged flow motion of cold water and reverse eddy current of the upper water body are reflected; the flow velocity distribution of them is basically the same; the calculated thickness of the submerged flow is in good agreement with the actual measured thickness, and the longitudinal flow velocity along the water depth direction is basically the same.

Figure 3 compares the variation of the outflow water temperature obtained from model calculation and that from actual measurement. The measured results show that the outflow water temperature began to decrease at t=14.6min, and decreased relatively fast at the beginning and then slowed down; the water temperature dropped to 19.4 °C at t=25.6min. The analysis of calculation results of the model and the actual measurement shows that time delay of outflow cooling is 2 minutes, and the degree of temperature drop of the outflow water is small; the temperature of the outflow is about 19.8 °C at stable state, about 0.4 °C higher than the actual situation

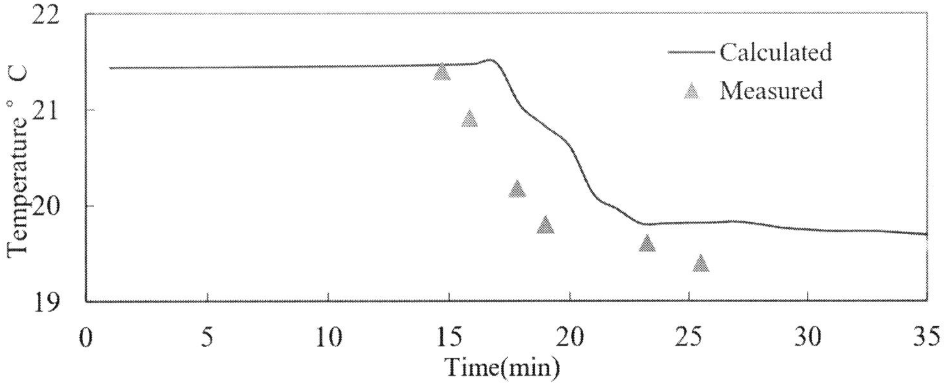


**Figure 3.** Outflow water temperature changes with time

Figures 4 show the calculation results of the turbid water thickness along the density-stratified flow with 38kg/m$^3$ sediment concentration.

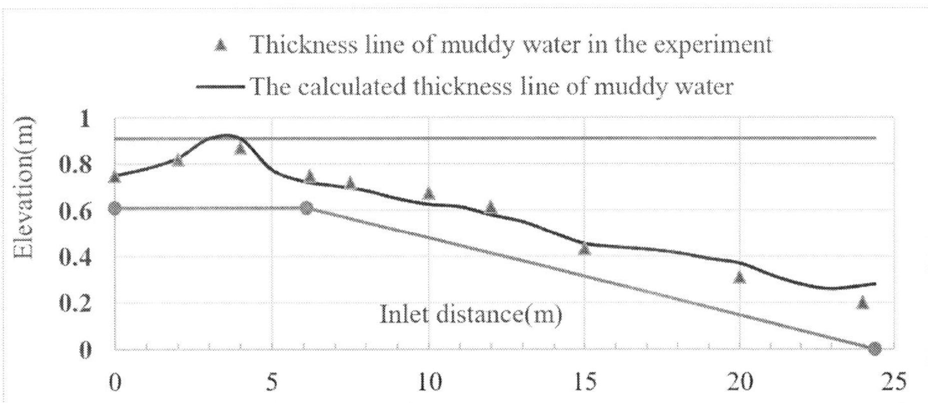


**Figure 4.** Turbid water thickness distribution along the sediment density-stratified flow (S=38kg/m$^3$)

The calculation results show that the density-stratified flow phenomenon due to density difference is obvious, and the thickness of the density-stratified flow is also significantly different under different sediment concentration. It can also be seen from the figure that under the boundary conditions of this experiment, the location of the density-stratified flow plunging point and plunging depth under different sediment concentrations are basically the same, and with the transfer of density-stratified flow, the difference of thickness of the turbid water shows a trend of gradual increase. It can be seen from the above calculation results that the three-dimensional turbulence model established in this study can simulate the flow pattern, water temperature and sediment transfer process well, and can provide good calculation effect.

## 6. Conclusion

The numerical relationship among vertical turbulence P/S $\sigma_{tz}$ turbulence and stratification feature parameters is established by analyzing the flux expression of turbulent stress/ flux algebraic mode. Corresponding density change rate parameters reflecting changes in temperature and sediment concentration are introduced to correct the vertical diffusion coefficient of turbulent. A three-dimensional turbulence model with multi-phase coupling of water flow, temperature and sediment is established on this basis. The model reflects the basic physical phenomena of density-stratified flow well. And the adopted finite volume method and SIMPLEC algorithm have the characteristics of good convergence, fast calculation speed and stable calculation, and solve the computational efficiency problem of three-dimensional turbulent density-stratified flow well. The numerical model proposed in this paper can simulate the density stratification and sediment stratification of reservoirs, and it has certain application value for the operation and management decision about reservoirs.

## Acknowledgments

This work was partially funded and supported by the National Natural Science Foundation of China (Grant Nos. 51479081) and the National Key Research and Development Program (2016YFC0402409-07).

## References

[1] Jin Haisheng, Zhang Shunong. Numerical simulation of laminar flow with temperature and salinity gradients [J].Journal of hydrodynamics research and development, 1991, (12): 72-76.
[2] Yang Ming, Liang Guoting, Lai Ruixun, et al. Principle and application of digital terrain generation based on surface interpolation technology [J].Journal of Water Conservancy, 2007, 38 (2): 221-225.
[3] Zhang Fangxiu et al. [R]. Three-dimensional mathematical model of water temperature in dam area of Xiaolangdi Reservoir. Yellow River Institute of Water Conservancy, 2009.

IOP Conf. Series: Materials Science and Engineering 472 (2019) 012081 doi:10.1088/1757-899X/472/1/012081

# Finite Element Analysis of the Brake Drum Based on ANSYS APDL

**Bin Zheng and Jingdong Zhang**

School of Transportation and Automobile Engineering, Panzhihua University. No. 4, New Hong international community, Airport Road, East District, Panzhihua, Sichuan. Email: 22198334@qq.com

**Abstract.** Brake drum is accompanied with vibration during it works, it is particularly important to research its static and dynamic characteristics. In this paper, the static analysis of the drum was carried out, and the displacement was computed by ANSYS APDL. Then, modal analysis of the drum was conducted, and the first six orders natural frequencies and vibration were computed. The results can offer references for the brake drum design and manufacture as well as improved the quality of the brake system.

## 1. Introduction

Braking system is a key part of automobile and the braking performance directly affects the safety of automobile driving. More and more researchers pay attention to the automobile brake system with the increasing of the automobiles. Static analysis of the brake drum has very important significance to improve the braking efficiency and stability of the automobile braking system. Many domestic experts and scholars conducted related research on the brake drum.

Liu established the three model of cam shaft by using ANSYS parametric design language, then the deformation at different positions and different angles is deduced and analyzed with the finite element method. The natural frequency, the vibration mode and the harmonic response of the cam shaft while grinding is analyzed. Zhang Jing-dong used ANSYS to obtain the first 12 natural frequencies and vibration shapes of a heavy-duty automobile brake drum and pointed out the probable resonance risk and hidden design danger. Xi presented a new method to research a kind of wheel brake of a heavy automobile by using virtual prototype technology.

Based on the above research, it concluded that the drum of brake drum is conceived, optimized and finite element analysis, but few parametric modeling and static analysis of the drum are researched. Because the static analysis of the drum has an important role, there is a need to establish parametric model through APDL and carried out the simulation. Static displacement and natural frequency of the drum were carried out by using ANSYS APDL which provided a theoretical basis for the drum deformation research.

## 2. Parametric Modelling of Drum Using APDL

Braking system is very important to a vehicle. If the brake system don not operate properly, the result can be disastrous. Brakes are actually energy conversion devices, which convert the kinetic energy of the car into thermal energy. There are two main types of brakes: drum brake and disc brake. The brake drum mainly includes the brake drum, brake shoes and friction lining. The brake process is that the leading shoe and trailing shoe can rotate with its axis of the eccentric shaft axis so that the friction lining contact with the brake drum to achieve braking purpose. The physical picture of the drum is shown as Figure 1.

**Figure 1.** Physical picture of the brake drum.

Based on the measured data, the APDL command flow is given to establish the model of the drum. The specific command flow is as follows.

```
/clear
/prep7
k, 1,0,120,0
k,2,0,175,0
k, 3,50,210,0
k,4,140,210,0
k, 5,140,220,0
k,6,240,220,0
k, 7,240,205,0
k,8,50,205,0
k, 9,15,175,0
k,10,15,120,0
k, 11,0,0,0
k,12,10,0,0
asel, none
a,1,2,3,4,5,6,7,8,9,10
cm, a-mid, area
vsel, none
et,1,solid 187
mpdata, ex, 1, , 1.3e11
mpdata, prxy, 1, , 0.25
vsee, , , , all
flst, 5, 4, 6, orde, 2
fitem, 5, 15
fitem, 5, -18
cm,y,volu
vesl, , , , p51x
cm, y1, volu
chkmsh,'volu''
cmsel,s,_y
vmesh,y1
mptemp, , , , , , , ,
mptemp, 1, 0
mpdata, ex, 1, , 1.3e11
mpdata, prxy, 1, , 0.25
```

By execute the above APDL command, the Figure 2 is obtained as following. The above ANSYS APDL command stream can be edited or modified by the text editing software, or it can be executed directly in ANSYS software.

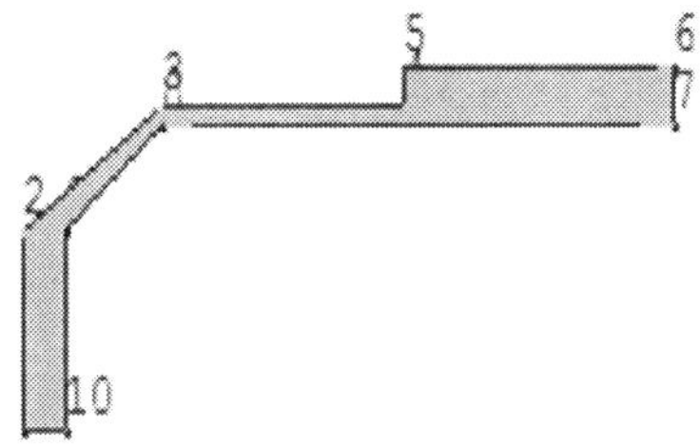


**Figure 2.** The surface model of the drum.

The Figure 3 gives the drum model that established by ANSYS APDL. During the modelling process, in order to facilitate meshing division and avoid the error of finite element calculation as well as improving the calculation speed, the physical model of drum is properly simplified as the following principles.

1) Delete all the features of the tiny structure, such as chamfer and fillet.
2) Ignore some structures whose thickness is less than 5 mm.
3) Fill the small threaded hole or tapped hole with the cylinder.

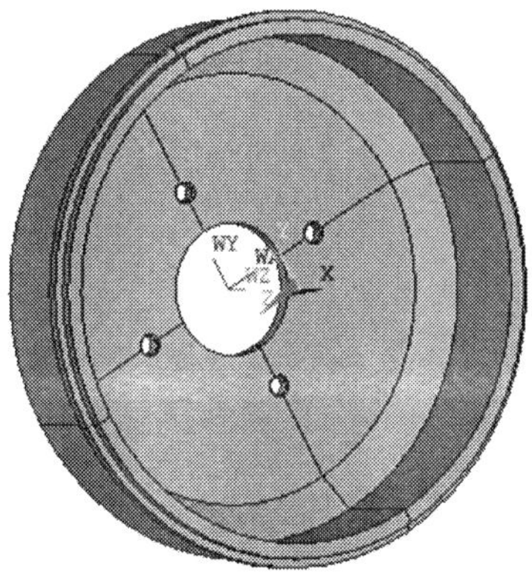


**Figure 3.** Three dimensional solid model of the drum.

When carrying out the mesh division of the drum, 3D solid element of 187 tetrahedron unit is used for creating finite element model, which includes 10 nodes. By using the free mesh division method to mesh the drum, the finite element model of the drum is created as shown in Figure 4. The relevant command stream of APDL is giving as following.

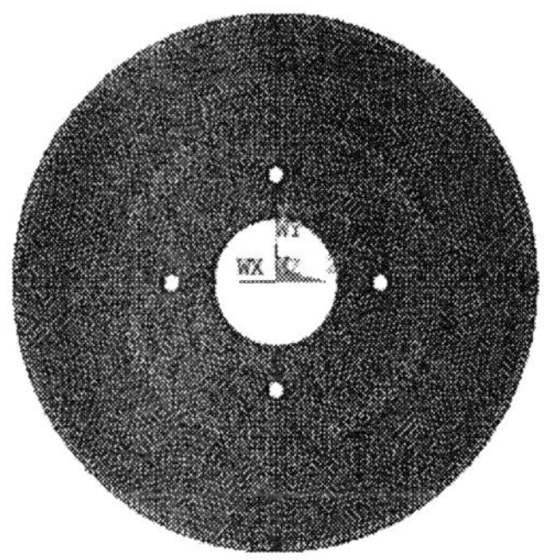


**Figure 4.** Finite element model of the drum

```
ET, 1, SOLID187
MPDATA, EX, 1, , 1.3e11
MPDATA, PRXY, 1, , 0.25
```

```
VSEL, , , , ALL
FLST, 5, 4, 6, ORDE, 2
FITEM, 5, 15
FITEM, 5, -18
CM, Y, VOLU
VSEL, , , , P51X
CM, Y1, VOLU
CHKMSH, 'VOLU'
CMSEL, S, _Y
VMESH, Y1
MPTEMP,,,,,,,,
MPTEMP, 1, 0
MPDATA, EX, 1, , 1.3e11
MPDATA, PRXY, 1, , 0.25
```

## 3. Static Analysis of the Drum

The function of static analysis is to solve the displacement and equivalent stress of the structure under the action of static load. The static analysis calculates the effect of a structure under a fixed load without considering the effects of inertia and damping. The following gives the APDL of calculation of static analysis.

```
FINISH
/SOL
/ANG, 1, -30, YS, 1
DA, P51X, ALL
SFA, P51X, 1, PRES, 3e6
APLOT
/STATUS, SOLU
SOLVE
```

The results show that the maximum displacement of the drum while the restraints of degree of freedom and surface loads of $3\times10^6$N are applied on the upper surfaces of the drum. From the figure 5, it can be seen that the overall displacement is 0.050743mm. It shows that the deformation of the drum is relatively large. Therefore, the following optimization of the drum should be taken into account the edge deformation of the drum. From the figure 6, it can be seen that the maximum displacement of the drum is 0.047224mm, which indicate that the distribution along X axis is relatively large. During the braking process, the friction between the brake shoe and the deformation of drum is large. In the design process of the brake drum, it should consider the reasonable structure to avoid large deformation and to ensure the service life of the brake and braking performance.

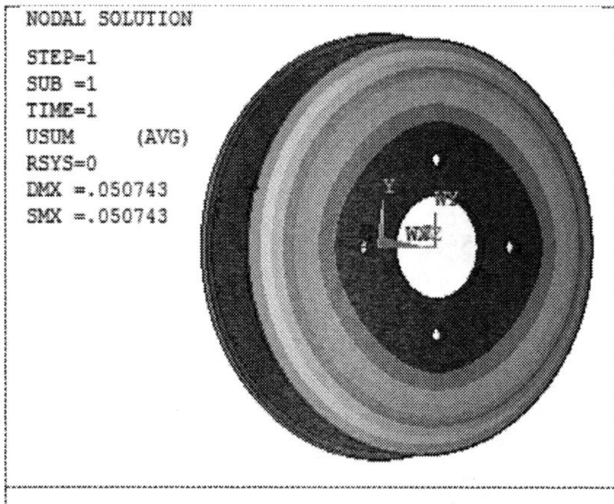


**Figure 5.** Overall displacement distribution of the drum.

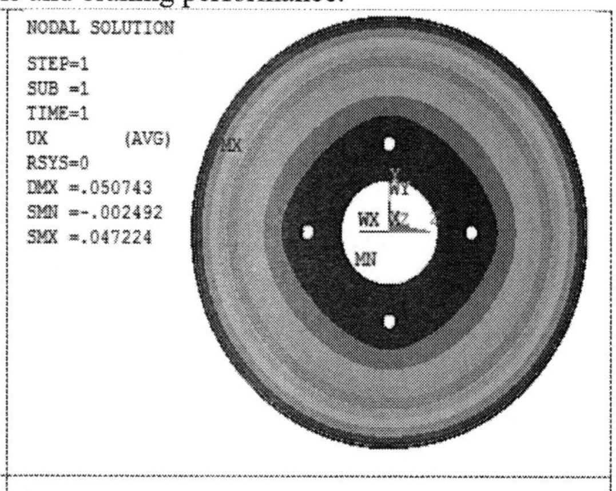


**Figure 6.** Displacement distribution along X axis of the drum.

From the figure 7, it can be seen that the maximum displacement is 0.023425mm; the displacement distribution along Y axis is relatively small. From the figure 8, it can be seen that the maximum displacement is 0.023489mm; the displacement distribution along Z axis is similar to Y axis.

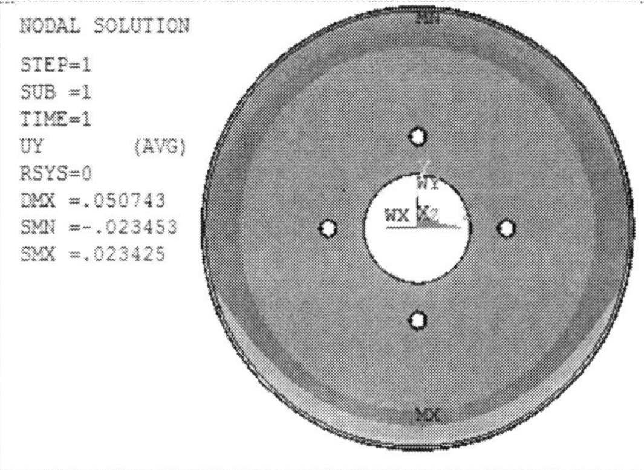


**Figure 7.** Displacement distribution along Y axis of the drum.

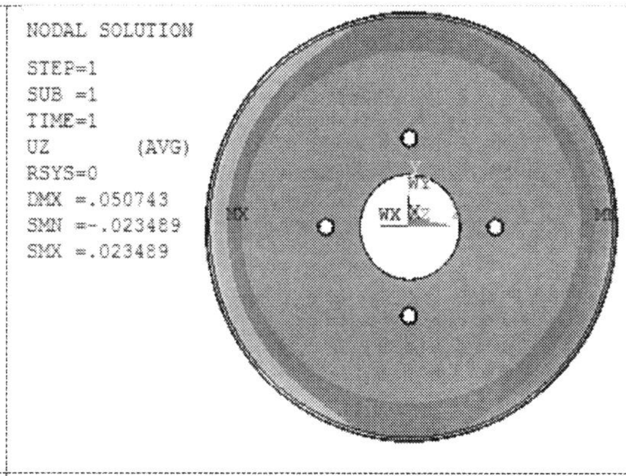


**Figure 8.** Displacement distribution along Z axis of the drum.

## 4. Modal Analysis of the Drum

While the drum works, it would suffer from the dynamic load and vibration disturbance. When the external excitation frequency is near to the natural frequencies of the drum, it would lead to resonate. The resonance will cause severe noise and fatigue damages of the drum. So, a modal analysis in the ANSYS software is applied to calculate the dynamic characteristics in the design stage. The modal analysis of the drum is performed and the first six natural frequencies are extracted. Figure 9 to figure 12 give the first four mode shapes and vibration of the drum. The following gives the APDL of calculation of static analysis.

```
FINISH
/clear
allsel,all
/solu
LSCLEAR, ALL
antype,2
modopt,lanb,10,0,999999,,,
mxpand,10,,,0
solve
/post1
result summary
plot ctrls--animate
read result
```

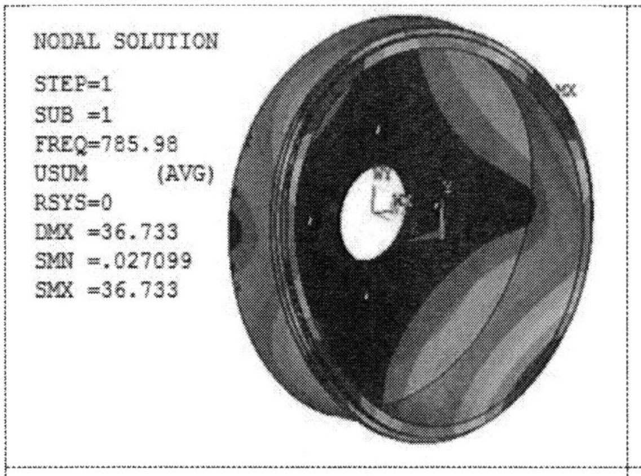


**Figure 9.** The first-order mode vibration of the drum.

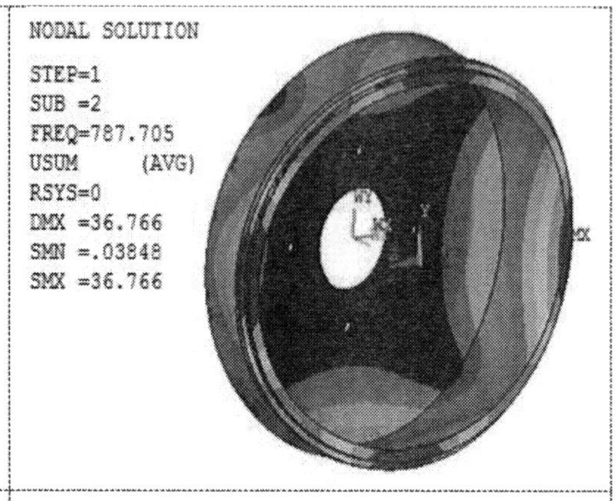


**Figure 10.** The second-order mode vibration of the drum.

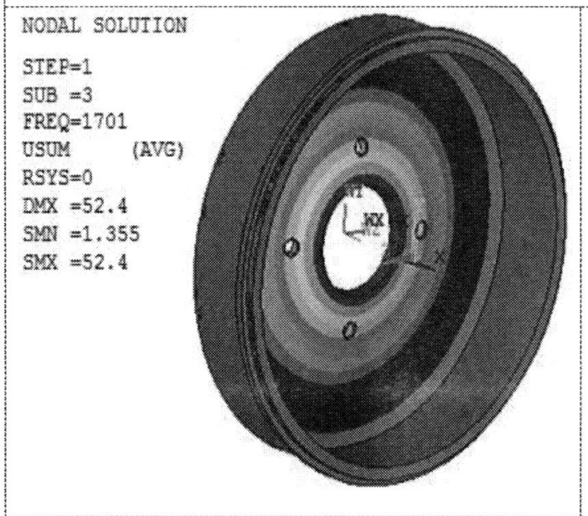


**Figure 11.** The three-order mode vibration of the drum.

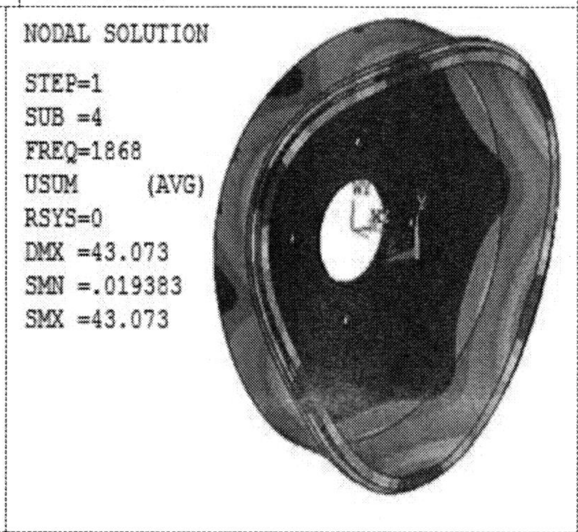


**Figure 12.** The four-order mode vibration of the drum.

As can be seen from the figure 9 to figure 12, the difference in the natural frequencies of the first four orders of the drum is relatively large. The main vibration mode of the drum in the low-order frequency is the vibration of the outer edge of the drum. The main vibration mode of the high-order frequency is being quadrilateral shape.

## 5. Conclusions

It is concluded that the use of APDL for three-dimensional modelling and finite element analysis can greatly reduce the workload and improve work efficiency. Model of the drum is established by using ANSYS APDL parametric design language. As the dimensions of the drum changed, it just modifies the relevant parameters of the APDL to establish new drum model. Static and modal analysis of the drum is carried out by using ANSYS APDL. Displacement, natural frequency and vibration mode are obtained. It is significant to improve the work efficiency and shorten the design cycle by using ANSYS APDL.

## Acknowledgment

This work was supported by educational commission of Sichuan province key project (18ZA0293), PhD Research Start-up Foundation of Panzhihua University (bkqj2017006), Education and teaching research and reform project of Panzhihua University (JJ1769).

**References**

[1] Zheng Bin, Zhang Jing-dong, Li Tang. Parametric Modelling and Finite Element Analysis of the Brake Drum Based on ANSYS APDL. 2017 3rd International Conference on Green Materials and Environmental Engineering, 2017, p 134
[2] Yang Sheng, Liu Shufen, Bai Heng. Application of APDL Language in Parameterized Modeling of ANSYS [J]. mechanical engineer, 2015, p 3
[3] Zheng Bin, Zhang Jing-dong. Modal analysis of the brake drum of heavy-scale automobile. 2017 IEEE 3rd Information Technology and Mechatronics Engineering Conference, 2017, p 984
[4] Guo Jie. Parameterized Design and Optimization Analysis of Drum Brake. Wuhan University of technology. P 221
[5] Zhang Jingdong, Zheng Bin, Yao Jie, et al. Modal analysis and optimization design on brake drum for a heavy-duty automobile. Bus & Coach Technology and Research, 2016, p1
[6] Ning Lianwang. ANSYS finite element analysis theory and development. [J]. Shanxi Science and Technology, 2008, p 3
[7] Xiao Jiang, Qin Houming, Kong Zhuo, et al. Calculation method of drum angle during drum brake's braking process. Heavy Truck, 2015, p 14

# Research Progress on Treatment Technology of Produced Water by Adsorption Method

**Xiangbin Liu** [1,2]

1. Daqing Oilfield Production Technology Institute, China
2. Northeast Petroleum University, China
E-mail: liuxiangbin@petrochina.com.cn

**Abstract.** Produced water is the largest by-product of oil production. Studying its high-efficiency treatment technology is one of the key issues in the sustainable development of oilfields. The adsorption has the characteristics of low cost, high efficiency and no secondary pollution, and has significant effects in the treatment of produced water. This paper introduces the adsorption technology. The research progress of the domestic and international adsorption treatment of oilfield produced water was summarized. The effects of different adsorbents on oilfield produced water treatment are briefly described, which points out a new research direction for the development of future adsorbent technology.

## 1. Introduction

As the amount of oil extraction continues to increase, the output of produced water in the oilfield is also increasing. The oil and gas industry will also have an adverse impact on the environment while producing oil. During the process of oilfield exploitation oilfield, the produced water is the water which is separated from produced liquid by the process of dewatering. It is the largest by-product of oil production in the oil field. The produced water in the oilfield contains a large amount of organic matter, inorganic salt ions, oils, corrosives and radioactive products. Its efflux will cause continuous pollution to the ecological environment, and its composition is extremely complicated [1]. Different types of oilfield produced water need to be treated with different treatment processes to achieve the re-injection criteria. Therefore, the treatment of oilfield produced water has brought challenges to the sustainable and stable production of oilfields, and its management has become an important issue for environmental supervision departments. China needs to deal with a large amount of oilfield produced water every year [2], and its output is increasing year by year. Most of the produced water in the oilfield is used to reinject the formation, and a small amount of produced water is directly discharged into the environment. The produced water used for reinjection of the stratum needs to reach the water quality index of the re-injection water in the oilfield, and the produced water that was directly discharged should also meet the criterion of wastewater discharge given by the environmental protection department. Therefore, the efficient treatment of oily wastewater has become a key issue that needs to be solved in the continuous exploitation of oil fields, and the discharge of produced water also poses a serious challenge to environmental protection.

At present, the main oilfield produced water treatment using some conventional processing technology, such as physical methods: gravity separation, centrifugal separation process, coalescence method, filtration method and membrane separation. Chemical methods: chemical demulsification method and chemical oxidation method. Physical and chemical method, salting-out method, coagulation and sedimentation method, and electrochemical method. Biological method: activated sludge method, biological membrane method, oxidation pond and anaerobic contact method. The method of combination between them. There are also many related review reports the research progress of the processing technology [3-4]. In recent years, membrane technology has become a research hotspot due to its characteristics of high efficiency and non-pollution. Among them, Xu Xiaodong, Lin Aiguo, et al. has summarized the application of membrane technology in oilfield produced water [5-6]. However, there are few reports about the research on adsorbent treatment of oilfield produced water. Therefore, this is summarized below.

## 2. Characteristics and Applications of Adsorbents

The adsorption method has become a hot spot in the research of produced water treatment technology because of its advantages of low cost, good removal effect and no secondary pollution. The main principle is to absorb the pollutants in the wastewater by utilizing the lipophilicity and chemisorption of the adsorbent. In addition, the treatment of some macromolecular organic pollutants that are difficult to remove by other methods is particularly significant. Therefore, the adsorption method has an irreplaceable role in the field of water treatment. Adsorbent treatment of heavy metal wastewater [7-9], printing and dyeing wastewater [10-13], chemical wastewater [14] and other wastewaters have been reported in many literatures, many of which have been successfully applied in industry. However, its application in oilfield wastewater treatment is still in the experimental stage. At present, the adsorbents used in oilfield produced water treatment are mainly divided into: clay adsorbents, carbonaceous adsorbents, adsorbents prepared by waste and other adsorbents.

### *2.1. Clay Adsorbent*

*2.1.1. Bentonite.* The main component of bentonite is montmorillonite, which is prone to interlayer peeling and expansion under the action of solvent. As a result, bentonite has a large surface area and thus has a strong adsorption capacity, which can effectively remove organic pollutants in wastewater. Moazed et al. [15] reported that organic bentonite treatment produced water from the Esteban well in Canada. The initial oil content of the produced water was 231 mg/L and the pH was 6.8. The experimental results showed that the oil removal rate of the produced water from the oil well treated with organic bentonite could reach 88-89%. Subsequently, Moazed et al. [16] also used the combined adsorbent of organic bentonite and anthracite to treat the same produced water, in which the mass ratio of organic bentonite and anthracite was 3:7, the average particle size of the combined absorbent was 2.1mm, and the porosity was 45%. It was found that after 5g of composite adsorbent was added and adsorbed for 15 minutes, the oil content in the produced water decreased dramatically from 231mg/L to less than 50mg/L. However, with the increase of adsorption time, the oil content in the produced water basically remains unchanged, indicating that the adsorbent had reached saturation state. It was also found that the oil removal rate increased with the increase of the amount of compound adsorbent.

*2.1.2. Vermiculite.* Vermiculite is a natural mineral, belonging to silicate, which is widely used in wastewater treatment because of its good ion-exchange and adsorption. Silva et al. [17] investigated the effect of vermiculite on the adsorption of crude oil in wastewater. The results showed that the adsorption of hydrated vermiculite on crude oil was poor, while expansion vermiculite and hydrophobic vermiculite had a good adsorption effect on crude oil. The experiment also found that for every 10% increase of the hydrophobic group of the hydrophobic vermiculite, the hydrophobic meteorite can increase the adsorption capacity by 50% compared with the expanded vermiculite. Mysore, etc on the basis of Silva, etc. [18] used expansion vermiculite and hydrophobic vermiculite as adsorbents to treat refinery wastewater. The results showed that the oil removal rate of refinery wastewater treated with expanded vermiculite as adsorbent was higher than that of hydrophobic vermiculite, but the oil removal rate of both was relatively low, with the oil removal rate of expanded vermiculite being 57%, while the oil removal rate of hydrophobic vermiculite was only 43%. Therefore, it can be found that vermiculite is less effective in treating oily wastewater from oil field.

### *2.2. Carbon Adsorbent*

*2.2.1. Powdered activated carbon.* Powdered activated carbon is mainly made of wood chips and husks as raw materials. After steam activation, it is refined and pulverized into micro-powder materials. It has the characteristics of large specific surface area and strong adsorption capacity, and is widely used in wastewater treatment. Khaled et al. [19] examined the oil removal effect of powdered activated carbon on oily wastewater. The experimental results showed that the mass of powdered activated carbon increased from 0.1g to 0.5g under the condition of initial oil content of 600mg/L and stirring time of 30min. The oil removal rate can be increased from 20% to 61%. When the initial oil content of the wastewater is 836mg/L, 0.5g of powdered activated carbon is added and stirred for 2h, the oil removal rate can reach 82.78%. And under the same conditions, 1g of powdered activated carbon is added and stirred for 4h and the oil removal rate was as high as 93.54%. In this experiment, they also found that the oil removal rate increases with the increase of dosage and contact time of powdered activated carbon, and decreases with the increase of initial oil content of wastewater.

*2.2.2. Carbon deposition.* Carbon deposition is the main component of particulate matter from diesel exhaust, which is of concern to researchers due to its low price and high oil removal efficiency. The literature [19] found that the oil removal effect of carbon deposition under the same conditions is better than that of powdered activated carbon. Because the porosity and surface area of carbon deposition are larger than that of powdered activated carbon and the oil removal rate of carbon deposition is more than 10% higher than that of powdered activated carbon. The experimental results show that the contact time, the initial oil content of the wastewater and the dosage of adsorbent all have an effect on the oil removal rate of carbon deposition, which is consistent with that of powdered activated carbon.

### *2.3. Adsorbent Prepared from Waste*

*2.3.1 Eggshell.* Muhammad et al. [20] used eggshells as biosorbents to remove dissolved oil and dispersed oil from oily wastewater. The surface morphology of the eggshell was characterized by scanning electron microscopy (SEM). It was found that the eggshell is a semi-permeable biofilm with complex porous structure, so it has strong adsorption in the natural state. The experimental results show that the eggshell can effectively remove the oily substances in the oily wastewater, and it is found that the oil removal rate will increase with the increase of the eggshell addition. When the initial oil content in oily wastewater is 194mg/L and the addition amount of eggshell is 200mg/L, the oil removal rate can reach 77.83%. And when the adding amount of egg shell is 1800mg/L, the oil removal rate can reach 100%. The result shows that the eggshell as adsorbent has a good application prospect. It not only has good adsorption characteristics, but also can turn waste into treasures, which is in line with the characteristics of the new adsorbent "cheap, high efficiency, no secondary pollution".

*2.3.2 Banana peel.* EI-Nafaty et al. [21] reported the use of banana peel to remove oily materials from oil produced water. The surface morphology of banana peel was characterized by SEM. It was found that there is a strong adsorption in the place where the layered material is deposited. And the charged oil droplets in the water will destroy the double and triple bonds in the banana peel structure to initiate the reaction and exchange their ions to neutralize the charge. It can eliminate the chargeability of the oil droplets, thereby reducing the repulsion between the oil droplets, which is conducive to the accumulation of oil droplets. The experimental results show that Banana peel has excellent oil removal effect. When the additive amount of banana peel is 10mg/L, the oil removal rate can reach 94.51%. And when the addition amount of banana peel is 50mg/L, the oil removal rate is 100%. At the same time, the effects of pH, adsorption time and temperature on oil removal were also analyzed in the literature.

### 2.4. *Other Adsorbents*

High oil-absorbing resin is a new kind of organic adsorbent, which is a functional polymer oil-absorbing material prepared by several oleophilic monomers through a certain degree of cross-linking polymerization. It has good heat resistance, cold resistance, and density less than water does not absorb water, oil absorption speed, not aging. In foreign countries, new high oil - absorbent resins have been widely studied and applied [22]. Li Feng, et al. [23] prepared copolymer high oil absorption resin by suspension polymerization with methacrylate as monomer. Under the condition of polymerization temperature 700°C, the mass ratio of the high oil absorption resin to the monomer is 65:20:15, and the dispersing agent is 0.5%, it was found that the high oil absorption resin oil absorption performance is the best, oil absorption ratio of 7.86g/g. After the experiment, the high oil absorbent resin and polypropylene were combined in a certain proportion to produce the high oil absorbent resin/polypropylene composite oil absorbent material. The experimental results show that the overall oil absorption performance is good. When the mass ratio of the high oil absorption resin is 5%, the adsorption material has the best overall performance, with oil absorption ratio of 11.5g/g, water absorption rate of 6.6% and oil retention rate of 96.3%. It is found that the high oil absorbing resin/polypropylene composite oil absorbing material has great application prospect in the recovery and treatment of oil floating on water surface.

Xin Shanshan, et al. [24] modified chitosan-acrylate composite resin was synthesized by suspension polymerization method. Chitosan was modified by A151, with factors analysis and orthogonal experiment. The result shows that, reaction power was 100W, time of reaction was 1h, the ratio of monomers was 1:1.5, BPO was 2.0% (wt), crosslinker was 1.5%(wt), PVA was 10%(wt), modified chitosan was 1.0%(wt). Adsorption capacity was found to be maximum at optimal condition. The as-prepared resin presented porous microspheres structures and the volume expansion rate reached 50 times.

## 3. Conclusion

The above research shows that the adsorbent method has the advantages of high oil removal rate, no secondary pollution and low treatment cost, compared with the oily wastewater treatment process commonly used in oil fields. However, due to its poor treatment effect on suspended solids, it is still only in the experimental research stage. The adsorbent has a good removal effect on the organic pollutants in the oily wastewater, so it can be considered to be applied to the pretreatment process of the oily wastewater.

## Acknowledgements

This work was financially supported by Postdoctoral researchers settled in Heilongjiang research funding, Study on the influence mechanism of clay particles on the stability of produced water from ASP flooding, LBH-Q17035.

## References

[1] Ebenzer T I, Georg Z C. Produced water treatment technologies [J]. International Journal of Low: Carbon Technologies, 2014, 9(13):1-21.
[2] Shen Chen. New progress of sewage treatment technology [M]. China Petrochemical Press, 2008:1-3.
[3] ZHANG Xue-jia, JI Wei, WANG Bao-hui, et al. WEN Yu-chen. Study on the Treatment Technology of Oilfield Produced Water [J]. Industrial Safety and Environmental Protection, 2007, 33(4):13-16.
[4] Zhou Weidong, Tong Deshui, Li Luopeng. Research and development of oilfield produced water treatment methods [J]. Industrial Water Treatment, 2008, 28(12):5-8.
[5] XU Xiao- dong, LI Ji, YUAN Xi- ming, et al. Application of Membrane Techniology in the Treatment of Oilfield Produced Water [J]. Journal of Filtration & Separation, 2005, 15(4):34-36.
[6] Lin Aiguo, Liu Peiyong, Liu Gang, et al. Progress of membrane separation technique in oil-bearingwater treatment in oil fields [J]. Industrial Water Treatment, 2006, 26(1):5-8.
[7] Wang Xiuli, Shang Yujun, Song dandan. Research progress in a new type of adsorbents

forremoving heavy metals from wastewater [J]. Industrial Water Treatment, 2014, 34(7):5-9.
[8] Deng Jingheng, Yu Kanping, Xiao Guoguang, et al. Research progress in the treatment of heavy metal wastewater by adsorption[J]. Industrial Water Treatment, 2014, 34(11):4-7.
[9] ZHANG Fan, LI Jing, TAN Jianhua, et al. Advance of the treatment of heavy metal wastewater by adsorption [J]. Chemical Industry and Engineering Progress, 2013, 32(11):2749-2756.
[10] WANG Kai-hua, LI Fu-quan, CHEN Sarina, et al. Research Review of Application of Algal Adsorbents in Dyeing Wastewater Treatment [J]. Water Pollution Control, 2016, 1:1-6.
[11] TANG Xiao-huan, MAO Yong, JIN Jing, et al. Latest Research Advance in Dye Wastewater Treatment with Adsorbent [J]. Acta Agriculturae Jiangxi, 2013, 25(7):116-119.
[12] Li Ping, Gong Lei. Adsorption Treatment of Dyeing Wastewater Using Fallen Leaves as Adsorbent [J]. Journal of University of Science and Technology (Natural Science Edition), 2012, 33(1):62-67.
[13] HE Lian-Fang, CHEN Cui-Jin, LIN Zuan-Tao, et al. Preparation and application of an excellent bioabsorbent by tetraethylenepentamine modified sugarcane bagasse for treatment of printing and dyeing wastewater [J]. Ecological, 2011, 30(4):441-445.
[14] XIE Tian, HAN Xiang-kui, WANG Xiao-ling, et al. Research on Advanced Treatment of Coal Chemical Wastewater by Granular Activated Carbon Adsorption Method[J]. Coal and Chemical Industry, 2015, 38(1):156-160.
[15] Moazed H T, Viraraghavan T. Removal of Oil from Water by Bentonite Organoclay[J]. Journal of Hazardous, Toxic, and Radioactive Waste, 2005, 9(2):130-134.
[16] Moazed H T, Viraraghavan T. Use of organo-clay/anthracite mixture in the separation of oil from oily waters [J]. Energy Sources, 2005, 27(1-2):101-112.
[17] Silva J,Umberto G,Marcus A,et al. Adsorption of crude oil on anhydrous and hydrophobized vermiculite[J]. Journal of Colloid and Interface Science, 2003, 260(2): 302-304.
[18] Mysore D,Viraraghavan T,Jin Y C. Treatment of oily waters using vermiculite [J]. Water Research, 2005, 39 (12): 2643-2653.
[19] Khaled O, Mona E S, Mohaned Y E K. Treatment of oil–water emulsions by adsorption onto activated carbon, bentonite and deposited carbon [J]. Egyptian Journal of Petroleum, 2011, 20(2):9-15.
[20] Muhammad I M, Ei-nafaty U A, Abdulsalam S, et al. Removal of oil from oil produced water using eggshell[J]. Civil and Environmental Research, 2012, 2(8):52-63.
[21] Ei-nafaty U A,Muhammad I M,Abdulsalam S. Biosorption and kinetic studies on oil removal from produced water using banana peel[J]. Civil and Environmental Research, 2013, 7(3):125-136.
[22] Zhang Chao. Synihesis and properties of modified acrylic ester compound oil absorption resin [D]. Jiangsu University, 2017.
[23] LI Feng, ZHANG Chao, LUAN Guo-hua, et al. Oil absorptive material based on high oil absorptive methylacrylate resin material and the polypropylene [J]. Journal of Hebei University of Technology, 2015, 44(01):45-49.
[24] Xin Shanshan. Synthesis and absorbency for organics of modified natural aterlal-acrylate composite sesins [D]. Yanshan University, 2014.

IOP Conf. Series: Materials Science and Engineering 472 (2019) 012083 doi:10.1088/1757-899X/472/1/012083

# Multi-Scale Full Waveform Inversion based on Curvelet Transform

**Helong Yang, Xue Chen *, Jiangwei Zhang and Daheng Guo**

Hebei GEO University, Shijiazhuang 050031, China.
E-mail: yanghlmail@163.com

**Abstract**. Full waveform inversion (FWI) is a rapidly developed inversion method in recent years, and the resolution is much higher than traditional inversion methods, such as tomography and migration velocity analysis. However, FWI is much dependent on starting model and low frequency information, and easy to fall into local minima because of cycle skipping issue. We proposed Curvelet transform multi-scale strategy to improve the effect of bad starting model by selecting different scale information of observed data during inversion. Test of frequency domain FWI on Marmousi model shows that the result of inversion has a better stability and convergence with Curvelet transform multi-scale method than traditional methods.

## 1. Introduction

FWI is an effective tool to obtain accurate velocity models with complex geological structures. Compared with conventional velocity model building method (eq. tomography, migration velocity analysis), the FWI could get a higher resolution result by taking full advantage of kinematics and kinetics information of the prestack seismic records [1, 2].

Successful application of FWI requires a good starting model to ensure that the modeled waveforms are kinematically less than half a cycle away from the recorded data. If one fails to provide a proper starting model, the gradient updates may fall down to the wrong direction and the cycle skipping could happen [3].

In this paper we proposed a new multi-scale FWI method combining with multi-scale characteristic of Curvelet transform. FWI goes from low frequency to high frequency to update the velocity. And at the different stages of FWI, select different scales of observed seismic data involved in the inversion according to the use of frequency. Thus simulated data can be better matched with observed data, and reduce the dependence on initial model of FWI.

## 2. Theory

### *2.1. Full wave form Inversion Theory*

FWI algorithm is an iterative process, and every iteration should carry out at least two times forward modeling (one for the incident wavefield and another for the back-propagated residual wavefield). And the cost of the inversion is highly depended on the forward modeling. In this paper, we choose the finite difference method for solving two dimensional frequency-domain acoustic wave equation:

$$\frac{\omega^2}{\kappa(x,z)}p(x,z,\omega)+\frac{\partial}{\partial x}\left(\frac{1}{\rho(x,z)}\frac{\partial p(x,z,\omega)}{\partial x}\right)+\frac{\partial}{\partial z}\left(\frac{1}{\rho(x,z)}\frac{\partial p(x,z,\omega)}{\partial z}\right)=-s(x,z,\omega), \tag{1}$$

where $p(x,z,\omega)$, $\kappa(x,z)$, and $s(x,z,\omega)$ denote the wavefield, bulk modulus, and source term,

respectively, $\omega$ is angle frequency and $\rho(x,z)$ denotes the density.
In the inversion problem, the objective function $C(m)$ is expressed as the difference between the real seismic data $d^{obs}$ and the simulated seismic data $d^{cal}$ ( Brossier, et al., 2010):

$$C(m)=\frac{1}{2}[d^{obs}-d^{cal}]^T[d^{obs}-d^{cal}]=\frac{1}{2}\left\|d^{obs}-d^{cal}\right\|_2^2, \quad (2)$$

where 'T' denotes conjugate transpose; $\|\cdot\|_2^2$ denotes $l2$ norm, and $m$ is geological model, such as subsurface velocity model.

The process of local optimization is to find the minimum value of the misfit function around the starting model $m_0$. In the framework of the Born approximation, we assume that the updated model $m$ can be expressed as the sum of starting model $m_0$ and the model update $\Delta m$:

$$m = m_0 + \Delta m \quad (3)$$

Conducting Taylor-Lagrange expansion for Equation (3) at $m_0$, and calculating derivative with respect to $m$ at both ends, we obtain a new equation expressed in matrix form:

$$\frac{\partial C(m)}{\partial m}=\frac{\partial C(m_0)}{\partial m}+\frac{\partial^2 C(m_0)}{\partial m^2}\Delta m. \quad (4)$$

Equation (3) has the minimum value when $\frac{\partial C(m)}{\partial m}=0$. Then the model update equals:

$$\Delta m=-[\frac{\partial^2 C(m_0)}{\partial m^2}]^{-1}\frac{\partial C(m_0)}{\partial m}, \quad (5)$$

where $\frac{\partial^2 C(m_0)}{\partial m^2}$ is Hessian matrix, and $\frac{\partial C(m_0)}{\partial m}$ is derivative of misfit function.

### 2.2. *Multi-Scale Full Waveform Inversion*

Recently Curvelet transform has been growing concerns because of its unique character. Curvelet transform was proposed by Emmanuel and David in 1999, developed from wavelet transform and overcome the inherent defects in the expression of image edge directional characteristics of wavelet transform. According to Curvelet transform theory, in signals containing noise, the larger Curvelet coefficients corresponds to valid signals, and whereas the noise. Therefore Curvelet transform can not only denoise the random noise, but also maintain valid signals. We can obtain Curvelet coefficients by the inner product of signal and the wavelet function:

$$C(j,l,k)=<f,\varphi_{j,l,k}>=\int_{R^2}f(x)\varphi_{j,l,k}(x)dx \quad (6)$$

Which $j$ is scale, $l$ is direction, $k$ is displacement, $f$ is signal, and $\varphi_{j,l,k}$ is wavelet function.

The Curvelet coefficients of an image can be divided into three different scale layers. The first layer is a matrix composed by low frequency coefficients, called coarse layer. The second layer is called detail layer, and is composed by middle frequency coefficients. And the third layer, fine layer, is composed by high frequency coefficients. With the increase of coefficient layer, the image is portrayed more and more sophisticated.

Multi-scale strategy is commonly used to improve the stability of FWI. In order to reduce the dependence on the initial model of FWI, this paper proposed a new multi-scale FWI method combining with multi-scale characteristic of Curvelet transform. FWI goes from low frequency to high frequency to update the velocity. And at the different stages of FWI, select different scales of observed seismic data involved in the inversion according to the use of frequency. That is, first extract large scale data of observed seismic data involved in inversion at the beginning stage of FWI, and gradually add smaller scale data with the increase of frequency and accuracy, as is shown in figure 2. Thus simulated data can be better matched with observed data, and reduce the dependence on initial model of FWI.

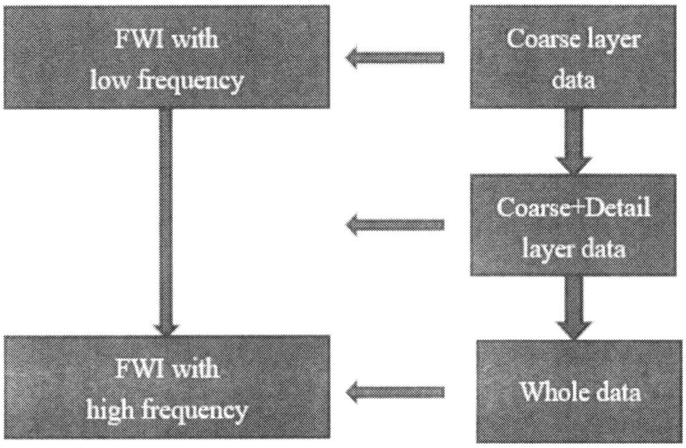


**Figure 1.** Schematic diagram of Curvelet transform multi-scale FWI.

## 3. Numerical Test

We test the multi-scale FWI method on a modified 2D Marmousi P-wave velocity model. We add a water layer above and don't update the water layer during the nonlinear procedure. Figure 3(a) and 3(b) are the true model and the starting model, respectively. The starting velocity model is a smoothed version of the true model. We use a Ricker wavelet with peak frequency of 12Hz. The source is located on the surface with the interval 24m. The grid size is 24m and sampling rate is 4ms. All the grid points on the surface acts as receivers. To illustrate the effectiveness of the new method, we reduce the accuracy of the starting model, as shown in figure 3(b). First extract coarse layer data of observed seismic data involved in inversion at the beginning stage of FWI, and gradually add detail layer data and fine layer data with the increase of frequency and accuracy. Figure 3(c) is the result of conventional FWI method, and figure 3(d), 3(e), 3(f) is the results of multi-scale FWI method. As we start the inversion with a relatively bad starting model, conventional FWI method fails to obtain an accurate result due to cycle skipping, while our new method obtains a more accurate result. By selecting different scale of seismic data involved in the inversion, simulated data can be better matched with observed data, thus the new method can reduce the dependence on starting model of FWI, and get a more accurate result. Figure 4 is the Kirchhoff migration results using the velocity results obtained by two FWI methods, respectively. From two migrated figures, we can see that more detail information of subsurface structure is recovered in multi-scale FWI result than that in conventional FWI result.

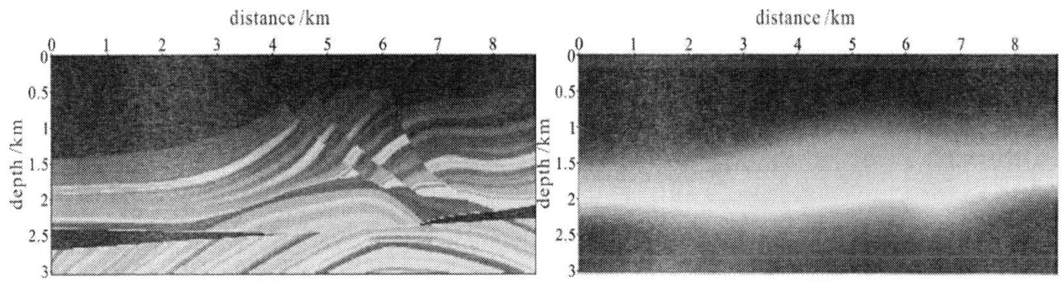


(a) (b)

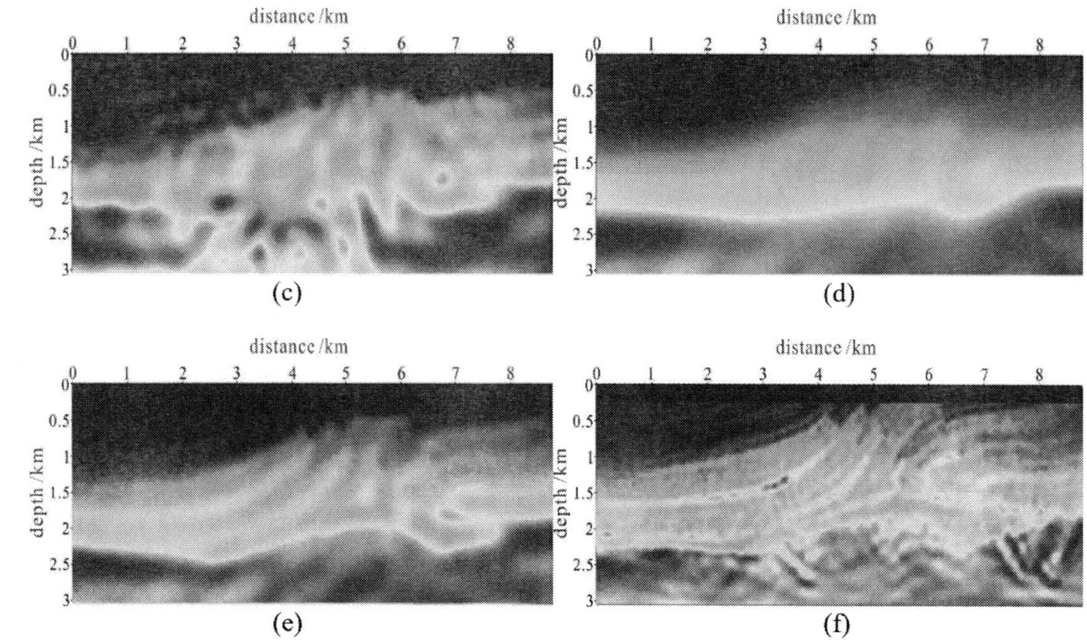


**Figure 2.** (a) True model, (b) starting model, FWI results of (c) traditional FWI method, (d) coarse scale data, (e) coarse + detail scale data, (f) whole data.

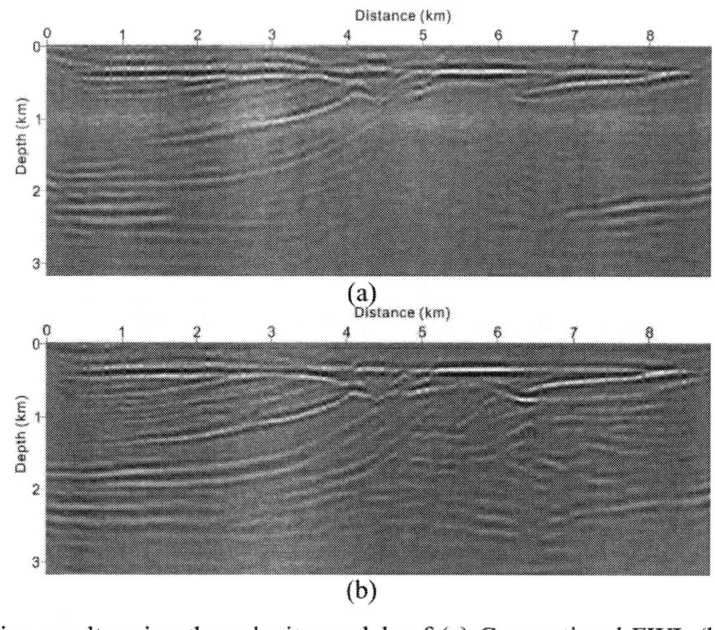


**Figure 3.** Migration results using the velocity models of (a) Conventional FWI; (b) multi-scale FWI method.

## 4. Conclusion

Cycle skipping is one of the main problems hindering FWI from practical application. To reduce the dependence on the starting model, we proposed a new multi-scale FWI method combining with multi-scale characteristic of Curvelet transform. FWI goes from low frequency to high frequency to update the velocity. And at the different stages of FWI, select different scales of observed seismic data involved in the inversion according to the use of frequency. Thus simulated data can be better matched with observed data, and reduce the dependence on starting model of FWI. Numerical examples on the

Marmousi model show that the new multi-scale method can mitigate the effect of cycle skipping problem and obtain a better result with a bad starting model than the conventional FWI method.

**Acknowledgments**
We would like to thank Scientific Research Foundation of Hebei GEO University (BQ 2017029) and Science and Technology Research Project of Hebei Higher Education Institution (QN 2018242) for the funding support.

**References**
[1] Lailly P 1983 the seismic inverse problem as a sequence of before stack migrations: Conference on inverse scattering: theory and application *Society for Industrial and Applied Mathematics, Philadelphia* 206-220.
[2] Pratt R 1999 Seismic waveform inversion in the frequency domain Part 1 Theory and verification in a physical scale model *Geophysics* **64 (3)** 888-901.
[3] Bunks C, Saleck F M and Zaleski S 1995 Multiscale seismic waveform inversion *Geophysics* **60(5)** 1457-1473.
[4] Brossier R, Operto S and Virieux J 2010 which data residual norm for robust elastic frequency-domain full waveform inversion *Geophysics* **75(3)** R37-R46.

IOP Conf. Series: Materials Science and Engineering 472 (2019) 012084 doi:10.1088/1757-899X/472/1/012084

# Improvement of Gas Drainage Efficiency of Low-Permeability Coal Seams Using Ultra-High Pressure Hydraulic Cutting Technique

**Yongjiang Zhang**[1,2]

1. State Key Laboratory of the Gas Disaster Detecting, Preventing and Emergency Controlling, Chongqing 400037, China;
2. China Coal Technology and Engineering Group Chongqing Research Institute, Chongqing 400039, China
E-mail: zhangyj_1026@163.com

**Abstract.** With the increase in the coal mining depth, the dynamic disasters of deep coal seam under the combined effects of high crustrals stress and high gas pressure seriously restrict the safety production of the coal mines. Moreover, it is very difficult to lower the gas content of the low-permeability coal seams. Even though numerous boreholes have been drilled inside the coal seam and gas drainage have been performed for a long period. High-pressure water jet cutting technology can effectively increase the exposure area of coal seams, improve the gas flow state in coal seams, so that the coal can be evenly and fully discharged in a certain range, and then the goal of rapid discharge and permeability enhancement can be achieved. After cutting the coal seam between 1m and 2m interval , a uniform pressure relief space is formed inside the boreholes to increase the permeability of the coal seam and improve the gas drainage efficiency The field test results shows that the equivalent diameter of boreholes with cutting is about 5 times that of boreholes without cutting. The gas drainage radius of boreholes with cutting is 2.8 times that of boreholes without cutting, which significantiy shortens the gas drainage time of the coal lane strip, and then attributes to relizing the rapid and safe excavation of coal seam roadway. This technique can provide reference significance for the modification of low-permeabilety coal seans in the similar condition area.

## 1. Introduction

The deep coal seams in Chinese coal mines show remarkable characteristics of "high stress, high pore pressure, high temperature and low permeability". At present, for coal seams without the protective layer mining conditions, the prevention and control of coal and gas outburst and coal rock gas dynamic disasters mainly use ground drilling and coal mine underground pre-pumping, which is mainly divided into straight drilling and penetrating drilling. Two ways. Ground drilling and mining technology has the advantages of large aperture, high extraction rate, large extraction radius, and good ground construction conditions, but the construction cost is high. Compared with drilling and drilling on the ground, the coal mine underground drilling and penetrating drilling are easier to implement, but most areas under the coal mine are because of the small diameter of drilling and the low permeability of coal seams. The gas volume is small and the gas concentration is low; increasing the diameter of drilling, improving the permeability of coal seams and reducing the stress state of coal strata is one of the commonly used methods to improve the effect of this measure. High-pressure hydraulic drilling cutting and unloading pressure enhancement technology, which is powered by water, scour and strip coal around boreholes and increase the number of cracks in coal, can greatly improve the flow state of

IOP Conf. Series: Materials Science and Engineering 472 (2019) 012084 doi:10.1088/1757-899X/472/1/012084

gas in coal seams and create favorable conditions for gas extraction. It can not only weaken or eliminate outstanding forces, but also greatly change the physical and mechanical properties of outstanding coal seams, so as to relieve pressure and prevent protrusions. It is an important development direction for coal mine to improve gas permeability and gas release ability.

## 2. Processof Ultra-High Pressure Hydraulic Cutting

In the case of soft coal seam, because of the soft coal seam, there is a common situation of drilling hole, and drilling hole causes the flow mode of gas extraction to change. From the radial gas flow that originally used drilling as the center line to the ball flow with drilling coal as the center of the ball, the drilling effect is affected, and the deep part of the coal seam is basically not affected by pumping. The two factors of low permeability and cave-hole in coal seam make it difficult for conventional drilling methods in soft and thick protruding coal seams to play a role.

According to the characteristics of soft coal seam through drilling, the best effect of strengthening pumping is the high pressure water discharge and increase penetration technology. The technique of water jet pressure reduction is mainly used to cut and flush the coal around the borehole to form a large area of pressure relief and penetration in the borehole to achieve the purpose of increasing the permeability.

The equipment can not only solve problem of permeability improved in layer-through borehole, but also solve the difficult problem of large area pressure releasing and permeability improved in bedding borehole. Schematic diagram of process of ultra-high pressure hydraulic cutting is shown in Fig 1.

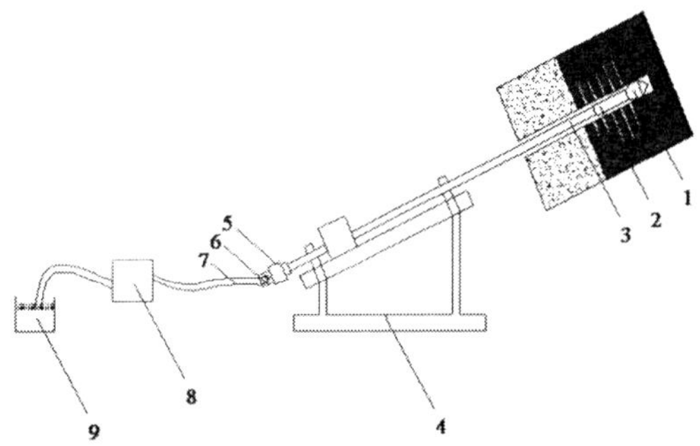


1-diamond bit for hydraulic cutting; 2- high / low voltage converter; 3-shallow helix drilling rod for hydraulic cutting; 4- drilling rig; 5-ultra-high rotating mio; 6-nipple; 7- ultra-high rubber tube; 8-ultra-high clean water pump; 9- water tank

**Figure 1.** Schematic diagram of ultra-high pressure hydraulic cutting

Technology and equipment is not only suitable for pressure releasing and permeability improved for bedding borehole of working face, layer-through borehole and uncovering coal in cross-cut, and rock burst prevention under high ground stress, high gas and low permeability, but also suitable for ultra-high pressure cutting in roof and working face roof caving partial. Cutting depth in bedding borehole can attain more than 80~100m, which can attain more than 80~120m in layer-through borehole. Cutting radius can attain 1.0~1.5m.

## 3. Application of Ultra-High Pressure Hydraulic Cutting

### *3.1. Basic Situation of Test Area*

Cutting test area of 1# coal seam is bulk mainly and flake little with developed fracture. Thickness of coal is 3.54~7.7m and the average is 6.6m. Gas content of 1# coal seam is 5.8~9.2m3/t and gas pressure is 0.8~1.4MPa, consistent coefficient (f) of coal is 0.5~0.9 and buried depth is 820m.

Using ultra-high pressure hydraulic cutting on 1415A tail entry bed plate tunnel to 1# coal roadway strip interval, cutting strike and inclination distance is 20m and 10m, type of bit is Φ113mm. Plan and profile of ultra-high pressure hydraulic cutting drilling are shown as Fig2 and Fig3.

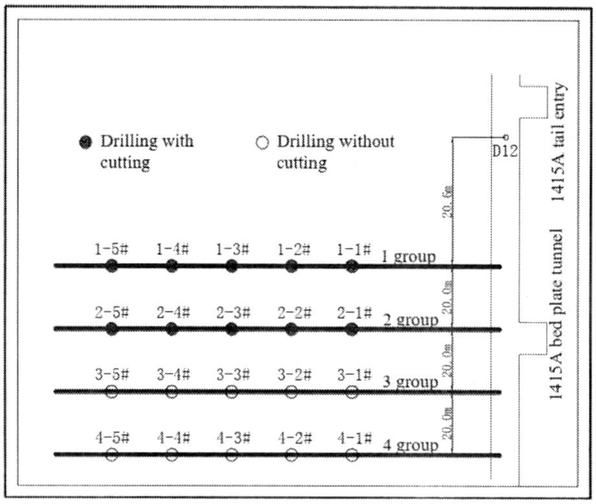


**Figure 2.** Plan of ultra-high pressure hydraulic cutting drilling

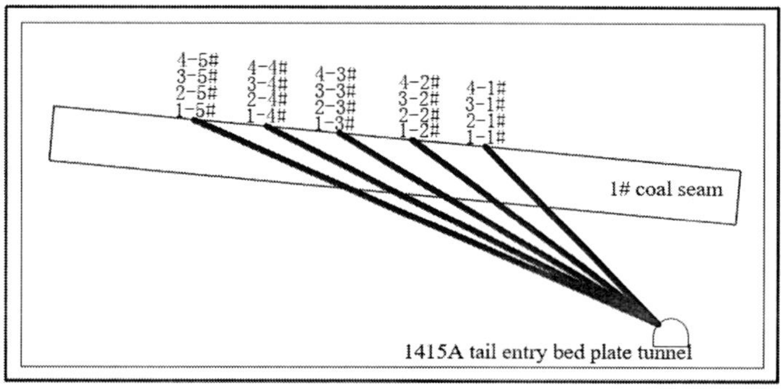


**Figure 3.** Profile of ultra-high pressure hydraulic cutting drilling

### *3.2. Effect Analysis for Pressure Releasing and Permeability Improved using Ultra-highPressure Hydraulic Cutting*

Comparative analysis of drainage effect between drilling with cutting and drilling without cutting is shown as Fig4 to Fig7.

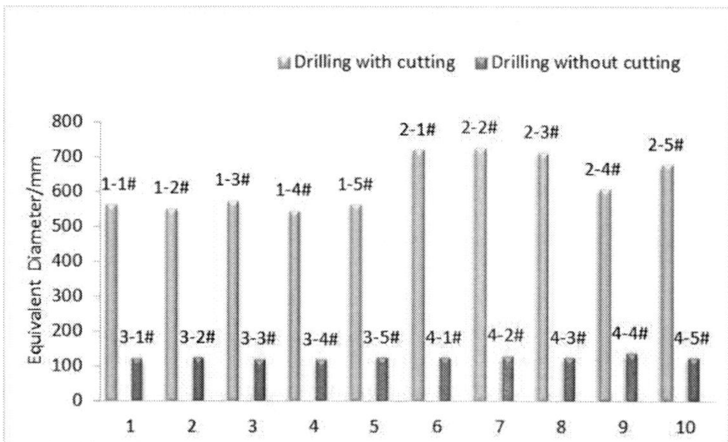


**Figure 4.** Contrast diagram of equivalent diameter between drilling with cutting and drilling without cutting

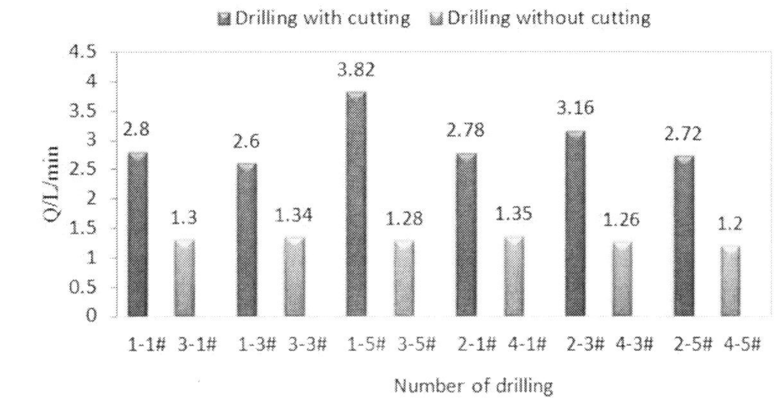


**Figure 5.** Contrast diagram of gas content between drilling with cutting and drilling without cutting

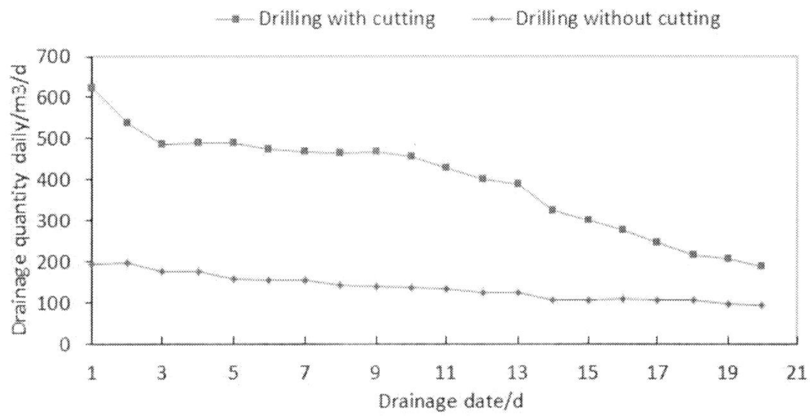


**Figure 6.** Contrast diagram of drainage quantity daily between drilling with cutting and drilling without cutting

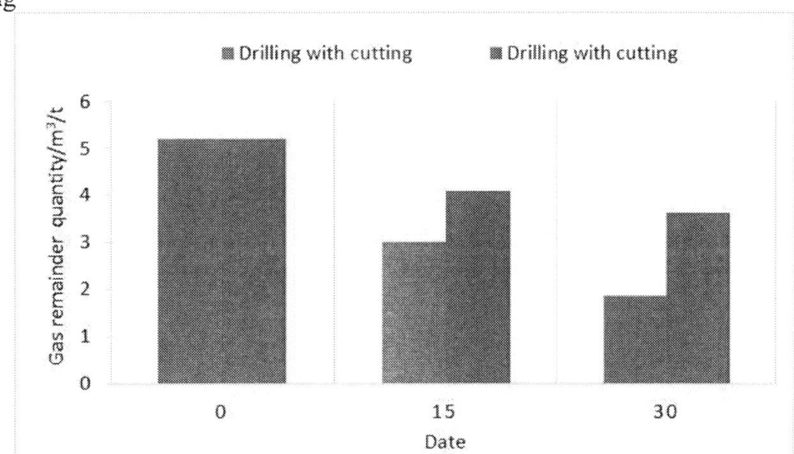


**Figure 7.** Contrast diagram of gas remainder quantity under the same drainage condition between drilling with cutting and drilling without cutting

Analyze gas drainage concentration and scalar comparatively on 1415A tail entry bed plate tunnel between drilling with cutting and drilling without cutting:

①After using ultra-high pressure hydraulic cutting, exposed area of coal seam was increased and equivalent diameter of drilling was enlarged. The average of equivalent diameter of drilling with cutting is 627mm and it of drilling without cutting is 124mm, equivalent diameter of drilling with cutting is 5 times that of drilling without cutting, gas flow state is improved in coal seam to create advantage for gas emission, and effect of cutting is significant.

②After using ultra-high pressure hydraulic cutting, the average of initial speed from borehole gas emission with cutting is 2.98L/min, and it of drilling without cutting is 1.29L/min, initial speed from borehole gas emission with cutting is 2.3 times that of drilling without cutting.

③After using ultra-high pressure hydraulic cutting, coal pressure is released full, gas drainage quantity during 20d of drilling with cutting is 7936m3 and it of drilling without cutting is 2645m3, gas drainage quantity of drilling with cutting is 3 times that of drilling without cutting and gas drainage radius is 2.8 times, which can reduce engineering quantity of drainage drilling.

④After using ultra-high pressure hydraulic cutting, gas content of drilling with cutting decreases about 38% and it of drilling without cutting decreases about 21% when draining 15d, gas content of drilling with cutting decreases about 53% and it of drilling without cutting decreases about 30% when draining 30d. Decreasing range of gas content of drilling with cutting is 1.8 times than that of drilling without cutting under the same condition, which can shorten gas pre-drainage time effectively.

## 4. Conclusions

(1) Using high pressure water jet to increase the exposed area and improve the gas flow state is an effective way to improve the gas extraction efficiency of coal seam.

(2) Using ultra-high pressure hydraulic cutting in 1415A tail entry bed plate tunnel, fractures of 1# coal seam was effectively increased, exposed area of coal seam was enlarged, desorption and flow of gas were improved and time of coal roadway strip gas pre-drainage is shortened by 1/3.

(3)Ultra-high pressure hydraulic cutting can make pressure in coal seam release uniform, change stress condition of coal seam, eliminate risks of high ground stress and gas dynamic phenomenon during coal seam tunnel excavating, which can realize fast and secure excavating.

## Acknowledgements

This work was financially supported by the national key research and development program of China (2017YFC0804206), Chongqing science and technology innovation project of social undertakings and people's livelihood guarantee (cstc2017shmsA90008); Key Technology Projects for Prevention and Control of Major Accidents in Safe Production (yangqi-0006-2017 AQ); Special fund for science and technology innovation and venture capital of China Coal Science and Technology Group Co., Ltd. (2018MS011).

## Reference

[1] Yuan Liang, Lin Baiquan and Yang Wei (2015) "Research Progress and Development Direction of Gas Control with Mine Hydraulic Technology in China Coal Mine,"Coal Science and Technology.J., (01): 45-49.

[2] Li Xiaohong, Wang Xiaochuan and Kang Yong, et al (2014) "Energy Characteristic and Dissipation in Transient Process of Hydraulic Cutting Seams System in Coal Seam," Journal of China Coal Society. J., (08): 1404-1408.

[3] Meng Xianzheng, Cao Jianjun and He Qin, et al (2009) "Study and Application on Hydraulic Highly Efficient Outburst Prevention Integration Technology," Mining Safety and Environmental Protection. J., (S1): 49-53.

[4] Chi Mingbo (2015) "Experimental Study on Hydraulic Cutting Seam Prevention and Cure of Xinzhouyao Mines Rock Burst,"Shan Xi: Tai Yuan University of Technology. D.

[5] Zheng Chunshan, Lin Baiquanand Yang Wei, et al (2014) "Hole-spraying Mechanism of Hydraulic Slotting Drilling and the Influence of Slotting Mode," Safety in Coal Mines. J., (01): 5-8+12.

[6] Yu Liya (2016) "Hydraulic Cutting Technology Control in High Gassy Seam,".Inner Mongolia Coal Economy. J., (01): 90-91+128.

[7] Wang Yong, Yang Bi andYao Jianda (2014)"Numerical Simulation of Process of Hydraulic Fracturing of Slotted Drilling," Coal Technology. J., (09): 45-47.
[8] Liang Yunpei (2007) "Study on the Mechanism of Coal Failure By Drilling of High Pressure Water Jetting," Shan Dong: Shandong: Shandong University Of Science And Technology. D.

IOP Conf. Series: Materials Science and Engineering 472 (2019) 012085 doi:10.1088/1757-899X/472/1/012085

# Preparation of GO-Based Iron-Containing and Adsorption of Cr in Water

**Qiaobing Wang, Tao Feng and Shuo Xiong**

College of Resources and Environment Engineering, Wuhan University of Science and Technology, Wuhan, Hubei 430081, China
Email: 514348251@qq.com

**Abstract.** The presence of Cr (VI) in the aqueous media is known to cause severe health issues in both humans and animals. Many technologies and especially adsorbents have been applied for its removal. In this work, Graphene oxide (GO) was prepared using the Hummers' method. A facile method was adopted to synthesize GO-Based Iron-Containing (GO-Fe) through the iron salt impregnation method. The GO-Fe was firstly applied as adsorbents for efficient elimination of Cr (VI) ions. Batch adsorption experiments showed that the adsorption of Cr (VI) on GO-Fe followed the pseudo-second-order kinetic model and Langmuir isotherm well, respectively. The absorption experiments showed that the GO-Fe have the largest adsorption capacity for Cr (VI) with the concentrations of $FeCl_2$ is 0.5mol/L. The GO-Fe exhibited excellent adsorption capacity of 31.24mg/g at pH 5.0, the contact time was 180 min, the dosage of GO was 0.5g/L and the initial concentrations of Cr (VI) was 40mg/L. This work highlighted the application of GO-Fe as novel and promising materials in the efficient elimination of Cr (VI) from contaminated water and industrial effluents in the environmental pollution management.

## 1. Introduction

Cr (VI) has been commonly used in a number of industrial processes, such as leather tanning, electroplating, mining and printing and dyeing [1]. It is quite dangerous to human and living things, leading to a variety of issues in the spectrum of health, include allergies, ulcers, and dermatitis. Cancer may be caused under certain conditions and may even lead to death in severe cases [2]. In order to reduce the risks associated with metal pollutants, it is essential that wastewaters be treated before being released to the environment [3]. Compared with other treatment technologies such as chemical precipitation [4], photocatalytic method [5] and ion exchange resin method [6], the adsorption procedure has attracted extensive attention due to its simplicity and low cost [7]. Numerous kinds of adsorbents such as activated carbon, fly ash and agricultural byproducts have been applied for the elimination of heavy metal ions from wastewater.

On the other hand, these materials have the disadvantage that the lack of model predictability and their incapability to elimination the contaminants in trace concentrations. Lack of specific surface area and the specific heavy metal adsorption capacity has also become the weakness of these materials. Therefore, the study on to find a new type of materials is of great concern.

Graphene oxide (GO) has excellent properties, the surface is rich in functional groups such as carboxyl, hydroxyl, ketone, and epoxy groups, which are favorable for the adsorption of heavy metal ions and provide a large number of active sites for functional modification [8]. As an ideal planar two-dimensional structure, a great specific surface area ($2630m^2 \cdot g^{-1}$), determined from the monolayer carbon structure [9]. In this sense, GO has great adsorption capacity for positively charged heavy metal ions in water, become the best nonporous absorbent. However, since the GO is negatively charged in aqueous solution, the adsorption of heavy metal such as cationic heavy metal is better, and the adsorption of Cr (VI) and other anionic pollutants in water is poor. Therefore, it is necessary to improve the adsorption performance through modified methods.

At present, it is a hot research direction to modify the iron content of the material by the iron salt impregnation method. A lot of work has been done about the synthesis of nanocomposites by iron salt impregnation. For example, Xiao *et al.*[ 10] use iron-containing salt impregnation method for preparing iron-loaded activated carbon for the elimination of As (III) from the aquatic environment. Li *et al.* [11] prepared an iron-containing activated carbon IOCACAC for Cr (VI) Adsorption in water. In continuation of the previous study, GO-Fe is prepared by iron-salt impregnation method with a simple method, low cost, and no secondary pollution. The adsorption and mechanism of Cr (VI) in aqueous solution with GO-Fe is discussed. On the one hand, the application of graphene oxide is enlarged. The scope, on the other hand, provides basic data and theoretical research on the treatment of chromium pollution.

## 2. Materials and Methods

### *2.1. Synthesis of Graphene Oxide*

All chemicals were of analytical grade and were used without any further purification. Graphene Oxide (GO) was synthesized according to the Hummers'method [12]. Flake graphite (3.0g) and $NaNO_3$ (0.75g) was dissolved in 98% concentrated $H_2SO_4$ (70ml) and then cooled in an ice bath (4°C for 30min), stirred for 30 minutes to disperse it. 9g of $KMnO_4$ was added to a three-necked flask and stirred for 2h, after the solution is purple-green, keep the bath temperature controlled at about 35°C and stirred the reaction for 1 hour, ultra-pure water (140ml) was added to solution then heated to a bath temperature (98°C for 40min), when the solution is golden yellow after hydrolysis, added appropriate amount of $H_2O_2$ until no bubbles produced, filtered, and washed with 5% HCl, ultrapure water and ethanol for several times until the solution neutral and then dried at 60°C for 24 h. The sample was eventually well maintained in a desiccator for future use.

### *2.2. Synthesis of GO-Based Iron-Containing*

0.1g of GO was added to a certain concentration (0.005ml/L, 0.01mol/L,0.025mol/L,0.05mol/L, 0.1mol/L) of $FeCl_2$ aqueous solution (100ml) and dipped for 24h at room temperature. A certain amount of NaClO solution was added to a mixture every 8h then put it in a 70°C electric heating oven to dry it for 24h, cooled to room temperature. Therefore, washed until there is no visible color and it is not red with potassium thiocyanate solution then continue to dry then get GO-Fe after grinding. The samples were characterized by SEM-EDS, FTIR and Zeta potentials.

### *2.3. Absorption Experiment*

In this experiment, the method of static adsorption was adopted. At room temperature, a certain concentration of heavy metal Cr (VI) solution was placed in a conical flask. The pH value of the solution was adjusted with sodium hydroxide and hydrochloric acid. Put a certain mass of adsorbent material in the solution, placed on a shaker for a certain period of time, and after the reaction was completed, the remaining heavy metal chromium ions in the solution were measured by flame atomic absorption spectrometry (FAAS). The adsorption capacity of the heavy metal ions Cr (VI) in the aqueous solution are calculated by the following formula (1).

$$Q_e = \frac{(C_0 - C_e) \times V}{W} \tag{1}$$

Where $C_0$ and $C_e$ are the initial and equilibrium Cr (VI) concentrations (mg/g). $Q_e$ is the quantity of Cr (mg/g) absorbed on unit mass of adsorbent at equilibrium. $V$ is the volume of the solution(ml), and $W$ is the mass of the adsorbent(g).

## 3. Results and Discussion

### *3.1. Material Characterization*

Fig.1shows the infrared spectrum of flake graphite powder, GO and GO-Fe. It can be seen from Fig.1, the absorption peak at 3444cm$^{-1}$ in the infrared spectrum of GO is the stretching vibration absorption peak of O-H. 1740cm$^{-1}$ is the stretching vibration absorption peak of COOH and C=O, 1645cm$^{-1}$ is the bending vibration absorption peak of C=C, and the obtained GO infrared spectrum results are similar to those reported in previous literature [13], it shows that the synthesized GO molecules contain oxygen groups such as carboxyl groups and hydroxyl groups. The newly emerged absorption peak at 590cm$^{-1}$ of the infrared spectral data of GO-Fe is the absorption peak of Fe-O of trivalent iron, this shows that Fe was successfully inserted into the graphene oxide and forming a new group.

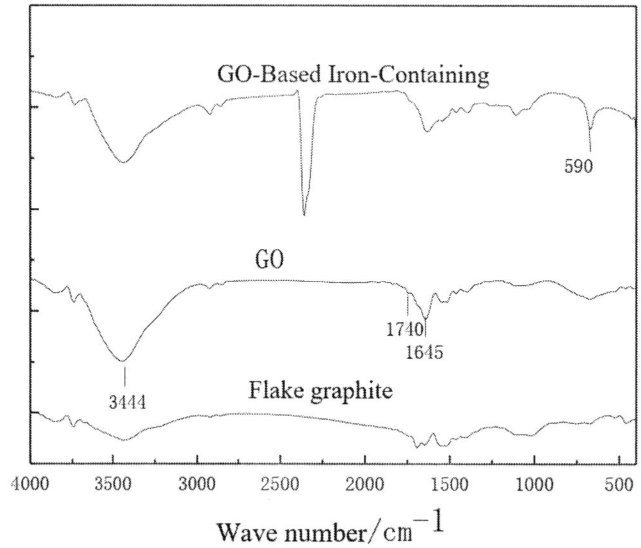


**Figure 1.** Infrared spectra of GO, GO-Fe and flake graphite

Fig.2 shows the SEM-EDS of GO and GO-Fe, spectral analysis of a little spectrum1 on the surface of the GO body indicated in Fig.2 (a). The result is as shown, it was found that GO is mainly composed of two elements: C and O, spectral analysis of the Spectrum3 of the surface of the GO-Fe body indicated in Fig.2 (b), the result is as shown; it was found that the content of GO-Fe is mainly composed of C, O, and Fe. The result of the insertion of iron into graphene oxide is consistent in the infrared spectra above.

Fig.3 shows Zeta potentials for GO and GO-Fe. It can be seen from Fig.3 that the Zeta potential is around -20mv at pH<7, when pH>7, the potential decreases slowly, when pH=10, its potential value is -47mv. When the pH is less than 7, the Zeta potential of GO-Fe is about 35mv, when the pH increases to 7, the potential value is 13mv, later when pH>7, the potential value of GO-Fe becomes negative, at pH=10, the potential value is -18mv. From this, it can be seen that GO is negatively charged in aqueous solution, and GO is positively charged under acidic conditions and negatively charged under alkaline conditions.

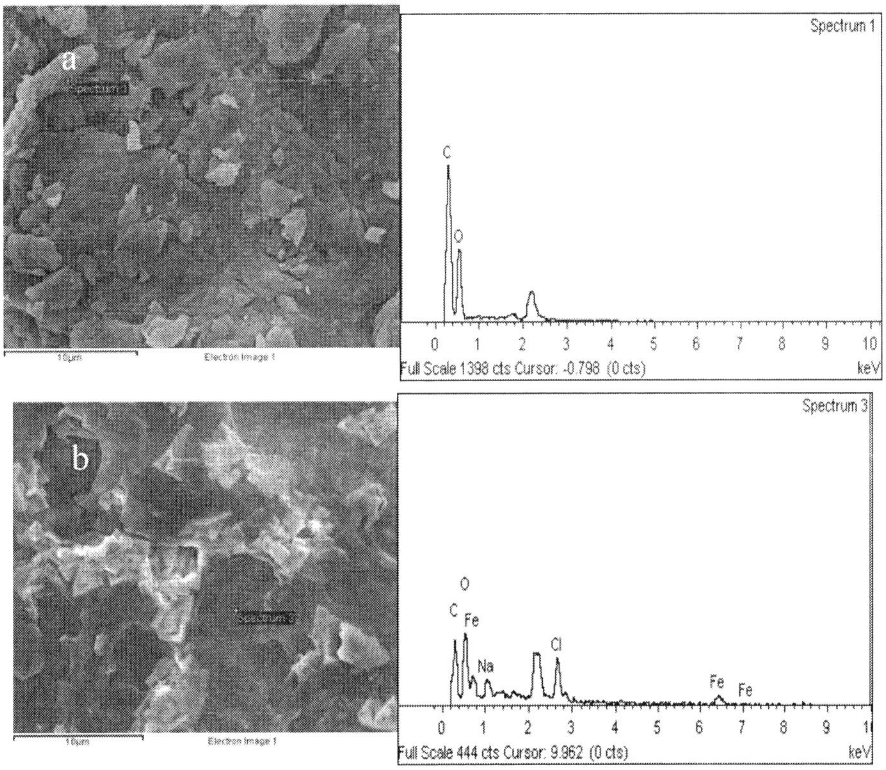


**Figure 2.** SEM-EDS of GO (a) and GO-Fe (b)

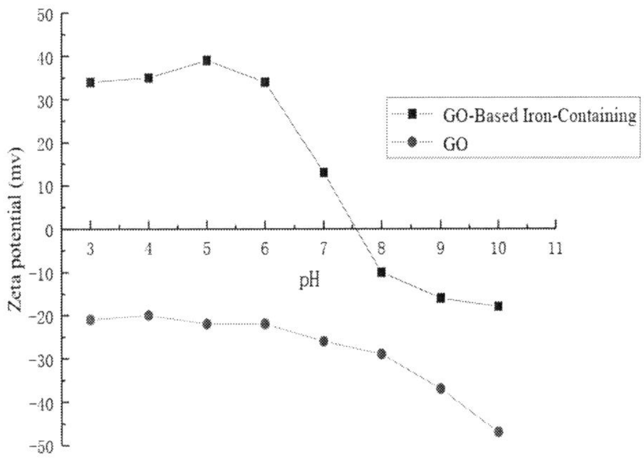


**Figure 3.** Zeta potentials of GO(a) and GO-Based Iron-Containing (b)

### *3.2. Effect of Iron Content on Adsorption of Cr(VI)*

GO(0.1g) was added to 100ml of a certain concentration of the $FeCl_2$ aqueous solution, prepared GO-Fe with different iron content, Cr(VI)adsorption capacity as an indicator, the effect of ferrous chloride concentration on the adsorption performance of GO-Fe was investigated. The initial concentration of Cr(VI)in aqueous solution is 10mg/L and adjust solution pH to 5.0 then shock absorption after 180min, using FAAS to test the concentration of Cr(VI)in solution, the adsorption capacity of GO-Fe with different iron contents was calculated. The results are shown in Fig.4.

At acidic conditions, the main species of Cr (VI) was $HCrO_4^-$ and its concentration kept almost unchanged. With the solution pH increased, $HCrO_4^-$ was then transformed into $CrO_4^{2-}$. It can be seen from Fig.4, as the concentration of $FeCl_2$ aqueous solution continues to increase, the adsorption capacity of GO-Fe gradually increases, when the concentration of $FeCl_2$ is increased to 0.05mol/L, the adsorption capacity of GO-Fe is maximized, when the concentration of $FeCl_2$ gradually increases, the adsorption capacity tends to be stable. This may be due to the limited active sites on graphene oxide, with the increasing concentration of the aqueous $FeCl_2$ solution, the active sites on graphene oxide are continuously occupied and reduced, after reaching the saturation state, and the adsorption effect tends to be stable. With the addition of $FeCl_2$, the adsorption capacity of GO-Fe is significantly increased, when the concentration of $FeCl_2$ was increased to 0.05mol/L, the adsorption capacity was 15.71mg/g. The reason for the analysis from Fig.1and Fig.3 is that GO contains a large number of oxygen-containing functional groups, its Zeta potential is negative, negatively charged in aqueous solution, Cr(VI) electrostatic repulsion, the Zeta potential of the GO-Fe is positive, positive charge in aqueous solution, and electrostatic attraction of Cr(VI). At the same time, GO-Fe contains Fe-O functional group, complexation with Cr (VI), therefore, the adsorption capacity of GO-Fe is significantly improved compared to GO. This experiment selected the best $FeCl_2$ concentration of 0.05mol/L.

### *3.3. Effect of pH on Cr(VI)Adsorption*

The mechanism of the synthesis and adsorption of GO-Fe is as follows: Added $FeCl_2$ to the graphene oxide solution then added NaClO to oxidize $Fe^{3+}$, it reacts with functional groups such as carboxyl and hydroxyl groups in graphene oxide. From the infrared data in Fig.5, a new group Fe-OH is formed in the GO-Fe. The Fe-OH group and the $HCrO_4^-$ in the aqueous solution adsorb Cr (VI) on the GO-Fe surface through complexation. At present, there is no unified conclusion about the model of Fe-OH group and Cr (VI)complexation. However, due to the easy desorption of adsorbed Cr (VI), it shows that the binding force between them is relatively weak and belongs to the characteristics of complexation in the outer region. Therefore, many researchers [14] believe that the surface complexation of Cr (VI) and Fe-OH groups belongs to the outer region complexation. The effect of pH on the adsorption of Cr (VI) on GO-Fe was investigated in the pH range of 2.0-6.0. The initial concentration of Cr (VI) in aqueous solution is 40mg/L, added adsorbent mass 50mg, the volume of the solution is 50ml, use hydrochloric acid to adjust pH, shock absorption after 180min at room temperature and determination of Cr (VI) concentration in solution by FAAS. The adsorption capacity of GO-Fe at different pH was calculated and the result is shown in Fig.5.

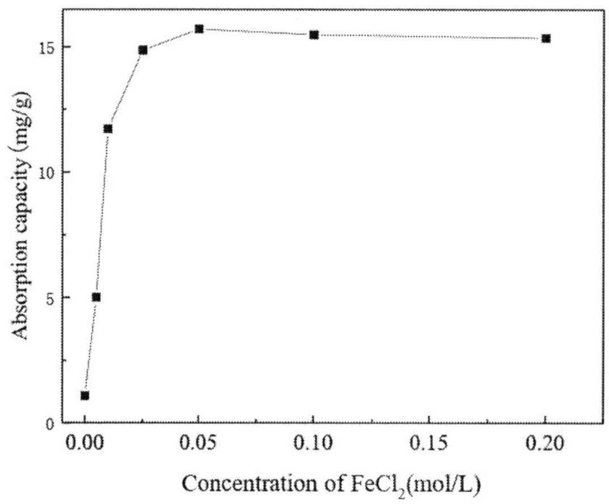


**Figure 4.** Effect of $FeCl_2$ concentration on Cr (VI) adsorption

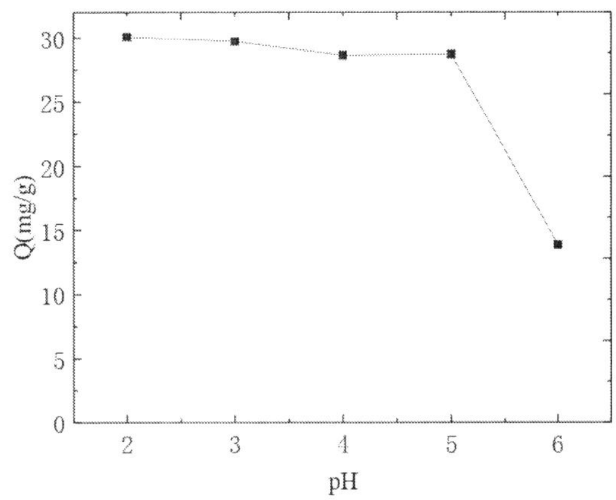


**Figure 5.** Effect of pH on the Adsorption of Cr (VI)

It can be seen from Fig.5, as the pH increases, the adsorption capacity of GO-Fe decreases, when the pH is 2-5, the adsorption capacity of GO-Fe basically stabilizes, when the pH increased to 6, the adsorption capacity of GO-Fe decreased. The possible reason is, at pH between 2-5, the Cr (VI) in the aqueous solution exists mainly in the form of $Cr_2O_7^{2-}$ and $HCrO_4^-$, with little change in concentration; When the pH is 5-7, the concentration of $Cr_2O_7^{2-}$ and $HCrO_4^-$ decreases and the concentration of CrO42- increases gradually; When the pH>7, the Cr (VI) in the aqueous solution exists mainly in the form of $CrO_4^{2-}$. When the pH is between 2 and 5, the concentration of $Cr_2O_7^{2-}$ and $HCrO_4^-$ remain almost the same. At the same time, we can see from Fig.3 that the Zeta potential of GO-Fe is basically unchanged, so the adsorption capacity of GO-Fe tends to be stable. When pH is 6, the concentration of $Cr_2O_7^{2-}$ and $HCrO_4^-$ decreases, the concentration of $CrO_4^{2-}$ increases, and the adsorption capacity of iron-bearing GO on $CrO_4^{2-}$ is relatively weak, so the adsorption capacity of iron-loaded GO decreases. At the same time, since the initial pH of the simulated Cr (VI) solution is close to 5.0, the optimum pH for the experiment is 5.0.

### *3.4. Adsorption Isotherm Analysis*

In order to study the adsorption isotherms of Cr (VI) on GO-Fe, different adsorption isotherm models were used for fitting. Langmuir and Freundlich isotherm adsorption models are two commonly used methods for the analysis of adsorption data. Langmuir isotherm adsorption line assumes adsorption is monolayer adsorption, Freundlich assumes that the adsorption is a multi-layer adsorption because the adsorbent adsorption site and the adsorbed metal ions in the solution exist in various forms. The isotherm constants for each model were calculated from the linearized Eqs. (2)- (4)
The Langmuir adsorption model is shown below:

$$C_e/Q_e = C_e/Q_0 + 1/(Q_0 b) \tag{2}$$

The Freundlich adsorption model equation is as follows:

$$lnQ_e = (1/n)lnC_e + lnk_f \tag{3}$$

Langmuir adsorption constant $R_L$ is an important parameter used to measure whether the adsorption system is effective if $R_L$ is between 0 and 1, it means that the reaction system is effective and the reaction is easy to proceed. The formula is as follows:

$$R_L=1/(1+bC_0) \tag{4}$$

Here, $C_e$ (mg/L) is the concentration of the adsorbate in solution after equilibrium is reached; $Q_e$ (mg/g) represents the equilibrium adsorption capacity of the adsorbent; $Q_0$ (mg/g) represents the maximum adsorption capacity of the adsorbent; $b$ (g/mg) indicates the adsorption strength (g/mg); $K$ (mg/g) represents the adsorption performance constant of the adsorbent; $1/n$ denotes the Freundlich constant related to the adsorption strength; $C_0$ (mg/L) represents the initial concentration of the solution.

**Table 1.** GO-Fe Adsorption Isothermal Fitting Parameters

| Adsorption materials | Freundlich adsorption model | | | Langmuir adsorption model | | | |
|---|---|---|---|---|---|---|---|
| | N | K | $R^2$ | $q_0$ | B | $R^2$ | $R_L$ |
| GO-Fe | 3.305 | 12.16 | 0.96 | 33.00 | 0.543 | 0.9974 | 0.035~0.27 |

From Table 1, the correlation of the Langmuir adsorption model for chromium (VI) adsorption on GO-Fe is 0.9974. The Freundlich adsorption model has a correlation of 0.96. The maximum adsorption capacity of Cr (VI) in wastewater is 33mg/g. Therefore, the adsorption of hexavalent chromium on the GO-Fe is a mono-layer adsorption on the homogeneous surface and the Langmuir adsorption constant $R_L$ obtained is between 0 and 1, which indicates that the adsorption of Cr (VI) by GO-Fe is easy to perform.

### *3.5. Mechanism Analysis*

Since Cr (VI) has a high positive charge and a small radius, there is no simple presence of $Cr^{6+}$ ions in solution. Cr (VI) exists mainly in the form of $Cr_2O_7^{2-}$, $HCrO_4^-$ and $CrO_4^{2-}$ in water. Their conversions are shown in Equations (5) and (6) below.

$$HCrO_4^- \rightarrow H^+ + CrO_4^{2-} \tag{5}$$

$$2HCrO_4^- \rightarrow H_2O + CrO_7^{2-} \tag{6}$$

It can be seen from the Zeta potential data in Fig.4. When the pH is between 2 and 5, the Zeta potential of GO is about -20mv. With the increase in pH, the Zeta potential decreases. When the pH is 10, the Zeta potential is -47mv. Because $Cr_2O_7^{2-}$ and $HCrO_4^-$ are both negatively charged and GO electrostatically repulsive, GO has poor adsorption of Cr (VI). When GO is modified by iron, when the pH is between 2 and 5, the Zeta potential of GO-Fe is about 35mv, and it is electrostatically attracted by $Cr_2O_7^{2-}$ and $HCrO_4^-$ and adsorbs Cr(VI)on the GO-Fe surface.

## 4. Conclusion

Under acidic conditions, the GO-Fe contains Fe-O groups and the surface is positively charged .The zeta potential of the iron-loaded GO is positive and $Cr_2O_7^{2-}$ and $HCrO_4^-$ in the aqueous solution are adsorbed on the surface of the material by electrostatic action.The GO-Fe exhibited excellent adsorption capacity of 31.24mg/g at pH 5.0, the concentrations of $FeCl_2$ is 0.5mol/L, the contact time was 180 min, the dosage of GO was 0.5g/L and the initial concentrations of Cr(VI) was 40mg/L.

## References

[1] Owlad M, Aroua M K, Daud WAW, et al. Removal of hexavalent chromium-contaminated water and wastewater: a review [J]. Water, Air, and Soil Pollution. 2009; 200(1-4): 59-77.
[2] Renfeng Huang, Xiaoguo Ma, Xin Li, Lihui Guo, Xiaowen Xie, Mengyuan Zhang, Jing Li. A novel ion-imprinted polymer based on graphene oxide-mesoporous silica nanosheet for fast and efficient removal of chromium (VI) from aqueous solution [J]. Journal of Colloid and Interface Science 514(2018)544-553.
[3] L Cui, Y. Wang, L. Gao, L. Hu, Q. Wei, B. Du, Removal of Hg (II) from aqueous solution by resin loaded magnetic β-cyclodextrin bead and graphene oxide sheet: synthesis, adsorption mechanism and separation properties, J. Colloid Interface Sci. 456(2015) 42-49.
[4] M. Erdem, F. Tumen, Chromium removal from aqueous solution by the ferrite process, J. Hazard. Mater. 109 (2004) 71–77.
[5] S. Luo, F. Qin, Y.A. Ming, H.P. Zhao, Y.L. Liu, R. Chen, Fabrication uniform hollow Bi2S3 nanospheres via Kirkendall effect for photocatalytic reduction of Cr(VI)in electroplating industry wastewater, J. Hazard. Mater. 340 (2017) 253–262.

[6] F. Gode, E. Pehlivan, Removal of Cr (VI) from aqueous solution by two Lewatitanion exchange resins, J. Hazard. Mater. 119 (2005) 175–182.
[7] Tayebeh Esfandiyari, Navid Nasirizadeh, Mohammad Dehghani, Mohammad Hassan Ehrampoosh, Graphene oxide based carbon composite as adsorbent for Hg removal: preparation, characterization, kinetics and isotherm studies. Chinese Journal of Chemical Engineering 25(2017) 1170-1175.
[8] Yang D, Velamakanni A, Bozoklu G, et al. Chemical analysis of graphene oxide films after heat and chemical treatments by X-ray photoelectron and Micro-Raman spectroscopy [J]. Carbon. 2009; 47(1): 145-152.
[9] G. Zhao, J. Li, X. Ren, C. Chen, X. Wang, Few-layered graphene oxide nanosheets as superior sorbents for heavy metal ion pollution management, Environ. Sci. Technol. 45 (2011) 10454-10462.
[10] Jing Xiao, KaiXun Tian, Yi Gao. Preparation of Iron-Containing Activated Carbon Adsorbent and Study on the Performance of Arsenic Removal (III) [J]. Industrial water treatment. 2012; 32(11): 28-32.
[11] Bing Li. Removal of Cr (VI) Adsorption by Granular Activated Carbon Loaded Iron Oxide (IOCGAC) [D]: Changsha: Hunan University.
[12] W.S. Hummers, R.E. Offeman, Preparation of graphitic oxide, J. Am. Chem. Soc. 80 (1958) 1339.
[13] Guo H-L, Wang X-F, Qian Q-Y, et al. A green approach to the synthesis of graphene nanosheets [J]. ACS nano. 2009; 3(9): 2653-2659.
[14] Aoki T, Munemori M. Recovery of chromium (VI) from wastewaters with iron (III) hydroxide-I: adsorption mechanism of chromium (VI) on iron (III) hydroxide [J]. Water Research, 1982, 16(6): 793-796.

IOP Conf. Series: Materials Science and Engineering 472 (2019) 012086 doi:10.1088/1757-899X/472/1/012086

# Design of Video Image Low-Pass Filtering Process Based on FPGA

**Xiaojing Zhang and Hua Wang**

Beijing Polytechnic. Communication University of China
Email: dky_zxj@sina.com;1014690221@qq.com

**Abstract**. A real-time video low-pass filter processing system is designed and implemented in this paper through VHDL language programming, applying Altera CycloneIII series FPGA as the main control chip. The changes in display quality of video signals of different bandwidths can be observed on the monitor by changing the bandwidths of the input video signals through low-pass filtering.

## 1.Introduction

The real-time video processing has been widely applied in many fields, such as video telephony, video conferences, monitoring, communications, and proved to be one of the research hotspots in the field of image processing. Field-Programmable Gate Arrays (FRGA) allows many image processing functions to be integrated on an FPGA chip, which has the distinct advantages of good flexible programmability, powerful parallel processing. In order to observe the display quality variation of different bandwidth video signals, this paper designs and implements a video image low-passing filter processing system based on FPGA .As reducing the bandwidths of the video signals only changes the horizontal resolutions of the video images, and the vertical resolutions of the video images are determined by the scanning parameters of the TV system and basically independent of the variation of the signal bandwidths, this paper only designs horizontal filters, while the design of 2-demensional filters is not required.

## 2.Overall System Design

The hardware platform of this system is mainly consistent of the following parts: the FPGA model EP3C55F484 used for real-time low-pass filtering of video, A/D chip converting the incoming analog video signals into digital ones to input FPGA, and D/A chip converting the processed digital video signals for ADV7171 to analog video which is displayed on the monitor at last. Besides, the FPGA connects SRAM cells of two 16*bit* data lines, used to store video data. The system board is shown in Figure1.

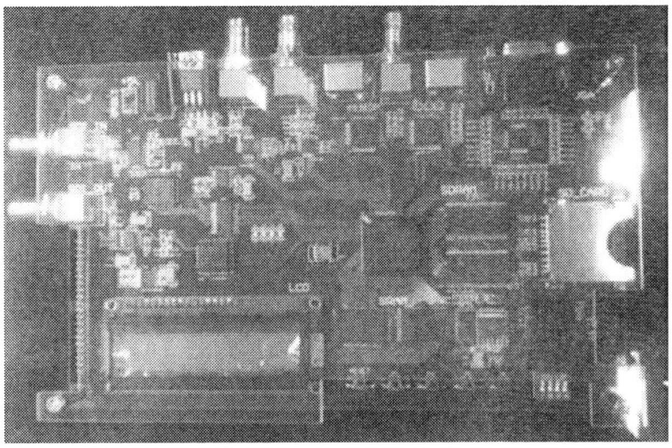

**Figure 1**. The System Board

*2.1.Signal Acquisition*

The camera is used to obtain optical images at the front end of the system and it's outputs are PAL composite video signals, completing the transition from the optical signals to the analog electrical signals. The analog electrical signals are then converted to digital signals by the TVP5150 video decoding chip. The FPGA is used to configure $I^2C$ for TVP5150 and the output format is selected as ITU-RBT.565, that is the $YCrCb$ signals of the $4:2:2$ sampling format with a clock frequency of$27MHz$.

*2.2.Field Decoding*

The complete PAL-system ITU-RBT656 data of one frame is divided into odd and even fields, the 23-311 rows are effective data of the even field, the 366-624 rows are effective data of the odd field, and the rest are the field control signals or invalid data. The first 288 bytes of each row are row control signals, with the first 4 bytes being EAV (Active Video End) signals and the last 4 bytes being SAV (Active Video Start) signals.

Both the EAV signals and the SAV signals have a 3-byte preamble 'FF, 00, 00', and the last 1 byte is 'XY'. The parity flag is 'F', the vertical blanking flag is 'V', the horizontal blanking flag is 'H' ,and the protection bits are 'P3,P2,P1,P0'. The extractions and detections of 'F','V' and 'H' are performed by continuously determining of 'FF, 00, 00' and 'XY'. If the numerical value of 'H' is 0, the signal should be SAV signal, and then judge the numerical value of 'F', if numerical value of 'F' is determined as 0,the current field is an odd field, and another case is that numerical value of 'F' is 1, then it is an even field. When the numerical value of 'F' changes from 1 to 0, it is the frame header of a frame. The next step is judging the numerical value of 'V', if it can be judged as 0, it is the field forward. Since we only process the valid data, the row is proved to be the valid data, only when the frame header is detected and the numerical values of 'H' and 'V' are 0.

*2.3.Frame Buffer*

This paper uses two blocks of SRAM for ping-pong storage to realize the conversion between processing and display,with storing in units of one frame of data. When the current frame is written to SRAM1, the previous frame stored in SRAM2 is being read.The ping-pong switch is only performed to read the data of SRAM1, after the frame of SRAM1 is stored.

The valid data in one frame are$720 \times 576$ pixels, and the write address pointer is located at $(0,78)$ at the beginning. 'FF,00' in the SAV of the valid data of the first row will be written on $(0,78)$,after the frame header is detected and the numerical values of 'H' and 'V' are judged as 0. The next '00' and 'XY' will be written on $(0,79)$,before the subsequent first valid pixel will be written on $(0,80)$.During the storage, if the numerical value of 'F' of the SAV is found to change from 0 to 1, which means that the odd field ends and the incoming data will be the even field data, the write address pointer will jump to$(1,78)$, and the coming stage is the storage of even field. When the numerical value of 'F' changes from 1 to 0, which is the new frame header, the write address pointer will reposition to (0,78), then cycle back and forth.

## 3.Video Low-Pass Filtering in the FPGA

### *3.1.The Design of Low-Pass Filter*

As the sensitivity of the human eyes to the changes of chromaticity is very low [1], this paper only performs low-pass filtering on the component of luminance signal Y. Since the clock frequency of the input signal is 27MHz and the input data is $YCbCr$ format,presenting the arrangement of a luminance signal and a chrominance signal, the pixel clock is 13.5MHz which means that the sampling frequency of the Y signal is 13.5MHz.The bandwidths of input video signals of the SD signals are generally not more than 6MHz so that this paper chooses to limit the bandwidths of the input video signals to within 3MHz, and the cutoff frequency of the corresponding digital low-pass filter is
wc = fc/fs*2pi = 3/13.5*2pi = 0.44pi

Thus a low-pass filter can be designed as shown in Figure2.

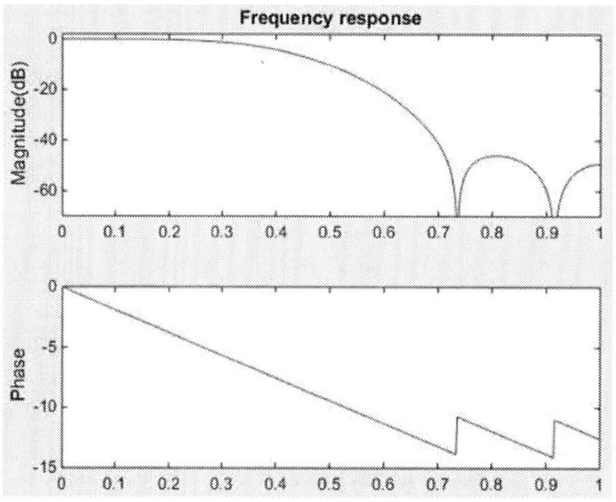


**Figure 2.** Frequency Characteristic of the Filter

### *3.2.The Low-Pass Filtering in the FPGA*

This paper selects to design and implement a 12-order FIR filter,resulting to 13 coefficients. The formula implemented by the FIR filter is

$$y(n) = \sum_{k=0}^{12} h(k)x(n-k) \tag{1}$$

'h(n)' of the formula above is the coefficient of the filter and 'x(n)' is the numerical value of the luminance Y in each row of pixels. 'y (n)' is the component of the luminance signal Y of pixels after filtering.

Filtering with this formula causes the output images to produce rightward translations relative to the input video images. For avoiding the problem of shifting the filtered images, this paper caches the input pixels before filtering them with a non-causal filter.

$$y(n) = \sum_{k=0}^{12} h(k)x(n-k+6) \tag{2}$$

As the shown equation above, the primary filtering requires 13 multiplications. Since the designed filter is a linear phase filter, with the coefficients being symmetric, a linear phase structure can be employed to improve the efficiency of calculations. The linear phase structure is shown in Figure3.

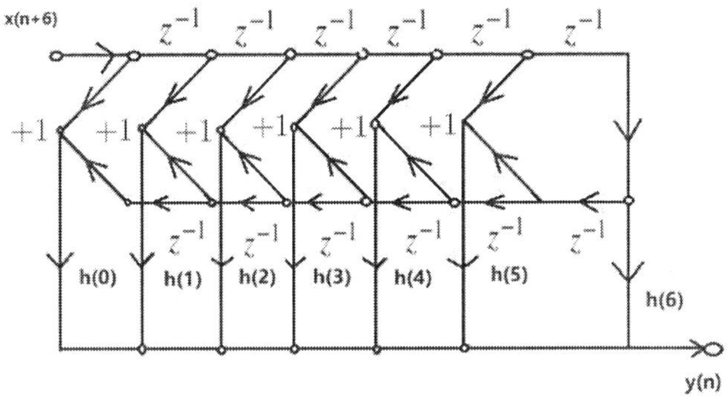


**Figure 3.** Linear Phase Structure

The corresponding calculation formula is

$$y(n) = \sum_{k=0}^{5} h(k)[x(n-k+6) + x(n-k-6)] + h(6)x(n) \quad (3)$$

The filtering requires only 7 multiplications at a time.

As the obtained coefficients of the FIR filter are all decimals, this paper shifts the coefficients of the filter to the left by 8 bits for amplification to realize the fractional multiplications. The result is shifted to the right by 8 bits to the original size, after the multiplications and accumulations of the filtering are completed. Thus this paper realizes the low-pass filtering processing of the component of the luminance signal Y.

**4.Experimental Result**

Through the steps above, the designs of each system module and the reduction processing of bandwidths of the camera acquisition video has been completed. The oscilloscope display of the video images before and after filtering is shown in Figure4 and Figure5. The input images are a multi-burst signals, including 0.5M, 1M, 2M, 4M, 4.8M and 5.8MHz signals.

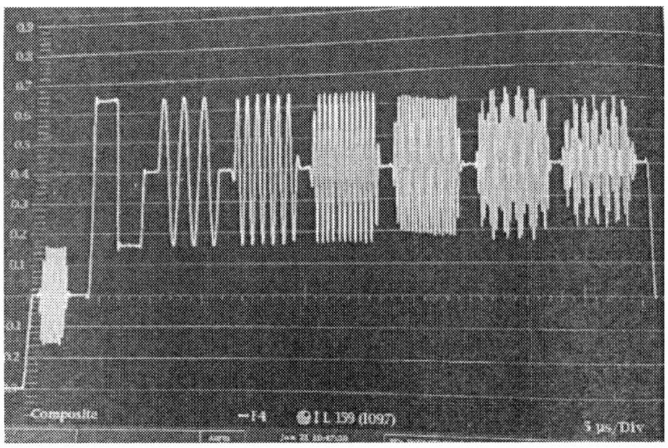


**Figure 4.** The Input of Multi-burst Signals

After low-pass filtering with a cutoff frequency of 3MHz,signals above 3MHz that are the 4MHz, 4.8MHz and 5.8MHz signals have been filtered out,only leaving 0.5MHz,1MHz and 2MHz signals shown in Figure 5.

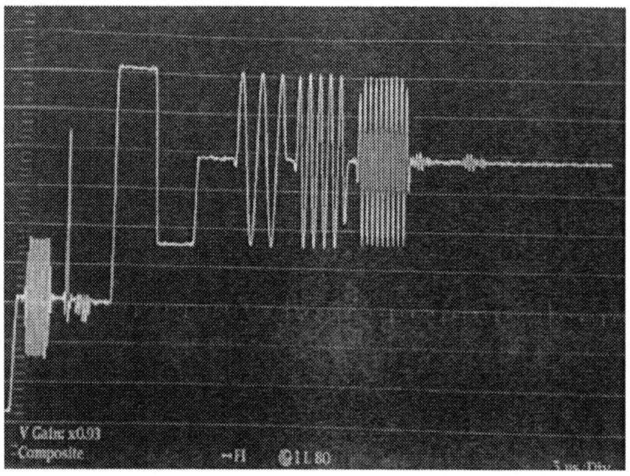

**Figure 5.** Filtered Multi-burst Signals

## 5.Conclusion

This paper designed and implemented low-pass filtering processing of video image signals based on FPGA.As the cutoff frequency of the filter can be different, such as 1MHz, 2MHz, 3MHz, 4MHz etc., which can be modified by the FIR coefficients, it is convenient to observe the changes in the display quality of video signals of different bandwidths. The whole system of this paper was divided into a field decoding module, a frame buffer (ping-pong storage) module and a low-pass filtering module. The experimental results had shown the feasibility of the method.

## References

[1] Xiuhua Jiang.2018.Modern TV Principles [M].Beijing: Higher Education Press.
[2] Jinming Wang, Shun Zhou.2010.Digital System Design and VHDL [M].Beijing: Publishing House of Electronic Industry.
[3] Hujun Zhang.2011.Implementation of video acquisition and output system based on FPGA [D].Dalian: Dalian Maritime University.
[4] Binjie Yu.2008.Design of full color LED synchronous display control system based on FPGA [D].Shanghai: Shanghai Jiao Tong University.
[5] Zhenyang Wu.2016.Digital signal processing [M].Beijing: Higher Education Press.

# Effect of Polypropylene Fiber on Properties of Aeolian-Sand Mortar

**Junhua Guo[1], Zhenghong Yang [1,2] and Chunyong Gao[1]**

[1]China Building Materials Academy, No.1 Guan Zhuang Dong Li, Chaoyang District, Beijing, China;
Email: guojunhua870108@163.com
[2]Tongji University, No. 1239, Siping Road, Shanghai, China;

**Abstract.** In this paper, the same consistency of eolian sand mortar was prepared by blending polypropylene fibers of 0, 0.3Kg/m$^3$, 0.7Kg/m$^3$, 1.0Kg/m$^3$, 1.3Kg/m$^3$ and 1.8Kg/m$^3$ respectively. Effects of fiber content on physical properties such as shrinkage, strength and crack resistance of aeolian sand dry mortar. The results show that polypropylene fiber can significantly improve the physical and mechanical properties of aeolian sand dry mortar. Polypropylene fiber content is within 1.3Kg/m$^3$, and with the increase of fiber content, the effect of aeolian sand mortar is obvious; when the content exceeds 1.3Kg/m$^3$, the shrinkage and mechanical properties are shrunk; the wind and sand accumulation mortar is aggregated. The suitable amount of propylene fiber is 1.3Kg/m$^3$. Under this dosage, polypropylene fiber can not only improve the construction performance of aeolian sand mortar, but also improve its basic mechanical properties.

## 1. Introduction

With the accelerated implementation of China's "One Belt, One Road" policy, infrastructure projects have also increased rapidly, and the demand for raw materials such as construction mortar is increasing, while the raw materials of related areas, especially construction sand resources, are extremely scarce. The systematic study of aeolian sand to prepare dry-mixed mortar can not only utilize the aeolian sand resources along the "Belt and Road"area, but also avoid over-exploitation of local natural river sand resources and protect the ecological environment.

In China, Inner Mongolia Agricultural University and Ningxia University have conducted preliminary research on aeolian sand as a fine aggregate of mortar and concrete. Ma Lei [1] and other studies pointed out that aeolian sand can be used as a fine aggregate to replace part of natural sand for the preparation of mortar; Dong Wei [2] and other studies have shown that aeolian sand can significantly improve the early strength of mortar after replacing some ordinary sand. The fine particles of aeolian sand cause the aeolian sand mortar to have low toughness and dry shrinkage, which is easy to crack. Therefore, improving the anti-drying performance is the focus of research on aeolian sand mortar. Polypropylene fiber has obvious toughening and anti-drying effect on mortar [3]. Therefore, this paper studies the effect of polypropylene fiber content on the physical and mechanical properties of aeolian sand mortar and the suppression of shrinkage.

## 2. Test

### *2.1. Material*

Cement: P·I42.5 benchmark cement; Polypropylene fibers: proportion-0.91 , diameter18-48μm, young's modulus 3.65 GPa, ultimate elongation rate 17%, length 6 mm. Aeolian sand: it comes from the taklamakan desert in xinjiang region, modulus of fineness 0.54, and the water content of dry sand is 0.14%. Its main components are shown in table 1. The micro morphology and particle size distribution are shown in Figure ure 1 and Figure 2.

**Table1.** Main comuositions of aeolian sand

| component | $SiO_2$ | $Al_2O_3$ | $Fe_2O_3$ | $Na_2O$ | CaO | $K_2O$ | MgO |
|---|---|---|---|---|---|---|---|
| Content/ % | 65.63 | 9.84 | 2.28 | 2.3 | 7.92 | 2.17 | 1.73 |

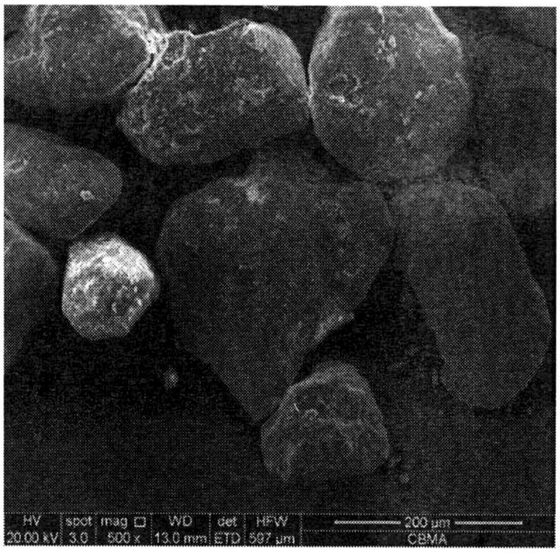

**Figure 1.** SEM image of aeolian sand

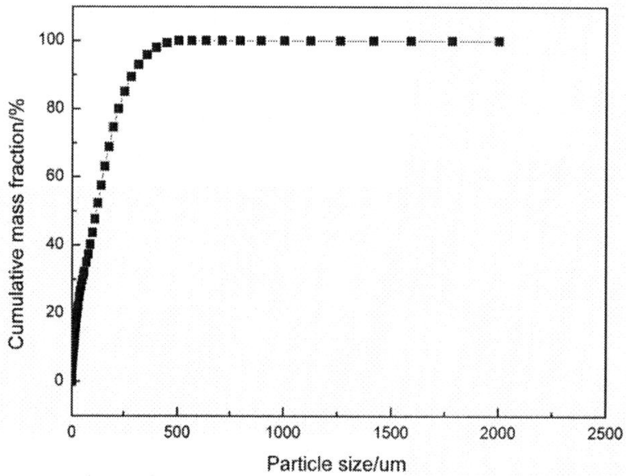


**Figure 2.** Particle size distributions of aeolian sand

It can be seen from Figure ure 1 that the aeolian sand particles are round, irregular in shape and with obviously edge angles. As shown in Figure ure 2, the aeolian sand particles are extremely fine and poorly graded, and their particle sizes are generally less than 500 μm.

*2.2. Test Method*

The general physical and mechanical properties of mortar were tested in accordance with "Method of testing cements-Determination of strength (ISO method) "and "Standard for test method of performance on building mortar".[4]

## 3. Test Results and Discussion

### *3.1. The Influence of the Content of Polypropylene Fiber on the Physical Mechanics and Dry Shrinkage Performance of Aeolian Sand Mortar*

The effect of adding polypropylene fiber of different quality on the physical and mechanical properties and dry shrinkage of aeolian sand mortar (the ratio of cement to sand is 1:2) is shown in table 2.

**Table 2.** The effect on physical and dry-shrinking performance of aeolian sand mortar with different contents PP-fiber

| Cement sand ratio | Fiber content $Kg/m^3$ | Denseness /*mm* | Segregation /*cm* | Water cement ratio | breaking strength/ *MPa* | compression strength/*MPa* | Tensile Strength/ *MPa* | shrinkage ratio/% |
|---|---|---|---|---|---|---|---|---|
| 1:2 | 0 | 81 | 1.5 | 0.688 | 4.61 | 22.88 | 2.72 | 0.763 |
| | 0.3 | 80 | 2.0 | 0.695 | 4.72 | 22.87 | 2.78 | 0.684 |
| | 0.7 | 83 | 2.1 | 0.704 | 4.81 | 21.78 | 2.88 | 0.630 |
| | 1.0 | 84 | 2.1 | 0.728 | 4.94 | 21.89 | 2.92 | 0.575 |
| | 1.3 | 83 | 2.3 | 0.715 | 5.36 | 21.56 | 2.95 | 0.525 |
| | 1.8 | 81 | 2.3 | 0.733 | 4.68 | 20.68 | 2.83 | 0.598 |

The fiber overlaps with each other to form a network system in aeolian sand mortar, resulting in decreased flowability of the mortar. Under the same denseness, polypropylene fiber will increase the water demand of mortar, and the delamination of sand mortar will increase accordingly.

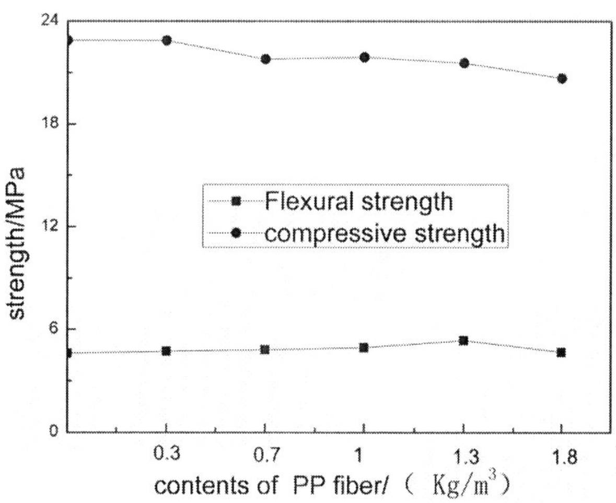


**Figure 3.** Compresive strength of aeolian sand mortar with different contents PP fiber

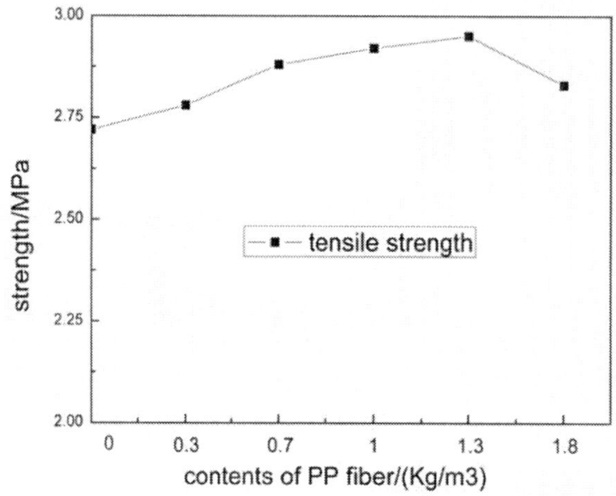


**Figure 4.** Tensile strength of Aeolian sandmortar with different contents PP fiber

It can be found from Figure 3 and Figure 4 that the compressive strength of mortar decreases gradually with the increase of the content of PP fiber. When the dosage is lower than 1.3kg /m3, the tensile strength and flexural strength of mortar increase slowly with the increase of fiber content.When the fiber content exceeds 1.3 $Kg/m^3$, the tensile and flexural strength of the mortar show decline. With the increase of fiber content, the water consumption of mortar increases, at the same time, excessive fiber agglomerates in mortar, so the strength of mortar decreases under the corresponding amount of fiber content.

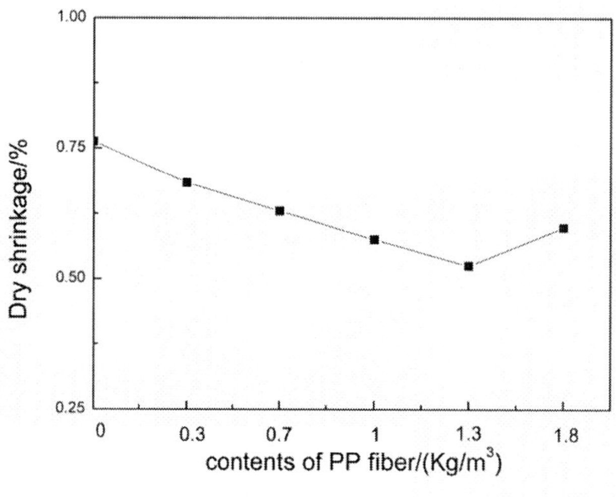


**Figure 5.** Dry shrinkage of aeolian sand mortar with different contents PP fiber

As shown in Figure 5, the dry shrinkage of sand slurry is large, and PP fiber has obvious inhibiting effect on the dry shrinkage of mortar. When the adding amount is $1kg/m^3$, the dry shrinkage rate of aeolian sand cement mortar is 24.6% lower than that of blank mortar without fiber. When $1.3kg/m^3$ fiber is added, the dry shrinkage rate of aeolian sand mortar reaches the minimum, this was 31.2% lower than undoped blank mortar. When the fiber content reaches $1.8kg/m^3$, because the fiber content is too large, the fiber agglomeration and dispersion are not uniform, so the shrinkage of mortar can not be effectively inhibited.

### 3.2. *The Influence of PP Fiber on the Cracking Properties of Aeolian Sand Mortar*

**Table 3.** Results of fracture on aeolian sand mortar

| cement-sand ratio | content ($Kg/m^3$) | crack length($mm$) | | | crack index | Relative cracking index |
|---|---|---|---|---|---|---|
| | | <1 | 1-2 | >2 | | |
| 1:2 | 0 | 70 | 65 | 0 | 100 | 1 |
| | 0.3 | 62 | 32 | 0 | 62 | 0.62 |
| | 0.7 | 86 | 0 | 0 | 43 | 0.43 |
| | 1.0 | 54 | 0 | 0 | 27 | 0.27 |
| | 1.3 | 43 | 0 | 0 | 22 | 0.22 |
| | 1.8 | 72 | 0 | 0 | 36 | 0.36 |

It can be seen from Figure 6 that the adding of PP fiber can lead to the decrease of crack length and number, meanwhile, the the number of long cracks reduce. When 1.3kg/m$^3$ fiber is added, the relative cracking index of mortar is 22, and the total number of cracks is 78% less than the blank mortar without fiber. In other words, the anti-cracking of eolian sand mortar can be significantly improved by adding polypropylene fiber. As the fiber content continues to increase, the fiber aggregate and dispersion are not uniform, resulting in an increase of total number of mortar cracks.

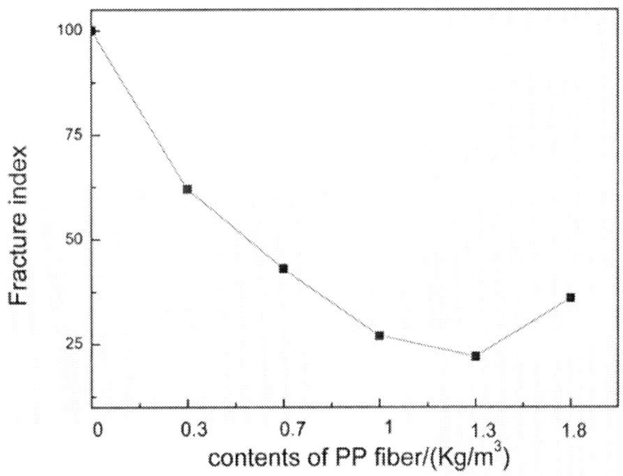


**Figure 6.** Fracture index of aeolian sand mortar with different contents PP fiber

### 3.3. *Influence Mechanism of Fiber on Mortar Properties*

The PP fiber is distributed uniformly in the mortar, which inhibits the sedimentation segregation of water and mortar aggregates, Reduces water evaporation, resulting in reducing the number of connecting holes of aeolian sand mortar. Meanwhile, after adding fibers, the fibers are kept on the surface of mortar, which slows down the evaporation rate of mortar water, improves the water-retaining performance of aeolian sand mortar, and further inhibits the dry shrinkage of mortar.

The PP fibers in aeolian sand mortar are randomly overlapping with each other, which have a significant inhibitory effect on mortar crack [5] [6] [7]. After adding PP fiber into the aeolian sand mortar, its tensile strength and toughness are improved, the total number of cracks is reduced, longer cracks are basically disappeared, and the tendency of crack refinement is obvious [8].

If the pp fiber is dispersed uniformly in aeolian sand mortar, the cracks in mortar encounters the PP fiber in expansion, and the presence of the fiber will force the crack to change its development direction or prevent the cracks from forming smaller micro-cracks, then the development trend of the mortar micro-cracks will be limited.

## 4. Conclusion

The fiber overlaps with each other to form a network system in aeolian sand mortar, resulting in decreased flowability of the mortar. Under the same denseness, polypropylene fiber will increase the water demand of mortar, and the delamination of sand mortar will increase accordingly. When $1.3kg/m^3$ fiber is added, the relative cracking index of mortar is 22, and the total number of cracks is 78% less than the blank mortar without fiber. The anti-cracking of eolian sand mortar can be significantly improved by adding PP fiber.

The dry shrinkage of sand slurry is large, and PP fiber has obvious inhibiting effect on the dry shrinkage of mortar. With the increase of fiber content, dry shrinkage of aeolian sand mortar first decreases and then increases.When the adding amount is 1kg/m3, the dry shrinkage rate of aeolian sand cement mortar is 24.6% lower than that of blank mortar without fiber. When $1.3kg/m^3$ fiber is added, the dry shrinkage rate of aeolian sand mortar reaches the minimum, this was 31.2 percent lower than undoped blank mortar.When the fiber content reaches $1.8kg/m^3$, because the fiber content is too large, the fiber agglomeration and dispersion are not uniform, so the shrinkage of mortar can not be effectively inhibited.

In conclusion, polypropylene fiber can improve the toughness of aeolian sand mortar, effectively inhibit the mortar's dry shrinkage, and also enhance the mortar's anti-cracking properties. The adding amount of pp fiber in aeolian sand mortar is appropriate to be $1.3kg /m^3$.

**Acknowledgments**

The authors thanks to the support of the National Key R&D Program of China (No. 2016YFC0701000)

## References

[1] Ma Lei, Liang Dong and Fan Xiaopeng 2012 Desert sand as a construction mortar fine aggregate feasibility study (low temperature construction technology, vol 34) chapter 4 pp9-10.
[2] Dong Wei and Shen Xiangdong 2013 Study on the fluidity and strength of cement mortar with different aeolian sand content (silicate Bulletin, vol32) chapter 9 pp1900-1904.
[3] Yuan Zhenyu, Wu Huimin and Yang Jianxi. 1999 Experimental study on the effect of polypropylene fiber on crack resistance of mortar (concrete and cement products, vol6) pp41-42.
[4] Ma Yiping andTan Muhua 2000 Study on physical and mechanical properties of polypropylene fiber cement-based composites(I)-anti-plastic dry shrinkage cracking performance (Journal of Building Materials vol3) chapter 1 pp48- 52.
[5] Sezan.O 2000 Investigation of vibration damping on polymer concrete with polyester resin.Concrete.(Cement & Research vol 30) pp 207-214
[6] Chung D D L.1996 Vibration damping admixture for cement. (Concrete & Cement Research vol 26) pp 73-84
[7] Fu Xuli 1998 improving the vibration damping capacity of cement (Materials Science vol 33) pp 17-21.
[8] Hillerborg A. 1985 the theoretical basis of a method to determine the fracture energy GF of concrete (Mater & Struck vol 18) pp 103-109.

# Finite Element Analysis of Overlying Soil Failure Based on Fault Dip Angle

**Tingting Guo [1, 2, a], Xiwei Xu [3] and Guihua Yu [3]**

1Shandong Institute of Earthquake Engineering, Jinan, China
2Earthquake Administration of Shandong Province, Jinan, China
3Key Laboratory of Active Tectonics and Volcano, Institute of Geology, CEA, Beijing, China
a gtt.yy@163.com

**Abstract.** When the strong earthquake occurs, the deformation response and rupture of the overlying soil are very complicated. In this paper, the influence of fault dip angle on the surface deformation and rupture of overlying soil is analyzed by means of finite element numerical simulation. When the vertical dislocation of the fault is about 3~5% of the thickness of the overlying soil layer, the surface rupture occurs only when the fault dip angle is 45°. When the vertical displacement of the fault increases to 10% of the thickness of the overlying soil layer, surface rupture occurs under three working conditions of 45°, 70° and 90° of fault dip angle. With the increase of the dip angle from 45°, 70° to 90°, the surface equivalent strain is gradually smaller, the deformation of the overlying soil is smaller, and the deformation width of the upper wall will change from about 2 times that of the lower wall to the same as the lower wall. The deformation and rupture of the overlying soil first begin with the fracture of the soil mass at the interface between the fault bedrock and the soil. With the increase of the dislocation quantity, the surface also will appear 1 breakpoint when the fault angle is 45° and 70°, the ground surface will appear 2 breakpoints when the fault angle is 90°, and at last overlying soil is broken through.

## 1. Introduction

Several historical strong earthquake damage phenomena indicate that the destruction of ground surface and the permanent large deformation caused by the earthquake faults make the destruction of the ground buildings, structures and lifeline engineering very serious. In recent years, the research personnel pay attention to the problems about the deformation and fracture of the overlying soil caused by fault dislocation. So the formation and development of overlying soil deformation and fracture caused by fault dislocation can further understand the failure mode of overlying soil and the failure mechanism of the near fault structure.

During the occurrence of strong earthquakes, the deformation and fracture of overlying soil are very complicated. Previous studies show that the deformation process and rupture characteristics of the overlying soil are affected by many factors. On the one hand, it is the factors of soil itself: such as different formation combinations [1], soil layer thickness [2-4], the physical and mechanical properties of different stratum soil and soil composition [5], weak intercalation or hard intercalation [3], etc. The two aspects are earthquake and its seismic fault factors, such as seismic amplitude, frequency spectrum characteristics and duration, seismic fault type, fault dip angles [6], fault displacement [7-8] and other factors. Based on the above research results, it is concluded that the fault dip angle has a great influence on the deformation and fracture of overlying soils, and it is worth further analysis and discussion.

Therefore, on the basis of previous research results, this paper uses finite element numerical simulation method to analyze the influence of fault dip angle on the surface deformation and fracture of the overlying soil layer, and discusses its influence law, which is used to provide reference basis for prediction of permanent deformation of the ground, active fault seismic hazard and risk and the seismic fortification of engineering structure.

## 2. Calculation Conditions

Using the plane strain finite element method, combined with the nonlinear constitutive relation of the soil, under the condition of fault vertical dislocation of 2m, different overlying soil thickness (60m, 40m, 20m), the influence of dip angle change on the deformation and rupture of the overlying soil layer and and its regularity analysis are analysed by changing fault dip angle. Nine operating conditions are calculated (Table 1).

**Table 1.** Calculation condition table

| working condition number | overlying soil thickness(m) | fault dip angle(°) | working condition number | overlying soil thickness(m) | fault dip angle(°) | working condition number | overlying soil thickness( m) | fault dip angle(°) |
|---|---|---|---|---|---|---|---|---|
| **NO.1** | 60 | 45 | **NO.4** | 40 | 45 | **NO.7** | 20 | 45 |
| **NO.2** | 60 | 70 | **NO.5** | 40 | 70 | **NO.8** | 20 | 70 |
| **NO.3** | 60 | 90 | **NO.6** | 40 | 90 | **NO.9** | 20 | 90 |

## 3. Establishment of Finite Element Model

### *3.1. Basic Assumptions*

In the model establishment, it is assumed that: 1 plane strain state; 2 considering the elastoplastic material of rock and rock, all adopting Drucker-Prager yield criterion; 3 the contacts between upper and lower disk of reverse fracture is assumed to be nonlinear contact.

### *3.2. Determination of Material Model*

*3.2.1.Original geological model.* The observation point of Xiaoyudong Town in Sichuan is taken as the original geological model. It is located in the northwestern edge of the Sichuan Basin, the heart of the Yujiang River. The lithology of the strata exposed on the west side of Xiaoyudong Street Road in the study area are:

Triassic (T3x): mainly argillaceous limestone sandstone with a thickness of 100m-120m.

Quaternary alluvial deposit (QAL+PL): covered in the Xujiahe Formation, with a thickness of 30-40m, consisting of sub-clay, sand and gravel.

According to field survey and comprehensive analysis of relevant literature [9], it is preliminarily determined that the fault dip angle is about 70 °.

*3.2.2.Determination of material model.* In the establishment of bedrock fault and overburden model, considering the nonlinear stress-strain characteristics of rock and soil, the Drucker-Prager elastoplastic model is applied to both bedrock and soil, and the Drucker-Prager yield criterion is followed.

The yield criterion is an approximation of the Mohr-Coulomb criterion to correct the Von Mises yield criterion, which includes an additional term in the Von Mises expression. Its equivalent stress expression is:

$$\sigma_e = 3\beta\sigma_m + \left[\frac{1}{2}\{S\}^T[M]\{S\}\right]^{\frac{1}{2}} \tag{1}$$

*3.2.3 Selection of material calculation parameters.* The model bedrock is argillaceous limestone, and there are Quaternary alluvial deposits and residual slopes on the surface. The main reference is to the "Engineering Geology Handbook" and a large number of previous numerical simulation experience. the mechanics parameters of the material is determined by engineering geology analogy method. The parameters are shown in Table 2.

**Table 2.** Parameter table of bedrock and geotechnical materials of the model

| | material Name | elastic modulus(Mpa) | Poisson's ratio | volumetric weight(KN/m$^2$) | cohesion (Mpa) | friction angle (°) |
|---|---|---|---|---|---|---|
| **bedrock** | argillaceous limestone | 2.0e4 | 0.2 | 23 | 0.15 | 30 |
| **overlying soil** | Quaternary alluvial deposits | 1.5e2 | 0.35 | 19 | 2.5e-2 | 15 |

### *3.3. Determination of Geometric Models*
A geometric calculation model is established by taking into account the boundary effects and the amount of calculation. The bottom of the model is 300m wide and the is 20m thick. In the different cases of fault vertical dislocation of 2m and overlying soil layers of 60m, 40m and 20m,fault dip angles are changed by 45°, 70° and 90°; and frictional nonlinear contact is adopted between the upper and lower faults. Taking vertical dislocation of 2m, overlying soil layer of 40m, and fault dip angle of 70° as an example, a geometric model is established and meshed as shown in Fig.1.

### *3.4. Boundary Conditions And Loading*

*3.4.1 Boundary conditions.* The upper boundary is the surface free surface, without constraint. The left boundary is free and unrestrained; the right boundary is fixed and completely bound; the footwall of bedrock is fixed and completely restrained.

*3.4.2. Loading.* As the near field of the fault is simulated, the dislocation displacement of bedrock at the A interface of the soil layer is consistent with the fault rupture of bedrock. In different working conditions, certain displacement values in X and Y directions were applied to the upper wall of bedrock in different models, that is, the movement of the reverse fault was simulated through upper and lower dislocation respectively.

## 4. Preliminary Analysis and Discussion of Finite Element Results

### *4.1. Finite Element Simulation Results of Deformation and Fracture of Overlying Soil Caused by Fault Dislocation*
Taking the model of overlying single soil layer on bedrock of thrust fault, with vertical dislocation of 2m, overlying soil thickness of 40m and fault dip angle of 45 degrees as example, the process and characteristics of overlying soil deformation and fracture caused by fault dislocation are preliminarily analyzed. In this model, the displacement in the X and Y directions (vertical dislocation 2m) is applied to the upper wall of bedrock, so that the upper and lower walls stagger to simulate the movement of thrust faults. Fig.2 (a) and (b) show the deformation cloud map of overlying soil layer and its 4-fold magnification. At the same time, the node horizontal displacement distribution diagram of overburden layer, the node vertical displacement distribution diagram of overburden layer, the node total displacement distribution diagram of overburden layer and the node equivalent strain distribution diagram of overburden layer are respectively drawn as shown in Fig 3-Fig.6 for further comparative analysis.

Through the calculation of a large number of conditions, it can be concluded that the deformation of the overlaying soil first will appear in the boundary surface of bedrock and soil, with the increment

IOP Conf. Series: Materials Science and Engineering 472 (2019) 012088 doi:10.1088/1757-899X/472/1/012088

of fault dislocation, scaling up of deformation, when the bedrock dislocation value reaches a certain value, the topsoil appear deformation, scaling down, the last through the reach soil (When the surface deformation exceeds 0.02, it is considered that surface rupture can be caused).

Next, through the calculation and comparison of the working conditions of different dip angles, the influence and rule of the variation of fault dip angle on the deformation and fracture of overlying soil are analyzed.

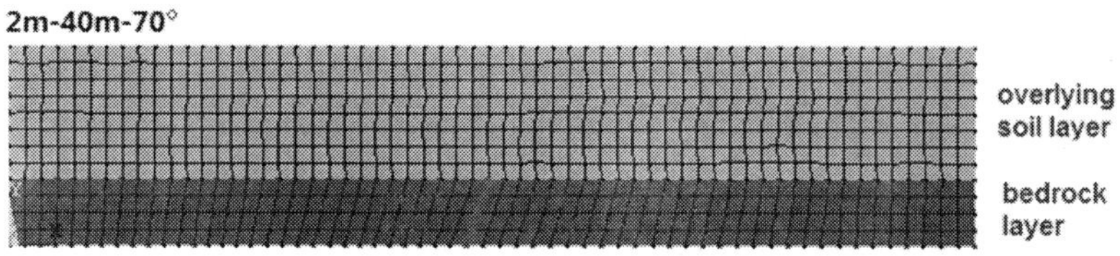


**Figure 1.** Geometric model and mesh division

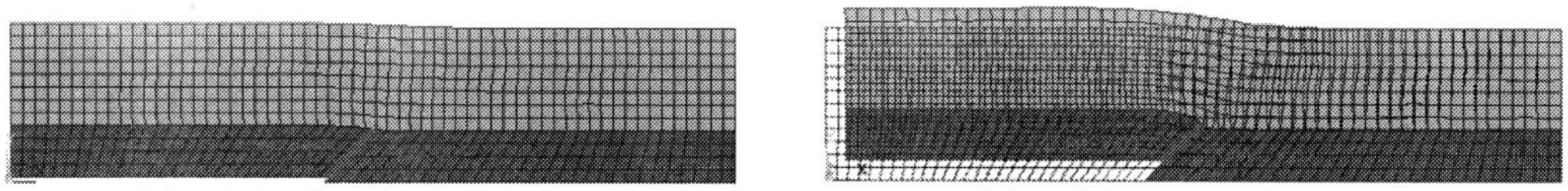

(a) original proportion (b) 4-fold magnification

**Figure 2.** Deformation cloud map of overlying soil

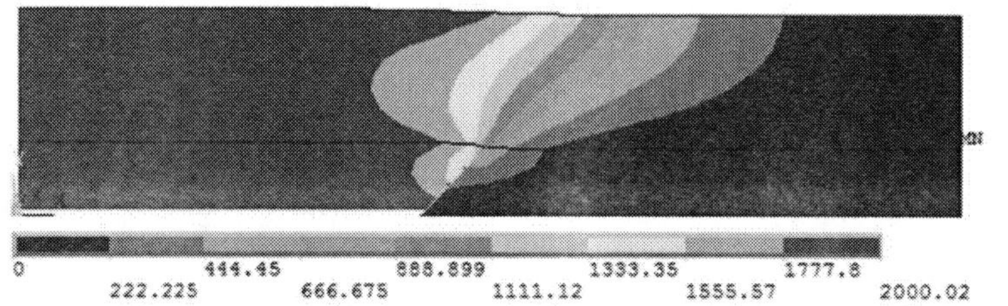


**Figure 3.** Horizontal displacement distribution of nodes in overlying soil (2m-40m-45°)

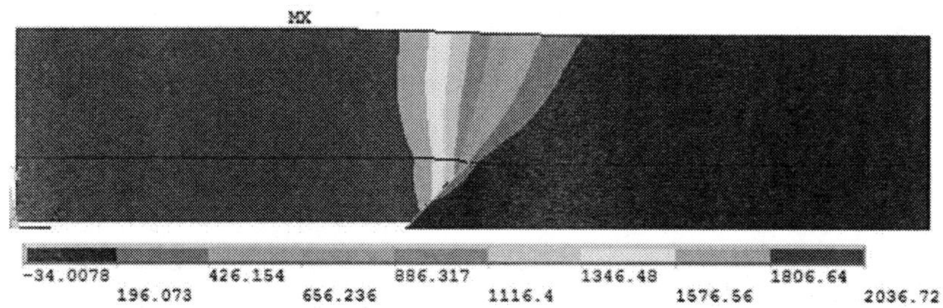


**Figure 4.** Vertical displacement distribution of nodes in overlying soil (2m-40m-45°)

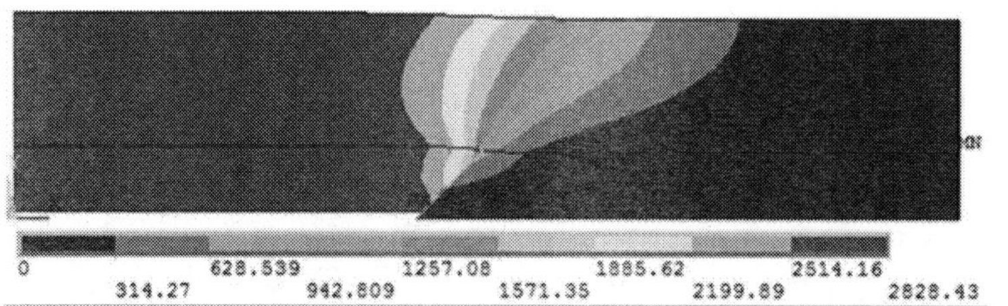


**Figure 5.** Total displacement distribution of nodes in overlying soil (2m-40m-45°)

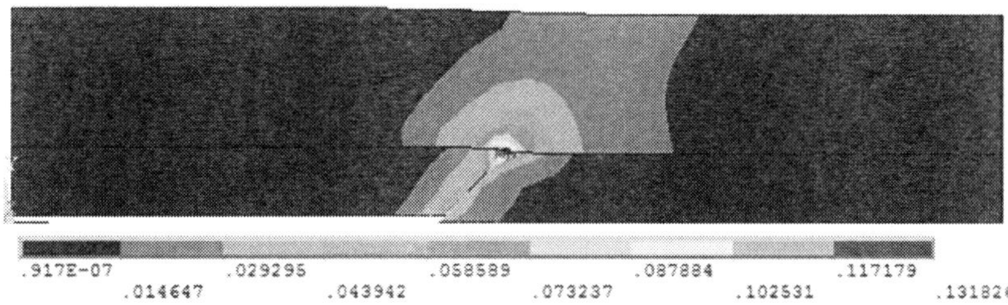


**Figure 6.** The equivalent strain distribution of nodes of overlying soil (2m-40m-45°)

### *4.2. Influence of Fault Dip Angle on the Deformation of Overlying Soil*

*4.2.1 Vertical dislocation amount of 2m and overlying soil thickness of 60m.* In the case of vertical dislocation of 2m and overlying soil thickness of 60m, fault dip Angle is increased in turn to 45 °, 70 °, 90°. The equivalent strain distribution of the overburden layer under three different working conditions is obtained as shown in Fig.7.

By comparison in Fig.7., it can be obtained:

①Under the condition of the thickness of overlying soil is 60m, when the vertical dislocation of the fault is about 3.3% of the thickness of the overlying soil, the overlying soils are deformed in different degrees under the above three working conditions. Among them, in the operation condition of the fault dip angle is 45 °, the surface is in the critical state of rupture, in the working condition of fault dip angle is 70 °, 90 °, neither the surface rupture occurred.

②As the fault dip angle increases from 45 °, 70 ° to 90 °, the surface of the equivalent strain is gradually become smaller, the overlaying soil deformation is smaller, which indicates that the amount of dislocation required for the same deformation is increased.

*4.2.2 Vertical dislocation amount of 2m and overlying soil thickness of 40m.* In the case of vertical dislocation of 2m and overlying soil thickness of 40m, fault dip angle is increased in turn to 45 °, 70 °, 90 °. The equivalent strain distribution of the overburden layer under three different working conditions is obtained as shown in Fig.8.

By comparison in Fig. 8, it can be obtained:

①Under the condition of the thickness of overlying soil is 40m, when the vertical dislocation of the fault is about 5% of the thickness of the overlying soil, the overlying soils are deformed in different degrees under the above three working conditions. Among them,

in the operation condition of the fault dip angle is 45 °, surface rupture occurred; in the working condition of fault dip angle is 70 °, 90 °, no surface rupture occurred, but compared to the operation condition of the overlying soil thickness of 60m, deformation increased obviously.

②As the fault dip angle increases from 45 °, 70 ° to 90 °, the surface of the equivalent strain is gradually become smaller, the overlaying soil deformation is smaller, which indicates that the amount of dislocation required for the same deformation is increased.

③According to the operation condition results which fault dip angle is 45 ° and 70 °(Fig.7, Fig.8), it can be seen: when the fault dip angle is 45 °, 70 °, first of all, the soil at the interface between the fault bedrock and the soil is fractured, and as the increase of dislocation amount(that is, the overlaying soil thickness decreases under same conditions), a rupture point appeared on the surface, and finally formed a transfixion fracture of the overlying soil.

④According to the results of different dip angle(Fig.7, Fig.8), it can be obtained: as the fault dip angle increases from 45 °, 70 ° to 90 °, the deformation width of the upper wall will change from about 2 times of the lower wall to close to the lower wall. When the fault dip angle is 45 °, as for the range of the overlaying soil deformation, upper wall of fault is greater than the lower wall; when the fault dip angle is 90 °, overlaying soil deformation is basic symmetric distribution.

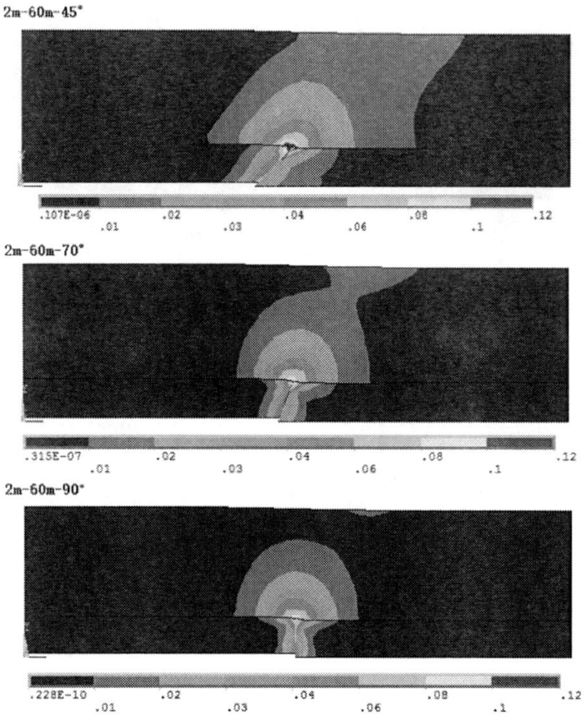


**Figure 7.** Comparison of equivalent strain distribution of overlying soil layer under different fault dip angles with 2 m vertical dislocation and 60 m thickness of overlying soil layer

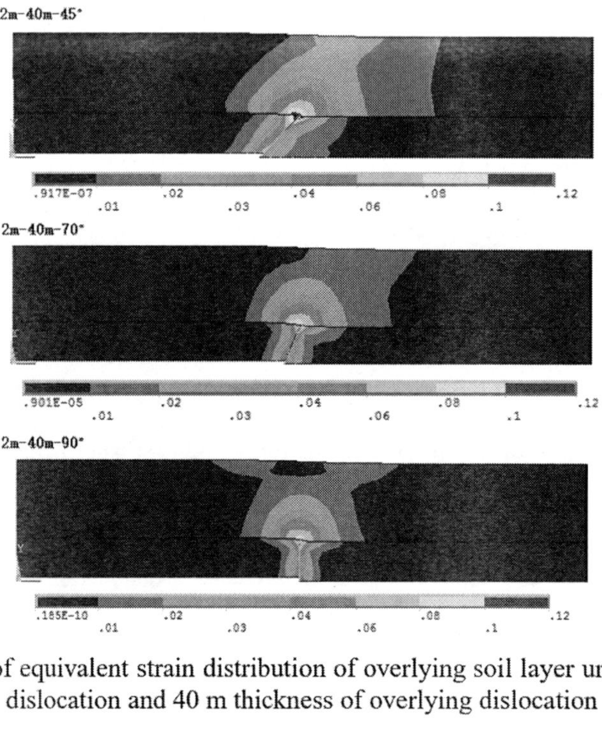


**Figure 8.** Comparison of equivalent strain distribution of overlying soil layer under different fault dip angles with 2 m vertical dislocation and 40 m thickness of overlying dislocation and 40 m thickness of overlying soil layer

*4.2.3 Vertical dislocation amount of 2m and overlying soil thickness of 20m.* In the case of vertical dislocation of 2m and overlying soil thickness of 20m, fault dip angle is increased in turn to 45 °, 70 °,

90 °. The equivalent strain distribution of the overburden layer under three different working conditions is obtained as shown in Fig.9.

By comparison in Fig. 9, it can be obtained:

①Under the condition of the thickness of overlying soil is 20m, when the vertical dislocation of the fault is about 10% of the thickness of the overlying soil, under the above three conditions, surface rupture all occurred.

②As the fault dip angle increases from 45 ° to 90 °, the surface of the equivalent strain is gradually become smaller, the overlaying soil deformation is smaller, which indicates that the amount of dislocation required for the same deformation is increased.

③As the fault dip angle increases from 45 ° to 90 °, the deformation width of the upper wall will change from about 2 times of the lower wall to close to the lower wall.

④According to the operation condition results which fault dip angle is 90 °(Fig.7, Fig.8, Fig.9), it can be seen: when the fault dip angle is 90 °, first of all, the soil at the interface between the fault bedrock and the soil is fractured, and as the increase of dislocation amount(that is, the overlaying soil thickness decreases under same conditions), two rupture points appeared on the surface(is different from one rupture point which dip angle is 45 °, 70 °), and finally formed a transfixion fracture of the overlying soil.

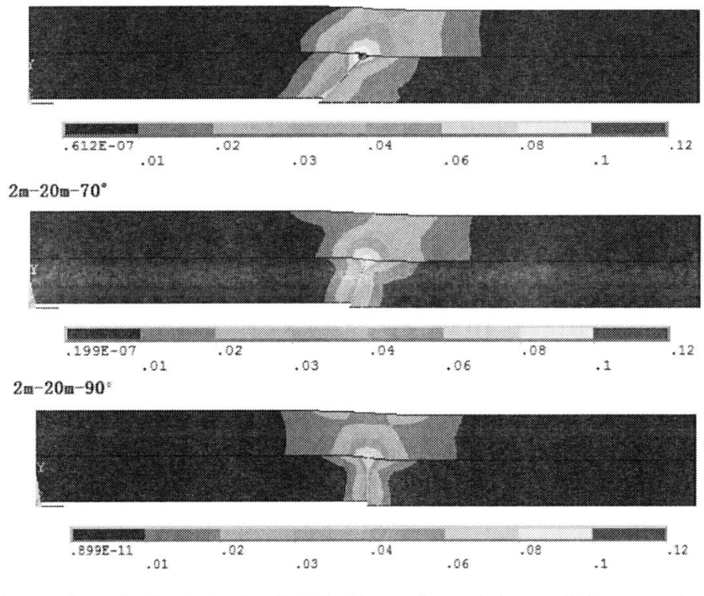


**Figure 9.** Comparison of equivalent strain distribution of overlying soil layer under different fault dip angles with 2m vertical dislocation and 20m thickness of overlying soil layer

## 5. Conclusions and Recommendations

In the process of strong earthquake, the deformation and fracture of overlying soil are very complicated. The deformation process and fracture characteristics of the overlying soil in the fault of earthquake are affected by the overlying soil thickness, soil property, fault type and fault dip angle. In this paper, the effect of fault dip angle on surface deformation and fracture of overlying soil is analyzed by means of finite element numerical simulation.

①When the amount of vertical fault dislocation is about 3-5% of the thickness of the overlying soil, only in the operation condition of the fault dip angle is 45 °, surface rupture occurred; in the working condition of fault dip angle is 70 °, 90 °, no surface rupture occurred. When the amount of vertical fault dislocation is about 10% of the thickness of the overlying soil, under the above three conditions(fault dip angle is 45°, 70°, 90°), surface rupture all occurred.

②As the fault dip angle increases from 45 °, 70 ° to 90 °, the surface of the equivalent strain is gradually become smaller, the overlaying soil deformation is smaller, which indicates that the amount of dislocation required for the same deformation is increased.

③As the fault dip angle increases from 45 °, 70 ° to 90 °, the deformation width of the upper wall will change from about 2 times of the lower wall to close to the lower wall. When the fault dip angle is 45 °, as for the range of the overlaying soil deformation, upper wall of fault is greater than the lower wall; when the fault dip angle is 90 °, overlaying soil deformation is basic symmetric distribution.

④The deformation and fracture of overlying soil first begins with the fracture of soil at the interface between fault bedrock and soil. As the increase of dislocation amount(that is, the overlaying soil thickness decreases under same conditions), when the fault dip angle is 45° and 70°, a rupture point appeared on the surface; when the fault dip angle is 90°, two rupture points appeared on the surface, finally a transfixion fracture of the overlying soil is formed.

⑤This paper is a preliminary study on the influence of fault dip angle on the surface deformation and fracture of overlying soil. In the future, the working condition of fault dip Angle can be increased to further refine the value of influencing factors, so as to further control the critical value of surface rupture.

**Acknowledgments**
This work was financially supported by Shandong Provincial Natural Science Foundation, China (ZR2014DQ022).

**Reference**

[1] Liu Shouhua, Dong Jincheng, Xu Guangming, et al. 2005. Influence on different overburden soils due to bedrock fracture [J]. Chinese Journal of Rock Mechanics and Engineering, 24(11):1868-1874.
[2] Cole DA, Lade PV. 1984. Influence zones in alluvium over dip-slip faults [J]. J Geotech Eng-ASCE, 110(5):599-616.
[3] Guo Endong, Shao Guangbiao, Bo Jingshan, et al. 2002. A method for earthquake ruptures analysis of overlying soil site [J]. Earthquake Engineering and Engineering Vibration, 22(5):122-126.
[4] Zhao Lei,Li Xiaojun,Huo Da.2006. Study on the Rupture Characteristics of the Overlaying Soil with Soft Interlayer Due to Fault Bedrock Dislocation [J]. Acta Seismologica Sinica, 28(5):523-528.
[5] Zhou Qing, Zhou Bengang, Ran Hongliu.2006. Comparative Study on Surface Rupture of Faults in Different Soil Mass [J]. Technology for Earthquake Disaster Prevention, 1(3):225-233.
[6] Astasopoulos I, Gazetas G, Bransby M, et a1. Fault rupture propagation through sand: finite-element analysis and validation through centrifuge experiments [J]. Geotechnical and geoenvironmental engineering, 2007, 133(8): 943-958.
[7] Taniyama and Watanabe, 2000. Deformation of sandy deposits by fault movement. Proc.12th WCEE (No. 2209), Vol. 6.
[8] Guo Endong, Feng Qimin, Bo Jingshan,et al. 2001. Seismic test of soil site rupture under fault displacements [J]. Earthquake Engineering and Engineering Vibration, 21(3):145-149.
[9] Li Yong, Huang Runqiu, A.L. Densmore, et al. 2009. Surface Rupture of Xiaoyudong Fault Caused by Wenchuan Earthquake and its Geological Significance [J]. Quaternary Sciences, 29(3):502-512.

# Anomaly Detection for Environmental Data Using Machine Learning Regression

**Fuqing Yuan and Jinmei Lu**

Department of Technology and Safety, UiT the Arctic University of Norway, 9037 Tromsø, Norway
Email: yuan.fuqing@uit.no

**Abstract.** Environmental data exhibits as huge amount and complex dependency. Utilizing these data to detect anomaly is beneficial to the disaster prevention. Big data approach using the machine learning method has the advantage not requiring the geophysical and geochemical knowledge to detect anomaly. This paper using the popular support vector regression (SVR ) to model the correlation between factors. From the residual of the regression, it develops a statistical method to quantify the extremity of some abnormal observed data. A case study is proposed to demonstrate the developed methods.

## 1. Introduction

Big data is a hot topic almost in all the engineering subjects [1]. The current challenges in engineering data crisis need a method can handle the data automatically to save analytic cost. In environmental engineering, the amount of data is unexceptionally huge. In Norway, the meteorological authorities collects all the temperature, humidity, wind, ice etc. data in their database, in order to utilize the data into helpful decision making, such as optimize the road traffic, give early operating warning for its big oil and gas industry, control its air traffic, the available data is a huge resources.

From the available to detect anomaly event such as abnormal high temperature, humidity, wind speed could herald nature disaster. In norther Norway, the temperature, wind speed correlates to the avalanche in the mountainous area. An abnormal temperature, eight too high or too low, could incur avalanche. On the road, unusual temperature in some seasons could also induce more traffic accidents. Since amount of data is huge and one are not able, perhaps not necessary to investigate the physical connection between the observed data and the correlated disaster. A method can correlate the environmental data and the disaster can benefit the society.

## 2. Methodology

Machine learning is key to solve big data problem. The development of machine learning is rapidly. For some machine learning, the abnormal pattern is detected based on the geometric distance between data sets. This method evaluates the similarity or dependency in terms of distances between data sets. This distance is not limited to the geometrical distance but also be the abstract distance, e.g. Euclidean distance, Riemannian distance, Mahalanobis distance, or Kullback-Leibler distance [2]. A simple machine learning method such as the K-Nearest Neighbor (KNN) is Euclidean distance based [3]. The Support Vector Machine (SVM) is Riemannian distance based [4]. The Artificial Neural Network (ANN) is diverse. In state of art, one can find all abstract distance based ANN.

The fundamental principle for most machine learning methods is general. In the problem this paper considering, the outlier or anomaly of some environmental measurement should be detected. For time-independent environmental factor, we can use machine learning method to define the boundary of the normal data, as shown in the left of Figure 1. For time-dependent factor, we can employ the machine learning regression model to remove the time effect. Thereafter we define the boundary for the normal data, the abnormal data then can be readily detected [5], as shown in the right of Figure 1. This paper uses the second one.

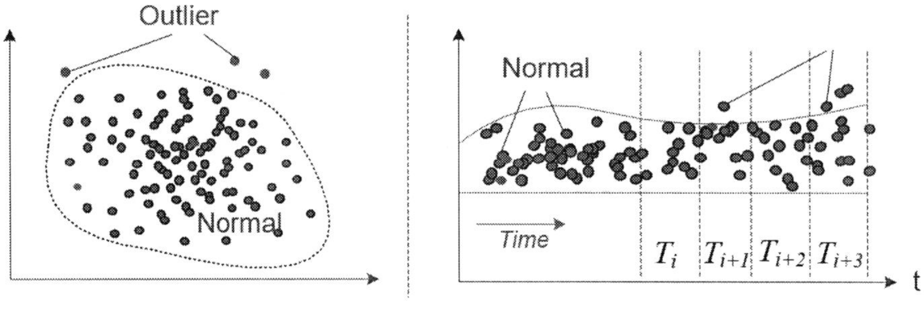


**Figure 1.** Approach to detect abnormal event

*2.1. Machine Learning Regression*

Support Vector Regression (SVR) is a nonlinear regression method. It can be realized by considering it as a support vector classifier. Suppose the desired regression function is $f(x)$. As shown in Figure 2, the data positioning in upper of the desired regression function $f(x)$ is considered as class 1, the data below $f(x)$ is considered as class 2. SVR converts the regression problem into a special classification problem. SVR uses soft margin to tolerate misclassification, i.e. the $\varepsilon$-insensitive loss function [6].

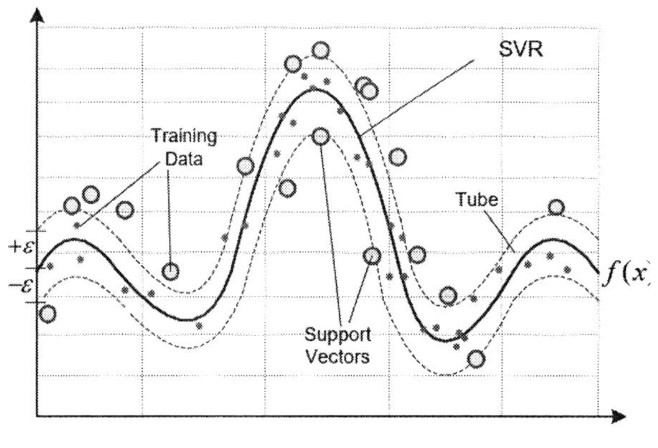


**Figure 2.** Support Vector Regression

Alike the support vector machine, the primal problem of SVR is to solve the problem:

$$\begin{aligned} &\min \frac{1}{2}\|w\|^2 + C\sum_{i=1}^{m}(\xi_i + \xi_i^*) \\ &s.t.\ f(x_i) - y_i \le \varepsilon + \xi_i \\ &\quad y_i - f(x_i) \le \varepsilon + \xi_i^* \\ &\quad \xi_i \ge 0, \xi_i^* \ge 0.\ i = 1,2,3,....m. \end{aligned} \tag{1}$$

Then introducing Lagrangian multipliers, a dual problem of Formal (1) is formatted as:

$$
\begin{aligned}
max \quad & W(\alpha,a^*) = -\varepsilon\sum_{i=1}^{m}(\alpha_i+\alpha_i^*)+\sum_{i=1}^{m}(\alpha_i^*-\alpha_i)y_i-\frac{1}{2}\sum_{j=1}^{m}(\alpha_i^*-\alpha_i)(\alpha_j^*-\alpha_j)\langle x_i,x_j\rangle \\
s.t. \quad & \alpha_i \geq 0,\ \alpha_j \geq 0; \\
& \alpha_i^* \leq C, \alpha_j^* \leq C; \\
& \sum_{i=1,j}^{m}(\alpha_i^*-\alpha_i)=0,\ i=1,2,....,m
\end{aligned} \tag{2}
$$

If we let the $<x_i,x_j>$ replaced by a kernel function $K<x_i,x_j>$. The desired function $f(x)$ is as follows:

$$
f(x)=\sum_{j=1}^{m}(\alpha_i^*-\alpha_i)K(x,x_j)+b \tag{3}
$$

The SVR is a black box method. In applying the method, we don't write the regression function as (3) explicitly, since it is not necessary as we solve the problem using software. In environmental engineering, it is not necessary to investigate the geophysical or geochemical background of these regression parameters $\alpha$ and b. There are data-driven and just adaptive to the data.

## 3. Case Study

The above mentioned approach is applied to detect the abnormal high or low temperature for environmental data collected in a Russian airport [7]. The data ranges from year 2016 to 2018. The January data is analysed in the case study. Original raw data for each day measured 8 times. We take the average of a day as this day's temperature. Totally 88 datasets is extracted from the raw data. Figure 3 shows screen shot of the original data.

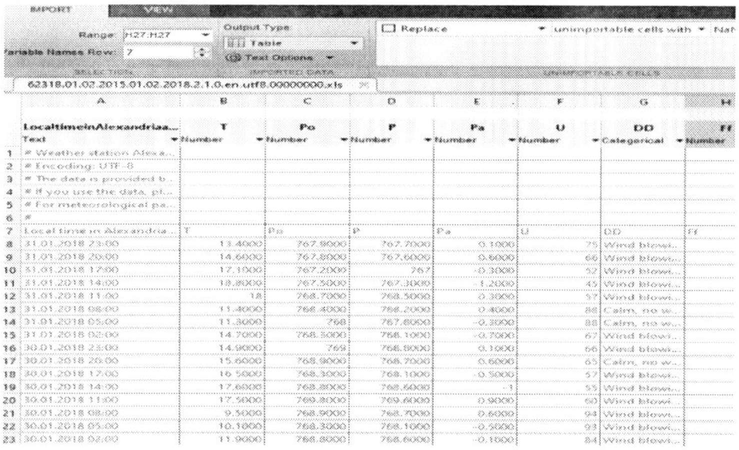

| | A | B | C | D | E | F | G | H |
|---|---|---|---|---|---|---|---|---|
| | LocaltimeinAlexandriaa... Text | T Number | Po Number | P Number | Pa Number | U Number | DD Categorical | Ff Number |
| 1 | # Weather station Alexa... | | | | | | | |
| 2 | # Encoding: UTF-8 | | | | | | | |
| 3 | # The data is provided b... | | | | | | | |
| 4 | # If you use the data, pl... | | | | | | | |
| 5 | # For meteorological pa... | | | | | | | |
| 6 | # | | | | | | | |
| 7 | Local time in Alexandria... | T | Po | P | Pa | U | DD | Ff |
| 8 | 31.01.2018 23:00 | 13.4000 | 767.9000 | 767.7000 | 0.1000 | 75 | Wind blowi... | |
| 9 | 31.01.2018 20:00 | 14.6000 | 767.8000 | 767.6000 | 0.6000 | 66 | Wind blowi... | |
| 10 | 31.01.2018 17:00 | 17.1000 | 767.2000 | 767 | -0.3000 | 52 | Wind blowi... | |
| 11 | 31.01.2018 14:00 | 18.8000 | 767.5000 | 767.3000 | -1.2000 | 45 | Wind blowi... | |
| 12 | 31.01.2018 11:00 | 18 | 768.7000 | 768.5000 | 0.3000 | 57 | Wind blowi... | |
| 13 | 31.01.2018 08:00 | 11.4000 | 768.4000 | 768.2000 | 0.4000 | 88 | Calm, no w... | |
| 14 | 31.01.2018 05:00 | 11.3000 | 768 | 767.8000 | -0.3000 | 88 | Calm, no w... | |
| 15 | 31.01.2018 02:00 | 14.7000 | 768.3000 | 768.1000 | -0.7000 | 67 | Wind blowi... | |
| 16 | 30.01.2018 23:00 | 14.9000 | 769 | 768.8000 | 0.1000 | 66 | Wind blowi... | |
| 17 | 30.01.2018 20:00 | 15.6000 | 768.9000 | 768.7000 | 0.6000 | 65 | Calm, no w... | |
| 18 | 30.01.2018 17:00 | 16.5000 | 768.3000 | 768.1000 | -0.5000 | 57 | Wind blowi... | |
| 19 | 30.01.2018 14:00 | 17.6000 | 768.8000 | 768.6000 | -1 | 55 | Wind blowi... | |
| 20 | 30.01.2018 11:00 | 17.5000 | 769.8000 | 769.6000 | 0.9000 | 60 | Wind blowi... | |
| 21 | 30.01.2018 08:00 | 9.5000 | 768.9000 | 768.7000 | 0.6000 | 94 | Wind blowi... | |
| 22 | 30.01.2018 05:00 | 10.1000 | 768.3000 | 768.1000 | -0.5000 | 93 | Wind blowi... | |
| 23 | 30.01.2018 02:00 | 11.9000 | 768.8000 | 768.6000 | -0.1000 | 84 | Wind blowi... | |

**Figure 3.** Screenshot of raw data

SVR is applied to model the correlation between the date and the average, i.e. the input for the SVR is date and the corresponding output for SVR is the temperature. It is a rather simple case with one dimensional input data. The penalty C is 1000000 and Epsilon is 0.5 for soft margin. Nonlinear Gaussian function is chosen as the kernel function with parameter 0.1. As the regression is an expression of kernel functions, we ignore the regression expression here. By using the regression function for the date, a figure is plot as shown in Figure 4. The predicted data is centred at the real data. The error plot in the right, it shows the error centred around 0 and roughly follows Normal distribution. It implies the obtained regression function can capture the information quite well, so that we can use them for abnormal detect.

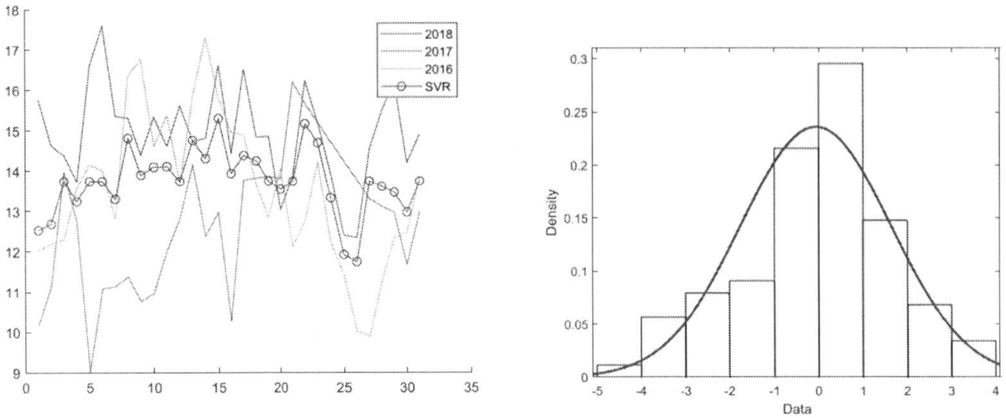


**Figure 4.** Predicted data and original data

We perform a residual analysis. The mean of the error is - 0.06234 and the standard deviation is 1.69. For a date with residual locates at extremes of the distribution can be considered as abnormal. As shown in Figure 5, on the right tail of the distribution, the date locates in the 2% area is abnormal.

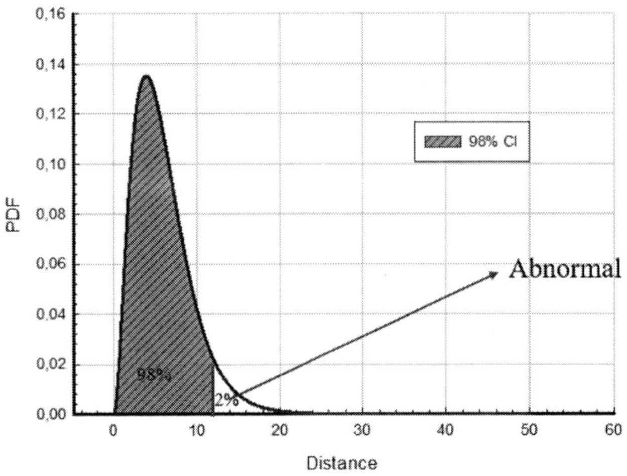


**Figure 5.** Abnormal Definition

For the data shown in Figure 3, if we defines the date locates within 1% of the extremum area as the abnormal date, we can find 2017.Jan. 05 is abnormal date. Above the level such as 0.999, we cannot find any abnormal for these three years. The corresponding extreme level and the abnormal dates are shown in Table 1. The extreme level 0.95 in the table are just for demonstration of the method. If we define the extreme level as 0.99, we found the 2017.Jan.5th has extreme temperature, i.e. it is an abnormal date.

**Table 1.** Abnormal Date

| Extreme Level | **Abnormal Date** |
|---|---|
| *0.95* | *2017.jan.05; 2017.jan.08; 2017.jan.16; 2016.jan. 27; 2018.jan.06.* |
| 0.99 | 2017.Jan. 05 |
| 0.999 | None |
| 0.9999 | None |
| 0.99999 | None |

## 4. Conclusions

The big data using machine learning method shows the feasibility for abnormal detect. SVR can successfully to capture the information of the correlation between date and the temperature. The abnormal date can be figured out from the residual analysis. One can generalize the approach to analyse humidity, wind speed or other environmental factors. The case study also shows the advantage of the machine learning method without requiring the geophysics or geochemical knowledge to detect the abnormal.

## Acknowledgement

We appreciate the support from the EU Interreg Min-north project. The project provides the original idea of the method and the finance for the research.

## References

[1] Jo J and Lee K W, "High-Performance Geospatial Big Data Processing System Based on MapReduce," *Isprs International Journal of Geo-Information,* vol. 7, Oct 2018.
[2] Patrick E A and Fattu J M, *Artificial intelligence with statistical pattern recognition*. Englewood Cliffs, N.J.: Prentice-Hall, Business and Professional Division, 1986.
[3] Theodoridis S and Koutroumbas K, *Pattern recognition*, 3rd ed. San Diego, CA: Academic Press, 2006.
[4] Amari S and Wu S, "Improving support vector machine classifiers by modifying kernel functions," *Neural Networks,* vol. 12, pp. 783-789, Jul 1999.
[5] Chen K Y, "Forecasting systems reliability based on support vector regression with genetic algorithms," *Reliability Engineering & System Safety,* vol. 92, pp. 423-432, Apr 2007.
[6] Schölkopf B and Smola A J, *Learning with kernels : support vector machines, regularization, optimization, and beyond*. Cambridge, Mass.: MIT Press, 2002.
[7] Weather archive in Alexandria (airport) [Online]. Available: https://rp5.ru/Weather_archive_in_Alexandria_(airport)

# Mechanical Properties of Riverbank Soil in the Jingjiang Reach and the Characteristics of Flow in Pocket Collapse Area

**Jun Wang[1], Zhaoxi Liu[2]* and Hongyan Yue[1]**

[1] Changjiang River Scientific Research Institute, Wuhan 430010, China
[2] Shihezi University, Shihezi 832000, China
* E-mail address: 985254641@qq.com

**Abstract.** Based on analyzing the field data and the experimental results, Mechanical properties of riverbank soil in the Jingjiang Reach and the characteristics of flow in the region of pocket collapse are discussed. It is calculated that riverbank of the Jingjiang Reach is characterized by a typical composite structure, covering the low liquid limit cohesive upper layer and the sandy lower layer. The cohesion of the cohesive soil increases first and then decreases with the increase of water content. The internal friction angle generally decreases with the increase of water content. The flow structure in pocket collapse area shows the intensive three-dimensional characteristics. The surface and vertical average velocity, the range and strength of the reflux increase with the increase of upstream flow discharge. Under different conditions of flow discharge, the vertical average velocity in pocket collapse area has a distributed feature of nearly linear or tiny oscillation along the relative direction of water depth. The research can provide the technical support for prevention measures of bank collapse.

## 1. Introduction

Bank failure is a main disaster in the Yangtze River. In 1998, more than 330 cases of large-scale collapse occurred in the middle and lower reaches of the Yangtze River, of which 17 occurred in the middle reach of the Yangtze River and 39 occurred in the lower reach [1]. And about 655 dangerous cases of bank failure occurred in the main stream of middle and lower reaches of Yangtze River during 2003 to 2010. In both 2002 and 2005, bank retreat occurred in Wencunjia reach more than 300m width. A continuous bank failure occurred at the right bank of the Jing98 section, which is located downstream of the apex of the Beimenkou bend, with the cumulative retreat width reaching 332m during the period from 2002 to 2010. In Oct.2012, a case of large-scale collapse, which was about 400m wide and 500m deep, occurred in the Zhenyang Reach. The damage caused by bank collapse is serious, especially pocket bank collapse has the characteristics of large scale, rapid development and great harm, which may decrease the stability of riverbank, the safety of embankments, the development and utilization of coastal line along the river and the safety of people's lives and property.

The Jingjiang Reach is located between Zhicheng and Chenglingji in the Middle Yangtze River, with a length of 347.2km. Bank failure often occurs in the Jingjiang Reach during the flood season and the dry season [2]. The types of bank failure can mainly be divided into the nest collapse (including general nest collapse, pocket bank collapse and isolated collapsing), strip collapsing and washed collapsing in the Yangtze River[1]. The three phases of fluvial process of the pocket collapse, including the riverbed change near shore, sharp scour and relative equilibrium. And the special factor of the forming the pocket bank failure is preliminarily studied in the flow and boundary condition [3].

In view of the previous researches, this paper studied mechanical properties of riverbank soil in the Jingjiang Reach by soil test and the flow structure of pocket collapse by the experiment of the flume model.

## 2. Soil Composition And Mechanical Properties of the Jingjiang Reach

Based on the investigation and sampling of the bank soil in the typical collapse area of the Jingjiang Reach, laboratory soil tests, including the soil composition, physical and mechanical properties of the cohesive soil, were determined by the Geotechnical Test Method (GB/T 50123-1999).

### *2.1. Soil Composition of the Bank*

Field exploration indicates that the riverbanks of the Jingjiang Reach are characterized by a typical composite structure, covering the cohesive upper layer and the non-cohesive lower layer, a thin sandy layer can be found between two cohesive layers at some sections. Figure 1 shows the vertical stratification structures of typical riverbank soils at Guan39 section and the failed bank profile at Jing178 section. The laboratory tests indicate that the cohesive upper soil layer of bank is mainly composed of clay and silt-clay, and the non-cohesive lower soil layer is mainly composed of fine sand and silt. The plasticity index $I_P$ of the cohesive upper layer ranges from 9.0 to 23.8, and liquid limit $\omega_L$ and plastic limit $\omega_P$ are both lower than 50%, so it is regarded that the riverbank of the Jingjiang Reach is a composite structure, covering the low liquid limit cohesive upper layer and the sandy lower layer, and the soil composition along the river bank is almost the same.

**Figure 1.** Failed bank profiles at typical sections (2018.10)

### *2.2. Mechanical Properties of the Bank Cohesive Soil*

The mechanical properties of the cohesive soil can often be represent by its shear strength and tensile strength, For the cohesive riverbank of the Jingjiang Reach, its shear strength which includes the parameters of cohesion ($c$) and the angles of internal friction ($\varphi$) was studied. The results of laboratory soil tests indicate that water content ranges from 30% to 43%, cohesion ranges from 5kPa to 22.2kPa and the internal friction angle ranges from 6.8°to 16.8°in nature.

According to the results of direct shear test, the relationship between water content and shear strength of cohesive soil (remolded) are shown in Figure 2, it is obvious that the cohesion of the cohesive soil increases first and then decreases with the increase of water content, and there is a critical water content, which makes the cohesion reach the peak value. The internal friction angle generally decreases with the increase of water content [4].

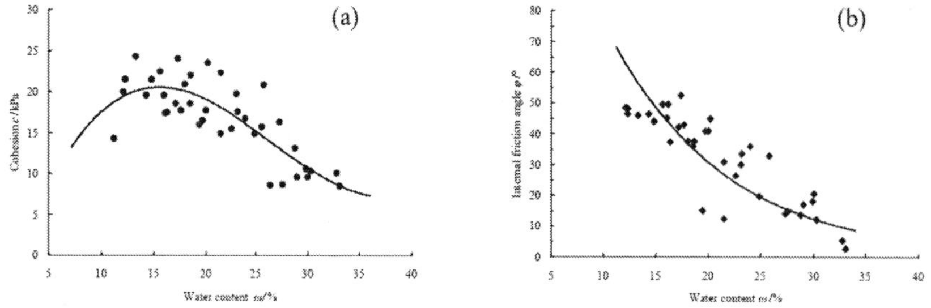


**Figure 2.** Relationships between shear strength parameters and water content: (a) cohesion and (b) internal friction angle.

## 3. Flume Model Test

### *3.1. Model Design*

The total length of the experiment flume is 49m. In order to simulate the natural pocket collapse of composite structure, the collapse process of the natural and composite riverbank is simulated by the different particle size of the new type of compound plastic sand (adding adhesive of the appropriate proportion 1% to the upper layer) through the indoor generalized model tests. The median size of the lower layer made of compound plastic sand is 0.2mm. The median size of the upper layer is 0.058mm.

### *3.2. Test Content*

The riverbank change and movement characteristics of flow and sediment in the made pocket collapse(thickness ratio of the upper layer to the lower layer is 2:1; the slope of the river bank is 1:1) are simulated under the different flow conditions(when the design water depth is 0.4m,the design discharge are $0.1m^3/s$ and $0.2m^3/s$). Horizontal width and vertical width ratio is 1:1, the entrance width is about 1m.

During the experiment, the incoming flow discharge is controlled by the electromagnetic flowmeter. The water level in the tail gate is controlled automatically by the WEL-II-type content gauge developed by the Wuhan University. The typical cross section topography is measured by the CY- I -type silting meter. The velocity is measured mainly by the 3D acoustic Doppler velocimeter (ADV), while the surface velocity is measured by the Rotary pulp type current meter.

## 4. Flow Characteristics in Pocket Collapse Area

### *4.1. Reflow Characteristics*

By the role of the reflux, the flow structure in the pocket collapse shows the intensive three-dimensional characteristics. Reflux in different strengths generate in the interior of the pocket collapse by the function of turbulent shear stress and gravity around the separation and interface of the water. Separation of flow and collapse zone boundaries is the immediate cause for the flow subregion of the primary and secondary streams, and turbulent shear stress which caused by lateral mixing of large-scale and small-scale vortex body in the interface of mainstream and reflux is the power source of the form of the reflux in the region of the pocket collapse. The water near the entrance in the region of the pocket collapse moves forward by the role the turbulent shear stress, following the mainstream. Then the water moves to the reflux area along the entrance downstream sidewall. The water level in the inside of the reflux area rises, resulting in the surface transverse slope. Water and sediment in the reflux area of the pocket collapse and that in mainstream area are going a mass, momentum and energy exchange through the entrance interface. Flow momentum exchange in both sides of mainstream and reflux cross interface not only include mixing of particles and eddy transferring, but also direct convective transfer of large-scale vortices. So it intensifies turbulent mixing of the water, which makes the flow more unstable in the region of the pocket collapse, and aggravates the exchange of material on both sides of the interface simultaneously.

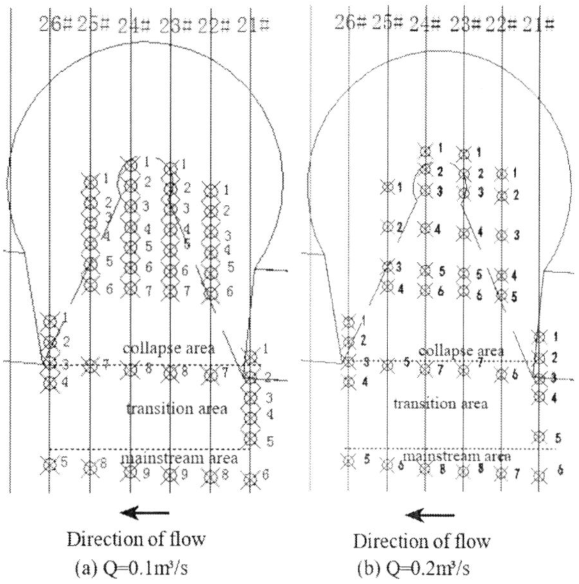


**Figure 3.** Layout of measurement of flow points of pocket bank collapse.

The reflux flow lines in the pocket collapse appear as multiple concentric circles of varying sizes when flow lines in general nest collapse are multiple irregular ovals and there are many independent of scale size varying vortex. Moreover, the water surface arise irregular distortions. Surface velocity and the range and intensity of the reflux increase with the increase of upstream discharge.

### *4.2. Velocity Distribution Characteristics*

As showed in Figure 4 and Table1, survey area can be divided into collapse area, transition area and mainstream area according to different measuring point location in the pocket collapse.

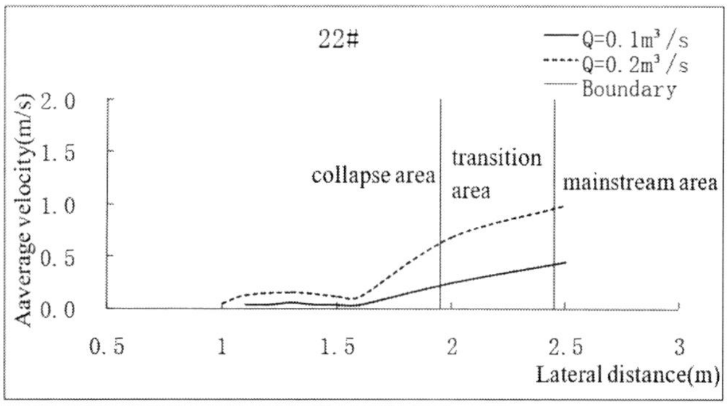


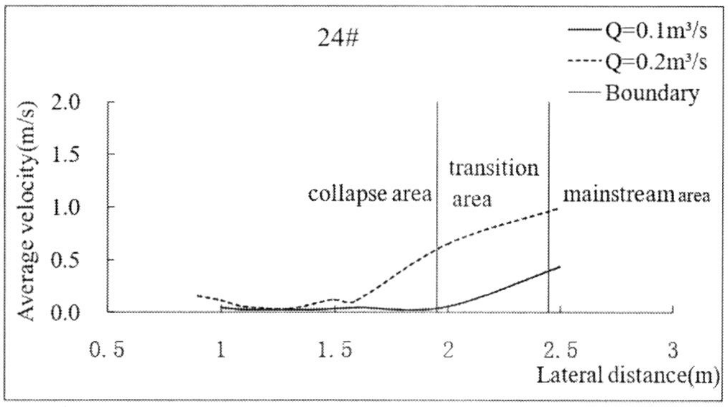

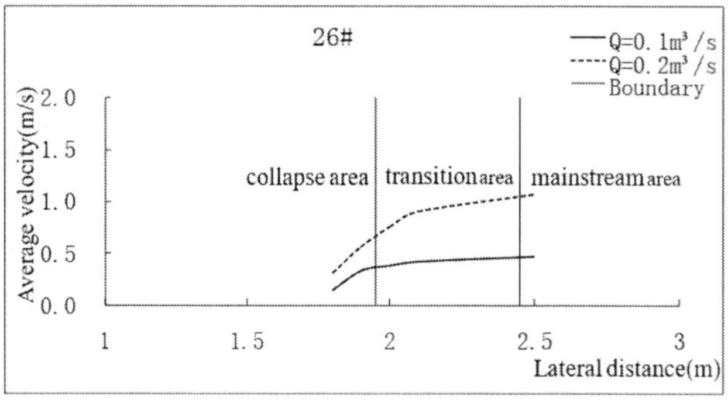


**Figure 4.** Distribution of average velocity of typical cross-sections of pocket bank collapse.

**Table 1.**Classification and arrangement of measuring points in pocket collapse area.

| Discharge($m^3/s$) | Collapse area | Transition area | Mainstream area |
|---|---|---|---|
| 0.1 | Measuring points except transition and mainstream area | 22#-7,23#-8,24#-8, 25#-7,26#-4 | 22#-8,23#-9,24#-9,25#-8,26#-5 |
| 0.2 | Measuring points except transition and mainstream area | 22#-6,23#-7,24#-7, 25#-5,26#-4 | 22#-7,23#-8, 24#-8,25#-6,26#-5 |

As showed in Figure 4, affected by flow instability in side wall of collapse area, separation zone of mainstream and reflux interface and recirculation zone, the regularity of lateral distribution form of offshore vertical average velocity of typical cross sections in the pocket collapse is not strong, and velocity distribution in the transverse direction is different. Flow velocity is in the range of 0.29m/s to 0.42m/s, and 0.76m/s to 1.00m/s, when discharge are $0.1m^3/s$ and $0.2m^3/s$, respectively. Consequently, the larger is the upstream flow, the greater is the average velocity in the pocket collapse. The average velocities of typical cross sections of the pocket collapse in collapse area are minimal, and the largest in mainstream area, while transition area is between the two.

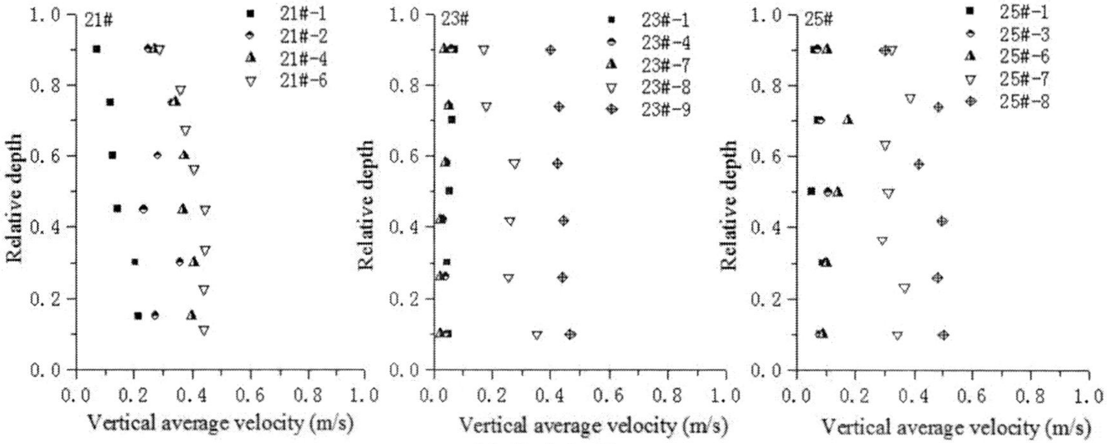


(a) $Q=0.1m^3/s$

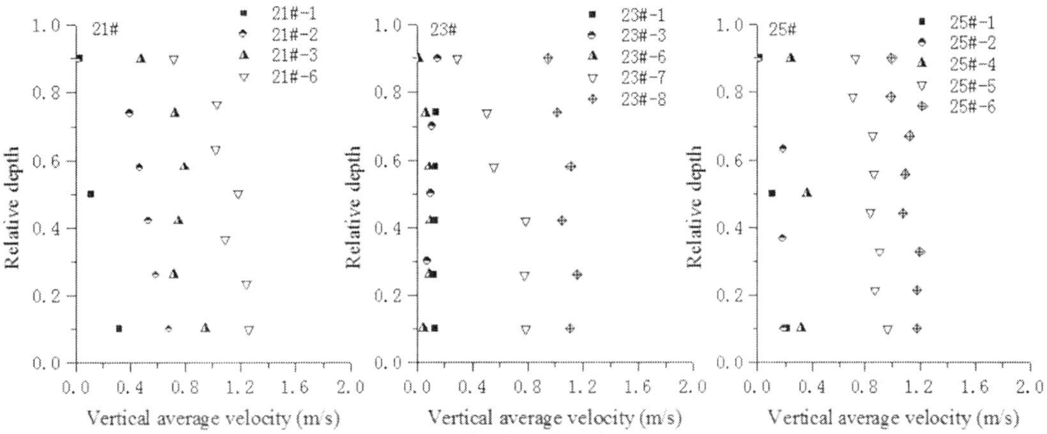


(b) Q=0.2m³/s

**Figure 5.** Distribution of vertical average velocity on typical cross-sections.

As seen in Figure 5, the vertical average velocity of various sections of the pocket collapse in collapse area are minimal, and the largest in mainstream area, while transition area is between the two. Under various discharge, mean velocity on a vertical of each measuring point is close to each other near the central area of collapse area, such as sections of 3#-1, 23#-4, 23#-7. And the velocities are significantly less than these in the transition region (23#-8) and in the mainstream area (23#-9). The average vertical velocity distribution characteristics are that the variation of vertical average velocity is small along the direction of the relative depth and the distribution is the approximate linear or small shocks. When flow increases to $0.2m^3/s$ from $0.1m^3/s$ , the law of vertical velocity distribution is similar but different in size, and the vertical average velocity distribution of various sections along the direction of the relative depth are more distributed.

## 5. Conclusions

(1)Riverbanks of the Jingjiang Reach are characterized by a typical composite structure, covering the low liquid limit cohesive upper layer and the sandy lower layer. The cohesion of the cohesive soil increases first and then decreases with the increase of water content. The internal friction angle generally decreases with the increase of water content.

(2)During the development of the pocket collapse, suffered by the reflux, the upstream and downstream of the entry both are scoured and collapsed seriously. The flow structure in the pocket collapse occur strong three-dimensional characteristic.

(3)With the increase of the upstream flow discharge, the surface and vertical average velocity both increase. The range and strength of the reflux are both bigger. Under the different flow discharge, the vertical average velocity distribution in the region of the pocket collapse has a distributed feature of nearly linear or tiny congestion along the relative direction of water depth. After the flow discharge increasing, it makes the vertical average velocity distribution in every measurement point more dispersed along the relative direction of water depth.

## Acknowledgements

Project supported by the National Natural Science Foundation of China (51479008) and basic scientific research fee of commonweal scientific research institutes at central government level (CKSF2016018/HL).

## References

[1] Yu W C, Lu J Y. 2008. Bank Erosion and Protection in the Yangtze River. China Water and Power Press, Beijing (in Chinese)

[2] Zong Q L, Xia J Q, Deng C Y and Xu Q X. 2013 J Modeling of the Composite Bank Failure Process Using BSTEM. Journal of Sichuan University (Engineering Science Edition) **45(03)** 69-78 (in Chinese)
[3] Yu W C. 2007 J. The Yangtze River. The Initial Research on the Forming Mechanism of the Pocket Bank failure in the middle and lower reaches **38** 156-159 (in Chinese)
[4] Xia J Q, Zong Q L, Deng S S, Xu Q X and Lu J Y. 2014 J Journal of Hydrology. Seasonal variations in composite riverbank stability in the Lower Jingjiang Reach, China **519** 3664-3673

# The Distribution of Heavy Metals in the Sediment of Low Tidal Flat, Eastern Chongming Island, China

**Jinmei Lu and Fuqing Yuan**

Department of Technology and Safety, UiT the Arctic University of Norway, 9037, Norway.
Email: jinmei.lu@uit.no

**Abstract.** The distribution of heavy metals in the tidal flat is of great significance on the estuary environment. This study aims to find the heavy metal distribution mechanism at low tidal flat in Yangtze Estuary, China. 4 sediment cores were collected at low tidal flat of eastern Chongming Island at four seasons spring, summer, autumn and winter. The concentrations of element Al, Cu, Cr, Fe, Mn, Pb, Rb, Zn, Zr in the sediment were analyzed. The vertical distribution of different element with depth were analyzed and the main factors that control the distribution of heavy metals in the sediments of low tidal flat were discussed. The results showed that the source of sediment and tidal hydrodynamics are the two main factors that controls the distribution of heavy metals in the sediment of low tidal flat. Zr in the sediments can be used as geochemical marker to reflect the erosion and deposition changes in tidal flat.

## 1. Introduction

Tidal flats, as wetland habitats, are an important ecosystem and of great ecological significance for humans. Estuarine flats provide vital feeding grounds for migrant and native birds as well as nursery zones for fish [1]. However due to increasing urban and industrial development in surrounding areas, more and more contaminants such as heavy metals are discharged into these ecologically sensitive areas. The tidal flats tend to become a sink or source for the heavy metals [1-3]. Heavy metals are adsorbed onto suspended particles, deposit on the bottom as sediments and accumulate in the sediments of tidal flats [3]. Sediments are regarded as an effective archive that can record heavy metal contamination [4].

Sediment pollution by heavy metals has become a critical problem in marine environment due to their toxicity, persistence, bioaccumulation and non-degradability in the environment [4-7]. Heavy metals in the sediments could significantly affect the health of marine ecosystems, and they can act as a source of heavy metals, imparting them into water and degrade water quality when the conditions of sedimentary environment change [4]. Therefore the study of heavy metal distribution in the sediments of estuary tidal flats is of great importance for the sustainable development of this region.

The Yangtze River is considered the largest and most important river in Asia [8]. It ranks ninth globally in terms of drainage area, and fourth in sediment flux (500 Mt/year before its decline in the 1970s) [8]. Among 470 million tons of silt carried by the Yangtze River annually, about half is deposited on Shanghai tidal flats [9]. Eastern Chongming tidal flat is a well-developed tidal flat and the largest intertidal zone of the Yangtze River estuary [9]. The study area is located on the eastern coast of Chongming island (Figure 1). The sediment in this tidal flat has obvious zonation and divided into high, middle and low tidal flat from land to sea. Our study is focus on the low tidal flat and study the heavy metal distribution mechanism in the sediment of this zone.

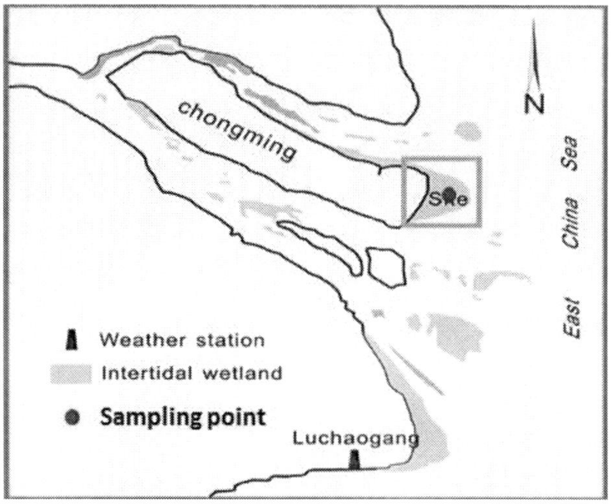


**Figure 1.** The study area and sampling site [9]

## 2. Materials and Methods

### *2.1. Sampling and Pretreatment*

Four sediment cores were collected with PVC pipe (ca. 40cm in length, 100mm in diameter) at low tidal flat during ebbing period (Figure 1). The sampling was performed during a year at four different seasons in January, April, July and September. The sampling and pretreatment follow the same procedure at that described in [9].

### *2.2. Major and Trace Element Analysis*

4g sediment sample was weighed and put onto a low pressure polyethylene base. The sample pretreatment and analysis methods is the same as that described in [9].

### *2.3. Standardization of Elements*

Grain size is an important parameter to affect the distribution of heavy metals. Natural transport processes of heavy metals mainly depend on the presence and transport of find-grained sediment, which is an efficient sink for heavy metals, with the capacity to concentrate and retain heavy metals [7]. Therefore before analyzing the temporal and spatial distribution of heavy metals in sediments, the concentrations of elements are normally standardized with Aluminum, one of the major chemical components of fine-grained sediments, to eliminate the effect of grain size on heavy metals content [9]. The standardization method is as follows:

$$\mathbf{C_{i\ normalised}} = \frac{\mathbf{C_i}}{\mathbf{C_{Al}}}$$

Where $\mathbf{C_{i\ normalised}}$, $\mathbf{C_i}$, $\mathbf{C_{Al}}$ indicate the content of element **i** after normalization, content of element **i** before normalization and content of Al before normalization.

## 3. Results and Discussions

The vertical distributions of heavy metals in the four seasons were analyzed and the results were shown in Figure 2-5. The distribution mechanisms of heavy metals in different seasons were discussed in the following section.

### *3.1. The Vertical Distribution of Heavy Metals in the Sediment in Spring*

In spring, Cr and Zr showed very consistent vertical distribution characteristics, which indicates that Cr mainly migrates with relatively course particles? In a depth deeper than 12cm, Fe and Mn showed similar distribution characteristics as that of Cr and Zr, which showed a relatively large peak at a depth of 16cm and a relatively small peak at a depth of 22cm. However, Cu, Pb, Rb and Zn exhibit an opposite vertical distribution from that of Zr and Cr. Since Zr mainly produced and enriched in heavy minerals, it is estimated that the vertical distribution of Fe and Mn in the sediments of low tidal flat in spring is primarily caused by the source of sediments. Fe and Mn are mainly enriched in the heavy minerals of course particles. While the distribution of Cr, Cu, Pb, Rb, and Zn are more affected by the grain size, and mainly attached to the fine sedimental particles and migrate with fine grain size particles. Due to the strong hydrodynamic effect in the low tidal flat, the sediment is under constant stirring. The enrichment of heavy metals in the redox boundary layer during early diagenesis is a relatively slow process. So if the sediment deposits at a relatively slow speed, heavy metals have sufficient time for post-deposition migration and form a peak at the redox boundary layer. In the contrary, if the sediment deposit at a fast speed, it is difficult to form a peak. Therefore the strong hydrodynamic conditions at low tidal flat are not conducive to the formation of redox boundary layer. So the vertical distribution of heavy metals in the sediment of low tidal flat is mainly affected by the source of sediment, and the post-migration after deposition process is weak.

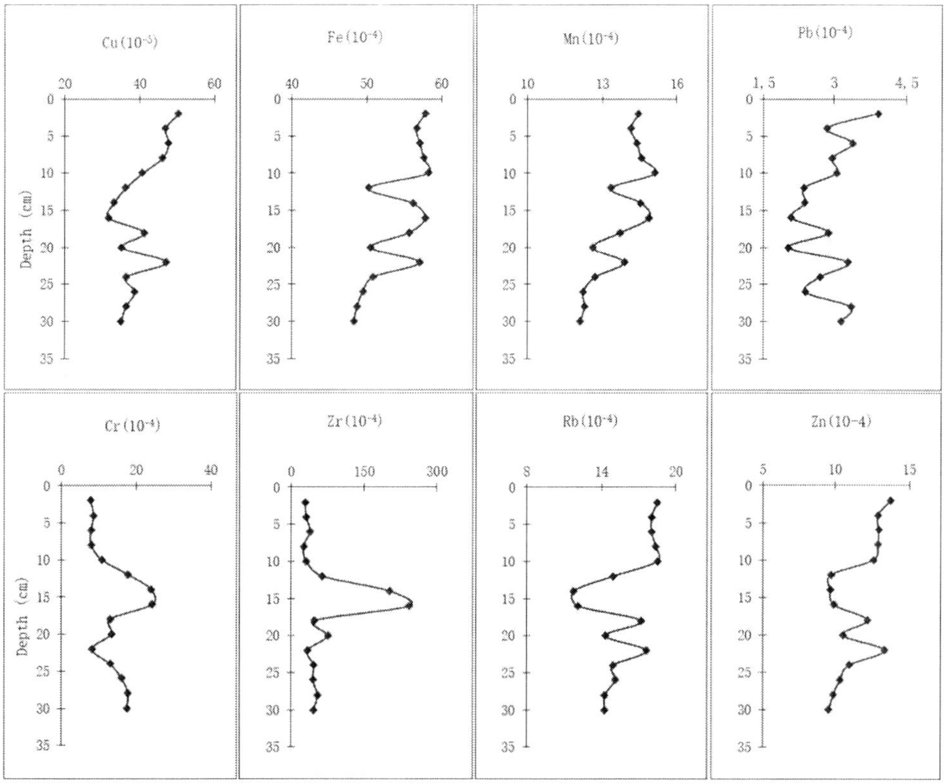


**Figure 2.** The vertical distributions of metals in the middle tidal flat in spring

### *3.2. The Vertical Distribution of Heavy Metals in the Sediment in Summer*

In summer, Fe, Mn and Cr, Zr showed similar vertical distribution in the sediment of low tidal flat at a depth deeper than 18cm. Zr is very stable in the sediment, and the post-migration is extremely weak. Therefore the similar distribution of Fe, Mn and Cr, Zr indicates that the distribution of Fe, Mn is mainly affected by the source of sediment. Fe and Mn are not in the form of oxides and are mainly present in the heavy minerals rich in Zr. The peak value moved from 16cm to about 26cm. This is mainly because the tidal flat is dominated by siltation process from spring to summer. The sedimentation depth is about 10cm. At a depth lower than 18cm, Cu, Pb, Zn, Fe and Mn exhibit an opposite distribution characteristic to Cr, which is especially obvious between the distribution of Cu and Cr. Zr indicates a relatively course sediment source and Cr migrates mainly with relatively course particles in the sediment. So the vertical distribution of heavy metals in the sediment of low tidal flat in the summer is mainly affected by the source of sediment and tidal flat siltation process.

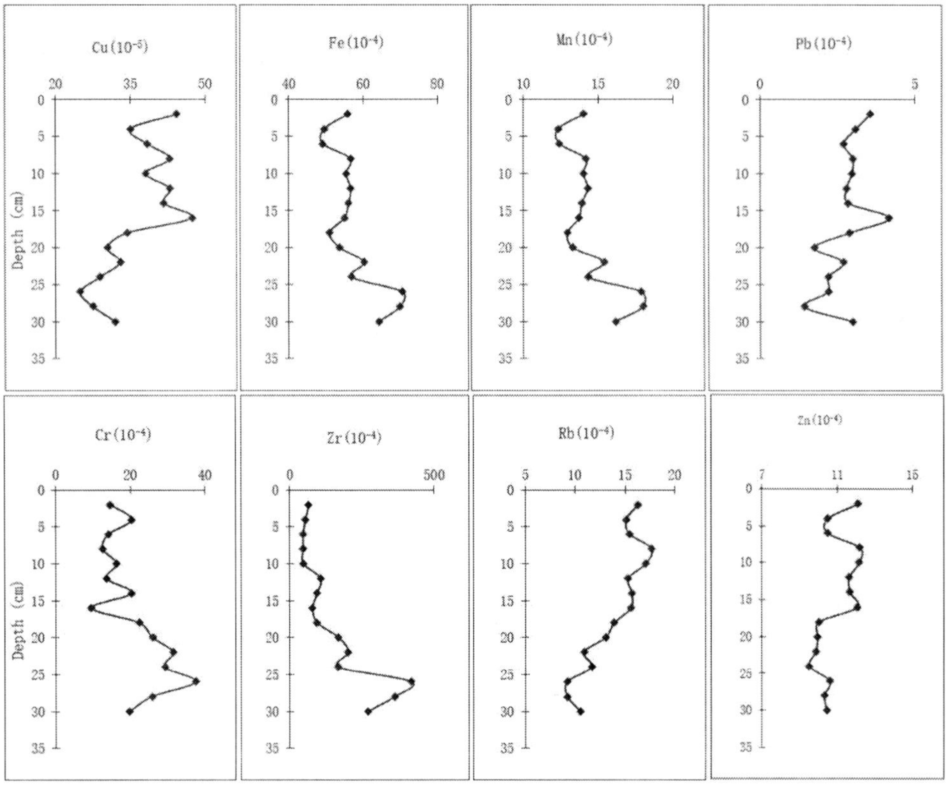


**Figure 3.** The vertical distributions of metals in the low tidal flat in summer

### *3.3. The Vertical Distribution of Heavy Metals in the Sediment in Autumn*

In autumn, Fe, Mn and Zr, Cr showed similar vertical distribution characteristics at a depth deeper than 14cm and both have a peak value at a depth of 16cm. This indicates that the tidal flat experiences mainly a scouring effect from summer to autumn and about 10cm of sediments are washed away from surface. From 14cm to surface layer, Zr and Cr showed little change, while Cu, Pb, Zn, Fe and Mn showed an increasing trend from 16cm to the surface. This may be the combined effect of sediment source and organic matter.

### *3.4. The Vertical Distribution of Heavy Metals in the Sediment in Winter*

From autumn to winter, the tidal flat is still dominated by scouring effect. The depth of scouring is about 10cm. The peak value moved from a depth of 16cm in autumn to about 6cm. From 14cm to the surface layer, the heavy metals generally showed a decreasing trend and enriched in the surface. This may be related to the sediment source and the hydrodynamic effect.

In summary, the distribution of heavy metals in the sediments of low tidal flat is mainly affected by the sediment source and the tidal hydrodynamic forces. The low tidal flat is under constant strong hydrodynamic effect and experiences intense erosion and siltation process throughout seasons. The average erosion and deposition depth is about 10cm. In winter and spring the siltation process dominates and in summer and autumn, the souring process dominates (Figure 6).

## 4. Conclusions

The distribution of heavy metals in the sediments of low tidal flat is mainly affected by the sediment source and the tidal hydrodynamic forces. The low tidal flat is under constant strong hydrodynamic effect and experiences intense erosion and siltation process throughout seasons. The average erosion and deposition depth is about 10cm. In winter and spring the siltation process dominates and in summer and autumn, the souring process dominates. The enrichment depth of Zr can be used as an indicator to indicate the erosion and deposition of tidal flat.

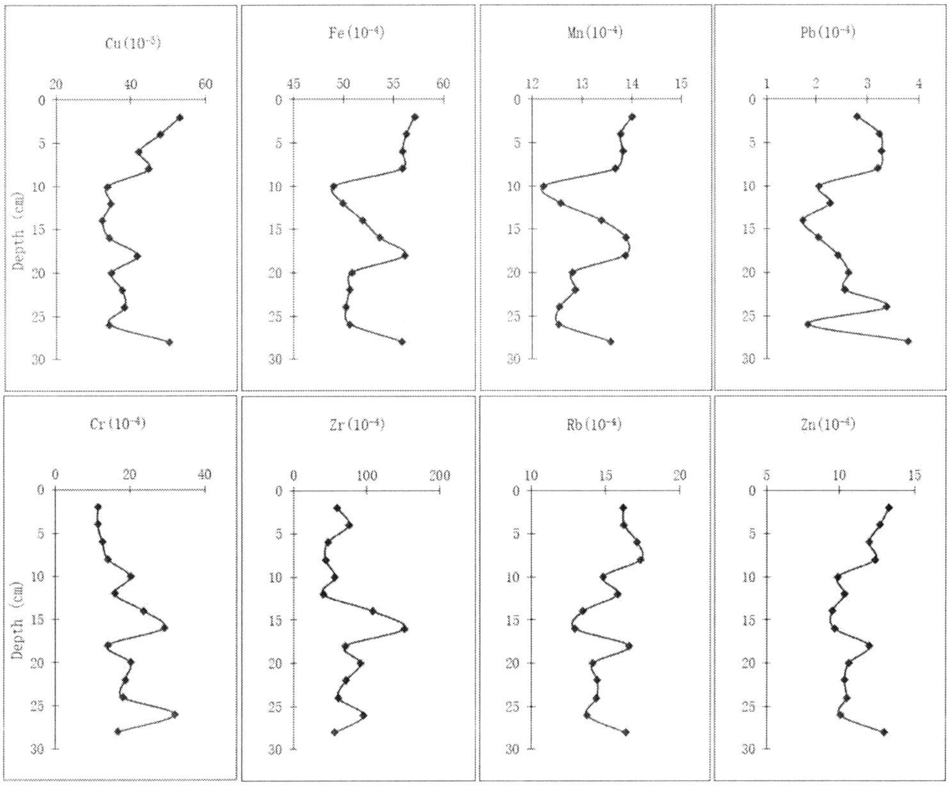


**Figure 4.** The vertical distributions of heavy metals in the low tidal flat in autumn

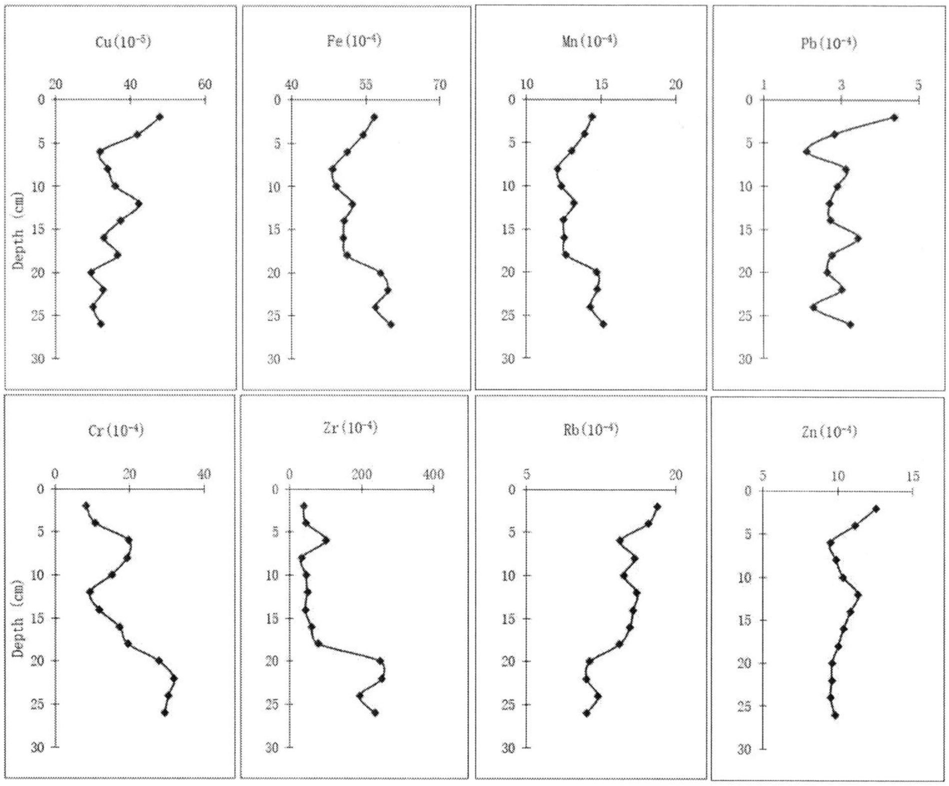


**Figure 5.** The vertical distributions of heavy metals in the low tidal flat in winter

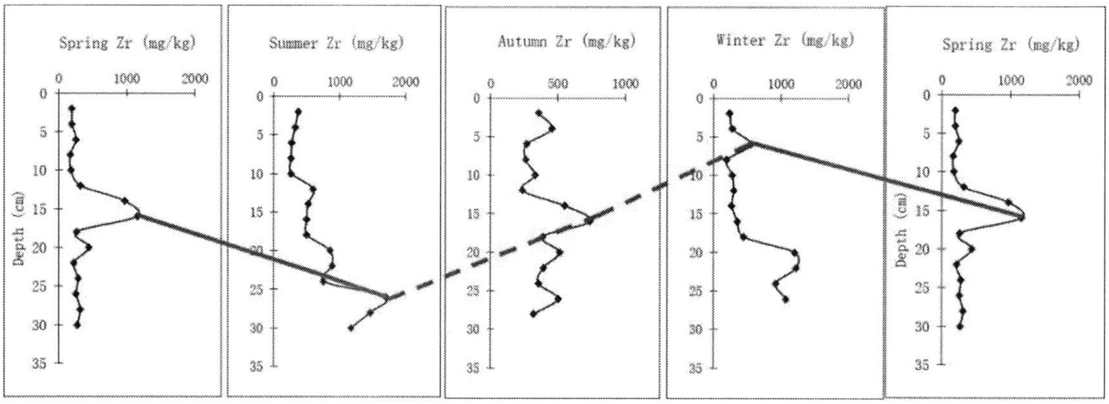


**Figure 6.** The erosion and siltation of low tidal flat in the four seasons

**Acknowledgement**

We appreciate the support from the Interreg EU Framework project MIN-NORTH Development, evaluation and optimization of measures to reduce the impact on the environment from mining activities in northern regions (Grant ID 20200531).

**References**

[1] S. E. Botte, R. H. Freije, and J. E. Marcovecchio, "Distribution of Several Heavy Metals in Tidal Flats Sediments within Bahia Blanca Estuary (Argentina)," (in English), *Water Air and Soil Pollution,* vol. 210, no. 1-4, pp. 371-388, Jul 2010.

[2] C. K. Na and H. J. Park, "Distribution of heavy metals in tidal flat sediments and their bioaccumulation in the crab Macrophthalmus japonicas in the coastal areas of Korea," (in English), *Geosciences Journal,* vol. 16, no. 2, pp. 153-164, Jun 2012.
[3] R. Zhang, F. Zhang, Y. J. Ding, J. R. Gao, J. Chen, and L. Zhou, "Historical trends in the anthropogenic heavy metal levels in the tidal flat sediments of Lianyungang, China," (in English), *Journal of Environmental Sciences,* vol. 25, no. 7, pp. 1458-1468, Jul 1 2013.
[4] H. Zhu, H. J. Bing, H. P. Yi, Y. H. Wu, and Z. G. Sun, "Spatial Distribution and Contamination Assessment of Heavy Metals in Surface Sediments of the Caofeidian Adjacent Sea after the Land Reclamation, Bohai Bay," (in English), *Journal of Chemistry,* 2018.
[5] F. H. Wesal, "Metal contamination in the sediments of Tidal Flat for Iraq coastal," *International journal of Science and Research (IJSR),* vol. 7, no. 3, pp. 120-125, 2016.
[6] F. Ye, X. P. Huang, D. W. Zhang, L. Tian, and Y. Y. Zeng, "Distribution of heavy metals in sediments of the Pearl River Estuary, Southern China: Implications for sources and historical changes," (in English), *Journal of Environmental Sciences,* vol. 24, no. 4, pp. 579-588, 2012.
[7] W. H. Gao, Y. F. Du, S. Gao, J. R. Ingels, and D. D. Wang, "Heavy metal accumulation reflecting natural sedimentary processes and anthropogenic activities in two contrasting coastal wetland ecosystems, eastern China," (in English), *Journal of Soils and Sediments,* vol. 16, no. 3, pp. 1093-1108, Mar 2016.
[8] S. L. Yang, K. H. Xu, J. D. Milliman, H. F. Yang, and C. S. Wu, "Decline of Yangtze River water and sediment discharge: Impact from natural and anthropogenic changes," (in English), *Scientific Reports,* vol. 5, Jul 24 2015.
[9] J. Y. Lu, F.; Zhang, F.; Zhao, Q., "The study on heavy metal distribution in the sediment of middle tidal flat in Yangtze Estuary, China," *Environmental Earth Sciences,* vol. 75, p. 557, 2016.

# Study on the Purification Effect of Clean Fresh Air on PM2.5 in the Bedroom

**Qian Yuan and Guojian Li**

School of Civil Engineering and Architecture, Zhejiang Sci-Tech University, Hangzhou, 310018, China;
E-mail: 389923440@qq.com

**Abstract.** In order to study the purification effect of clean fresh air on PM2.5 in bedroom, the concentration distribution of PM2.5 in bedroom was numerically simulated using RNG k-e turbulence model based on DPM discrete phase model. In the case of severe haze, clean fresh air is introduced to meet the needs of indoor personnel. Aiming at two common types of airflow in the bedroom (same side/opposite side upper and lower back) and two different air supply speeds, studied the purification effect of fresh air on PM2.5. The results showed that there was no significant difference in the decrease rate of PM2.5 in the respiratory surface under the two inlet velocities. In the first 600s, the concentration of PM2.5 decreased by 75% with the inlet and outlet were on the same side, while the PM2.5 concentration decreased by 63% with the inlet and outlet were on the opposite side. The PM2.5 concentration distribution was more uniform when the inlet and outlet were on the same side than the opposite side. It is concluded that the air supply position has a greater influence on the purification effect than the air inlet velocity when the fresh air volume is constant. When the inlet and outlet were on the same side the introduction of the clean fresh air in the bedroom is more effective in purifying the PM2.5 than the opposite side.

## 1. Introduction

With the influence of the outdoor smog and people's life activities, indoor particle pollution becomes a big problem [1] which affects people's physical and mental health seriously, and become one of the major components of indoor pollution. When the indoor air of fine particles of pollutants to a certain concentration, it will pose a great threat to human health and even threaten people's lives [2-5]. According to the current relevant research data at home and abroad show that PM2.5 concentration in the building are generally high, so the control of indoor particulate matter without delay.

At present, most residential buildings are equipped with mechanical ventilation to improve the indoor environment. However, the mechanical ventilation used in the buildings is not equipped with a particle-specific filtering device [9]. Although the air-handling equipment also has some filtering effects, for fine particles, its purification effect fails to meet people's needs. Bedroom is the essential resting place for people, it is very important to control the particles in the bedroom. Currently, two main methods to remove indoor particles are air purification and fresh air systems [10]. Fresh air systems can better meet people's needs in terms of hygiene requirements. However, at present, the research on indoor particulate matter mainly focuses on the correlation research of indoor and outdoor particulate matter [6-8]. There is insufficient research on the purification of indoor PM2.5 after introducing fresh air into treatment.

The purification effect of clean fresh air on PM2.5 in bedroom is studied. Under the condition of guaranteeing the fresh air quantity needed by indoor personnel, the purification efficiency of clean fresh air on PM2.5 and the distribution of indoor PM2.5 are studied.

## 2. Theoretical analysis

Since the PM2.5 distribution is mainly affected by the airflow organization, and the influence factors of the airflow organization are mainly the air supply position and parameters. In this paper, the fresh air is introduced to meet the fresh air demand of the indoor personnel, not bear the indoor heat load. Therefore, does not consider the influence of supply air temperature for the research, the influence of the air supply position and the air supply speed on the PM2.5 distribution was studied.

This study is typical of gas-solid two-phase flow. The RNG k-ε model was applied to simulate the airflow. For the PM2.5 in a bedroom, due to the volume fraction of discrete items is less than 10%, a Discrete Phase Model (DPM) was used for simulating the PM2.5 injection and dispersion.

## 3. Model

Taking the resident bedroom in typical residential buildings as the simulation object, the size of bedroom is 3.6 m x 4.8 m x 2.6 m, the bed is 2 m x 2 m x 0.5 m, the table is 0.6 m x 1.5 m x 0.8 m, the wardrobe is 2 m x 0.5 m x 2 m, The layout is shown in figure 1.

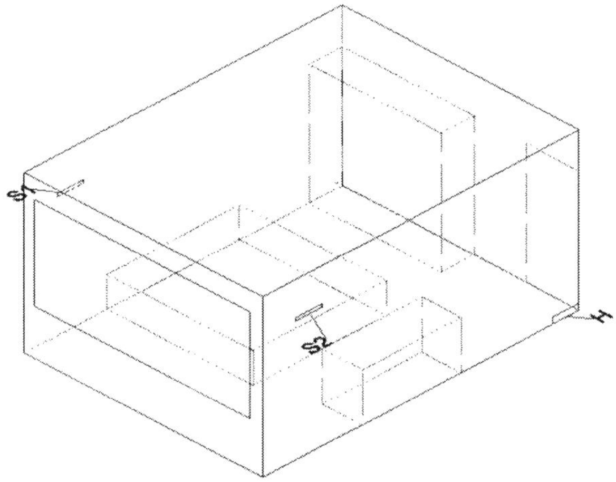


**Figure 1.** Bedroom and vents set simplified diagram

Taking into account the convenience of the introduction of fresh air in the outdoor, the position of the fresh air supply port is shown in figure 1. There are two settings of S1 and S2. Since the exhaust vent is usually not provided in the bedroom, the air is discharged from the door seam, and the door slit is simplified here as the outlet as shown in figure 1, and the size is 0.4 m × 0.1 m.

## 4. Condition Setting

The amount of fresh air is determined by the number of indoor air changes and the amount of fresh air required by personnel. The number of air changes in the bedroom is 1/h, the required air volume is 40 $m^3/h$. The number of indoor personnel is 2; the total fresh air required is 60 $m^3/h$, so the introduction of fresh air volume is 60 $m^3/h$.

Indoor PM2.5 distribution is mainly affected by air distribution. In this paper, the air distribution is changed by changing the position or inlet velocity. The position of air supply is shown in figure 1. Referring to the size of wall-mounted air conditioner and air purifier in bedroom, two kinds of air inlets with sizes of 0.4 m × 0.05 m and 0.2 m × 0.05 m are selected in this paper, and the indoor air inlet velocities are calculated to be 0.83 m/s and 1.67 m/s, respectively. According to the location of the inlet and the inlet velocity, the following four simulation conditions are designed.

**Table 1**. Simulation Case Setting Table

| Case | Inlet and outlet setting | Air volume $Q/(m^3 \cdot h^{-1})$ | Inlet velocity $v/(m \cdot s^{-1})$ |
|---|---|---|---|
| 1 | opposite side | 60 | 0.83 |
| 2 | opposite side | 60 | 1.67 |
| 3 | same side | 60 | 0.83 |
| 4 | same side | 60 | 1.67 |

When the fog is severe, the outdoor PM2.5 concentration is 200 μg/m$^3$, and the high-efficiency filter is used in the fresh air treatment equipment. The purification efficiency of PM2.5 is 90%, that is, the PM2.5 concentration in the air supply is 20 μg/m$^3$. The height of the bed is 0.5 m, the height of the respiration surface is 0.6 m when sleeping, in the simulation study, z = 0.6 were taken as the monitoring surface.

## 5. Simulation Results

### *5.1. Flow field and PM2.5 Concentration Distribution under Different Working Conditions*

*5.1.1. Case 1: Set inlet and outlet on the different sides, the inlet velocity is 0.83 m/s.* It can be seen from the velocity profile in figure 2 that the airflow distribution is complicated, and multiple airflow vortices appear on the monitoring surface. Figure 3 shows the specific distribution of the PM2.5 concentration on the monitoring surface. As can be seen from figure 2 and figure 3, the PM2.5 concentration distribution is affected by the airflow field. PM2.5 concentration is relatively high in the area with low airflow velocity, while PM2.5 concentration is relatively low in the area with high airflow velocity.

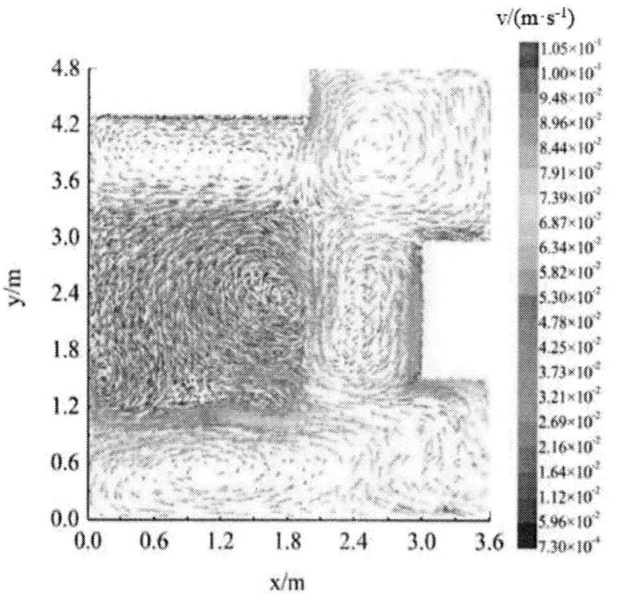


**Figure 2.** Plane z = 0.6 velocity profile

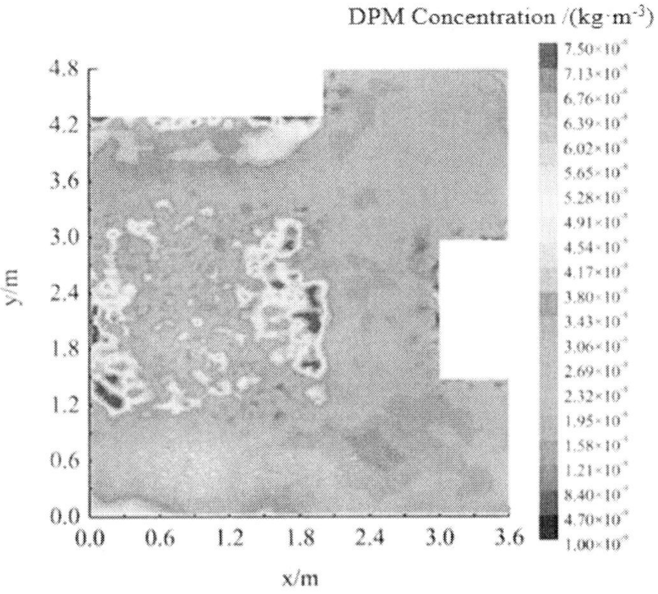


**Figure 3.** 1000s plane z = 0.6 PM2.5 concentration profile

*5.1.2. Case 2: Set inlet and outlet on the different sides, the inlet velocity is 1.67 m/s.* Comparing with the velocity profile in figure 2 and figure 4, it can be seen that under the premise of unchanged supply air position and fresh air volume, there is no obvious difference in the distribution law of the flow field when adjusting the inlet velocity. As the distribution of PM2.5 is affected by the air distribution, the air velocity increases, and the distribution of PM2.5 changes accordingly. As can be seen from figure 5, the PM2.5 concentration above the bed is significantly lower than case 1.

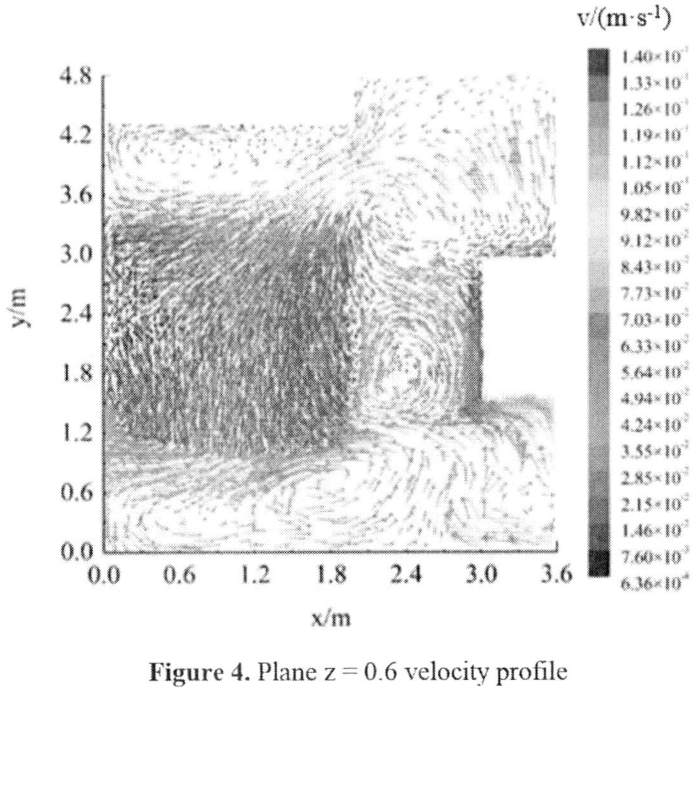


**Figure 4.** Plane z = 0.6 velocity profile

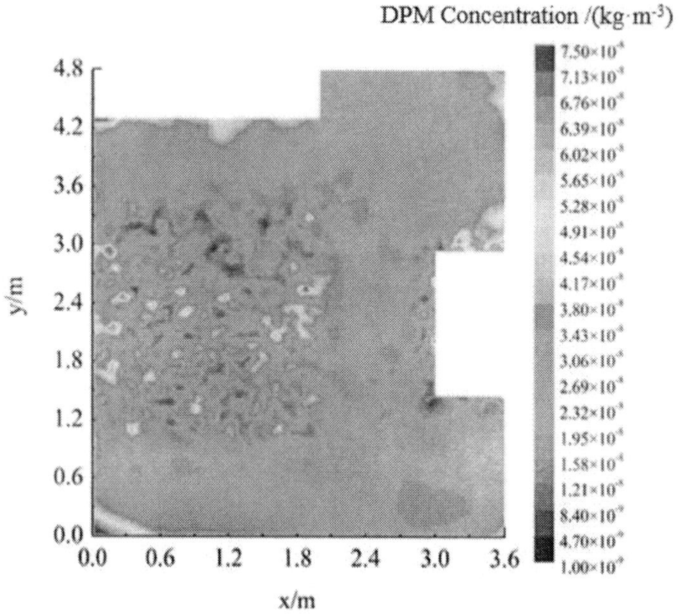


**Figure 5.** 1000s plane z = 0.6 PM2.5 concentration profile

5.1.3. *Case 3: Set inlet and outlet on the same sides, the inlet velocity is 0.83m/s.* Figure 6 shows the velocity distribution of the monitoring surface at the speed of 0.83 m/s from S2. On the whole monitoring surface, the vortices mainly appear in the corners, and the velocity of the air flow above the bed is relatively high. Figure 7 shows the specific distribution of PM2.5 on the monitoring surface. The PM2.5 concentration in the breathing surface of the bed was greatly reduced, and the PM2.5 concentration distribution on the respiratory surface was relatively uniform.

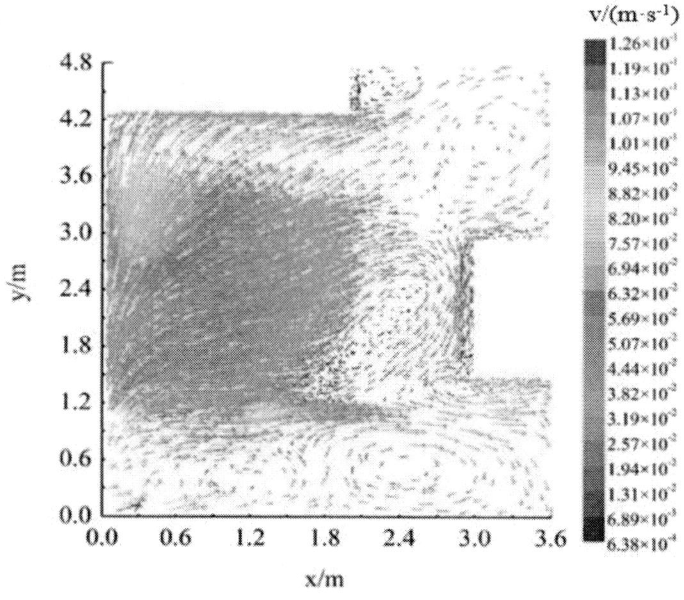


**Figure 6.** Plane z = 0.6 velocity profile

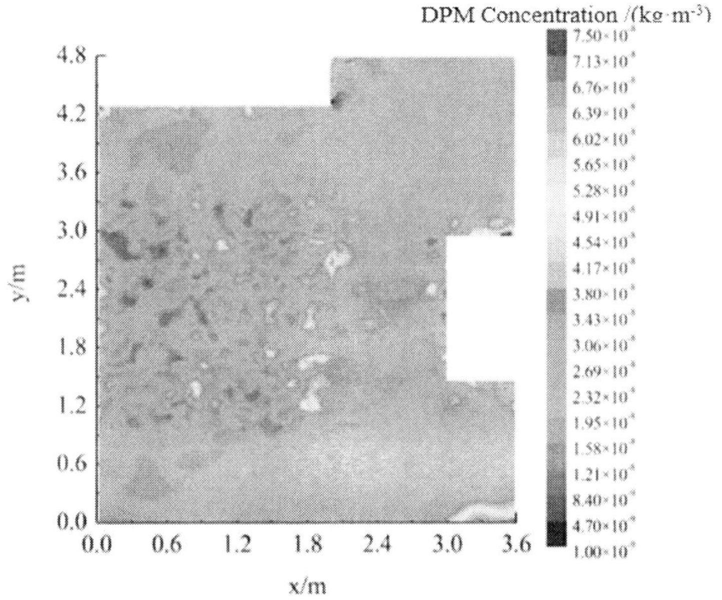


**Figure 7.** 1000s plane z = 0.6 PM2.5 concentration profile

*5.1.4. Case 4: Set inlet and outlet on the same sides, the inlet velocity is 1.67m/s.* Comparing the velocity profile of figure 6 and figure 8, it can be seen that the distribution law of the airflow velocity is almost the same, but after increasing the inlet velocity, the airflow velocity on the bed is increased as a whole. Figure 9 shows the specific distribution of PM2.5 on the monitoring surface. The overall PM2.5 concentration on the bed is slightly lower than other areas.

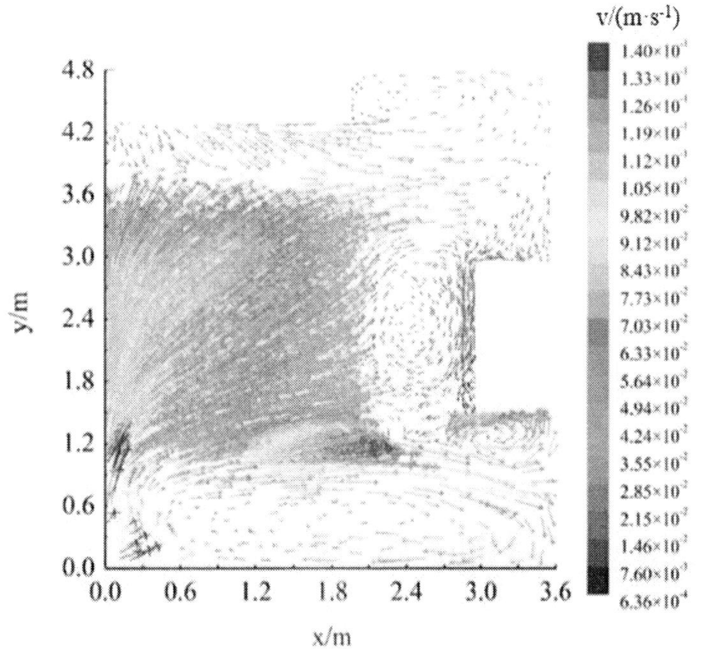


**Figure 8.** Plane z = 0.6 velocity profile

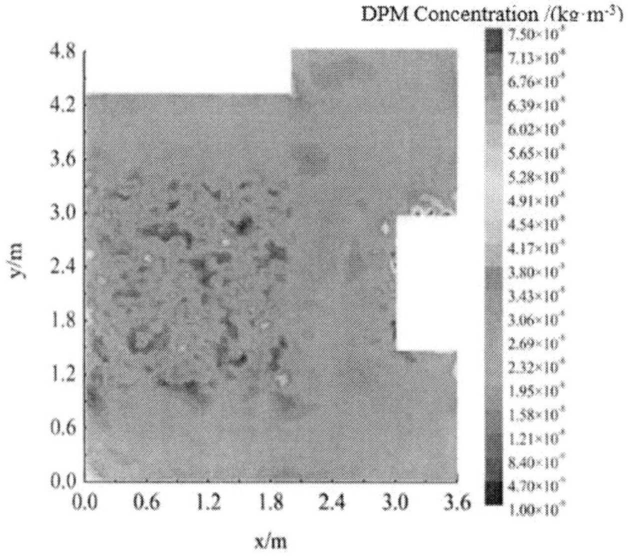


**Figure 9.** 1000s plane z = 0.6 PM2.5 concentration profile

### *5.2. Changes in PM2.5 Concentration*

Figure 10 shows the curve of PM2.5 average concentration varying with time at different working conditions. It can be seen that the introduction of fresh air has a good purification effect on indoor PM2.5.

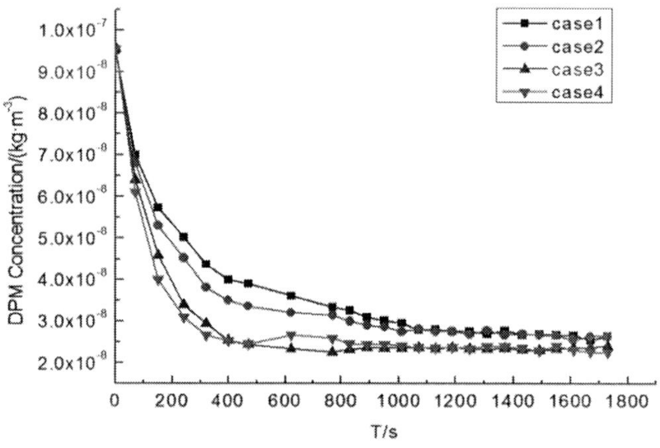


**Figure 10.** Plane z = 0.6 PM2.5 average concentration curve

It can be seen from figure 10 that there is not much difference between the case 1 and the case 2, and the changes of the case 3 and case 4 curves are the same basically. In the first 600 s the concentration of PM2.5 in the monitoring surface of case 1 and case 2 decreases by 57% and 63% respectively, while the concentration of PM2.5 in the monitoring surface of case 3 and case 4 decreases by 74% and 75% respectively. After that, slowly descend until dynamic equilibrium is reached.

It can be obtained that the adjustment of inlet velocity has little effect on the purification efficiency of PM2.5, and the position of the vent has a great influence on the purification efficiency, and the purification efficiency of the same side air supply to PM2.5 is higher than that of the opposite side.

## 6. Summary

Considering the needs of personnel and energy consumption, this paper selects the minimum fresh air volume required for indoor purification of indoor PM2.5. Because of the different purification conditions of clean fresh air to PM2.5 under different setting conditions, Aiming at two common types of airflow in the bedroom and two different inlet velocities, the purification effect of fresh air on PM2.5 was studied. The research results are as follows:

a) When the fresh air volume is constant and the vent position is the same, the change of the inlet velocity has no significant effect on the purification effect.

b) In the bedroom, when the fresh air volume and the inlet velocity is the same, the change of the vent position has a great influence on the purification effect. When the inlet and outlet were on the same side the introduction of the clean fresh air in the bedroom is more effective in purifying the PM2.5 than the opposite side.

c) In the bedroom, when using the minimum clean fresh air supply, the indoor PM2.5 will fall about 12% faster with the inlet and outlet were on the same side than the opposite side in the first 600s.

**Acknowledgments**
This research was supported in part by the Natural Science Foundation of Zhejiang Province under grant number LY18E080032

**References**

[1] He K, Yang F, Ma Y, et al. The characteristics of PM 2.5 in Beijing, China [J]. *Atmospheric Environment*, 2001, 35(29): 4959-70.

[2] Sorensen M, Daneshvar B, Hansen M, et al. Personal PM2.5 exposure and markers of oxidative stress in blood [J]. *Environmental Health Perspectives*, 2003, 111(2): 161-66.

[3] Chen R, Li Y, Ma Y, et al. Coarse particles and mortality in three Chinese cities: the China Air Pollution and Health Effects Study (CAPES) [J]. *Science of the Total Environment*, 2011, 409(23): 4934-44448.

[4] Weichenthal S, Villeneuve J P, Burnett T R, et al. Long-term exposure to fine particulate matter: association with nonaccidental and cardiovascular mortality in the agricultural health study cohort [J]. *Environmental Health Perspectives*, 2014, 122(6): 609-15.

[5] Cakmak S, Dales R, Kauri M L, et al. Metal composition of fine particulate air pollution and acute changes in cardiorespiratory physiology [J].*Environmental Pollution*, 2014, 189(12): 208-14.

[6] Chen Chun, Zhao Bin. Review of relationship between indoor and outdoor particles: I/O ratio, infiltration factor and penetration factor [J]. *Atmospheric Environment*, 2011, 45(2): 275-88.

[7] Han Y, Qi M, Chen Y, et al. Influences of ambient air PM 2.5 concentration and meteorological condition on the indoor PM 2.5 concentrations in a residential apartment in Beijing using a new approach [J]. *Environmental Pollution*, 2015, 205(553): 307-14.

[8] Massey D, Kulshrestha A, Masih J, et al. Seasonal trends of PM 10 , PM 5.0 , PM 2.5 & PM 1.0 in indoor and outdoor environments of residential homes located in North-Central India[J]. *Building & Environment*, 2012, 47(1): 223-31.

[9] Adair D, Jaeger M. Integration of computational fluid dynamics into a fluid mechanics curriculum [J]. *Computer Applications in Engineering Education*, 2014, 22(1): 131-41.

[10] Zhou Yu, Deng Ye-lin, Wu Peng, et al. The effects of ventilation and floor heating systems on the dispersion and deposition of fine particles in an enclosed environment [J]. *Building & Environment*, 2017, 125: 192-205.

# PIV Experimental Study on Flow Field near Lateral jet under the Condition of Vegetation

**Sufen Teng[1], Minquan Feng[2,*] and Kailin Chen[3]**

[1,2,3] State Key Laboratory of Eco-hydraulic in Northwest Arid Region of China, Xi'an University of Technology, Xi'an 710048, China
Email: sfteng@stu.xaut.edu.cn, Email: mqfeng@xaut.edu.cn,
Email: klchen@stu.xaut.edu.cn;
*Corresponding author: Minquan Feng

**Abstract.** The high-frequency PIV was applied to measure the flow field near the lateral jet under the condition of rigid vegetation and flexible vegetation. Both the Single row vegetation and broadleaf grass were analyzed. The results showed that different vegetation arrangements have different effects on the jet flow field. The vegetation arrangement and vegetation resistance cause significant changes. The changes in flow velocity with changes in water depth displayed "S" type and anti-"S" type distributions, and the flow velocity of free layer was approximately logarithmic. Due to vegetation resistance and jet pressure differences, the lateral jet trajectory in water with vegetation also have different distributions. The lateral jet trajectory in water with broadleaf grass was more likely to bend than in water with rigid vegetation Compared to the influence of two vegetation arrangements on the water flow near the jet, the flexible vegetation has a better deceleration effect on the flow field near the lateral jet.

## 1. Introduction

Maintaining healthy rivers and promoting the sustainable use of water resources is an important issue for water resources managers and researchers. Plants not only directly affect the transport of sediment, pollutants, and nutrients in rivers, but they also facilitate the transportation of bedrock and suspended sediment in rivers. Plants also change the movement and diffusion of pollutants in a river. Under the action of water flow, plants will have a certain degree of bending deformation and fluctuation as a response to the water flow. Therefore, how to maximize the beneficial effects of aquatic vegetation on the river, while inhibiting its adverse effects, are essential to the ecological restoration and flood control of rivers.

A jet is a turbulent shear stream that plays an important role in crossflow. In terms of jet flow characteristics, a large number of studies were conducted by Fischer et al [1~5], showed that the axial velocity, concentration attenuation law, jet penetration height and jet trajectory equation. According to the different methods used by Fischer et al [6] or Metftah et al [7], the bottom jet was used to enter the open channel containing vegetation. Ben [8] analyzed the influence of emergent rigid vegetation on a jet going into an open channel. In the same way, Malcangio [9] showed that the effect rigid vegetation has on a circular whole floating jet, and with the decrease of the flow velocity in the crossflow, the penetration height and dilution of the jet increase significantly. Meftah [10] indicated that vertically dense jets within a crossflow, the flow field, jet trajectory, turbulence intensity, turbulent kinetic energy, turbulent length scale, and diffusion coefficient.

In order to study the interaction between the jet and vegetation from different angles, we focuses on two vegetation conditions: flow velocities, and the rest of this paper is organized as follows: Section 2 presents experimental facility and experimental conditions (such as the detailed laboratory flume measurements of the instantaneous velocity were obtained using a Particle image velocimeter (PIV). Section 3 presents an evaluation of the scheme's performance in two cases (the variation of the average flow-field distribution of a lateral jet in an open channel; the longitudinal jet flow velocity along a vertical line; the trajectory of a transverse jet; Section 4 provides the conclusions.

## 2. Experiment Equipment and Conditions

### *2.1. Sink Device*

The experiment consists of a water supply device, steady flow device, test water tank, and water return device (as shown in Fig.1). The main channel is constructed of Plexiglas's; the coordinate system is defined by the center of the jet orifice as the coordinate origin.

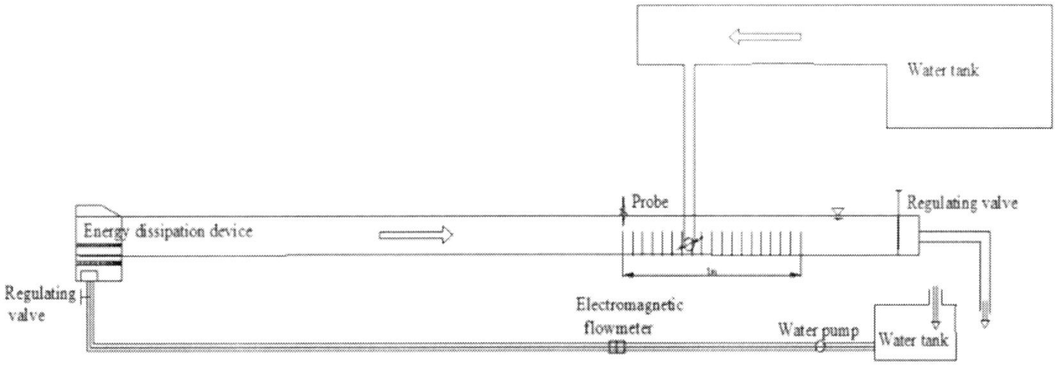


**Figure 1.** Sketch of the sink.

### *2.2. Experiment Equipment*

Experiments were carried out in a smooth horizontal rectangular flume in the National Key Laboratory of Ecological and Water Conservancy of Northwest Arid Region of Xi'an University of Technology. The glass flume was 7.3 m long, 0.3 m wide and 0.25 m deep. In order to ensure the stability of the flow, a trash rack, a glass ball, and a floating row were arranged at the head of the sink. The side row jet hole was set at a distance of 2.6 m from the inlet of the water tank. The diameter of the jet hole was 0.01 m, which was represented with $D$, and it was installed at 0.1 m from the bottom of the water tank. Fig. 1 showed the experimental setup and coordinate system. The center of the jet orifice was considered to be the origin for the coordinate system. The $x$-axis was the streamwise direction, with $x = 0$ at the leading edge of the flume. The $y$-axis was the span wise direction and $y = 0$ was the edge of the flume. The $z$-axis was the vertical direction and $z = 0$ was the channel bed. The hollow glass sphere acts as the tracer particle D =10 μm.

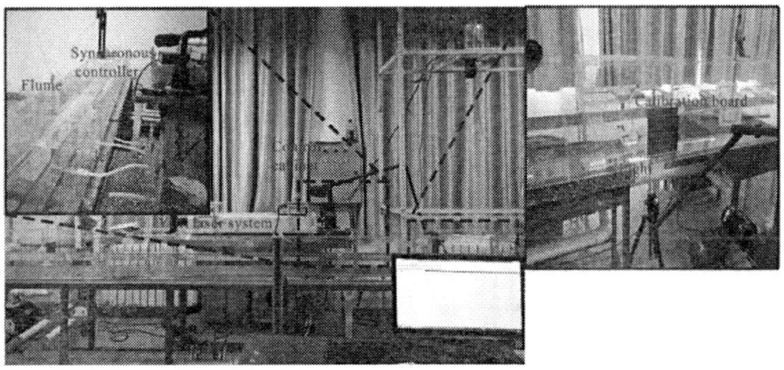

**Figure 2.** PIV measurement system and photo area

### *2.3. Experiment Conditions*

The initial parameters and experimental conditions were illustrated in Table 1. *H* was the ambient flow depth. *Re* was the Reynolds number for the jet. $U_a$ was the main velocity. $U_0$ was the jet velocity. *R* was the velocity ratio. *X* was the lateral spacing, while *Y* was the vertical spacing.

**Table 1.** Parameters of the experiments

| Run | condition | $Q$(m$^3$/h) | $H$(m) | $Ua$(m/s) | $U_0$(m/s) | R | Re | $X$(m) | $Y$(m) |
|---|---|---|---|---|---|---|---|---|---|
| V1J1 | Single row | 6 | 0.15 | 0.037 | 0.35 | 9.46 | 3500 | 0.03 | 0.02 |
| V2J1 | Broadleaf grass | 6 | 0.15 | 0.037 | 0.35 | 9.46 | 3500 | 0.01 | 0.02 |

### *2.4. Vegetation Materials and Configuration*

The transparent Plexiglas's rod was used to simulate rigid vegetation. The plant diameter was 8cm *and* the height was 10 cm. After inserting the place, the height is 9.2 cm. The Plexiglas's rod was fixed to a PVC board that was 100 cm long, 30 cm wide and 0.05 cm thick. The schematic diagram as follows.

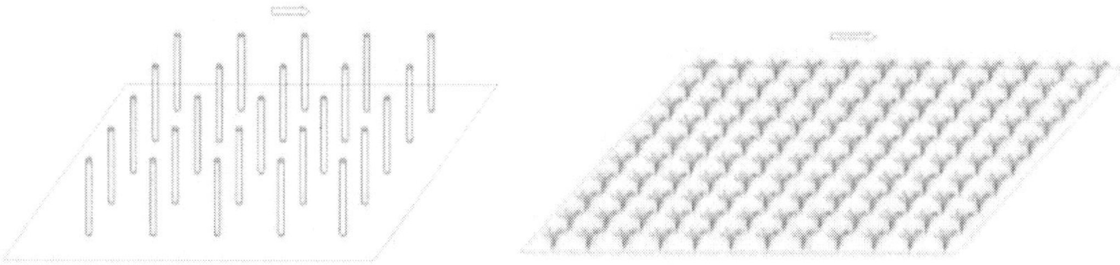

(a) Single row vegetation arrangement (b) Broadleaf grass arrangement.

**Figure 3.** Schematic diagram of vegetation arrangement

## 3. Results and Discussion

### *3.1. Distribution of Average Flow Field in a Transverse Jet under the Condition of Vegetation*

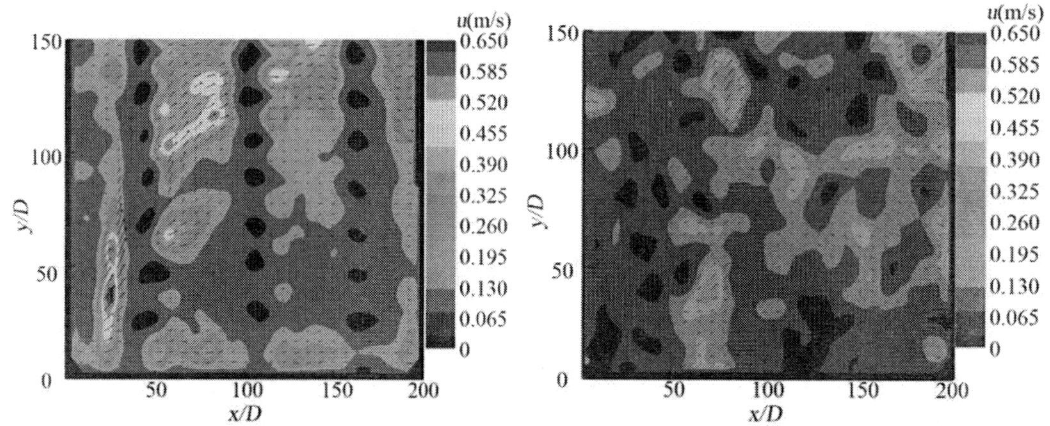


(a) vector map of $u_{xz}$ for run $V_1J_1$ (b) vector map of $u_{xz}$ for run $V_2J_1$

**Figure 4.** Time-average flow rate vector (z/D = 5)

Figure 4a showed that with single-row vegetation ($V_1J_1$), the maximum jet velocity of $u_{max}$ = 0.065 m/s was reached at the jet nozzle. Due to the flow around the vegetation, after the jet passed through the vegetation area, the jet velocity drastically deceased. There was a reverse vortex of varying shapes and sizes, and the velocity approached zero. The velocity gradually increased after the jet passed beyond the vegetation. Figure 4b showed that with broadleaf grass ($V_2J_1$), the velocity of the flexible vegetation was more complicated. The velocity of the flexible vegetation was weaker than rigid vegetation.

### 3.2. Vertical Flow Velocity Distribution of a Transverse Jet along a Vertical Line with Vegetation

The average flow velocity distribution along the vertical line had obvious regional characteristics. To highlight the significant effects of a cylindrical array of vegetation on the jet flow, we compared some vertical profiles of the dimensionless velocity $u/u_a$ between jets discharged into the rigid vegetation channels ($V_1J_1$ ) and into the flexible vegetation channels ($V_2J_1$) at different downstream positions $x/D$ (seen Figure 5).

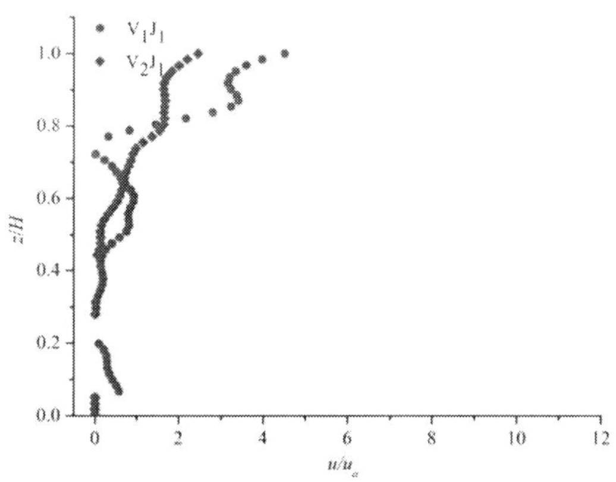


(a) Velocity distribution along the vertical at x/D = 50.

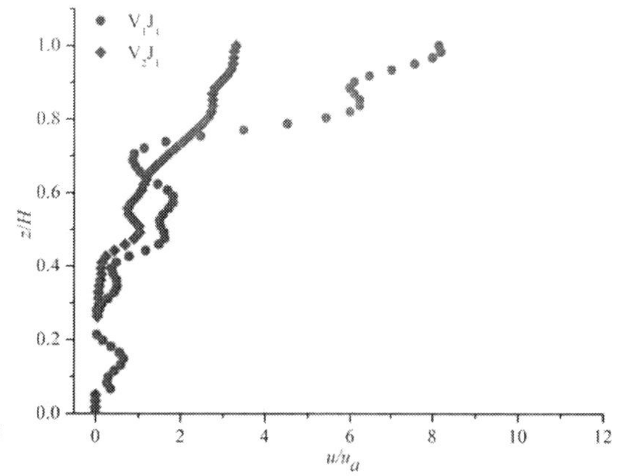


(b) Velocity distribution along the vertical at x/D = 80.

**Figure 5.** Vertical flow velocity distribution of the lateral jet at the different positions $x/D$ under different conditions.

In Figure 5, $u/u_a$ represents the flow ratio and the ordinate z/H represents the water depth ratio. Figure 5a showed the water depth ratio of $z/H = 0$ to $z/H = 0.5$ for the $V_1J_1$ condition with a single row vegetation. The longitudinal velocity changed drastically in the vertical direction and the velocity distribution showed obvious S and anti-S types. For the $V_2J_1$ conditions, the longitudinal flow velocity was approximately linear. In summary, there were many factors affecting the velocity distribution, such as roughness, jet resistance coefficient, drag coefficient *Cd* and infiltration height.

### 3.3. Transverse Jet Motion Trajectory with Rigid vegetation and Flexible Vegetation

The jet trajectory, which was the jet entered the cross-flow, was an important parameter. Ben, Meftah [11] used the dimensionless *D* and the dimensionless $r_jD$ to analyze jet trajectory changed. With vegetation and without vegetation, the study was conducted to analyze the changes in the transverse jet trajectory. To further analyze the jet trajectory, Kam[12] proposed an alternative scaling, normalizing the coordinates with *D* that was, $x/D$ and $y/D$, and went on to obtain a power-law of the jet trajectories in the form of

$$\frac{y}{D} = A\left(\frac{x}{D}\right)^B \tag{1}$$

In Equation (1), A and B were experimental coefficients. Margason[13] reviewed several correlations and concluded that much of the data could be collapsed by normalizing coordinates with the product $r_jD$, leading to a simple power-law trajectory in the form of

$$\frac{y}{r_j D} = A\left(\frac{x}{r_j D}\right)^B \tag{2}$$

A power-law similar to Equation (2) was also proposed by Pratte[14]. It was based on their dimensional analysis for a jet of $r_j D$ ranging from 5 to 35.New. [15]. Indicated that some researchers affirmed that the jet trajectories were best scaled with $r_j D$, instead of D or $r_j^2 D$.

Figure 6 showed the jet trajectories for different conditions. In Figure 6a, *x*/*D* represented the water flow ratio, and *y*/*D* represented the river width ratio. In Figure 6b, x/ $r_j D$ represented the water flow ratio, y/$r_j D$ represented the river width ratio, and $r_j$ represented the ratio of the ambient flow rate to the jet flow rate. $V_1J_1$ represented the lateral jet trajectory with rigid vegetation, and $V_2J_1$ represented the lateral jet trajectories with flexible vegetation.

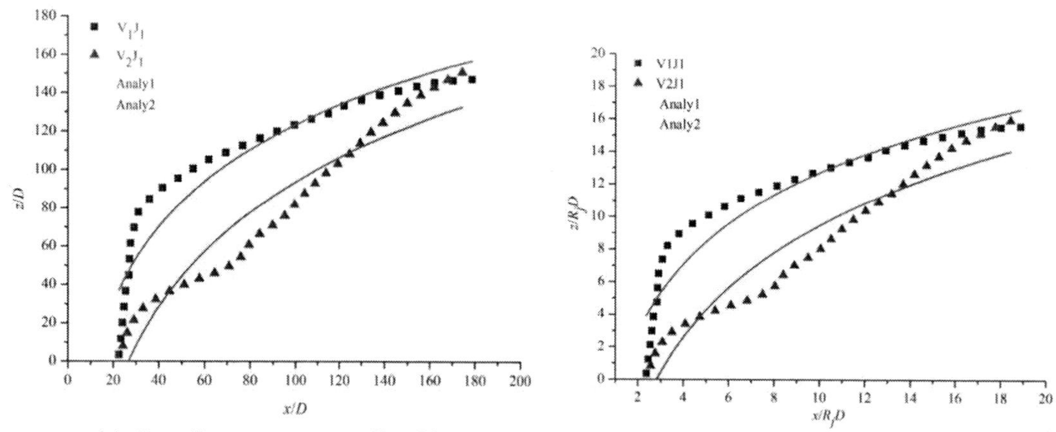


(a) Coordinates are normalized by *D* (b) Coordinates are normalized by $r_j D$

**Figure 6.** Jet axis motion trajectory

As shown in Figure 6a, by comparing the scatter points with the fitting curves of the jet trajectories under two conditions, it was found that different vegetation arrangements resulted in different curves of the jet trajectories. The curve of Analy1 fitted well with that of $V_1J_1$ under with single row vegetation. Under the effect of flexible vegetation, the trend of the jet trajectory curves was consistent and the fitting curves were also very similar. Under the action of flexible vegetation, the scatter points of $V_2J_1$ fluctuated greatly with the curve of Analy2.

## 4. Conclusion

The PIV flow rate measurement system can quickly, accurately, and effectively acquire the time-averaged and turbulent flow field characteristics near a lateral jet, and then has no interference to analyzing the flow field at a high resolution. The main conclusions are as follows:

According to the analysis of the vertical flow velocity, the flow velocity in the inner region of the vegetation and near the bottom of the trough was reduced. The flow velocity along the water depth was of S and anti-S types, by comparing the jet trajectories under different conditions, it was found that jet penetration was significantly greater. And the jet trajectory was more curved for the flexible vegetation arrangement.

## Acknowledgement

This work was funded by the National Natural Science Foundation (Fund No. 51679191); in particular, it was funded by the Xi'an University of Technology Doctoral Innovation Fund (Fund No. 310-252071507).

## References

[1] Fischer, Hugo B., E. John List, Robert C. Y. Koh, Jörg Imberger, and Norman H. Brooks. 1979. 'Chapter 6 – Mixing in Reservoirs', *Mixing in Inland & Coastal Waters*: 148-228.
[2] HUAI Wen xin, NA Yu tong, HUANG Ji zhong, TONG Han yi, JIANG Guo qiang.2002. An experimental study of vertical round jets in shallow water ,Advances In Water Science, 01):26-30.
[3] ZHANG ChaJiao, MA Jian, CA Qin and YANG Dejun, 2006, Simulation of vertical round turbulent jets in shallow water[J], Journal of Zhejiang University (Engineering Science), (12):2163-2167.
[4] Xiao Yang, Lei Ming, Li Kaijie, Liu Guiping, and Yan Jing. 2012. Experimental study on flow characteristics of porous jets in cross-flows, Advances in Water Science, 23: 390-95.
[5] LI Wenbin, GU Jie, YAN Cuiping, WANG Yuan. Experimental study on jet trajectory and diffusion characteristics in crossflow based on PIV[J]. Hydrodynamic Research and Progress (Series A), 2018, 33(02): 207-215 .
[6] Meftah, Mouldi Ben, Francesca De Serio, and Michele Mossa. 2014. 'Hydrodynamic behavior in the outer shear layer of partly obstructed open channels', Physics of Fluids, 26: 1624-35.
[7] Meftah, M. Ben, and M. Mossa. 2013. 'Prediction of channel flow characteristics through square arrays of emergent cylinders', Physics of Fluids, 25: 13-25.
[8] Ben, Meftah M, Serio F De, D Malcangio, M Mossa, and A. F. Petrillo. 2015. 'Experimental study of a vertical jet in a vegetated crossflow', Journal of Environmental Management, 164: 19-31.
[9] Malcangio, D., M. Ben Meftah, G. Chiaia, F. De Serio, M. Mossa, and A. Petrillo. 2016. "Experimental studies on vertical dense jets in a crossflow." In The International Conference on Fluvial Hydraulics, 890-96.
[10] Meftah, Mouldi Ben, and Michele Mossa. 2018. 'Turbulence Measurement of Vertical Dense Jets in Crossflow', Water, 10.
[11] Ben, Meftah M, Serio F De, D Malcangio, M Mossa, and A. F. Petrillo. 2015. 'Experimental study of a vertical jet in a vegetated crossflow', Journal of Environmental Management, 164: 19-31.
[12] Kamotani, Yasuhiro, and Isaac Greber. 1971. 'Experiments on a Turbulent Jet in a Cross Flow', Aiaa Journal, 32: 1451-60.
[13] Margason, R. J. 1968. 'The path of a jet directed at large angles to a subsonic stream', N. A. S. A., TN. D. -4919.
[14] New, T. H., T. T. Lim, and S. C. Luo. 2006. 'Effects of jet velocity profiles on a round jet in cross-flow', Experiments in Fluids, 40: 859-75.
[15] Pratte, B. P. 1967. 'Profiles of the Round Turbulent Jet in a Crossflow', Journal of the Hydraulics Division Proc of the Asce, 93: 53-64.

# Discussion on the Methods to Make the Military Highway Unimpeded in Plateau

**Yunhong Zhang, Wei Qiao and Hai Wang**

Army Military Transportation University, Tianjin 300161, China
E-mail: zhangyunhong2005@126.com

**Abstract.** This paper analyzed the types and feature of disaster which threatened the safety of military highway in plateau, and discussed the engineering methods to keep the military highway unimpeded from the disaster of mud-rock flow, landslide, sand-sliding, and damaged by flood, and the management methods.

## 1. Introduction

Most border areas on the plateau of China have an average altitude of more than 3, 500 m, cold climate, lack of oxygen along the route; sparsely populated areas, poor water transportation conditions, and railway and air only cover a small part of the region, and cannot form an integrated three-dimensional transportation system. The high and steep hills and steep, tortuous and dangerous military highways serve as a bridge and link between guard posts and troops. The only support for personnel transportation and material support is also an important basis for maintaining national security, an important prerequisite for realizing national military strategy, and an important link in transforming war potential into war strength. Therefore, strengthening the construction of military highways on the plateau and keeping them unobstructed play a vital role in enhancing the ability of border troops to cope with various security threats and accomplish diversified military tasks.

## 2. The Types, Causes and Characteristics of Disasters Affecting Border Military Highways

The particularity of the region where the plateau military highway lies leads to the fact that the quantity and quality of highway are seriously lower than the overall level of national highway. Most military highways are classified as off-road, with an average altitude of 3,500 m. Some need to climb over Snow Mountains above 5,000 m, and some need to travel through fast-flowing glaciers. A considerable part of the roads are ruts and natural roads on earth and stone, with few safety protection facilities and low ability to withstand natural disasters [1].Due to the complex and special natural conditions such as geological geomorphology, hydrology and climate, etc. along the military highways in the gaoyuan area, the disasters are not only of various types, but also widely distributed and frequent. The damages to the highways are easy to destroy and difficult to repair. The main types of disasters that directly threaten road traffic safety include debris flow, landslide, sand removal and water destruction.

### *2.1. Debris Flow Disaster [2]*

Debris flow can be divided into glacier debris flow, rain - flood debris flow and glacier - rain - flood debris flow. Rain-flood debris flow is mainly debris flow induced by rainfall runoff, which is characterized by high occurrence frequency and high risk. The occurrence of debris flow is caused by intense erosion, mixing and transport of loose solid materials in valley slope caused by rainstorm runoff in summer. Glacier debris flow is formed by the hydrodynamic conditions of a large number of ice deposits and glacial lake outburst flood, glacier and snowmelt water. Glacier-rain flood debris flow is a hydrodynamic condition of glacier snow melt water and rainfall. The scale of disaster increases with the increase of catchment area.

### 2.2. Landslide

The landslide is widely distributed in the plateau area, which has the characteristics of serious road obstruction, difficulty and high cost. Usually, the larger landslide distribution is relatively concentrated in the cross compound of geological structure and the active canyon area of neotectonic movement. In tg-p due to early quaternary glacier accumulation formation of loose accumulation platform, poor stability, combined with rainfall, slope extremely easily, or in some loose, broken on both sides of the hill and valley under the fluvial erosion, leading edge collapse, the slope deformation and failure gradually development and form a series of landslide slope upward, in a highway landslide, the landslide sliding out direction is perpendicular to the line to more, directly to the slitting line, and landslide will occupy river, creating block and a serious threat to the downstream area road, personnel safety.

### 2.3. Sliding Sand

Sand sliding disaster is a process in which sand and gravel slide to the foot of slope under the action of gravity and accumulate in the foot of slope into cone slope. The material composition of sand-sliding breaking is mainly composed of grains of sand and mixed with gravel. When the slope is unstable, it reaches a new equilibrium state in the form of sand flow. Because of the continuous supply of sand grains, sand sliding activity has the characteristics of frequent activity, which seriously affects the safety of the highway passing the slope foot.

### 2.4. Water Damage

As the frontier defense military highway of the plateau moves towards the multi-along river and stream extension line, the roadbed, bridge and culvert foundation and bank protection of the highway have been washed and submerged by the summer flood for a long time, resulting in the damage of highway structures and the interruption of routes, which has a great impact on the military transport. Some sections of the sichuan-tibet line can reach a length of 220.8 m/km. The main causes of the water damage are as follows: first, due to topography, narrow river valley, easy to plug, large river longitudinal slope, river erosion on the bank slope serious erosion; The second is that some highway subgrade elevation is lower than flood level or the return water level formed due to debris flow, landslide and other obstructed river channels, resulting in road flooding. Flood water level, flow rate, flow rate, sand content and other changes in a large range, sudden outbreak, instantaneous flood volume, etc., are all difficulties to ensure the smooth flow of border military highways on the plateau.

## 3. Technical Methods for Prevention and Cure of Major Highway Diseases

### 3.1. Prevention Programs and Countermeasures of Debris Flow

For roads built in debris flow drainage area or debris flow hazard area, the following strategies can be adopted to prevent and control them. First, measures can be taken when selecting the route. Second, the use of Bridges, tunnels and other structures across serious disease zone, but the cost of building Bridges and tunnels is relatively high; In addition, construction of structures such as drainage, rectification and blockage are also engineering measures often adopted in response to debris flow disasters.

When determining the route direction for preventing and controlling debris flow on military highways, the intersection of roads and debris flow gullies should be controlled, various prevention and control measures should be compared technically and economically, and the most favorable plan should be selected, which will be reduced to the minimum, so that the debris flow can pass safely and guarantee the smoothness of military roads.

### *3.2. Prevention and Treatment of Landslide*

Since the occurrence of landslide is mainly the movement of soil under gravity, the prevention and control of landslide should proceed from this fundamental reason. First is to avoid landslide, in the selection of road line, avoid landslide frequent location; The second is to control landslide, such as the line can not avoid landslide prone areas, must take control measures, mainly from two directions: improve anti-sliding force, reduce sliding force. The reinforcement method such as setting up supports, adding anti-slide pile, anchor and anchor cable can be adopted to improve the anti-sliding force. To reduce the sliding force, the methods of decompression and cutting of soil above the landslide can be adopted. Planting suitable vegetation on the slope, removing surface water and draining groundwater, and reducing human disturbance on the slope can enhance the anti-sliding force and reduce the sliding force. In addition, in highway operation, monitoring of landslide prone areas is an important means to prevent landslide.

### *3.3. Technical Measures and Countermeasures for Sand Slope Prevention*

There is no unique treatment method for sand sliding disaster at home and abroad. Most of the methods are combined with landslide control and sand sliding control. Controlling sand source is the most fundamental way to control sand sliding, which is the most ideal and difficult to implement. Secondly, sand fixation can be adopted by planting vegetation and solidifying the surface of sand slope. The relatively direct and effective method is to use the route to avoid winding, or to build the sand discharge aqueduct, baffle wall, sand protection shelter and other sand discharge methods.

### *3.4. Preventive Measures against Water Damage [3]*

The main purpose of the project is to protect the roadbed from erosion. According to the flow structure and mechanism, the protection of roadbed can be divided into direct protection and indirect protection. Direct protection is to strengthen the slope foot or foundation directly and improve its anti-scour ability. The main protective engineering measures include slope protection, retaining wall, retaining foundation, riprap, precast concrete, geotextile, stone cage, etc. Indirect protection is to construct the ding dam, along the dam and other projects or renovate the river course, change the flow structure of the river course, make the flow deviate from the protected river bank, pier or change the erosion section into the silt section, so as to achieve the purpose of protection. This kind of engineering encroaches on the river course, changes the river course flow greatly, has certain influence on the upstream and downstream even the opposite bank, accordingly these structures are also subjected to the strong scour of the water flow. Even if the dam end of the ding dam and the diversion dam is destroyed in the flood, the safety of the roadbed or bridge will not be immediately threatened. In fact, the protection purpose of the roadbed or bridge has been achieved, and it can be repaired after the flood season.

## 4. Management Means to Ensure the Smooth Military Highway of Border Defense on the Plateau

As a military road, use and management of the department in daily work, should be carefully in combination with the practical situation of sections of their jurisdiction, actively and effectively eliminate the factors of highway diseases, enhance the durability of highway facilities and the ability to resist natural disasters, in addition to the necessary engineering technology means, in the construction of highway management should also actively explore scientific and reasonable method.

### *4.1. Building a Good Military Highway is the Material Basis for Ensuring the Highway Unimpeded*

The construction of military highway usually includes the planning, evaluation, decision-making, design, construction, completion acceptance, putting into production or delivery, etc., all the work must follow the sequence of work in the whole construction process, which cannot be reversed at will, but can be reasonably crossed. The national defense of the military highway on the plateau is of great significance. The special region, harsh natural environment and complicated links and procedures of the project construction must be paid close attention to all aspects of the project construction to ensure the maximum efficiency of the frontier military highway. Is a military road planning, road planning must again after full investigation and research of decision-making, the line of line selection must meet countries, border guards and defense transportation under the major premise of optimizing circuit to avoid all kinds of natural disasters, according to the facts in can't avoid, also will do the least cross section. Secondly, the natural condition survey, the detailed investigation of the geological, hydrological and other natural conditions around the preliminarily planned route, is the most important basis for determining the technical means and methods of highway construction. Three is the construction process management, to determine the scientific and reasonable construction method, in construction process to design in strict accordance with the requirements of construction, can't let already facing all sorts of natural threats road add unnecessary human losses, at the same time, the construction of highway management and supervision of party must strict supervision, control of construction quality, for the plateau of the national defense highway traffic build quality.

### *4.2. Do Well in Daily Maintenance and Keep the Highway in the Best Condition*

In order to maintain the best use status of military highways on the plateau and constantly improve the road rate, it is necessary to firmly set up the idea of "three-point construction and seven-point maintenance" and make unremitting efforts for daily maintenance. First, a scientific maintenance plan should be formulated. The maintenance amount of military highway in the plateau area is large, but the corresponding personnel is less, and the annual suitable maintenance time is short. Therefore, scientific and reasonable maintenance schedule, efficient use of maintenance power and adequate maintenance equipment are all necessary means to improve the maintenance efficiency. Second, the conservation measures should be seriously implemented, and then the scientific planning plan can only be fully implemented.3 it is to the rational use of maintenance funds, compared with the condition of local roads, military highway maintenance funds for plateau is lower, in the face of such situation, the financial support of border troops should actively strive for higher level units, at the same time its also want to improve their capital utilization, to ensure that all maintenance funds with maximum efficiency to use on the plateau military highway maintenance.

### *4.3. Strengthen the Capacity Building of Rapid Emergency Repair to Ensure the Timely Recovery of Damaged Highways*

Although the border military highways that have been constructed and maintained to the maximum extent are unimpeded, road damage still occurs frequently in border areas of the plateau. In order to guarantee the needs of military transport in time of war and emergency, when roads are damaged by enemies or blocked or interrupted by natural disasters, they must be repaired in time. In order to ensure the smooth road recovery as soon as possible, first of all, it is necessary to establish the emergency repair team with strong business capacity, whose personnel have the relevant knowledge and technology of engineering, communication, transportation, repair, medical treatment, etc.; With a strong personnel team, it is also necessary to deploy complete emergency repair equipment and equipment that can meet the construction and operation requirements of the plateau region, as well as emergency personnel and equipment, and improve efficient emergency rescue and command schemes to enable people and things to do their best to ensure that damaged roads can be repaired as soon as possible.

### *4.4. Adhering to Military-Civilian Integration Development and Comprehensively Strengthening Military Highway Construction and Management*

Military and civilian integration development is an important and long-term road for national defense transportation construction. For the construction and management of plateau military highways in difficult conditions, it is more important for the military and civilian to jointly build and manage them, and make use of civilian manpower and material resources to make up for their own deficiencies. Therefore, from the construction of border defense military highway in the plateau, to daily maintenance, emergency repair and access, can give play to the advantages of local forces; fully mobilize civil resources, so that civil resources can be added to the construction of national defense communications. For the joint construction and co-management of military highways on the plateau, sound laws and regulations and a reasonable civil resource compensation mechanism are the most basic guarantee for safeguarding the rights and interests of both sides and mobilizing the enthusiasm of civil resources. However, there are still some deficiencies in the construction of laws and regulations on the civil transport capacity compensation mechanism in China. Therefore, strengthening the construction of laws and regulations in the border area of the plateau is also an important work to guarantee the smoothness of border military highways.

## References

[1] Wang shuangjie. Development of highway construction technology in the permafrost region on the qinghai-tibet plateau [J]. Science in China 2009(02)
[2] Wang shaungjie. Study on the technology for highway construction and engineering practices in permafrost regions [J]. Sciences in cold and arid regions 2009(05)
[3] Zhang bin. Integrated evaluation on highway engineering geological hazard in mountainous area of Enshi, Hubei [J]. Wuhan university journal of natural sciences, 2006(04)

# The Impact of Electrified Railway on Oil Depot by Military Railway

**Wei Qiao, Yunhong Zhang and Yang Wu**

Army Military Transportation University, Tianjin 300161, China
E-mail:626893334@qq.com

**Abstract.** Railway electrification transformation brought us transport efficiency, due to its high pressure, strong electromagnetic radiation characteristics also give the military oil depot brings great danger risk. The paper analysis the impact of electrified railway on oil depot by special military railway from the main line electrification, the military Line electrification, the length of special military railway , position , earth conductivity and the potential difference between earth and railway six respects. To provide a basis for doing the protection of the depot electrification project.

## 1. Introduction

Electrified railway has the characteristics of large traction quality, fast running speed, low transportation cost, reduction of environmental pollution and superior technical and economic performance. From 1961, China's first electrified railway baofeng railway was officially delivered and operated, and by the end of 2012, the operating mileage of electrified railway in China reached 48,000 km, ranking the first in the world. By the end of the 12th five-year plan for railway development, China's railway operating mileage will reach about 120,000 kilometers and the electrification rate will reach over 60%.With the development of China's railway electrification, most of the oil depot special line located along the railway has been electrified. Due to the characteristics of the electrified railway itself, the 27.5KV plus or minus 10% high pressure line on the contact network is directly introduced into the reservoir area, which poses a threat to the safety of oil depot, and also has a certain impact on the completion of the combat readiness and normal supply support tasks of the troops. In order to guarantee the safety of all kinds of war equipment under electrification, we must understand the influence mechanism of electrified railway on oil depot. This paper expounds the influence of electrification line on oil depot from the following aspects through field research.

## 2. Influence of Military Special Line Electrification on Oil Depot [1]

Because the electrified railway USES the contact network -- the rail circuit to supply power to the electric locomotive, the rail ACTS as a conductor in the power supply circuit, the electric locomotive picks up the flow through the contact network and returns back to the substation through the rail, forming a complete power supply circuit. In the rail - earth loop, there is leakage resistance between the track bed composed of sleeper and ballast and the rail, so the rail backflow through the leakage resistance generates rail - to - ground potential. The backflow of the rail through the leakage resistance causes the current flowing into the ground to increase the ground potential in the area.

Ac electric traction section. Traction current is via rail - earth regression substation, as shown in the figure below:

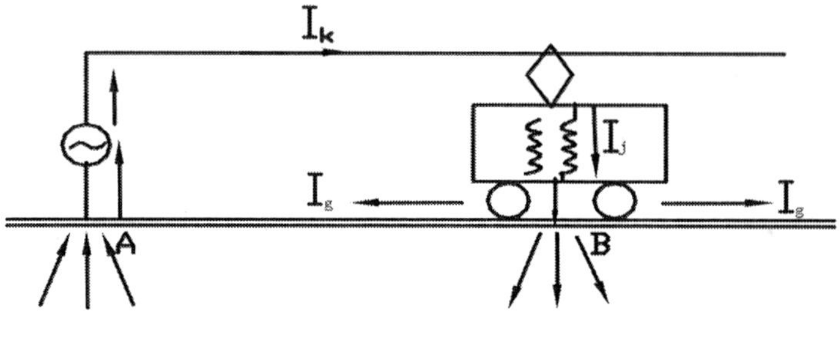


**Figure 1.** Schematic diagram of traction current reflux ( $I_K$ Represents the current in the contact network; $I_g$ Represents the current in the rail; $I_1$ Represents the Ground current)

The regression current enters the ground from point B via the rail, which increases the potential of the rail to the ground. The current distribution in the rail is as follows:

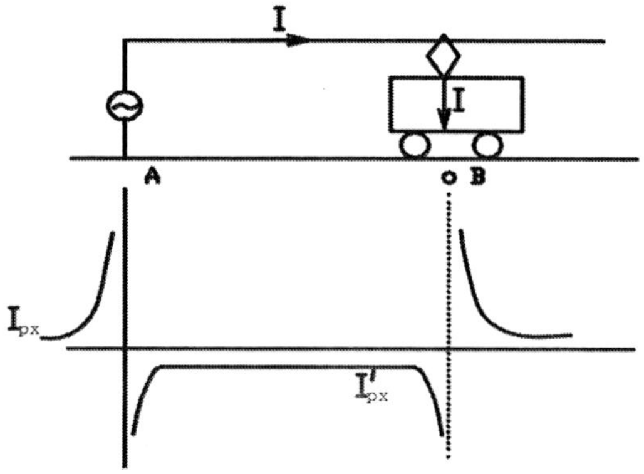


**Figure 2.** Rail regression current distribution diagram ( $I_{px}$, $I'_{px}$ Represents the regression current in orbit)

In the figure, the rail current in AB section is the vector sum of regression current and induction current, while the rail current outside AB section is regression current. That is to say, when the electric locomotive runs on the military special line of the electrified railway, the rail current between the electric locomotive and the substation is the vector sum of the rail return current and induced current, while the rail from the locomotive to the end of the military special line is the return current. According to the variation trend of rail current in the figure, the rail current is the largest at the ground end of the substation and the position of the locomotive.

Regression current in rail: $I_{dPX} = \frac{I}{2} e^{-\gamma_p x}$ [2]

Where: $I_{px}$ is the regression current in orbit (A); $I_1$ Represents the incoming current (A); $\gamma_p$ Represents the propagation constant (1/Km) of the "rail - earth" loop here.

It can be seen that the rail current changes along the x value in an exponential curve. The higher the value of x, the greater the rail potential to ground, and the smaller the rail potential to ground. X represents the distance between the electric locomotive and the loading and unloading platform of the oil depot.

When the electric locomotive is located near the loading and unloading platform of the special military line, x=0 indicates that the rail current at the loading and unloading oil platform in the reservoir area where the locomotive is located is the largest, and the current in the rail and the rail have the highest potential to the ground, which is the most harmful to the dangerous goods warehouse, especially the oil depot.

## 3. The Influence of Positive Line Electrification on Oil Depot

The rise of the ground potential of the rail of military special line is caused by the superposition of the conduction component of the backflow of the rail of the positive line electrified railway and the induction component of the contact network current on the rail.

### *3.1. Conduction Component*

The special line in the reservoir area is generally shorter, and there are many cases that are obliquely close to or perpendicular to the direct line railway. Therefore, the conduction component of the backflow of the positive line rail is the main reason for the increase of the potential of the special line rail to the ground.

The electric locomotive on the direct line railway is in continuous operation, which is generally divided into three situations: the locomotive is located on the left side of the contact point, the locomotive is located on the contact point, and the locomotive is located on the right side of the contact point.

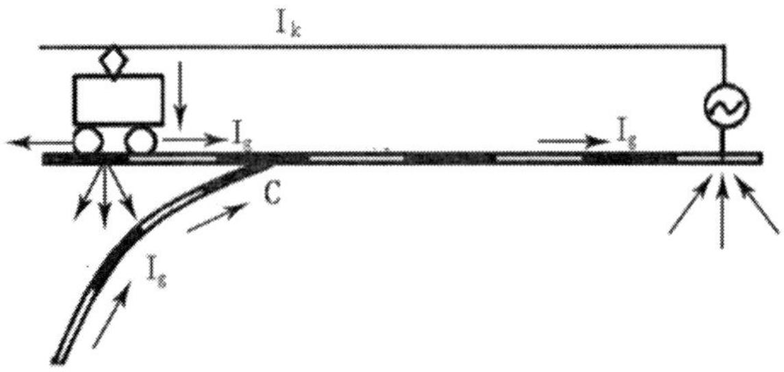


**Figure 3.** Track current distribution diagram of locomotive at the left side of point C ( $I_K$ Represents the current in the contact network; $I_g$ Represents the current in the rail)

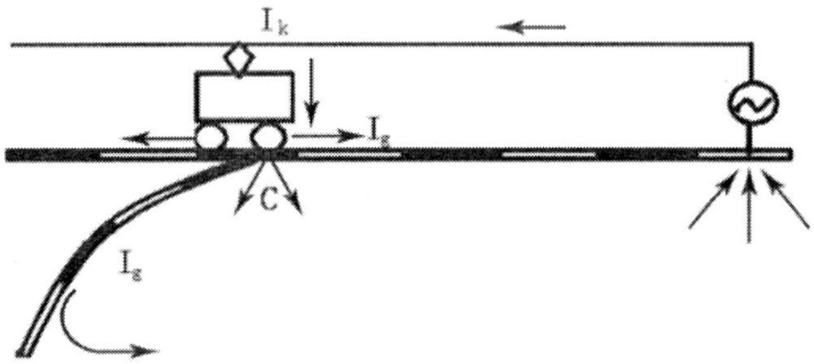


**Figure 4.** Track current distribution diagram of locomotive at point C ( $I_K$ represents the current in the contact network; $I_g$ Represents the current in the rail)

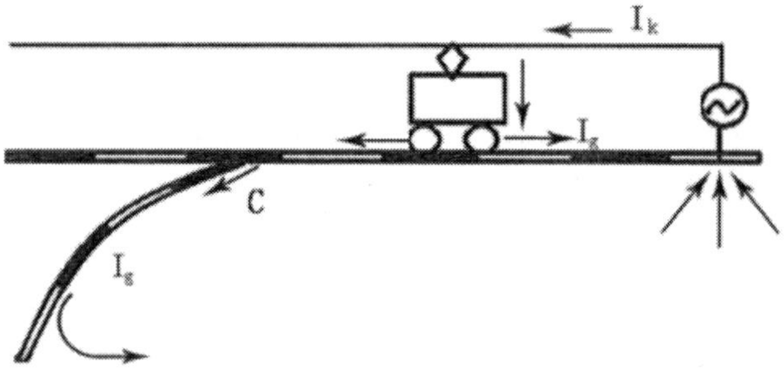


**Figure 5.** Track current distribution diagram of locomotive on the right side of point C ( $I_K$ represents the current in the contact network; $I_g$ Represents the current in the rail)

According to the analysis of these three working conditions, as long as there is a locomotive running on the line, there is a current flowing through the dedicated line, but the direction and composition of the current are different. When the locomotive is at point C, the regression current shunt to the dedicated line is larger. When the locomotive is located to the right of point C, the current in the dedicated line is mainly the regression current, and there is no induced current in the contact network. The potential and current located at point C in the positive rail can be transmitted to the end of the special line into the ground along the special line and return to the substation via the earth.

### *3.2. Induction Component*

The induced voltage on the rail of the railway military special line is due to alternating magnetic field generated by alternating current on the contact network. Through the mutual inductance M between the contact network and the rail, induced electromotive force is generated in the rail by the changing magnetic field, and thus induced current is formed in the rail, which is superimposed and propagated along the rail. This is the mutual sense between the contact network and the rail. The schematic diagram is as follows:

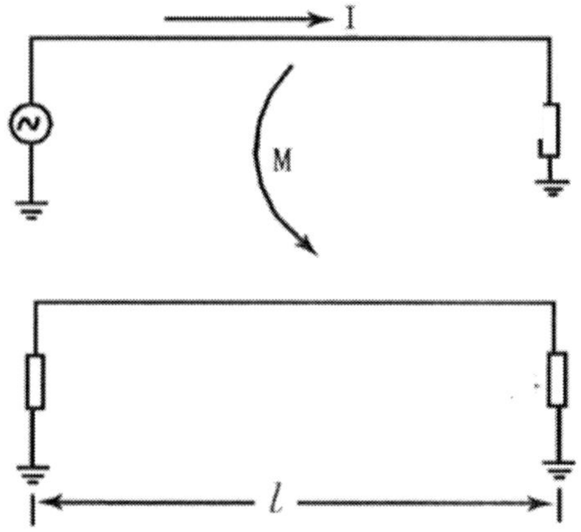


**Figure 6.** Contact network - rail mutual inductance diagram (I: Represents the current in the contact network; M: Represents the mutual inductance between the contact network and the rail; L: Denotes the length of contact network or rail)

According to the principle of mutual inductance, the induction current on the rail is related to the length of induction, the frequency of contact network current, the conductivity of the rail, the grounding resistance and the earth conductivity.

### 3.3. Ground Potential and Current of Rail

Based on the above analysis of the rail return current and induced current, the rail ground potential and current of the military special line can be obtained by vector sum of the induction component and conduction component. The rail ground potential and rail reflux of the military special line are:

$$\overline{U}_X = \overline{U}_{mpx} + \overline{U}_{dpx}$$

Where: $\overline{U}_x$ represents the ground potential (V) of the rail of the special military line; $\overline{U}_{mpx}$ Represents the induction voltage component (V); $\overline{U}_{dpx}$ Represents the conduction voltage component (V)

$$\overline{I}_X = \overline{I}_{mpx} + \overline{I}_{dpx}$$

Where: $\overline{I}_x$ represents ground current (A) from rail of military special line; $\overline{I}_{mpx}$ Represents the induced current component (A); $\overline{I}_{dpx}$ Represents the conduction current component (A)

Because there is leakage resistance between the rail and the ground, the current in the rail takes conduction component as the main component, induction component is generally small, sometimes can ignore.

## 4. The Influence of Length of Military Special Line on Oil Depot

Trunk line electrification railway's influence on the oil depot is associated with the length of the special military line, took the flow when the trunk line electrification railway electric locomotive, the branch line rail rail as the trunk line, the line on the rail there are rail backflow, the line current on the rail by stray current in the land along the rail assembled, exponential curve.That is, the longer the military special line, the smaller the rail potential to ground at the end of the military special line, and the shorter the rail potential to ground at the end of the military special line.

Through field tests, it is found that the length of military special line over 2 km is generally less affected. When the main line passes through the truck, the ground potential of the rail is relatively small when it is tested near the loading and unloading platform in the reservoir area, which is generally below 2 V.If the distance between the military special line and the loading and unloading oil platform is within 2 km from the joint gauge point, the influence is relatively large, and the actual test results show that the rail has high ground potential, and the maximum measured value can reach 18.6v.

Influence of relative position of military special line and direct line on oil depot

In the actual test of the ground potential of the steel rail at the end of the military special line when the main line passes the electrified locomotive, the following situations will occur in different oil depot:

(1) when the locomotive has not reached the oil depot test point, the rail potential to the ground gradually increases; when the locomotive reaches the oil depot test point, the rail potential to the ground is the highest; when the locomotive is far away, the rail potential to the ground gradually decreases.

(2) when the locomotive has not reached the test point of oil depot, the rail's potential to the ground does not increase significantly. When the locomotive reaches the test point of oil depot, the rail's potential to the ground is not the highest.

(3) when the locomotive has not reached the test point of oil depot, the ground potential of the rail starts to rise, and gradually decreases when it reaches the highest level. When the locomotive passes the test point of oil depot, the rail potential is not high.

(4) the potential distribution on the rail of the military special line, the potential of the contact point is high, and it gradually decays in line with the exponential curve, and decays to the minimum when it reaches the end of the military special line.

(5) the potential distribution on the rail of the military special line, the terminal potential of the military special line is the highest, and it gradually decreases in line with the exponential curve.

The above situation will occur in different oil depot, which is caused by the difference of the military special line railway, the main line railway, the loading and unloading oil platform of the oil depot, the oil storage area and the mutual location. Electrified railway power supply section length range from 10 km to a few tens of kilometers, the railway, winding catenary above railway erection, when the locomotive in the power supply section, rail by rail - earth loop flow back to the substation, flows through the earth return current caused by the stray current, the earth if by the way these stray current flows through a terminal area, will result in the reservoir of ground potential rise, after the reservoir of stray current intensity, direction and other factors will have different effect to rail to earth potential, through the reservoir area the stronger private sidings rail backflow of stray current, the higher the private rail to earth potential will be. If the flow direction is from the end of the military special line in the reservoir area to flow back to the substation, the ground potential of the rail at the end of the reservoir area will be higher than the potential of the joint. If the direction of the return current is from the joint point to the loading and unloading oil platform at the end of the military special line, the potential at the joint point is the highest, and the oil platform at the end of the military special line decreases gradually in line with the exponential curve.

The mutual position of railway military special line, main line, oil depot loading and unloading platform, oil storage area and so on has different influence on the flow direction of the return stray current in the ground.

The influence of the earth conductivity on oil depot

The earth conductivity is an important parameter in mutual inductance, the traction current in electrified railway catenary will on surrounding parallel with metal conductor produced by the principle of electromagnetic induction and the longitudinal electromotive force and inductive current, the size of the induced current is associated with mutual inductance, mutual inductance and the size of the associated with the size of the earth conductivity, can say, induction voltage and current is the dependent variable, the earth conductivity is the independent variable. The smaller the earth conductivity value indicates that the soil resistivity is high and the soil conductivity is poor, the higher the induction voltage will be. On the contrary, the higher the earth conductivity value indicates that the soil resistivity is small and the soil conductivity is good, the smaller the induction voltage will be. The earth conductivity has an influence on the stray current in the ground. The earth conductivity is small, the soil resistivity is high, the soil conductivity is poor, the potential gradient of stray current in the soil is large, the ground potential increases, and conversely the ground potential decreases.

Rail to earth potential are also associated with the earth conductivity, the conductivity is small, high soil resistivity, the conductivity of the soil is poor, the rail current to ground leak resistance will also increase, the rail current is not easy to leak to the ground, cause rail current is big, high potential, rail for small vice rail current, rail to earth potential.

Influence of rail on ground potential difference of military special line

The rail to ground potential difference, which is the voltage difference between rail and ground, is one of the key points for protecting electrified railway. Common rail to ground potential can reach dozens of volts, high speed and heavy duty railway can reach several hundred volts, when rail to ground potential reaches 60 V or above, it will cause harm to the human body, and will also cause damage to the signal equipment on the rail or other equipment associated with the rail.

Electrification railway private sidings military in the use of electric locomotive shunting operation, when walking on electric locomotive in the special line, the current rail can reach tens of amperes to hundreds of ampere, rail to earth potential can reach dozens of volts to hundreds of V, the harm of it mainly depends on when the voltage reaches a certain degree, will produce is the spark discharge phenomenon, the general rail to earth potential to reach a minimum of 2 V in contacts with other metal objects, it is obvious of edm, this for military railway private sidings to operate near the line workers, line on the use of precision instrument is very dangerous.

In military warehouse loading and unloading operations of dangerous goods, especially in the process of oil and other liquid dangerous goods warehouse handling, loading and unloading crane tube with pot mouth close distance, crane tube when loading and unloading oil and fuel tank mouth of the inevitable happens contacts, and both of the potential difference of different, very likely spark, and the dangerous goods such as fuel in the tank mouth site volatile concentration is very big, when the oil vapor concentration reaches explosion limit, rail and crane tube potential difference is 0.7 V, spark energy just 0.15 mm focal, a serious threat to private sidings and warehouse security [3].

With the continuous development of railway electrification in China and the continuous advancement of railway electrification transformation of military special line, it is bound to produce hidden danger to the oil depot of military special line. Therefore, we should start from the influence of electrified railway on military special line, reduce the potential of rail to ground, reduce the rail reflux in the rail, and reduce the potential difference between rail and other metal objects, so as to eliminate the potential risk of electric spark, and provide guarantee for the safety of the loading and unloading operation of military special line oil depot under electrification.

**Reference**

[1] Li zhuxuan, Shen changli. Railway military special line [G].Beijing: PLA press, 2011.11.
[2] Electrification engineering bureau of ministry of railways. Electromagnetic influence and protection of ac electrified railway [G].Tianjin: tianjin education press, 1990 (12).249-250.
[3] TB/T 2832-1997 protective standards for the impact of ac electrified railway on oil (gas) pipelines.

# Simultaneous Reduction and Covalent Combining of Tetraethylenepentamine on Graphene Oxide

**Qunxiang Ren, Xiaoyan Shen, Jun Zhang and Jiayun Liu**

Department of Chemistry, School of Basic Medical Sciences, Shenyang Medical College, Shenyang 110034, PR China
Email: renqunxiang@163.com

**Abstract.** [Objective] To prepare tetraethylenepentamine functionalized reduced graphene oxide (TEPA-Gn) by a one-step process. [Method] Tetraethylenepentamine (TEPA) was grafted with graphene oxide (GO) under mild condition of 70°C and in the presence of sodium hydroxide which acted as a catalyzer. [Results] The structural analysis of TEPA-Gn by FTIR and XPS showed that tetraethylenepentamine was attached to the six-membered ring of graphene by forming a covalent bond with an epoxy group in GO. Raman spectrum showed that the disorder of TEPA-Gn increased. [Conclusion]The tetraethylenepentamine was both a linker agent and a reducing agent in the reaction of preparing TEPA-Gn.

## 1. Introduction

Graphene has some excellent performance, such as high specific surface area, excellent electrical conductivity, outstanding mechanical properties and high thermal conductivity [1-6]. The van der Waals force between the graphene sheets is very large, and it is easy to cause aggregation, making it difficult to dissolve in water and common organic solvents, which has caused great difficulty for further research and application of graphene. The introduction of functional groups onto the graphene surface by chemical modification is the main way to solve its dispersion. There are oxygen-containing groups on the surface of GO, like hydroxyl group, epoxy group, carbonyl group, carboxyl group and ester group. These oxygen-containing groups can easily react with compounds containing amino group, carboxyl group, and isocyanate group, so as to realize covalent modification of GO [7-9]. The thickness of the modified GO is higher than that of graphene. At the same time, the conjugated structure of GO is damaged, resulting in a significant reduction in its electrical and mechanical properties [10-11]. However, some of the conjugated structures can be recovered by chemical reduction to improve their performance.

In this work, the graphene oxide was simultaneously modified and reduced with amino acid-rich tetraethylenepentamine under the catalysis of sodium hydroxide to prepare tetraethylenepentamine functionalized reduced graphene oxide with good dispersion performance. The reaction condition was mild and the operation process was simple without hydrazine hydrate or sodium borohydride reduction steps.

## 2. Experimental

### *2.1. Reagents*

Tetraethylenepentamine, graphite powder, $H_2SO_4$, $KMnO_4$, $NaNO_3$, HCl, NaOH and absolute ethanol were purchased from Sinopharm Chemical Reagent Co., Ltd., and all the above reagents were of analytical grade.

### *2.2. Apparatus*

DF-101S collector thermostatic magnetic stirrer produced by Gongyi Yuhua of China, ESCALAB250 multi-functional surface analysis system of Thermo VG Company of USA, 2000 type Fourier transform infrared spectrometer produced by American PE Company and the DXR Raman spectrometer produced by Thermo Fisher Company of the United States.

### *2.3. Preparation of TEPA-Gn*

The graphene oxide was prepared using the improved Hummers method [12]. 25 mg of graphene oxide was dissolved in 25 mL of deionized water, and after 1 hour of sonication, a homogeneous dispersing solution was formed. 3 ml of a 0.2 mol/L NaOH solution was added dropwise with stirring and after ultrasonication for 20 minutes, 25 μL of tetraethylenepentamine was slowly dropped under stirring. The solution was stirred at 70 °C for 20 hours, centrifuged by a centrifuge, and washed three times with deionized water. The product after centrifugation was redispersed in deionized water and no precipitation occurred within three months.

## 3. Results and Discussion

### *3.1. FTIR spectra of GO and TEPA-Gn*

As shown in Figure 1a, GO has a large number of hydrophilic functional groups. The broad absorption peak at 3397 $cm^{-1}$ corresponds to the stretching vibration of the hydroxyl group in graphene oxide, the absorption peak at 1731 $cm^{-1}$ is attributed to the stretching vibration of the carbonyl group, and the absorption peak at 1623 $cm^{-1}$ is derived from the stretching vibration of the aromatic ring C=C. The absorption peaks at 1050 $cm^{-1}$ and 853 $cm^{-1}$ are symmetric and asymmetric stretching vibration peaks of the epoxy group, In the IR spectrum of TEPA-Gn, the absorption peaks at 1050 $cm^{-1}$ and 853 $cm^{-1}$ disappear because the amino group in tetraethylenepentamine reacted with the epoxy group in graphene oxide. The absorption peak at the 1731 $cm^{-1}$ also disappears, which can be attributed to the reduction of graphene oxide. Several new absorption peaks appeared in the FTIR spectrum of TEPA-Gn because of the introduction of tetraethylenepentamine. The attribution of these peaks can be divided as follows: N–H stretching vibration absorption peak (3407 $cm^{-1}$), N–H bending vibration peak (1583 $cm^{-1}$), and C–N stretching vibration absorption peaks (1351 $cm^{-1}$ and 1221 $cm^{-1}$). The results of IR spectra show that graphene oxide reacted with tetraethylenepentamine and was reduced to form tetraethylenepentamine functionalized reduced graphene oxide.

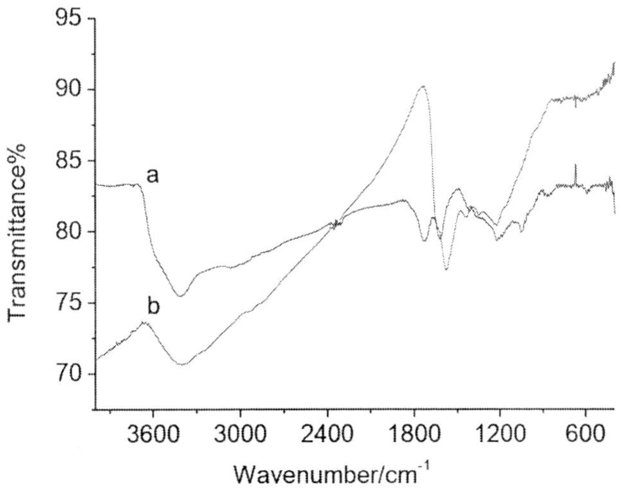


**Figure 1.** FTIR spectra of GO (a) and TEPA-Gn (b)

IOP Conf. Series: Materials Science and Engineering 472 (2019) 012096 doi:10.1088/1757-899X/472/1/012096

*3.2. XPS characterization of GO and TEPA-Gn*

The XPS full spectra of GO and TEPA-Gn are shown in Figure 2. There are only C1s (284 eV) peaks and O1s (532 eV) peaks in the full XPS spectrum of GO while a new N1s peak(399 eV) appears in the full XPS spectrum of TEPA-Gn, indicating that tetraethylenepentamine has been grafted on graphene oxide. The intensity ratio of C/O in TEPA-Gn is higher than that of GO, which is due to the removal of oxygen-containing groups after the reaction of tetraethylenepentamine and graphene oxide under the presence of sodium hydroxide.

Figure 3 is the C1s high-resolution XPS spectra of GO and TEPA-Gn. The combined states of C in graphene oxide are C-C and C=C (284.5 eV), C-OH (286.6 eV) and C=O (288.1 eV), respectively [13]. Compared with graphene oxide, the content of C-OH and C=O in TEPA-Gn is significantly reduced, indicating that most of the oxygen-containing groups have been reduced. And the presence of a new carbon-bonded C-N (285.5 eV) in TEPA-Gn much more confirms the combination of graphene oxide and tetraethylenepentamine, which is consistent with the results confirmed by FTIR spectra and XPS full spectra.

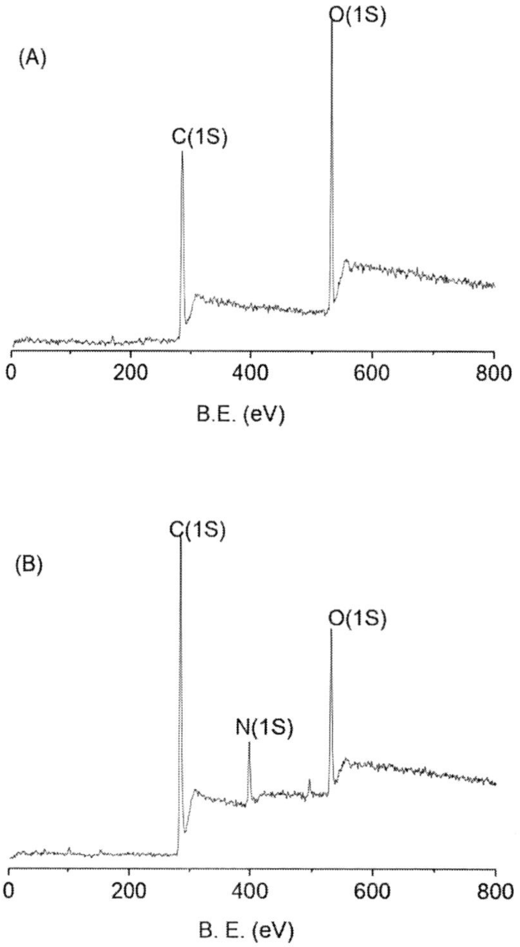


**Figure 2.** XPS spectra of GO (A) and TEPA-Gn (B)

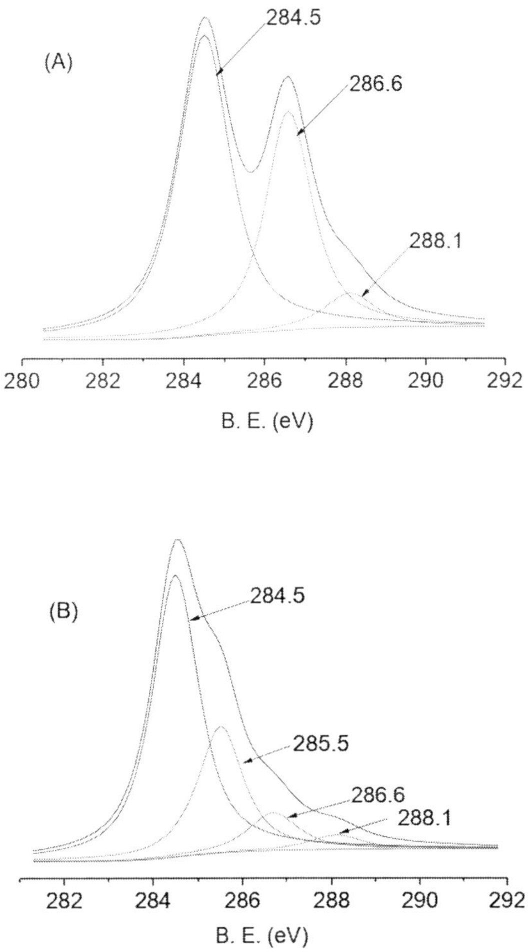


**Figure 3.** Peak deconvolution of C (1s) of GO (A) and TEPA-Gn(B)

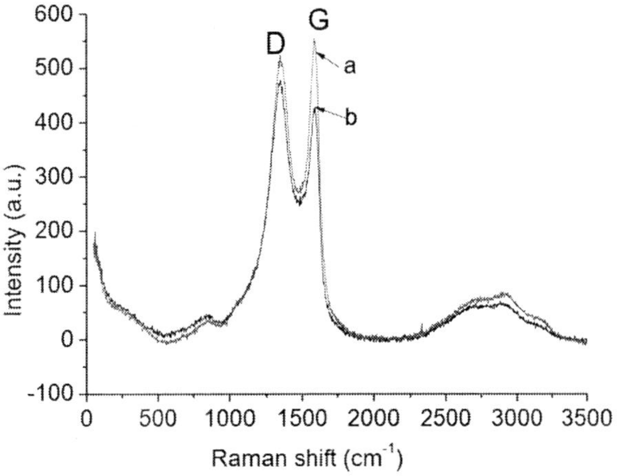


**Figure 4.** Raman spectra of GO (a) and TEPA-Gn (b).

### *3.3. Raman spectra of GO and TEPA-Gn*

Raman scattering is a method that can directly observe the interaction of electrons and phonons without damage, it has high sensitivity to electronics and crystallographic structure so it has been widely used for structural detection of carbon materials [14]. Graphene is composed of symmetrical carbon-carbon covalent bonds, and its structure can be detected by Raman spectroscopy even with slight changes. Raman spectrum of the graphene oxide shows two sharp peaks. The G peak is located in around 1590 $cm^{-1}$ produced by the stretching motion of the $sp^2$ hybrid orbital. And the D peak is located in around 1350 $cm^{-1}$ which represents $sp^2$ hybridization defects. The $sp^2$ hybridization defects are caused by crystal defects and disorder induction. Generally, the disorder degree of graphene is measured by the intensity ratio of D peak to G peak ($I_D/I_G$), and the larger the ratio, the higher the degree of disorder of graphene. The $I_D/I_G$ ratio of GO was 0.92; while the $I_D/I_G$ of TEPA-Gn prepared in this study was 1.07, which shows that the degree of disorder of graphene oxide was increased because of the introduction of tetraethylenepentamine.

## 4. Conclusion

The tetraethylenepentamine was both a linker agent and a reducing agent in the reaction of preparing TEPA-Gn. Sodium hydroxide can catalyze the ring-opening reaction of tetraethylenepentamine and graphene oxide epoxy, and at the same time, it can reduce graphene oxide. The mild reaction conditions avoided the aggregation of the TEPA-Gn. The TEPA-Gn can further participate in the reaction because of its good dispersion and active amino group, which creates conditions for further utilization of functionalized graphene.

## Acknowledgments

This study was funded by the Natural Science Funds of Liaoning Province (20170540876) and Science Research Funds of Shenyang Medical College (20161010).

## References

[1] Lee C G, Wei X D, Kysar J W, Hone J, et al. Science, 2008, 321: 385-388.
[2] Seol J H, Jo I, Moore A L, et al. Science, 2010, 328(5975):213.
[3] Sun Z, Chang H. Acs Nano, 2014, 8(5): 4133-4156.
[4] Sykes E C. Nature Chemistry, 2009, 1 (3): 175-176.
[5] Cinti S, Arduini F. Biosensors & Bioelectronics, 2016, 89 (Pt 1): 107-122.
[6] Wan S, Bi H, Zhou Y, et al. Carbon, 2017, 114: 209-216.
[7] Hou S, Su S, Kasner M L, et al.Chemical Physics Letters, 2010, 501 (1–3): 68-74.
[8] Liu H, Liu C, Peng S, et al. Carbohydrate Polymers, 2018, 182 (118): 52–60.
[9] Choi E Y, Han T H, Hong J. J. Mater. Chem., 2010, 20(10): 1907-1912.
[10] Moon I K, Lee J, Ruoff R S, et al. Nat. Commun., 2010, 1(6),73.
[11] Liu L, Zhang J, Zhao J, et al. Nanoscale, 2012, 4(19), 5910-5916.
[12] Zaaba N I, Foo K L, Hashim U, et al. Procedia Engineering, 2017, 184: 469-477.
[13] Akhavan O. Carbon, 2015, 81 (1): 158-166.
[14] Mafra D L, Kong J, Sato K. Phys. Rev. B, 2012, 86(19):273-278.

IOP Conf. Series: Materials Science and Engineering 472 (2019) 012097 doi:10.1088/1757-899X/472/1/012097

# Design of Air Floating EGR Water Treatment Device

**Fangyong Wang[1*] and Zhanguang Wang[2]**

1. College of Engineering & Technology, Northeast Forestry University, Harbin, Heilongjiang, China, 150001
2. College of Power &Engineering, Harbin Engineering University, Harbin, Heilongjiang, China, 150001

Email: wfy166@163.com

**Abstract.** EGR is an important technical means to reduce NOx emissions from marine diesel engines, but EGR washing waste liquid contains a large amount of pollutants that must be treated before discharging the hull. On January 1, 2016, the IMO Tier III regulations have been implemented and the marine EGR system will be more widely used. Therefore, it is imperative to study the ship EGR plant wastewater treatment system. In this paper the author studied the IMO ship washing wastewater discharge regulations, and introduced the principle, composition, classification, overall layout and causes of wastewater generation of EGR devices. At the same time, the author analyzed the components of wastewater EGR system wastewater pollutants, and compared emission regulations. For the requirements of water quality, various technologies for treating EGR wastewater pollutants were studied. Based on various technologies, the author designed an air-floating wastewater treatment system to treat wastewater. The overall treatment plan, design ideas, structure and size design, and process flow of the device were analyzed and calculated in detail.

## 1. Introduction

Since 2000, the International Maritime Organization (IMO) has revised the emission regulations several times. The latest ship emission regulations, IMO Tier III, have been implemented on January 1, 2016. Due to the demanding requirements, the latest technology must be adopted to meet NOx emission reduction standards. EGR technology is the cutting-edge technology invented by MAN in Germany. The principle of EGR is to return some of the exhaust gas back to the combustion chamber to reduce the concentration of O2, thereby reducing the temperature of the combustion chamber to achieve the purpose of suppressing NOx formation. This technology can achieve 80% NOx reduction [1]. While as we all know, two-stroke low-speed marine diesel engines use HFO which can cause large number of SOx and PM. That means the recirculated exhaust gas must be treated to remove SOx and PM. The requirements for the low-speed marine diesel EGR system wastewater treatment plant are also increasing. The wastewater generated by the low-speed diesel engine EGR device is also a kind of ship wastewater. However, there are very few studies on the ship low-speed diesel engine EGR device wastewater treatment system [2-4]. The low-speed marine EGR plant wastewater treatment is only recovered by simple cyclone separation or aeration oxidation. The EGR wastewater treated by this process can only reduce the turbidity value and COD value in the EGR plant wastewater, and can't fully degrade the polycyclic aromatic hydrocarbons (PAHs) and sulfites in the EGR wastewater. Under increasingly stringent environmental requirements, it does not meet the IMO's latest regulations for the discharge of washing water from ship exhaust systems [5-6]. Therefore, it is necessary to improve the existing equipment and processing technology or to develop a new EGR wastewater treatment unit.

## 2. Material and Methods

### *2.1. Aeration Technology*

Aeration is a mechanical process in which sodium sulfite is treated by intense contact with air and water in order to dissolve oxygen in the atmosphere while discharging unwanted volatiles and gases from the water out of the water. Aeration promotes the exchange of substances between liquid and gas. The oxygen in the air is transferred to the water by aeration technology, and the oxygen is transferred to the liquid phase.

The aeration method is divided into three types: blast aeration, mechanical aeration and new aeration.

Blast aeration is also called compressed air aeration, and the device includes an aerator and an aeration fan. The air compressor uses a pipe to transfer air to an air diffusing device installed from the bottom of the water tank, which is bubbled out by the diffusing device and finally dissolved in water. The mechanical aeration is rotated by a mechanical impeller installed at the bottom of the tank, and the waste water is vigorously stirred to integrate oxygen in the air. In the past few years, there have been many new types of aeration methods, including deep well aeration, oxygen-enriched or pure oxygen aeration, and aeration technology to assist other biological treatment methods. Pure oxygen aeration is a method similar to blast aeration to inject oxygen-enriched or pure oxygen into the water. The upper part of the aeration tank is provided with a sealing cover, which is beneficial to improve oxygenation efficiency. Biofilm carrier fluidized bed aeration and biological contact oxidation are aeration methods still under development.

### *2.2. Coagulation Technology*

The methods for treating the suspended matter in the wastewater include filtration, coagulation, centrifugal separation, and a gravity sedimentation method. Among them, the filtration method is only suitable for the advanced treatment of wastewater; the gravity sedimentation method is suitable for the treatment of suspended solids with large particle size; therefore, the suspension in the wastewater is treated by centrifugal separation or coagulation air flotation. The essence of the coagulation air floatation method is to rapidly separate the suspended matter which has been precipitated from the wastewater with water by the air flotation method on the basis of the coagulation method.

### *2.3. Air Floatation Technology*

In the low-speed diesel engine EGR wastewater treatment technology, according to the bubble size and the form of action, the air floatation method can be roughly divided into four categories, namely induced air flotation method, dissolved air flotation method, chemical air floatation method and electrolytic air floatation method [19]. The commonly used methods are the jet air flotation method and the impeller air flotation method in the air flotation method, and the pressurized dissolved air flotation method in the dissolved air flotation method, wherein the dissolved air flotation method is most widely used. The dissolved air flotation method includes a dissolved gas vacuum floatation method and a dissolved gas pressurized air floatation method. The working principle of the impeller air floatation method is that the impeller that rotates at high speed attracts a large amount of air, the air that is caught in the impeller is continuously crushed, and becomes a large number of tiny bubbles, and finally a large amount of pollutants adhere to the bubbles; the work of the jet air float method The principle is: the ejector rapidly sprays water and creates a negative pressure inside the nozzle suction chamber. Therefore, a large amount of gas is sucked into the chamber. When the water passes through the mixing section at high speed, the gas carried by the gas is cut into small bubbles, and finally the contaminants are attached thereto. on.

The characteristics of the air floatation method: the scum can be completely removed by the scraper, and the water content is low, and the operation is convenient; in the process of air floatation, a large amount of oxygen is dissolved in the water, and a protection mechanism is easily formed; the equipment and the process flow are simple, and the management Convenient and economical; there is a lot of dissolved air in the water, which can be applied to a variety of water quality.

### *2.4. Activated Carbon Adsorption Technology*

The methods for treating polycyclic aromatic hydrocarbons include microbial degradation, ultrasonic degradation, ultraviolet photolysis, and natural degradation. Among them, the microbial degradation method and the natural degradation method take a long time and a slow speed, and can not be matched and applied in the entire desulfurization system. The ultrasonic degradation method and the Fenton reagent oxidation method have high cost, complicated operating system, large floor space, and complicated operating system, which are not suitable for the ship environment. Ultraviolet photolysis is easy to produce other toxic substances, and the degradation efficiency is not high. Therefore, the best method for treating polycyclic aromatic hydrocarbons in wastewater is adsorption, and the adsorbent can be used with the lowest cost, most widely used, and recyclable activated carbon.

## 3. Air Floating Wastewater Treatment Device

According to the water quality research results, the pollutants to be treated in the EGR plant wastewater of the low-speed diesel engine are sulfite, suspended solids and polycyclic aromatic hydrocarbons. Aeration technology is used to treat sulfite in EGR wastewater, flocculation technology is used to treat suspended solids in EGR wastewater, and activated carbon adsorption technology is used to treat polycyclic aromatic hydrocarbons in EGR wastewater. From the above, we can see that the comprehensive scheme of air-floating wastewater treatment is as follows:

- EGR wastewater is first discharged into the waste tank for storage, and the waste tank is opened every six hours to discharge the waste water into the waste gas tank.
- EGR wastewater is oxidized to sulphate in water by aeration in aeration tank.
- EGR wastewater is completely mixed with coagulant and flocculant in the mechanical coagulation tank to form floc.
- EGR wastewater enters the air flotation tank for air flotation treatment and the air bubbles and flocs generated by the air flotation device are combined to form scum, and the scum is discharged through the scraper.
- The water after coagulation and air flotation treatment is discharged into an activated carbon filter device to remove polycyclic aromatic hydrocarbons from wastewater.
- The final treated water enters the detection device. If the concentration of each component meets the IMO standard, it is discharged into the discharge port. If it does not meet the standard, it is discharged into the waste container and treated again.

The air-floating wastewater treatment device is composed of four parts: an aeration device, a coagulation air flotation device, an activated carbon adsorption device, and a detection device.

In this paper, the author selected blower aeration as aeration device, selected advection air flotation cell, and pressurized dissolved air flotation for wastewater treatment. The selected coagulant and flocculant are PAC and PAM.

Table 1 shows the main equipment of air floating water treatment device.

**Table 1.** Main equipment of air floating water treatment device

| Number | Equipment | parameter | Amount |
|---|---|---|---|
| 1 | air compressor | Q=0.1 $m^3$/min | 1 |
| 2 | aeration tank | 2200×2200×2200 | 1 |
| 3 | Backflow pump | Q=2 $m^3$/h<br>H=3m | 2 |
| 4 | air tank | Φ250×1000 | 1 |
| 5 | Waste tank | 2000×2000×2000 | 1 |
| 6 | water supply pump | Q=2 $m^3$/h<br>H=3m | 2 |
| 7 | Digital constant flow pump | 3~6000ml/h | 3 |
| 8 | dosing tank | 420×320×280 | 3 |
| 9 | Activated carbon adsorption unit | Φ600×600 | 1 |
| 10 | water collecting pipe | Φ50 | 1 |
| 11 | clean water tank | 500×500×600 | 1 |
| 12 | flotation tank | 1600×1600×1600 | 1 |
| 13 | Slag scraper | 0.1KW | 1 |
| 14 | Stirring motor | 0.3KW | 1 |
| 15 | Dissolved gas releasing device | 2 $m^3$/h | 1 |
| 16 | Mechanical reaction cell | 500×500×600 | 1 |
| 17 | flowmeter | 10 $m^3$/h | 3 |
| 18 | monitoring unit | 400×400×400 | 1 |
| 19 | one-way throttle valve | 10 $m^3$/h | 1 |
| 20 | Air blower | m3/h | 1 |

## 4. Results

In this paper the author firstly described the comprehensive scheme and design ideas of the air-floating wastewater treatment device, and then comprehensively calculated and analyzed the selection design and size structure of each device and component, and determines the following parameters.: the volume of the waste tank is 8m3, the length of the waste tank is 2000mm, the width is 2000mm, the height is 2000mm; the volume of the aeration tank is 9.92m3, the aeration tank is 2200mm long, the width is 2200mm, the height is 2200mm, the total air volume of the aeration fan is 24m3/h; the diameter of the dissolved gas tank is 250mm, the height is 1000mm, the length of the air floating pool is 1600mm, the width is 1600mm, height is 600mm, mechanical mixing pool and clear water tank are 500mm long, width is 500mm, height is 600mm, air compressor ventilation

0.48m3/h; the volume of the activated carbon adsorption container is 0.32m3, the diameter is 600mm, the height is 600mm; the length of the chemical tank is 420mm, the width is 320mm, the height is 280mm, there are three dosing boxes, respectively, PAC, PAM And NaOH;

The detection device has a length of 400 mm, a height of 400 mm and a width of 400 mm.

## References

[1] ingjiang Yuan, Shitang Tong, Suoqi Zhao et al. Adsorption of olycyclic aromatic hydrocarbons from water using petroleum coke-derived porous carbon[J]. Journal of hazardous material, 2010, 181(1/3):1112-1120.
[2] Ulrich Dohle. MTU solutions for meeting future exhaust emissions regulation[C]. CIMAC, 2010, Paper No. 284.

[3] Mingfa Yao. Experimental Study on the Effects of EGR and Octane Number of PRE Fuel on Combussion and Emission Characteristics of HCCI Engines[C]. SAE Paper 2005-01-0174, 2005.
[4] Masanori Higashida. Newly Developed Combined EGR & WEF System to comply with IMO Regulation Tier 3 for Two-stroke Diesel Engine[C]. CIMAC.2013.Paper No.200.
[5] Jorhan Kaltoft. Development of integrated EGR system for two-stroke diesel engines[C]. CIMAC.2013.Paper No. 219.
[6] MAN DIESEL &TURBO. Emission Project Guide; 8-33.

IOP Conf. Series: Materials Science and Engineering 472 (2019) 012098 doi:10.1088/1757-899X/472/1/012098

# Design of Mixed Scrubbing and Desulfurization System for a Marine Engine

**Fangyong Wang[1*] and Hao Guo[2]**

1.College of Engineering & Technology, Northeast Forestry University, Harbin, Heilongjiang, China, 150001
2.College of Power &Engineering, Harbin Engineering University, Harbin, Heilongjiang, China, 150001
Email: wfy166@163.com

**Abstract.** Along with the rapid development of the shipbuilding industry, the control of $SO_x$ in the exhaust gas discharged from marine power plants has become increasingly strict. Unlike NOx emissions, sulphur oxide are derived from fuels, and economic and practical considerations need to be considered while looking for alternative fuels and low-sulfur fuels. Therefore, post-treatment of emissions is the best means to control $SO_x$. In order to meet the new IMO emission control standards, a more stringent exhaust gas desulfurization system should be designed. In this paper, the design of the mixed ship exhaust gas desulfurization system is studied. The absorption principle and the respective characteristics of the seawater method open desulfurization system are analyzed. The spray tower was selected as the desulfurization absorption equipment, and the seawater was used as the main absorbent to calculate the size data of the spray tower, the spray device and the size of the gas inlet and outlet. A set of specific exhaust gas desulfurization systems was designed for the design of liquid handling equipment, piping and pumps.

## 1. Introduction

With the development of ships industry, the proportion of $SO_x$ from marine power plant accounting for total global emissions increases year by year, with enable the usage of high-sulfur heavy fuel oils (HFO) [1]. In order to prevent atmospheric pollution from ships, according to the regulations of International Maritime Organization (IMO), from 2015 beginning, for the ships sailing in sulfur controlled area (SECA) the sulfur content of fuel oil shall not exceed 0.1%wt; by 2020, the global worldwide sailing ship in non-SECA, the fuel sulfur content shall not exceed 0.5%wt. In accordance with the comparative analysis of ships SOx emissions performance technologies, ships exhaust gas emissions after-treatment technologies are the most economic and feasible [2-3]. Therefore, it is necessary to improve the existing equipment and processing technology or to design a new mixed marine tail gas washing and desulfurization system.

The wet scrubbing system is more widely used than the dry scrubbing system. It is the main desulfurization system currently researched and applied in foreign countries. At present, Wärtsilä, Clean Marine, MAN, Aalborg, etc. have mastered their core technologies in this respect [4-5]. After February 2012, Wärtsilä's exhaust scrubber desulfurization system incorporates Hamworthy's technology, and several vessels have already installed or ordered the ship's exhaust scrubber desulfurization system developed by Wärtsilä.

## 2. Reaction Mechanism

When the desulfurization system starts to operate, the initial absorbent gradually contacts and reacts with $SO_2$, and the pH of the absorption liquid continuously decreases. When the pH value of the absorbent is lower than the limit value and the required desulfurization efficiency cannot be satisfied, the system automatically discharges the portion. The absorbent is simultaneously added with a fresh absorbent to automatically adjust the pH to maintain a high desulfurization efficiency. During the whole absorption process in the scrubbing tower, various reversible and irreversible chemical reactions occur in the contact zone. The chemical reaction process is very complicated, and water dissociation and sulfite mainly occur during the period. Oxidation produces sodium sulfite, sodium bisulfite and a small amount of sodium sulfate.

1. Absorption reaction

$SO_2(g)+H_2O \rightarrow SO_2(l)+H_2O$(mass transfer)

$SO_2+H_2O \rightarrow H_2SO_3$ (dissolved)

$SO_2+H_2O \rightarrow H^++HSO_3^-$(ionization)

$H_2SO_3 \rightleftharpoons H^++HSO_3^-$(ionization)

2. Neutralization reaction

The sodium hydroxide absorbent maintains a certain pH value, and will neutralize the product formed by the absorption of sulfur dioxide in the flue gas in the previous step, and the neutralized reaction in the spray tower after the neutralization process is again sprayed in the desulfurization closed cycle. The tower absorbs sulfur dioxide, and the following is the chemical equation for the neutralization reaction.

$NaOH \rightarrow Na^++OH^-$

$2NaOH+H_2SO_3 \rightarrow Na_2SO_3+2H_2O$

$Na_2SO_3+ SO_2(l)+ H_2O \rightarrow 2NaHSO_3$

$NaOH+NaHSO_3 \rightarrow Na_2SO_3+H_2O$

$Na^++HSO_3^- \rightarrow NaHSO_3$

$2Na^++CO_3^{2-} \rightarrow Na_2CO_3$

$H^++CO_3^{2-} \rightarrow H_2O+CO_2\uparrow$

3. Oxidation reaction

The absorbed part of $HSO_3^-$ is oxidized in the absorption zone of the absorption tower with oxygen in the flue gas. More of the part of $HSO_3^-$ is completely oxidized in the slurry tank below the spray tower and the air. The reaction equation is as follows:

$HSO_3^-+1/2O_2 \rightarrow HSO_4^-$

$HSO_4^- \rightleftharpoons H^++SO_4^{2-}$

## 3. Hybrid Desulfurization Tower Structure Design

### *3.1. Exhaust Gas Properties*

The model selected in this paper is the MAN 6S50 diesel engine as the supporting diesel engine model. When designing the exhaust gas mass flow, the maximum exhaust gas flow should be fully considered. After data inquiry, the mass flow rate of the exhaust gas is finally determined to be 82500 kg/h.

The absorption target of this paper is the exhaust gas produced after the combustion of the fuel with a sulfur content of 3.5%. After the test, the concentration of $SO_2$ in the exhaust gas after combustion of the fuel with a sulfur content of 3.5% was measured and further measured. Below are some other data of the exhaust gas.

Exhaust gas density: $1.293kg/m^3$,

Smoke molar mass: 29g/mol,

Smoke viscosity: $2\times10^{-5}$Pas.

### *3.2. Spray Tower Structure Size Calculation*

The structural size calculation of the spray tower mainly consists of two parts, one is the tower diameter calculation of the spray tower, and the other part is the tower height calculation of the spray tower. The diameter of the spray tower can be calculated from the exhaust gas intake flow rate and the empty tower gas velocity as follows:

$$D=\sqrt{\frac{4V_s}{\pi u}} \quad (1)$$

Where: *D* - a diameter, m.

Volume - flow of exhaust gas under *Vs* operating conditions, $m^3/h$.

*U* - gas velocity of mixed gas, m/s.

In the design process, the rationality of the design and the operability of the processing should be considered. Therefore, the final design of the spray tower diameter is 2.5m.

### *3.3. Spray Tower Height Calculation*

The height of the spray tower is composed of many sections. The design is mainly for the height of the absorption zone, the height of the defogging zone, the height of the slurry pool and the height of the gas inlet pipe. The rest can be increased or shortened more or less according to the actual situation of the design. .

(1) Design of the absorption zone of the spray tower

The flue gas containing sulfur dioxide passes through the spray tower to average the total sulfur dioxide absorption in the tower to the inner volume of the tower within the height of the absorption zone, which is the average volumetric load of the absorption tower, and the value is expressed as the average volumetric absorption rate. The symbol $\xi$ indicates.

$$\xi=3600\times\frac{64}{22.4}*\frac{273}{273+t}u*y_1\eta/h \quad (2)$$

Where: $\eta$- $SO_2$ absorptivity,

*h* - Height of absorption zone in H tower,

*u*- gas velocity u(m/s),

$Y_1$- the molar fraction of the solute in the gas entering the tower.

According to the design experience, the average volumetric absorption rate $\xi$ should be selected between 5.5 and 6.5kg/ ($m^3$.s). Take $\xi$=6.0kg/ ($m^3$.s).

$$h=\frac{3600\times\frac{64}{22.4}\times\frac{273}{273+75}\times4.0\times1.12\times10^{-3}\times97.2\%}{6.0}=5.86 \quad (3)$$

(2) Design of the height of the defogging zone of the spray tower

The mist eliminator designed in this paper has two-stage defogger. The first-stage demister and the uppermost spray layer are 1m, and the distance between the first and second demister is 1.5m. The secondary demister and the upper spray tower is one meter, so the final demister height design height $h_2$=3.5m.

(3) Design of spray tower pulp pool height

In the design process of the slurry tank, the volumetric capacity should be designed from the volume, so that a certain amount of liquid is stored at the bottom of the bottom of the slurry. Therefore, the capacity of the slurry tank $V_1$ can be designed according to the ratio of the liquid to gas ratio *L/V* and the residence time of the slurry. To determine, the formula is as follows:

$$V_1=\frac{L}{V}\times V_N\times t_1 \quad (4)$$

Where: *L/V* - operating liquid to gas ratio,

$V_N$ - Fume gas standard state wet volume,

$t_1$ - slurry residence time.

$$V_1 = \frac{7.38\times\frac{65893.27}{3600}\times 120}{1000} = 16.21\text{m}^3 \tag{5}$$

$$S = \pi D^2 = 5.826\times\pi = 18.30\text{m} \tag{6}$$

$$h_3 = \frac{V}{S} = \frac{16.21}{18.30} = 0.886m \tag{7}$$

The result is rounded upwards, and the final designed pulp pool height $h_3$=0.9m.

(4) Design of spray tower flue gas inlet height

The height of this part should be consistent with the caliber of the inlet flue gas pipeline, and should be appropriately increased on the basis of the known flue caliber. The final design of the gas inlet height $h_4$=1.1m.

Based on the above four parts analysis, it is concluded that the overall height of the spray tower is mainly composed of the four parts, the height of the absorption zone, the height of the defogging zone, the height of the slurry pool and the inlet height of the flue gas.

$$H = h_1 + h_2 + h_3 + h_4 \tag{8}$$

Finally, the overall height of the spray tower is $H$=11.5m.

When the spray tower flue gas inlet device is designed, the velocity of the gas entering the spray tower is about 10-20m/s. Generally speaking, the higher the gas entering the tower speed can promote the absorption of the absorbent more effectively and improve the overall spray tower. Desulfurization efficiency, so this design takes gas into the tower speed of 20m/s. When designing the seawater liquid inlet, it should be calculated according to the designed liquid velocity. According to the pipe material standard, the liquid velocity in the pipe is generally between 1.8m/s and 2.4m/s. In this design, the inlet speed of the absorbent is 2m/s.

## 4. Results

The ship hybrid exhaust gas desulfurization system is an effective measure to control the emissions of SOx. In this paper, the design and research of the mixed exhaust gas desulfurization system is carried out. The emission control of the exhaust gas of the 10MW model is studied, and a set of scrubbing desulfurization system is designed, the conclusion is as follows:

1. The material balance between the seawater method and the sodium-alkali method is obtained. The designed liquid-gas ratio of the seawater method is 7.38L/$m^3$, and the designed liquid-gas ratio of the sodium-alkali method is 2.8 L/$m^3$, which is calculated according to the liquid-gas ratio. The scrubber has a diameter of 2.5m.

2. The height of scrubber is divided into four parts, which are calculated as the height of the absorption zone 6.0m, the height of the defogging zone is 3.5m, the height of the pulp pool is 0.9m, the inlet height of the flue gas is 1.1m, and the overall height of the spray tower is calculated to be 11.5m.

3. The spray device is selected as a pressure atomizing nozzle, the flue gas flow rate is 3.5m/s, the seawater liquid inlet pipe diameter is 0.3m, and the NaOH solution inlet pipe diameter is 0.23m.

## References

[1] Cordoba, Patricia Status of Flue Gas Desulphurization (FGD) systems from coal-fired power plants: Overview of the physic-chemical control processes of wet limestone FGDs [J]. Elsevier journal, 2015.
[2] Caillahua, Mariella Cortez, Moura, Francisco Jose. Technical feasibility for use of FGD gypsum as an additive setting time retarder for Portland cement [J]. Elsevier journal, 2018.
[3] Yu, Jicheng, Lu Jia,Kang Yong. Removal of sulfate from wet FGD wastewater by co-precipitation with calcium hydroxide and sodium aluminate [J]. Energy&Fuels, 2015.
[4] MAN. Emission Project Guide, 2017.6.
[5] Wartsila. Wartsila-Environmental Product Guide, 2017.

# AUTHOR INDEX

# AUTHOR INDEX

# AUTHOR INDEX

**Institute of Physics**
Dirac House, Temple Back
Bristol BS1 6BE UK

ISSN: 1757-8981
ISBN 978-1-5108-8220-1